# RANDOM SEAS AND DESIGN
# OF MARITIME STRUCTURES

3rd Edition

# ADVANCED SERIES ON OCEAN ENGINEERING

Series Editor-in-Chief
**Philip L- F Liu** (*Cornell University*)

*Published*

Vol. 20 The Theory and Practice of Hydrodynamics and Vibration
*by Subrata K. Chakrabarti* (Offshore Structure Analysis, Inc., Illinois, USA)

Vol. 21 Waves and Wave Forces on Coastal and Ocean Structures
*by Robert T. Hudspeth* (Oregon State Univ., USA)

Vol. 22 The Dynamics of Marine Craft: Maneuvering and Seakeeping
*by Edward M. Lewandowski* (Computer Sciences Corporation, USA)

Vol. 23 Theory and Applications of Ocean Surface Waves
Part 1: Linear Aspects
Part 2: Nonlinear Aspects
*by Chiang C. Mei* (Massachusetts Inst. of Technology, USA),
*Michael Stiassnie* (Technion–Israel Inst. of Technology, Israel) and
*Dick K. P. Yue* (Massachusetts Inst. of Technology, USA)

Vol. 24 Introduction to Nearshore Hydrodynamics
*by Ib A. Svendsen* (Univ. of Delaware, USA)

Vol. 25 Dynamics of Coastal Systems
*by Job Dronkers* (Rijkswaterstaat, The Netherlands)

Vol. 26 Hydrodynamics Around Cylindrical Structures (Revised Edition)
*by B. Mutlu Sumer and Jørgen Fredsøe* (Technical Univ. of Denmark,
Denmark)

Vol. 27 Nonlinear Waves and Offshore Structures
*by Cheung Hun Kim* (Texas A&M Univ., USA)

Vol. 28 Coastal Processes: Concepts in Coastal Engineering and Their
Applications to Multifarious Environments
*by Tomoya Shibayama* (Yokohama National Univ., Japan)

Vol. 29 Coastal and Estuarine Processes
*by Peter Nielsen* (The Univ. of Queensland, Australia)

Vol. 30 Introduction to Coastal Engineering and Management (2nd Edition)
*by J. William Kamphuis* (Queen's Univ., Canada)

Vol. 31 Japan's Beach Erosion: Reality and Future Measures
*by Takaaki Uda* (Public Works Research Center, Japan)

Vol. 32 Tsunami: To Survive from Tsunami
*by Susumu Murata* (Coastal Development Inst. of Technology, Japan),
*Fumihiko Imamura* (Tohoku Univ., Japan),
*Kazumasa Katoh* (Musashi Inst. of Technology, Japan),
*Yoshiaki Kawata* (Kyoto Univ., Japan),
*Shigeo Takahashi* (Port and Airport Research Inst., Japan) and
*Tomotsuka Takayama* (Kyoto Univ., Japan)

Vol. 33 Random Seas and Design of Maritime Structures, 3rd Edition
*by Yoshimi Goda* (Yokohama National University, Japan)

*For the complete list of titles in this series, please write to the Publisher.

Advanced Series on Ocean Engineering — Volume 33

# RANDOM SEAS AND DESIGN OF MARITIME STRUCTURES

## 3rd Edition

**Yoshimi Goda**

Yokohama National University, Japan

 **World Scientific**

NEW JERSEY · LONDON · SINGAPORE · BEIJING · SHANGHAI · HONG KONG · TAIPEI · CHENNAI

*Published by*

World Scientific Publishing Co. Pte. Ltd.

5 Toh Tuck Link, Singapore 596224

*USA office:* 27 Warren Street, Suite 401-402, Hackensack, NJ 07601

*UK office:* 57 Shelton Street, Covent Garden, London WC2H 9HE

**British Library Cataloguing-in-Publication Data**
A catalogue record for this book is available from the British Library.

**RANDOM SEAS AND DESIGN OF MARITIME STRUCTURES, 3rd Edition**
**Advanced Series on Ocean Engineering, Vol. 33**

ISBN-13 978-981-4282-39-0
ISBN-10 981-4282-39-1
ISBN-13 978-981-4282-40-6 (pbk)
ISBN-10 981-4282-40-5 (pbk)

Printed by FuIsland Offset Printing (S) Pte Ltd. Singapore

# Preface to the Third Edition

As the title tells, the book deals with random sea waves on their statistical nature and interaction with maritime structures. Why do we need to introduce the random nature of sea waves into engineering practice? It is because the randomness is the fundamental characteristics of sea waves and its neglect leads to gross errors in engineering assessment. The concept of random seas has already been accepted by the majority of coastal engineers and researchers. The International Standards ISO 21650 *Actions from Waves and Currents on Coastal Structures* issued in 2007 adopts the wave spectrum and statistics as the fundamental tool for assessing wave actions. The Proceedings of the 5th International Conference on Coastal Structures in 2007 and the 31st International Conference on Coastal Engineering in 2008 contain many papers employing random waves in theory, laboratory tests, and engineering applications.

This is the third edition of the book first published in 1985 by University of Tokyo Press. The second edition was published in 2000 by World Scientific as one of *Advanced Series on Ocean Engineering*. The book has been evolved from its Japanese edition in 1977 published by the Kajima Institute Publishing Company, which was subsequently revised in 1990 and 2007. The book has been enlarged through successive revisal works.

The book is composed of four parts. Part I is for practicing engineers by providing design tools to deal with random seas. Part II is for graduate students to serve as a textbook of random wave theory. Part III deals with the methodology of extreme wave analysis for selection of design wave height. Part IV is a new addition to the present edition, which overviews beach morphological problems from the viewpoint of coastal managers and field engineers who are rather unacquainted with research works. It emphasizes a historical view of a coast under study and the necessity of field validation data whenever one tries to make prediction of future morphological changes.

The content of the book has been enlarged quite much. In addition to Part IV, two chapters are newly added on the topics of the probabilistic design of breakwaters and the 2-D computation of wave transformation with random breaking and nearshore currents. The total pages increased from 443 in the second edition to 708 in the present edition. The number of references including duplication increased from 316 to 718. Forty new figures and fourteen new tables were introduced.

One feature of the book is inclusion of the information available only in Japanese. Among 718 references, 150 papers were written in Japanese. They are referred to in various chapters of the book but appear mainly in Chapters 3 to 7 that deals with wave actions on structures. I believe that the Japanese experience and technology would make good contribution to the progress of coastal engineering in the world.

Nowadays a great many number of research works and field reports are presented in various conferences and technical journals. It is almost impossible for any one to make appropriate review of all recent accomplishments. I have picked up papers that attracted my attention rather subjectively and tried to convey what is useful for practitioners and future researchers. If I failed in referring to important accomplishments, I must request the kindness of the readers for pardoning my failure.

Similarly as the previous editions, many colleagues and friends helped me greatly by providing me with valuable information and data. From the Port and Airport Research Institute, Dr. Ken-ichiro Shimosako gave advices on the expected sliding distance of caisson breakwaters, Dr. Katsuya Hiramaya clarified my questions on the Boussinesq equation, Dr. Takeshi Nagao gave a lecture on reliability design method, and Dr. Tetsuya Hiraishi and Dr. Yoshiaki Kuriyama provided me with many references related to the subjects in the present book. Mr. Tadashi Tamada of IDEA Corporation kindly provided me with his data set and diagrams of wave overtopping rates of inclined seawalls, which are reproduced as Figs. 5.8 to 5.10 in Sec. 5.2.3. Prof. Kyung-Duck Suh of Seoul National University in Korea kindly sent me a meticulous corrigenda to the previous edition, which have duly been corrected.

As for Part IV, Professor Robert G. Dean, Dr. Yoshiaki Kuriyama, and Dr. Atilla Bayram kindly read the manuscript and advised me a number of corrections and revisions. Especially I am very grateful for Prof. R. G. Dean for his meticulous review of the draft and many excellent advices.

I am indebted to many more friends and express my thanks for their kind corporation extended to me.

Yoshimi GODA
Yokosuka, Japan
November 2009

# Preface to the Second Edition

This is the second edition of the book first published in 1985 by University of Tokyo Press, which was an enlarged English edition of my book in Japanese published in 1977 by the Kajima Institute Publishing Company. The Japanese edition was revised in 1990 with the addition of new material, especially of a new chapter on the statistical analysis of extreme waves. In the present edition, further revisions have been made to update the book's content.

Additions to the first English edition are the new sections on the relationships between wave statistics and spectrum, the longshore currents induced by random waves, the motions of ships at mooring, the tests with multidirectional random waves, the advanced theories of directional spectral estimates, and the new chapter on extreme wave statistics. Several sections have been thoroughly rewritten, such as those on wave grouping, sampling variability of wave parameters, and numerical simulation of random wave profiles.

The objective of this book is twofold: to provide practicing engineers with design tools to deal with random seas (Part I), and to serve as a textbook of random wave theory for graduate students (Part II). The very warm response by many readers to the first English edition seems to reflect the success in achieving the objective. Because of the limited number of printed copies of the first edition, there was quite a large demand for the publication of the second edition. With the recommendation of Professor Philip L-F Liu of Cornell University, World Scientific Publishing Company succeeded in getting the transfer of the publishing copyright from the University of Tokyo Press in 1996. The present edition has been newly typeset from old text and additional new manuscript, the work of which took some time.

The major portion of the book is based on my research work at the Port and Harbour Research Institute, Ministry of Transport, Japan, where I worked from 1957 to 1988. Many staff members of my research group there assisted me in carrying out a number of laboratory experiments and field data analysis. I am very grateful for their dedicated efforts. Many research topics were brought to me by many government engineers in charge of port construction and coastal protection. They taught me what problems in the field were in need of rational engineering solutions.

The publication of the first English edition was suggested by Dr Nicholas C Kraus, presently at the Waterways Experimental Station, US Army Corps of Engineers. He critically reviewed the first manuscript, clearing up ambiguities and correcting many idiomatic expressions. The success of the first edition owed much to him. My wife, Toshiko, patiently allowed me to spend many off-duty hours at home for study and writing of the first and second editions of this book. I would like to conclude this preface with gratitude to her.

Yoshima GODA
Yokosuka, Japan
November 1999

# Preface to the First Edition

Our understanding of sea waves has grown considerably since the appearance of Minikin's celebrated book *Winds, Waves and Maritime Structures* in 1950. In particular, the random nature of sea waves has become much clearer, and sea waves are now described and analyzed by means of statistical theories. The ocean wave spectrum, for example, is presently a working tool of oceanographers and researchers in coastal and offshore engineering. However, engineering application of the random wave concept is as yet limited to a rather small number of researchers and engineers. This book appears to be the first attempt to present a systematic treatment and applications of the concept of random sea waves from the perspective of the engineer.

The original edition of this book was published in Japanese in October 1977 by Kajima Institute Publishing Company. I am pleased to say that the Japanese version was greeted with enthusiasm by a large number of coastal and harbor engineers in Japan. The present edition is a revision and translation of the Japanese edition.

The book is separated into two parts. Part I is mainly addressed to practicing engineers who are looking for immediate answers to their daily problems. This part consists of Chapters 1 to 7. Part II, consisting of Chapters 8 to 10, is directed toward researchers, engineers and graduate students who wish to learn the fundamental theory of random sea waves in order to carry out further developments in coastal and offshore engineering. It is hoped that the readership of Part II will not be limited to those who are civil engineers by training, but will also include naval architects, physical oceanographers and scientists in other disciplines who are interested in the complexity of random sea waves. It is assumed that the reader has a basic understanding of small amplitude wave theory, a description of which can be found in most textbooks on coastal engineering and physical oceanography.

For practicing engineers, Part II may be somewhat difficult, but perseverance will be rewarded by a glimpse of the theoretical foundation upon which the engineering application of the random wave concept is constructed. However, an understanding of Part II is not necessary, because Part I is self-contained. For those who are pursuing subjects at the forefront of knowledge, in contrast, the discussions and recommendations presented in Part I may sound subjective, or even dogmatic. It should be remembered, however, that engineers often face situations in which they must design structures and produce drawings for construction with minimal background information on the site and limited reliable theoretical machinery for calculation. I have tried to provide guidelines for engineers — guidelines I believe to be the best available at present — in the hope of helping those who are tackling various problems related to sea waves. With advances in knowledge, some of the solutions and recommendations in this book will become obsolete and be replaced by new ones. But so long as this book serves for today's needs, its purpose will be judged to be fulfilled.

Major revisions made for the English edition are the rewriting of Chapter 4 on breakwaters and the addition of new material in Chapters 9 and 10, based on recent studies. Chapter 4 introduces Japan's vertical breakwaters and the procedures for their design, which are probably not well known to engineers outside Japan. The new material in Chapter 9 concerns the theory of wave grouping and the analysis of wave nonlinearity. In Chapter 10, discussions of the directional wave spectrum calculated by means of the maximum-likelihood method and a simulation technique for two-dimensional waves were added.

The material presented in this book mainly derives from research work conducted by the members of the Wave Laboratory of the Port and Harbour Research Institute, Ministry of Transport, Japan, which I headed from 1967 to 1978. I sincerely wish to acknowledge the efforts of the dedicated staff members of the Wave Laboratory in the successful accomplishment of many research projects. I am also indebted to many fellow government engineers who brought to my attention various stimulating problems of a difficult and urgent nature. Often they supported the research projects financially.

The publication of this English edition was suggested and initial arrangements were made by Dr N. C. Kraus of the Nearshore Environment Research Center, Tokyo. I am very grateful to him for the above and for his critical review of the manuscript. As the first reader of the English edition, he helped to clarify ambiguities and to correct many idiomatic expressions. The staff members of the International Publications Department, University of Tokyo Press, were helpful and supportive throughout the publication process. Kajima Institute Publishing Company, publishers of the Japanese-language book, generously lent the illustration plates for use in this edition. I am also grateful to my wife, who patiently allowed me to spend many off-duty hours at home for study and writing of this book. Finally, I wish to acknowledge the financial support of the Ministry of Education, Science and Culture, Japan, for translation of the text.

Yoshimi GODA
Yokosuka, Japan
May 1984

# Contents

*Preface to the Third Edition*                                          v

*Preface to the Second Edition*                                         ix

*Preface to the First Edition*                                          xi

**Part I. Random Sea Waves and Engineering Applications**

1. Introduction                                                        3

   1.1   Waves in the Sea . . . . . . . . . . . . . . . . . . . . . . .   3
   1.2   Overview of Historical Development of Random Waves
        Applications . . . . . . . . . . . . . . . . . . . . . . . . . .   5
   1.3   Outline of Design Procedures Against Random
        Sea Waves . . . . . . . . . . . . . . . . . . . . . . . . . . .   10
        1.3.1   Wave Transformation . . . . . . . . . . . . . . . .   10
        1.3.2   Methods of Dealing with Random Sea Waves . . .   13

2. Statistical Properties and Spectra of Sea Waves                     19

   2.1   Random Wave Profiles and Definitions of Representative
        Waves . . . . . . . . . . . . . . . . . . . . . . . . . . . . .   19
        2.1.1   Spatial Surface Forms of Sea Waves . . . . . . . .   19
        2.1.2   Definition of Representative Wave Parameters . .   21
   2.2   Distributions of Individual Wave Heights and Periods . .   24
        2.2.1   Wave Height Distribution . . . . . . . . . . . . .   24
        2.2.2   Relations Between Representative Wave Heights .   27
        2.2.3   Distribution of Wave Period . . . . . . . . . . .   30
   2.3   Spectra of Sea Waves . . . . . . . . . . . . . . . . . . . . .   31
        2.3.1   Frequency Spectra . . . . . . . . . . . . . . . . .   31
        2.3.2   Directional Wave Spectra . . . . . . . . . . . . .   37

2.4    Relationship Between Wave Spectra and Characteristic
       Wave Dimensions . . . . . . . . . . . . . . . . . . . . . .    48
       2.4.1    Relationship Between Wave Spectra and Wave
                Heights . . . . . . . . . . . . . . . . . . . . . . .    48
       2.4.2    Relationship Between Wave Spectra and Wave
                Periods . . . . . . . . . . . . . . . . . . . . . . .    53
2.5    Long-Period Waves Accompanying Wind Waves and
       Swell . . . . . . . . . . . . . . . . . . . . . . . . . . . .    54

3.  Generation, Transformation and Deformation of Random
    Sea Waves                                                          63

    3.1    Simplified Forecasting Method of Wind Waves and
           Swell . . . . . . . . . . . . . . . . . . . . . . . . . .    63
           3.1.1    Simplified Forecast of Wave Height and Period . .    63
           3.1.2    Simplified Forecast of Swell Height and Period . .   66
           3.1.3    Relationship Between Significant Height and
                    Period of Wind Waves and Swell . . . . . . . . .     67
    3.2    Wave Refraction . . . . . . . . . . . . . . . . . . . . .    68
           3.2.1    Introduction . . . . . . . . . . . . . . . . . . .    68
           3.2.2    Refraction Coefficient of Random Sea Waves . . .     70
           3.2.3    Computation of Random Wave Refraction by
                    Means of the Energy Balance Equation . . . . . .     75
           3.2.4    Wave Refraction on a Coast with Straight, Parallel
                    Depth-Contours . . . . . . . . . . . . . . . . . .    78
    3.3    Wave Diffraction . . . . . . . . . . . . . . . . . . . . .    80
           3.3.1    Principle of Random Wave Diffraction Analysis .      80
           3.3.2    Diffraction Diagrams of Random Sea Waves . . .       82
           3.3.3    Random Wave Diffraction of Oblique Incidence .       89
           3.3.4    Approximate Estimation of Diffracted Height by
                    the Angular Spreading Method . . . . . . . . . .     92
           3.3.5    Application of Regular Wave Diffraction Diagrams     95
    3.4    Equivalent Deepwater Wave . . . . . . . . . . . . . . . .    96
    3.5    Wave Shoaling . . . . . . . . . . . . . . . . . . . . . .    99
    3.6    Wave Deformation Due to Random Breaking . . . . . . .       102
           3.6.1    Breaker Index of Regular Waves . . . . . . . . .    102
           3.6.2    Hydrodynamics of Surf Zone . . . . . . . . . . .    104
           3.6.3    Wave Height Variations on Planar Beaches . . . .    117
           3.6.4    Prediction of Random Wave Breaking Process on
                    Beaches of Complicated Profiles . . . . . . . . .    128

3.7 Reflection of Waves and Their Propagation and
     Dissipation . . . . . . . . . . . . . . . . . . . . . . . . . . . 131
     3.7.1 Coefficient of Wave Reflection . . . . . . . . . . . 131
     3.7.2 Propagation of Reflected Waves . . . . . . . . . . 133
     3.7.3 Superposition of Incident and Reflected Waves . . 137
3.8 Spatial Variation of Wave Height Along Reflective
     Structures . . . . . . . . . . . . . . . . . . . . . . . . . . . . 139
     3.8.1 Wave Height Variation Near the Tip of a
           Semi-Infinite Structure . . . . . . . . . . . . . . . 139
     3.8.2 Wave Height Variation at an Inward Corner of
           Reflective Structures . . . . . . . . . . . . . . . . 141
     3.8.3 Wave Height Variation Along an Island
           Breakwater . . . . . . . . . . . . . . . . . . . . . . 144
3.9 Wave Transmission of Breakwaters and Low-Crested
     Structures . . . . . . . . . . . . . . . . . . . . . . . . . . . . 146
     3.9.1 Wave Transmission of Coefficient of Composite
           Breakwaters . . . . . . . . . . . . . . . . . . . . . . 146
     3.9.2 Wave Transmission Coefficient of Low-Crested
           Structures . . . . . . . . . . . . . . . . . . . . . . . 148
     3.9.3 Propagation of Transmitted Waves in a Harbor . 153

4. Design of Vertical Breakwaters                              161
4.1 Overview of Vertical and Composite Breakwaters . . . . . 161
4.2 Wave Pressures Exerted on Upright Sections . . . . . . . 168
     4.2.1 Overview of Development of Wave Pressure
           Formulas . . . . . . . . . . . . . . . . . . . . . . . . 168
     4.2.2 Goda Formulas of Wave Pressure Under a Wave
           Crests . . . . . . . . . . . . . . . . . . . . . . . . . . 170
     4.2.3 Impulsive Breaking Wave Pressure and
           Its Estimation . . . . . . . . . . . . . . . . . . . . . 180
     4.2.4 Sliding of Upright Section by Single Wave Action 185
4.3 Preliminary Design of Upright Sections . . . . . . . . . . 188
     4.3.1 Stability Condition for an Upright Section . . . . 188
     4.3.2 Stable Width of Upright Section . . . . . . . . . . 190
4.4 Several Design Aspects of Composite Breakwaters . . . . 194
     4.4.1 Wave Pressure Under a Wave Trough . . . . . . . 194
     4.4.2 Uplift on a Large Footing . . . . . . . . . . . . . . 197
     4.4.3 Wave Pressure on Horizontally-Composite
           Breakwaters and Other Special Breakwaters . . . 198
     4.4.4 Comments on Design of Concrete Caissons . . . . 200

4.5    Design of Rubble Mound Foundation of Composite
       Breakwaters . . . . . . . . . . . . . . . . . . . . . . . . . . .   201
       4.5.1   Dimensions of Rubble Mound  . . . . . . . . . . .   201
       4.5.2   Foot-Protection Blocks  . . . . . . . . . . . . . .   202
       4.5.3   Protection Against Scouring of the Seabed in Front
               of a Breakwater  . . . . . . . . . . . . . . . . . .   206

5.  Design of Coastal Dikes and Seawalls                              211

    5.1    Random Wave Run-Up on Coastal Dikes and Seawalls  . .   211
           5.1.1   Run-Up Height by Standing Waves at Vertical
                   Wall . . . . . . . . . . . . . . . . . . . . . . . . . . .   211
           5.1.2   Run-Up Height on Smooth Slopes and Coastal
                   Dikes . . . . . . . . . . . . . . . . . . . . . . . . . .   212
    5.2    Wave Overtopping Rate of Coastal Dikes and Seawalls  . .   216
           5.2.1   Overtopping Rate by Random Sea Waves . . . . .   216
           5.2.2   Mean Rate of Wave Overtopping at Vertical and
                   Block-Mound Seawalls  . . . . . . . . . . . . . .   218
           5.2.3   Mean Rate of Wave Overtopping at Coastal Dikes
                   of Plane Slope . . . . . . . . . . . . . . . . . . .   225
           5.2.4   Unified Formulas for Wave Overtopping Rate of
                   Vertical and Inclined Seawalls  . . . . . . . . . .   229
    5.3    Influence of Various Factors on Wave Overtopping Rate  .   233
    5.4    Tolerable Rate of Wave Overtopping and Determination of
           Crest Elevation . . . . . . . . . . . . . . . . . . . . . . . .   238
           5.4.1   Design Principles for the Determination of Crest
                   Elevation . . . . . . . . . . . . . . . . . . . . . .   238
           5.4.2   Tolerable Rate of Wave Overtopping  . . . . . . .   239
           5.4.3   Examples of Determining Crest Elevation of
                   Seawalls  . . . . . . . . . . . . . . . . . . . . . .   241
    5.5    Additional Design Problems Related to Seawalls  . . . . .   246

6.  Probabilistic Design of Breakwaters                               253

    6.1    Uncertainty of Design Values . . . . . . . . . . . . . . .   253
           6.1.1   Overview . . . . . . . . . . . . . . . . . . . . . .   253
           6.1.2   Examples of Uncertainty of Design Parameters for
                   Breakwater Design  . . . . . . . . . . . . . . . .   255
           6.1.3   Uncertainty of Offshore Significant Wave Height  .   258

6.2    Reliability-Based Design of Breakwater . . . . . . . . . .   260
       6.2.1   Classification of Reliability-Based Design
               Method . . . . . . . . . . . . . . . . . . . . . . .   260
       6.2.2   Evaluation of External Safety by Level II
               Method . . . . . . . . . . . . . . . . . . . . . . .   261
       6.2.3   Design of Breakwaters with Partial Factor
               System . . . . . . . . . . . . . . . . . . . . . . .   270
6.3    Performance-Based Design of Breakwaters . . . . . . . .   274
       6.3.1   Outline of Performance-Based Design Method  . .   274
       6.3.2   Performance-Based Design with Expected Sliding
               Distance Method . . . . . . . . . . . . . . . . . .   276
       6.3.3   Vertical Breakwater Design with Modified Level I
               Method . . . . . . . . . . . . . . . . . . . . . . .   285

7. Harbor Tranquility                                               291

7.1    Parameters Governing Harbor Tranquility . . . . . . . .   291
7.2    Estimation of the Probability of Wave Height Exceedance
       Within a Harbor . . . . . . . . . . . . . . . . . . . . . .   294
       7.2.1   Estimation Procedure . . . . . . . . . . . . . . .   294
       7.2.2   Joint Distribution of Significant Wave Height,
               Period and Direction Outside a Harbor . . . . . .   296
       7.2.3   Selection of the Points for the Wave Height
               Estimation . . . . . . . . . . . . . . . . . . . . .   298
       7.2.4   Estimation of Wave Height in a Harbor Incident
               Through an Entrance . . . . . . . . . . . . . . . .   298
       7.2.5   Estimation of Waves Transmitted Over a
               Breakwater . . . . . . . . . . . . . . . . . . . . .   300
       7.2.6   Estimation of the Exceedance Probability of Wave
               Height Within a Harbor . . . . . . . . . . . . . .   301
       7.2.7   Estimation of Storm Wave Height in a Harbor . .   304
7.3    Graphical Solution of the Distribution of Wave Height in
       a Harbor . . . . . . . . . . . . . . . . . . . . . . . . . .   305
7.4    Some Principles for Improvement of Harbor Tranquility  .   309
7.5    Motions of Ships at Mooring . . . . . . . . . . . . . . .   314
       7.5.1   Modes and Equations of Ship Motions . . . . . . .   314
       7.5.2   Ship Mooring and Natural Frequency of Ship
               Mooring System . . . . . . . . . . . . . . . . . .   317
       7.5.3   Time-Domain Analysis of Moored Ships . . . . . .   319
       7.5.4   Some Remarks on Ship Mooring . . . . . . . . . .   321

7.6     Allowable Ship Movements and Mitigation of Mooring
        Troubles . . . . . . . . . . . . . . . . . . . . . . . . . . . .   322
        7.6.1   Allowable Movements of Moored Ships  . . . . . .   322
        7.6.2   Mitigation of Mooring Troubles  . . . . . . . . . .   323

8.  Hydraulic Model Tests with Random Waves                           331

    8.1     Similarity Laws and Scale Effects . . . . . . . . . . . . .   331
            8.1.1   Selection of Model Scales with the Froude
                    Similarity Law . . . . . . . . . . . . . . . . . . . .   331
            8.1.2   Possible Scale Effects in Model Tests  . . . . . . .   333
    8.2     Necessity of Hydraulic Model Tests with Random Waves .   335
    8.3     Generation of Random Waves in Test Basins  . . . . . . .   337
            8.3.1   Random Wave Generator . . . . . . . . . . . . .   337
            8.3.2   Preparation of Input Signal to the Generator . . .   340
            8.3.3   Input Signals to a Multidirectional Random Wave
                    Generator  . . . . . . . . . . . . . . . . . . . . . .   343
            8.3.4   Non-Reflective Wave Generator  . . . . . . . . . .   344
            8.3.5   Other Topics on Wave Generation in Test Flumes   344
    8.4     Model Tests Using Multidirectional Radom Waves  . . . .   347
    8.5     Some Remarks on Execution of Random Wave Tests . . .   348
            8.5.1   Number of Test Runs and Their Durations . . . .   348
            8.5.2   Calibration of Test Waves  . . . . . . . . . . . . .   349
            8.5.3   Resolution of Incident and Reflected Waves in a
                    Test Flume . . . . . . . . . . . . . . . . . . . . . .   350
            8.5.4   Statistical Variability of Damage Ratio of Armor
                    Units . . . . . . . . . . . . . . . . . . . . . . . . .   351

**Part II. Statistical Theories of Random Sea Waves**

9.  Description of Random Sea Waves                                   357

    9.1     Profiles of Progressive Waves and Dispersion Relationship   357
    9.2     Description of Random Sea Waves by Means of Variance
            Spectrum  . . . . . . . . . . . . . . . . . . . . . . . . . .   360
    9.3     Stochastic Process and Variance Spectrum . . . . . . . .   362

10.  Statistical Theory of Irregular Waves                             369

    10.1    Distribution of Wave Heights . . . . . . . . . . . . . . .   369
            10.1.1  Envelope of Irregular Wave Profile . . . . . . . .   369
            10.1.2  The Rayleigh Distribution of Wave Heights . . . .   371
            10.1.3  Probability Distribution of Largest Wave Height .   376

10.2   Wave Grouping . . . . . . . . . . . . . . . . . . . . . . . . 379
     10.2.1   Wave Grouping and Its Quantitative Description . 379
     10.2.2   Probability Distribution of Run Length for
            Uncorrelated Waves . . . . . . . . . . . . . . . . . 382
     10.2.3   Correlation Coefficient Between Successive Wave
            Heights . . . . . . . . . . . . . . . . . . . . . . . . 383
     10.2.4   Theory of Run Length for Mutually Correlated
            Wave Heights . . . . . . . . . . . . . . . . . . . . 388
10.3   Distribution of Wave Periods . . . . . . . . . . . . . . . . 391
     10.3.1   Mean Period of Zero-Upcrossing Waves . . . . . . 391
     10.3.2   Marginal Distribution of Wave Periods and Joint
            Distribution of Wave Heights and Priods . . . . . 393
10.4   Maxima of Irregular Wave Profiles . . . . . . . . . . . . . 402
10.5   Nonlinearity of Sea Waves . . . . . . . . . . . . . . . . . . 407
     10.5.1   Nonlinearity of Surface Elevation . . . . . . . . . 407
     10.5.2   Effects of Wave Nonlinearity on Characteristic
            Wave Heights and Periods . . . . . . . . . . . . . 414
     10.5.3   Nonlinear Components of Wave Spectrum . . . . 419
10.6   Sampling Variability of Sea Waves . . . . . . . . . . . . . 424

11.   Techniques of Irregular Wave Analysis             433

11.1   Statistical Quantities of Wave Data . . . . . . . . . . . . . 433
     11.1.1   Analysis of Analog Data . . . . . . . . . . . . . . 433
     11.1.2   Analysis of Digital Data . . . . . . . . . . . . . . 435
11.2   Frequency Spectral Analysis of Irregular Waves . . . . . . 440
     11.2.1   Theory of Spectral Analysis . . . . . . . . . . . . 440
     11.2.2   Spectral Estimate with Smoothed Periodograms . 447
11.3   Directional Spectral Analysis of Random Sea Waves . . . 453
     11.3.1   Relation Between Directional Spectrum and
            Covariance Function . . . . . . . . . . . . . . . . . 454
     11.3.2   Estimate of Directional Spectrum with a Wave
            Gauge Array . . . . . . . . . . . . . . . . . . . . . 456
     11.3.3   Estimate of Directional Wave Spectra with a
            Directional Buoy or with a Two-Axis Current
            Meter . . . . . . . . . . . . . . . . . . . . . . . . . 464
     11.3.4   Advanced Theories of Directional Spectrum
            Estimates . . . . . . . . . . . . . . . . . . . . . . . 467

11.4　Resolution of Incident and Reflected Waves of Irregular
　　　Profiles . . . . . . . . . . . . . . . . . . . . . . . . . . . 472
　　　11.4.1　Measurement of the Reflection Coefficient in a
　　　　　　Wave Flume . . . . . . . . . . . . . . . . . . . . . 472
　　　11.4.2　Measurement of the Reflection Coefficient of
　　　　　　Prototype Structures . . . . . . . . . . . . . . . 476
11.5　Numerical Simulation of Random Sea Waves and
　　　Numerical Filters . . . . . . . . . . . . . . . . . . . . . 478
　　　11.5.1　Principles of Numerical Simulation . . . . . . . . 478
　　　11.5.2　Methods of Numerical Simulation . . . . . . . . 479
　　　11.5.3　Pseudo-Random Number Generating Algorithm . 484
　　　11.5.4　Numerical Filtering of Wave Record . . . . . . . 485

12. 2-D Computation of Wave Transformation with Random
　　Breaking and Nearshore Currents　　　　　　　　　　491

12.1　Overview of Numerical Computation Models for 2-D Wave
　　　Transformations . . . . . . . . . . . . . . . . . . . . . . 491
12.2　Outline of Phase-Averaged Type Wave Transformation
　　　Models . . . . . . . . . . . . . . . . . . . . . . . . . . . 494
12.3　Outline of Phase-Resolving Type Wave Transformation
　　　Models . . . . . . . . . . . . . . . . . . . . . . . . . . . 497
12.4　Wave Transformation Analysis with PEGBIS Model . . . 501
12.5　Outline of Numerical Computation of Nearshore
　　　Currents . . . . . . . . . . . . . . . . . . . . . . . . . . 511
12.6　Prediction of Wave Setup and Longshore Currents on
　　　Planar Beaches . . . . . . . . . . . . . . . . . . . . . . 520

**Part III. Statistical Analysis of Extreme Waves**

13. Statistical Analysis of Extreme Waves　　　　　　　　　537

13.1　Introduction . . . . . . . . . . . . . . . . . . . . . . . . 537
　　　13.1.1　Data for Extreme Wave Analysis . . . . . . . . . 537
　　　13.1.2　Distribution Functions for Extreme Data
　　　　　　Analysis . . . . . . . . . . . . . . . . . . . . . . 539
　　　13.1.3　Characteristics of Selected Distribution
　　　　　　Functions . . . . . . . . . . . . . . . . . . . . . 542
　　　13.1.4　Return Period and Return Value . . . . . . . . . 545
　　　13.1.5　Spread Parameter of Distribution Functions . . . 546

13.2 Estimation of Best-Fitting Distribution Function . . . . . 549
  13.2.1 Selection of Plotting Position . . . . . . . . . . . 549
  13.2.2 Estimation of Return Values with the Least
         Squares Method . . . . . . . . . . . . . . . . . . . 553
  13.2.3 Selection of Most Probable Parent Distribution . 559
  13.2.4 Rejection of Distribution Function . . . . . . . . 563
13.3 Confidence Interval of Return Value . . . . . . . . . . . 568
  13.3.1 Statistical Variability of Samples of Extreme
         Distributions . . . . . . . . . . . . . . . . . . . . 568
  13.3.2 Confidence Interval of Parameter Estimates . . . 570
  13.3.3 Confidence Interval of Return Value . . . . . . . 571
13.4 Several Topics on Extreme Wave Statistics . . . . . . . . 579
  13.4.1 Treatment of Mixed Populations . . . . . . . . . 579
  13.4.2 Encounter Probability . . . . . . . . . . . . . . . 581
  13.4.3 *L*-Year Maximum Wave Height and Its Confidence
         Interval . . . . . . . . . . . . . . . . . . . . . . . 582
13.5 Design Waves and Related Problems . . . . . . . . . . . 586
  13.5.1 Database of Extreme Waves and Its Analysis . . . 586
  13.5.2 Selection of Design Wave Height and Period . . . 590

**Part IV. Waves and Beach Morphology**

14. Coastline Change and Coastal Reconnaissance                599

14.1 Introduction . . . . . . . . . . . . . . . . . . . . . . . . 599
14.2 Overview of Historical Coastline Change . . . . . . . . . 600
  14.2.1 Geological View of Coastline Change . . . . . . . 600
  14.2.2 Geological Features of Sandy Coast . . . . . . . . 603
  14.2.3 Natural Process of Shoreline Change . . . . . . . 607
  14.2.4 Examples of Estimated Rate of Littoral Sediment
         Transport . . . . . . . . . . . . . . . . . . . . . . 609
14.3 Anthropogenic Influence on Coastal Morphology . . . . . 613
  14.3.1 Outline . . . . . . . . . . . . . . . . . . . . . . . 613
  14.3.2 Typical Cases of Significant Shoreline Recession by
         Anthropogenic Influence . . . . . . . . . . . . . . 614
  14.3.3 Patterns of Shoreline Changes Caused by
         Structure Construction on Sandy Coast . . . . . . 618
14.4 Coastal Reconnaissance for Beach Protection Project . . . 624
  14.4.1 Collection and Analysis Documents Before Coastal
         Reconnaissance . . . . . . . . . . . . . . . . . . . 624

14.4.2   Field Inspection Works . . . . . . . . . . . . . . .   629

14.4.3   Guidelines for Search of Sediment Supply and
         Longshore Transport Direction . . . . . . . . . .   635

15. Prediction and Control of Shoreline Evolution          643

15.1   Introduction . . . . . . . . . . . . . . . . . . . . . . . .   643

15.2   State of the Art of Beach Morphological Prediction . . . .   644

15.2.1   Overview of Studies on Sediment Movement  . . .   644

15.2.2   Problems Inherent to Beach Morphological
         Prediction . . . . . . . . . . . . . . . . . . . . . . .   648

15.2.3   Sediment Suspension Rate in Surf Zone . . . . . .   652

15.2.4   Fall Velocity and Equilibrium Beach Profile . . . .   655

15.3   Overview of Beach Morphology Models  . . . . . . . . .   657

15.3.1   Prediction Formulas for Longshore Sediment
         Transport  . . . . . . . . . . . . . . . . . . . . . . .   657

15.3.2   Shoreline Change Models . . . . . . . . . . . . .   661

15.3.3   Numerical Models for 3-D Beach Deformation  . .   665

15.3.4   Quantitative Assessment of Sediment
         Impoundment by Groin . . . . . . . . . . . . . . .   667

15.4   A Case Study of Shoreline Change Prediction by One-Line
       Model . . . . . . . . . . . . . . . . . . . . . . . . . . . .   670

15.5   Overview of Shore Protection Facilities . . . . . . . . . . .   676

*Appendix*                                                   687

*Author Index*                                               693

*Subject Index*                                              699

Part I

# Random Sea Waves
# and Engineering Applications

# Chapter 1

# Introduction

## 1.1 Waves in the Sea

Engineers build various types of maritime structures. Breakwaters and quaywalls for ports and harbors, seawalls and jetties for shore protection, and platforms and rigs for the exploitation of oil beneath the seabed are some examples. These structures must perform their functions in the natural environment, being subjected to the hostile actions of winds, waves, tidal currents, earthquakes, etc. To ensure their designated performance, we must carry out comprehensive investigations in order to understand the environmental conditions. The investigations must be as accurate as possible so that we can rationally assess the effects of the environment on our structures.

Waves are the most important phenomenon to be considered among the environmental conditions affecting maritime structures, because they exercise the greatest influence. The presence of waves makes the design procedure for maritime structures quite different from that of structures on land. Since waves are one of the most complex and changeable phenomena in nature, it is not easy to achieve a full understanding of their fundamental character and behavior.

Waves have many aspects. They appear as the wind starts to blow, grow into mountainous waves amid storms and completely disappear after the wind ceases blowing. Such changeability is one aspect of the waves. An observer on a boat in the offshore region easily recognizes the pattern of wave forms as being made up of large and small waves moving in many directions. The irregularity of wave form is an important feature of waves in the sea. However, upon reaching the shore, an undulating swell breaks as individual waves, giving the impression of a regular repetition. Yuzo

3

Yamamoto, in his novel *Waves*, sees an analogy between successive waves and a son's succession to the father.

The generation of waves on a water surface by wind and their resultant propagation has been observed throughout history.[a] However, the mathematical formulation of the motion of water waves was only introduced in the 19th century. In 1802, Gerstner, a mathematician in Prague, published the trochoidal wave theory for waves in deep water, and in 1844, Airy in England developed a small amplitude wave theory covering the full range of water depth from deep to shallow water. Thereafter, in 1847, Stokes gave a theory of finite amplitude waves in deep water, which was later extended to waves in intermediate-depth water. This solution is now known as the Stokes wave theory. The existence of a solitary wave which has a single crest and propagates without change of form in shallow water was reported by Russell in 1844. Its theoretical description was given by Boussinesq in 1871 and Rayleigh in 1876. Later, in 1895, Korteweg and de Vries derived a theory of permanent periodic waves of finite amplitude in shallow water. This is now known as the cnoidal wave theory.

Thus, the fundamental theories of water waves were established by the end of the 19th century. Nevertheless, several decades had passed before civil engineers were able to make full use of these theories in engineering applications. An exception is the theory of standing wave pressure derived in 1928 by Sainflou,[2] an engineer at Marseille Port. Sainflou's work attracted the attention of harbor engineers soon after publication; his pressure formula was adopted in many countries for the design of vertical breakwaters. It should be mentioned, however, that it was during the Second World War when the mathematical theory and engineering practice was successfully combined together. This led to the formation of the discipline of coastal engineering, which can be said to have begun with the wave forecasting method introduced by Sverdrup and Munk;[3] this later evolved into the more sophisticated S–M–B method, the calculation of wave diffraction by a breakwater developed by Penney and Price,[4] and other milestone developments.

In proposing the foundation for the present S–M–B method, Sverdrup and Munk clearly understood that sea waves are composed of large and small waves. They introduced the concept of the *significant wave*, the height of which is equal to the mean of the heights of the highest one-third waves in a wave group, as representative of a particular sea state. Therefore, the

---

[a]The following historical overview of the study of water waves is based on the literature listed by Lamb.[1]

significant wave concept was based upon the understanding of sea waves as a random process. However, the significant wave, expressed in terms of a single wave height and wave period, is sometimes misunderstood by engineers to represent waves of constant height and period. The theory of monochromatic waves and experimental results obtained from a train of regular waves have been directly applied to prototype problems in the real sea on the belief that the regular waves correspond exactly to the significant wave.

As early as in 1952, a group of American oceanographers, headed by Pierson,[5] took the first step in recognizing the irregularity of ocean waves as a fundamental property and incorporating this fact in the design process. The so-called P–N–J method[6] of wave forecasting, often compared with the S–M–B method, introduced the concept of wave spectrum as the basic tool for describing wave irregularity. The generation and development of wind waves, the propagation of swell and wave transformation near the shore were all explained in detail via the concept of wave spectrum. Although the spectral concept was readily accepted by oceanographers at an early stage, coastal and harbor engineers with the exception of a few researchers considered it too complicated. Hence, the introduction of spectral computation techniques into the design process for coastal structures was much delayed. The situation began to change gradually since the 1970s. In the next subsection, an overview is given on the development of random wave concept and its engineering applications in the field of coastal engineering, which are discussed in detail in the chapters to follow in the present book.

## 1.2   Overview of Historical Development of Random Wave Applications

(A) *Statistical description of random seas*

At the early stage of coastal engineering, sea waves were measured with pressure gauges mounted on the seabed. The amplitudes of individual oscillations of wave pressure were converted to surface amplitudes with the small amplitude wave theory by using the respective wave periods. Statistical distributions of wave heights and periods thus obtained were investigated eagerly. For example, Putz[7] reported in 1952 that the mean ratio of the significant to the mean wave heights is 1.57, that of the one-tenth to the significant wave height is 1.29, and the wave height distribution can be approximated with the Pearson III type distribution. In the same year, however, Longuet-Higgins[8] verified the applicability of the Rayleigh distribution to sea waves and presented the theoretical relationships among various wave heights. Since then, the Rayleigh distribution of wave heights has been widely accepted.

The basic assumption of the Rayleigh distribution is that the wave spectrum is confined in a narrow frequency band. The spectrum of sea waves is broad-banded as evidenced through many field observations, however, and the question on the wave height distribution of broad-band spectrum arose soon. Most of field data yielded the wave height distribution close to the Rayleigh so long as individual waves are defined with the zero-upcrossing or zero-downcrossing method regardless of the spectral shape. The first approach to the question of wave height distribution under broad-band spectra was made by the author[9] in 1970 by means of the numerical simulation of wave profiles under various spectral shape. The simulation work confirmed the above finding through field observations.

The distribution of wave period is discussed in the form of the joint distribution with the wave height; the marginal distribution of wave period alone excites little interest from the practical viewpoint. Longuet-Higgins[10] presented the first theory of the joint distribution of wave height and period in 1975, followed by several researchers. Though the period distribution is affected by the spectral shape, the periods of individual large waves are clustered around the significant wave period.

## (B) *Wave transformation analysis with wave spectrum*

Until the late 1960s, coastal engineers used to analyze wave transformations and actions by means of the regular waves, the height and period of which are equal to the significant wave height and period, respectively. Waves are thought to come from a single direction. Computer drawing of wave rays for wave refraction analysis was quite common. A breakthrough was made by Karlsson[11] in 1969, who introduced a numerical computation method for the spectral wave shoaling and refraction. The method was an application of the spectral wave forecasting method of the first generation, which had began to replace the S–M–B method in the late 1960s.

Karlsson's method, which is based on the energy balance equation, was immediately applied for an actual harbor planning by Nagai *et al.*[12] in 1974 and the method soon became a design tool among Japanese coastal engineers. Contrarily, American and European engineers did not seem to appreciate the spectral method at that time, probably because of unfamiliarity with the concept of directional wave spectrum. Measurement and analysis of directional wave spectra had been promoted by oceanographers since the late 1950s, and coastal engineers joined in the efforts from the 1970s.

In 1975, Mitsuyasu *et al.*[13] proposed a formulation of the directional spreading function based on a number of detailed ocean measurements. Goda and Suzuki[14] immediately adopted it as the standard functional form of the directional spectrum for engineering applications under the name of the Mitsuyasu-type directional spreading function. They made use of this directional spectrum for computing and presenting a full set of directional random diffraction diagrams. Goda *et al.*[15] presented several sets of these diagrams at the 16th International Conference on Coastal Engineering in 1978, but it did not awake the response of the audience.

The paper on the laboratory measurements of the refraction and diffraction of directional waves over an elliptical shoal by Vincent and Briggs[16] in 1989 seems to have arisen the interests of American and European engineers on directional spectral waves. A number of numerical schemes for directional spectral wave transformations have been presented since then. Holthuijsen *et al.*[17] developed the SWAN model in 1993 for the transformation of directional random waves in presence of currents. The model is a shallow water version of the spectral wave forecasting model of the third generation, WAM.

### (C) *Random wave breaking and surf zone hydrodynamics*

While regular waves in a laboratory flume break at a fixed location with almost the same height, random waves break at various locations over quite a distance, which is called the surf zone. At a beach having sand bars offshore, the outer bar usually defines the edge of the surf zone. Otherwise, an observer needs to define the area of the surf zone somewhat subjectively.

Quantitative evaluation of wave decay within the surf zone was initiated by Collins[18] in 1970, who truncated the Rayleigh distribution of wave height beyond the height exceeding the breaking limit of regular waves. Truncation of the wave height distribution yielded a decrease of the total wave energy, thus approximating the wave decay process within the surf zone. Several modifications of this model followed, including the author's model[19] in 1975 to be discussed in Sec. 3.6.

In 1978, Battjes and Janssen[20] proposed another modeling approach with estimation of the percentage of breaking waves and evaluation of energy dissipation with the analogy of a bore in hydraulic jump. Though predictability of the model is limited to the variation of the root-mean-square wave height only, it can be applied to beaches of arbitrary bathymetry and thus it has been used as an engineering tool by European engineers.

The third approach to wave energy dissipation by breaking was proposed by Dally *et al.*[21] in 1985, who assumed the rate of energy dissipation being proportional to the difference between the energy level of the local waves and that of the stable waves after breaking when the former exceeds the latter. The model was initially developed for regular waves, but it has been applied to individual random waves by several researchers. One of such models will be presented in Sec. 12.4.

The surf zone is characterized with the spatial variation of local mean water level, called the *wave setup* and *wave setdown*, and the generation of nearshore currents, both of which are induced by the spatial gradients of radiation stresses associated with the wave momentum fluxes. Different behaviors of regular and random waves with respect to wave breaking bring forth quite different results of the local mean water level and nearshore currents. Although Battjes[22] demonstrated such differences in 1972, the majority of coastal engineers have been late in recognizing the difference. Even in the 2000s, there are some papers dealing with wave setup and nearshore currents induced by regular waves. This aspect will be discussed in Sec. 12.3.

### (D) *Wave actions on structures*

It has been the established practice that offshore structures for oil exploitation are designed against the wave of maximum height among individual waves, because a single action of the largest wave loading may causes collapse of a structure. Composite breakwaters with upright sections are also designed against the maximum wave. The author's wave pressure formulas[23,24] presented in 1973 has been utilized for composite breakwater designs in Japan since the late 1970s. Interests in caisson breakwaters were arisen in Europe after the collapse of a deepwater mound breakwater at Sines Port in 1978, and a number of joint research projects on the design methodology of caisson breakwaters have been carried out under the EU science and technology sponsorship. Goda's formulas have served as a guiding reference in these studies.

Before the late 1970s, the stability of armor units of mound breakwaters had been tested by using regular waves. A question of how to convert the regular wave height in the tests to the representative height of random waves such as the significant height or the highest one-tenth height was hotly debated by researchers and practitioners. Wide spread of irregular wave machines in laboratory flumes since the late 1970s changed the situation by making irregular wave tests as the standard procedure for breakwater stability analysis. The test results are now presented in terms of the significant wave height. Sometimes the 2% exceedance wave height is used to describe the test results when there is deviation in the wave height distribution from the Rayleigh.

The crest elevation of a coastal dike is often designed such that it is equal to or higher than the 2% exceedance level of wave run-up. The practice was introduced by Dutch engineers when they designed the reclamation dike of the North East Polder in 1936 within the Ijsselmer that was created by closing the Zeidel Sea in 1932.[25] Use of the 2% run-up height for coastal dike design seems to be common in European countries.

Coastal dikes and seawalls are also designed against the mean discharge of overtopping under the design wave and water level conditions. In 1968, Tsuruta and Goda[26] introduced a concept of the expected wave overtopping rate, which synthesized the data of regular wave overtopping rate by averaging them with the weight of the probability density function of wave heights under the Rayleigh distribution. Then, Goda *et al.*[27] presented the design diagrams of the expected wave overtopping rate of vertical seawalls and sloped seawalls with concrete block mounds based on series of irregular wave tests in 1975. They were published in Japanese and remained unknown outside Japan until the publication of the first edition of the present book in 1985. In UK, Owen[28] presented the first overtopping formula in 1980 based on irregular wave tests. Since then, many studies on irregular wave overtopping have been carried out in various countries.

(E) *Methodology of laboratory tests using random waves*

Reproduction of random waves in laboratory basins requires a wave generator of the servo-control system that drives the wave paddle according to the control signal. Such irregular wave machines were first introduced in ship testing tanks in the early 1960s, and soon hydraulic laboratories in UK and Denmark equipped their test basins with such machines for harbor tranquility tests. Hydraulic laboratories in other countries also followed the trend, and irregular wave tests gradually became a standard procedure in coastal engineering studies.

One of the early problems encountered in irregular wave tests was the multireflection of waves between a model under test and the wave paddle. A regular wave test can be carried out by finishing measurements before the first wave reflected by the model and re-reflected by the paddle would reach the model again. An irregular wave test must take measurements of a few hundred waves and thus has to accept the multireflected waves in the measurement records. A solution was given by Goda and Suzuki[29] in 1976, who introduced the algorithm for resolving the incident and reflected waves from the wave records at two neighboring stations. The method was immediately accepted as one of the standard test procedures in hydraulic laboratories in the world.

Laboratory generation of multidirectional random waves was initiated by Salter[30] in the late 1970s, when he worked for development of a wave power device made of rotational floats. Many hydraulic laboratories adopted Salter's concept and developed their own multidirectional random wave generators for their test basins. By the early 1990s, such devices became the standard laboratory equipment in many countries. Advancement of random wave studies and progress in reliable design methods against random seas owe to the improvement in laboratory testing methods with random wave generators since the 1970s.

## 1.3    Outline of Design Procedures Against Random Sea Waves

### 1.3.1    *Wave Transformation*

A prerequisite for the reliable estimation of wave actions on maritime structures is a detailed understanding of how waves transform during their propagation toward the shore, after they have been generated and developed by the wind in the offshore region. The various types of wave transformations are schematically shown in Fig. 1.1.[31]

First, wind waves become swell when they move out of the generating area. The height of the swell gradually decreases with distance as it propagates (② in Fig. 1.1). When wind waves and swell encounter an island or a headland during their propagation in deep water, they are diffracted and penetrate into the area behind the obstacle (④ in Fig. 1.1).

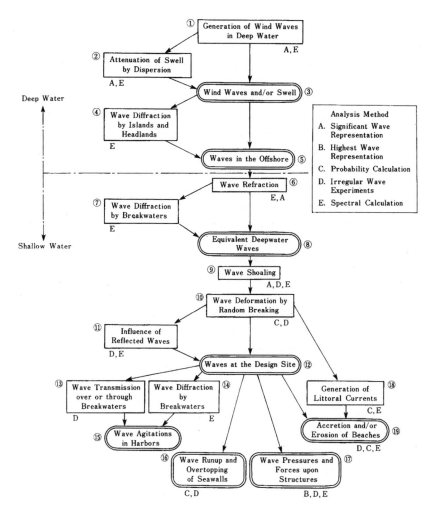

Fig. 1.1 Flow of the transformations and actions of sea waves with suggested methods for their calculation.[7]

When waves enter an area of depth less than about one-half of their wavelength, they are influenced by the seabed topography. These waves are called *intermediate-depth water waves*. Where the water is shallower than about one-twentieth a wavelength, the waves are called *long waves*. For the sake of simplicity, both intermediate-depth water waves and long waves may be classified together and called *shallow water waves*; this terminology is employed in the present chapter. Waves in an area having a depth greater than about one-half of a wavelength are called *deepwater waves* (⑤ in Fig. 1.1). In a shallow bay or estuary, wind-generated waves may become shallow water waves during their process of development. In such a case, wave forecasting or hindcasting needs to take into account the effect of water depth on the wave development.

Having propagated into a shallow region, waves undergo *refraction* by which the direction of wave propagation, as well as the wave height, varies according to the seabed topography (⑥ in Fig. 1.1). Concerning waves inside a harbor, the phenomenon of *diffraction* by breakwaters is the governing process (⑦ in Fig. 1.1). Although not shown in Fig. 1.1, wave attenuation due to bottom friction and other factors may not be neglected in an area of relatively shallow water which extends over a great distance with a very gentle inclination in the sea bottom. For convenience, the change in wave height due to wave refraction, diffraction and attenuation is often incorporated into the concept of *equivalent deepwater waves* (⑧ in Fig. 1.1). Equivalent deepwater waves are assigned the same wave period as deepwater waves, but the wave height is adjusted to account for the spatial change due to wave refraction, diffraction and attenuation.

Waves propagating in a shallow region gradually change their height as a result of the change in the rate of energy flux due to the reduction in water depth, even if no refraction takes place. This is the phenomenon of *wave shoaling* (⑨ in Fig. 1.1). When waves reach an area of depth less than a few times the significant wave height, waves of greater height in a wave train begin to break one by one and the overall wave height decreases as the wave energy is dissipated. This is the wave deformation caused by *random wave breaking* (⑩ in Fig. 1.1).

Waves arriving at the site of a proposed structure (⑫ in Fig. 1.1) will experience the above transformations and deformations of wave refraction, diffraction, shoaling, breaking, etc. If a long extension of vertical breakwater is already located in the adjacent water area, or if the design site is within a harbor, the influence of waves reflected from neighboring structures must be added to that of the waves arriving directly from offshore (⑪ in Fig. 1.1). These wave transformations and deformations will be discussed in detail in Chapter 3. Once the characteristics of the waves at the site have been estimated, design calculations can be made according to the nature of the problem. For example, the problem of harbor tranquility requires the analysis of waves transmitted over or through breakwaters (⑬

in Fig. 1.1), waves diffracted through harbor entrances (⑭ in Fig. 1.1) and waves reflected within a harbor. These will be discussed in Chapter 7.

The planning of seawalls and coastal dikes to stand against storm waves requires the estimation of wave run-up and wave overtopping rate. Technical information for such problems is usually provided by conducting a scale model test in the laboratory or a set of design diagrams based on laboratory data. In such cases, the flow of calculation takes a jump from ⑧ to ⑯ in Fig. 1.1, because these data are prepared using the parameter of equivalent deepwater waves by directly incorporating the effects of transformations ⑨ and ⑩ into the data. Problems related to wave run-up and overtopping are discussed in Chapter 5.

In the design of breakwaters, the magnitude of the wave pressure is the focal point that requires an appropriate selection of calculation formulas. Chapter 4 discusses the formulas for wave pressure and their applications. It is remarked here that littoral drift, which is associated with beach erosion and accretion, is closely related to wave deformation by breaking and is induced by the resultant longshore currents (⑱ in Fig. 1.1). Prediction of longshore currents is discussed in Chapter 12, while beach morphology problems are dealt with in Chapters 14 and 15.

In the process of calculating the wave transformations and deformations described above, the significant wave height and period are used as indices of the magnitude of random sea waves. These parameters are converted to those of the highest waves or other descriptive waves whenever necessary, as in the case of a wave pressure calculation. Thus, at first sight, this procedure may look the same as the conventional design procedure, for which the significant wave is regarded as a train of regular waves. In the present treatise, however, the effects of wave randomness are accounted for the estimation of the respective wave transformations, and the resultant height of the significant wave after such transformation often takes a value considerably different from that obtained with the regular wave approach.

### 1.3.2 *Methods of Dealing with Random Sea Waves*

At present, the following five methods are available to deal with the transformation and action of random sea waves:

(A) Significant wave representation method
(B) Highest wave representation method

(C)  Probability calculation method

(D)  Irregular wave test method

(E)  Spectral calculation method

The *significant wave representation method* makes use of a train of regular waves with height and period equal to those of the significant wave
as representative of random sea waves. Transformations of sea waves are
estimated with the data of regular waves on the basis of theoretical calculation and/or laboratory experiments. The method had has widely been
employed in the field of coastal engineering since the introduction of the
significant waves as the basis of the S–M–B method for wave forecasting. It
has the merits of easy understanding and simple application, but it also has
the demerit of containing a possibly large estimation error, depending on
the type of wave phenomenon being analyzed. Diffraction, to be discussed
in Sec. 3.3, is an example: the wave height behind a single breakwater may
be estimated to be less than a third of the actual height if the diffraction
diagram of regular waves is directly applied. The design of a steel structure
in the sea is another example in which the maximum force exerted by individual waves is the governing factor. If the structure is designed against a
regular wave equal to the significant wave, the structure will most probably
fail under the attack of waves higher than the significant wave height when
the design storm hits the site.

The danger of underestimating wave forces through the use of significant
waves was well understood at the early stages of construction of offshore
structures such as oil drilling rigs. It is an established practice to use a train
of regular waves of height and period equal to those of the highest wave,
and to design structures against this train of regular waves. This is called
the *highest wave representation method* herein. The method is mainly used
for structural designs.

In contrast to the above examples, the phenomenon of diffraction is
quite sensitive to the characteristics of the wave spectra, especially to the
directional spreading of wave energy. In the analysis of diffraction, refraction and wave forces upon a large isolated structure such as oil storage tank
in the sea, calculation is made for individual components of the directional
spectrum. The resultant total effect is estimated by summing the contributions from all spectral components. This is called the *spectral calculation
method*.

The rate of wave overtopping of a seawall and the sliding of a concrete
caisson of a vertical breakwater differ from the previous examples in the

sense that the cumulative effect of the action of individual waves of random nature is important. The probability distribution of individual wave heights and periods is the governing factor in this type of problems. The phenomena of irregular wave run-up and wave deformation by random breaking belong to the same category. The calculation of these cumulative wave effects may be called the *probability calculation method*.

If a large wind–wave flume or a wave basin with a random wave generator is available, wave transformations and wave action on structures can be directly investigated by using simulated random water waves. This is the *irregular wave test method*. When the first edition of this book was published in 1985, the reproduction of directional random waves in model basins was possible only at a limited number of hydraulic laboratories in the world. As described in Sec. 1.2(E), many laboratories are now equipped with multidirectional random wave generators and become capable of carrying out model tests of various problems, including random wave refraction and diffraction. It may be said that the majority of hydraulic model tests related to sea waves are presently carried out with random waves, and wave tests with regular waves are mostly reserved for fundamental research purposes.

In Fig. 1.1, the symbols A to E indicate the analysis method appropriate to the respective phenomenon. As can be seen, the problems related to random sea waves must be solved by selecting the appropriate calculation method among the five, A to E, to obtain a safe and rational design. None of the five methods can be used alone to treat all problems concerning sea waves. This stems from the complicated nature of waves in the real sea. In the following chapters, the above methods of analyzing the various wave phenomena are presented and discussed.

## References

1. H. Lamb, *Hydrodynamics*, 6th edn., Chap. IX (Cambridge Univ. Press, 1932).
2. G. Sainflou, "Essai sur les digues maritimes verticales," *Annales de Ponts et Chaussées* **98** (4) (1928).
3. H. U. Sverdrup and W. H. Munk, *Wind, Sea, and Swell; Theory of Relations for Forecasting*, U.S. Navy Hydrographic Office, H. O. Publ. No. 601 (1947).
4. W. G. Penney and A. T. Price, "Diffraction of sea waves by breakwaters," *Directorate of Miscellaneous Weapons Development, Tech. History* No. 26, Artificial Harbours, Sec. 3-D (1944).
5. W. J. Pierson, Jr., J. J. Tuttell and J. A. Woolley, "The theory of the refraction of a short-crested Gaussian sea surface with application to the

Northern New Jersey Coast," *Proc. 3rd Conf. Coastal Engrg.*, Cambridge, Mass. (1952), pp. 86–108.

6. W. J. Pierson, Jr., G. Neumann and R. W. James, *Practical Methods for Observing and Forecasting Ocean Waves by Means of Wave Spectra and Statistics*, U.S. Navy Hydrographic Office, H. O. Pub. No. 603 (1955).

7. R. R. Putz, "Statistical distributions for ocean waves," *Trans. Amer. Geophys. Union* **33** (5) (1952), pp. 685–692.

8. M. S. Longuet-Higgins, "On the statistical distributions of sea waves," *J. Marine Res.* **XI** (3) (1952), pp. 245–265.

9. Y. Goda, "Numerical experiments on wave statistics with spectral simulation," *Rept. Port and Harbour Res. Inst.* **9** (3) (1970), pp. 3–57.

10. M. S. Longuet-Higgins, "On the joint distribution of the periods and amplitudes of sea waves," *J. Geophys. Res.* **80** (18) (1975), pp. 2688–2694.

11. T. Karlsson, "Refraction of continuous ocean wave spectra," *Proc. Amer. Soc. Civil Engrs.*, **95** (WW4) (1969), pp. 471-490.

12. K. Nagai, T. Horiguchi, and T. Takai, "Computation of directional spectral deepwater waves propagating into shallow water area," *Proc. 21st Japanese Conf. Coastal Engrg.* (1974), pp. 437–448 (*in Japanese*).

13. H. Mitsuyasu, F. Tasai, T. Suhara, S. Mizuno, M. Ohkusu, T. Honda and K. Rikiishi, "Observation of the directional spectrum of ocean waves using a cloverleaf buoy," *J. Phys. Oceanogr.* **5** (1975), pp. 750–760.

14. Y. Goda and Y. Suzuki, "Computation of refraction and diffraction of sea waves with Mitsuyasu's directional spectrum," *Tech. Note of Port and Harbour Res. Inst.* **230** (1975), 45p. (*in Japanese*).

15. Y. Goda, T. Takayama and Y. Suzuki, "Diffraction diagrams for directional random waves," *Proc. 16th Int. Conf. Coastal Engrg.*, Hamburg, ASCE (1978), pp. 628–650.

16. C. L. Vincent and M. J. Briggs, "Refraction-diffraction of irregular waves over a mound," *J. Waterways, Port, Coastal and Ocean Engrg.* **115** (2) (1989), pp. 269–284.

17. L. H. Holthuijsen, N. Booji and R. C. Ris, "A spectral wave model for the coastal zone," *Proc. 2nd Int. Symp. on Ocean Wave Measurement and Analysis*, New Orleans, ASCE (1993), pp. 630–641.

18. J. I. Collins, "Probabilities of breaking wave characteristics," *Proc. Int. 12th Conf. on Coastal Engrg.*, Washington, D.C., ASCE (1970), pp. 399–414.

19. Y. Goda, "Irregular wave deformation in the surf zone," *Coastal Engineering in Japan* **18** (1975), pp. 13–26.

20. J. A. Battjes and J. P. F. M. Janssen, "Energy loss and set-up due to breaking of random waves," *Proc. 16th Int. Conf. Coastal Engrg.*, Hamburg, ASCE (1978), pp. 1–19.

21. W. R. Dally, R. G. Dean and R. A. Darlymple, "Wave height variation across beaches of arbitrary profile," *J. Geophys. Res.* **90** (C6) (1985), pp. 11917–11927.

22. J. A. Battjes, "Setup due to irregular wave," *Proc. 13th Int. Conf. Coastal Engrg.*, Vancouver, ASCE (1972), pp. 1993–2004.

23. Y. Goda, "A new method of wave pressure calculation for the design of composite breakwaters," *Rept. Port and Harbour Res. Inst.* **12** (3) (1973), pp. 31–69 (*in Japanese*).

24. Y. Goda, "New wave pressure formulae for composite breakwater," *Proc. 14th Int. Conf. Coastal Engrg.*, Copenhagen, ASCE (1974), pp. 1702–1720.

25. EA (UK), ENW (NL) and KFKI (DE), *EurOtop Wave Overtopping of Sea Defenses and Related Structures — Assessment Manual*, June 2007, Section 5.2.1.

26. S. Tsuruta and Y. Goda, "Expected discharge of irregular wave overtopping," *Proc. 11th Conf. Coastal Engrg.*, London, ASCE (1968), pp. 833–852.

27. Y. Goda, Y. Kishira and Y. Kamiyama, "Laboratory investigation on the overtopping rate of seawalls by irregular waves," *Rept. Port and Harbour Res. Inst.* **14** (4) (1975), pp. 3–44 (*in Japanese*).

28. N. W. Owen, "Design of seawalls allowing for wave overtopping," *Report No. EXS 924, HR Wallingford*, United Kingdom (1980).

29. Y. Goda and Y. Suzuki, "Estimation of incident and reflected waves in random wave experiments," *Proc. 15th Int. Conf. Coastal Engrg.*, Hawaii, ASCE (1976), pp. 828–845.

30. S. H. Salter, "Absorbing wave-makers and wide tanks," *Proc. Conf. on Directional Wave Sptectra Applications*, Univ. California, Berkeley, ASCE (1981), pp. 185–202.

31. Y. Goda, "Irregular sea waves for the design of harbour structures (integrated title)," *Trans. Japan Soc. Civil Engrs.* **8** (1976), pp. 267–271.

# Chapter 2

# Statistical Properties and Spectra of Sea Waves

## 2.1 Random Wave Profiles and Definitions of Representative Waves

### 2.1.1 *Spatial Surface Forms of Sea Waves*

Photograph 2.1 exemplifies the random nature of sea waves. It is a picture of the sea surface taken at a slanted angle, when a breeze is generating these wind waves. Sunbeams, which are reflected everywhere, produce patches of glitter. These reflected beams clearly show that there are many small and large wavelets moving in various directions. However, the waves as a whole are moving from right to left, following the direction of the wind. Photograph 2.2 is another example, showing laboratory waves generated by the wind blowing over the water surface. This photograph shows the profiles of the waves through a glass pane of the wind-wave flume, as well as a view of the surface.

Photo 2.1  Glittering on the sea surface caused by wind waves.

Photo 2.2  Wind waves in a wind water tunnel.

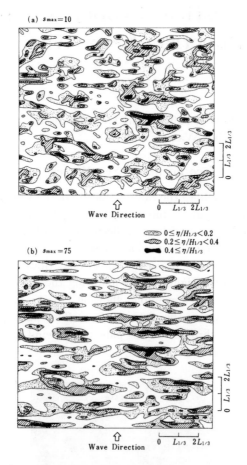

Fig. 2.1   Computer simulation of surface elevation contours of random waves.[1]

Such surface forms of sea waves, as discussed above, can be depicted by means of a contour map of the surface elevation. The result will be similar to that shown in Fig. 2.1, though it is not the topography of a real sea surface but the output of a computer simulation using the directional wave spectrum (Goda[1]) which is to be discussed in Sec. 2.3. The notation $s_{max}$

in the figure captions refers to the degree of wave energy spreacding with respect to direction, which will be described in Sec. 2.3.2. Figures 2.1(a) and (b) may be taken as representative examples of wind waves and swell, respectively. In the legend, the symbol $\eta$ denotes the water surface elevation above the mean water level. Dots, hatching and shadowing indicate the magnitude of the elevation while the blank areas designate the water surface below the mean water level. The actual surface topography of sea waves, given by an analysis using aerial stereophotogrametry, closely resembles the example in Fig. 2.1. The wave pattern in Fig. 2.1 also indicates that wave crests do not have a long extent, but instead consist of short segments. Because of this feature, waves in the sea, especially wind waves, are termed as *short-crested waves*.

In the theoretical treatment of water waves, their surface forms are often represented by sinusoidal functions. But such sinusoidal wave forms are observable only in the laboratory; sinusoidal waves are never found in the natural environment as single wave forms.

### 2.1.2 *Definition of Representative Wave Parameters*

In the profile shown in Photo 2.2, four waves are visible through the glass pane. If we were to trace such a wave profile over a much longer distance, we would get a wave profile undulating in an irregular manner. A longitudinal cross section of the water surface depicted in Fig. 2.1 will also yield a similar wave profile of irregular shape. Both profiles could be plotted against the horizontal distance.

On the other hand, a strip chart record from a wave gauge in the sea yields a wave form as shown in Fig. 2.2. In this example, the horizontal axis gives the elapsed time from the start of the recording. It is rather difficult

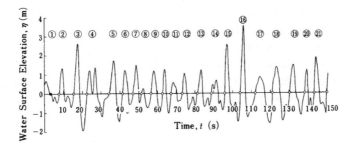

Fig. 2.2   Example of a wave record.

to define individual waves appearing in such irregular wave records; in fact, there is no absolute method of definition. However, the customary practice in wave analysis is to utilize either the *zero-upcrossing method* or the *zero-downcrossing method* as the standard techniques for defining waves.

We will now briefly describe these methods. First, the mean level of the water surface is deduced from the surface record and defined as the zero line. Next, a search is made for the point where the surface profile crosses the zero line upward. That point is taken as the start of one individual wave. Following the ups and downs of the irregular surface profile, a search is continued to find the next zero-upcrossing point after the surface profile has once gone below the zero line. When the next zero-upcrossing point is found, it defines the end of the first wave and the start of the second wave. The distance between the two adjacent zero-upcrossing points defines the wave period if the abscissa is the time, whereas an apparent wavelength[a] is defined if the abscissa is the horizontal distance. The vertical distance between the highest and lowest points between the adjacent zero-upcrossing points is defined as the wave height, disregarding small humps which do not cross the zero line.

In the case of zero-downcrossing method, the points where the surface profile crosses the zero line downward are taken as the starting and ending points of individual waves. The difference between the two definitions is mainly conceptual; that is, whether the wave height is defined using a crest and the following trough or using a crest and the preceding trough. Statistically, they are equivalent except in the surf zone where the conspicuous forward tilting of surface profiles tends to yield the zero-downcrossing wave period slightly shorter than the zero-upcrossing wave period. However, no difference appears between the zero-upcrossing and zero-downcrossing wave height[3] (refer to Sec. 10.5 for further discussion).

Application of the zero-upcrossing method to the sample record of Fig. 2.2 yields twenty-one individual waves. Their heights and periods as read from the record are listed in Table 2.1 in the order of their appearance. The fourth column indicates the order beginning with the highest wave. In field wave observations, a standard procedure is to take records of about 100 consecutive waves, and thus a long list for the first to one-hundredth wave heights and periods similar to that in Table 2.1 must be made by

---

[a]Pierson[2] states that the wavelength defined in this way is an apparent one, being shorter than the wavelength derived from the wave period and water depth based on the small amplitude wave theory. However, the average wavelength measured in Fig. 2.1 is approximately equal to the small amplitude wavelength.

Table 2.1   Wave heights and periods read from Fig. 2.2.

| Wave number | Wave height $H$ (m) | Wave period $T$ (s) | Order number $m$ | Wave number | Wave height $H$ (m) | Wave period $T$ (s) | Order number $m$ |
|---|---|---|---|---|---|---|---|
| 1 | 0.54 | 4.2 | 21 | 12 | 1.95 | 8.0 | 15 |
| 2 | 2.05 | 8.0 | 12 | 13 | 1.97 | 7.6 | 14 |
| 3 | 4.52 | 6.9 | 2 | 14 | 1.62 | 7.0 | 18 |
| 4 | 2.58 | 11.9 | 8 | 15 | 4.08 | 8.2 | 3 |
| 5 | 3.20 | 7.3 | 4 | 16 | 4.89 | 8.0 | 1 |
| 6 | 1.87 | 5.4 | 17 | 17 | 2.43 | 9.0 | 9 |
| 7 | 1.90 | 4.4 | 16 | 18 | 2.83 | 9.2 | 7 |
| 8 | 1.00 | 5.2 | 20 | 19 | 2.94 | 7.9 | 6 |
| 9 | 2.05 | 6.3 | 13 | 20 | 2.23 | 5.3 | 11 |
| 10 | 2.37 | 4.3 | 10 | 21 | 2.98 | 6.9 | 5 |
| 11 | 1.03 | 6.1 | 19 | | | | |

reading off the values from a strip chart record. Nowadays, almost all the data are recorded in digital form, and the above process is executed by a computer.

Based on such a long list of height and period data, the following four kinds of representative waves are defined:

(a) *Highest wave*: $H_{max}$, $T_{max}$. This refersto the wave having the height and period of the highest individual wave in a record. The quantities are denoted as $H_{max}$ and $T_{max}$, respectively. In the case of the data in Table 2.1, the 16th wave is the highest wave, with $H_{max} = 4.9$ m and $T_{max} = 8.0$ s.

(b) *Highest one-tenth wave*: $H_{1/10}$, $T_{1/10}$. The waves in the record are counted and selected in descending order of wave height from the highest wave, until one-tenth of the total number of waves is reached. The means of their heights and periods are calculated and denoted as $H_{1/10}$ and $T_{1/10}$, respectively. In the example of Table 2.1, the 16th and 3rd waves are selected for this definition, yielding the values of $H_{1/10} = 4.7$ m and $T_{1/10} = 7.5$ s. An imaginary wave train having height and period of $H_{1/10}$ and $T_{1/10}$ is defined as the highest one-tenth wave.

(c) *Significant wave*, or *highest one-third wave*: $H_{1/3}$, $T_{1/3}$. For this representative wave the waves in the record are counted and selected in descending order of wave height from the highest wave, until one-third of the total number of waves is reached. The means of their heights and periods are calculated and denoted as $H_{1/3}$ and $T_{1/3}$, respectively. In the example of Table 2.1, the 16th, 3rd, 15th, 5th, 21st, 19th and 18th waves are selected for this definition, yielding the values of $H_{1/3} = 3.6$ m and $T_{1/3} = 7.8$ s. An imaginary wave train having height and period equal to $H_{1/3}$ and $T_{1/3}$ is defined as the significant wave or the highest one-third wave. The height $H_{1/3}$ is often called the significant wave height, and the period $T_{1/3}$ is called the significant wave period.

(d) *Mean wave*: $\overline{H}, \overline{T}$. The arithmetic means of the heights and periods of all waves in a record are calculated and denoted as $\overline{H}$ and $\overline{T}$, respectively. The example in Table 2.1 has mean values of $\overline{H} = 2.4$ m and $\overline{T} = 7.0$ s. An imaginary wave having height and period equal to $\overline{H}$ and $\overline{T}$ is defined as the mean wave.

Among the above definitions of representative waves, the significant wave is most frequently used. The height and the period of wind waves and swell, including the results of wave hindcasting, usually refer to the significant waves unless otherwise specified.

## 2.2    Distributions of Individual Wave Heights and Periods

### 2.2.1    *Wave Height Distribution*

Among the various statistical properties of random waves in the sea, the distribution of individual wave heights will be examined first. Figure 2.3 is an example of the histogram of wave heights obtained from a long wave record, from which the wave profile shown in Fig. 2.1 was extracted. The ordinate $n$ is the number of waves in the respective class of wave height $H$. The wave heights in this example are distributed in a wide range from 0.1 to 5.5 m, with the mode in the class $H = 1.57$ to 2.10 m (a uniform class interval of $\overline{H}/4 = 0.525$ m is used).

The histogram of wave heights of a wave record containing about 100 waves usually exhibits a rather jagged shape because of the relatively small sample size. However, we can obtain a smoother distribution of wave heights by assembling many wave records with the wave heights normalized by the mean heights of the respective records, and by counting the relative frequencies of the normalized wave heights in their respective classes. Figure 2.4 presents a typical result of such a manipulation (Goda and Nagai[4]) in which the ordinate is the relative frequency, $n/N_0$ ($N_0$ is the total number of waves), divided by the class interval of the normalized wave height, $\Delta(H/\overline{H})$, so that the area under the histogram is equal to unity.

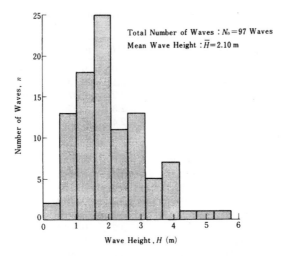

Fig. 2.3   Example of a histogram of wave heights.

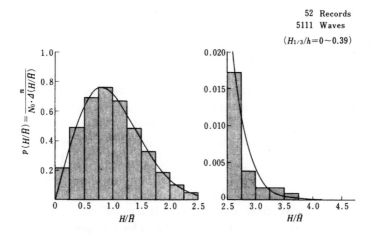

Fig. 2.4   Normalized histogram of wave heights.[4]

The smoothed histogram of normalized wave heights shown in Fig. 2.4 suggests the existence of some theoretical distribution function. In fact, the so-called *Rayleigh distribution* given by Eq. (2.1) and drawn with the solid line in Fig. 2.4 has been proposed for the distribution of individual wave heights:

$$p(x) = \frac{\pi}{2} x \exp\left[-\frac{\pi}{4} x^2\right] \; : \; x = \frac{H}{\overline{H}}. \tag{2.1}$$

Table 2.2  Probability density and exceedance probability of the
Rayleigh distribution.

| $H/\overline{H}$ | $p(H/\overline{H})$ | $P(H/\overline{H})$ | $H/\overline{H}$ | $p(H/\overline{H})$ | $P(H/\overline{H})$ |
|---|---|---|---|---|---|
| 0 | 0 | 1.0000 | 2.0 | 0.1358 | 0.0432 |
| 0.1 | 0.1559 | 0.9222 | 2.1 | 0.1033 | 0.0313 |
| 0.2 | 0.3044 | 0.9691 | 2.2 | 0.0772 | 0.0223 |
| 0.3 | 0.4391 | 0.9318 | 2.3 | 0.0567 | 0.0157 |
| 0.4 | 0.5541 | 0.8819 | 2.4 | 0.0409 | 0.0108 |
| 0.5 | 0.6454 | 0.8217 | 2.5 | 0.0290 | 0.00738 |
| 0.6 | 0.7104 | 0.7537 | 2.6 | 0.0202 | 0.00495 |
| 0.7 | 0.7483 | 0.6806 | 2.7 | 0.0138 | 0.00326 |
| 0.8 | 0.7602 | 0.6049 | 2.8 | 0.0093 | 0.00212 |
| 0.9 | 0.7483 | 0.5293 | 2.9 | 0.0062 | 0.00135 |
| 1.0 | 0.7162 | 0.4559 | 3.0 | 0.0040 | 0.00085 |
| 1.1 | 0.6680 | 0.3866 | 3.1 | 0.0026 | 0.00053 |
| 1.2 | 0.6083 | 0.3227 | 3.2 | 0.0016 | 0.00032 |
| 1.3 | 0.5415 | 0.2652 | 3.3 | 0.0010 | 0.00019 |
| 1.4 | 0.4717 | 0.2145 | 3.4 | 0.0006 | 0.00011 |
| 1.5 | 0.4025 | 0.1708 | 3.5 | 0.0004 | 0.000066 |
| 1.6 | 0.3365 | 0.1339 | 3.6 | 0.0002 | 0.000038 |
| 1.7 | 0.2759 | 0.1033 | 3.7 | 0.0001 | 0.000021 |
| 1.8 | 0.2219 | 0.0785 | 3.8 | 0.0001 | 0.000012 |
| 1.9 | 0.1752 | 0.0587 | 3.9 | 0.0000 | 0.000065 |
| 2.0 | 0.1358 | 0.0432 | 4.0 | 0.0000 | 0.000035 |

The function $p(x)$ represents the probability density; that is, the probability
of a normalized wave height taking an arbitrary value between $x$ and $x +$
$dx$ is given by the product $p(x)\,dx$. The *Rayleigh distribution* commonly
appears in the field of statistics. The ordinate of the plot in Fig. 2.4,
$n/(N_0 \cdot \Delta x)$, is an approximation to $p(x)$. The integral of $p(x)$ over the
range $x = 0$ to $x = \infty$ represents the probability of $x$ taking a value
between 0 and $\infty$. By the definition of probability, the integral over the
full range should give unity and thus $p(x)$ is normalized.

Table 2.2 lists the values of $p(x)$ calculated for selected values of $x$. The
function $P(x)$ listed in the third and sixth columns gives the probability of
a particular wave height exceeding a prescribed value. It is given by

$$P(x) = \int_x^\infty p(\xi)\,d\xi = \exp\left[-\frac{\pi}{4}x^2\right]. \qquad (2.2)$$

The Rayleigh distribution was originally derived by Lord Rayleigh in the
late 19th century to describe the distribution of the intensity of sounds emit-
ted from an infinite number of sources. In 1952, Longuet-Higgins[5] demonstrated that this distribution is also applicable to the heights of waves in the

sea. Since then, the Rayleigh distribution has been universally employed to describe wave heights. Strictly speaking, Longuet-Higgins only verified the applicability of the Rayleigh distribution for irregular waves which have very small fluctuations in the individual wave periods and whose heights exhibit a beat-like fluctuation. However, real sea waves usually exhibit fairly wide fluctuations in the individual wave periods. As of yet, no exact theory, which is applicable to real waves in the sea characterized by certain fluctuations in wave periods, has been proposed.

Examinations of real wave records have already indicated a slight departure of the actually occurring wave height distribution from the Rayleigh distribution; the degree of departure depends on the frequency spectrum of sea waves (refer to Sec. 2.4.1 for a detailed discussion). Nevertheless the Rayleigh distribution provides a good approximation to the distribution of individual wave heights which are defined by the zero-upcrossing and zero-downcrossing methods. This is true not only for wind waves or and swell individually, but also for the combined sea state of wind waves and swell propagating simultaneously. This seems to be one of the virtues of the zero-upcrossing and zero-downcrossing methods of wave definitions.

### 2.2.2 *Relations Between Representative Wave Heights*

If we adopt the Rayleigh distribution as an approximation to the distribution of individual wave heights, representative wave heights such as $H_{1/10}$ and $H_{1/3}$ can be evaluated by the manipulation of the probability density function (refer to Sec. 10.1 for detail). Thus, we have

$$H_{1/10} = 1.27H_{1/3} = 2.03\overline{H}, \quad H_{1/3} = 1.60\overline{H}. \tag{2.3}$$

These results represent the mean values of a number of wave records ensembled together. Individual wave records containing only 100 waves or so may give noticeable departures from these mean relations. Figures 2.5 and 2.6 show the relative frequencies of the wave height ratio $H_{1/10}/H_{1/3}$ and $H_{1/3}/\overline{H}$, respectively, based on an examination of 171 observed wave records.[4] The ratio of $H_{1/10}/H_{1/3}$ is found to lie in the range from 1.15 to 1.45, while the ratio $H_{1/3}/\overline{H}$ is distributed from 1.40 to 1.75. The overall means are 1.27 for $H_{1/10}/H_{1/3}$ and 1.59 for $H_{1/3}/\overline{H}$, which are quite close to the theoretical predictions of Eq. (2.3).

The relation between $H_{\max}$ and $H_{1/3}$ can also be derived from the Rayleigh distribution (refer to Sec. 10.1.3 for a detailed discussion). However, the basic nature of $H_{\max}$ is such that individual wave records having

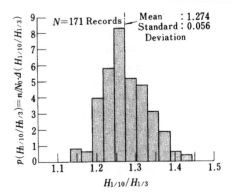

Fig. 2.5   Histogram of the wave height ratio $H_{1/10}/H_{1/3}$.[4]

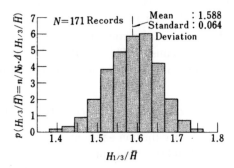

Fig. 2.6   Histogram of the wave height ratio $H_{1/3}/\overline{H}$.[4]

the same value of $H_{1/3}$ contain different values of $H_{\max}$, because $H_{\max}$ refers to the height of one wave which happens to have the greatest wave height in a particular wave train. For example, the relative frequency of the ratio $H_{\max}/H_{1/3}$ for 171 wave records is shown in Fig. 2.7;[4] the ratio is broadly spread in the range between 1.1 to 2.4. The ratio of $H_{\max}/H_{1/3}$ is affected by the number of waves in a record. This will be discussed in connection with Eqs. (2.4) to (2.6). The solid and dashed-dot lines indicate the theoretical probability density function of $H_{\max}/H_{1/3}$ for wave numbers of 50 and 200, respectively. The records, which contain from 55 to 198 waves per record exhibit a distribution of $H_{\max}/H_{1/3}$ bounded by the two theoretical curves.

The deterministic prediction of particular values of $H_{\max}$ for individual wave trains is impossible. Although it has little meaning to attempt to

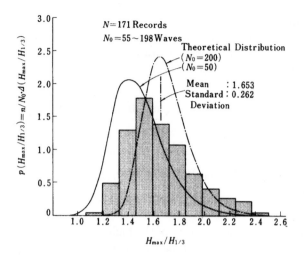

Fig. 2.7   Histogram of the wave height ratio $H_{\max}/H_{1/3}$.[4]

derive a single value of $H_{\max}$ for one storm wave condition, based on the Rayleigh distribution, a probability density can be reasonably defined for the ratio of $H_{\max}/H_{1/3}$. The most probable value, or the mode of distribution, is a function of the number of waves in a wave train or wave record. This is given by

$$(H_{\max}/H_{1/3})_{\text{mode}} \simeq 0.706\sqrt{\ln N_0}\,, \qquad (2.4)$$

where $N_0$ refers to the number of waves.

The arithmetic mean of $H_{\max}/H_{1/3}$ is greater than the most probable value, as seen from the skewed shape of the curves in Fig. 2.7. The mean is given approximately as

$$(H_{\max}/H_{1/3})_{\text{mean}} \simeq 0.706\left[\sqrt{\ln N_0} + \gamma/(2\sqrt{\ln N_0})\right]\,, \qquad (2.5)$$

where $\gamma$ is Euler's constant having the value of $0.5772\ldots$.

Furthermore, we can define a quantity $(H_{\max})_\mu$ whose probability of exceeding is $\mu$; that is, integration of a probability density of the largest wave height beyond $(H_{\max})_\mu$ yields the probability of $\mu$. The height $(H_{\max})_\mu$ is given by

$$\frac{(H_{\max})_\mu}{H_{1/3}} \simeq 0.706\sqrt{\ln\left[\frac{N_0}{\ln 1/(1-\mu)}\right]}\,. \qquad (2.6)$$

### Example 2.1

A sea state with $H_{1/3} = 6.0$ m continues for the duration of 500 waves. Calculate the mode and arithmetic mean of $H_{max}$ as well as the value of $H_{max}$ with $\mu = 0.01$.

### Solution

We have $\sqrt{\ln 500} = 2.493$ and $\sqrt{\ln[500/\ln(1/0.990)]} = 3.289$. Therefore,

$$(H_{max}/H_{1/3})_{mode} \simeq 0.706 \times 2.493 \times 6.0 = 10.6 \text{ m},$$

$$(H_{max}/H_{1/3})_{mean} \simeq 0.706 \times [2.493 + 0.5772/(2 \times 2.493)] \times 6.0 = 11.1 \text{ m},$$

$$(H_{max}/H_{1/3})_{0.01} \simeq 0.706 \times 3.289 \times 6.0 = 13.9 \text{ m}.$$

The nondeterministic property of the highest wave causes inconvenience as well as uncertainty in the design of maritime structures. It is an inevitable consequence, however, of the random nature of sea waves. The value of $H_{max}$ should be estimated based upon consideration of the duration of storm waves and the number of waves, and by allowing some tolerance for a range of deviation. The prediction generally employed falls within the range

$$H_{max} = (1.6 \sim 2.0)H_{1/3}, \tag{2.7}$$

in which the particular final value is chosen by consideration of the reliability of the estimation of the design storm waves, the accuracy of the design formula, the importance of the structure, the type and nature of the possible structural failure, and other factors. In the design of offshore structures, $H_{max} = 2.0H_{1/3}$ or a higher value is often employed. For the design of vertical breakwaters, the author[6] has proposed the use of the relation $H_{max} = 1.8H_{1/3}$.

### 2.2.3 *Distribution of Wave Period*

The periods of individual waves in a wave train exhibit a distribution narrower than that of wave heights, and the spread of periods lies mainly in the range of 0.5 to 2.0 times the mean wave period. However, when wind waves and swell coexist, the period distribution becomes broader. In some cases, the period distribution is bi-modal, with two peaks corresponding to the mean periods of the wind waves and swell. Thus the wave period does not exhibit a universal distribution law such as the Rayleigh distribution in the case of wave heights.

Nevertheless, it has been empirically found that the representative period parameters are interrelated. From the analysis of field wave data, the following results have been reported[4]:

$$\left. \begin{array}{l} T_{\max} = (0.6 \sim 1.3)T_{1/3}, \\ T_{1/10} = (0.9 \sim 1.1)T_{1/3}, \\ T_{1/3} = (0.9 \sim 1.4)\overline{T}. \end{array} \right\} \tag{2.8}$$

The above indicates the range of variations. The average values for many wave records can be summarized as

$$T_{\max} \simeq T_{1/10} \simeq T_{1/3} \simeq 1.2\overline{T}. \tag{2.9}$$

The ratio of $T_{1/3}$ to $\overline{T}$, indicated in Eq. (2.9), provides only a guideline because its value is affected by the functional shape of frequency spectrum analyzed from a wave record (refer to Sec. 2.4.2 for a detailed discussion).

Equations (2.8) and (2.9) reflect the characteristics of the joint distribution of wave heights and periods. Waves of smaller heights in a wave record often have shorter periods, whereas waves of heights greater than the mean height do not show a correlation with the wave period. Thus, the overall mean period is shorter than the mean periods of the higher waves (refer to Sec. 10.3 for a detailed discussion).

## 2.3  Spectra of Sea Waves

### 2.3.1  *Frequency Spectra*

(A) *General*

The concept of spectrum can be attributed to Newton, who discovered that sunlight can be decomposed into a spectrum of colors (red to violet) with the aid of a prism.[7] The spectrum indicates how the intensity of light varies with respect to its wavelength. The concept is based on the principle that white light consists of numerous components of light of various colors (wavelengths). The technique of decomposing a complex physical phenomenon into individual components has been applied in many physical problems.

Sea waves, which at first sight appear very random, can be analyzed by assuming that they consist of an infinite number of wavelets with different frequencies and directions. The distribution of the energy of these wavelets when plotted against the frequency and direction is called the *wave spectrum*. More precisely, the wave energy distribution with respect to the frequency alone, irrespective of wave direction, is called the *frequency*

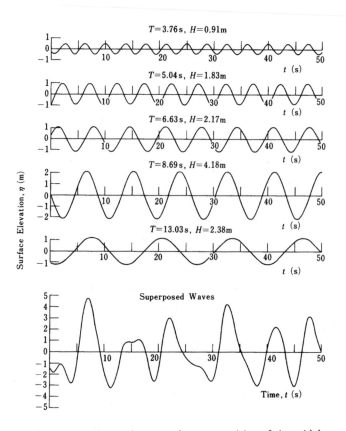

Fig. 2.8   Simulation of irregular waves by superposition of sinusoidal waves.

*spectrum*, whereas the energy distribution expressed as a function of both frequency and direction is called the *directional wave spectrum*.

Figure 2.8 gives an example of an irregular wave profile which was constructed by adding five sinusoidal waves (component waves) of different heights and periods. Although the irregularity of the wave profile is not remarkable in this example, we can obtain quite irregular profiles similar to those of real sea waves by increasing the number of component waves. The inverse process is also possible, and irregular wave profiles as shown in Fig. 2.2 can be broken into a number of component waves (refer to Sec. 11.2 for the actual computation procedure). The way in which the component waves are distributed is expressed by plotting the component wave energy, or the square of the component amplitude, against the frequency of component waves. The irregular profile in Fig. 2.8, for example, is represented

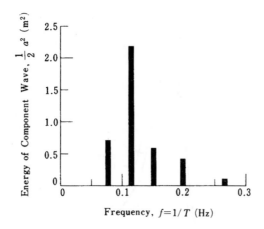

Fig. 2.9  Spectral representation of superposed waves.

by the diagram with five bars shown in Fig. 2.9. In the case of sea waves, the energy distribution manifests itself as a continuous curve because there exist an infinite number of frequency components. The wave record which included the profile in Fig. 2.2 was used to calculate the frequency spectrum shown in Fig. 2.10. Such a continuous spectrum is called the *frequency spectral density function* in more precise language, and it has units of m²·s, or units of similar dimensions.

The wave spectrum in Fig. 2.10 indicates that the wave energy is spread in the range of about $f = 0.05 \sim 0.4$ Hz, or equivalently $T = 2.5 \sim 20$ s, even though the significant wave period is 8 s. The figure also indicates that the wave energy is concentrated around the frequency $f_p \simeq 0.11$ Hz, which is slightly less than the frequency $f = 0.125$ Hz corresponding to the significant wave period.

## (B) *Frequency spectrum of sea waves*

The characteristics of the frequency spectra of sea waves have been fairly well established through analyses of a large number of wave records taken in various waters of the world. The spectra of fully developed wind waves, for example, can be approximated by the following standard formula:

$$S(f) = 0.257 H_{1/3}^2 T_{1/3}^{-4} f^{-5} \exp[-1.03(T_{1/3}f)^{-4}]. \qquad (2.10)$$

The dash-dot line in Fig. 2.10 is the result of fitting Eq. (2.10) with the values of the significant wave height and period of the record. Although

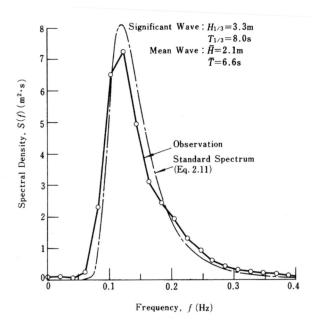

Fig. 2.10 Example of spectrum of sea waves.

some difference is observed between the actual and standard spectra, partly because of the shallow water effect in the wave record which was taken at the depth of 11 m, the standard spectrum describes the features of the actual spectrum quite well.

Equation (2.10) is based on the proposal by Bretschneider[8] with adjustment of the coefficients by Mitsuyasu,[9] and it is called the *Bretschneider–Mitsuyasu spectrum*. There are several other expressions used as standard forms of the frequency spectrum, e.g., those of Pierson and Moskowitz,[10] Mitsuyasu,[11] Hasselmann *et al.*,[12] Ochi and Hubble,[13] etc. The formula of Pierson and Moskowitz includes the wind velocity as the principal parameter, because it was derived for the purpose of wave prediction in the ocean. But its expression can be converted into an equation equivalent to Eq. (2.10) by introducing the wave height and period, because the functional dependence with respect to frequency is the same.

The coefficients in Eq. (2.10) have been so set by Mitsuyasu to yield the relation of $T_p \simeq 1.05T_{1/3}$, which was derived by his field measurements ($T_p$ denotes the wave period corresponding to the frequency $f_p$ at the spectral peak, or $T_p = 1/f_p$). However, a number of later measurements suggests the relation of $T_p \simeq 1.1T_{1/3}$ as more appropriate for wind waves. With further adjustment on the relation of the significant wave height and total wave energy, the following alternative formula has been proposed by the author[14] for the frequency spectrum of wind waves:

$$S(f) = 0.205 H_{1/3}^2 T_{1/3}^{-4} f^{-5} \exp[-0.75(T_{1/3}f)^{-4}]. \qquad (2.11)$$

The spectrum of Eq. (2.11) may be called the *modified Bretschneider–Mitsuyasu spectrum*.

Equations (2.10) and (2.11) are applied for the wind waves fully developed in the ocean. Wind waves rapidly developed in a relatively restricted water by very strong wind usually exhibit the spectral peak much sharper than that given by Eqs. (2.10) and (2.11). This feature has been taken up by Hasselmann *et al.*[12] in their proposal of wave spectrum, which was based on the results of the joint wave observation program in the North Sea (referred to as JONSWAP). The *JONSWAP spectrum* includes the wind speed as the parameter for the purpose of wave forecasting, but it can be rewritten in approximate form in terms of the parameters of wave height and period as follows[14]:[b]

$$S(f) = \beta_J H_{1/3}^2 T_p^{-4} f^{-5} \exp[-1.25(T_pf)^{-4}] \gamma^{\exp[-(T_pf-1)^2/2\sigma^2]}, \qquad (2.12)$$

in which

$$\beta_J = \frac{0.0624}{0.230 + 0.0336\gamma - 0.185(1.9 + \gamma)^{-1}} [1.094 - 0.01915 \ln \gamma], \qquad (2.13)$$

$$T_p \simeq T_{1/3}/[1 - 0.132(\gamma + 0.2)^{-0.559}], \qquad (2.14)$$

$$\sigma = \begin{cases} \sigma_a : f \le f_p, \\ \sigma_b : f \ge f_p, \end{cases} \qquad (2.15)$$

$\gamma = 1 \sim 7$ (mean of 3.3), $\sigma_a \simeq 0.07$, $\sigma_b \simeq 0.09$.

The JONSWAP spectrum is characterized by a parameter $\gamma$ which is called the *peak enhancement factor*; this controls the sharpness of the spectral peak. For $\gamma = 1$, Eq. (2.12) reduces to Eq. (2.11). For $\gamma = 3.3$, which is the mean value determined for the North Sea, the peak value of the spectral density function becomes 2.1 times higher than that of Eq. (2.11) for the same significant wave height and period.

---

[b]The original spectral formula was integrated over the range $f = 0$ to $f = 6f_p$ for various values of $\gamma$. The integrated values were correlated to $H_{1/3}$ of simulated wave records together with Eq. (2.33) and Tables 2.3 and 2.4 to appear later, and the coefficient of proportionality was empirically determined as a function of $\gamma$.

In the shallow water area, the frequency spectrum tends to exhibit the attenuation of the spectral density at the high frequency region more gradual than that of the spectrum described by Eqs. (2.10) to (2.12). Some research works, on the other hand, deal with a spectrum with the high frequency tail attenuating rapidly. The flexibility in spectral shape can be realized by using the Wallops spectrum proposed by Huang et al.[15] The Wallops spectrum has been rewritten by the author[14] in terms of the parameters of wave height and period as follows:

$$S(f) = \beta_W H_{1/3}^2 T_p^{1-m} f^{-m} \exp\left[-\frac{m}{4}(T_p f)^{-4}\right], \qquad (2.16)$$

in which

$$\beta_W = \frac{0.0624 m^{(m-1)/4}}{4^{(m-5)/4}\,\Gamma[(m-1)/4]}\left[1 + 0.7458(m+2)^{-1.057}\right], \qquad (2.17)$$

$$T_p \simeq T_{1/3}/[1 - 0.283(m - 1.5)^{-0.684}]. \qquad (2.18)$$

For $m = 5$, Eq. (2.16) reduces to Eq. (2.11). The symbol $\Gamma(\cdot)$ in Eq. (2.17) denotes the Gamma function.

Another frequency spectrum dedicated to shallow water waves is the *TMA spectrum* proposed by Bouws et al.,[16] which is based on field measurements in water of finite depth. According to the formulation by Tucker,[17] the TMA spectrum is expressed as follows:

$$S(f) = S_J(f) \cdot \phi(kh) \;:\; \phi(kh) = \frac{\tanh^2 kh}{1 + 2kh/\sinh 2kh}, \qquad (2.19)$$

where $S_J(f)$ denotes the JONSWAP spectrum of Eq. (2.12). The value of the function $\phi(kh)$ decreases as the relative depth $kh$ decreases. The decrease implicitly expresses the process of wave decay by depth-limited breaking.

Some researchers in US and Europe employ the TMA spectrum as the target spectrum to be reproduced in random wave tests, because the waves in front of the wave paddle are shallow water waves in general. However, the shallow water wave spectrum contains a considerable amount of secondary frequency components produced by nonlinear wave interactions among the linear spectral components as will be discussed in Sec. 10.5.3. When linear waves having the TMA spectrum are generated by the wave paddle, laboratory waves would induce the nonlinear interactions much stronger than the prototype waves while propagating from the area of wave generation to the test site. It is advised to check the degree of nonlinear wave interactions at the test location if the TMA spectrum is employed as the target spectrum.

(C) *Frequency spectrum of swell and spectrum of combined sea state*

The spectrum of swell is transformed from that of wind waves through its propagation over a long distance after it has left the wave generating area. The process of velocity dispersion (first discussed by Pierson *et al.*[18]) takes place during the swell propagation, because the low frequency wave components propagate faster than the high frequency components. The swell observed at a fixed station has a spectrum restricted to a narrow frequency range. Thus, the swell spectrum exhibits a peak much sharper than that of wind waves.

According to the analysis of swell which was generated off New Zealand, propagated over the distance of some 9000 km to the Pacific coast of Costa Rica and still maintained the significant wave height of about 3 m (Goda[19]), the swell spectral peaks were equivalent to the JONSWAP spectra of Eq. (2.12) with $\gamma = 8 \sim 9$, and to the Wallops spectra of Eq. (2.16) with $m = 8 \sim 10$, on the average. Thus, the swell spectrum for engineering applications may be approximated by the JONSWAP spectra of Eq. (2.12) with the peak enhancement factor being chosen between $\gamma = 3 \sim 10$, depending on the distance traveled.

Actual wave spectra usually exhibit some deviations from these standard forms. In particular, when swell coexists with wind waves, a secondary peak appears at the frequency corresponding to the representative period of swell or wind waves, depending on their relative magnitudes. In some cases, not only bi-modal but also tri-modal frequency spectra can be observed. The standard spectra proposed by Ochi and Hubble[13] can represent bi-modal spectra by means of 11 patterns which were developed from a data base of 800 ocean wave spectra. When the representative heights and periods of wind waves and swell are given *a priori*, the spectrum of the resultant sea state can be estimated by linearly superposing the standardized spectra of wind waves and swell.

## 2.3.2 *Directional Wave Spectra*

(A) *General*

Sea waves cannot be adequately described by using the frequency spectrum alone. Irregular waves specified solely by the frequency spectrum, if viewed from above in a manner similar to Fig. 2.1, would appear as so-called long-crested waves which have straight and parallel crestlines. The patterns of the wave crests, as shown in Fig. 2.1, imply the existence of many component waves propagating in various directions. The concept of directional spectrum is therefore introduced to describe the state of superimposed directional components. The directional spectrum represents the distribution of wave energy not only in the frequency domain but also in direction (angle $\theta$). It is generally expressed as

$$S(f, \theta) = S(f) G(\theta|f), \qquad (2.20)$$

where $S(f, \theta)$ is the directional wave spectral density function or simply the directional wave spectrum, and $G(\theta|f)$ is the *directional spreading function*, alternatively called the *spreading function*, the angular distribution function, or the directional distribution.

The function $G(\theta|f)$, which represents the directional distribution of wave energy in direction, has been found to vary with frequency. Therefore, the function $G(\theta|f)$ contains the frequency variable $f$. The directional spreading function carries no dimensions[c] and is normalized as

$$\int_{-\pi}^{\pi} G(\theta|f)\, d\theta = 1. \qquad (2.21)$$

Thus the frequency spectrum $S(f)$ gives the absolute value of the wave energy density, while the function $G(\theta|f)$ represents the relative magnitude of directional spreading of wave energy.

(B) *Directional spreading function of the Mitsuyasu-type*

Knowledge of the directional distribution of the energy of sea waves is still limited because of the difficulty in making reliable field measurements. Techniques of field measurements and analyses will be discussed in Sec. 11.3. Only a few reports which give the results of measurements of directional spectra of ocean waves are available. Therefore, the establishment of a standard functional form for the directional wave spectrum has not been achieved yet, in contrast to the case of frequency spectrum. Nonetheless, Mitsuyasu *et al.*[20] have proposed the following function on the basis of their detailed field measurements with a special cloverleaf-type instrument buoy, as well as other available data:

$$G(\theta|f) = G_0 \cos^{2s}\left(\frac{\theta - \theta_0}{2}\right), \qquad (2.22)$$

where $\theta$ is the azimuth and $\theta_0$ denotes the azimuth of the principal wave direction. In this expression, $G_0$ is a constant introduced to satisfy the

---

[c]However, the directional spreading function can be thought of as having the units of inverse angle, such as rad$^{-1}$.

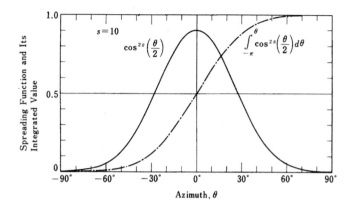

Fig. 2.11   Example of directional spreading function.

condition of Eq. (2.21); i.e.,

$$G_0 = \left[ \int_{\theta_{min}}^{\theta_{max}} \cos^{2s}\left(\frac{\theta - \theta_0}{2}\right) \right]^{-1}, \tag{2.23}$$

and $s$ is a parameter related to the frequency as further discussed below. If the range of the angle is $\theta_{min} = -\pi$ and $\theta_{max} = \pi$, the constant $G_0$ becomes

$$G_0 = \frac{1}{\pi} 2^{2s-1} \frac{\Gamma^2(s+1)}{\Gamma(2s+1)}, \tag{2.24}$$

where $\Gamma(\cdot)$ denotes the Gamma function. For example, by setting $s = 10$, $G_0$ becomes about 0.9033, and the directional spreading function is calculated as shown by the solid line in Fig. 2.11 for the case of $\theta_0 = 0°$. The cumulative value of $G(\theta|f)$ from $\theta = -\pi$ is also shown in Fig. 2.11 as the dash-dot line. From this cumulative distribution of $G(\theta|f)$, it is observed that about 85% of the wave energy is contained in the angular range of $\pm 30°$.

The directional spreading function formulated by of Mitsuyasu *et al.*[20] has the features that the parameter $s$ representing the degree of directional energy concentration takes a peak value around the frequency of the spectral peak, and that the value of $s$ decreases as the value of the frequency moves away from that of the spectral peak toward both lower and higher frequencies. That is to say, the directional spreading of wave energy is narrowest around the spectral peak frequency. Although the original proposal of Mitsuyasu *et al.* relates the spreading parameter $s$ to the wind speed $U$, Goda and Suzuki[21] have rewritten the original equation into the following

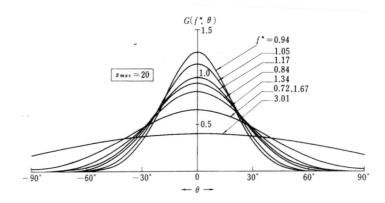

Fig. 2.12   Example of Mitsuyasu-type spreading function for $s_{\max} = 20$.

form by introducing the peak value of $s$, denoted as $s_{\max}$, as the principal parameter for the purpose of engineering applications:

$$
s = \begin{cases} (f/f_p)^5\, s_{\max} & : f \leq f_p, \\ (f/f_p)^{-2.5}\, s_{\max} & : f \geq f_p. \end{cases} \tag{2.25}
$$

Here, $f_p$ denotes the frequency at the spectral peak and is related to the significant wave period as $f_p = 1/(1.05 T_{1/3})$ for the case of Bretschneider–Mitsuyasu frequency spectrum. The relation between $f_p$ and $T_{1/3}$ for other frequency spectra can be obtained from Eqs. (2.14) and (2.18). Figure 2.12 illustrates the directional spreading function at various frequencies $f^* = f/f_p$ for $s_{\max} = 20$. The range of the energy distribution is set as $-\pi/2 \leq \theta \leq \pi/2$ and the normalization constant $G_0$ is evaluated by numerical integration for each directional spreading function.

### (C)  *Estimation of the spreading parameter $s_{\max}$*

The degree of directional spreading of wave energy greatly affects the extent of wave refraction and diffraction. This will be discussed in Chapter 3. Thus, the estimation of the value of the parameter $s_{\max}$ requires careful study of the nature of the waves at the design site. The observations by Mitsuyasu *et al.*[20] have shown that the peak value $s_{\max}$ increases as the parameter $2\pi f_p U/g$, which represents the state of wind wave growth. They introduced the relation

$$
s_{\max} = 11.5\,(2\pi f_p U/g)^{-2.5}, \tag{2.26}
$$

where $U$ denotes the wind speed.

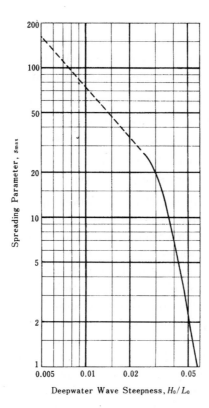

Fig. 2.13   Relationship between spreading parameter and deepwater wave steepness.[21]

Acccording to Wilson's formulas[22] for the growth of wind waves to be introduced in Sec. 3.1, the decrease of $2\pi f_p U/g$ can be associated with the decrease of deepwater wave steepness $H_0/L_0$. By utilizing the formulas, $s_{max}$ can be related to $H_0/L_0$ by means of Eq. (2.26). Figure 2.13 shows the result of such a calculation.[21] The portion of the result in the range $H_0/L_0 < 0.026$, where Wilson's formula is not applicable, is an extrapolation using the slope of the curve drawn with the solid line. Qualitatively speaking, the inverse proportionality between $s_{max}$ and $H_0/L_0$ in the range of swell is expected, as the pattern of wave crests should become long-crested as the swell propagates. The computer simulated crest patterns of Fig. 2.1 also demonstrates the tendency of long-crestedness with increase of $s_{max}$.

The curve in Fig. 2.13 is an estimation of the mean relationship between $s_{max}$ and $H_0/L_0$. Actual wave data would show a wide scatter around the curve of Fig. 2.13, as Wilson's formula itself represents the mean relation of wind wave growth derived from data with a fairly large scatter. Consideration of such scatter in the data and other factors leads to the recommendation of the following values of $s_{max}$ for engineering applications:[21]

$$
\left.
\begin{array}{lll}
\text{(i) Wind waves:} & s_{max} = 10\,, \\
\text{(ii) Swell with short decay distance:} & s_{max} = 25\,, \\
\text{\ \ \ \ (with relatively large wave steepness)} & \\
\text{(iii) Swell with long decay distance:} & s_{max} = 75\,. \\
\text{\ \ \ \ \ (with relatively small wave steepness)} &
\end{array}
\right\} \quad (2.27)
$$

Values in Eq. (2.27) as well as the relationship shown in Fig. 2.13 are to be applied to deepwater waves. In water of finite depth, where structures are to be built, waves have transformed under the effect of wave refraction which result in longer wave crests and reduced dispersion in wave directions. The directional wave spectrum also transforms accordingly. The degree of wave transformation by refraction depends on the bathymetry of the seabed. In an area where the seabed topography can be represented with straight, parallel depth-contours, the variation of the directional spreading function can be treated by means of an apparent increase in the value of $s_{max}$, as shown in Fig. 2.14.[21] In this figure, $(\alpha_p)_0$ denotes the angle of incidence of the deepwater waves and $L_0$ appearing in the abscissa is the length of the deepwater waves corresponding to the significant wave period; it is calculated as $L_0 = 1.56\,T_{1/3}^2$ in the units of meters and seconds. As the

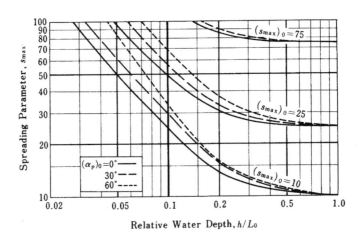

Fig. 2.14   Estimation of spreading parameter $s_{max}$ in shallow water area.[21]

effect of the incident angle $(\alpha_p)_0$ on $s_{max}$ is seen to be small in Fig. 2.14, this figure would appear to be applicable to a seabed of general topography as a reasonable approximation.

**Example 2.2**

Estimate the value of $s_{max}$ when wind waves of $H_{1/3} = 6$ m and $T_{1/3} = 10$ s reach water of depth $h = 15$ m.

**Solution**

The deepwater wavelength corresponding to the significant wave period is $L_0 = 156$ m, and the deepwater wave steepness is fairly large, taking the value $H_0/L_0 \simeq 0.04$. The value of $s_{max}$ in deepwater is thus estimated as $s_{max} = 10$. Because $h/L_0 \simeq 0.096$ at the specified water depth, the spreading parameter $s_{max}$ is estimated to increase to $25 \sim 35$ according to Fig. 2.14.

### (D) *Cumulative distribution curve of wave energy*

The characteristics of the directional wave spectrum can also be expressed from the viewpoint of the directional distribution of total wave energy. The cumulative relative energy $P_E(\theta)$ is defined for this purpose as follows[21]:

$$P_E(\theta) = \frac{1}{m_0} \int_{\theta_0 - \pi/2}^{\theta} \int_0^\infty S(f, \theta) \, df \, d\theta \,, \qquad (2.28)$$

where $\theta$ is the azimuth, $\theta_0$ is the azimuth of the principal wave direction, and $m_0$ denotes the representative value of the total wave energy and is given by

$$m_0 = \int_0^\infty \int_{\theta_0 - \pi/2}^{\theta_0 + \pi/2} S(f, \theta) \, df \, d\theta \,. \qquad (2.29)$$

The range of integration in the azimuth is set as $[\theta_0 - \pi/2, \theta_0 + \pi/2]$ because the wave components moving in the direction opposite to the principal wave direction are discarded in most designs of maritime structures.

Figure 2.15 shows the calculated result for the cumulative distribution of relative wave energy with respect to direction.[21] The frequency spectrum of Eq. (2.10), the directional spreading function of Eq. (2.22) and the spreading parameter of Eq. (2.25) were employed to obtain Fig. 2.15. The cumulative distribution curves for $s_{max} = 5, 10, 25$ and $75$ are given, as well as the curve for the direction spreading function of SWOP, which will be discussed in the next section.

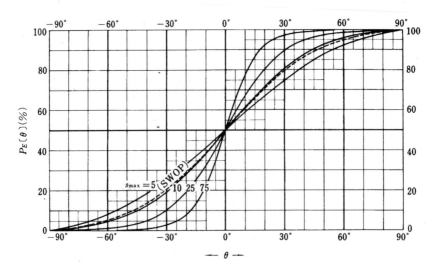

Fig. 2.15  Cumulative distribution of relative wave energy with respect to azimuth from principal wave direction.[21]

### Example 2.3

Estimate the ratio of wave energy contained in the range of $\pm 15°$ from the principal wave direction for wind waves with $s_{\max} = 10$.

### Solution

By reading off the values of the cumulative distribution curves of $s_{\max} = 10$ in Fig. 2.15 at $\theta = 15°$ and $\theta = -15°$, we obtain

$$\Delta E = P_E(15°) - P_E(-15°) = 0.67 - 0.33 = 0.34 \,.$$

That is, 34% of the total energy is contained in the range between $\pm 15°$.

The above example may appear to be in disagreement with the cumulative distribution curve of Fig. 2.11. However, $s_{\max} = 10$ represents the peak value and the overall mean of $s$ for the full frequency range remains low, while Fig. 2.11 represents the cumulative curve of $G(\theta|f)$ at a single frequency.

### (E) *Other directional spreading functions*

When Pierson *et al.*[23] proposed the P-N-J wave forecasting method based on the directional spectral concept, they adopted the $\cos^2 \theta$-type spreading function based on the conceptual suggestion by Arthur.[24] Though it was not validated by field measurements, it was simple to use and its extended version of the following form is sometimes used:

$$
G(\theta|f) \equiv G(\theta) = \begin{cases} \dfrac{2l!!}{\pi(2l-1)!!} \cos^{2l}(\theta - \theta_0) & : \ |\theta - \theta_0| \le \dfrac{\pi}{2}, \\[3mm] 0 & : \ |\theta - \theta_0| > \dfrac{\pi}{2}, \end{cases} \qquad (2.30)
$$

where $2l!! = 2l \cdot (2l-1) \cdots 4 \cdot 2$, $(2l-1)!! = (2l-1) \cdot (2l-3) \cdots 3 \cdot 1$. Pierson et al.[23] used the above function with $l = 1$ for the analysis of directional wave refraction.

The $\cos^{2l}$-type spreading function of Eq. (2.30) disregards the frequency-dependency of directional spreading as expressed by Eq. (2.25) of the Mitsuyasu-type function. It is an approximate expression of how the wave energy is distributed over the azimuth as a whole.

An ambitious effort called the Stereo Wave Observation Project (SWOP) was undertaken in the 1950s, which took stereo-photographic pictures of sea surface with two airplanes flying in parallel and analyzed the directional wave spectra from the sea surface contours reproduced from the stereo-photographs.[25] The project produced the first resolution of the directional spreading function of sea waves, which is called the SWOP function of the following:

$$
G(\theta|\omega) = \frac{1}{\pi} \left\{ 1 + \left( 0.50 + 0.82 \exp\left[ -\frac{1}{2}\left(\frac{\omega}{\omega_0}\right)^4 \right] \right) \cos 2(\theta - \theta_0) \right.
$$
$$
\left. + 0.32 \exp\left[ -\frac{1}{2}\left(\frac{\omega}{\omega_0}\right)^4 \right] \cos 4(\theta - \theta_0) \right\} : \ |\theta| \le \frac{\pi}{2}, \qquad (2.31)
$$

in which $\omega = 2\pi f$, $\omega_0 = g/U_{5.0}$ and $U_{5.0}$ denotes the wind speed at an elevation of 5.0 m above the sea surface. This directional spreading function is similar to the Mitsuyasu-type in that the wave components in the high frequency region exhibit broad spreading of wave energy.

Since the late 1980s, American researchers began to use the following wrapped normal spreading function, which is a representation of the Fourier series expression of directional spectrum and proposed by Borgman,[26] when they simulate directional random waves in multidirectional wave basins.

$$
G(\theta|f) = \frac{1}{2\pi} + \frac{1}{\pi} \sum_{n=1}^{N} \exp\left[ -\frac{(n\sigma_\theta)^2}{2} \right] \cos n(\theta - \theta_0), \qquad (2.32)
$$

in which $\sigma_\theta$ denotes the directional spreading standard deviation defined by the following:

$$\sigma_\theta^2(f) = \int_{\theta_{\min}}^{\theta_{\max}} (\theta - \theta_0)^2 \, G(\theta|f) \, d\theta \,. \tag{2.33}$$

The number of the serial terms $N$ in Eq. (2.32) is taken at some large figure to assure the convergence of the series. When using the wrapped normal spreading function, the directional spreading standard deviation $\sigma_\theta$ should vary with the frequency. However, many tests with directional random wave have employed some constant values over the full frequency range such that $\sigma_\theta = 30°$ for waves with wide directional spreading and $\sigma_\theta = 10°$ for waves with narrow directional spreading.

When Ewans[27] analyzed the directional spectrum of swell that had traveled across the Indian Ocean and reached the west coast of the Northern Island of New Zealand in 2001, he employed a Fourier series expression of the spreading function. He presented his swell data in the form of the wrapped normal distribution of Eq. (2.32) and found that $\sigma_\theta$ is the smallest with the value of $10°$ at the spectral peak frequency and its value increases as the frequency moves away from the peak frequency in both the lower and higher frequency ranges. This frequency dependency of directional spreading of swell is common with the Mitsuyasu-type spreading function of Eq. (2.25). The value of $\sigma_\theta = 10°$ is equivalent to $s = s_{\max} \simeq 65$ according to Eq. (2.35) to appear later. Thus, Ewan's data provides a supporting evidence to the proposed value of $s_{\max} = 75$ for swell in Eq. (2.27).

(F) *Interrelations between various spreading functions*

Because the functional forms of directional spreading differ among the Mitsuyasu-type of Eqs. (2.22) to (2.25), the $\cos^{2l}$-type of Eq. (2.30), and the wrapped normal distribution type of Eq. (2.32), direct comparison between them is difficult. Nevertheless, calculation of the directional spreading standard deviation $\sigma_\theta$ of Eq. (2.33) enables mutual conversion possible. Goda[28] has calculated $\sigma_\theta$ for the directional spreading functions mentioned above with the results as introduced below.

For the $\cos^{2l}$-type spreading function, $\sigma_\theta$ is given by

$$\sigma_\theta \simeq \frac{180°}{\pi} \left( \frac{2}{2.2 + 4l} \right)^{1/2} \,. \tag{2.34}$$

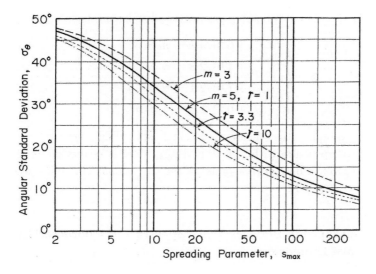

Fig. 2.16 Design diagram of directional spreading standard deviation of the Mitsuyasu-type directional spectrum.[28]

For the spreading function of $\cos^{2s}(\theta/2)$-type, $\sigma_\theta$ is calculated as below.

$$\sigma_\theta \simeq \frac{180°}{\pi} \left(\frac{2}{1+s}\right)^{1/2}. \tag{2.35}$$

In case of the Mitsuyasu-type spreading function, the spreading parameter varies with the frequency. For evaluation of the overall directional spreading, the directional spectral density is integrated with respect to the frequency so that the following overall directional spreading is defined:

$$\overline{G}(\theta) = \frac{G_0}{m_0} \int_0^\infty S(f) \cos^{2s}\left(\frac{\theta - \theta_0}{2}\right) df. \tag{2.36}$$

The cumulative relative energy $P_E(\theta)$ defined by Eq. (2.28) is equivalent to the integration of $\overline{G}(\theta)$ with respect to the azimuth $\theta$. Calculation of the directional spreading standard deviation $\sigma_\theta$ of the Mitsuyasu-type function is made by inserting $\overline{G}(\theta)$ into Eq. (2.33) and using the JONSWAP and Wallops frequency spectra for $S(f)$. The result of calculation is shown in Fig. 2.16.

The variation of $\sigma_\theta$ shown in Fig. 2.16 was empirically formulated as a function of $s_{\max}$. By substituting the resultant value of $\sigma_\theta$ into Eq. (2.35), the mean directional spreading parameter $\bar{s}$ has been estimated as below.[28]

$$\bar{s} = \begin{cases} (s_{\max} + 6.3)/4.1 & : \quad m = 3\,, \\ (s_{\max} + 4.0)/2.8 & : \quad \gamma = 1.0 \text{ or } m = 5\,, \\ (s_{\max} + 1.0)/2.1 & : \quad \gamma = 3.3\,, \\ (s_{\max} + 0.2)/1.7 & : \quad \gamma = 10\,. \end{cases} \tag{2.37}$$

For example, the mean directional spreading parameter of the JONSWAP spectrum with $\gamma = 10$ and $s_{\max} = 75$ is estimated as $\bar{s} = 44.2$ by Eq. (2.37), which corresponds to the directional spreading standard deviation of $\sigma_\theta \simeq 12°$.

## 2.4  Relationship Between Wave Spectra and Characteristic Wave Dimensions

### 2.4.1  *Relationship Between Wave Spectra and Wave Heights*

The representation of sea waves with characteristic height and period parameters and the spectral description of sea waves reflect two aspects of the same physical phenomenon, though they look quite different. It is possible to relate the two representations to each other. For example, spectral information for engineering applications can be obtained from the data of the significant wave by adopting one of the frequency spectra of Eqs. (2.10) to (2.16) and by combining the directional spreading function of Eq. (2.22) or other spreading functions.

It is also possible to estimate the heights and periods of representative waves from the wave spectrum. First, the representative value of the total wave energy $m_0$ is obtained by integrating the directional wave spectrum in the full frequency and directional ranges as given by Eq. (2.29). The integral, which has the units of m$^2$, cm$^2$, or similar dimensions, is by definition of the wave spectrum equal to the variance of the surface elevation. Thus,

$$m_0 = \overline{\eta^2} = \lim_{t_0 \to \infty} \frac{1}{t_0} \int_0^{t_0} \eta^2 dt\,. \tag{2.38}$$

The root-mean-square (rms) value of the surface elevation is then given by

$$\eta_{\text{rms}} = \sqrt{\overline{\eta^2}} = \sqrt{m_0}\,. \tag{2.39}$$

This rms value bears a certain relationship to the heights of representative waves when the wave height follows the Rayleigh distribution. In particular,

$$H_{1/3} \simeq H_{m0} = 4.004\eta_{\text{rms}} = 4.004\sqrt{m_0}, \qquad (2.40)$$

in which the notation of $H_{m0}$ is introduced to specify the significant wave height being estimated from the spectral information.

The proportionality between the significant wave height and the rms surface elevation has been confirmed by many wave observation data taken throughout the world. The coefficient of proportionality is slightly different from that derived via the Rayleigh distribution. The significant wave height is better expressed by the mean relationship $H_{1/3} \simeq 3.8\sqrt{m_0}$ for deep water.[29] However, with decrease in water depth, the coefficient of proportionality tends to increase to a value of 4.0 or greater, owing to wave nonlinearity as will be discussed in Sec. 10.5.2.[30]

Decrease of the proportionality coefficient reflects a narrowing of wave height distribution from that of the Rayleigh. Forristall,[31] for example, proposed the following function for the probability of exceedance for storm waves generated by hurricanes:

$$P(\xi) = \exp[-\xi^{-2.126}/8.42] \; : \; \xi = H/\eta_{\text{rms}}. \qquad (2.41)$$

With the above equation, the statistical prediction of $H_{\max}$ becomes slightly smaller than those given by Eqs. (2.4) to (2.7). The coefficient of proportionality in Eq. (2.40) also becomes smaller than 4.0 (see Sec. 10.1.2 (B) for details).

To clarify the definition of the significant wave height, the IAHR Working Group on Wave Generation and Analysis[32] has proposed to use the notation of $H_{m0}$ when estimated by $m_0$ or $H_\sigma$ when estimated by $\eta_{\text{rms}}$: the significant wave height calculated from the zero-upcrossing or zero-downcrossing method is recommended to be denoted as $H_{1/3,\,\text{u}}$ or $H_{1/3,\,\text{d}}$. The latter differentiation of $H_{1/3,\,\text{u}}$ and $H_{1/3,\,\text{d}}$ is not meaningful, because they are statistically the same.[3] A confusion in naming may be avoided by calling $H_{m0}$ as the *spectral* significant wave height and $H_{1/3}$ as the *statistical* significant wave height. When the technical term of "significant wave height" is used in the present book, it refers to the statistical one of $H_{1/3}$ unless otherwise stated.

Such a departure of the wave heights from the Rayleigh distribution is due to the spread of wave spectra over a wide frequency range, contrary to the assumption of the narrow-banded spectrum in the derivation of the

Table 2.3  Mean values of wave height ratios for the JONSWAP-type and Wallops-type spectra.

| Wave Height Ratio | Wallops-type Spectrum | | | | JONSWAP-type Spectrum | | | Rayleigh Distribution |
|---|---|---|---|---|---|---|---|---|
| | $m=3$ $\kappa=0.32$ | $m=5$ $\kappa=0.39$ | $m=10$ $\kappa=0.56$ | $m=20$ $\kappa=0.75$ | $\gamma=3.3$ $\kappa=0.55$ | $\gamma=10$ $\kappa=0.71$ | $\gamma=20$ $\kappa=0.80$ | |
| $[H_{max}/\eta_{rms}]^*$ | 0.89* | 0.93* | 0.96* | 0.96* | 0.93* | 0.94* | 0.93* | 1.00* |
| $H_{1/10}/\eta_{rms}$ | 4.66 | 4.78 | 4.91 | 5.01 | 4.85 | 4.92 | 4.96 | 5.090 |
| $H_{1/3}/\eta_{rms}$ | 3.74 | 3.83 | 3.90 | 3.95 | 3.87 | 3.91 | 3.93 | 4.004 |
| $\overline{H}/\eta_{rms}$ | 2.36 | 2.45 | 2.49 | 2.50 | 2.46 | 2.48 | 2.50 | 2.507 |
| $[H_{max}/H_{1/3}]^*$ | 0.96* | 0.97* | 0.98* | 0.98* | 0.97* | 0.96* | 0.95* | 1.00* |
| $H_{1/10}/H_{1/3}$ | 1.247 | 1.248 | 1.263 | 1.268 | 1.253 | 1.259 | 1.261 | 1.271 |
| $H_{1/3}/\overline{H}$ | 1.584 | 1.565 | 1.561 | 1.577 | 1.573 | 1.576 | 1.577 | 1.597 |

Note: (1) The values listed in this table are based on the results of numerical simulations of wave profiles with given wave spectra in the frequency range of $f = (0.5 \sim 6.0)f_p$.

(2) Figures with the mark * refer to the ratio to the theoretical values based on the Rayleigh distribution.

(3) $\kappa$ value is calculated by Eq. (2.42).

Rayleigh distribution. The effect of spectral shape on the wave height distribution has been examined by the author[14] by means of the Monte Carlo simulation technique. Table 2.3 lists the results of numerical simulations for the mean values of various wave height ratios for the JONSWAP-type spectrum of Eq. (2.12) and the Wallops-type spectrum of Eq. (2.16). In the case of the JONSWAP-type spectrum, the wave height ratios gradually increase toward those of the Rayleigh distribution as the peak enhancement factor $\gamma$ becomes large. In the case of the Wallops-type spectrum, approach to the wave height ratios of the Rayleigh distribution is realized with the increase of the exponent $m$. The ratio $H_{1/3}/\eta_{\rm rms}$ is about 3.8 when $\gamma = 1$ or $m = 5$, which is in good agreement with the result of field measurement data.

Departure of the wave height distribution from the Rayleigh is controlled by the *spectral shape parameter* defined by the following[33]:

$$\kappa(\overline{T})^2 = \left|\frac{1}{m_0}\int_0^\infty S(f)\cos 2\pi f\overline{T}, df\right|^2 + \left|\frac{1}{m_0}\int_0^\infty S(f)\sin 2\pi f\overline{T}\, df\right|^2 .$$
(2.42)

The parameter $\kappa$ was first introduced by Rice[34] in his classical paper on random noise in 1944 (see Sec. 10.2 for details).

Figure 2.17 shows the variation of the wave height ratio $H_{1/3}/\eta_{\rm rms}$ of field data with the spectral shape parameter $\kappa(T_{01})$, where the mean period is evaluated from frequency spectra as $\overline{T} = T_{01} = m_0/m_1$, where $m_0$ and $m_1$ refer to the zeroth and first spectral moments, respectively. Legends to the symbols refer to the location names with the number of wave records analyzed. Each symbol is plotted at the position of the mean values with the horizontal and vertical lines equivalent to twice the standard deviations.

The dotted curve represents the empirical relationship derived from the results of numerical simulation studies and is empirically expressed as below.

$$H_{1/3}/\eta_{\rm rms} = 3.459 + 1.353\,\kappa - 1.385\,\kappa^2 + 0.5786\,\kappa^3 .$$
(2.43)

Field data exhibit large scatter owing to the statistical variability due to small sample sizes (around 100 waves) of wave records, but they generally follow the trend of the simulation results. The data marked "Kochi-infra" are the records of infragravity waves, having been reconstructed by means of the inverse FFT method by using the Fourier amplitudes in the frequency less than 1/30 Hz. Their spectra showed almost uniform energy density over the frequency range. The statistical significant wave height of the waves with such flat spectra indicates the relationship of $H_{1/3} = (3.3 \sim 3.6)\eta_{\rm rms}$,

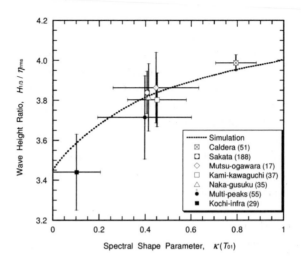

Fig. 2.17   Variation of wave height ratio $H_{1/3}/\eta_{\rm rms}$ versus spectral shape parameter $\kappa(T_{01})$.[33]

being much smaller than that of $H_{1/3} = 4.004\eta_{\rm rms}$ under the Rayleigh distribution.

The Bretschneider–Mitsuyasu spectrum of Eq. (2.10) has been adjusted so as to yield the result $m_0 = 0.0624H_{1/3}^2$ when it is integrated in the range $f = 0$ to $f = \infty$. In the case of Eqs. (2.11), (2.12) and (2.16), adjustments have been made on the basis of the relationship between $H_{1/3}$ and $\eta_{\rm rms}$ listed in Table 2.3. Thus, the frequency spectra can be constructed by means of Eqs. (2.10) to (2.16) when the significant wave height and some representative wave period are specified.

Inversely, the estimation of significant wave height can always be made from the given information of wave spectrum, by first evaluating the zeroth spectral moment $m_0$ with Eq. (2.38), then converting it to the rms surface elevation $\eta_{\rm rms}$ with Eq. (2.39), and finally utilizing the relationship of Eq. (2.40). Such operations for the wave height estimation from spectral information become necessary in the analysis of wave refraction, diffraction, etc., in which the transformation of the directional wave spectrum is principally computed. A slight adjustment to the relationship of Eq. (2.40) using the result of Table 2.3 is possible, but will not be practical in the computation of wave transformations because the wave spectrum changes its shape through the transformations. The heights of representative waves other than the significant wave height can be derived by means of Eqs. (2.3) to (2.6).

## 2.4.2 Relationship Between Wave Spectra and Wave Periods

According to the statistical theory of random waves (see Sec. 10.3), the mean wave period defined by the zero-upcrossing method is given by the zeroth and second moments of the frequency spectrum as follows:

$$\overline{T} = \sqrt{m_0/m_2} \ : \ m_2 = \int_0^\infty f^2 S(f)\, df. \qquad (2.44)$$

This relationship is often utilized when period information is sought from data of the wave spectrum. However, there have been reports that Eq. (2.44) produces a period shorter than the period directly counted on wave profiles (about 83% on the average, according to Ref. 4), though some other reports indicate quasi-equality between the two mean periods.[29] One reason is the different performances of wave recording instruments, and another is the effect of wave nonlinearity, in that actual sea waves are not simply composed of an infinite number of infinitesimal independent wavelets but are also accompanied by some phase-locked harmonic components (cf. Sec. 10.5.3).

Since the late 1990s, the mean period $T_{m-1,0}$ defined by the following equation is becoming popular among European engineers:

$$T_{m-1,0} = m_{-1}/m_0 \ : \ m_{-1} = \int_0^\infty f^{-1} S(f)\, df. \qquad (2.45)$$

The mean period $T_{m-1,0}$ is not much affected by the high frequency part of the spectrum. It also yields a smooth variation when wind waves gradually develop at the background of swell presence and the spectral peaks corresponding to wind waves surpasses that of swell. If the spectral peak period is employed as the representative period, its value would jump from the swell period to the wind wave period at a certain stage of wave development. Avoidance of a sudden change of the representative period seems to be one reason favoring $T_{m-1,0}$. The ratio of $T_{m-1,0}/T_p$ has been calculated for the Wallops- and JONSWAP-type spectra[35] and is listed in the row below that of $T_{1/3}/T_p$ in Table 2.4. As seen there, $T_{m-1,0}$ is almost the same as $T_{1/3}$ for single-peaked spectra. A few comparisons have also been made for double-peaked spectra with the data of numerically simulated wave profiles, yielding a near equality between them.[35] Therefore, the representative period $T_{m-1,0}$ may be called the *spectral significant wave period* and can be used as the alternative to $T_{1/3}$ when the latter information is missing.

Table 2.4 is the result of a Monte Carlo simulation study with various wave spectral shapes.[14] As the spectral peak becomes sharp, the differences

Table 2.4   Mean values of wave period ratios for the JONSWAP-type and Wallops-type spectra.

| Wave Period Ratio | Wallops-type Spectrum | | | | JONSWAP-type Spectrum | | |
|---|---|---|---|---|---|---|---|
| | $m = 3$ | $m = 5$ | $m = 10$ | $m = 20$ | $\gamma = 3.3$ | $\gamma = 10$ | $\gamma = 20$ |
| $T_{1/10}/T_p$ | 0.82 | 0.89 | 0.93 | 0.96 | 0.93 | 0.96 | 0.97 |
| $T_{1/3}/T_p$ | 0.78 | 0.88 | 0.93 | 0.96 | 0.93 | 0.97 | 0.98 |
| $T_{m-1,0}/T_p$ | 0.77 | 0.86 | 0.93 | 0.97 | 0.90 | 0.94 | 0.96 |
| $\overline{T}/T_p$ | 0.58 | 0.74 | 0.89 | 0.95 | 0.80 | 0.87 | 0.91 |
| $T_{\max}/T_{1/3}$ | 1.07 | 0.99 | 0.99 | 0.99 | 0.99 | 0.99 | 0.99 |
| $T_{1/10}/T_{1/3}$ | 1.06 | 1.00 | 0.99 | 1.00 | 1.00 | 1.00 | 1.00 |
| $T_{1/3}/\overline{T}$ | 1.35 | 1.19 | 1.06 | 1.02 | 1.16 | 1.11 | 1.09 |

Note: The values listed in this table are based on the results of numerical simulations of wave profiles with given wave spectra in the frequency range of $f = (0.5 \sim 6.0)f_p$.

between various wave period parameters become small and these periods approach the peak period $T_p$. In the wave transformation analysis for refraction, diffraction, etc., the information on the period parameters other than $T_p$ is often sought for. For this purpose, the relationship between $T_{m-1,0}$ computed by Eq. (2.45) and the input period parameter, such as $T_{1/3}$, is first investigated for the input wave spectrum. This relationship is utilized for correlating $T_{m-1,0}$ computed from the spectrum after transformation to other period parameters of interest.

## 2.5   Long-Period Waves Accompanying Wind Waves and Swell

### (A)   *General*

Wind waves generated and developed under strong storm winds may have the significant height over 15 m with the significant period exceeding 18 s. When such well-developed storm waves propagate as swell over long distance, their period become elongated to 20 s or more; such swell provides board surfers with best opportunity to display their skill. However, no wave groups with the period longer than 30 s have ever been recorded, even though the spectrum of large waves in shallow water usually exhibit an appreciable amount of energy at the frequency range less than $T_p/2$.

When we watch the wave run-up on a beach minutely, we can detect irregular ups and down of the maximum run-up point with the period of one to several minutes. It manifests a presence of a slow, irregular fluctuation of the mean water level around the shoreline, which is called the *surf beats*. The phenomenon was first reported by Munk[36] in 1949, and confirmed by Tucker[37] and Yoshida[38] in 1950. Surf beats are one aspect of the long-period waves or low-frequency waves, which are sometimes regarded as same as surf beats.

Long-period waves are also called the *infragravity waves* in general. In the present book, the waves with the period range of around 30 to 300 s are mainly referred to as the long-period waves. Slow fluctuations of the sea water level with the period longer than 600 s or 10 minutes are usually called as "seiche" or "secondary undulations of oceanic tides," which are a kind of resonant oscillations pertaining to respective topographic features of embayment, harbor basins, and others. Such resonant oscillations are often excited by tsunamis or minute barometric oscillations. Hibino and Kajiura[39] proved that large seiches in Nagasaki Bay, which sometimes exceed 2 m in height and are locally called "Abiki," are induced by barometric oscillations over the East China Sea and grow in height by resonances.

### (B) *Bounded versus free long-period waves*

Long-period waves generally follow the growth and decay of wind waves and/or swell (both are hereafter called as the sea waves) on the whole. Among a group of irregular waves propagating in the sea, the mean water level of the area where high waves come together is found to be lower than the rest, while that of the area with low waves is higher. The difference among the local water levels is caused by the spatial variation of the radiation stresses associated with the wave momentum flux, which is proportional to the square of wave height; the strong stress under high waves pushes down the water level, while the weak stresses under low waves pulls up the water level.

The spatial variation of the local water level can be regarded as the waves of long wavelengths, which propagate together with the group of original irregular waves, being bounded by the envelope of the original wave group. In this sense, the long waves are called the *group-bounded long waves*. The propagation speed of the group-bounded long waves is same as that of the original wave group and much slower than the group velocity corresponding to the period of the long waves. Generation and development of the group-bounded long waves are analyzed with the nonlinear wave interaction theory to be discussed in Sec. 10.5.3. The height of the group-bounded long waves is proportional to the square of the representative height of the original wave group, such as the significant wave height, and it decreases as the directional spreading of the original wave group increases.

When the original wave group encounters with a cape or breakwaters, waves are subject to the diffraction process and propagate into the sheltered area with the lessened height. Because of the decrease in the binding power of the radiation stresses in the sheltered area, the long waves partly begin to behave as free waves. Within the surf zone, the envelope structure of the original wave group is destroyed by the depth-limited breaking process, resulting in evolution of group-bounded long waves into *free long waves*. These are qualitative explanation of the mechanism for generation of free long waves, but quantitative description of the mechanism is incomplete yet. Free long waves have very small steepness, and thus they are almost fully reflected from natural beaches toward the offshore. Upon beaches of mild slopes, there are formed the cross-shore standing wave systems of long waves in most cases.

Separation of the group-bounded and free long waves requires a certain effort. Figure 2.18 has been presented by Sekimoto *et al.*,[40] who analyzed the wave records measured at the depth of 15 m off Kashiwa-zaki Coast, Niigata Prefecture in Japan. They separated the spectral density below and above the threshold frequency at $f_{cut} = 1/20$ Hz, and summed up the wave energies of long-period and sea waves to evaluate the zeroth moments of respective waves, i.e.

$$\text{Long-period waves} \; : \; m_{0,L} = \int_0^{f_{\text{cut}}} S(f)\,df\,, \qquad (2.46)$$

$$\text{Sea waves} \; : \; m_{0,S} = \int_{f_{\text{cut}}}^\infty S(f)\,df\,. \qquad (2.47)$$

Then they estimated the significant wave height of long-period waves as $G_{1/3} = 4.0\sqrt{m_{0,L}}$ and that of sea waves as $H_{1/3} = 4.0\sqrt{m_{0,S}}$. The data in Fig. 2.18 indicate that the height of the long-period waves is proportional to the height of sea waves when the latter is small, while it tends to be proportional to the square of the height as the sea waves becomes large. The two diagonals with the legend "Goda (1975)" represents the empirical prediction with Eq. (3.46) in Sec. 3.6.2, while the two steep lines with the legend "Nonlinear Interaction Theory" have been calculated with the theory in Sec. 10.5.3. The result of Fig. 2.18 suggests the dominance of group-bounded long waves as sea waves become large.

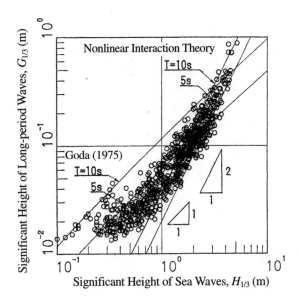

Fig. 2.18 Example of observed relationship between the heights of long-period waves and sea waves after Sekimoto *et al.*[40]

(C) *Empirical prediction of long-period wave height*

The generation mechanism and behavior of long-period waves are not fully understood yet. From the practical point of views, efforts have been made to correlate the long-period wave height with the height and period of sea waves, because long-period waves often hinder safe ship mooring and efficient cargo handling operations in harbors. Several empirical formulas have been proposed for respective harbors where measurements were made, but it is difficult to synthesize them into some unified form because of differences in relative dominance of the bounded waves over free waves, characteristic features of measurement sites, and other factors.

When attention is focused on the group-bounded long waves, their spectra and heights can be computed with the nonlinear interaction theory. Bowers,[41] for example, carried out such calculations and gave the following approximation formula for the case of unidirectional sea waves:

$$H_B = 0.074 H_{1/3}^2 T_p^{\,2}/h^2 \,, \qquad (2.48)$$

where $H_B$ denotes the significant height of the long-period waves as defined by $H_B = 4.0\sqrt{m_{0,L}}$ with Eq. (2.46). The wave heights $H_B$ and $H_{1/3}$ and the depth $h$ are given the units of m, while the period $T_p$ is expressed with the units of s.

Kato and Nobuoka[42] also computed the bounded long wave height numerically and derived three empirical formulas for the cases of the directional spreading parameter being $s_{max} = 10, 100$, and $1000$; the height of bounded long waves was expressed as proportional to some powers of Ursell's number. The formulas were further simplified by Nakai[43] as in the following:

$$H_B = 0.014 H_{1/3}^2 T_{1/3}{}^2 s_{max}{}^{0.26}/h^2. \qquad (2.49)$$

When the directional spreading parameter is set $s_{max} = 1,260$ and the peak period is converted as $T_p = 1.1 T_{1/3}$, the term of $0.014 s_{max}$ in Eq. (2.49) yields a constant of $0.074$, which is the same as Eq. (2.48). When $s_{max} = 10$ with the relation of $T_p = 1.1 T_{1/3}$ is employed, however, the term of $0.014 s_{max}$ in Eq. (2.49) decreases to a value of $0.021$, which is $1/3.5$ of the former case. Thus, Eq. (2.49) exemplifies the effect of directional spreading on the bounded long-period waves.

The frequency spectra of long-period waves are mostly flat without concentration at particular frequency. If there appear some spectral peaks, a confidence interval check should be made because the interval is usually quite wide owing to low degrees of freedom in spectral analysis of low frequency waves. In some coastal areas, long-period waves may persistently exhibit spectral peaks at several fixed frequencies that correspond to the resonant frequencies of the water areas. For analysis of the motions of vessel moored at berths, Hiraishi[44] has proposed a standard spectrum of long-period waves, which is characterized with uniform spectral density at the low frequency range.

The significant height of long-period waves shown in Fig. 2.18 and Eqs. (2.48) and (2.49) are all the spectral significant wave height. When the profiles of long-period waves are recorded directly with some digital filters attached to the recording instrument or the profiles are recovered by means of the inverse FFT method, the statistical analysis with the zero-upcrossing method would yield the statistical significant wave height $H_{1/3,L}$ much smaller than the spectral one $H_{m0,L}$ as in the following:

$$H_{1/3,L} = (3.3 \sim 3.6)\eta_{rms\,L} = (0.82 \sim 0.90) H_{m0,L}. \qquad (2.50)$$

The above relationship is based on the experience of the data analysis of the infragravity waves at Kochi, shown in Fig. 2.17.

# References

1. Y. Goda, "A proposal of systematic calculation procedures for the transformations and actions of irregular waves," *Proc. Japan Soc. Civil Engrs.* **253** (1976), pp. 59–68 (*in Japanese*).

2. W. J. Pierson, Jr., "An interpretation of the observable properties of sea waves in term of the energy spectrum of the Gaussian record," *Trans. American Geophys. Union* **35**(5) (1954), pp. 747–757.

3. Y. Goda, "Effect of wave tilting on zero-crossing wave heights and periods," *Coastal Engineering in Japan* **29** (1986), pp. 79–90.

4. Y. Goda and K. Nagai, "Investigation of the statistical properties of sea waves with field and simulation data," *Rept. Port and Harbour Res. Inst.* **13**(1) (1974), pp. 3–37 (*in Japanese*).

5. M. S. Longuet-Higgins, "On the statistical distributions of the heights of sea waves," *J. Marine Res.* **IX**(3) (1952), pp. 245–266.

6. Y. Goda, "New wave pressure formulae for composite breakwaters," *Proc. 14th Int. Conf. Coastal Engrg.* (Copenhagen, ASCE) (1974), pp. 1702–1720.

7. For example, "Light," *Encyclopaedia Britannica* **14** (1964), p. 59.

8. C. L. Bretschneider, "Significant waves and wave spectrum, *Ocean Industry* (Feb. 1968), pp. 40–46.

9. H. Mitsuyasu, "On the growth of spectrum of wind-generated waves (2) – spectral shape of wind waves at finite fetch," *Proc. Japanese Conf. Coastal Engrg.* (1970), pp. 1–7 (*in Japanese*).

10. W. J. Pierson, Jr. and L. Moskowitz, "A proposed spectral form for fully developed wind seas based on the similarity law of S. A. Kitaigorodskii," *J. Geophys. Res.* **69**(24) (1964), pp. 5181–5190.

11. H. Mitsuyasu, "On the growth of the spectrum of wind-generated waves (1)," *Rept. Res. Inst. Applied Mech., Kyushu Univ.* **XVI**(55) (1968), pp. 459–482.

12. K. Hasselmann *et al.*, "Measurements of wind-wave growth and swell decay during the Joint North Sea Wave Project (JONSWAP)," *Deutsche Hydr. Zeit Reihe A* (8°) **12** (1973).

13. M. K. Ochi and E. N. Hubble, "On six-parameter wave spectra," *Proc. 15th Int. Conf. Coastal Engrg.* (Hawaii, ASCE) (1976), pp. 301–328.

14. Y. Goda, "Statistical variability of sea state parameters as a function of a wave spectrum," *Coastal Engineering in Japan* **31**(1) (1988), pp. 39–52.

15. N. E. Huang, S. R. Long, C.-C. Tung, Y. Yuan and L. F. Bliven, "A unified two-parameter wave spectral model for a general sea state," *J. Fluid Mech.* **112** (1981), pp. 203–224.

16. E. Bouws, H. Gunther, W. Rosenthal and C. Vincent, "Similarity of the wind wave spectrum in finite depth water," *J. Geophys. Res.* **90**(C1) (1985), pp. 975–986.

17. M. J. Tucker, "Nearshore waveheight during storms," *Coastal Engineering* **24** (1995), pp. 111–136.

18. W. J. Pierson, Jr., G. Neumann and R. W. James, *Practical Methods for Observing and Forecasting Ocean Waves by Means of Wave Spectra and Statistics*, U.S. Navy Hydrographic Office, H. O. Pub. 603 (1955).

19. Y. Goda, "Analysis of wave grouping and spectra of long-travelled swell," *Rept. Port and Harbour Res. Inst.* **22**(1) (1983), pp. 3–41.

20. H. Mitsuyasu, F. Tasai, T. Suhara, S. Mizuno, M. Ohkusu, T. Honda and K. Rikiishi, "Observation of the directional spectrum of ocean waves using a cloverleaf buoy," *J. Physical Oceanogr.* **5**(4) (1975) pp. 750–760.

21. Y. Goda and Y. Suzuki, "Computation of refraction and diffraction of sea waves with Mitsuyasu's directional spectrum," *Tech. Note of Port and Harbour Res. Inst.* **230** (1975), 45p (*in Japanese*).

22. B. W. Wilson, "Numerical prediction of ocean waves in the North Atlantic for December, 1959," *Deutche Hydr. Zeit* **18**(3) (1965), pp. 114–130.

23. W. J. Pierson, Jr., J. J. Tuttle and J. A. Wooley, "The theory of the refraction of a short-crested Gaussian sea surface with application to the northern New Jersey coast," *Proc. 3rd Conf. Coastal Engrg.* (Cambridge, Mass., 1952), pp. 86–108.

24. R. W. Arthur, "Variability in direction of wave travel in ocean surface waves," *Ann. New York Acad. Sci.* **51**(3) (1949), pp. 511–522.

25. For example, B. Kinsman, *Wind Waves* (Prentice-Hall, Inc., 1965), p. 401 and pp. 460–471.

26. L. E. Borgman, "Directional spectrum estimation for the $S_{xy}$ gauges," *Tech. Rept., Coastal Engrg. Res. Center, USAE Waterways Exper. Station* (Vicksburg) (1984), pp. 1–104.

27. K. C. Ewans, "Directional spreading in ocean swell," *Proc. Int. Symp. WAVES 2001* (ASCE) (2002), pp. 517–529.

28. Y. Goda, "A comparative review on the functional forms of directional wave spectrum," *Coastal Engineering Journal* **41**(1) (1999), pp. 1–20.

29. For example, Y. Goda, "A review on statistical interpretation of wave data," *Rept. Port and Harbour Res. Inst.* **18**(1) (1979), pp. 5–32.

30. Y. Goda, "A unified nonlinearity parameter of water waves," *Rept. Port and Harbour Res. Inst.* **22**(3) (1983), pp. 3–30.

31. G. Z. Forristall, "On the statistical distribution of wave heights in a storm," *J. Geophys. Res.* **83**(C5) (1978), pp. 2353–2358.

32. The IAHR Working Group on Wave Generation and Analysis. "List of seastate parameters," *J. Waterway, Port, Coastal, and Ocean Engrg.*, ASCE **115**(6) (1989), pp. 793–808.

33. Y. Goda and M. Kudaka, "On the role of spectral width and shape parameters in control of individual wave height distribution," *Coastal Engineering Journal* **49**(3) (2007), pp. 311–335.

34. S. O. Rice, "Mathematical analysis of random noise," 1944, reprinted in *Selected Papers on Noise and Stochastic Processes* (Dover Pub., 1954) pp. 132–294.

35. Y. Goda, "A performance test of nearshore wave height prediction with CLASH datasets," *Coastal Engineering* **56** (2009), pp. 385–399.

36. W. H. Munk, "Surf beats," *Trans. Amer. Geophys. Union* **30**(6) (1949), pp. 849–854.

37. M. J. Tucker, "Surf beats: sea waves of 1 to 5 minute period," *Proc. Roy. Soc., London, Series A* **202** (1950), pp. 565–573.

38. K. Yoshida, "On the ocean wave spectrum, with special reference to the beat phenomenon and the '1-3 minute waves'," *J. Oceangra. Soc. Japan* **6**(2) (1950), pp. 49–56.

39. T. Hibino and K. Kajiura, "Origin of the Abiki phenomena (a kind of seiche) in Nagasaki Bay," *J. Oceanogr. Soc. Japan* **38**(3) (1981), pp. 172–182.

40. T. Sekimoto, T. Shimizu, Y. Kubo and S. Imai, "Field investigation on generation and propagation of surf beats outside and inside a harbor," *Proc. Coastal Engrg., JSCE* **37** (1990), pp. 86–90 (*in Japanese*).

41. E. C. Bowers, "Low frequency waves in intermediate water depths," *Coastal Engineering 1992* (*Proc. 23rd Int. Conf.*, Venice, ASCE) (1992), pp. 832–845.

42. H. Kato and H. Nobuoka, "Properties of the second order waves in numerical simulation for coastal area (2)," *Ann. J. Coastal Eng., JSCE* **52** (2005), pp. 136–140 (*in Japanese*).

43. K. Nakai, "Analysis of the relationship between bound waves and total infragravity waves using observed wave data," *Ann. J. Civil Eng. in the Ocean, JSCE* **22** (2006), pp. 151–156 (*in Japanese*).

44. T. Hiraishi, "Applicability of standard energy spectrum density for long period waves," *Tech. Note of Port and Harbour Res. Inst.* **934** (1975), 45p (*in Japanese*).

## Chapter 3

# Generation, Transformation and Deformation of Random Sea Waves

### 3.1 Simplified Forecasting Method of Wind Waves and Swell

#### 3.1.1 *Simplified Forecast of Wave Height and Period*

Waves in the sea are generated by winds and their growth is governed by the *wind speed*, the *fetch* (distance over which winds blow), and the *wind duration*. The current practice of wave forecasting is the numerical computation of directional wave spectrum by computers. There are several numerical models, among which WAM is an open-source program of spectral wave forecasting methods of the third generation (see Komen *et al.*[1] for details). When calibrated with instrumental measurement records, the forecast may attain the accuracy of $\pm 10\% \sim \pm 20\%$. However, it requires a large quantity of input data on the temporal and spatial variations of wind fields as well as the service of specialists for computation.

On the other hand, the S-M-B method essentially uses a simplified wind field of a constant wind speed $U$ over a fixed fetch length $F$ for a fixed duration $t$. Even though the accuracy is not as good as the spectral method, the S-M-B method can provide a first order estimate of the wave dimensions for conceptual design works. The method makes use of a set of nondimensional relationships between wave height, wave period, wind speed, and fetch length. The relationships are often called the *fetch diagrams* when expressed in graphical forms. The formulas for the relationships have been developed by a number of researchers, even though they yield somewhat different predictions of wave generation; see Sec. 4.2.4.6 of *Rock Manual*[2] for details. Within the group of the S-M-B method, the author regards Wilson's formulas most suitable for practical applications.

In 1965, Wilson[3] summarized a large number of wave observation data into the following relationship:

$$\frac{gH_{1/3}}{U_{10}^2} = 0.3 \left\{ 1 - \left[ 1 + 0.004 \left( \frac{gF}{U_{10}^2} \right)^{1/2} \right]^{-2} \right\}, \tag{3.1}$$

$$\frac{gT_{1/3}}{U_{10}} = 8.61 \left\{ 1 - \left[ 1 + 0.008 \left( \frac{gF}{U_{10}^2} \right)^{1/3} \right]^{-5} \right\}, \tag{3.2}$$

where $g$ is the acceleration of gravity, $H_{1/3}$ denotes the significant wave height, $T_{1/3}$ is the significant wave period, $U_{10}$ represents the wind speed at the elevation of 10 m above the sea surface, and $F$ is the fetch length.

Figures 3.1 and 3.2 are the dimensional representation of Eqs. (3.1) and (3.2) by Goda.[4]

The growth of wind waves are limited by the minimum wind duration, $t_{\min}$. It has been estimated through numerical calculation by Goda[4] as follows:

$$\frac{t_{\min}U}{F} = 43 \left( \frac{gF}{U_{10}^2} \right)^{-0.27}. \tag{3.3}$$

When $t_{\min}$ is counted by hours, $U$ by m/s, and $F$ by km, then Eq. (3.3) is rewritten as

$$t_{\min} = 1.0\, F^{0.73} U_{10}^{-0.46}. \tag{3.4}$$

Alternatively, the minimum fetch length $F_{\min}$ (km) necessary for full wave growth under a given wind duration $t$ (h) is expressed as follows:

$$F_{\min} = 1.0\, t^{1.37} U_{10}^{0.63}. \tag{3.5}$$

### Example 3.1

Estimate the significant height and period of wind waves generated under the condition of $U = 25$ m/s, $F = 350$ km, and $t = 10$ h.

### Solution

The minimum duration is calculated as $t_{\min} = 16.4$ h by Eq. (3.4), while the minimum fetch is $F_{\min} = 310.4$ km by Eq. (3.5). Thus, wave growth is governed by the wind duration of $t = 10$ h. By using $U = 25$ m/s and $F_{\min} = 310.4$ km, we can have the estimate of $H_{1/3} = 7.4$ m and $T_{1/3} = 10.3$ s.

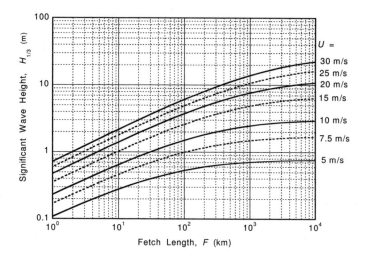

Fig. 3.1   Prediction curves of significant wave height $H_{1/3}$.[4]

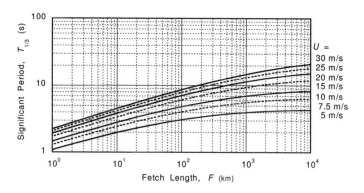

Fig. 3.2   Prediction curves of significant wave period $T_{1/3}$.[4]

Calculation of the significant wave height and period for various wind speeds and fetch lengths by Wilson's formulas has yielded the relationship between them as shown in Fig. 3.3. For a generated height of significant waves, the significant period depends on the wind speed, becoming shorter with increase in speed. When approximation for the mean relationship is made by neglecting differences due to wind speed, the following formula for wind waves is obtained[4]:

$$T_{1/3} \simeq 3.3 \, (H_{1/3})^{0.63} \, . \tag{3.6}$$

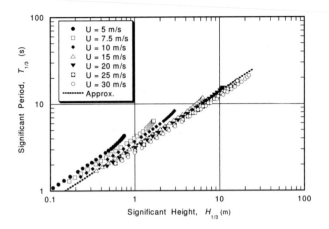

Fig. 3.3   Correlation between significant height and period of wind waves according to Wilson's formulas.[4]

### 3.1.2   *Simplified Forecast of Swell Height and Period*

Wind waves continue to grow within the fetch area, but they become swell and begin to decay after they leave the fetch area. Swell may travel over a distance of several thousand kilometers and cause damage on structures. The author[5] once identified a swell propagation over 7,000 to 9,000 km from the southeastern sea of New Zealand to the Pacific coast of Costa Rica, Central America in May 1981. It was estimated that the storm waves off New Zealand were generated by the mean wind speed of $\overline{U}_{10} = 22$ m/s over the fetch area of $F \simeq 2,000$ km with the result of $H_{1/3} = 10 \sim 11$ m and $T_{1/3} = 13 \sim 15$ s. When the swell traveled over the distance of $D \simeq 6,800$ km along the great circle of the earth to Costa Rica, its height was reduced to $H_{1/3} \simeq 3.1$ m and its period increased to $T_{1/3} \simeq 17$ s. The swell caused a partial damage on the rubble mound break-water of Caldera Port and accelerated the northward longshore sediment transport in the adjacent beaches.

The author[6] had another occasion to identify a swell propagation over some 4,500 km from the southwestern offshore of Australia to Mále Island, Maldive in early April, 1987. A graphical wave hindcasting method by means of Wilson's fetch diagrams produced an estimate of $H_{1/3} = 9.4$ m and $T_{1/3} = 12.3$ s over the fetch of about 1,800 km on the 7th of April. In the night of the 10th to the morning of the 11st of April, Mále Island was inundated by more than 1 m above the mean sea level, and people thought the flooding as caused by a storm surge. However, it was caused by the water overtopping the outer reef by the swell of about 3 m high and 16 s in the period.

For a quick estimation of swell height and period, Bretschneider[7, 8] presented in 1968 the following empirical formulas based on various field data and some theoretical consideration:

$$\frac{(H_{1/3})_D}{(H_{1/3})_F} = \left[ \frac{0.4F_{min}}{0.4F_{min} + D} \right]^{1/2} , \tag{3.7}$$

$$\frac{(T_{1/3})_D}{(T_{1/3})_F} = \left[ 2.0 - \frac{(H_{1/3})_D}{(H_{1/3})_F} \right]^{1/2} , \tag{3.8}$$

in which the suffixes $F$ and $D$ denotes the quantities at the ends of the fetch and decay areas, respectively, $F_{min}$ is the minimum fetch length, and $D$ is the swell travel distance. The above two cases of swell propagation at Costa Rica and Maldive approximately fit to the relationships of Eqs. (3.7) and (3.8).

### 3.1.3 Relationship Between Significant Height and Period of Wind Waves and Swell

Figure 3.4 gives the approximate range of the significant height and period of wind waves and swell according to the relationships of Eqs. (3.6) and (3.8). The diagram is composed of two regions marked as "zone of no waves" and "swell zone," being divided by a gray thick line marked as "wind waves." The group of thin curves represent the constant values of the deepwater wave steepness with the wavelength $L_0$ being calculated by using the significant wave period $T_{1/3}$. If $L_0$ is calculated with $T_p$ with the result of $L_{0,p} \simeq 1.2L_{0,1/3}$, then the value of wave steepness is reduced to about 80%.

The line of "wind waves" shows the relationship between $H_{1/3}$ and $T_{1/3}$ based on Eq. (3.6). The line should be regarded to have a certain range of variation, because Eq. (3.6) is an approximation as indicated in Fig. 3.3. At an early stage of wind wave growth, the wave steepness may be as large as 0.06 if the significant wave height is less than 2 m. As waves grow in height and period, the wave steepness decreases toward the asymptotic value of 0.026.

The area above the line of "wind waves" is the "zone of no waves," in which no combination of $H_{1/3}$ and $T_{1/3}$ in this zone is possible physically. Existence of such no wave zone is confirmed in the joint height and period

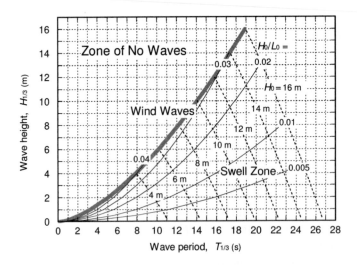

Fig. 3.4   Relationship between significant wave height and period of wind waves and swell.

tables of wave climate statistics. The dotted lines in the "swell zone" indicate the elongation of wave period with the decay in swell height according to Eq. (3.8). For example, wind waves developed to the height of 10 m with the period of 14 s have the steepness of 0.033. As the waves propagate as swell and the height is decreased to 5 m, the swell will have the period of 17 s and the steepness of 0.011.

## 3.2   Wave Refraction

### 3.2.1   *Introduction*

In an area where the water depth is greater than about one-half of the wavelength, i.e., a region of deep water, waves propagate without being affected by the seabed topography. When waves enter into a region of shallower water, however, the direction of wave propagation gradually shifts and the wave crestlines are bent into the pattern of the depth contours of the sea bottom. This process is the phenomenon of water wave refraction, which is analogous to that of light and sound waves, and is produced by the spatial variation of propagation speed.

Figure 3.5 exhibits a crest pattern of young swell approaching the shoreline of a planar coast with the uniform slope of 1 on 100 from the depth 20 m to 0.1 m. This has been obtained by numerical simulation of spatial wave profiles by assuming the directional spreading parameter of $s_{\max} = 25$

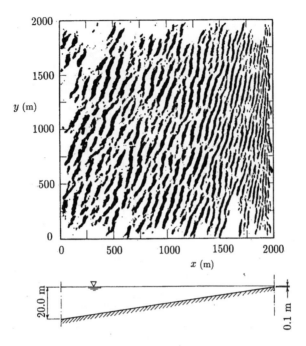

Fig. 3.5   Crest pattern of young swell being refracted in shoaling waters of planar beach $(T_p = 8.01 \text{ s})$.[1]

and the deepwater incident angle of $30°$.[9] As waves approach the shoreline, wavelengths are shortened and waves become long-crested, relative to the local wavelength.

The variation of the wave direction by refraction for the case of regular waves can be estimated graphically by hand or by numerical computation with a digital computer.[a] Figure 3.6 shows an example of the variation of wave direction analyzed by means of the graphical method for regular waves with a period of 12 s incident from SSE to a hypothetical water area. The solid lines with arrows are the wave rays, indicating the direction of wave propagation. Such a diagram of wave rays is called a *wave refraction diagram*. Although the distance between rays is constant in deep water (or in water of uniform depth), the rays converge and diverge as they approach the shoreline, depending on the local bottom features.

Because the flux of wave energy bounded by two rays is conserved if wave energy dissipation is negligible, the variation in ray separation distance

---

[a]See Refs. 10 to 13 for details of the techniques of refraction analysis for regular waves.

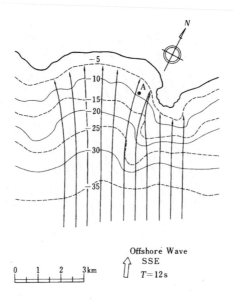

Fig. 3.6 Refraction diagram of regular waves.

means that the wave energy density varies inversely to the ray distance. Because the wave energy is proportional to the square of the wave height, the variation in wave height due to refraction is given by the following equation:

$$\frac{H}{H_0} = \sqrt{\frac{b_0}{b}} = K_r, \tag{3.9}$$

where $b$ denotes the distances between the two wave rays in the region of interest and $b_0$ is the distance between the same two rays in deep water. The wave height ratio $K_r$ is called the *wave refraction coefficient*. In the example of Fig. 3.6, the refraction coefficient at Point A is estimated to be $K_r = 0.94$ for waves of $T = 12$ s incident from SSE.

### 3.2.2 *Refraction Coefficient of Random Sea Waves*

The above refraction coefficient corresponds to regular waves with constant period and single direction of wave propagation. The variation in the heights of real sea waves due to refraction is not necessarily well represented by a refraction coefficient for regular waves. As discussed in Chapter 2, waves in the real sea are composed of an infinite number of components having different frequencies and directions. Therefore, the variation in the heights of sea waves is determined by the contributions from all components, each component wave undergoing the process of refraction individually. The fundamental equation for the estimation of the refraction coefficient of random sea waves is given by

$$(K_r)_{\text{eff}} = \left[ \frac{1}{m_{s0}} \int_0^\infty \int_{\theta\min}^{\theta\max} S(f,\theta)\, K_s^2(f)\, K_r^2(f,\theta)\, d\theta\, df \right]^{1/2} , \qquad (3.10)$$

in which

$$m_{s0} = \int_0^\infty \int_{\theta\min}^{\theta\max} S(f,\theta)\, K_s^2(f)\, d\theta\, df . \qquad (3.11)$$

The subscript "eff," standing for "effective," is used to denote quantities related to random waves, as opposed to regular waves. In the above, $S(f,\theta)$ denotes the directional wave spectrum, $K_s(f)$ is the wave shoaling coefficient to be introduced in Sec. 3.5, and $K_r(f,\theta)$ stands for the refraction coefficient of a component wave (i.e., a regular wave) with frequency $f$ and direction $\theta$. In actual calculations, the integration is replaced by a summation.

A simple way to make an approximate estimate of the refraction coefficient of random sea waves is to use the following equation:

$$(K_r)_{\text{eff}} = \left[ \sum_{i=1}^{M} \sum_{j=1}^{N} (\Delta E)_{ij}\, (K_r)_{ij}^2 \right]^{1/2} , \qquad (3.12)$$

for which it is assumed that the effect of shoaling is negligible.

The term $(\Delta E)_{ij}$ in the above equation denotes the relative energy of component waves with $i$th frequency and $j$th direction, when the frequency range of the random sea waves is divided into segments running from $i = 1$ to $M$ and the directional range is divided into segments running from $j = 1$ to $N$. That is,

$$(\Delta E)_{ij} = \frac{1}{m_0} \int_{f_i}^{f_i + \Delta f_i} \int_{\theta_j}^{\theta_j + \Delta\theta_j} S(f,\theta)\, d\theta\, df , \qquad (3.13)$$

in which

$$m_0 = \int_0^\infty \int_{\theta\min}^{\theta\max} S(f,\theta)\, d\theta\, df . \qquad (3.14)$$

In actual calculations, representative frequencies and directions of component waves must be selected. If the frequency spectrum of the Bretschneider–Mitsuyasu type as expressed by Eq. (2.10) in Sec. 2.3.1 is

Table 3.1   Representative periods of component waves for refraction analysis.

| Numbers of Component Waves | $T_i/T_{1/3}$ | | | | | | |
|:---:|:---:|:---:|:---:|:---:|:---:|:---:|:---:|
| | $i=1$ | $i=2$ | $i=3$ | $i=4$ | $i=5$ | $i=6$ | $i=7$ |
| 3 | 1.16 | 0.90 | 0.54 | – | – | – | – |
| 4 | 1.20 | 0.98 | 0.81 | 0.50 | – | – | – |
| 5 | 1.23 | 1.04 | 0.90 | 0.76 | 0.47 | – | – |
| 7 | 1.28 | 1.11 | 1.00 | 0.90 | 0.81 | 0.69 | 0.43 |

employed, the division of the frequency range can be made so as to equalize the wave energy in each frequency interval; note that Eq. (2.10) in Sec. 2.3.1 is integrable in closed form with respect to frequency. Such a division reduces the calculation time for the random refraction coefficient. The representative frequency in each interval is best determined as the mean of the second spectral moment of each interval so that the variation of wave period by refraction can be estimated with minimum error (because the mean wave period, Eq. (2.44), is given by the second moment of the frequency spectrum). The formula for the representative frequency of the second spectral moment of each interval has been given by Nagai[14]:

$$f_i = \frac{1}{0.9T_{1/3}} \left\{ 2.912M \left[ \Phi\left(\sqrt{2\ln\frac{M}{i-1}}\right) - \Phi\left(\sqrt{2\ln\frac{M}{i}}\right) \right] \right\}^{1/2},$$

(3.15)

in which $\Phi(t)$ is the error function defined by

$$\Phi(t) = \frac{1}{\sqrt{2\pi}} \int_0^t e^{-x^2/2}\, dx.$$

(3.16)

The representative frequencies by Eq. (3.15) have been converted to the wave periods listed in Table 3.1.

For a rapid computation, the median frequency (the frequency which bisects the area of the wave spectrum in each interval) may be employed. It is given by

$$f_i = \frac{1}{0.9T_{1/3}} \left\{ \frac{0.675}{\ln[2M/(2i-1)]} \right\}^{1/4} = \frac{1.007}{T_{1/3}} \left\{ \ln[2M/(2i-1)] \right\}^{-1/4}.$$

(3.17)

When the representative frequencies are selected by either one of the above two methods, the relative energy of the component waves may be approximated with the following:

$$(\Delta E)_{ij} = \frac{1}{M} D_j.$$

(3.18)

Table 3.2   Ratio of wave energy in each direction to the total energy.

| Direction of Component Waves | 16-Point Bearing $s_{max}$ | | | 8-Point Bearing $s_{max}$ | | |
|---|---|---|---|---|---|---|
| | 10 | 25 | 75 | 10 | 25 | 75 |
| 67.5° | 0.05 | 0.02 | 0 | – | – | – |
| 45.0° | 0.11 | 0.06 | 0.02 | 0.26 | 0.17 | 0.06 |
| 22.5° | 0.21 | 0.23 | 0.18 | – | – | – |
| 0° | 0.26 | 0.38 | 0.60 | 0.48 | 0.66 | 0.88 |
| −22.5° | 0.21 | 0.23 | 0.18 | – | – | – |
| −45.0° | 0.11 | 0.06 | 0.02 | 0.26 | 0.17 | 0.06 |
| −67.5° | 0.05 | 0.02 | 0 | – | – | – |
| Total | 1.00 | 1.00 | 1.00 | 1.00 | 1.00 | 1.00 |

The quantity $D_j$ represents the ratio of the wave energy in each direction to the total energy. It is read off from the cumulative distribution curves of relative wave energy shown in Fig. 2.15 in Sec. 2.3.2(D). For a directional division with 16 or 8 point bearings, values of $D_j$ have been prepared as listed in Table 3.2. If the directional range of component waves is limited to less than $\pm 90°$, owing to the topographic conditions at the site of interest, the values of $D_j$ should be linearly increased so that the summation in the range of possible wave approaches will give unity. When the frequency division of the wave spectrum is different from that listed in Table 3.1, the energy ratio $(\Delta E)_{ij}$ must be computed by integrating the frequency spectrum and by evaluating the ratio of wave energy in each frequency interval.

The above assignment of relative wave energy by Eq. (3.18) and Table. 3.2 ignores the fundamental nature of the directional wave spectrum by neglecting the degree of directional spreading of wave energy with variation in the frequency. This technique should be reserved for situations when the weighted summation of Eq. (3.12) is done manually.

As an explanatory example, the refraction coefficient at the point A in Fig. 3.6 will be calculated for random sea waves. By employing three divisions in the frequency range taken from Table. 3.1, the wave periods of the component waves are obtained as $T_1 = 14$ s, $T_2 = 11$ s, and $T_3 = 6.5$ s for $T_{1/3} = 12$ s. The directions of the component waves are chosen from a 16-point bearing within the range of $\pm 90°$ around the principal direction of SSE. The waves are assumed to be swell of relatively large wave steepness so that the spreading parameter $s_{max}$ is set at the value of 25. The initial stage

Table 3.3    Example of random wave refraction analysis.

| Direction of Component Waves | $K_r$ | | | $\sum K_r^2$ | $D_j$ | $\frac{D_j}{M} \sum K_r^2$ |
|---|---|---|---|---|---|---|
| | 14 s | 11 s | 6.5 s | | | |
| E | 0.69 | 0.60 | 0.65 | 1.259 | 0.02 | 0.008 |
| ESE | 0.90 | 0.77 | 0.76 | 1.981 | 0.06 | 0.040 |
| SE | 1.07 | 1.11 | 0.95 | 3.280 | 0.23 | 0.251 |
| SSE | 1.11 | 0.86 | 0.95 | 2.874 | 0.38 | 0.364 |
| S | 0.64 | 0.78 | 0.99 | 1.998 | 0.23 | 0.153 |
| SSW | 0.84 | 0.95 | 1.02 | 2.649 | 0.06 | 0.053 |
| SW | 0.72 | 0.62 | 0.76 | 1.480 | 0.02 | 0.010 |

$\sum \frac{D_j}{M} \sum K_r^2 = 0.879$, $(K_r)_{\text{eff}} = 0.938$.

of the estimation process is to draw refraction diagrams and to estimate the refraction coefficient at the point A for all 21 component waves (three wave periods and seven directions). The result is listed in the 2nd to 4th columns of Table 3.3. The second step is to calculate the sum $\sum_{i=1}^{M} K_r^2$ and to complete the 5th column of Table 3.3. Also, the wave energy ratio $D_j$ is to be read from Table 3.2 for $s_{\text{max}} = 25$ and listed in the 6th column of Table 3.3. Then the products of the entries in the 5th and 6th columns divided by $M = 3$ are written in the 7th column. By summing the entries in the 7th column and by taking the square root, the refraction coefficient of random sea waves in this example is obtained as $(K_r)_{\text{eff}} = 0.94$. The refraction coefficient for the principal deepwater wave direction of SE or S can also be obtained by shifting the entry for $D_j$ at the respective principal directions. The result becomes $(K_r)_{\text{eff}} = 0.94$ for SE and $(K_r)_{\text{eff}} = 0.89$ for S.

For comparison, the refraction coefficient corresponding to regular waves for the wave direction SSE is $K_r = 0.94$, which happens to be the same as the refraction coefficient found above for random sea waves from SSE. But the refraction coefficients of regular waves incident from SE and S are 1.10 and 0.70 respectively, which are seen to differ greatly from the value 0.94 found for the direction SSE. It is rather difficult, however, to believe that the wave conditions after refraction are so different for a change in the wave direction of only $\pm 22.5°$. The process of taking a weighted mean in Eq. (3.12) produces smoothing of the erratic values of the refraction coefficient as obtained for regular waves. Thus, by introducing the concept of directional wave spectrum, we obtain more realistic values for the refraction coefficient for random sea waves than those obtained for regular waves.

The prevailing direction of waves after undergoing refraction may be taken as the direction which has the maximum value in the 7th column of Table 3.3. In the example of Table 3.3, the offshore wave direction which contains the greatest wave energy density after refraction is SSE. By referring to the refraction diagram of regular waves, the prevailing direction of the refracted waves at Point A is estimated to be N165°.

### 3.2.3 Computation of Random Wave Refraction by Means of the Energy Balance Equation

In addition to the above method of superposing refracted component waves, the refraction of random sea waves can be analyzed by numerically solving the equation for the balance of the flux of wave energy as described by the directional spectrum of waves in water of variable depth and topography (e.g., Refs. 15 to 17), which is due to Karlsson.[15] The fundamental equation, which is called the *energy balance equation*, takes the form

$$\frac{\partial}{\partial x}(Sv_x) + \frac{\partial}{\partial y}(Sv_y) + \frac{\partial}{\partial \theta}(Sv_\theta) = 0, \qquad (3.19)$$

where $S$ denotes the directional wave spectral density and $v_x, v_y$ and $v_\theta$ are given by

$$\left.\begin{aligned}
v_x &= c_g \cos\theta, \\
v_y &= c_g \sin\theta, \\
v_\theta &= \frac{c_g}{c}\left(\frac{\partial c}{\partial x}\sin\theta - \frac{\partial c}{\partial y}\cos\theta\right).
\end{aligned}\right\} \qquad (3.20)$$

The symbols $c$ and $c_g$ denote the phase and group velocities, respectively, and they are calculated with the following formulas for waves with period $T$ and length $L$ in water of depth $h$:

$$\left.\begin{aligned}
c &= \frac{L}{T} = \frac{g}{2\pi}T \tanh\frac{2\pi h}{L}, \\
c_g &= \frac{1}{2}\left[1 + \frac{4\pi h/L}{\sinh(4\pi h/L)}\right]c.
\end{aligned}\right\} \qquad (3.21)$$

The method has been applied to the case of wave refraction occurring at a spherical shoal as shown in Fig. 3.7, with a diameter of 40 m and water depth of 5 m at its top, set in water of uniform depth of 15 m.[17] The distributions of wave height and period by random wave refraction are shown in Fig. 3.8 for waves with $T_{1/3} = 5.1$ s. The directional wave spectrum was

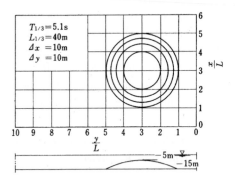

Fig. 3.7   Shape of spherical shoal.[17]

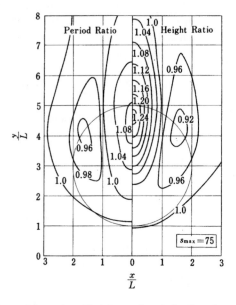

Fig. 3.8   Distributions of the ratios of heights and periods of random waves on a spherical shoal.[17]

assumed to be the Bretschneider–Mitsuyasu frequency spectrum combined with the Mitsuyasu-type spreading function having $s_{max} = 75$. The right side of Fig. 3.8 gives the variation of refracted wave height, while the left side shows the variation of wave period. The transformation of random sea waves is generally accompanied by some change in wave period because the directional spectrum varies through the wave transformation. Figure 3.8 is one such example.

The refraction of regular waves over this shoal has been solved by Ito *et al.*[18] with their numerical technique of wave propagation analysis. The result of the distribution of wave height is presented in Fig. 3.9. As seen in this figure, the refraction of regular waves often produces a significant spatial variation in the wave height. The computation of wave refraction using spectral components of various directions and frequencies brings forth the effect of smoothing such large spatial variations. Vincent and Briggs[19] have investigated the pattern of wave height behind an elliptical shoal in a laboratory for both regular and directional random waves. They reported that the most significant factor affecting the wave height distribution was the amount of directional spread.

Strictly speaking, waves over a shoal are affected not only by the refraction but also by the diffraction phenomenon, especially when wave caustics is formed by the crossing of wave rays over a shoal. A number of numerical schemes have been developed to solve such refraction-diffraction problems for regular waves (see Ref. 20). Even though the energy balance equation method of Eqs. (3.19) and (3.20) cannot take the diffraction effect into account, it provides a reasonable estimate of wave height for directional random waves around a shoal, or over a sea bottom topography of such complexity that the conventional refraction analysis of regular waves will give the crossing of wave rays. It is especially so when the directional

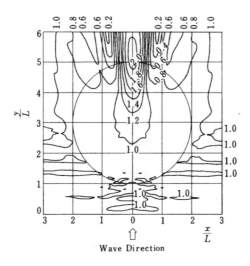

Fig. 3.9 Distribution of the height ratio of regular waves on a spherical shoal (after Ito *et al.*[18]).

angular spread of incident spectrum is broad, as indicated by O'Reilly and Guza.[21]

### 3.2.4    *Wave Refraction on a Coast with Straight, Parallel Depth-Contours*

For the case of coastal water with straight, parallel depth-contours, the variation in the ray direction and the refraction coefficient of the component waves can be obtained analytically. Then, the computation of the refraction of random sea waves is rather easily made with the superposition method. The refraction coefficient of random sea waves and the variation in their predominant wave directions have been obtained as shown in Figs. 3.10 and 3.11, respectively (Goda and Suzuki[17]).

The computation was made with the frequency and directional components of $M = N = 36$ for the directional spectrum, using the Bretschneider–Mitsuyasu frequency spectrum and the Mitsuyasu-type spreading function.

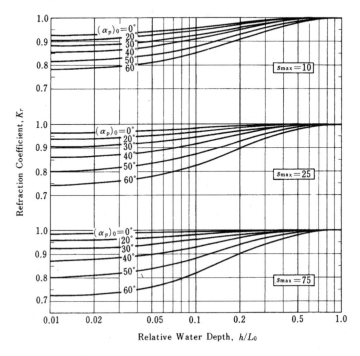

Fig. 3.10   Refraction coefficient of random sea waves on a coast with straight, parallel depth-contours.[17]

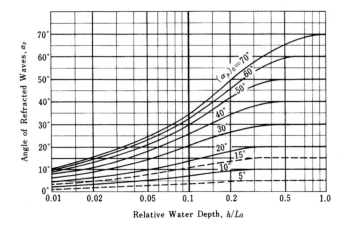

Fig. 3.11   Variation of predominant direction of random sea waves due to refraction on a coast with straight, parallel depth-contours.

The wavelength $L_0$ appearing in the abscissa of Figs. 3.10 and 3.11 is the deepwater wavelength corresponding to the significant wave period. The parameter $(\alpha_p)_0$ denotes the angle between the principal direction of the deepwater waves and the line normal to the depth-contours.

The refraction coefficient of random sea waves varies with the value of $s_{max}$, but the difference is less than several percent. The predominant direction of refracted waves, which was defined by Nagai[22] as the direction corresponding to the highest directional spectral density, is little affected by the value of $s_{max}$.

It should be noted in Fig. 3.10 that even normally incident waves experience a decrease in refraction coefficient as they propagate into shallow water. This is caused by refraction of the obliquely incident component waves on both sides of the principal direction of normal incidence, because waves described by means of a directional wave spectrum always contain component waves with directions different from the principal direction.

### Example 3.2

Describe the condition of refracted waves at the water depths of 20 and 10 m, when swell with height of 2 m and period of 12 s is incident at an angle of 40° to a coast with straight, parallel depth-contours.

### Solution

Referring to Eq. (2.27) in Sec. 2.3.2 with the information on swell condition, the spreading parameter $s_{max}$ is set at 75. Since the

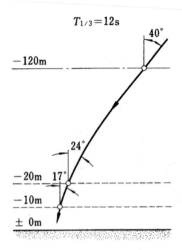

Fig. 3.12   Change in wave direction.

deepwater wavelength corresponding to $T_{1/3} = 12$ s is $L_0 = 225$ m (refer to Table A.3 in the appendix), the relative water depth is $h/L_0 = 0.089$ at $h = 20$ m. The refraction coefficient is $K_r = 0.92$ by Fig. 3.10 and the predominant direction is $\alpha_p = 24°$ by Fig. 3.11. At the water depth of $h = 10$ m, $K_r = 0.90$ and $\alpha_p = 17°$ with $h/L_0 = 0.044$. The variation in predominant wave direction is illustrated in Fig. 3.12.

## 3.3   Wave Diffraction

### 3.3.1   *Principle of Random Wave Diffraction Analysis*

When water waves encounter an obstacle such as a breakwater, island, or headland during propagation, they pivot about the edge of the obstacle and move into shadow zone of the obstacle. This phenomenon is called the diffraction of water waves, and it is common with the other wave motions of sound, light, and electromagnetic waves. The spatial distribution of diffracted wave height of regular waves in uniform depth can be computed by means of the Sommerfeld solution based on velocity potential theory. The results are compiled as diagrams showing the distribution of the ratio of diffracted to incident wave heights, which and are called *diffraction diagrams*. Conventional diagrams given in many references (e.g., Ref. 23) have been prepared for regular waves with constant period and single directional component. Direct application of such conventional diagrams to real situations is not recommended, because they can lead to erroneous results.

The diffracted heights of real sea waves should be computed as follows by introducing the directional wave spectrum:

$$(K_d)_{\text{eff}} = \left[ \frac{1}{m_0} \int_0^\infty \int_{\theta\,\text{min}}^{\theta\,\text{max}} S(f,\theta) \, K_d^2(f,\theta) \, d\theta \, df \right]^{1/2} , \qquad (3.22)$$

where $(K_d)_{\text{eff}}$ denotes the *diffraction coefficient of random sea waves* (i.e., the ratio of diffracted to incident heights of significant or other representative waves), $K_d(f,\theta)$ is the diffraction coefficient of component (regular) waves with frequency $f$ and direction $\theta$, and $m_0$ is the integral of the directional spectrum specified by Eq. (3.14).

The validity of the computation of random wave diffraction with Eq. (3.22) has been confirmed through simultaneous wave observations inside and outside the storm-surge breakwater at Nagoya Port.[24] Capacitance wave gauges were set at the point A outside and the point B inside the harbor as shown in Fig. 3.13. An example of the frequency spectra at both points is presented in Fig. 3.14. Because the breakwater is of the caisson type and reflects incident waves almost completely, the wave energy observed at the point A is considered to be the sum of incident and reflected wave energy. By taking one-half of the energy at the point A as that of the incident waves, the height and period of the incident waves were estimated as $H_{1/3} = 0.46$ m and $T_{1/3} = 2.8$ s.

The wave direction was judged to be from SW on the basis of the observed wind direction. The spectrum of the diffracted waves was calculated for this wave condition, and the resultant spectrum was in good agreement with the observed spectrum at the point B. If the diffraction is calculated with the theory of regular waves, the diffraction coefficient becomes

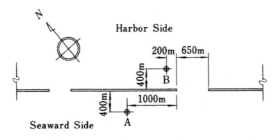

Fig. 3.13   Location of wave observation stations.

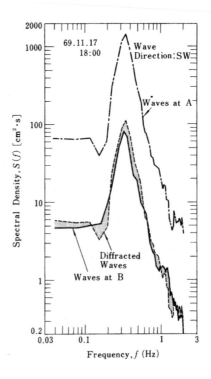

Fig. 3.14   Example of frequency spectrum of diffracted waves observed in the field.[24]

$K_d \simeq 0.07$ for the condition of $x/L \simeq 20$ and $y/L \simeq 31$ with the wavelength of $L \simeq 12$ m corresponding to the significant wave period. In terms of the spectral density, the theory of regular wave diffraction yields only 3% of the observed value.

### 3.3.2   *Diffraction Diagrams of Random Sea Waves*

Diffraction diagrams of random sea waves have been computed with Eq. (3.22), and they are shown in Figs. 3.15 to 3.19 (Goda *et al.*[24]). The directional wave spectrum employed is the combination of the Bretschneider–Mitsuyasu frequency spectrum and the Mitsuyasu-type spreading function. The integrations in Eq. (3.22) were replaced by summations over 10 frequency intervals by Eq. (3.15) and 20 to 36 directional intervals of equal spacing ($\Delta\theta = 9°$ to $5°$). Figure 3.15 pertains to diffraction by a semi-infinite straight breakwater, whereas Figs. 3.16 to 3.19 are for diffraction through the opening between two semi-infinite straight breakwaters with

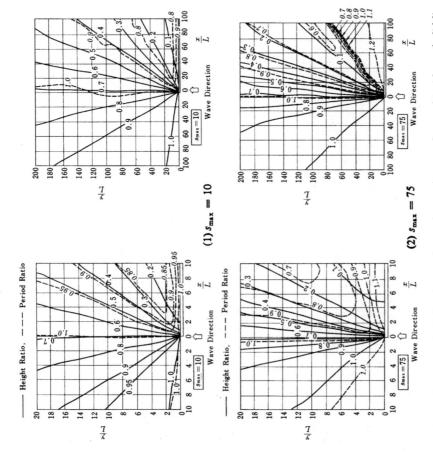

Fig. 3.15 Diffraction diagrams of a semi-infinite breakwater for random sea waves of normal incidence (solid lines for wave height ratio and dash lines for wave period ratio).[24]

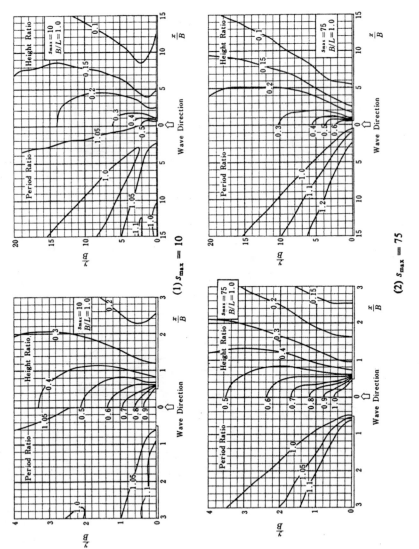

Fig. 3.16   Diffraction diagrams of a breakwater opening with $B/L = 1.0$ for random sea waves of normal incidence.[24]

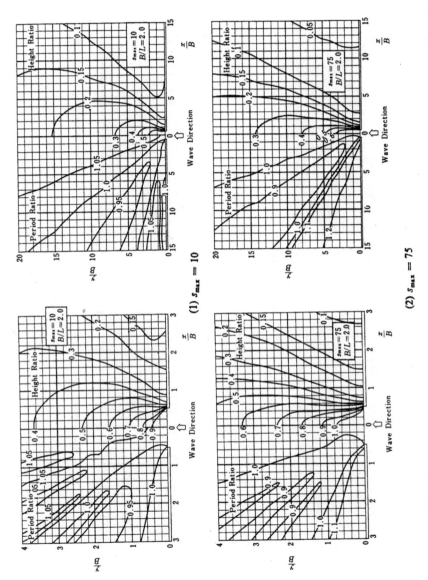

Fig. 3.17   Diffraction diagrams of a breakwater opening with $B/L = 2.0$ for random sea waves of normal incidence.[24]

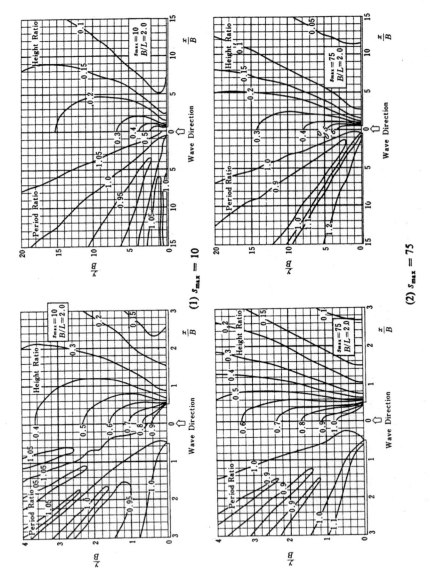

Fig. 3.18    Diffraction diagrams of a breakwater opening with $B/L = 4.0$ for random sea waves of normal incidence.[24]

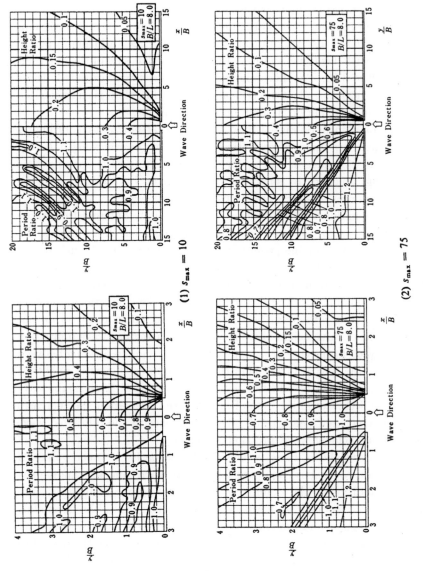

Fig. 3.19 Diffraction diagrams of a breakwater opening with $B/L = 8.0$ for random sea waves of normal incidence.[24]

the opening width being 1, 2, 4 and 8 times the wavelength corresponding to the significant wave periods. The wave direction is normal to the axis of the breakwaters. Each figure is composed of four diagrams for two values of $s_{max}$ (10 and 75) in the near and distant areas.

This set of random wave diffraction diagrams shows not only variations in wave height but also variations in wave period. The variation in period is indicated with dash lines in Fig. 3.15 and on the left side of the diagrams of Figs. 3.16 to 3.19. The diffraction of random sea waves, especially those with a frequency-dependent directional spreading function, is characterized by changes in wave period in addition to wave height.

Another comment on Figs. 3.16 to 3.19 is that the horizontal coordinates are normalized by the opening with $B$ instead of the wavelength $L$. By doing so, the difference in $(K_d)_{eff}$ for different values of the opening ratio $B/L$ appears much to be smaller than that given by plotting with the coordinates normalized by wavelengths.

The random diffraction coefficient values are quite different from those of regular waves. For example, the diffraction coefficient of random waves along the boundary of the geometric shadow (or the straight line drawn from the tip of the breakwater parallel to the wave direction) takes the value of about 0.7, while regular wave diffraction theory gives a diffraction coefficient of about 0.5. The difference between the predictions increases in the sheltered area behind the breakwater, and would result in an underestimation of wave height there if diagrams for regular wave diffraction were employed. In the case of wave diffraction through an opening between breakwaters, the spatial variation of the diffraction coefficient is smoothed to some extent by the introduction of wave directionality. That is, the wave height ratio decreases in the area of direct wave penetration and increases in the sheltered area. As a result, the dependence of the diffracted wave height on the incident wave direction decreases.

As discussed above, there is considerable disparity between values of the diffraction coefficient computed for regular waves and for random sea waves. In addition to the example of Fig. 3.14, another example of wave diffraction from the field is shown in Fig. 3.20, illustrating the dissimilarity between regular and random wave. The data were taken by the Akita Port Construction Office, the First District Port Construction Bureau, Ministry of Transport (presently Ministry of Land, Infrastructure, Transport and Tourism) of Japan (Irie[25]). Figure 3.20 compares the ratio of wave height inside to outside a single breakwater, measured with wave recorders of the inverted echo-sounder type, with calculated values of the diffraction

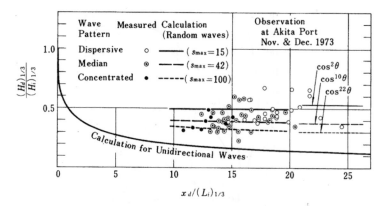

Fig. 3.20   Measured diffraction coefficients for single breakwater at Akita Port in comparison with calculation.[25]

coefficient. The curves with the legends $s_{max} = 15$, 42 and 100 denote the results for random wave diffraction, while the continuous curve at the bottom represents the calculation for regular wave diffraction. It is clear from this figure that the calculation of random waves is in reasonable agreement with the (widely scattered) data, whereas the calculation for regular waves yields inadequately small values of the diffraction coefficient and does not describe the field data at all.

In 1995, Briggs *et al.*[26] carried out laboratory tests of wave diffraction by a semi-infinite breakwater for both regular and directional random waves. Clear differences between traditional regular wave diffraction and random wave diffraction were observed in the test results. Briggs *et al.* concluded that directional spreading is very important and should be considered in diffraction analysis of engineering problems.

When using random diffraction diagrams, the effect of wave refraction upon the parameter $s_{max}$ must be taken into account, because most breakwaters are built in relatively shallow water compared to the predominant wavelength, and the directional wave spectrum has been transformed from that of deepwater waves. The change in $s_{max}$ can be estimated with the curves in Fig. 2.14 in Sec. 2.3.2.

### 3.3.3   *Random Wave Diffraction of Oblique Incidence*

In most situations, waves will arrive at an oblique angle to a breakwater. In the case of wave diffraction by a semi-infinite breakwater, problem of

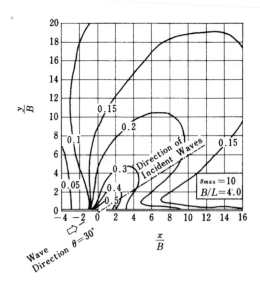

Fig. 3.21  Diffraction diagram of random sea waves of oblique incidence.

oblique incidence can be treated by rotating the axis of the breakwater in the diffraction diagrams of Fig. 3.15 while keeping the wave direction and the coordinate axes at their original positions. This technique produces some error when the angle between the principal direction of wave approach and the line normal to the breakwater exceeds ±45°.

On the other hand, for waves diffracted through an opening in a breakwater, the axis of the diffracted waves or the line connecting the points of deepest penetration of contour lines deviates slightly toward the line normal to the breakwaters as demonstrated in Fig. 3.21. The angle of deviation varies depending on the angle between the wave direction and the line parallel to the breakwaters, the relative opening $B/L$, and the spreading parameter $s_{\max}$. From the analysis of several diffraction diagrams for obliquely incident waves, the deviation angle of the axis of the diffracted waves has been estimated as listed in Table 3.4.

Unless the diffraction diagrams of random sea waves are directly computed with the aid of a computer, the diagrams in Figs. 3.16 to 3.19 for normal incidence must be utilized to estimate wave heights behind breakwaters. When the waves are obliquely incident, the direction of wave approach should be shifted by the amount of the angle of deviation listed in Table. 3.4, and the apparent opening width as viewed from the shifted wave direction should be employed instead of the actual width of the opening.

Table 3.4  Deviation angle of diffracted waves through a breakwater opening for obliquely incident waves.

| $s_{max}$ | $B/L$ | $\Theta = 15°$ | $\Theta = 30°$ | $\Theta = 45°$ | $\Theta = 60°$ |
|---|---|---|---|---|---|
| | | Deviation Angle $\Delta\Theta$ | | | |
| | 1.0 | 37° | 28° | 20° | 11° |
| 10 | 2.0 | 31° | 23° | 17° | 10° |
| | 4.0 | 26° | 19° | 15° | 10° |
| | 1.0 | 26° | 15° | 10° | 6° |
| 75 | 2.0 | 21° | 11° | 7° | 4° |
| | 4.0 | 15° | 6° | 4° | 2° |

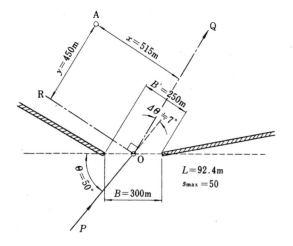

Fig. 3.22  Layout of breakwaters and calculation point.

## Example 3.3

Estimate the diffraction coefficient at Point A for the layout of the breakwaters shown in Fig. 3.22 when wind waves with $T_{1/3} = 10$ s are incident from a direction which makes an angle of 50° with the line connecting the two breakwater heads. The water depth is $h = 10$ m.

## Solution

The spreading parameter $s_{max}$ is set at the value of 10 in deep water, from the condition of wind waves. However, as the site of the breakwaters is shallow, the parameter $s_{max}$ must have increased there.

By referring to Fig. 2.14 in Sec. 2.3.2, we estimate the parameter $s_{max} = 50$ for $h/L_0 = 0.064$. The wavelength corresponding to the significant wave period at the water depth of $h = 10$ m at the entrance is read as $L = 92.3$ m from Table A.2 in the Appendix. Next, the deviation angle of the axis of the diffracted waves needs to be estimated. For the opening ratio $B/L = 3.2$ and angle between the incident waves and breakwaters of $\Theta = 50°$, the deviation angel is interpolated as $\Delta\Theta \simeq 14°$ for $s_{max} = 10$ and $\Delta\Theta \simeq 4°$ for $s_{max} = 75$. Therefore, $\Delta\Theta \simeq 7°$ may be taken for $s_{max} = 50$.

When applying the diffraction diagram, the opening width is an apparent one as viewed from the direction $\overrightarrow{OQ}$ which is the predominant direction after diffraction. In this example, it is measured as $B' = 250$ m, giving $B'/L = 2.7$. By taking $\overrightarrow{OR}$ as the $x$-axis and $\overrightarrow{OQ}$ as the $y$-axis, the coordinates of the point A are read as $x = 515$ m and $y = 450$ m, or $x/B' = 2.1$ and $y/B' = 1.8$. By employing the diffraction diagrams with $B/L = 2.0$ as the ones with the opening ratio closest to the problem at hand, the diffraction coefficient is obtained as $(K_d)_{eff} = 0.26$ for $s_{max} = 10$ and $(K_d)_{eff} = 0.17$ for $s_{max} = 75$. Therefore, the interpolated diffraction coefficient for $s_{max} = 50$ is $(K_d)_{eff} \simeq 0.20$. The diffraction diagram for regular waves would yield the much smaller value of $K_d \simeq 0.14$ for this problem.

Caution should be exercised whenever there are vertical quay walls or some reflective structure within the water area behind a breakwater. In such cases, the reflection of the diffracted waves needs to be included in the estimation of the wave height behind the breakwater (cf. Sec. 7.3).

### 3.3.4    *Approximate Estimation of Diffracted Height by the Angular Spreading Method*

Diffraction diagrams of random sea waves for breakwaters are also applicable to the problem of wave diffraction by islands and headlands. Another method of estimating the effect of wave diffraction by large topographic barriers is to utilize the cumulative distribution curves of total wave energy discussed in Sec. 2.3.2, though this method should be considered as giving only an approximate estimate. The method is called the *angular spreading method*.

For the angular spreading method, it is assumed in the fundamental equation, Eq. (3.22), that $K_d = 0$ in the geometric shadow zone and $K_d = 1$ in the illuminated region. Then it will be understood that Eq. (3.22) takes a form similar to Eq. (2.28) for $P_E(\theta)$ by setting $K_d = 0$ or 1 depending on the azimuth. In practice, the ratio of total wave energy penetrating directly to the point of interest is estimated by means of Fig. 2.15 in Sec. 2.3.2, and the wave height ratio is calculated by taking the square root. The error introduced by setting the diffraction coefficient equal to either 0 or 1 is not large, as errors in the estimated diffraction coefficient around the boundary of the geometric shadow mostly cancel out among the many component waves from various directions. This method is based on the same principle as that of swell attenuation due to angular spreading as formulated in the Pierson–Neumann–James method for wave forecasting,[27] and it may be called the angular spreading method. In 1966, Hom-ma *et al.*[28] applied the same basic method to estimate the effect of wave sheltering by Sado Island on the Niigata Coast in Japan, though they employed the cosine square law for the directional spreading function.

A limitation on this method is that the dimensions of the barriers must be sufficiently large, on the order of several tens of wavelengths or greater; otherwise the error introduced by simplifying the diffraction coefficient may not be negligible.

### Example 3.4

Estimate the wave height ratio at Point O which is partly sheltered by the headland P sketched in Fig. 3.23.

### Solution

The angle between the geometric shadow line $\overline{OP}$ and the principal direction of wave approach of wave approach is read on the sketch as $\theta_1 = 17°$. Among the component waves, those in the range of $\theta = 17°$ to $90°$ are assumed to be blocked by the headland so as not to reach Point O. If the waves are assumed to be wind waves with $s_{max} = 10$, the cumulative energy ratio at $\theta_1 = 17°$ is $P_E(17°) \simeq$ 0.685 by Fig. 2.15. This means that 68.5% of the total wave energy will reach Point O and the wave height ratio is therefore estimated as

$$K_d \simeq \sqrt{0.685} = 0.83 \,.$$

If the waves are assumed to be swell with $s_{max} = 75$, then $P_E(17°) \simeq$ 0.89 and $K_d \simeq 0.94$.

### Example 3.5

Estimate the wave height ratio at Point O which is sheltered by the island $\overrightarrow{PQ}$ shown in Fig. 3.24.

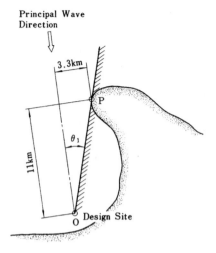

Fig. 3.23   Wave diffraction by a headland.

## Solution

Referring to Fig. 3.24, the angles $\theta_1$ and $\theta_2$ are obtained as $-27°$ and $31°$, respectively. Thus, the ratio of wave energy blocked by the island to the total incident energy is estimated for the case of wind waves (by assuming $s_{max} = 10$) as

$$\Delta E = P_E(31°) - P_E(-27°) = 0.82 - 0.22 = 0.60.$$

Therefore the ratio of the wave height at Point O to the incident height is estimated to be

$$K_d = \sqrt{1 - \Delta E} = \sqrt{1 - 0.60} = 0.63.$$

Although the above calculation gave a single wave height ratio, the actual situation is that waves are arriving at Point O from two different directions, from the left end and the right end of the island. The ratios of the respective wave heights to the incident height are estimated as follows:

Wave group from the left side

$$(K_d)_1 = \sqrt{P_E(-27°) - P_E(-90°)} = \sqrt{0.22} = 0.47.$$

Wave group from the right side

$$(K_d)_2 = \sqrt{P_E(90°) - P_E(31°)} = \sqrt{1 - 0.82} = 0.42.$$

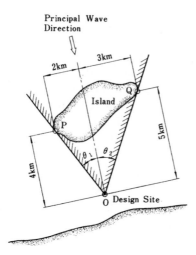

Fig. 3.24   Wave diffraction by an island.

When information on the wave direction as well as the wave height is important, such as in the case of harbor planning with the entrance at Point O, it is best to consider that there are two wave groups with height ratios of 0.47 and 0.42 incident from different directions instead of assuming a single wave group with a height ratio of 0.63.

In actual situations, waves will be refracted somewhat around the edges of islands and headlands, unless these are formed of cliffs which drop steeply into the sea. Because of such wave refraction effects, some energy from component waves incident from outside the geometric boundary will reach the design site. Therefore, from the viewpoint of the practicing engineer, the angle $\theta_1$ in Fig. 3.23 should be set slightly larger than the geometric angle, while the angles $\theta_1$ and $\theta_2$ in Fig. 3.24 should be set slightly smaller. There is no guideline, however, for the amount of adjustment to be made. Refraction diagrams for the region of the tip of the island or headland may provide some clues. The analysis of random wave refraction described in Sec. 3.2 is an alternative approach to such a problem.

### 3.3.5   *Applicability of Regular Wave Diffraction Diagrams*

As explained so far, diffraction diagrams prepared for regular waves yield values of the diffraction coefficient differing considerably from those of actual wave diffraction in the sea, and thus, they are not recommended for practical applications in general. The main cause of the discrepancy is the directional spreading of random sea waves. In other words, use of regular wave diffraction diagrams in real situations requires very narrow directional spreading of wave energy. Such a situation is realized at a harbor located at the inner end of a narrow bay when it is subject to swell penetration, and also for the case of secondary diffraction by inner breakwaters of waves which have already been diffracted by outer breakwaters. Another example is the diffraction of swell by a series of short detached breakwaters parallel to the shoreline to be built in order to promote formation of salients. As such breakwaters are usually built in water only several meters deep, the approaching swell has undergone the full effect of wave refraction; therefore its directional spread of energy has become very narrow. These are cases in which diffraction diagrams for regular waves are applicable.

## 3.4   Equivalent Deepwater Wave

The analysis of wave transformation is often facilitated by introducing the concept of the *equivalent deepwater wave*, which is listed as item ⑧ in Fig. 1.1 in Sec. 1.3. This wave is a hypothetical one devised for the purpose of adjusting the heights of waves which may have undergone refraction, diffraction and other transformations, so that the estimation of wave transformation and deformation can be more easily carried out when dealing with complex topographies. The height and period of the equivalent deepwater wave are defined by

$$H_0' = K_d K_r (H_{1/3})_0 , \quad T_{1/3} = (T_{1/3})_0 , \tag{3.23}$$

where

$H_0'$ : equivalent deepwater wave height (corresponding to the significant wave),

$(H_{1/3})_0 = H_0$ : deepwater significant wave height,

$(T_{1/3})_0$ : significant wave period of deepwater waves,

and $K_r$ and $K_d$ denote the coefficients of random wave refraction and diffraction, respectively.

For example, if the deepwater waves have height $(H_{1/3})_0 = 5$ m and period $(T_{1/3})_0 = 12$ s, and the changes in wave height by refraction and diffraction at a particular point are given by $K_r = 0.92$ and $K_d = 0.83$, the equivalent deepwater wave is defined by $H_0' = 3.8$ m and $T_{1/3} = 12$ s, and subsequent calculations are done with these values.

In a water area where the average slope of the sea bottom is very gentle and in which the zone of shallow depth continues for a great distance, wave attenuation due to bottom friction may not be negligible. In such a case, the equivalent deepwater wave height is estimated by multiplying a coefficient of wave height attenuation $K_f$ on the right side of Eq. (3.23). For details on the wave height attenuation coefficient $K_f$, reference is made to Bretschneider and Reid[29] and others (e.g., Ref. 30). The period of the equivalent deepwater wave is generally regarded as equal to the deepwater significant wave period, but in reality the significant wave period may vary during wave transformation, as in the sheltered area behind a breakwater.

The idea of adjusting the deepwater wave height in consideration of wave refraction effect appeared in *Shore Protection Planning and Design* by Beach Erosion Board[31] in 1961 or earlier. The symbol of $H_0'$ was introduced as being equal to $K_r H_0$ but without any name. It was later called the *unrefracted* deepwater wave height in *Shore Protection Manual.*[32] In the early 1960s, Japanese coastal engineers gave the name of *equivalent* deepwater wave height with the definition of Eq. (3.23).

The concept of equivalent deepwater wave was introduced in order to relate the phenomena of wave breaking, run-up, overtopping and other processes to the characteristics of the deepwater waves. Various processes of wave deformation and wave action are investigated through experiments in laboratory wave flumes, and data sets from many experiments are available. These investigations are usually carried out in flumes of uniform width. On the other hand, waves in the sea experience a variation in the distance between wave rays due to refraction, as shown in Fig. 3.25, and the wave height near the shore varies from place to place even though the deepwater wave height is constant. By introducing imaginary wave rays represented by dashed lines and assigning spatially varying equivalent deepwater wave heights, we can make full use of the wave flume data on breaking and other phenomena.

When waves are measured in water of the depth $h_1$, the significant wave height $(H_{1/3})_1$ needs to be converted to the deepwater wave height $(H_{1/3})_0$ by using the following formula:

$$(H_{1/3})_0 = (H_{1/3})_1 / [(K_r)_1 (K_s)_1], \qquad (3.24)$$

where $(K_r)_1$ is the refraction coefficient at $h_1$ and $(K_s)_1$ is the shoaling coefficient at $h_1$ to be discussed in the next section.

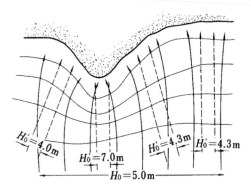

Fig. 3.25   Explanatory sketch of equivalent deepwater waves undergoing refraction.

### Example 3.6

In water of the depth $h = 20$ m, the significant wave of $H_{1/3} = 4.2$ m and $T_{1/3} = 8.0$ s was observed. The depth-contours around there are nearly straight and parallel to the shoreline. Waves are incident to the depth-contours with the angle of $22°$. Estimate the deepwater significant height of this wave and the equivalent deepwater wave height for a structure at the depth of $h_2 = 7$ m.

### Solution

The deepwater wavelength for this wave is $L_0 = 99.8$ m and the relative depth is $h_1/L_0 = 0.200$. By referring to Fig. 3.11 in Sec. 3.2.4, the deepwater incident angle corresponding to $\alpha_p = 22°$ is estimated as $(\alpha_p)_0 = 24°$. Then, the refraction coefficient at $h/L_0 = 0.20$ is estimated as $(K_r)_1 = 0.96$ by referring to Fig. 3.10 and assuming $s_{max} = 10$. For evaluation of the shoaling coefficient, we need the wavelength at the depth $h = 20$ m, which is obtained as $L_1 = 88.7$ m from Table A-2 in the Appendix. By using Eq. (3.25) in Sec. 3.5, we obtain $(K_s)_1 = 0.964$ for $h_1/L_1 = 0.225$. Thus, the deepwater significant wave height is estimated as

$$(H_{1/3})_0 = 4.2/[0.96 \times 0.964] = 4.54 \text{ m}.$$

At the depth of $h_2 = 7$ m, the wavelength is $L_2 = 61.4$ m and the relative depth is $h_2/L_2 = 0.114$. The refraction coefficient is estimated as $(K_r)_2 = 0.94$. Thus, the equivalent deepwater wave height at $h_2 = 7$ m is obtained as

$$H_0' = 4.54 \times 0.94 = 4.27 \text{ m}.$$

The concept of equivalent wave height is restricted to use in connection with the significant wave as the representative wave of a wave group; it is not applied to other characteristic waves such as the highest wave or the mean wave.

## 3.5   Wave Shoaling

As waves propagate in a channel of gradually decreasing depth but of constant width, the wavelength and celerity decrease as approximately given by the theory of small amplitude waves. At the same time, the wave height also changes. The change in wave height due to varying depth is called *wave shoaling*.

The cause of the variation in wave height is the variation in the speed of energy propagation; i.e., the group velocity, with the water depth. For small amplitude waves with a single period, the variation in wave height due to the wave shoaling effect is calculated by the following equation:

$$K_{si} = \frac{H}{H_0'} = \sqrt{\frac{(c_g)_0}{c_g}} = \left[\left(1 + \frac{2kh}{\sinh 2kh}\right) \tanh kh\right]^{-1/2}$$

$$= \left[\tanh kh + kh(1 - \tanh^2 kh)\right]^{-1/2} , \qquad (3.25)$$

where

$K_s$ : shoaling coefficient by small amplitude wave theory,

$c_g$ : group velocity given by Eq. (3.21),

$(c_g)_0$ : group velocity in deep water $= \frac{1}{2}c_0 = 0.78T\,(\mathrm{m/s})$,

$k$ : wavenumber $= 2\pi/\mathrm{L}$,

$L$ : wavelength, and

$h$ : water depth.

For real sea waves, some modifications to the shoaling coefficient given by Eq. (3.25) are necessary. One reason is the effect of the energy distribution in the frequency domain as expressed through the frequency spectra, and another is the effect of the finite amplitude of the individual waves. The former can be evaluated by computing the wave shoaling coefficient at various frequency components in the wave spectrum and by summing the results in a manner similar as in the calculation of random wave refraction and diffraction. This procedure produces a kind of smoothing of the variation of the shoaling coefficient with respect to the *relative water depth* (ratio of water depth to wavelength). For example, the minimum value of the shoaling coefficient becomes $(K_s)_{min} = 0.937$ by introducing the frequency spectrum (Goda[33]) whereas it is $(K_s)_{min} = 0.913$ for regular waves. The difference is on the order of 2% to 3%, which may be neglected in practical design procedures.

The second effect by finite wave amplitude, which is called *nonlinear wave shoaling*, can be calculated by making use of various theories of finite amplitude waves. Iwagaki and Sakai,[34] for example, have presented a design diagram for the estimation of the shoaling coefficient. Shuto,[35] on the other hand, has formulated the variation of wave height occurring in relatively shallow water with a set of fairly simple equations. In terms of the shoaling coefficient, these equations are rewritten as[33]:

$$\left.\begin{array}{ll} K_s = K_{si} & : h_{30} \leq h\,, \\[2mm] K_s = (K_{si})_{30}\left(\dfrac{h_{30}}{h}\right)^{2/7} & : h_{50} \leq h < h_{30}\,, \\[2mm] K_s(\sqrt{K_s} - B) - C = 0 & : h < h_{50}\,, \end{array}\right\} \qquad (3.26)$$

in which $K_{si}$ denotes the shoaling coefficient for a small amplitude wave as given by Eq. (3.25), $h_{30}$ and $(K_{si})_{30}$ are the water depth satisfying Eq. (3.27) below and the shoaling coefficient for that depth, respectively, $h_{50}$ is the water depth satisfying Eq. (3.28), and $B$ and $C$ are constants given by Eq. (3.29):

$$\left(\frac{h_{30}}{L_0}\right)^2 = \frac{2\pi}{30}\frac{H_0'}{L_0}(K_{si})_{30}\,, \qquad (3.27)$$

$$\left(\frac{h_{50}}{L_0}\right)^2 = \frac{2\pi}{50}\frac{H_0'}{L_0}(K_s)_{50}\,, \qquad (3.28)$$

$$B = \frac{2\sqrt{3}}{\sqrt{2\pi H_0'/L_0}}\frac{h}{L_0}\,, \quad C = \frac{C_{50}}{\sqrt{2\pi H_0'/L_0}}\left(\frac{L_0}{h}\right)^{3/2}\,, \qquad (3.29)$$

where $L_0$ denotes the wavelength in deep water, $(K_s)_{50}$ is the shoaling coefficient at $h = h_{50}$ and $C_{50}$ is a constant given by

$$C_{50} = (K_s)_{50}\left(\frac{h_{50}}{L_0}\right)^{3/2}\left[\sqrt{2\pi\frac{H_0'}{L_0}(K_s)_{50}} - 2\sqrt{3}\frac{h_{50}}{L_0}\right]\,. \qquad (3.30)$$

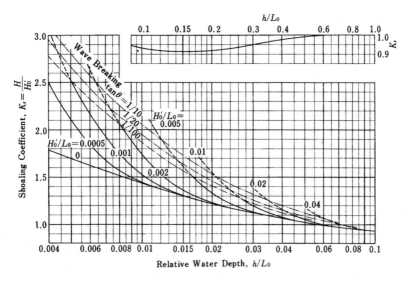

Fig. 3.26  Diagram of nonlinear wave shoaling.[33]

In the computation of the shoaling coefficient, the water depths $h_{30}$ and $h_{50}$ satisfying Eq. (3.27) and (3.28) are first solved for by an iterative method, and then it can be determined in which range the water depth at the design site is located. If the water depth is less than $h_{50}$, the third equation of Eq. (3.26) must be solved by a numerical method.

Figure 3.26 presents the shoaling coefficient including the finite amplitude effect calculated by Goda,[33] based on Shuto's theory. The shoaling coefficient in the upper right corner corresponds to water of relative depth $h/L_0$ greater than 0.09, where it is assigned the same value as that of small amplitude waves.

A closed form of the nonlinear shoaling coefficient has been presented by Iwagaki *et al.*[36] based on another nonlinear wave theory. Kweon and Goda[37] employed their functional form and adjusted the coefficients to approximate the curves shown in Fig. 3.26 as in the following:

$$K_s = K_{si} + 0.0015 \left( \frac{h}{L_0} \right)^{-2.87} \left( \frac{H_0'}{L_0} \right)^{1.27} . \tag{3.31}$$

**Example 3.7**

Estimate the wave height at water depth of 8 m for an equivalent deepwater wave of $H_0' = 4.5$ m and $T_{1/3} = 12$ s incident to a coast with a uniform slope of 1/10.

**Solution**

As the deepwater wavelength of this wave is $L_0 = 225$ m, we have

$$H_0'/L_0 = 4.5/225 = 0.0200 \quad \text{and} \quad h/L_0 = 8/225 = 0.036.$$

Therefore, by entering Fig. 3.22 and reading off the curve for $H_0'/L_0 = 0.02$ we obtain

$$K_s = 1.24, \quad \text{thus} \quad H_{1/3} = K_s H_0' = 1.24 \times 4.5 = 5.6 \text{ m}.$$

If the wave height is assumed to be very small in the above example, the shoaling coefficient becomes $K_s = 1.09$ by the curve for $H_0'/L_0 = 0$ in the figure, thus differing by about 12% from the previous result. The approximate formula of Eq. (3.31) yields $K_s = 1.23$, which is close to 1.24 in the above.

If the bottom slope is much gentler, waves begin to break before they reach the depth of 8 m, and the wave height becomes smaller as a result of attenuation by breaking. The dash-dot curves with the legend "wave breaking" indicate the boundary beyond which the attenuation of $H_{1/3}$ exceeds 2%. In such shallow zones, the wave height should be estimated by taking into account the phenomenon of random wave breaking as described in the next section.

It should be cautioned that the nonlinear shoaling phenomenon exemplified in Fig. 3.26 is not accompanied by any increase in the wave energy density. It is caused by deformation of wave profiles with sharpening of wave crests and flattening of wave troughs. Such a deformed profile can have a height larger than that of a sinusoidal profile under the same potential energy, yielding the nonlinear shoaling coefficient apparently larger than the linear shoaling coefficient. The nonlinear shoaling process is important for evaluation of wave actions on structures, but it should not be incorporated into the computation of radiation stresses and related phenomena, which are proportional to the linear wave energy.

## 3.6  Wave Deformation Due to Random Breaking

### 3.6.1  *Breaker Index of Regular Waves*

We can easily observe that a train of regular waves in a laboratory flume undergoes shoaling over a sloping bottom and breaks at a certain depth. The location at which waves break is almost fixed for regular waves, and there is a distinct difference between the oscillatory wave motion before breaking and the turbulent wakes with air entrainment after breaking. The terminology *wave breaking point, depth* and *height* is employed to denote the location, water depth, and height of wave breaking, respectively. The expression "limiting breaker height" is sometimes also used, in the sense of the upper limit of progressive waves physically attainable at a certain water depth for a given wave period. The ratio of limiting breaker height to water depth, which is called the *breaker index*, depends on the bottom slope and the relative water depth. The author[38] has compiled a number of laboratory results in a form of design diagram for four beach slopes in 1970, and then re-expressed it in the following empirical function in 1974[39]:

$$\frac{H_b}{h_b} = \frac{A}{h_b/L_0} \left\{ 1 - \exp\left[ -1.5\frac{\pi h}{L_0} \left( 1 + 15\tan^{4/3}\theta \right) \right] \right\} \quad : \ A = 0.17, \quad (3.32)$$

where $\theta$ denotes the angle between the sea bottom and the horizontal plane, and thus $\tan\theta$ represents the bottom slope.

However, Rattanapitikon and Shibayama[40] pointed out in 2000 that Eq. (3.32) has a tendency of overprediction for steep slope. Accordingly, the constant of 15 to be multiplied to $\tan\theta$ is reduced to 11 as in the following[41]:

$$\frac{H_b}{h_b} = \frac{A}{h_b/L_0} \left\{ 1 - \exp\left[ -1.5\frac{\pi h}{L_0} \left( 1 + 11\tan^{4/3}\theta \right) \right] \right\} \quad : \ A = 0.17. \quad (3.33)$$

This modification caused to reduce the breaker index for the slope of 1/10 by 11%, but the reduction for the slope of 1/50 remains at 2% only.

Even though the breaker index is expressed with functional forms such as Eq. (3.33), the laboratory data of regular wave breaking always exhibit a certain range of scatter. Figure 3.27 exhibits such scatter of data for the beach slope of 1/20 with various symbols corresponding to different authors. The solid curve represents the prediction by Eq. (3.33), while the curves of dashes and dots correspond to the upper limit of 115% and lower limit of 87%, respectively. As commented by Smith and Kraus,[42] such scatter of laboratory data should be considered inherent to the breaking process of water waves. Thus, the breaker height should be treated as a stochastic variable. The coefficient of variation is 11% for the slope of 1/20 shown in Fig. 3.27, while it rises to 14% for the slope of 1/10 but falls to about 6% for the slope of 1/100.

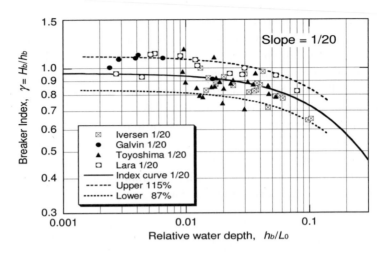

Fig. 3.27  Comparison of new breaker index formula with laboratory data on beach slope of 1/20.[41]

### 3.6.2  *Hydrodynamics of Surf Zone*

(A) *Incipient breaking of random waves and surf zone*

As opposed to the breaking of a regular wave train, actual sea waves, in particular wind waves, exhibit a wide spatial spread in the region of breaking; some waves break far from shore, some at an intermediate distance, and others approach quite near to the shoreline before they break. In coastal waters, therefore, wave breaking takes place in a relatively wide zone of variable water depth, which is called the *wave breaking zone* or the *surf zone*. Within the surf zone, the percentage of breaking waves gradually increases toward the shoreline, while representative wave heights such as the significant height gradually decrease. Only for the situation of swell incident to a coast with a single, well-developed longshore bar, we are able to define the unambiguous outer edge of the surf zone, as being located at the offshore side of the longshore bar. Otherwise, it is difficult to define the location at which random waves begin to break.

Nevertheless, Kamphuis[43] proposed the concept of *incipient breaking* of random waves. He measured the variation of significant wave height from the offshore to the shoreline in the laboratory and drew two curves outside and inside the surf zone. The curve of wave height variation outside the surf zone represented wave shoaling, while the curve inside the surf zone represented wave decay by breaking. He extended the two curves to find the

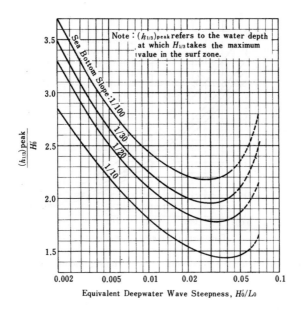

Fig. 3.28   Index curves for the water depth at which the significant wave height takes the maximum value in the surf zone.[45]

crossing point, and defined the point as the location of incipient breaking. When he applied the breaker index formula of Eq. (3.33), he found that the condition of incipient breaking can be expressed by changing the value of the constant $A$ from 0.17 to 0.12. Li *et al.*[44] also found the value of $A = 0.12$ as the condition such that several large waves among a train of random waves begin to break. Thus,

$$\frac{H_b}{L_0} = 0.12 \left\{ 1 - \exp\left[ -1.5 \frac{\pi (h_b)_{\text{incipient}}}{L_0} (1 + 11 \tan^{4/3}\theta) \right] \right\} . \qquad (3.34)$$

In 1975, the author presented a diagram for the depth $(h_{1/3})_{\text{peak}}$ at which the significant wave height takes the maximum value within the surf zone and another diagram for the peak significant height $(H_{1/3})_{\text{peak}}$ based on his random wave breaking model (Goda[45]); they have been evaluated from a set of design diagrams for estimation of $H_{1/3}$ within the surf zeone, which are shown in Sec. 3.6.3. The two diagrams are reproduced here as Figs. 3.28 and 3.29. The symbol $H_0'$ denotes the equivalent deepwater wave height defined in Sec. 3.4.

The curves of $(h_{1/3})_{\text{peak}}/H_0'$ in Fig. 3.28 can be approximated with the following empirical formulas for the ranges of $0.002 \leq H_0'/L_0 \leq 0.05$ and

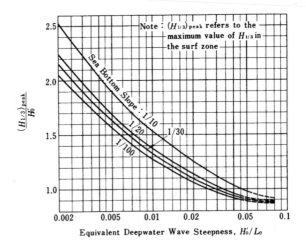

Fig. 3.29   Index curves for the maximum value of the significant wave height in the surf zone.[45]

$0.01 \leq \tan\theta \leq 0.1$:

$$(h_{1/3})_{\text{peak}}/H_0' = a_0 + a_1 x + a_2 x^2 + a_3 x^3 \ : \ x = \ln(H_0'/L_0), \qquad (3.35)$$

$$\left.\begin{aligned}
a_0 &= -4.773 - 5.156y - 0.5463y^2, \\
a_1 &= -1.5504 - 1.9915y - 0.2095y^2 \ : \ y = \ln(\tan\theta), \\
a_2 &= -0.0580 - 0.2534y - 0.0252y^2, \\
a_3 &= \ \ \ 0.0105 - 0.0041y - 0.002y^2.
\end{aligned}\right\} \qquad (3.36)$$

Similarly the curves of $(H_{1/3})_{\text{peak}}/H_0'$ in Fig. 3.29 can be approximated with Eq. (3.36) of the following for the same ranges:

$$(H_{1/3})_{\text{peak}}/H_0' = b_0 + b_1 x + b_2 x^2 \ : \ x = \ln(H_0'/L_0), \qquad (3.37)$$

$$\left.\begin{aligned}
b_0 &= 0.8119 - 0.1470y - 0.0136y^2, \\
b_1 &= 0.1509 - 0.0711y - 0.0070y^2 \ : \ y = \ln(\tan\theta), \\
b_2 &= 0.1127 + 0.0186y + 0.00227y^2.
\end{aligned}\right\} \qquad (3.38)$$

By combining the two diagrams of Figs. 3.28 and 3.29, Goda[41] has prepared a diagram for the breaker index of random waves corresponding to the peak height of significant wave in the surf zone with the result shown in Fig. 3.30. It may be regarded as the incipient breaker index. The curves of the incipient breaker index correspond to the constant value of $A = 0.11 \sim 0.13$ when fitted to Eq. (3.33). The constant value is slightly small

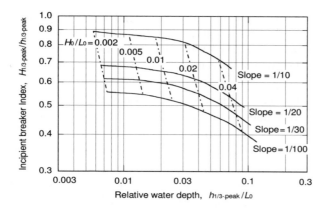

Fig. 3.30 Incipient breaker index of random waves.[41]

for waves with large steepness breaking on a gentle slope, while it is slightly large for waves with small steepness breaking on a steep slope. However, the constant value is nearly the same as that of $A = 0.12$ in Eq. (3.34). These constant values are for the incipient breaking condition only. As waves progress toward the shoreline within the surf zone, the ratio of significant height to water depth gradually increases.

### Example 3.8

Estimate the maximum value of significant wave height and the water depth of its appearance when the equivalent deepwater wave of $H_0' = 2.0$ m and $T_{1/3} = 14$ s comes on a planar beach with the uniform slope of $1/50$.

### Solution

The deepwater wavelength of this wave is $L_0 = 306$ m, the wave steepness is $H_0'/L_0 = 2.0/306 = 0.0065$, and the beach slope is $\tan \theta = 0.02$. Thus, the variables in Eqs. (3.35) and (3.36) become $x = \ln(0.0065) = -5.036$ and $y = \ln(0.02) = -3.912$. By substituting these variables into Eqs. (3.35) and (3.36), we have the following result:

$$a_0 = 7.036,\ a_1 = 3.034,\ a_2 = 0.5476,\ a_3 = 0.0235 :$$
$$(h_{1/3})_{\text{peak}}/H_0' = 2.64 \Rightarrow (h_{1/3})_{\text{peak}} = 5.3m\,,$$

$$b_0 = 1.179,\ b_1 = 0.322,\ b_2 = 0.0746 :$$
$$(H_{1/3})_{\text{peak}}/H_0' = 1.45 \Rightarrow (H_{1/3})_{\text{peak}} = 2.9m\,.$$

When the depth of $(h_{1/3})_{\text{peak}}$ = 5.3 m is substituted into Eq. (3.34), we have an estimate of $H_b$ = 3.0 m, which is slightly larger than $(h_{1/3})_{\text{peak}}$ = 2.9 m above. Such a difference comes from a slight deviation of the constant $A$ from the mean value of 0.12.

The boundary lines of the 2% attenuation of the significant wave height shown in Fig. 3.26 and also in Figs. 3.38 to 3.41 approximately correspond to the constant value of $A = 0.11$ in Eq. (3.33). At the water depth of the peak significant wave height, therefore, several percent of large waves must have been broken. As the waves further propagate toward the shore, the percentage of broken waves increases while the reduced energy flux retains nonbreaking waves of smaller heights. The ratio of the significant height to the local water level gradually rises toward the shore, as expressed by the increase of the constant value $A = 0.12$ to around 0.15. Except for the area near the shoreline where the wave setup is conspicuous, the significant wave height on genetle beaches rarely exceeds 0.6 times the water depth.

A representative wave height mainly used in Europe and US other than $H_{1/3}$ is the *root-mean-square wave height* defined as below.

$$H_{\text{rms}} = \sqrt{\frac{1}{N} \sum_{n=1}^{N} H_i^2} = \sqrt{8m_0} = \sqrt{8}\eta_{\text{rms}}. \tag{3.39}$$

Sallenger and Holman[46] have reported that the upper limit of the root-mean-square wave height is $0.3 \sim 0.5$ times the water depth under a strong influence of the beach slope, e.g. $\gamma_{\text{rms}} \equiv H_{\text{rms}}/h = 3.2 \tan\theta + 0.32$. They have estimated the wave heights from the spectra of orbital velocities measured with the electromagnetic current meters by means of the linear theory. It is likely that the surface profiles directly measured would have produced somewhat larger heights.

(B) *Distribution of individual wave heights within the surf zone*

Individual wave heights defined by the zero-upcrossing method in relatively deep water approximately follow the Rayleigh distribution as discussed in Sec. 2.2.1. When waves propagate into shallow water, the process of nonlinear wave shoaling enhances large waves more than small waves. The wave height distribution becomes wider than the Rayleigh with a long extended right-hand tail. As waves enter the surf zone, however, large waves begin to break first, and then medium-size waves break. The wave height distribution becomes narrower than the Rayleigh, with the upper tail truncated by the breaking height controlled by the water depth. Figure 3.31 illustrates the change of the functional shape of the wave height distribution when irregular waves generated in a laboratory flume propagate on a uniform beach with the slope of $1/10$.

As listed in the top of Fig. 3.31, the equivalent deepwater wave with $H_0' = 10.3$ cm and $(T_{1/3})_0 = 1.24$ s was generated at the constant depth of $h = 50$ cm and propagated on a fixed bed slope with $\tan\theta = 1/10$. Waves were measured at the section of constant depth and at the three locations on the slope with the depth of $h = 15$, $10$, and $6$ cm. The mean wave height $\overline{H}$ varied from $6.1$, $6.9$, $6.8$, and $4.5$ cm, respectively. The four diagrams in Fig. 3.31 show the wave height distribution in the form of probability density function, which is normalized with $\overline{H}$. The theoretical curves that have been computed with the author's model in 1975 (see Sec. 3.6.3) agree well with the laboratory data at the four locations.

Though the measurements at much shallower locations were not made, the measurements if made would have produced a widening of wave height distribution toward the Rayleigh owing to re-generation of nonbreaking waves, irregular fluctuations of mean water level (surf beats), and other factors. Laboratory data obtained by Hamm and Peronnard[47] and another by Ting[48] exhibit such a trend of returning to the Rayleigh in very shallow water.

Changes in the functional shape of the wave height distribution are also observed as the variations of the ratio among representative wave heights. Figure 3.32 shows the variation of the wave height ratios $H_{2\%}/H_{1/3}$ and $H_{\mathrm{rms}}/H_{1/3}$ in the laboratory tests against the relative water depth $h/H_0$. The former ratio is based on the test by Smith *et al.*,[49] while the latter ratio is due to Ting.[48] The symbol $H_{2\%}$ represents the threshold height of 2% exceedance and is equal to $1.4H_{1/3}$ when the wave height distribution is the Rayleigh, while $H_{\mathrm{rms}}$ is equal to $0.706H_{1/3}$ for the Rayleigh. The dashed lines are the prediction by the directional random wave breaking model named the PEGBIS by Goda,[50] which will be described in Sec. 12.4. The model predicts that the ratio $H_{2\%}/H_{1/3}$ goes down from 1.4 to around 1.2 at $h/H_0 \simeq 1.5$ and rises up to 1.35 near the shoreline, while the ratio $H_{\mathrm{rms}}/H_{1/3}$ goes up from 0.70 to 0.80 around $h/H_0 \simeq 0.6$ and returns to 0.70 near the shoreline. Though the laboratory data, especially of $H_{2\%}/H_{1/3}$, exhibit behaviors somewhat different from the model prediction, the tendency of the wave height distribution narrowing in the middle of the surf zone and re-widening near the shore is the same with the model.

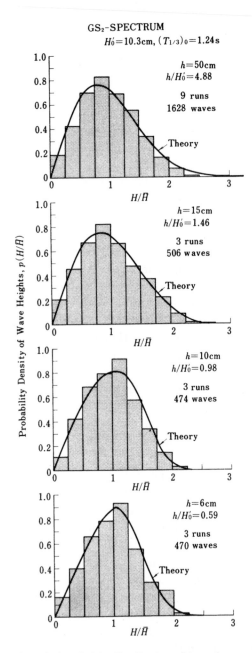

Fig. 3.31   Deformation of wave height distribution of irregular waves in a laboratory flume with the slope of 1 to 10.[45]

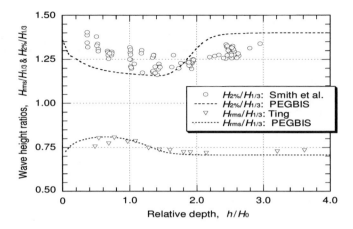

Fig. 3.32    Laboratory data of variations of wave height ratios $H_{2\%}/H_{1/3}$ and $H_{\rm rms}/H_{1/3}$ across the surf zone.[41]

Figure 3.33 is another evidence of the narrowing and re-widening of the wave height distribution across the surf zone for the case of field data. The data of the three field campaigns at Ajigaura Beach, Duck85, and Super-duck have been analyzed for the ratios of $H_{1/10}/H_{1/3}$ and $H_{\rm rms}/H_{1/3}$ by the author.[41] The data at Ajigaura Beach, Ibaragi Prefecture in Japan were reported by Hotta and Mizuguchi,[51,52] while the data at the Duck85 campaign were reported by Ebersole and Hughes.[53] The data at the SuperDuck campaign was given to the author through the courtesy of Dr. S. A. Hughes. Hotta and Mizuguchi mobilized twelve motion-picture cameras set on an observation pier extended perpendicularly from the shore and took pictures of sea surface around surveyor's poles erected every 2 m in the sea. The sea surface elevations in the frames of motion-pictures were later digitized manually and produced the time series data of wave profiles. Ebersole and Hughes applied the same technique through the collaboration of Dr. Hotta and called the technique as the *photopole method.*.

The field data approximately follow the predictions by the PEGBIS model. The wave height ratio $H_{1/10}/H_{1/3}$ goes down to around 1.15 at $h/H_0 \simeq 1.6$ from the value of 1.27 outside the surf zone, while the ratio $H_{\rm rms}/H_{1/3}$ rises to 0.8 around $h/H_0 \simeq 1.0$. Both the ratios show the tendency of returning to the values of the Rayleigh distribution near the shoreline. Large scatter of data is caused by the statistical variability owing to small numbers (70 to 100) of waves per record. According to the information on the statistical variability of wave data to be presented in

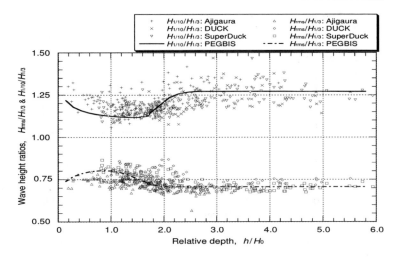

Fig. 3.33   Field data of variations of wave height ratios $H_{1/10}/H_{1/3}$ and $H_{\rm rms}/H_{1/3}$ across the surf zone.[41]

Sec. 10.6, the wave height ratio $H_{1/10}/H_{1/3}$ is estimated to have the coefficient of variation of some 4%, which yields a 2-sigma confidence interval of $\Delta(H_{1/10}/H_{1/3}) \simeq 0.12$.

The degree of the changes of the wave height distribution across the surf zone depends on the wave steepness, the beach slope and other factors in addition to the relative water depth. There have been some efforts to model the wave height distribution in the surf zone with the combination of the Rayleigh and some Weibull distributions, but no composite distribution model has yet succeeded in predicting the re-widening of the wave height distribution near the shoreline.

### (C) *Spatial variation of mean water level and its irregular fluctuations*

Within the surf zone, the mean water level varies locally with lowering in the middle and then linearly rising toward the shore. This phenomenon is induced by the existence of a stress acting on the water due to the presence of wave motion, called *radiation stress*. The radiation stress is associated with the momentum flux accompanying wave propagation, and its magnitude is proportional to the square of wave height. When the wave height varies due to shoaling and breaking during wave propagation from the offshore toward the shoreline, the magnitude of the radiation stress also varies. Around the outer edge of the surf zone, the wave height is large and the radiation stress is strong. As waves progress toward the shore, the wave height is gradually lessened and so is the radiation stress. The cross-shore variation of the radiation stress induces an inclination of the local mean water level. When the amount of the variation of the local mean water level from the still water level is denoted by $\bar{\eta}$, it can be evaluated by numerically integrating the following differential equation from deep water toward the shoreline (Longuet-Higgins[54]:)

$$\frac{d\bar{\eta}}{dx} = -\frac{1}{(\bar{\eta}+h)} \frac{d}{dx} \left[ \frac{1}{8}\overline{H^2} \left( \frac{1}{2} + \frac{2kh}{\sinh 2kh} \right) \right], \qquad (3.40)$$

where $x$ is taken offshore from the shoreline and $\overline{H^2}$ denotes the mean square of the heights of random water waves.

The variation in the mean water level modifies the local water depth, which determines the breaker height. Because the individual wave heights are controlled by the breaker height, the variations in the mean water level and the wave height distribution must be solved simultaneously. Figure. 3.34 presents the result of a computation of the mean water level variation on a bottom with a uniform slope of $1/100$ based on the Goda model in 1975.[b] Waves with low steepness produce a slight depression in the mean water level, which is called *wave setdown*, in the range $h/H_0' = 2$ to 4. In the range of $h/H_0' < 2$, the mean water level rises almost linearly. The rise of mean water level near the shore is called *wave setup*. If the wave setup and setdown are calculated with the regular wave theory, the both are evaluated excessively large.

As indicated in Fig. 3.34, the amount of wave setup at the shoreline is on the order of $0.1H_0'$. Such wave setup can be detected in marigrams. The First District Port Construction Bureau of the Ministry of Transport (presently Ministry of Land, Infrastructure and Transport) of Japan[55] has verified the existence of wave setup of about $0.1H_0'$ by comparing the recorded tide curves with wave data. In case of storm surges, a considerable amount of wave setup may contribute to the increase of the surge height depending on the track of a typhoon/cyclone.

Good prediction of the amount of wave setup requires use of a reliable random wave breaking model. Figure 3.35 presents a result of the computation of the wave setup at the shoreline as denoted by $\zeta$ for the coast with straight, parallel depth-contours (Goda[56]), which has been obtained

---

[b]As will be discussed in Sec. 12.6, the computation of the mean water level should take into consideration the contribution of surface rollers. The result shown in Fig. 3.34 is the case in which the surface roller effect is neglected.

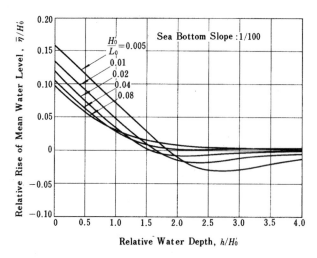

Fig. 3.34  Variation of mean water level due to shoaling and breaking of random sea waves.[45]

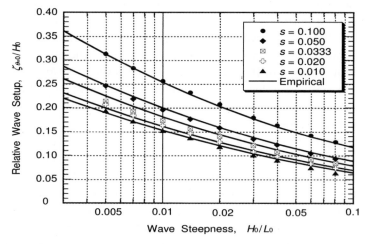

Fig. 3.35  Relative wave setup at the shoreline by waves of normal incidence at the coast with straight, parallel depth-contours.[56]

by using the PEGBIS model for directional random waves developed by the author.[50] (see Sec. 12.4 for details.) The notation $s$ in the legend refers to the beach slope. The symbols in Fig. 3.35 denote the numerically computed values, while the four curves represent the estimates with empirically fitted formula of the following for the wave setup at the shoreline in case of normal incidence, i.e., $\zeta_{\theta_0=0}$:

$$\frac{\zeta_{\theta_0=0}}{H_0} = A_0 + A_1 \ln(H_0/L_0) + A_2 \left[\ln(H_0/L_0)\right]^2 , \tag{3.41}$$

$$\left.\begin{array}{rl} A_0 = & 0.0063 + 0.768\,s , \\ A_1 = & -0.0083 - 0.011\,s , \\ A_2 = & 0.00372 + 0.0148\,s . \end{array}\right\} \tag{3.42}$$

The wave setup by obliquely incident waves is estimated with the following equation:

$$\zeta = \zeta_{\theta_0=0} \left(\cos\theta_0\right)^{0.545+0.038\,\ln(H_0/L_0)} , \tag{3.43}$$

where $\theta_0$ denotes the offshore wave incident angle.

### Example 3.9

Estimate the amount of wave setup at the shoreline when the deepwater waves with $H_0 = 2.5$ m, $T_{1/3} = 12$ s are incident with the angle of $\theta_0 = 30°$ to a a planar beach with the uniform slope of $1/50$.

### Solution

The deepwater wavelength is $L_0 = 225$ m and the wave steepness is $H_0/L_0 = 2.5/225 = 0.0111$. The coefficients $A_0, A_1$, and $A_2$ are calculated with Eq. (3.40) for $s = 0.02$ as follows:

$$A_0 = 0.02166, \ A_1 = -0.00852, \ A_2 = 0.004016 :$$
$$\zeta_{\theta_0=0}/H_0' = 0.141 \Rightarrow \zeta_{\theta_0=0} = 0.353 \text{ m} .$$

The above estimate is adjusted for oblique incidence with Eq. (3.43) with the result of the following:

$$\zeta = 0.353 \times \left(\cos 35°\right)^{0.374} = 0.328 \approx 0.33 \text{ m} .$$

The rise of the mean water level at the actual coast has been investigated by Yanagishima and Katoh[57] through field measurements. They analyzed one-year record of mean water level at an ocean observation pier located at the Hazaki Coast, Ibaraki, Japan. The astronomical tide level was subtracted from the observed mean water level and the effects of barometric pressure and wind setup were corrected. The remaining amount of mean water level rise was correlated with the significant wave height and period measured at an offshore station nearby. The average beach slope around the observation pier was 1 on 60, and the equivalent deepwater wave steepness varied from 0.01 to 0.04. For this range of data, Yanagishima and Katoh derived the following estimation formula of the normalized wave setup:

$$\overline{\eta}/H_0' \approx 0.052\,(H_0'/L_0)^{-0.2}\,. \tag{3.44}$$

In comparison with their estimation formula, Eq. (3.41) tends to predict slightly higher setup for low wave steepness and slightly lower setup for high wave steepness. However, Eq. (3.41) provides the prediction equivalent to their field data on the average. During the field investigation by Yanagishima and Katoh, the rise of mean water level at times of storm waves was also observed at the tide station of Kashima Harbor, which is located about 15 km north of the Hazaki observation pier. However, the amount of mean water level rise was about one-half of the rise at Hazaki. The rise of mean water level in a harbor seems to be affected by the surrounding topography, breakwater layout, location of tide gage, and others.

The wave setup shown in Fig. 3.35 and predicted by Eq. (3.41) refer to the value at the still-water shoreline. The rise of the mean water level continues onshore over the swash zone of sandy beaches on which waves run up. The final rise of the mean water level on the swash zone denoted by $\zeta_s$ is higher than the setup $\zeta$ at the still-water shoreline. Hanslow and Nielsen[58] have proposed the following estimation formula for $\zeta_s$ on the basis of their field measurements on the New Southwales coast in Australia:

$$\frac{\zeta_s}{(H_0)_{\mathrm{rms}}} = \begin{cases} 0.45\,[(H_0)_{\mathrm{rms}}/L_0]^{-0.5}\tan\theta_F : \tan\theta_F > 0.06\,, \\ 0.048\,[(H_0)_{\mathrm{rms}}/L_0]^{-0.5} \qquad : \text{all data}\,, \end{cases} \tag{3.45}$$

where $(H_0)_{\mathrm{rms}}$ is the root-mean-square offshore wave height and $\tan\theta_F$ is the beachface slope.

Another source contributing to the variation in mean water level is the phenomenon of surf beats; i.e., the irregular fluctuations in the mean water level with period of several to a few tens of times the period of the incoming surface waves. The phenomenon attracted the attention of oceanographers in the early 1950s as discussed in Sec. 2.5. Although the amplitude of surf beat is on the order of 10% of the surface wave amplitude in water of 10 m depth or so, it may reach more than 30% of the deepwater wave amplitude near the shoreline, as reported in field observations.[45] An example of a surf beat profile will be shown in Fig. 11.7 in Sec. 11.5.4. The following is an empirical expression for estimating the amplitude of surf beat within the surf zone:[45]

$$\xi_{\mathrm{rms}} = \frac{0.01 H_0'}{\sqrt{\dfrac{H_0'}{L_0}\left(1 + \dfrac{h}{H_0'}\right)}}\,, \tag{3.46}$$

where $\xi_{rms}$ denotes the root-mean-square value of the profile of the surf beat and $H'_0$ refers to the equivalent deepwater wave height as defined in Sec. 3.4

Equation (3.46) was derived by empirical fitting to some field data without theorectical background. It is recommended to make a survey of recent theory/field data on surf beats when more information is needed.

### 3.6.3  *Wave Height Variations on Planar Beaches*

(A) *Goda's model in 1975*

The author[33, 45] presented a numerical model for wave deformation by random breaking in 1975, which was validated with laboratory test data and wave measurement data at Sakata Port in Japan. Based on the model, design diagrams were elaborated for estimation of the variations of the maximum wave height $H_{max}$ and the significant wave height $H_{1/3}$ from the offshore to the shoreline, as presented in clause (B) of this section. The diagrams have been utilized by many practitioners as reproduced in various engineering manuals in the world. The model was constructed with the assumptions sketched in Fig. 3.36.

First, the distribution before wave breaking is assumed to have a Rayleigh distribution as shown in Fig. 3.36(1). The abscissa $x$ is a nondimensional wave height normalized with a reference height $H^*$.

Among the waves obeying that distribution, those with height exceeding the breaking limit will break and cannot occupy their original position in the wave height distribution. The breaking limit for random sea waves should be allowed a range of variation because even a regular wave train exhibits some fluctuation in breaker height, and a train of random sea waves would show a greater fluctuation owing to the variation of individual wave periods and other characteristics. Therefore, wave breaking is assumed to take place in the range of relative wave height from $x_2$ to $x_1$ with a probability of occurrence which varies linearly between the two boundaries (Fig. 3.36(2)). The upper limit $x_1$ is set by assigning the constant value $A = 0.18$ in Eq. (3.32) for the breaker index of regular waves, while the lower limit is set with $A = 0.12$.

With this assumption, the portion of waves which is removed from the original distribution due to the process of breaking is represented by the zone of slashed lines shown in Fig. 3.24(3). The broken waves do not lose all of their energy but retain some. Because little information is available on

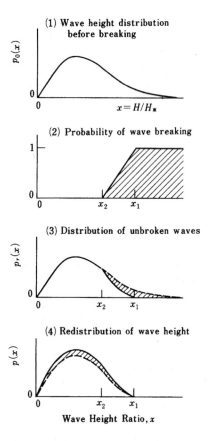

Fig. 3.36   Explanatory sketch of the model of random waves breaking.[45]

the heights of individual random waves after breaking, they are assumed to be distributed in the range of nondimensional wave heights between 0 and $x_1$ with a probability proportional to the distribution of unbroken waves. With this model, the wave height distribution within the surf zone is expressed as shown in Fig. 3.36(4), where the area with slashed lines represents the heights of attenuated waves after breaking.

When the range of wave breaking is computed by Eq. (3.32) with $A =$ 0.12 to 0.18, the local water depth is adjusted by taking into account the wave setdown and setup induced by the radiation stresses. Contribution of the surf beats to local water depth is probabilistically represented by assigning the normal distribution to the water level variation due to the surf beats.

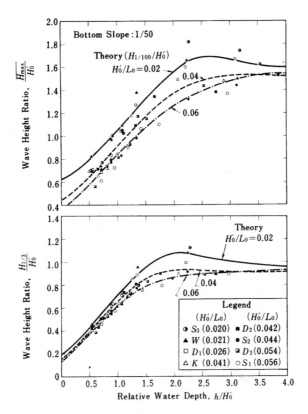

Fig. 3.37  Variations of the maximum and significant wave heights in a laboratory flume with the slope of 1 on 50.[45]

The deepwater wavelength $L_0$ is calculated with the significant wave period when computing the breaker height. That is to say, an irregular wave train with a single period is employed without consideration for frequency spectrum or directional spreading of wave energy.

The theoretical curves in Fig. 3.31 in Sec. 3.6.2 have been computed with this model, which exhibits good agreement with laboratory data of individual wave height distributions. As discussed in Sec. 3.6.1, Eq. (3.32) had a tendency of overestimation on steep slopes and a new breaker index formula of Eq. (3.33) has been proposed to remedy this tendency. However, the random wave breaking model itself does not seem to require any modification as demonstrated in the example of Fig. 3.31.

The attenuation in the characteristic heights of irregular waves by breaking was measured in a laboratory flume.[33] As shown in Fig. 3.37, the

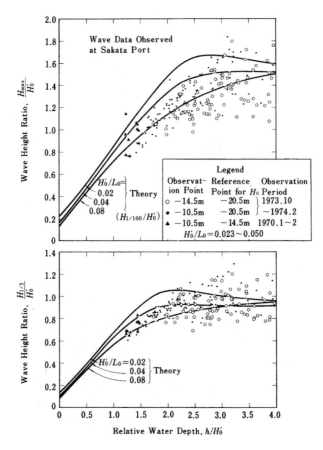

Fig. 3.38   Variations in the maximum and significant wave heights observed at Sakata Port.[45]

variations in the various wave heights are gradual. The experiments were done in a range of wave steepness with different wave spectral forms as indicated by the various symbols. Each data point represents the arithmetic mean of three runs with different input wave profiles of about 200 waves long. The wave height variation predicted by the random breaking model is seen to describe the laboratory data well.

Figure 3.38 presents a comparison of the computation to field data on the wave height variation near a coast. The data were taken at Sakata Port by the First District Port Construction Bureau of Japan[18]; three wave recorders were employed simultaneously at the depths 20, 14 and 10 m below the datum level. The equivalent deepwater wave height $H_0'$ was estimated from the data taken with the deepest recorder by correcting for the effects of wave shoaling and refraction. Field wave measurements by their very nature exhibit large scatter due to the sampling variability originating from wave irregularity itself (refer to Sec. 10.6), and this is one such typical example. Although the wave attenuation seems to be slightly larger in the data than given by the prediction as a whole, the pattern of the wave height variation in the field data is well predicted with the random wave breaking model.

### (B) *Diagrams for the estimation of wave height in the surf zone*

The wave height changes shown in Figs. 3.37 and 3.38 are expressed in terms of the parameter of equivalent deepwater wave steepness. This is because the effect of wave steepness enters quite strongly in the wave breaking process; waves of large steepness begin to break before they attain further appreciable increase in wave height by shoaling. Also, the slope of the bottom is important in regular or random wave breaking; waves approaching a coast of steep slope do not break until quite near to the shore.

By taking the wave steepness and the bottom slope as the principal parameters, changes in the maximum and significant wave heights have been computed with the random wave breaking model discussed previously. The results are shown in Figs. 3.39 to 3.42 for bottom slopes of 1/10, 1/20, 1/30 and 1/100. The maximum wave height $H_{\max}$ is set as equal to the highest one-250th wave $H_{1/250}$, i.e., the mean of the heights of the waves included in 1/250 of the total number of waves, counted in descending order of height from the highest wave. This definition yields the approximate relation of $H_{\max} \simeq 1.8 H_{1/3}$ outside the surf zone. Also, each figure contains a dash-dot curve denoted "attenuation less than 2%." In the zone to the right of this dash-dot curve, the attenuation in wave height due to wave breaking is less than 2% and the variation in wave height can be estimated by the shoaling coefficient presented in Fig. 3.26.

As an example of the usage of the diagrams, the variation in wave height from the offshore area to the shoreline has been estimated for swell with $H_0' = 4.5$ m and $T_{1/3} = 12$ s approaching a coast with uniform bottom slope of 1/10 or 1/100. By entering Figs. 3.26, 3.39 and 3.42, the wave heights at various water depths are estimated as listed in Table 3.5. Although $H_0'$ is taken as constant in this example, it varies in actual situations because of wave refraction and other phenomena, and the values of $H_0'$ and $H_0'/L_0$ must be evaluated at the respective water depths. According to Table 3.5,

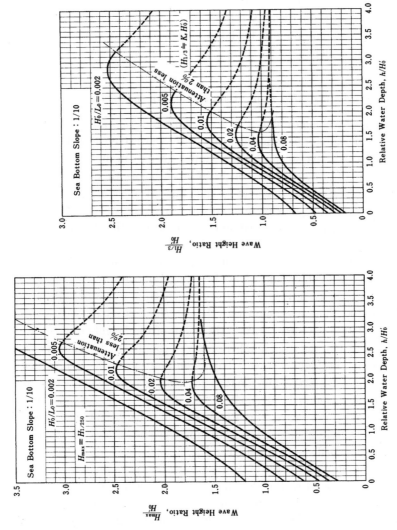

Fig. 3.39    Diagrams for the estimation of wave heights in the surf zone (sea bottom slope of 1/10). [45]

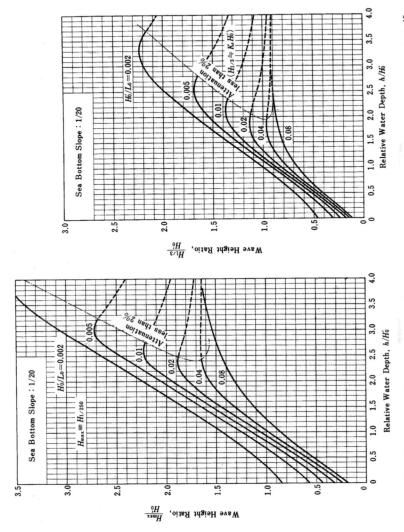

Fig. 3.40   Diagrams for the estimation of wave heights in the surf zone (sea bottom slope of 1/20).[45]

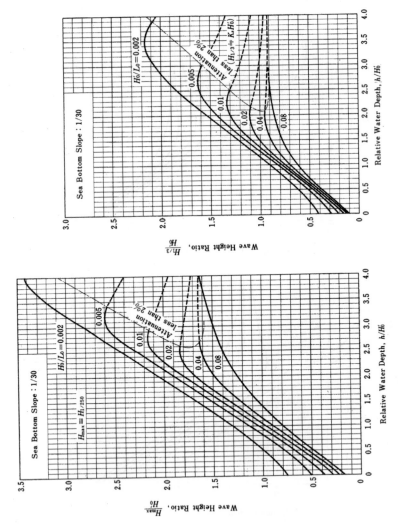

Fig. 3.41   Diagrams for the estimation of wave heights in the surf zone (sea bottom slope of 1/30).[45]

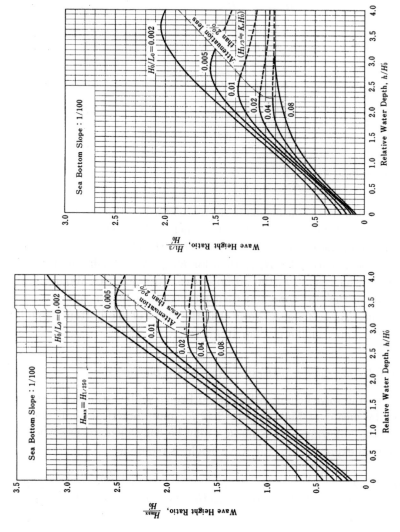

Fig. 3.42   Diagrams for the estimation of wave heights in the surf zone (sea bottom slope of 1/100).[45]

Table 3.5  Variation of wave height due to shoaling and random wave breaking.

$$- H_0' = 4.5 \text{ m}, T_{1/3} = 12 \text{ s} -$$

| Water Depth | Sea Bottom Slope 1/10 | | Sea Bottom Slope 1/100 | |
|---|---|---|---|---|
| $h$ (m) | $H_{\max}$ (m) | $H_{1/3}$ (m) | $H_{\max}$ (m) | $H_{1/3}$ (m) |
| 100 | 7.9 | 4.4 | 7.9 | 4.4 |
| 50 | 7.5 | 4.1 | 7.5 | 4.1 |
| 20 | 7.6 | 4.2 | 7.6 | 4.2 |
| 10 | 8.9 | 5.0 | 7.4 | 4.8 |
| 8 | 9.0 | 5.6 | 6.3 | 4.5 |
| 6 | 7.5 | 5.5 | 5.0 | 3.7 |
| 4 | 5.6 | 4.3 | 3.6 | 2.7 |
| 2 | 3.8 | 2.7 | 2.3 | 1.6 |
| 0 | 2.2 | 1.3 | 1.2 | 0.7 |

Note: $H_{\max} \equiv H_{1/250}$

the significant wave height begins to decrease at a water depth between 10 to 8 m on a coast with a slope of 1/100, whereas $H_{\max}$ starts to decrease at a water depth between 20 and 10 m. Thus, the effect of random wave breaking appears earlier for $H_{\max}$ than for $H_{1/3}$. As a result, the ratio of $H_{\max}/H_{1/3}$, which is 1.8 in the offshore area, goes down to about 1.3 at some depth in the surf zone and then recovers to a higher value toward the shoreline. Table 3.5 also shows that the effect of the bottom slope is noticeable in water shallower than 10 m.

## (C) *Formulas for wave height estimation within the surf zone*

The variation in wave height within the surf zone can be estimated without difficulty by means of Figs. 3.39 to 3.42 for most practical applications. However, the following approximation may be convenient to use when a mathematical expression for the wave height is required:

$$H_{1/3} = \begin{cases} K_s H_0' & : h/L_0 \geq 0.2, \\ \min\{(\beta_0 H_0' + \beta_1 h), \beta_{\max} H_0', K_s H_0'\} & : h/L_0 < 0.2, \end{cases} \quad (3.47)$$

$$H_{\max} \equiv H_{1/250}$$

$$= \begin{cases} 1.8 K_s H_0' & : h/L_0 \geq 0.2, \\ \min\{(\beta_0^* H_0' + \beta_1^* h), \beta_{\max}^* H_0', 1.8 K_s H_0'\} & : h/L_0 < 0.2. \end{cases} \quad (3.48)$$

Table 3.6 Coefficients for approximate estimation of wave heights within surf zone.[45]

| Coefficients for $H_{1/3}$ | Coefficients for $H_{max}$ |
|---|---|
| $\beta_0 = 0.028(H_0'/L_0)^{-0.38} \exp[20 \tan^{1.5} \theta]$ | $\beta_0^* = 0.052(H_0'/L_0)^{-0.38} \exp[20 \tan^{1.5} \theta]$ |
| $\beta_1 = 0.52 \exp[4.2 \tan \theta]$ | $\beta_1^* = 0.63 \exp[3.8 \tan \theta]$ |
| $\beta_{max} = \max\{0.92, 0.32(H_0'/L_0)^{-0.29}$ | $\beta_{max}^* = \max\{1.65, 0.53(H_0'/L_0)^{-0.29}$ |
| $\times \exp[2.4 \tan \theta]\}$ | $\times \exp[2.4 \tan \theta]\}$ |

Note: max $[a, b]$ gives the larger of $a$ or $b$.

The symbol $\min\{a, b, c\}$ stands for the minimum value among $a, b$ and $c$. The coefficients $\beta_0$, $\beta_1$, ... have been formulated as listed in Table 3.6.

**Example 3.10**

Estimate the significant wave height at the depth of $h = 8$ m by means of Eq. (3.47), when swell with $H_0' = 6$ m and $T_{1/3} = 15$ s is attacking a coast with a slope of $1/50$.

**Solution**

Since the deepwater wavelength is $L_0 = 351$ m and the wave steepness is $H_0'/L_0 = 6/351 = 0.0171$, the $\beta$ coefficients are calculated as follows:

$$\beta_0 = 0.028 \times (0.0171)^{-0.38} \times \exp[20 \times (0.02)^{1.5}] = 0.139 \,,$$

$$\beta_1 = 0.52 \times \exp[4.2 \times 0.02] = 0.566 \,,$$

$$\beta_{max} = \max\{0.92, 0.32 \times (0.0171)^{-0.29} \times \exp[2.4 \times 0.02]\}$$

$$= \max\{0.92, 1.093\} = 1.093 \,.$$

As for the shoaling coefficient in Fig. 3.26, the point corresponding to $h/L_0 = 0.023$ and $H_0'/L_0 = 0.0171$ appears to be located above the dash-dot curve corresponding to $\tan \theta = 1/100$, indicating a large attenuation of wave height due to breaking. Nevertheless, by extrapolating the curves of $K_s$, the shoaling coefficient is estimated as $K_s \simeq 1.7$. Thus, the significant wave height is estimated as

$$H_{1/3} = \min\{(0.139 \times 6.0 + 0.566 \times 8.0), 1.093 \times 6.0, 1.7 \times 6.0\}$$

$$= \min\{5.36, 6.56, 10.2\} \simeq 5.4 \text{ m} \,.$$

Equations (3.47) and (3.48) may give a difference of several percent in the estimated heights compared to those read in Figs. 3.39 to 3.42. For waves of steepness greater than 0.04, the formulas yield a significant height in excess of 10% compared to that from the diagrams around the water depth at which the value of $H_{1/3} = \beta_0 H_0' + \beta_1 h$ becomes equal to the value of $H_{1/3} = \beta_{\max} H_0'$. A similar difference also appears for the case of $H_{\max}$. Waves of large steepness may have a discontinuity in the height $H_{\max}$ estimated with Eq. (3.26) at the boundary $h/L_0 = 0.2$. Caution should be exercised in applying Eqs. (3.47) and (3.48) in regard to such differences and discontinuity.

Applicability of the formulas of Eqs. (3.47) and (3.48) has been tested against the large database of CLASH project (De Rouck *et al.*[59]). Laboratory measured data of significant wave heights in shallow water were compared with the heights estimated from the input in deeper locations by means of Eq. (3.47) for 1214 cases from 29 datasets, which covered a wide range of beach slope from 1 on 8 to 1 on 1000. The ratio of the estimated to the measured heights had the overall mean of 1.106 with the standard deviation of 0.155, indicating a tendency of slight overestimation.[60]

A unique feature of the formulas of Eqs. (3.47) and (3.48) is a prediction of finite heights of $H_{1/3} = \beta_0 H_0'$ and $H_{\max} = \beta_0^* H_0'$ at the shoreline where the initial water depth is zero. The presence of finite wave heights there is due to the increase in actual water depth owing to the effects of wave setup and surf beats. The motion of water around the shoreline, however, is more intensive than that of ordinary wave motion corresponding to the height estimated above at the increased water depth, because of the uprushing and downrunshing surging motion of water there. Therefore, the estimated wave height at $h = 0$ should be regarded as an apparent one which does not adequately represent the magnitude of the wave action. Evaluation of the weight of armor stone at the shoreline with the above estimated wave height by means of Hudson's formula, for example, would be poor practice because it would underestimate the required mass of the stone.

### 3.6.4 *Prediction of Random Wave Breaking Process on Beaches of Complicated Profiles*

The author's random wave breaking model in 1975 as previously described has been employed in many engineering applications. One shortcoming of this model is that the sea bottom should have a uniform slope. For coasts with nonuniform slopes, a compromise is made by defining a local slope and utilizing it in computation. However, it is difficult to apply this model to beaches of complicated profiles such as those having bars and troughs.

In addition to that model by the author, a number of random wave breaking models have been proposed and are being utilized in numerical analysis of wave transformations. The earliest model seems to have been proposed by Collins[61] in 1970 who truncated the Rayleigh distribution of wave height beyond the depth-controled breaker height. It was an approach based on deformation of the probability density function (*pdf*) of wave heights. Similar approaches were taken by Battjes,[62] Kuo,[63] and Goda,[33, 45] though the method of deforming *pdf* differed each other.

Then Battjes and Janssen[64] came up with the analogy of wave energy dissipation by breaking to the bore in hydraulic jumps. Their model with a revised breaker index by Battjes and Stive[65] has been widely utilized by European engineers. Thornton and Guza[66] have proposed a modified version for application to the field observation data at the California coast.

Another approach to evaluation of wave energy dissipation by breaking was proposed by Dally *et al.*[67] who introduced the concept of energy saturation at a stable wave height in the surf zone. The concept originally for regular waves was expanded to random waves by Larson and Kraus[68] and Goda.[50] While the *pdf* deformation model is applicable for planar beaches only, both the bore-type and Dally-type models can deal with random wave breaking on beaches of complicated profiles.

In addition to the above mentioned models, there are several more models for wave deformation by random breaking. Most of these models aim at predicting the root-mean-square wave height $H_{\mathrm{rms}}$ only without examining the changes in the *pdf* of wave heights; exceptions are the Goda model in 1975 and Goda's PEGBIS model[50] to be discussed in Sec. 12.4. Each model employs different criterion for incipient breaking, some model does not introduce the wave setup/setdown in calculation of local water depth, and many models assume uni-directional, single-frequency irregular waves without consideration for directional spectral characteristics. Because of such differences, performance of these random wave breaking models differs considerably.

The author[69] has made an exercise of comparison of the wave heights within the surf zone predicted by the models mentioned above except for those by Collins[61] and Dally *et al.*[67] Table 3.7 lists the characteristics of the seven models employed in the exercise.

The exercise was made for prediction of the root-mean-square wave height $H_{\mathrm{rms}}$ on a planar beach with the slope of 1 on 50 when the deepwater waves with $H_0 = 2.0\,\mathrm{m}$ and $T = 8\,\mathrm{s}$ are approaching with the deepwater

Table 3.7   Main characteristics of the seven models for random wave-breaking.[69]

| Models | Energy Dissipation Mechanism | pdf of Broken Waves | Factors Affecting Breaking Wave Height | | | | | |
|---|---|---|---|---|---|---|---|---|
| | | | $H_b/h$ | $h/L$ | Beach Slope | Range Allowance | Wave Setup | Wave Spectrum |
| Battjes[62] | pdf Deformation | Delta function at breaker height | Yes | Yes | No | No | Yes | No |
| Kuo and Kuo[63] | Ditto | Remove and adjust the remainder | Yes | No | No | No | No | No |
| Goda[45] | Ditto | Ditto | Yes | Yes | Yes | Yes | Yes | No |
| Battjes and Janssen[64] | Bore model | Delta function at breaker height | Yes | Yes | No | No | Yes | No |
| Thornton and Guza[66] | Ditto | Adjust with weight function | Yes | No | No | No | No | No |
| Larson and Kraus[68] | Dally model | Delta function at breaker height | Yes | No | No | No | Yes | No |
| Goda[50] (PEGBIS) | Ditto | Remove and adjust the remainder | Yes | Yes | Yes | Yes | Partially | Yes |

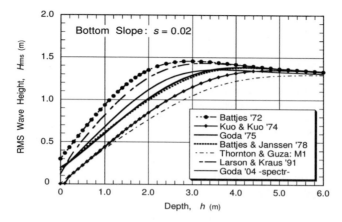

Fig. 3.43 Wave height variations in the surf zone on planar beaches with slope of $s = 1/50$ computed by seven random wave-breaking models for $(H_{1/3})_0 = 2.0\,\text{m}$, $T = 8.004\,\text{s}$, and $\theta_0 = 30°$.[56]

incident angle of $\theta_0 = 30°$. The results of predicted wave heights are shown in Fig. 3.43.

As exhibited in Fig. 3.43, the predicted values of $H_{\text{rms}}$ differ greatly among the seven models; they differ more than two fold in water shallower than $2\,\text{m}$. Difference in predicted wave heights directly affects the computed velocity of nearshore currents as demonstrated by the author.[69] When selecting a random wave breaking model in solving coastal engineering problems, due caution should be taken for the characteristics of the model under consideration.

## 3.7 Reflection of Waves and Their Propagation and Dissipation

### 3.7.1 *Coefficient of Wave Reflection*

When waves are reflected by a structure, the reflected waves cause increased agitation of the water in front of the structure, or they may propagate some distance to become a source of disturbance in an otherwise calm area of water. Thus, it is desirable to suppress the reflection of waves as much as possible. It is a problem of energy dissipation as well as estimation of the propagation of the reflected waves.

Concerning wave energy dissipation, the degree of wave reflection needs to be quantified. For this purpose, the coefficient of wave reflection $K_R$,

Table 3.8  Approximate values of reflection coefficients.

| Structural Type | Reflection Coefficient |
| --- | --- |
| Vertical wall with crown above water | $0.7 \sim 1.0$ |
| Vertical wall with submerged crown | $0.5 \sim 0.7$ |
| Slope of rubble stones (slope of 1 on 2 to 3) | $0.3 \sim 0.6$ |
| Slope of wave-dissipating concrete blocks | $0.3 \sim 0.5$ |
| Vertical structure of energy dissipating type | $0.3 \sim 0.8$ |
| Natural beach | $0.05 \sim 0.2$ |

which is the ratio of reflected wave height $H_R$ to incident height $H_I$, is generally employed; i.e.,

$$K_R = H_R/H_I .\tag{3.49}$$

Both the incident and reflected wave heights refer to the significant heights unless otherwise stated.

*Reflection coefficients* for most structures are usually estimated by means of laboratory model tests, because a theoretical analysis is not feasible for wave reflection associated with partial wave breaking by structures. Approximate values of reflection coefficients as reported in various sources are listed in Table 3.8. The range in coefficients for a vertical wall depends on the degree of wave overtopping, and it increases as the crown elevation increases. For sloped mounds and natural beaches, the reflection coefficient is inversely proportional to the steepness of the incident waves and the upper bounds correspond to swell of long period. Seelig and Ahrens[70] gave empirical formulas to estimate the reflection coefficient for beaches, revetments and rubble mound breakwaters, based on a large amount of laboratory data including that of irregular waves.

For sloping-type coastal structures, Zanuttigh and van der Meer[71] have proposed the following estimation formula for the wave reflection coefficient:

$$K_R = \tanh\left[a\left(\frac{\tan\alpha_s}{\sqrt{H_{m0}/L_{m-1,0}}}\right)^b\right],\tag{3.50}$$

where $\alpha_s$ denotes the angle between the slope surface and the horizontal, $H_{m0}$ refers to the spectral significant wave height (Eq. (2.40)) at the structure toe, and $L_{m-1,0}$ is the deepwater wavelength corresponding to the spectral significant wave period $T_{m-1,0}$ (Eq. (2.45)). The optimum values of the constants are $a = 0.16$ and $b = 1.43$ for smooth slopes, while they are $a = 0.12$ and $b = 0.87$ for slopes covered with rubble stones or wave-dissipating concrete blocks.

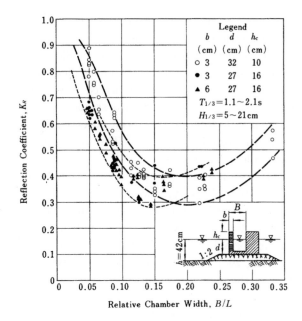

Fig. 3.44  Reflection coefficient for a perforated-wall caisson breakwater for irregular waves (after Tanimoto *et al.*[72])

For a vertical wall of the energy dissipating type, such as a perforated wall, the reflection coefficient is governed by the shape of the structure, the width of the energy dissipating section of the structure relative to the incident wavelength, and other factors. Therefore, individual scale model tests are required to estimate the reflection coefficient for these types of structures for a given wave condition. Figure 3.44 is an example of early model tests performed by Tanimoto *et al.*[72] It gives the reflection coefficient for irregular waves incident on a concrete caisson with a front wall having circular perforations. The reflection coefficient has a minimum value of $(K_R)_{min} \simeq 0.3$ at the relative cell width of $B/L \simeq 0.15$ and it rises above 0.7 when the relative cell width becomes less than 0.05. There have been published many model test reports on the reflection coefficient of energy-dissipating type coastal structures, and the readers are recommended to make a literature survey when designing such a structure.

### 3.7.2  *Propagation of Reflected Waves*

Each component wave of a random sea is assumed to be reflected at an angle equal to the angle of incidence and to continue to propagate in that direction, as in the theory of geometrical optics (see Fig. 3.45). A possible departure from geometric wave reflection is the case of long-period waves of large amplitude incident at a large angle to a reflective structure in a shallow water area. In such a case, waves being incident nearly parallel to the structure may not produce clear reflected waves but instead form gradually swelling crests which run along the structure. This phenomenon is called *Mach-stem reflection* and has been observed on the occasion of tsunami attacks, as reported by Wiegel.[73] The wave crests running along the structure are called *stem waves*.

The stem waves can also appear for ordinary waves. Mase *et al.*[74] reported their observation of the stem waves along a straight wall built on a 3-D model bathymetry when random waves with the directional spreading parameter of $s_{max} = 75$ or $\infty$ (uni-directional) are incident with the angle of 70° to the normal (the angle between the wave direction and the wall alignment being 20°). The height of the stem waves increased along the wall. However, the stem waves did not appear when the directional spreading parameter was $s_{max} = 25$.

Another aspect of reflected waves is that they have a finite length along their crest lines, because reflective structures such as caisson breakwaters or quay walls have limited extent. Therefore, reflected waves disperse during propagation away from the source of reflection in a manner similar to the phenomenon of wave diffraction. The dispersion of the reflected waves can be analyzed by means of the theoretical solution for wave reflection by an island barrier,[75,76] or by the numerical integration method for simulating wave propagation (Tanimoto *et al.*[77]).

As an engineering approximation, the spatial distribution of reflected wave height can be estimated by making use of diffraction diagrams for an opening between breakwaters. The idea is to treat the zone of wave reflection as the opening between fictitious breakwaters, as sketched in Fig. 3.46 and to apply a diffraction diagram accordingly. The direction of the imaginary incident waves is set at the mirror image of the actual waves at the reflective boundary. (In applying the diffraction diagram, a rotation of the axis of the diffracted waves specified in Table 3.4 in Sec. 3.3.3 must be made.) Then the amount of wave height attenuation of the reflected waves can be taken as equal to the diffraction coefficient obtained from the diagram thus applied.

The essential point in this method is to employ diffraction diagrams of random sea waves. This is because directional spectral characteristics must be taken into account in every problem involving two-dimensional

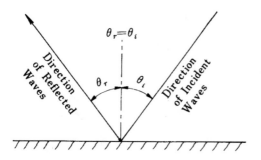

Fig. 3.45   Direction of reflected waves.

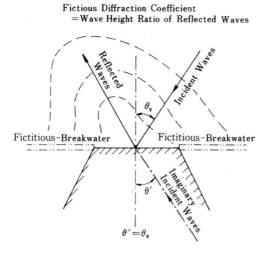

Fig. 3.46   Estimation of propagation of reflected waves by means of fictitious breakwaters.

propagation of sea waves. Likewise, if the problem is solved either analytically or numerically, the solution for a single direction of incidence with one wave period is not sufficient. The computation must be done for many wave components representative of the directional wave spectrum, and the results must be combined with the weight of the directional energy density so as to yield a meaningful answer to the problem.

Sometimes a reflective structure extends over such a great distance as to make inapplicable the diffraction diagram for a breakwater opening. In that case, one can apply diffraction diagrams for a semi-infinite breakwater or use the angular spreading method with the cumulative distribution curve

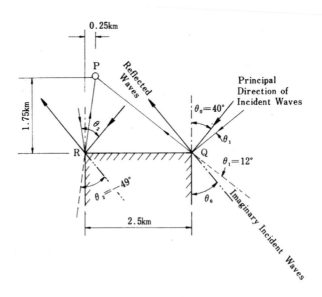

Fig. 3.47   Estimation of the height of reflected waves by the angular spreading method.

of total wave energy with respect to the azimuth from the principal wave direction.

### Example 3.11

Wind waves are incident at an angle of 40° to vertical wall $\overline{QR}$ of length 2.5 km as shown in Fig. 3.47. Estimate the height of the reflected waves at Point P, assuming that the reflection coefficient of the vertical wall is 1.

### Solution

First, the lines $\overline{PQ}$ and $\overline{PR}$ are drawn and the angles that the extensions of these lines make with the direction of the imaginary incident waves (mirror images of the actual waves) are measured. The angles are $\theta_1 = 12°$ and $\theta_2 = -49°$ in the example of Fig. 3.47. These angles are equal to the azimuths from the principal wave direction of the component waves which are reflected from the edges of Q and R and which propagate straight to Point P. Thus, the angle between $\theta_2$ and $\theta_1$ represents the azimuth of the component waves reflected from an arbitrary point between Q and R to reach Point P.

Next, the energy of the reflected waves reaching Point P is estimated as being equal to the energy of the component waves having an azimuth between $\theta_2$ and $\theta_1$. The ratio of this energy to the total energy is estimated with the cumulative distribution curve for $P_E(\theta)$ shown in Fig. 2.15 in Sec. 2.3.

By using the curve pertaining to $s_{\max} = 10$ for wind waves, we obtain $\Delta E = P_E(\theta_1) - P_E(\theta_2) = 0.63 - 0.08 = 0.55$. Therefore, the ratio of reflected to incident wave heights at Point P is obtained as the square root of the above energy ratio, or $\sqrt{0.55} = 0.74$.

If the structure has a reflection coefficient less than 1, the above estimate for the dispersion effect of reflected waves should be multiplied by the reflection coefficient of the structure to yield the final estimate of the reflected wave height reaching the point of interest. Another cause of the attenuation of reflected waves is energy dissipation by an adverse wind. We sometimes observe such a phenomenon when waves are reflected from a vertical wall; some of the wave crest are blown out by the opposing wind. The reflected waves lose additional energy through violent interaction with the incident waves. However, the rate of wave attenuation by adverse wind is not known quantitatively. It is only known empirically that wind waves of large steepness are quickly attenuated, whereas swell of low steepness continues to propagate over a long distance with only minor dissipation. In harbor planning and design, it will be on the safe side to disregard the attenuation of wave height by an adverse wind. If the water area in a harbor is quite broad and the harbor is to be designed against locally-generated wind waves, the effective value of the reflection coefficient of vertical walls within the harbor may be reduced to the level of say 80% as a measure of incorporating the effect of an adverse wind.

### 3.7.3 *Superposition of Incident and Reflected Waves*

The effect of reflected waves on harbor agitation and structural design is rather complicated, because not only the wave height but also the wave direction enters into the problem. When only the total wave height is of interest, however, it can be estimated by the principle of summation of energy components as in the following:

$$H_S = \sqrt{H_I^2 + (H_R)_1^2 + (H_R)_2^2 + \cdots}, \tag{3.51}$$

where $H_S$ denotes the significant height of the superposed waves, and $(H_R)_1, (H_R)_2, \ldots$ represent the significant heights of the reflected waves originating from various reflective sources. Equation (3.51) is not applicable in the immediate vicinity of structures because of the fixed phase relation between incident and reflected waves. But the phase interference almost cancels out among the various components of random sea waves if

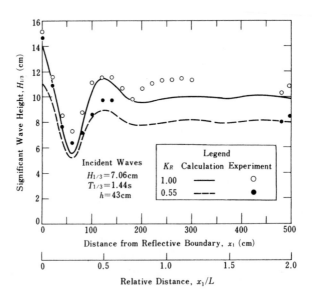

Fig. 3.48   Spatial variation of irregular standing wave height.[78]

the distance from the reflective structure is more than one wavelength or so, and Eq. (3.51) yields a reasonable estimate of the wave height.

The applicability of Eq. (3.51) is demonstrated in Fig. 3.48, which shows the spatial variations of significant wave height in front of model structures in a laboratory flume.[78] The incident waves are irregular trains with frequency spectra of the Bretschneider–Mitsuyasu type. Open circles represent the data for a fully reflective vertical wall, and filled circles represent the data for a model structure having a reflection coefficient of 0.55. The solid and dash lines indicate wave heights estimated from the frequency spectra of superposed wave systems, which were obtained by calculating the amplitudes of standing waves at each location for various frequency components of the incident wave spectrum. Although the significant height of the superposed waves exhibits a fluctuation corresponding to the nodes and antinodes of standing waves in the vicinity of the structures, as seen in Fig. 3.48, the fluctuations decrease rapidly with distance from the structure and the wave height approaches an asymptotic value. In fact, Eq. (3.51) predicts such an asymptotic value for points far from a reflective structure.

The theoretical basis of Eq. (3.51) is Eq. (2.40) in Sec. 2.4.1, which states that the significant wave height is proportional to the square root of the total wave energy, irrespective of the spectral shape.

## 3.8 Spatial Variation of Wave Height Along Reflective Structures

### 3.8.1 *Wave Height Variation Near the Tip of a Semi-Infinite Structure*

When waves are incident to a structure, a standing wave system is formed at its front. If the structure reflects the incident waves completely and the wave system is one-dimensional, the standing wave height at the front wall becomes twice the incident height. In actual situations, however, the standing wave height along the structure undulates, because there usually exists a two-dimensional effect owing to the finite extent of the structure. In addition, the plan shape of the structure may not be in straight lines.

First, the wave height around the tip of a structure is dominated by diffraction at the tip. Figure 3.49 exhibits such an example, which shows the spatial variation of wave height in the front as well as at the rear of a semi-infinite breakwater of perfect wave reflection when waves of a regular train are incident normally.[79] Although a standing wave system is seen to be formed in front of the breakwater, the wave height along the anti-nodal line at $y/L = 0, 0.5, 1.0, \ldots$ exceeds twice the incident wave height at some locations, whereas the wave height along the nodal lines remains at some finite value.

The ratio of the wave height along a breakwater to the incident height is generally calculated with the following formula, when waves approach the breakwater at an angle $\Theta$ between their direction of propagation and the face line of the breakwater (Ito and Tanimoto[80]:)

$$\frac{H_S}{H_I} = \sqrt{(C + S + 1)^2 + (C - S)^2}, \qquad (3.52)$$

where $C$ and $S$ stand for the *Fresnel integrals*:

$$C = \int_0^u \cos\frac{\pi}{2}t^2 dt, \quad S = \int_0^u \sin\frac{\pi}{2}t^2 dt \; : u = 2\sqrt{\frac{2x}{L}}\sin\frac{\Theta}{2}, \qquad (3.53)$$

in which $x$ denotes the distance from the breakwater tip and $L$ is the wavelength.

Figure 3.50 shows the distribution of wave height along a breakwater as computed with Eq. (3.52) for the case of waves with period 10 s normally incident to a breakwater located in a water depth of 10 m. The solid line represents regular waves of period $T = 10$ s, and the dash line presents the result for long-crested random waves with $T_{1/3} = 10$ s. The random waves were assigned the Bretschneider–Mitsuyasu frequency spectrum of

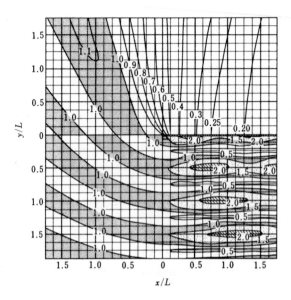

Fig. 3.49  Variation of the height of regular waves around a semi-infinite breakwater (after Morihira and Okuyama.[79])

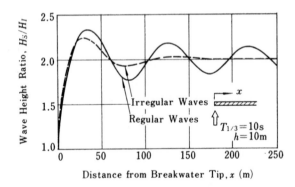

Fig. 3.50  Variation of wave height in front of a semi-infinite breakwater.

Eq. (2.10) in Sec. 2.3.1 without any directional spreading (all component waves normally incident to the breakwater). The wave height ratio of random waves was computed with Eq. (3.22) in Sec. 3.3.1 by replacing the factor $K_d$ with the ratio $H_S/H_I$ given by Eq. (3.52). The computed wave height shows undulations around the line $H_S/H_I = 2.0$. The first peak takes a value of 2.34 for regular waves and 2.25 for random waves. A large undulation repeats several times for the regular waves, but it diminishes rapidly after the second peak for the random waves. If directional spreading is introduced, the undulation in the wave height will be further reduced.

The undulation in wave height along a breakwater produces a spatial variation in the wave pressure exerted on the breakwater. Because the waves diffracted into the back side of the breakwater are shifted in phase from the standing waves at the front of the breakwater, the variation in total wave force exerted on a unit length of breakwater is greater than the variation in the wave height itself. For example, the first peak of wave force shows an increase of more than 20% compared with the standing wave force on an infinitely long breakwater. Surveys of concrete caissons of vertical breakwaters which have slid due to wave action often reveal evidence of an undulation in the sliding distance. Ito and Tanimoto[80] have explained this phenomenon as the effect of the waves diffracted from the tip of the breakwater and called it *meandering damage* to breakwaters.

A similar undulation in wave height is predicted in the vicinity of the corner of reclaimed land protected by straight, reflective seawalls. The wave height distribution for such a configuration can be analyzed with the theory of Mitsui and Murakami.[81]

### 3.8.2 *Wave Height Variation at an Inward Corner of Reflective Structures*

An inward corner of a reflective wall is sometimes formed when an extension of a vertical breakwater is made with an outward bend to the offshore, or when a short jetty is attached at the end of reclaimed land at an angle of bend toward the offshore. Waves incident to such a configuration are reflected toward the inward corner from the two reflective walls, and a pronounced increase in wave height can be observed at the corner. By denoting the angle between the two walls as $\beta$, as sketched in Fig. 3.51, the ratio of wave height at the corner to the incident height is given by the following formula, provided the reflective walls at both sides are sufficiently long:

$$\frac{H_S}{H_I} = \frac{2\pi}{\beta}, \tag{3.54}$$

where $\beta$ is expressed in radians. The formula is applicable to random sea waves described with directional wave spectra as well.

The distributions of the height of regular waves along both reflective walls usually exhibit large fluctuations, because of interference between the

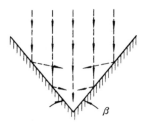

Fig. 3.51   Wave concentration at an inward corner.

incident waves and the two reflected waves. Depending on the direction
of wave incidence and the angle between the two walls, the wave height
at some point along the walls may become greater than the height at the
corner. In the case of random sea waves, the spatial fluctuation in wave
height is mostly smoothed out because of the wide spread in the frequency
and directional energy distributions. The wave height by Eq. (3.54) gives
the maximum value along the two reflective walls for the case of random
sea waves. Kobune and Osato[82] have made a series of computations for the
distribution of the wave height along two reflective walls for several values
of the angle between them for random sea waves having the Bretschneider–
Mitsuyasu frequency spectrum and Mitsuyasu-type spreading function with
a specific value of $s_{\max} = 75$. They proposed the following approximate
method for estimating the wave height for such geometry:

(i) The wave height ratio at the inward corner is estimated with Eq. (3.54).
(ii) The distribution of the height of reflected waves along the wings of the
reflective walls is estimated by the technique introduced in Sec. 3.7.2,
as sketched in Fig. 3.52. Then the height of these waves reaching the
line of the other wing is read from the estimated wave height distribution, and the overall wave height along the other wing is estimated
with Eq. (3.51).
(iii) The distances at which the wave height ratio takes the minimum value
of 2.0 are estimated for both wings by the following:

$$\frac{x_{\min}}{L} = 0.16 \exp\left[1.05 \tan\frac{\alpha}{2}\right], \qquad (3.55)$$

where $\alpha$ denotes the angle between the direction of wave approach and
the face line of the reflective wall along which $x_{\min}$ is to be determined.
(iv) The wave height distribution between the corner and the point
$x = x_{\min}$ is assumed to vary linearly from the value obtained with
Eq. (3.54) at $x = 0$ and the value 2.0 at $x = x_{\min}$.

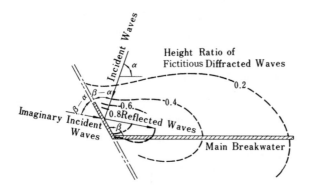

Fig. 3.52 Estimation of reflected wave height.

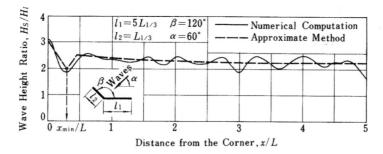

Fig. 3.53 Distribution of wave height along a main breakwater with an auxiliary break-water (after Kobune and Osato).[82]

(v) In the range $x > x_{min}$, the mirror image of the line of wave height distribution estimated in Step (iv) is drawn about the axis of symmetry at $x = x_{min}$. Then the resultant wave height ratio is compared with the value estimated in Step (ii), and the lesser of the two values is to be adopted as an estimation of the wave height ratio.

Figure 3.53 illustrates the above method of wave height estimation for an auxiliary reflective-type breakwater of length $L_{1/3}$ attached at an angle of $\beta = 120°$ to a main breakwater of the reflective-type with a length of $5\,L_{1/3}$. Random sea waves are incident to this structure with an angle of approach of $\alpha = 60°$ to the main breakwater. In Fig. 3.53, the result of the approximate estimation is denoted by the dash line, which can be compared with the result of a numerical computation with the directional wave spectrum, denoted by the solid line. Although some small differences are observed between the two estimations, the approximate method does represent the pattern of the wave height quite well.

### 3.8.3   *Wave Height Variation Along an Island Breakwater*

If the length of a single isolated breakwater in the offshore (hereafter called an island breakwater) is on the order of a few wavelengths of the incident waves or less, the distribution of the wave height along the breakwater will show considerable variation owing to the effects of the waves diffracted from both tips of the breakwater. Figure 3.54 presents the result of a computation of the height at the front and the rear of an island breakwater obtained with the solution of the velocity potential in problem (Goda and Yoshimura[83]). The incident waves are regular trains with angles of approach of 30° (dash line) and 90° (solid line). When waves are obliquely incident, the wave height gradually increases toward the down-wave side from the up-wave side of the breakwater, probably because some distance is needed before the reflected waves can achieve full height.

Such a variation in the distribution of wave height modifies the wave force acting on the structure, as in the case of the tip of a semi-infinite breakwater. An island breakwater experiences a greater variation in wave force than a semi-infinite breakwater, because waves diffracted behind the island breakwater may be just opposite in phase to the waves in front of the breakwater while still maintaining appreciable height. If the waves arrive in regular trains, the total wave force per unit length of breakwater may reach 1.8 times that of an infinitely long vertical breakwater (one-dimensional standing wave system), though the spatial variation in total

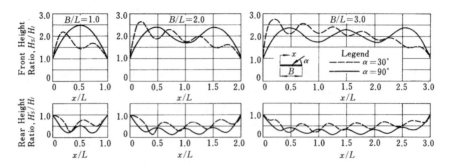

Fig. 3.54  Distributions of wave height along the front and rear of an island breakwater.[83]

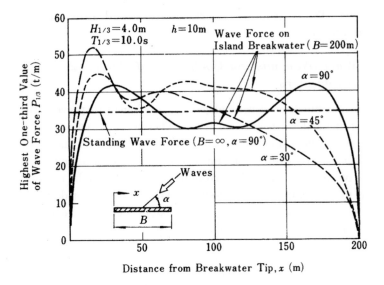

Fig. 3.55 Calculated distribution of wave force along an island breakwater due to irregular waves.[83]

wave force along an island breakwater is reduced to some extent by the effect of the wave spectral characteristics for random sea waves.

Figure 3.55 is an example of the distribution of total force of irregular waves along an island breakwater when the frequency spectral characteristics are included in the computation, but with no directional spreading of wave energy considered.[83] The breakwater is of the upright type and 200 m in length, erected in water 10 m deep and subject to the attack of irregular waves with $H_{1/3} = 4$ m, $T_{1/3} = 10$ s, and $L_{1/3} = 92.3$ m from several directions. The total wave force is estimated from the force spectrum, computed at each frequency by using the theory of small amplitude standing wave force. The mean of the highest one-third wave force is calculated from the integral of the force spectrum by means of Eq. (2.40) in Sec. 2.4.1; the coefficient of proportionality 4.0 has been changed to 2.0 for this case because the wave force refers to the height of the crest and not to the distance between crest and trough. In comparison with the standing wave force on an infinitely long breakwater, denoted by the horizontal line of heavy dash-dots in Fig. 3.55, the island breakwater is observed to experience an increase in total wave force by 20% to 50%, depending on the angle of wave approach.

## 3.9 Wave Transmission at Breakwaters and Low-Crested Structures

### 3.9.1 *Wave Transmission Coefficient of Composite Breakwaters*

The principal function of a breakwater is obviously to prevent the penetration of incident waves into a harbor. Therefore, waves passing through the gaps/voids in a breakwater and waves generated on the leeside of a breakwater by wave overtopping must be reduced to a minimum. Complete stoppage of these passing and overtopping waves may not be recommended, however, in consideration of excessive construction cost for doing so.

The wave reduction function of breakwaters is generally expressed with the *wave transmission coefficient* $K_T$, which is defined as

$$K_T = H_T/H_I, \tag{3.56}$$

where $H_T$ and $H_I$ denote the significant height of transmitted and incident waves, respectively.

Wave transmission at vertical and composite breakwaters is mainly the result of waves generated in the lee of the breakwater by the impact of the fall of the overtopping water mass. The water mass temporarily raises the water surface behind the breakwater, then it is pulled down by gravity, and oscillations of water surface starts; it is a kind of Cauchy–Poisson waves. The principal parameter governing the wave transmission coefficient in this case is the ratio of the crest elevation of the breakwater to the incident wave height, i.e. relative crest elevation $h_c/H_I$. Figure 3.56 shows a compilation of the author's data from laboratory tests[84] with regular waves. The data of the transmission coefficient $K_T$ were grouped by the ratio of the water depth above the foundation mound $d$ to the water depth at the toe $h$. The four curves in Fig. 3.56 for the relative mound depth $d/h = 0, 0.3, 0.5,$ and $0.7$ have been drawn through a scatter amounting to $\pm 0.1$ in the absolute value of $K_T$.

The diagram of Fig. 3.56 is applicable to irregular waves also, as shown by the example of some tests with irregular waves[85] in Fig. 3.57. The curve drawn through the data points of $K_T$ for irregular waves was taken from Fig. 3.56 for regular waves. Figure 3.57 also displays the data for the reflection coefficient. Alberti *et al.*[86] have carried out a series of irregular wave tests on wave transmission over vertical breakwaters and proposed

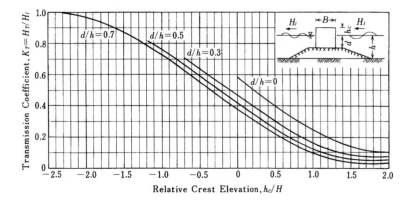

Fig. 3.56   Wave transmission coefficient for a vertical breakwater.[84]

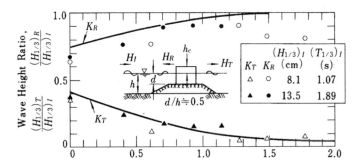

Fig. 3.57   Example from irregular wave tests on wave transmission and reflection coefficients for a vertical breakwater.[85]

the following formula:

$$K_T = \exp\left[-\left(1.14 + 1.16\frac{h_c}{H_{si}}\right)\right] \quad : \quad \frac{h_c}{H_{si}} > -0.2 , \qquad (3.57)$$

where $h_c$ denotes the crest elevation above the mean water level and $H_{si}$ is the significant height of incident waves. Equation (3.57) yields the transmission coefficient slightly smaller the curves of $d/h = 0.7$ in Fig. 3.56.

For the wave transmission coefficient of horizontally-composite breakwaters, i.e., those protected with a frontal mound made of wave-dissipating concrete blocks such as Tetrapods, Kondo and Sato[87] have proposed the simple expression of the following:

$$K_T = 0.3\left(1.1 - \frac{h_c}{H_I}\right) \quad : \quad 0 \le \frac{h_c}{H_I} \le 0.75 . \qquad (3.58)$$

Figure 3.56 and Eq. (3.58) are approximately applicable to individual waves in an irregular train of waves. This implies that the distribtuion of transmitted wave heights immediately behind an emerged breakwater has a range wider than the Rayleigh distribution discussed in Sec. 2.2, because individual waves of large height have a low value of $h_c/H$ and a large value of $K_T$ compared to the mean wave. However, the distribution approaches to the Rayleigh as the overtopped waves propagate over a certain distance.

Waves transmitted by overtopping tend to have shorter periods, because the impact of the falling water mass often generates harmonic waves with periods of one-half and one-third the incident wave period. Thus, the frequency spectra of waves transmitted over vertical and composite breakwaters show conspicuous enhancement of spectral energy in the high frequency range, while the spectral peak density is reduced to some extent, as demonstrated by van der Meer and de Waal.[88] Alberti et al.[86] have reported a decrease of the mean wave period by up to 50% while the specral peak period remains unchanged. They have presented an empirical relationship for the change of the significant wave period as $(T_{1/3})_T/(T_{1/3})_I = 1.2K_T + 0.28$.

### 3.9.2   *Wave Transmission Coefficient of Low-Crested Structures*

The terminology of *low-crested structures* has appeared in the early 2000s, which mostly refers to shore-parallel, low-crested breakwaters for protection of sandy beaches. They are expected to reduce, not eliminate, the wave energy exerted upon the beach to a certain degree; a key factor in their design is the allocation of the amount of wave energy to be transmitted to the beach. Lamberti et al.[89] summarizes European experiences of low-crested structures for coastal management. These structures have often been called detached breakwaters or reef breakwaters; the latter is applied when the whole structure is built with single size of stone or concrete blocks. A low-crested structure may have either an emerged or a submerged crest. Hereinafter, the acronym of LCS is employed for low-crested structures.

#### (A) *Wave transmission through LCS*

Early studies of wave transmission coefficient of LCS go back to the 1960s in Japan, where the detached breakwater system had been adopted at many coasts for beach protection. Iwasaki and Numata[90] carried out a series of experiments on waves passing through rubble-mound breakwaters and

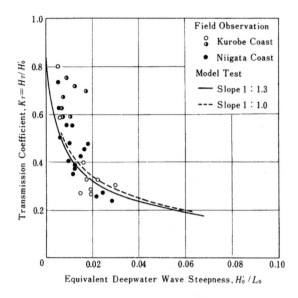

Fig. 3.58 Wave transmission coefficient of a mound breakwater constructed with Tetrapods (after Hattori).[91]

formulated the wave transmission coefficient as a function of the wave steepness in 1969. Dependence of the transmission coefficient on the wave steepness was also confirmed by Hattori,[91] who compared his laboratory test curves with some data from field observations (Tominaga and Sakamoto[92] and Katayama et al.[93]), as shown in Fig. 3.58.

In 1975, Numata[94] proposed a unified formula for wave transmission through porous breakwaters composed of wave-dissipating concrete blocks by synthesizing various laboratory and field data. His formula is expressed as below.

$$(K_T)_{\text{thru}} = 1/\left[1 + C\left(\frac{H_I}{L}\right)^{0.5}\right]^2 \ : \ C = 1.135\left(\frac{B_{\text{SWL}}}{D}\right)^{0.65}, \qquad (3.59)$$

where $B_{\text{SWL}}$ is the breakwater width measured at the still water level and $D$ is the height of concrete blocks. The notation $(K_T)_{\text{thru}}$ stands for the transmission coefficient by waves passing through structures.

(B) *Wave transmission over LCS*

As for waves transmitted over LCS, Goda and Ahrens[95] has proposed the following formulas by empirical fitting to the curves in the design diagram of wave transmission coefficient derived by Tanaka.[96]

$$\left.\begin{aligned}(K_T)_{\text{over}} &= \max\left\{0, (1 - \exp[a\,(h_c/H_I - F_0)])\right\}, \\ a &= 0.248\exp[-0.384\ln(B_{\text{eff}}/L_0)],\end{aligned}\right\} \quad (3.60)$$

where $(K_T)_{\text{over}}$ denotes the transmission coefficient of waves passing over LCS, $B_{\text{eff}}$ is the effective width of LCS, and $F_0$ represents an approximate limit of dimensionless wave runup, which is evaluated by the following:

$$F_0 = \begin{cases} 1.0 & : D_{\text{eff}} = 0, \\ \max\{0.5, \min(1.0, H_I/D_{\text{eff}})\} & : D_{\text{eff}} > 0, \end{cases} \quad (3.61)$$

where $D_{\text{eff}}$ denotes the effective diameter of the materials composing LCS. It is taken as equal to the median diameter $D_{50}$ for rubble stone and $(M/\rho)^{1/3}$ for concrete blocks with $M$ being the mass and $\rho$ being the density.

An artifice is made for setting the effective width $B_{\text{eff}}$ of LCS as in the following, depending on the crest height relative to the still water level:

$$\left.\begin{aligned}&\text{LCS with emerged crest :} \quad B_{\text{eff}} = \text{width at still water level}, \\ &\text{LCS with zero freeboard :} \quad B_{\text{eff}} = (9 \times \text{crest width} + \text{bottom width})/10, \\ &\text{LCS with submerged crest :} \ B_{\text{eff}} = (4 \times \text{crest width} + \text{bottom width})/5.\end{aligned}\right\}$$
$$(3.62)$$

That is to say, the width of the breakwater with zero freeboard is measured at the level of 10% below the crest, while that of the submerged breakwater is measured at the level of 20% below the crest. The artifice of Eq. (3.62) was introduced empirically so as to incorporate the effect of the gradient of the front and rear slopes of LCS on wave transmission and to yield overall agreement between the laboratory data and the prediction.

### (C) Overall transmission coefficient of LCS

When the structure's crest is high and little wave overtopping occurs, the wave transmission takes place through the void spaces among stone and blocks. The transmission coefficient can be evaluated with Eq. (3.59). When the structure is submerged, the wave transmission is mainly governed by breaking and frictional dissipation over the crest and slope surfaces of the structure. The transmission coefficient can be evaluated with Eqs. (3.60) and (3.61). A slightly emerged structure is subject to both energy dissipation through the void spaces and wave decay by breaking and frictional dissipation. For such a structure, the concept of energy summation is applied by referring to the approach by Wamsley and Ahrens.[97] The resultant formula becomes as follows:

$$(K_T)_{\text{all}} = \min\left\{1.0, \sqrt{(K_T)_{\text{over}}^2 + (K_h)^2 (K_T)_{\text{thru}}^2}\right\}:$$
$$K_h = \min\{1.0, \ h_t/(h + H_I)\}, \qquad (3.63)$$

where $h_t$ and $h$ denotes the total height of LCS and the water depth, respectively. The factor $K_h$ is introduced under the assumption that the runup height can be approximately equal to $H_I$ and the contribution of waves passing through LCS to the energy of transmitted waves is proportional to the ratio of the LCS height to the height of wave runup above the sea bottom. The upper bound of $(K_T)_{\text{all}} = 1.0$ is imposed in Eq. (3.63), because there is a possibility that the sum of the energies of the waves passing over and through a LCS may exceed 1.0 when the waves incident to a porous and submerged LCS is very small in height. When evaluating the transmission coefficient of waves passing through LCS, i.e., $(K_T)_{\text{thru}}$, with Eq. (3.61), the block height $D$ is replaced with the effective diameter $D_{\text{eff}}$. For a composite LCS with the armor layer and the core, $D_{\text{eff}}$ may be evaluated by taking the weighted mean diameter by using the respective cross-sectional areas as the averaging weights.

### Example 3.12

A reef-type LCS composed of wave-dissipating concrete blocks with the effective diameter of $D_{\text{eff}} = 2.5\,\text{m}$ is planned at the water of $h = 6$ m under the design tide level. The equivalent deepwater wave is $H_0' = 6$ m and $T_{1/3} = 10$ s, and the beach slope at the site is $s = 1/60$. The crest of LCS is submerged by 1 m or $h_c = -1.0\,\text{m}$, the crest width is $B_0 = 15\,\text{m}$ and the front and rear slopes have the inclination of 1 on 1.5. Estimate the wave transmission coefficient of this LCS and examine the effect of widening the width to $B_0 = 30\,\text{m}$ on the transmission coefficient.

### Solution

First, the significant wave height incident to the structure must be estimated. By using the simplified formula of Eq. (3.47) and referring to Example 3.11, the incident wave height is calculated as $H_I = 3.95\,\text{m}$. The wavelength is $L_0 = 156\,\text{m}$ and $L = 73.6\,\text{m}$, and thus the wave steepness is $H_I/L = 0.0537$. With the artifice of Eq. (3.62), the effective width becomes $B_{\text{eff}} = 18$ and $33\,\text{m}$ for $B_0 = 15$ and $30\,\text{m}$, respectively.

For the wave transmission through LCS, the constant in Eq. (3.59) is calculated as

$$C = 1.135 \times (18/2.5)^{0.65} = 4.1 \quad \text{for } B_0 = 15\,\text{m}.$$

Thus, by using Eq. (3.59)

$$(K_T)_{\text{thru}} = 1/[1 + 4.1 \times 0.0537^{0.5}]^2 = 0.267.$$

For the wave transmission over LCS, the constant $a$ and the nominal runup height $F_0$ are calculated as

$$a = 0.248 \times \exp[-0.384 \times \ln(18/156)] = 0.568,$$

$$F_0 = \max\{0.5 \ \min(1.0, 3.95/2.5)\} = 1.0.$$

Thus, by using Eq. (3.60)

$$(K_T)_{\text{over}} = \max\{0, \ (1 - \exp[0.568 \times (-1.0/3.95 - 1.0)])\} = 0.509.$$

The total height of LCS is $h_t = 6.0 - 1.0 = 5.0\,\text{m}$ and the factor in Eq. (3.60) becomes $K_h = 5.0/(6.0 + 3.95) = 0.503$. Therefore, the overall transmission coefficient is obtained by using Eq. (3.61) as

$$(K_T)_{\text{all}} = \sqrt{0.509^2 + 0.503^2 \times 0.267^2} = 0.526.$$

By expanding the crest width to 30 m, the above values are changed as below.

$$C = 6.07, \ (K_T)_{\text{thru}} = 0.173, \ a = 0.450, \ (K_T)_{\text{over}} = 0.391, \ (K_T)_{\text{all}} = 0.400.$$

## (D) Validation of new prediction formulas for wave transmission coefficient

The new set of prediction formulas has been tested against a large number of laboratory data. Figure 3.59 shows comparison of the laboratory transmission coefficient of 851 data and the predicted coefficient values. Because of versatile test conditions and configurations of model LCS in the data sets, a certain scatter of data would be inevitable. The determination coefficient between $(K_T)_{\text{pred}}$ and $(K_T)_{\text{obs}}$ is calculated as $r^2 = 0.865$. The arithmetic mean of the difference between $(K_T)_{\text{pred}}$ and $(K_T)_{\text{obs}}$ is 0.0185, while the standard deviation of the difference is 0.0879. See Ref. 95 for details of laboratory data.

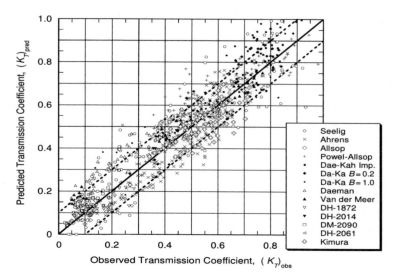

Fig. 3.59    Comparison of laboratory tested and predicted transmission coefficients.[95]

The new formulas have also been applied for the field data of wave transmission coefficient of submerged LCSs at Itoigawa, West Niigata, and Yugawara Coasts in Japan.[98] Predicted coefficients agreed well with the observed data in consideration of wide scatter of data inherent in field measurements. However, it was found necessary to reduce the constant value of $C = 1.135$ in Eq. (3.59) to $C = 0.6$ to yield good agreement for the case of West Niigata Coast, probably because of the reduction of the drag coefficient of 20-ton Tetrapods in the prototype situations.

### 3.9.3    *Propagation of Transmitted Waves in a Harbor*

Presently no reliable information is available on the subject of propagation and dispersion of waves transmitted to the lee of a breakwater. It is believed that for rubble and block mound breakwaters, the transmitted waves arising mainly from permeation would continue to propagate in a pattern similar to that of the incident waves, but with much reduced wave heights. For vertical breakwaters, where wave transmission by overtopping is predominant, the generation of transmitted waves would take place intermittently at short distances here and there, because incident waves in the sea are random and their wave crests are discontinuous. Thus, the transmitted waves would take the pattern of dispersing crests of concentric circles, with the generating sources positioned at various points along the breakwater. There is the possibility that frequent wave overtopping may occur at a particular section of a breakwater, with frequent appearance of resultant transmitted waves, owing to the layout of the breakwater. Because a general description of such individual situations is not feasible, it is necessary to assume in the analysis of harbor agitation that the transmitted waves with heights described as above would propagate in a pattern similar to that of the incident waves.

There is a report[99] of two-dimensional laboratory tests on obliquely incident waves to a vertical breakwater that indicates a decrease in the wave transmission coefficient by several percent compared to the case of normal incidence. Also, the direction of propagation of the transmitted waves by overtopping was observed to be deflected slightly toward the line normal to the breakwater. This can be understood as the effect of generation of short-period waves by overtopping, which propagate with a celerity less than that of the incident waves, and a process similar to wave refraction over varying water depth must have taken place.

## References

1. G. J. Komen, L. Cavaleri, M. Donelan, K. Hasselman, S. Hasselman and P. A. E. M. Janssen, *Dynamics and Modelling of Ocean Waves*, (Cambridge Univ. Pres., 1994), 532p.
2. CIRIA, CUR, and CETMEF, *The Rock Manual. The use of rock in hydraulic engineering* (2nd edition) CIRIA C683, London (2007).
3. B. W. Wilson, "Numerical prediction of ocean waves in the North Atlantic for December, 1959," *Deutsche Hydrographische Zeit* **18**(3) (1965), pp. 114–130.
4. Y. Goda, "Revisiting Wilson's formulas for simple wind-wave prediction," *J. Waterway, Port, Coastal, and Ocean Engrg.* **129**(2) (2003), pp. 93–95.
5. Y. Goda, "Analysis of wave grouping and spectra of long-travelled swell," *Rept. Port and Harbour Res. Inst.* **22**(1) (1983), pp. 3–41.
6. Y. Goda, "Report on environmental conditions and related problems at Mále Port (Contract No. CAS/A/88-003)," *Asian Development Bank, Maldives/Malé Port Development Project* (1988), 34p.
7. C. L. Bretschneider, "Decay of wind generated waves to ocean swell, Fundamentals of ocean engineering? Part 8," *Ocean Industry* (March 1968), pp. 36–39.
8. C. L. Bretschneider, "Decay of ocean waves, Fundamentals of ocean engineering? Part 8b," *Ocean Industry,* (April 1968), pp. 45–50.
9. Y. Goda, "Transformation of wave crest pattern in shoaling water," *Proc. 23rd Int. Conf. Coastal Engrg.* (Venice, ASCE, 1992), pp. 199–211.
10. U.S. Army Coastal Engineering Research Center, *Shore Protection Manual* (Third Edition), U.S. Govt. Printing Office (1977), Vol. I, pp. 2–69~2–75.
11. A. T. Ippen (ed.), *Estuary and Coastline Hydrodynamics,* (McGraw-Hill, Inc., 1966), pp. 260–263.

12. G. M. Griswold, "Numerical calculation of wave refraction," *J. Geophys. Res.* **68**(10) (1963), pp. 1715–1723.
13. W. S. Wilson, "A method for calculating and plotting surface wave rays," *U.S. Army Corps of Engrs., Coastal Engrg. Res. Center, Tech. Memo.* (17) (1966), 57p.
14. K. Nagai, "Diffraction of the irregular sea due to breakwaters," *Coastal Engineering in Japan* **15** (1972), pp. 59–69.
15. T. Karlsson, "Refraction of continuous ocean wave spectra," *Proc. ASCE* **95** (WW4) (1969), pp. 437–448.
16. K. Nagai, T. Horiguchi and T. Takai, "Computation of the propagation of sea waves having directional spectra from offshore to shallow water," *Proc. 21st Japanese Conf. Coastal Engrg.* (1974), pp. 349–253 (*in Japanese*).
17. Y. Goda and Y. Suzuki, "Computation of refraction and diffraction of sea waves with Mitsuyasu's directional spectrum," *Tech. Note of Port and Harbour Res. Inst.* **230** (1975), 45p (*in Japanese*).
18. Y. Ito, K. Tanimoto and S. Yamamoto, "Wave height distribution in the region of ray crossings — Application of the numerical analysis method of wave propagation," *Rept. Port and Harbour Res. Inst.* **11**(3) (1972), pp. 87–109 (*in Japanese*).
19. C. L. Vincent and M. J. Briggs, "Refraction-diffraction of irregular waves over a mound," *J. Waterway, Port, Coastal and Ocean Engrg.* **115**(2) (1989), pp. 269–284.
20. Y. Goda "Directional wave spectrum and its engineering applications," *Adv. Coastal and Ocean. Engrg.* (Ed. P. L.-F. Liu) **3** (World Scientific, 1997), pp. 67–102.
21. W. C. O'Reilly and R. T. Guza, "Comparison of spectral refraction and refraction-diffraction wave models," *J. Waterway, Port, Coastal and Ocean Engrg.* ASCE **117**(3) (1991), pp. 195–215.
22. K. Nagai, "Computation of refraction and diffraction of irregular sea," *Rept. Port and Harbour Res. Inst.* **11**(2) (1972), pp. 47–119 (*in Japanese*).
23. U.S. Army Coastal Engineering Research Center, *loc cit* Ref. 10, pp. **2**–83 ∼ **2**–109.
24. Y. Goda, T. Takayama and Y. Suzuki, "Diffraction diagrams for directional random waves," *Proc. 16th Int. Conf. Coastal Engrg.* (Hamburg, ASCE, 1978), pp. 628–650.
25. I. Irie, "Examination of wave deformation with field observation data," *Coastal Engineering in Japan* **18** (1975), pp. 27–34.
26. M. J. Briggs, E. T. Thompson and C. L. Vincent, "Wave diffraction around breakwater," *J. Waterway, Port, Coastal and Ocean Engrg.* **121**(1) (1995), pp. 23–35.
27. W. J. Pierson, Jr., G. Neumann and R. W. James, *Practical Methods for Observing and Forcasting Ocean Waves by Means of Wave Spectra and Statistics* (U.S. Navy Hydrographic Office, H.O. Pub. No. 603) (1955).
28. M. Hom-ma, K. Horikawa and Y. T. Chau, "Sheltering effects of Sado Island on wind waves off Niigata Coast," *Coastal Engineering in Japan* **9** (1966), pp. 27–44.

29. C. L. Bretschneider and R. O. Reid, "Modification of wave height due to bottom friction, percolation, and refraction," *U.S. Army Corps of Engrs., Beach Erosion Board, Tech. Memo.* (45) (1954).

30. O. H. Shemdin *et al.*, "Mechanisms of wave transformation in finite-depth water," *J. Geophys. Res.* **85**(C9) (1980), pp. 5012–5018.

31. Beach Erosion Board, Office of the Chief of Engineers, Department of the Army, Corps of Engineers, *Shore Protection Planning and Design*, Technical Report **4** (1961), p. 50.

32. U.S. Army Coastal Engineering Research Center, *loc cit* Ref. 10, p. **2**–121.

33. Y. Goda, "Deformation of irregular waves due to depth-controlled wave breaking," *Rept. Port and Harbour Res. Inst.* **14**(3) (1975), pp. 59–106 (*in Japanese*).

34. Y. Iwagaki and T. Sakai, "On the shoaling of finite amplitude waves (2)," *Proc. 15th Japanese Conf. Coastal Eng.* (1968), pp. 10–15 (*in Japanese*).

35. N. Shuto, "Nonlinear long waves in a channel of variable section," *Coastal Engineering in Japan* **17** (1974), pp. 1–12.

36. Y. Iwagaki, K. Shiota and H. Doi, "Shoaling and refraction of finite amplitude waves," *Proc. 28th Japanese Conf. Coastal Eng.* (1981), pp. 99–103 (*in Japanese*).

37. H. M. Kweon and Y. Goda, "A parametric model for random wave deformation by breaking on arbitrary beach profiles," *Proc. 25th Int. Conf. Coastal Eng.* (Orlando, ASCE) (1996), pp. 261–274.

38. Y. Goda, "A synthesis of breaker indices," *Trans. Japan Soc. Civil Engrs.* **2**(2) (1970), pp. 227–230.

39. Y. Goda, "New wave pressure fromulae for composite breakwaters," *Proc. 14th Int. Conf. Coastal Engrg.* (Copenhagen, ASCE) (1974), pp. 1702–1720.

40. W. Rattanapitikon and T. Shibayama, "Verification and modification of breaker height formulas," *Coastal Engineering Journal* **42**(4) (2000), pp. 389–406.

41. Y. Goda, "Reanalysis of regular and random breaking wave statistics," *Coastal Engineering Journal* **52**(1) (2010), pp. 71–106.

42. E. R. Smith and N. C. Kraus, "Laboratory study of wave-breaking over bars and artificial reefs," *J. Waterway, Port, Coastal, and Ocean Engrg.* **117**(4) (1991), pp. 307–325.

43. J. W. Kamphuis, "Incipient wave breaking," *Coastal Engineering* **15** (1991), pp. 185–203.

44. Y. C. Li, Y. Yu, L. F. Cui and G. H. Dong, "Experimental study of wave breaking on gentle slope," *China Ocean Engineering* **14**(1) (2000), pp. 59-67.

45. Y. Goda, "Irregular wave deformation in the surf zone," *Coastal Engineering in Japan* **18** (1975), pp. 13–26.

46. A. H. Sallenger and R. A. Holman, "Wave energy saturation on a natural beach of variable slope," *J. Geophys. Res.* **90**(C6) (1985), pp. 11,939–11,944.

47. L. Hamm and C. Peronnard, "Wave parameters in the nearshore: A clarification," *Coastal Engineering* **32** (1997), pp. 119–135.

48. F. C. K. Ting, "Laboratory study of wave and turbulence velocities in a broad-band irregular wave surf zone," *Coastal Engineering* **43** (2001), pp. 183–208.

49. G. Smith, I. Wallast, I. and M. R. A. van Gent, "Rock slope stability with shallow foreshore," *Coastal Engineering 2002 (Proc. 28th Int. Conf.)* (Cardiff, Wales, World Scientific) (2003), pp. 1524–1536.

50. Y. Goda, "A 2-D random wave transformation model with gradational breaker index," *Coastal Engineering Journal* **46**(1) (2004), pp. 1–38.

51. S. Hotta and M. Mizuguchi, "A field study of waves in the surf zone," *Coastal Engineering in Japan* **23** (1980), pp. 59–79.

52. S. Hotta and M. Mizuguchi, "Statistical properties of field waves in the surf zone," *Proc. 33rd Japanese Coastal Eng. Conf.* (JSCE, 1986), pp. 154–157 (*in Japanese*).

53. B. A. Ebersole and S. A. Hughes, "DUCK85 photopole experiment," *US Army Corps of Engrs., WES, Misc. Paper* (CERC-87-18) (1987) pp. 1–165.

54. M. S. Longuet–Higgins and R. W. Stewart, "Radiation stress and mass transport in gravity waves, with application to 'surf beats,'" *J. Fluid Mech.* **13** (1962), pp. 481–504.

55. Niigata Investigation and Design Office, First District Port Construction Bureau, Ministry of Transport, "On the winter storm waves of 1971," *NID Note* (45–6) (1971), pp. 93–107 (*in Japanese*).

56. Y. Goda, "Wave setup and longshore currents induced by directional spectral waves: Prediction formulas based on numerical computation results," *Coastal Engineering Journal.* **50**(4) (2008), pp. 397–440.

57. S. Yanagishima and K. Katoh, "Field observation on wave set-up near the shoreline," *Proc. 22nd Int. Conf. Coastal Engrg.* (Delft, ASCE) (1990), pp. 95–108.

58. D. Hanslow and P. Nielsen, "Shoreline set-up on natural beaches," *J. Coastal Res.*, Special Issue No. 15 (1993), pp. 1–10.

59. J. De Rouck, J. W. van der Meer, N. W. H. Allsop, L. Franco and H. Verhaeghe, "Wave overtopping at coastal structures: Development of a database toward up-graded prediction model," *Coastal Engineering 2002 (Proc. 28th Int. Conf.*, Cardiff, Wales, World Scientific) (2003), pp. 2140–2152.

60. Y. Goda, "A performance test of nearshore wave height prediction with CLASH datasets," *Coastal Engineering*, **56** (2009), pp. 220–229.

61. J. I. Collins, "Probabilities of breaking wave characteristics," *Proc. 12th Int. Coastal Eng. Conf.*, (Washington, D.C., ASCE) (1970), pp. 399–414.

62. J. A. Battjes, "Setup due to irregular wave," *Proc. 13th Int. Conf. Coastal Enrg.* (Vancouver, ASCE) (1972), pp. 1993–2004.

63. C. T. Kuo and S. T. Kuo, "Effect of wave breaking on statistical distribution of wave heights," *Proc. Civil Eng. Ocean*, (ASCE, 1974), pp. 1211–1231.

64. J. A. Battjes and J. P. F. M. Janssen, "Energy loss and set-up due to breaking of random waves," *Proc. 16th Int. Conf. Coastal Eng.*, (Hamburg, ASCE) (1978), pp. 1–19.

65. J. A. Battjes and M. J. F. Stive, "Calibration and verification of a dissipation model for random breaking model," *J. Geophys. Res.* **90**(C5) (1985), pp. 9155–9167.

66. E. B. Thornton and R. T. Guza, "Transformation of wave height distribution," *J. Geophys. Res.* **88**(C10) (1983), pp. 5925–5938.

67. W. R. Dally, R. G. Dean and R. A. Darlymple, "Wave height variation across beaches of arbitrary profile," *J. Geophys. Res.*, **90**(C6) (1985), pp. 11,917–11,927.

68. M. Larson and N. C. Kraus, "Numerical model of longshore current for bar and trough beaches," *J. Waterway, Port, Coastal, and Ocean Eng.*, **117**(4) (1991), pp. 326–347.

69. Y. Goda, "Examination of the influence of several factors on longshore current computation with random waves," *Coastal Engineering*, **53**(2-3) (2006), pp. 157–170.

70. W. N. Seelig and J. P. Ahrens, "Estimation of wave reflection and energy dissipation coefficients for beaches, revetments, and breakwaters," *U.S. Army. Corps of Engrs., Coastal Engrg. Res. Center, Tech. Paper* (81–1) (1981), 40p.

71. B. Zanuttigh and J. W. van der Meer, "Wave reflection from coastal structures," *Coastal Engineering 2006 (Proc. 30th Int. Conf.)* (San Diego, World Scientific, 2007), pp. 4337–4349, also B. Zanuttigh and J. W. van der Meer, "Wave reflection from coastal structures in design conditions," *Coastal Engineering* **55**(10) (2008), pp. 771–779.

72. T. Tanimoto, S. Haranaka, S. Takahashi, K. Komatsu, M. Todoroki and M. Osato, "An experimental investigation of wave reflection, overtopping and wave forces for several types of breakwaters and seawalls," *Tech. Note of Port and Harbour Res. Inst.* **246** (1976), 38p (*in Japanese*).

73. R. L. Wiegel, *Oceanographical Engineering* (Prentice-Hall, Inc., 1964), pp. 72–75, p. 194.

74. H. Mase, T. Memita, M. Yuhi, T. and Kitano, "Stem waves along vetical wall due to random wave incidence," *Coastal Engineering,* **44**(4) (2002), pp. 339–350.

75. Y. Goda, T. Yoshimura and M. Ito, "Reflection and diffraction of water waves by an insular breakwater," *Rept. Port and Harbour Res. Inst.* **10**(2) (1971), pp. 3–52 (*in Japanese*).

76. H. Mitsui, Y. Kawamura and K. Komatsu, "On the wave height distribution along coastal structures of uneven alignments (6th Report)," *Proc. 22nd Japanese Conf. Coastal Engrg.* (1975), pp. 103–107 (*in Japanese*).

77. K. Tanimoto, K. Kobune and K. Komatsu, "Numerical analysis of wave propagation in harbours of arbitrary shape," *Rept. Port and Harbour Res. Inst.* **14**(3) (1975), pp. 35–58 (*in Japanese*).

78. Y. Goda and Y. Suzuki, "Estimation of incident and reflected waves in random wave experiments," *Proc. 15th Int. Conf. Coastal Engrg.* (Hawaii, ASCE) (1976), pp. 828–845.

79. M. Morihira and Y. Okuyama, "Presentation of wave diffraction diagrams obtained through a digital computer," *Tech. Note of Port and Harbour Res. Inst.* **21** (1965), 45p (*in Japanese*).

80. Y. Ito and K. Tanimoto, "Meandering damage of composite type breakwaters," *Tech. Note of Port and Harbour Res. Inst.* **112** (1971), 20p (*in Japanese*).

81. H. Mitsui and H. Murakami, "On the wave height distribution along coastal structures of uneven alignments (2nd Report)," *Proc. 14th Japanese Conf. Coastal Engrg.* (1967), pp. 53–59 (*in Japanese*).

82. K. Kobune and M. Osato, "A study of wave height distribution along a breakwater with a corner," *Rept. Port and Harbour Res. Inst.* **25**(2) (1976), pp. 55–88 (*in Japanese*).

83. Y. Goda and T. Yoshimura, "Wave force computation for structures of large diameter, isolated in the offshore," *Rept. Port and Harbour Res. Inst.* **10**(4) (1971), pp. 3–52 (*in Japanese*).

84. Y. Goda, "Re-analysis of laboratory data on wave transmission over breakwaters," *Rept. Port and Harbour Res. Inst.* **8**(3) (1969), pp. 3–18.

85. Y. Goda, Y. Suzuki and Y. Kishira, "Some experiences in laboratory experiments with irregular waves," *Proc. 21st Japanese Conf. Coastal Engrg.* (1974), pp. 237–242 (*in Japanese*).

86. P. Alberti, T. Bruce, L. and Franco, "Wave transmission behind vertical walls due to overtopping," *Breakwaters, coastal structures and coastlines* (*Proc. Int. Conf.* in 2001, London) (Thomas Telford, 2002), pp. 269–280.

87. H. Kondo and I. Sato, "A study on the required elevation of breakwater crown," *Hokkaido Development Bureau, Civil Engineering Institute, Monthly Report* **117** (1964), pp. 1–15 (*in Japanese*).

88. J. W. van der Meer and J. P. de Waal, "Wave transmission: spectral changes and its effects on run-up and overtopping," *Coastal Engineering 2000* (*Proc. 26th Int. Conf.*, Sydney, ASCE) (2001), pp. 2156–2168.

89. A. Lamberti, R. Archetti, M. Kramer, D. Paphitis, C. Mosso and M. Di Risio, "European experience of low crested structures for coastal management," *Coastal Engineering,* **52** (2005), pp. 841–866.

90. T. Iwasaki and A. Numata, "A study on wave transmission coefficient of permeable breakwater," *Proc. 16th Japanese Conf. Coastal Eng.* (JSCE, 1969), pp. 329-334 (*in Japanese*).

91. M. Hattori, "Coastal development and wave control," *Lecture Series on Hydraulic Engineering,* 75-B2, Hydraulics Committee, Japan Soc. Civil Engrs. (1975), pp. B2-1 ∼ B2-24 (*in Japanese*).

92. M. Tominaga and T. Sakamoto, "Field measurements of wave attenuation by offshore breakwaters," *Proc. 18th Japanese Conf. Coastal Engrg.* (1971), pp. 149–154 (*in Japanese*).

93. T. Katayama, I. Irie and T. Kawakami, "Effects of Niigata Coast offshore breakwaters on beach protection and wave attenuation," *Proc. 20th Japanese Conf. Coastal Engrg.* (1973), pp. 519–524 (*in Japanese*).

94. A. Numata, "Experimental study on wave attenuation by block mound breakwaters," *Proc. 22nd Japanese Conf. Coastal Eng.* (1975), pp. 501–505 (*in Japanese*).

95. Y. Goda and J. P. Ahrens, "New formulation of wave transmission over and through low-crested structures," *Coastal Engineering 2008* (*Proc. 31th Int. Conf.*, Hamburg, World Scientific) (2009), pp. 3530–3541.

96. N. Tanaka, "Wave deformation and beach stabilization capacity of wide-crested submerged breakwaters," *Proc. 23rd Japanese Conf. Coastal Eng.* (1976), pp. 152–157 (*in Japanese*).

97. T. V. Wamsley and J. P. Ahrens, "Computation of wave transmission coefficient at detached breakwaters for shoreline response modeling," *Coastal Structures 2003* (*Proc. Int. Conf.*, Portland, Oregon, ASCE) (2003), pp. 593–695.

98. Y. Goda, H. Yoshida, K. Hachisuka and K. Kuroki, "New practical formulas for wave transmission coefficient of low-crested structures and their applications to field sites," *J. Japan Soc. Civil Engrs.* **65-B**(1) (2009), pp. 56–69 (*in Japanese*).

99. Niigata Investigate and Design Office, First District Port Construction Bureau, Ministry of Transport, "Several problems on breakwaters related to the safety of harbors (III)," *Proc. 14th District Symp. Port Construction Works* (1976) 76p. (*in Japanese*).

# Chapter 4

# Design of Vertical Breakwaters

## 4.1 Overviews of Vertical and Composite Breakwaters

(A) *Definition of breakwater types*

Breakwaters are generally classified as either mound breakwaters or vertical breakwaters. Functionally, mound breakwaters dissipate the energy of incident waves by forcing them to break on a slope, and thus they do not produce appreciable reflection. *Vertical breakwaters*, on the other hand, reflect the incident waves without dissipating much wave energy. A third category of composite or mixed type breakwaters was sometimes used to denote the type of breakwaters that function as mound breakwaters when the tide level is low and as vertical breakwaters when the tide level is high. (Larras.[1]) However, this definition is not employed nowadays because few composite breakwaters have rubble mounds high enough to emerge at a low tide level. According to ISO 21650 *Actions from Waves and Currents on Coastal Structures*,[2] a vertical breakwater is a structure of rectangular or nearly rectangular cross-section having a vertical or nearly vertical front wall extending directly from the seabed or built on top of a thin bedding layer. A *composite breakwater* is a combined structure of a main body of rectangular or nearly rectangular cross-section placed on a rubble mound that is submerged at all tidal levels. For the sake of simplicity, the present chapter uses the term of vertical breakwaters inclusive of composite breakwaters except for the cases specifically referring to the latter.

(B) *Historical development of Japanese vertical breakwaters*

In Western countries, vertical breakwaters are usually built in water of sufficient depth where wave breaking is not a problem (Ito[3]). But in the modern history of harbor construction in Japan, vertical breakwaters have been built primarily to withstand breaking waves. The construction of mound breakwaters has been confined to quite shallow water where construction work by barges is difficult. The scarcity of mound breakwaters is partly attributed to the young geological formation of the Japanese islands and to the humid climate, which made rock fragile. Therefore it is difficult to quarry sufficient amounts of large-sized rocks. This tendency to favor vertical breakwaters has been reinforced by many successful experiences in the construction of vertical breakwaters in Japan.

The first vertical breakwaters in the modern age of Japan are those at Yokohama Port, which were designed by H.S. Palmer (a retired British Major-General) and built from 1890 to 1896. One of the sections is shown in Fig. 4.1(a). They were redesigned with solid concrete blocks for the whole upright section after damage by a storm in 1902. A vertical breakwater consisting of solid concrete blocks was first built at Otaru Port in Hokkaido by I. Hiroi (a professor of civil engineering at Hokkaido University and later at the University of Tokyo) from 1897 to 1907. Figure 4.1(b) shows its cross-section. He took as examples the breakwaters at Karachi, Madras, and Colombo. After that, breakwaters with reinforced concrete caissons were introduced at Kobe Port for the first time in 1911, and then utilized for the island breakwater at Otaru Port in 1912. Because the wave conditions are much severer at Otaru than Kobe, the breakwater, as shown in Fig. 4.1(c), was quite sturdy. The caissons weighed 883 tons and were filled with poured concrete. Caisson breakwaters soon became the major structural type of breakwater for use in withstanding rough seas. Figure 4.1(d) shows the first breakwater of Onahama Port in Fukushima Prefecture. In this case, precast concrete blocks were placed in the caisson cells to save work time. Another noteworthy point is that the engineer-in-charge designed it against a maximum possible breaker height of about 8.5 m at a water depth of 9.7 m below the high water level. Many vertical breakwaters in Japan have been designed and built against breaking waves, such as the one in this example.

The filler material for the caisson cells has gradually evolved from concrete to gravel and then to sand. One of the early examples of a sand-filled breakwater is the outer breakwater at Yokohama Port, which was designed by S. Samejima (a distinguished government civil engineer at the time) and built from 1928 to 1943. As shown in Fig. 4.1(e), the caissons were very sturdy, with an outer wall 60 cm thick. The interiors of the caissons were filled with a mixture of sand and cobble to give the greatest possible density. Filling of caissons with sand was first done in water areas where

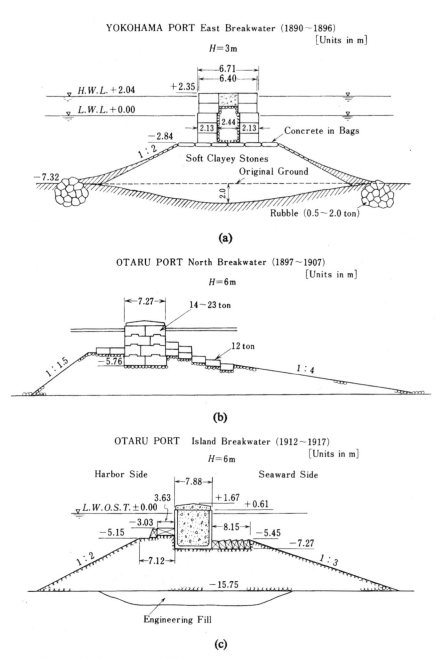

(a)

(b)

(c)

Fig. 4.1(a–c) Historical development of vertical breakwaters in Japan.

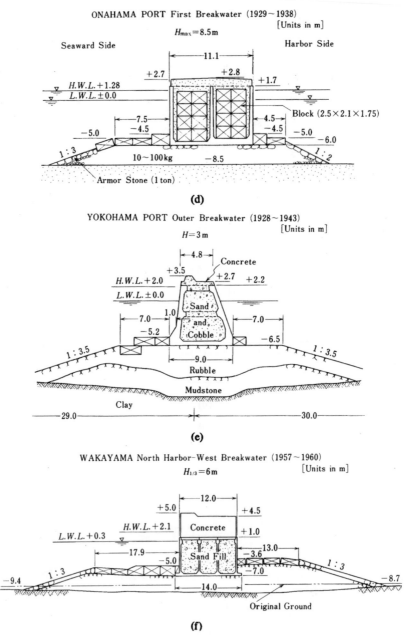

Fig. 4.1(d–f) Historical development of vertical breakwaters in Japan (continued).

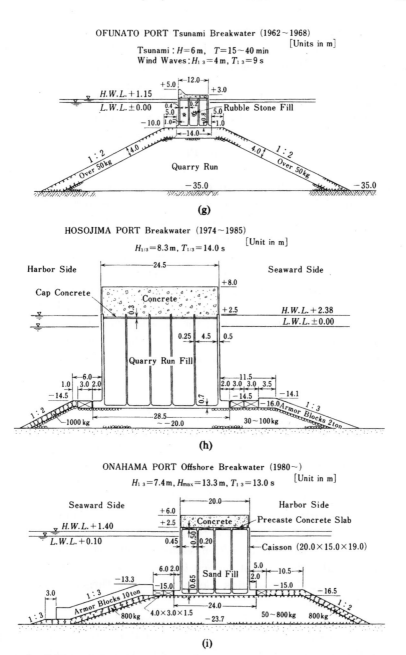

Fig. 4.1(g–i) Historical development of vertical breakwaters in Japan (continued).

the wave conditions were relatively mild, but sand filler soon became popular in areas of rough seas as well. This was particularly so in the period shortly after World War II, when cement was scarce because of war damage to production facilities.

Figure 4.1(f) is one such breakwater built after World War II. Because of soft ground conditions, the breakwater was given a wide rubble foundation to counteract circular slip of the ground. Figure 4.1(g) is the tsunami breakwater at Ofunato Port in Iwate Prefecture, built 35 m deep into the water. One of the largest breakwater caissons is found in the breakwater at Hosojima Port in Miyazaki Prefecture, shown in Fig. 4.1(h). It has a mass of nearly 5,000 Mg (ton), launched from a floating dock. Figure 4.1(i) presents a typical cross section of contemporary Japanese vertical breakwaters having been designed with the Goda wave pressure formulas.

The largest breakwater in the world is the tsunami breakwater at Kamaishi Port, Iwate Prefecture, Japan, construction of which was completed in 2008. The construction site has the maximum water depth of 63 m, and specially-shaped caissons with a mass of 16,000 Mg (ton), each are set on top of a rubble mound foundation at the elevation 25 m below the datum. Tanimoto and Goda[4] provide some details of this breakwater. In addition, Takahashi[5] describes various aspects of vertical breakwater designs in Japan.

### (C) Failure modes of vertical breakwaters

Harbor engineers of many countries appear to be rather unfamiliar with the design of vertical breakwaters; therefore the present chapter is dedicated to the introduction of Japanese design formulas and some of the design principles. Figure 4.2 is provided to acquaint the reader with the terminology associated with vertical breakwaters, illustrating the various parts of a vertical breakwater which appear in the present chapter. As seen in the examples of Fig. 4.1, the upright section is mostly built with concrete caissons filled with sand. The term of *caisson breakwater* is often used to denote such kind of vertical breakwaters. Concrete crown is the superstructure of the upright section.

Figure 4.3 shows the possible modes of major failures of vertical breakwaters by wave actions. These are *sliding* and *overturning* of the upright section as well as *geotechnical failure* of the rubble mound foundation and/or original seabed. However, the failure mode of overturning rarely materializes itself in reality because the stability of the rubble mound

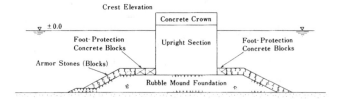

Fig. 4.2   Idealized typical section of a vertical breakwater.

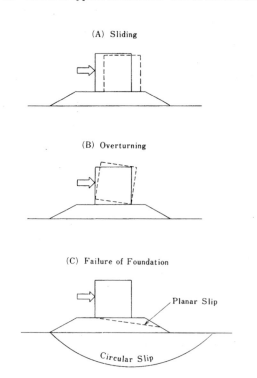

Fig. 4.3   Modes of major failure of vertical breakwaters.

foundation is destroyed before the overturning takes place, judging from the damage experience of Japanese breakwaters, the number of which counts more than ten thousands.

In addition to the failures shown in Fig. 4.3, displacement of individual foot-protection blocks and armor stones may also occur under wave actions. The sandy seabed in front of a vertical breakwater may be scoured by oscillatory flow by waves and the scour may induce the erosion of the foundation toe. A vertical breakwater may also be subject to settlement due to washing out of seabed sand beneath the rubble foundation if the countermeasures are not adequate. The subsequent sections mainly discuss the subjects of wave pressure, design of upright sections, and design of rubble mound foundation.

## 4.2    Wave Pressures Exerted on Upright Sections

### 4.2.1    *Overview of Development of Wave Pressure Formulas*

In parallel with the history of the construction of vertical breakwaters themselves, the formulas for the wave pressure exerted on upright sections have a long history of their own.[a] As early as 1842, Stevenson[6] measured wave pressure with a special gauge which he invented. Then from 1890 to 1902, Gaillard[7] made several series of wave pressure measurements and proposed a formula for estimating wave pressure on the basis of water particle velocities. In Japan, Hiroi[8] also measured wave pressures with a modified version of Stevenson's gauge. Although he obtained pressures as high as $350 \text{ kN/m}^2$, he correctly regarded such a high pressure as a local phenomenon and did not relate it directly to the stability of upright sections. Instead, he considered wave pressure as similar to the pressure of a water jet and derived the following formula, which was published in 1919:

$$p = 1.5\rho g H \,. \tag{4.1}$$

The wave pressure $p$ is assumed to act uniformly over the full height of an upright section, or to an elevation of 1.25 times the wave height above the still water level, whichever is less. The symbol $\rho$ in Eq. (4.1) denotes the density of sea water $(= 1{,}030 \text{ kg/m}^3)$, $g$ is the acceleration of gravity, and $H$ is the incident wave height. In situations where information on the design wave is unreliable, Hiroi recommended taking 0.9 times the water depth as the design wave height. Hiroi's formula was apparently intended for calculating the pressure caused by breaking waves, and it has been so used in breakwater design in Japan for more than 60 years.

A pressure formula for standing waves was introduced by Sainflou[9] in 1928, and it acquired immediate acceptance by harbor engineers throughout the world. Before 1979, Japanese harbor engineers have employed a dual

---

[a]For the content of this paragraph, the author gratefully acknowledges the treatise by Ito.[3]

system of wave pressure formulas, using both Hiroi's formula for breaking waves and Sainflou's formula for nonbreaking waves. Although Minikin[10] proposed a formula for breaking wave pressure in 1950, based partly on Bagnold's laboratory data,[11] Minikin's formula is rarely employed in actual breakwater design because of the excessive values which it predicts. Until the 1960s even Sainflou's formula was infrequently used in Japan, because design waves are relatively large and most vertical breakwaters built at the time did not reach to a water depth where design waves could be safely regarded as nonbreaking.

In the 1960s, extensive port development in Japan, keeping pace with national economic development, resulted in new circumstances in the application of wave pressure formulas. That is to say, many new breakwaters had to be built over long stretches from the shoreline to water deeper than 20 m to accommodate very large bulk carriers. At some point, the pressure formula for the design of these breakwaters had to be switched from that pertaining to breaking waves to nonbreaking waves. At that boundary point of the applicability of the two pressure formulas, the predicted wave pressure abruptly changes by more than 30%. A slight modification of the design wave height immediately moves the location of the cutoff between the two formulas, and the design section of a vertical breakwater has, in principle, to be changed accordingly. Such a situation is considered too artificial, and it is hard to convince engineers of the appropriateness of the dual formula system for wave pressure on breakwaters.

Another problem with the formulas of Hiroi and Sainflou was the ambiguity of the wave height to be used in design works. With advances in instrumental wave recording, engineers came to realize the complexity of sea waves and began to question which height, $H_{1/3}$, $H_{1/10}$, or $H_{max}$, should be substituted into the wave pressure formulas.

To resolve the above predicament, based on hydraulic model experiments, in 1966 Ito *et al.*[12] proposed a single formula covering both breaking and nonbreaking wave pressures, including the effect of the presence of a rubble mound foundation. At the same time, he specified $H_{max}$ as the height to be employed in his formula. Later, in 1973, the author extended Ito's work through the use of much more laboratory data (Goda and Fukumori[13]) as well as theoretical considerations (Goda and Kakizaki[14]). He also examined many cases of sliding and nonsliding of vertical breakwaters, and proposed a new set of wave pressure formulas for upright sections of vertical breakwaters.[15] With a later modification by Tanimoto *et al.*[16] to account for the effect of oblique wave approach, these formulas have been employed as the standard formulas for the design of vertical and composite breakwaters in Japan since 1979.

### 4.2.2    *Goda Formulas of Wave Pressure Under Wave Crests*

The wave pressure formulas proposed by the author for the design of vertical breakwaters assume the existence of a trapezoidal pressure distribution along a vertical wall, as shown in Fig. 4.4, regardless of whether the waves are breaking or nonbreaking. In this figure, $h$ denotes the water depth in front of the breakwater, $d$ the depth above the armor layer of the rubble foundation, $h'$ the distance from the design water level to the bottom of the upright section, and $h_c$ the crest elevation of the breakwater above the design water level. The wave height for the pressure calculation and other factors are specified below.

### (A)  *Design wave*

The highest wave in the design sea state is to be employed. Its height is taken as $H_{\max} = 1.8H_{1/3}$ seaward of the surf zone, whereas within the surf zone the height is taken as the highest of random breaking waves $H_{\max}$ at the location at a distance $5H_{1/3}$ seaward of the breakwater. The latter height $H_{1/3}$ is to be estimated with the random wave breaking model described in Sec. 3.6 at the depth of the location of the breakwater.

The period of the highest wave is taken as that of significant wave, as in Eq. (2.9) of Section 2.2.3; i.e., $T_{\max} = T_{1/3}$; the latter may be replaced by $T_{m-1,0}$ (see Sec. 2.4.2).

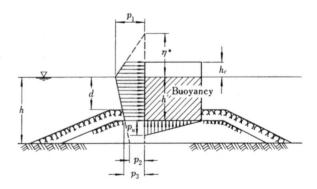

Fig. 4.4    Distribution of wave pressure on an upright section of a vertical breakwater.

(B) *Elevation to which the wave pressure is exerted*

The wave pressure is assumed to act to the elevation below:

$$\eta^* = 0.75(1 + \lambda_1 \cos \beta)H_{\max}, \tag{4.2}$$

in which $\beta$ denotes the angle between the direction of wave approach and a line normal to the breakwater. The wave direction should be rotated by an amount of up to 15° toward the line normal to the breakwater from the principal wave direction. This *directional adjustment* is made in view of the uncertainty in the estimation of the design wave direction. The practice of adjusting the wave direction up to 15° has been employed in the design of Japanese breakwaters ever since Hiroi proposed his wave pressure formula. The factor $\lambda_1$ has been introduced to deal with breakwaters of special configurations other than breakwaters made of simple upright sections. For the latter, the factor is set at $\lambda_1 = 1$.

(C) *Wave pressure on the front of a vertical wall*

The pressure intensities are to be estimated as below:

$$p_1 = \frac{1}{2}(1 + \cos \beta)(\alpha_1 \lambda_1 + \alpha_2 \lambda_2 \cos^2 \beta)\rho g H_{\max}, \tag{4.3}$$

$$p_2 = \frac{p_1}{\cosh(2\pi h/L)}, \tag{4.4}$$

$$p_3 = \alpha_3 p_1, \tag{4.5}$$

in which

$$\alpha_1 = 0.6 + \frac{1}{2}\left[\frac{4\pi h/L}{\sinh(4\pi h/L)}\right]^2, \tag{4.6}$$

$$\alpha_2 = \min\left\{\frac{h_b - d}{3h_b}\left(\frac{H_{\max}}{d}\right)^2, \frac{2d}{H_{\max}}\right\}, \tag{4.7}$$

$$\alpha_3 = 1 - \frac{h'}{h}\left[1 - \frac{1}{\cosh(2\pi h/L)}\right], \tag{4.8}$$

where $\min\{a, b\}$ refers to the smaller one of $a$ or $b$, and $h_b$ is the water depth at the location at a distance $5H_{1/3}$ seaward of the breakwater.

The above pressure intensities are assumed not to change even if wave overtopping takes place. The value of the coefficient $\alpha_1$ can be read off of Fig. 4.5, and the value of $1/\cosh(2\pi h/L)$ for $\alpha_3$ is obtained from Fig. 4.6. The symbol $L_0$ in both figures denotes the wavelength corresponding to the significant wave period in deep water. The symbol $\lambda_2$ denotes a correction factor that depends on the structural type of breakwaters, and it is set at $\lambda_2 = 1$ for the standard breakwaters with simple upright sections.

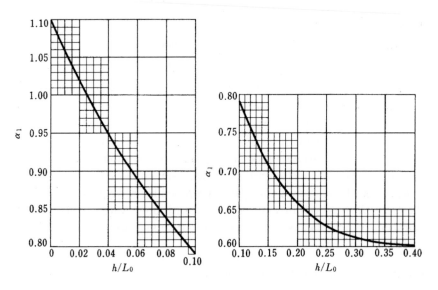

Fig. 4.5   Calculation diagrams for the parameter $\alpha_1$.[13]

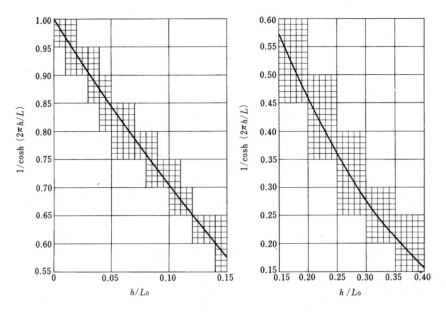

Fig. 4.6   Calculation diagrams for the factor of $1/\cosh(2\pi h/L)$.[13]

(D) *Buoyancy and uplift pressure*

The buoyancy is to be calculated for the displacement volume of the upright section in still water below the design water level, and the uplift pressure acting on the bottom of the upright section is assumed to have a triangular distribution with toe pressure $p_u$ given by Eq. (4.9) below, and with a heel pressure of zero.

$$p_u = \frac{1}{2}(1 + \cos\beta)\alpha_1\alpha_3\lambda_3\rho g H_{\max}, \qquad (4.9)$$

where the symbol $\lambda_3$ denotes a correction factor that depends on the structural type of breakwaters, and it is set at $\lambda_3 = 1$ for the standard breakwaters with simple upright sections. Both the buoyancy and uplift pressure are assumed to be unaffected by wave overtopping.

(E) *Background for wave pressure formulation*

Adoption of the wave height $H_{\max}$ in the above pressure formulas is taken after the proposal by Ito.[12] It is based on the principle that a breakwater should be designed to be safe against the single wave with the largest pressure among storm waves. As discussed in Sec. 2.2.2, the height $H_{\max}$ is a probabilistic quantity. But to avoid possible confusion in design, a definite value of $H_{\max} = 1.8H_{1/3}$ is recommended in consideration of the performance of many prototype breakwaters as well as with regard to the accuracy of the wave pressure estimation. Certainly there remains the possibility that one or two waves exceeding $1.8H_{1/3}$ will hit the site of the breakwater when storm waves equivalent to the design condition attack. But the distance of sliding of an upright section, if it were to slide, would be very small. It should be remarked, however, that the prescription $H_{\max} = 1.8H_{1/3}$ is a recommendation and not a rule. The design engineer can use his judgment in choosing another value, such as $H_{\max} = 1.6H_{1/3}$, $H_{\max} = 2.0H_{1/3}$, or some other value.

The recommended design water depth as discussed above is based on recognition of the fact that the greatest wave pressure is exerted not by waves just breaking at the site, but by waves which have already begun to break at a short distance seaward of the breakwater, midway through the plunging distance. Although this distance depends on the wave conditions and other factors, a single criterion of $5H_{1/3}$ has been adopted in consideration of some laboratory data on breaking wave pressures, for the sake of convenience.

The value of the *wave pressure coefficient* $\alpha_1$ in the pressure intensity $p_1$ at the still water level has been empirically determined on the basis of laboratory data. The formula for the coefficient $\alpha_1$ represents the mean tendency of wave pressure in that it increases with the wave period; its functional representation does not carry any theoretical significance. The simple functional form of the coefficient $\alpha_2$ represents the tendency of the pressure to increase with the height of the rubble foundation. The increase of wave pressure due to the presence of a rubble foundation may be regarded as result of the change in behavior of waves from nonbreaking to breaking, although actual waves never exhibit such marked changes. With this consideration, the reduction factor $\cos^2 \beta$ for the effect of oblique wave attack is multiplied to the coefficient $\alpha_2$ in addition to the general reduction factor $0.5(1 + \cos \beta)$. The coefficient $\alpha_3$ was derived based on the simplifying assumption of a linear pressure variation between $p_1$ and $p_2$ along a vertical wall.

Theoretically, the intensity of uplift pressure $p_u$ at the toe of an upright section should be the same as the front pressure $p_3$. It was judged, however, that the uplift pressure would be assigned too great a value if $p_u$ were set equal to $p_3$, in view of the performance of prototype breakwaters and other considerations. Thus, $p_u$ is set as given by Eq. (4.9). Although the uplift pressure is not appreciably affected by the occurrence of wave overtopping, as evidenced in the laboratory data reported by the author,[13] a very low crown elevation of breakwater is expected to result in some reduction in the uplift pressure.

(F) *Calculation formulas for total wave pressure and uplift*

With the above formulas for the wave pressure, the *total wave pressure* and its moment around the bottom of an upright section (see Fig. 4.7) can be calculated with the following equations:

$$P = \frac{1}{2}(p_1 + p_3)h' + \frac{1}{2}(p_1 + p_4)h_c^*, \tag{4.10}$$

$$M_p = \frac{1}{6}(2p_1 + p_3)h'^2 + \frac{1}{2}(p_1 + p_4)h'h_c^* + \frac{1}{6}(p_1 + 2p_4)h_c^{*2}, \tag{4.11}$$

in which

$$p_4 = \begin{cases} p_1(1 - h_c/\eta^*) & : \eta^* > h_c, \\ 0 & : \eta^* \leq h_c, \end{cases} \tag{4.12}$$

$$h_c^* = \min\{\eta^*, h_c\}. \tag{4.13}$$

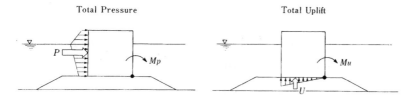

Fig. 4.7   Definition sketch of total pressure and uplift as well as their moments.

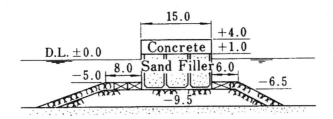

Fig. 4.8   Cross-section of a vertical breakwater (units in meters).

The *total uplift pressure* and its moment around the heel of the upright section (see Fig. 4.7) are calculated with

$$U = \frac{1}{2}p_u B \,, \tag{4.14}$$

$$M_U = \frac{2}{3}UB \,, \tag{4.15}$$

where $B$ denotes the width of the bottom of the upright section.

### Example 4.1

Calculate the wave pressure, uplift pressure, and their moments produced by waves of the following characteristics incident on the upright section of the vertical breakwater shown in Fig. 4.8.

Waves : $H_0' = 6.3$ m, $T_{1/3} = 11.4$ s, $\beta = 15°$,

Tide level : W.L. = +0.6 m,

Sea bottom slope : $\tan\theta = 1/100$.

### Solution

(1) Water depth and crest elevation

$$h = 10.1 \text{ m}, \quad h' = 7.1 \text{ m}, \quad d = 5.6 \text{ m}, \quad h_c = 3.4 \text{ m}.$$

(2) Wavelength and wave height

$L_0 = 202.7$ m (by Table A $-$ 3 in the appendix, or by $L_0 = 1.56\ T^2$),

$H_0'/L_0 = 0.031$,    $h/L_0 = 0.050$,

$H_{1/3} = 5.8$ m (estimated with Eq. (3.25) in Sec. 3.5.5
   using $h = 10.1$ m),

$h_b = 10.1 + 5 \times 5.8 \times \dfrac{1}{100} = 10.4$ m,

$H_{max} = 8.0$ m (estimated by Eq. (3.26) in Sec. 3.5.5
   with $h_b = 10.4$ m).

(3) Coefficients for the wave pressure

$\alpha_1 = 0.920$ (by Fig. 4.5),

$\alpha_2 = \min \left\{ \dfrac{10.4 - 5.6}{3 \times 10.4} \times \left( \dfrac{8.0}{5.6} \right)^2 , \dfrac{2 \times 5.6}{8.0} \right\} = \min \{0.314, 1.40\}$

$= 0.314$,

$1/\cosh(2\pi/L) = 0.847$ (by Fig. 4.6),

$\alpha_3 = 1 - \dfrac{7.1}{10.1} \times (1 - 0.847) = 0.892$.

(4) Maximum elevation of the wave pressure

$\cos \beta = 0.966$,    $0.5 \times (1 + 0.966) = 0.983$,

$\eta^* = 0.75 \times (1 + 0.966) \times 8.0 = 11.8$ m.

(5) Pressure components

$p_1 = 0.983 \times [0.920 + 0.314 \times (0.966)^2] \times 1030 \times 9.8 \times 8.0$

$= 96.3$ kN/m$^2$,

$p_3 = 0.892 \times 96.3 = 85.9$ kN/m$^2$,

$p_4 = 96.3 \times \left( 1 - \dfrac{35}{11.8} \right) = 68.6$ kN/m$^2$,

$p_u = 0.983 \times 0.920 \times 0.892 \times 1030 \times 9.8 \times 8.0 = 65.1$ kN/m$^2$.

(6) Total pressure and uplift

$$h_c^* = \min\{11.8, 3.4\} = 3.4 \text{ m},$$

$$P = \frac{1}{2} \times (96.3 + 85.9) \times 7.1 + \frac{1}{2} \times (96.3 + 68.6) \times 3.4$$

$$= 927 \text{ kN/m},$$

$$U = \frac{1}{2} \times 65.1 \times 15.0 = 488 \text{ kN/m}.$$

(7) Moment of wave pressure

$$M_p = \frac{1}{6} \times (2 \times 96.3 + 85.9) \times 7.1^2 + \frac{1}{2} \times (96.3 + 68.6) \times 7.1$$

$$\times 3.4 + \frac{1}{6} \times (96.3 + 2 \times 68.6) \times 3.4^2 = 4780 \text{ kN} \cdot \text{m/m}.$$

(8) Moment of uplift pressure

$$M_U = \frac{2}{3} \times 488 \times 15.0 = 4880 \text{ kN} \cdot \text{m/m}.$$

### (G) *Accuracy of the Goda wave pressure formulas*

The reliability of the calculation method for the wave pressure on a vertical breakwater is judged by the accuracy of the prediction of breakwater stability. The fundamental data for that judgment are the records of the sliding of breakwaters under storm waves as well as the records of breakwaters which withstood the attacks of high waves without experiencing any damage. The author[15] has made a survey of the performance of prototype breakwaters under waves with heights nearly equal to or greater than the heights of the design waves.

With a collection of cases for 21 breakwaters which experienced sliding (with sliding distances of 0.1 m to several meters) and 13 nondamaged breakwaters, the stability of these breakwaters was examined by means of the conventional pressure formulas of Sainflou and Hiroi, Goda's new formulas, and Minikin's formula. The significant wave heights listed in the reports of the breakwater performance were assumed to refer to the estimated heights of the incident waves at the locations of the breakwaters without consideration of the random wave breaking effect. From these heights, the equivalent deepwater heights were calculated as $H_0' = H_{1/3}/K_s$

and the wave heights were then corrected for the effect of random wave breaking.

The stability of prototype breakwaters was examined by using the safety factor against sliding of the upright section. This safety factor is defined as the ratio of the sliding resistance of an upright section against the wave force (see Eq. (4.36) in Sec. 4.3.1). If the safety factor is less than 1, the upright section is thought to slide under wave action.

Figure 4.9 illustrates the safety factors against sliding for the examined breakwaters, calculated by means of the conventional pressure formulas and the new formulas.[b] Open circles denote cases of breakwaters which did not slide, filled circles represent cases of sliding, and half-filled circles denote a case judged to be at the threshold condition. The letters A to R next to the circles refer to the associated harbors (the harbor names are given in Ref. 15). If the wave conditions were exactly known and the estimation of the wave pressure accurate, all the breakwaters which slid would have had safety factors less than 1.0, whereas those which did not slide would have had safety factors equal to or greater than 1.0. In other words, the filled circles would lie below the line S.F. = 1.0, whereas the open circles would lie above the line S.F. = 1.0, thus showing a clear distinction between the cases of sliding and nonsliding.

The actual trend of the calculated breakwater stability is not so clear. In particular, the analysis with the conventional pressure formulas as shown in Fig. 4.9(a) has yielded a rather ambiguous mixture of open and filled circles. For example, the breakwater at E Harbor did not slide though it had a safety factor of 0.67, whereas the breakwater at J Harbor slid even though the safety factor was 1.49. In the case of Goda's new formulas in Fig. 4.9(b), a separation between open and filled circles can be observed with a boundary around S.F. = 0.9 ~ 1.0, though some mixing still remains. Application of Minikin's formula to this set of prototype breakwater performance data (the wave height for the calculation being taken as the significant height) yielded the poorest prediction of breakwater stability; some breakwaters remained in position with estimated safety factors of less than 0.4, while some others slid with safety factors of more than 2.0. Thus, Minikin's formula is found to be impractical, and the

---

[b]Results previously presented by the author[15] did not properly incorporate the effect of random wave breaking. Also, the effect of oblique wave incidence was estimated in a slightly different manner. Thus, the results shown here, based on increased knowledge, differ to a certain extent from the previous results.

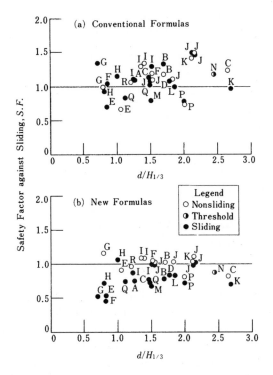

Fig. 4.9   Examination of the safety factor of prototype breakwaters against sliding.

accuracy of Goda's new formulas is judged superior to that of the conventional pressure formulas.

Table 4.1 gives a summary of the safety factor against sliding calculated by the conventional formulas and the Goda formulas. The mean and standard deviation of the safety factor are calculated separately for the cases of slid and non-slid breakwaters. The conventional method with the combination of the Saiflou and Hiroi formulas yields the difference of only 0.05 between the mean values of the slid and non-slid cases with the standard deviation of 0.21 to 0.24. On the other hand, the Goda formulas yield the difference of 0.21 with the standard deviation of 0.10 to 0.17, indicating their superiority in the predictability of wave actions on breakwaters.

The Goda formulas have a tendency of overestimation, however, as indicated by the mean safety factor of 0.998 for non-slid cases. The overestimating tendency has been pointed out by several later studies. The probabilistic design method to be discussed in Chapter 6 can take into

Table 4.1 Mean and standard deviation of safety factor against sliding of the slid and non-slid breakwaters.

| Breakwaters | Cases | Sainflou and Hiroi | Goda |
|---|---|---|---|
| Non-slid ones | 13 | 1.149 (0.243) | 0.998 (0.101) |
| Slid ones | 21 | 1.096 (0.218) | 0.786 (0.171) |

Note: Figures under the parentheses refer to the standard deviation.

account such bias and variability in the prediction of wave pressure for breakwater design.

### 4.2.3 Impulsive Breaking Wave Pressure and Its Estimation

(A) *Mechanism of impulsive pressure generation*

The wave pressure calculated with Eqs. (4.3) to (4.8) rarely exceeds the intensity of the hydrostatic pressure corresponding to the wave height, i.e., $1.0\rho g H_{max}$. As demonstrated in the laboratory experiments performed by Bagnold[11] and many other researchers, as well as by the results of field measurements, a gauge embedded in a vertical wall will record very high pressures due to breaking waves. This pressure may rise to more than ten times the hydrostatic pressure corresponding to the wave height, though its duration will be very short. Such an abnormally high breaking wave pressure is called the *impulsive breaking wave pressure*. The study on this problem began in the late 1930s in U.K. and France, and it has been revived in Europe since the 1990s using very large wave flumes and theoretical analysis.

An impulsive pressure is exerted on a vertical wall when an incident wave begins to break in front of the wall and collides with it, having the wave front being almost vertical as shown in Fig. 4.10. The impinging wave loses its forward momentum in the short time during which the collision takes place. The forward momentum is converted into an impulse which is exerted on the vertical wall. By denoting the forward momentum of the breaking wave per unit width as $M_v$, the total impulsive pressure on the wall as $P_I$, and its duration as $\tau$, the momentum equation for the present situation becomes as follows:

$$\int_0^\tau P_I dt = M_v. \tag{4.16}$$

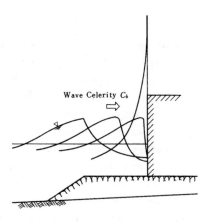

Fig. 4.10   Profiles of a breaking wave colliding with a vertical wall.

To estimate the magnitude of this momentum, let us consider the case of a water mass in the form of a semi-circular cylinder with a diameter of $H_b$ advancing with the speed of the wave $C_b$. Thus, we have

$$M_v \simeq \frac{\pi \rho g}{8g} C_b H_b^2 , \qquad (4.17)$$

where $g$ denotes the acceleration of gravity ($= 9.8$ m/s$^2$).

If it is assumed that the impulsive pressure increases linearly at the start of the collision ($t = 0$) to a maximum value at $t = \tau$, and then reduces to zero for $t > \tau$, the peak value of the impulsive pressure is obtained as

$$(P_I)_{\max} \simeq \frac{\pi \rho g C_b H_b^2}{4g\tau} . \qquad (4.18)$$

This model predicts that the impulsive pressure is inversely proportional to its duration. Thus, the impulsive pressure can attain a very large value when the front face of the breaking wave takes the form of a vertical flat plane and collides with the vertical wall over very short time duration. However, it is found in laboratory experiments that the front face of an impinging breaking wave is always curved and a small amount of air is entrapped at the instant of collision. As suggested by Bagnold[11] and formulated in a theoretical model by Takahashi and Tanimoto,[17] the entrapped air acts to dampen the impulsive pressure and prevents it from becoming abnormally high.

(B) *Effect of impulsive pressure on breakwater stability*

The impulsive breaking wave pressure of high intensity does not cause immediate threat to the stability of breakwaters, however. The rubble mound foundation and the ground around a prototype breakwater will be elastically deformed under the application of an impulsive breaking wave pressure, and this softens the wave impact on the upright section. The author had a personal experience of observing a peculiar motion of the superstructures of two adjacent breakwater caissons at Niigata West Port in the late 1960s such that they moved each other along an elliptical orbit with an horizontal stroke of about 0.15 m and a vertical stroke of about 0.1 m when swell of about 3 m was obliquely incident to the breakwater.

A computation of such *elastic deformation of the foundation*[18] indicates that the effective wave pressure in causing an upright section to slide is at most twice to thrice the hydrostatic pressure corresponding to the wave height, when the wave pressure is averaged over the exposed area of wall. The degree of absorbing the wave impact depends on the spring constant of the foundation or the modulus of ground reaction. When a vertical breakwater is built on a rock foundation with a thin rubble mound, the modulus would be high and the impulsive breaking wave pressure would directly threaten the breakwater stability. Thus, an appropriate evaluation of the ground reaction modulus is indispensable for analysis of the actions of impulsive pressure on vertical breakwaters. Lamberti and Martinelli,[19] for example, have measured the oscillatory motions of a caisson of the Genoa Breakwater which was given an impact by hitting of a tugboat and estimated the ground reaction modulus.

Even though the effective strength of impulsive breaking wave pressure on prototype breakwaters is not as high as those measured in laboratories, the impulsive pressure can cause sliding of caissons when the conditions are unfavorable. Japan has a case of steady sliding of caissons during a stormy winter at one port (Takahashi *et al.*[20]). There are also several cases of the breakage of a part of the front wall of caisson which was hit by impulsive breaking wave forces and the washing out of fill sand through the holes (Takahashi *et al.*[20] and Tsuda and Takayama[21]).

(C) *Situations at which breaking waves may induce impulsive forces*

Breaking waves acting on vertical and composite breakwaters induce impulsive forces at certain limited situations only. Impulsive forces are induced when waves are nearly normally incident to the breakwater with the incident angle being less than 20° and either one of the following conditions are met:

(a) The rubble mound foundation has a broad berm at a relatively high elevation (see the coefficient of impulsive pressure to appear later), or

(b) The sea bottom has the slope steeper than 1/50 and the equivalent deepwater wave steepness is less than about 0.03.

A vertical breakwater with a low crest elevation will not be in danger of impulsive pressure even though the conditions stated above are met. The threshold elevation seems to be about 0.3 times the wave height for ordinary vertical breakwaters. However, an impulsive pressure may be exerted on a vertical breakwater with a low crest elevation if the rubble mound is quite large and if the incident waves break on its slope in the form of surging breakers.

Some caisson breakwater is provided with a large sloped mound of wave-dissipating concrete blocks in its front so as to reduce wave reflection, overtopping, and/or wave forces; such structure is sometimes called a *horizontally-composite breakwater*. When the mound maintains the crest elevation at the level equal to or higher than that of the upright section, the mound can deter the impulsive breaking wave pressure by deforming the advancing wave front. However, if the mound crest is located at the level much lower than that of the upright section because of subsidence after construction or at the transition location to the section of no mound, incoming waves can run up the mound and collide with the exposed part of vertical wall. The impact of colliding waves may cause structural damage on the wall and lead to the loss of the breakwater stability.

The danger of the generation of impulsive breaking wave pressure can be judged to a certain extent with the conditions stated above. It is admitted however that the criteria are of a rather qualitative nature, and many cases may fall in the border zone. This uncertainty is inevitable because the phenomenon is affected by many factors in a complex and delicate manner. Therefore, it is advisable to resort to hydraulic model testing whenever there remains the suspicion of the generation of impulsive breaking wave pressure.

(D) *Coefficient of impulsive breaking wave pressure*

The Goda formulas of wave pressures themselves do not address to the problem of impulsive breaking wave pressure, though the coefficient $\alpha_2$ of

Eq. (4.7) represents the effect of the presence of rubble mound foundation on the pressure intensity. To remedy the inadequacy, Takahashi *et al.*[22] have proposed a set of formulas to estimate the magnitude of impulsive pressure on the basis of a systematic laboratory test on the sliding of model upright sections by Tanimoto *et al.*[23] The proposal is to rewrite Eq. (4.3) as in the following so as to include the term of impulsive pressure in the coefficient $\alpha_2$:

$$p_1 = \frac{1}{2}(1 + \cos\beta)(\alpha_1 + \alpha^* \cos^2\beta)\rho g H_{\max} \quad : \quad \alpha^* = \max\{\alpha_2, \alpha_I\}, \quad (4.19)$$

where the coefficient $\alpha^*$ is taken as the larger one of $\alpha_2$ of Eq. (4.7) or the following newly-defined coefficient $\alpha_I$ for impulsive breaking wave pressure:

$$\alpha_I = \alpha_{IH}\,\alpha_{IB}. \quad (4.20)$$

The coefficients $\alpha_{IH}$ and $\alpha_{IB}$ are evaluated with the following set of equations, in which $B_M$ denotes the berm width:

$$\alpha_{IH} = \min\{H/d, 2.0\}, \quad (4.21)$$

$$\alpha_{IB} = \begin{cases} \cos\delta_2/\cosh\delta_1 & : \delta_2 \le 0, \\ 1/(\cosh\delta_1 \cosh^{1/2}\delta_2) & : \delta_2 > 0, \end{cases} \quad (4.22)$$

$$\delta_1 = \begin{cases} 20\delta_{11} & : \delta_{11} \le 0, \\ 15\delta_{11} & : \delta_{11} > 0, \end{cases} \quad (4.23)$$

$$\delta_2 = \begin{cases} 4.9\delta_{22} & : \delta_{22} \le 0, \\ 3.0\delta_{22} & : \delta_{22} > 0, \end{cases} \quad (4.24)$$

$$\delta_{11} = 0.93\left(\frac{B_M}{L} - 0.12\right) + 0.36\left(0.4 - \frac{d}{h}\right), \quad (4.25)$$

$$\delta_{22} = -0.36\left(\frac{B_M}{L} - 0.12\right) + 0.93\left(0.4 - \frac{d}{h}\right). \quad (4.26)$$

The *coefficient of impulsive pressure* $\alpha_I$ has been so set that it will have the maximum value 2.0 at the conditions of $B_M/L = 0.12$, $d/h = 0.4$ and $H/h \ge 2.0$. When applying Eqs. (4.19) and (4.20) for breakwaters in relatively deep water, use of the water depth at site may yield quite a large value of impulsive pressure coefficient $\alpha_I$, making an excessive assessment of impulsive breaking wave pressure. Shimosako and Osaki[24] have recommended to use an apparent water depth of $h = 2H_{max}$ when the water depth at site $h$ exceeds $2H_{max}$.

**Example 4.2**

Check the possibility of the occurrence of impulsive breaking wave pressure for the upright section shown in Example 4.1.

**Solution**

The upright section has the values of $B_M = 8.0$ m, $d = 5.6$ m, and $h = 10.1$ m. The wavelength $L$ for the period 11.4 s at the depth 10.1 m is 107.5 m. The impulsive pressure coefficient $\alpha_I$ is evaluated by Eqs. (4.25) to (4.30) as below.

$$\delta_{11} = 0.93 \times (8.0/107.5 - 0.12) + 0.36 \times (0.4 - 5.6/10.1) = -0.0980,$$
$$\delta_{22} = -0.36 \times (8.0/107.5 - 0.12) + 0.93 \times (0.4 - 5.6/10.1) = -0.127,$$
$$\delta_1 = 20 \times (-0.0980) = -1.960,$$
$$\delta_2 = 4.9 \times (-0.127) = -0.622,$$
$$\alpha_{IB} = \cos(-0.622)/\cosh(-1.960) = 0.813/3.62 = 0.225,$$
$$\alpha_{IH} = \min \{8.0/5.6, 2.0\} = \min \{1.43, 2.0\} = 1.43,$$
$$\alpha_I = 1.43 \times 0.225 = 0.320.$$

The pressure coefficient $\alpha_2$, has been evaluated as 0.314 in Example 4.1. This value is slightly smaller than that of $\alpha_I = 0.320$ in the above, and thus the coefficient $\alpha^*$ is assigned the value 0.320. However, the pressure intensity calculated by Eq. (4.24) is much lower than that of impulsive pressure, and it can be judged that the upright section of Example 4.1 is not subject to the action of impulsive breaking wave pressure.

### 4.2.4  *Sliding of Upright Section by Single Wave Action*

An engineer in charge of designing breakwaters has to confirm the stability of his breakwater against the design wave as will be discussed in Sec. 4.3. Nowadays however, he is also asked to estimate the magnitude of deformation or possible damage on his breakwater in the event when extraordinary waves exceeding the design condition may attack the design site. It is an approach of the performance-based design method to be discussed in Sec. 6.3. The information he has to provide is the possibility if the breakwater may slide or not and the total amount of sliding distance during an extraordinary storm event. The basic data to answer this question is the sliding distance of an upright section by a single wave.

Estimation of the *sliding distance* of a composite breakwater was first presented by Ito *et al.*[12] in 1966. Efforts for obtaining more reliable information continued in Japan since then, which resulted in the proposal in 1994 by Shimosako and Takahashi,[25] who approximated the wave force with a triangular pulse of relatively short duration for estimation of the sliding distance. Then Tanimoto *et al.*[26] proposed a modification to it for the case in which the peak of triangular wave force is not so large and the succeeding pulsating force yields a certain contribution to the sliding distance; Tanimoto *et al.* introduced a pulsating force of sinusoidal form to be considered simultaneously with the triangular pulse force in sliding calculation. Presently the new method by Shimosako and Takahashi[27] with introduction of modification by Tanimoto *et al.* is employed as the standard procedure for breakwater design in Japan.

According to Tanimoto *et al.*,[26] the temporal variation of horizontal wave force is approximated as shown in Fig. 4.11 by taking a larger one of the pulsating force component $P_1(t)$ or the impulsive force component $P_2(t)$. That is,

$$P(t) = \max\{P_1(t), P_2(t)\}. \tag{4.27}$$

The two component wave forces are defined as below.

$$P_1(t) = \gamma_P \, (P_1)_{\max} \, \sin \frac{2\pi t}{T}, \tag{4.28}$$

$$P_2(t) = \begin{cases} \dfrac{2t}{\tau_0} \, (P_2)_{\max} & : \; 0 \le t \le \dfrac{\tau_0}{2}, \\[2mm] 2\left(1 - \dfrac{t}{\tau_0}\right)(P_2)_{\max} & : \; \dfrac{\tau_0}{2} \le t \le \tau_0, \\[2mm] 0 & : \; \tau_0 < t, \end{cases} \tag{4.29}$$

$$\gamma_P = 1 - \frac{\pi}{(P_1)_{\max}} \int_{t_1}^{t_2} \left[ P_2(t) - (P_1)_{\max} \sin \frac{2\pi t}{T} \right] dt$$

$$\text{for} \;\; P_2(t) - (P_1)_{\max} \sin \frac{2\pi t}{T} \ge 0, \tag{4.30}$$

where $(P_1)_{\max}$ denotes the peak value of the horizontal wave force component when the wave pressure $p_1$ is calculated by using only the coefficient $\alpha_1$ in Eq. (4.19), $(P_2)_{\max}$ stands for the peak value of the force component when $p_1$ is calculated with $\alpha^*$ only, $T$ is the wave period, $\tau_0$ expresses the duration of impulsive force component, the time $t_1$ is set at 0, and $t_2$ refers to the time when $P_2$ becomes smaller than $P_1$.

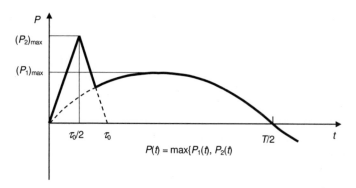

Fig. 4.11 A model of temporal variation of horizontal wave force (after Tanimoto et al.[27]).

The constant $\gamma_P$ defined by Eq. (4.30) has been introduced for determining the intensity of pulsating wave force in consideration of the impulse of impulsive force component. Shimosako and Takahashi[27] have given the duration of the impulsive force component as follows:

$$\tau_0 = k\,\tau_{0,F}\,, \tag{4.31}$$

$$\tau_{0,F} = \begin{cases} \left(0.5 - \dfrac{H}{8h}\right) T & : \; 0 < \dfrac{H}{h} \leq 0.8\,, \\[2mm] 0.4\,T & : \; 0.8 < \dfrac{H}{h}\,, \end{cases} \tag{4.32}$$

$$k = [(\alpha^*)^{0.3} + 1]^{-2}\,. \tag{4.33}$$

The upright section of composite breakwater under the action of the wave force expressed above makes a sliding motion with the horizontal acceleration, the magnitude of which is linearly proportional to the difference between the wave force and the frictional resistance, while the former exceeds the latter. When the former becomes equal to or less than the latter, the upright section stops the sliding motion. The impulsive wave pressure coefficient $\alpha^*$ of Eq. (4.19) inclusive of the impulsive coefficient $\alpha_I$ in Eqs. (4.20) to (4.26) has been derived by Takahashi et al.[19,21] on the basis of the performance of model breakwaters rested on rubble mounds. Therefore, the impulsive wave component $P_2$ represents the effective shear in consideration of the elastic deformation of the rubble mound foundation built on the seabed ground. The analysis of the sliding motion can be carried out by assuming the upright section being rested on a rigid floor under the action of the wave force model of Fig. 4.11. The displacement of the center of gravity of the upright section denoted with $x_G$ can be obtained by numerically integrating twice the following equation of motion:

$$\left(\frac{W_a}{g} + M_a\right) \ddot{x}_G = P - F_R \; : \; M_a \simeq 1.086 \, \rho_w \, h'^2 \, , \; F_R = f[(W_a - W_w) - U] \, ,$$
$$(4.34)$$

where $W_a$ is the weight of the upright section in the air, $M_a$ the added mass of the upright section moving in the water, $P$ the horizontal wave force, $F_R$ the frictional resistance against sliding of the upright section, $\rho_w$ the density of sea water, $f$ the frictional coefficient between the upright section and the rigid floor, $W_w$ the weight of water displaced by the upright section in still water (equal to the buoyancy), and $U$ the uplift force.

If the contribution of the pulsating force component is small, the horizontal wave force may be represented by the triangular pulse of $P_2$ alone, and the following closed form solution for the sliding distance by Shimosako and Takahashi[25] is available.

$$S = \frac{g\tau_0^2 (F_S - fW_e)^3 \, (F_S + fW_e)}{8fW_a W_e F_S^2} \; : \; F_S = P_{\max} + U, \; W_e = W_a - W_w \, .$$
$$(4.35)$$

The peak wave force $P_{\max}$ in Eq. (4.35) denotes the total force by Eq. (4.10) with the wave pressure calculated by Eq. (4.19) using the pressure coefficients $\alpha_1$ and $\alpha^*$.

## 4.3 Preliminary Design of Upright Sections

### 4.3.1 *Stability Condition for an Upright Section*

The upright section of a vertical breakwater must be designed to be safe against sliding and overturning. At the same time, the bearing capacity of the rubble mound foundation and the seabed should be examined to ascertain that they remain below the allowable limit. Presently, the stability of a vertical breakwater is examined either by the reliability or the performance-based design methods to be discussed in Chapter 6. As for the preliminary design, however, the method using the total safety factor may be employed. The safety factors against sliding and overturning of an upright section under wave action are defined by the following:

$$\text{Against sliding}: \quad \text{S.F.} = \frac{f\,(Mg - U)}{P} \, , \qquad (4.36)$$

$$\text{Against overturning}: \quad \text{S.F.} = \frac{Mgt - M_U}{M_p} \, , \qquad (4.37)$$

where $M$ denotes the mass of the upright section per unit extension in still water, $f$ the coefficient of friction between the upright section and the rubble mound, and $t$ the horizontal distance between the center of gravity and the heel of the upright section.

In the design of vertical and composite breakwaters in Japan, the *safety factors* against sliding and overturning have been set as equal to or larger than than 1.2. The *coefficient of friction* between concrete and rubble stones is usually taken as 0.6.

The *bearing capacity* of the foundation is to be analyzed by means of the methodology of geotechnical engineering for eccentric inclined loads. At sites where the seabed consists of a dense sand layer or soil of good bearing capacity, however, a simplified technique of examining the magnitude of the heel pressure is often employed. For this method, it is assumed that a trapezoidal or triangular distribution of bearing pressure exists beneath the bottom of the upright section, and the largest bearing pressure at the heel $p_e$ is calculated as

$$
p_e = \begin{cases} \dfrac{2W_e}{3t_e} & : t_e \leq \dfrac{1}{3}\,B\,, \\[3mm] \dfrac{2W_e}{B}\left(2 - 3\dfrac{t_e}{B}\right) & : t_e > \dfrac{1}{3}\,B\,, \end{cases} \tag{4.38}
$$

in which

$$
t_e = \frac{M_e}{W_e}\,, \quad M_e = Mgt - M_U - M_p\,, \quad W_e = Mg - U\,. \tag{4.39}
$$

The bearing pressure at the heel is to be kept below the value of 400 to 500 kN/m$^2$, but recent breakwater designs are gradually increasing this limit to 600 kN/m$^2$ or greater, with advancement of breakwater construction sites into deeper water and with increases in the weight of upright sections.

### Example 4.3

Examine the stability of the upright section of the breakwater shown in Example 4.1, assuming its mass is $M_a = 342.9$ Mg(ton)/m.

### Solution

The weight of the upright section in still water is calculated as

$$
W = Mg = (342.9 - 1.03 \times 15 \times 7.1) \times 9.8 = 2285 \text{ kN/m}\,.
$$

Thus, the safety factors are

Against sliding :     $\text{S.F.} = \dfrac{0.6 \times (2285 - 488)}{927} = 1.16$ ,

Against overturning : $\text{S.F.} = \dfrac{2285 \times 0.5 \times 15.0 - 4880}{4780} = 2.56$ .

Therefore, the upright section has sufficient stability against overturning, but the safety factor for stability against sliding does not have a sufficient margin.

Next, the bearing pressure at the heel is examined as

$$W_e = 2285 - 488 = 1797 \text{ kN/m} ,$$

$$M_e = 2285 \times 0.5 \times 15.0 - 4880 - 4780 = 7478 \text{ kN} \cdot \text{m/m} ,$$

$$t_e = \frac{7478}{1797} = 4.16 \text{ m} < \frac{1}{3}B = 5.0 \text{ m} ,$$

$$p_e = \frac{2 \times 1797}{3 \times 4.16} = 288 \text{ kN/m}^2 .$$

This value is far less than the allowable limit, and there would be no problem concerning the bearing capacity of the foundation.

### 4.3.2   *Stable Width of Upright Section*

The wave pressure exerted on an upright section of a vertical breakwater is approximately proportional to the height of the waves incident to the breakwater, but it is also controlled somewhat by the wave period, the sea bottom slope, the shape and dimensions of the rubble mound foundation, and other factors. The Goda wave pressure formulas described in the preceding sections are characterized by the incorporation of these factors. In 1976, Tanimoto et al.[16] calculated the minimum width of the upright section of a breakwater required for various combinations of design conditions such as the wave height, wave period, water depth, mound thickness and bottom slope. They utilized a breakwater of the shape shown in Fig. 4.12, and searched for the minimum width satisfying the conditions of safety factors against sliding and overturning equal to or greater than 1.2. The density of the upright section was taken to be $\gamma' = 1.1$ Mg(ton)/m$^3$ (concrete caisson filled with sand) for the submerged portion, $\gamma = 2.1$ Mg(ton)/m$^3$ for the portion of the caisson above the still water level, and $\gamma = 2.3$ Mg(ton)/m$^3$ for the concrete crown. The crest elevation was set at a height of 0.6

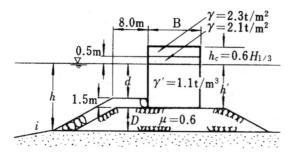

Fig. 4.12 Specification of vertical breakwater for trial calculation of the required caisson width (after Tanimoto *et al.*[16]).

times the significant wave height.[c] The coefficient of friction was taken as $f = 0.6$.

Because the impulsive pressure coefficient $\alpha^*$ was introduced by Takahashi *et al.*[22] in 1994, the original computation by Tanimoto *et al.*[16] needs to be revised. Figures 4.13 to 4.16 present the result of new computation with introduction of $\alpha^*$. First, Fig. 4.13 shows the variation in the required design caisson width for a given water depth above the mound in the range of $d = 0 \sim 28$ m for several wave conditions, for a constant thickness of the rubble mound of $D = 3$ m. The water depth at the breakwater location is $h = d + 4.5$ m. The bottom slope was taken to be $1/100$. The wave period was selected so as to keep the steepness of the equivalent deepwater waves in the range of $H_0'/L_0 = 0.037 \sim 0.040$ for wave heights of $H_0' = 3 \sim 10$ m. The wave height $H_{\max}$ for breakwater design varies with the water depth due to the effect of random wave breaking, but the effect of wave refraction was not taken into account.

The variation of the caisson width has two peaks. The first sharp peak around $d \simeq 3$ m corresponds to the maximum value of the impulsive breaking wave pressure coefficient $\alpha_I$. Dominance of $\alpha_I$ over $\alpha_2$ in Eq. (4.20) continues up to $d = 4.5$ m for the case of $H_0' = 10$ m. The second mild peak around $d = 17$ m for the case of $H_0' = 10$ m is caused by the artifice of Eq. (3.48) in Sec. 3.6.3(C) for the estimation of $H_{\max}$, used for the sake of convenience in the calculation. The peak would be more rounded if $H_{\max}$ were estimated by means of Fig. 3.42. The decrease of the caisson width at

---

[c]This criterion for specifying the crest elevation is used in breakwater design in Japan in situations where a small amount of wave overtopping and resultant wave transmission is tolerated. If very little wave overtopping is to be allowed, the crest elevation is set higher.

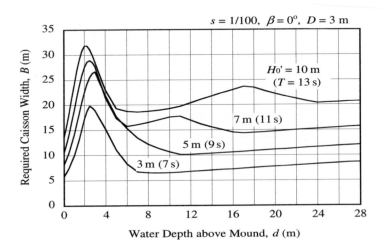

Fig. 4.13 Variation of the required caisson width for different incident wave heights: water depth is $h = d + 4.5$ m.

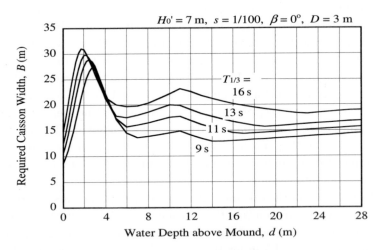

Fig. 4.14 Variation of the required caisson width for different incident wave periods: water depth is $h = d + 4.5$ m.

the right of the second peak is due to the decrease of the pressure coefficient $\alpha_1$ by increase of the relative water depth $h/L$ as expressed in Eq. (4.6). The increase of the caisson width at the further right zone indicates the dominance of the safety requirement against overturning over the sliding safety.

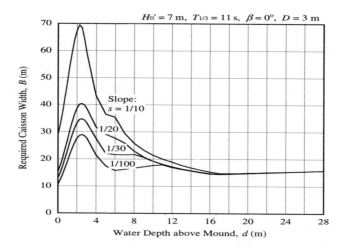

Fig. 4.15 Variation of the required caisson width for different sea bottom slopes: water depth is $h = d + 4.5$ m.[16]

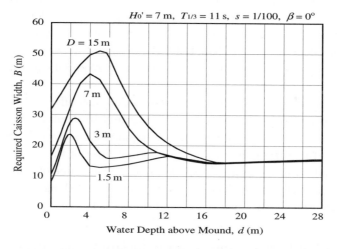

Fig. 4.16 Variation of the required caisson width for different thicknesses of the rubble mound: water depth is $h = d + D + 1.5$ m.

Next, Fig. 4.14 shows the effect of the wave period on the required caisson width. In this figure the wave period varies from $T_{1/3} = 9$ s to 16 s, and the equivalent deepwater wave height is kept constant at $H_0' = 7$ m. The required caisson width is seen to increase as the wave period increases. This is partly due to the increase in the coefficient $\alpha_1$ with decrease in relative water depth $h/L$. The increase in required caisson width is also affected by the process of random wave breaking in that the attenuation of $H_{max}$ is less for long-period waves than for short-period waves. Because of the effect of wave period on the wave pressure and the resultant required caisson width, careful examination of the design wave is required not only for the wave height but also for the wave period.

Several laboratory tests specifically performed to determine the wave pressure on vertical walls have demonstrated that the breaking wave pressure increases as the sea bottom slope becomes steeper. Figure 4.15 shows the effect of bottom slope on the required caisson width as calculated with the Goda wave pressure formulas. In the region of the breaker zone, an increase in the required caisson width with steepening of the bottom slope is clearly evident. This effect is introduced through the pressure formulas, not directly in the pressure itself, but indirectly via the term for the design wave height $H_{max}$. This happens because, first, the limiting height of the breaking waves increases as the bottom slope becomes steeper; second, the proposed technique of estimating $H_{max}$ to be the height at a distance $5H_{1/3}$ seaward from the breakwater enhances the effect of slope on the design value of $H_{max}$ to be employed in the pressure calculation. In an area where the slope is steeper than $1/30$, the design of a vertical breakwater should be done very carefully, as the increase in required caisson width is significant.

The height of the rubble mound foundation also affects the resultant wave pressure exerted on a structure. Figure 4.16 displays the effect of the thickness of a rubble mound on the required caisson width. Since the abscissa is the water depth $d$ above the armor layer of the rubble mound, the original water depth $h$ corresponding to a given value of $d$ depends on the mound thickness $D$. The effect of impulsive breaking wave pressure appears conspicuously when the mound thickness is large.

## 4.4    Several Design Aspects of Composite Breakwaters

### 4.4.1    *Wave Pressure Under a Wave Trough*

When the trough of an incident wave makes contact with a vertical wall, the pressure exerted on the wall becomes less than the hydrostatic pressure under the still water level. As a result, the vertical wall experiences a net pressure directed offshore. Such a pressure may govern the stability of an upright section against sliding seaward, the structural design of the front walls of concrete caissons, etc.

The problem of wave pressure under a wave trough, in particular that of breaking waves, has not been examined in detail. But as far as the pressure of standing waves is concerned, the author[14] has prepared a set of diagrams shown in Figs. 4.17 to 4.19. These diagrams are based on theoretical calculations using finite amplitude standing waves, with modifications introduced on the basis of laboratory data. Figure 4.17 gives the total offshore-directed pressure under a wave trough, Fig. 4.18 shows its lever arm length, and Fig. 4.19 gives the magnitude of the bottom pressure at the time of the greatest offshore-directed total pressure. The symbol $w_0$ in the ordinates of these figures stands for the specific weight of sea water $(= \rho_w g)$.

Figure 4.19, for example, indicates that the negative pressure near the bottom can become quite large even under the condition of quasi-deepwater waves, if the wave height is sufficiently large. This is caused by the appearance of second and fourth harmonic pressure components of appreciable amplitude due to wave nonlinearity. The harmonic pressure components act uniformly from the water surface to the sea bottom with the phases opposite to that of the primary component.

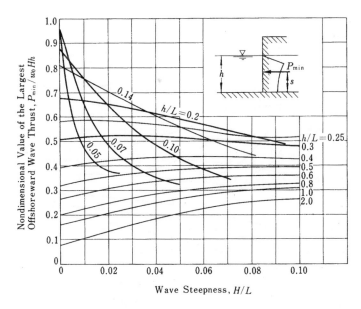

Fig. 4.17  Calculation diagram for the total pressure of standing waves under a wave trough.[14]

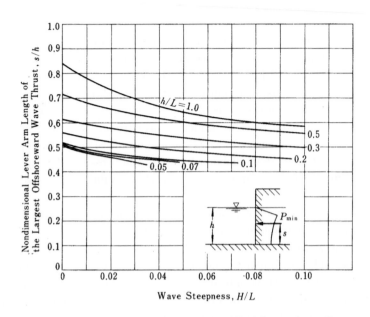

Fig. 4.18   Calculation diagram for the lever arm length of the total standing wave pressure under a wave trough.[14]

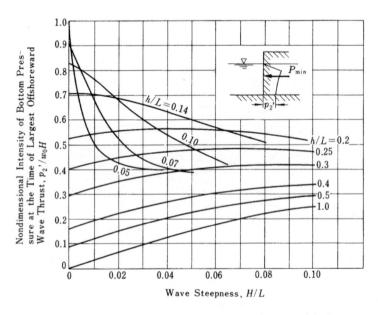

Fig. 4.19   Calculation diagram for the bottom pressure at the time of the largest seaward wave thrust during wave trough.[14]

In comparison with the onshore-directed pressure under a wave crest, the pressure under a trough is seen to become larger than that under a crest if the relative water depth (ratio of water depth to wavelength) exceeds about 0.25. Under such a condition, an upright section may slide toward the offshore during heavy storm waves. In fact, Tanimoto et al.[28] have demonstrated that a model of the tsunami-protection breakwater at Kamaishi Port at the water depth of $h = 60$ m with the mound depth of $d = 25$ m slid both onshore and offshore over almost the same distance when it was exposed to the irregular waves with the period corresponding to $T_{1/3} = 12$ s.

### 4.4.2 *Uplift on a Large Footing*

Some breakwater caisson is so designed to have footings extended from its bottom to reduce the toe pressure and/or for other purposes. The width of a footing is mostly 0.5 to 1.0 m, and the uplift is calculated by neglecting the presence of such a narrow footing. Recently however, some of breakwater caissons have been built with the hybrid structure composed of steel frames and reinforced concrete. It is not difficult to design a large footing for such hybrid caissons, though the onshore side footing is subject to large bending moments and shears by the reaction force from the foundation mound. Careful consideration is required for the moment distribution and stresses in the footing and the caisson wall for designing such a large footing.

The offshore side footing is subject to both the downward wave pressure and the upward uplift as shown in Fig. 4.20. The uplift is regarded to have a triangular distribution between the toes of the offshore and onshore side footings. On the basis of a series of laboratory model tests, Esaki et al.[29,30] have proposed the following formulas for estimation of the downward wave pressure $p_c$ and the uplift pressure $p_{ue}$ at the front of the offshore side footing:

$$p_c = p_5 \times \min\{0.7, \, [0.7\exp(-20x/L) + 0.3]\}, \qquad (4.40)$$

$$p_{uc} = p_u \times [0.7\exp(-11b/L) + 0.3], \qquad (4.41)$$

where $p_5$ denotes the wave pressure on the front wall at the elevation of the top of the footing, $p_u$ is the uplift pressure calculated by Eq. (4.9), $x$ stands for the distance from the front of the footing, $b$ expresses the extended width of the footing, and $L$ refers to the wavelength.

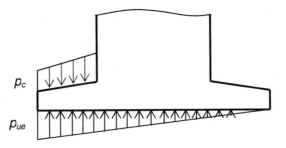

Fig. 4.20   Distributions of downward wave pressure and uplift exerted on a large footing and caisson.

### Example 4.4

A caisson breakwater is designed in the depth of $h = 13.0$ m under the design tide level with the caisson height of $h' = 11.0$ m below the water level. The design wave has the dimensions of $H_{1/3} = 3.5$ m, $H_{\max} = 6.3$ m, and $T_{1/3} = 7.3$ s. The offshore side footing will have the extension width of $b = 4.0$ m and the thickness of $d_F = 2.5$ m at the junction to the caisson wall. Calculate the intensities of the downward wave pressure $p_c$ and the uplift pressure $p_{uc}$.

### Solution

The wave length is calculated ass $L = 68.8$ m at the given conditions, and the pressure coefficients are obtained as $\alpha_1 = 0.700$, $\alpha_2 = 0.019$ and $\alpha_3 = 0.627$. The junction between the footing and the front wall is located at the level of $h_F = h' - d_F = 11.0 - 2.5 = 9.5$ m. Because the intensities of the wave pressure are $p_1 = 45.6$ kN/m$^2$ and $p_3 = 28.6$ kN/m$^2$, a linear pressure distribution between the water surface and the caisson bottom yields the pressure intensity of $p_5 = 32.5$ kN/m$^2$ at the junction level of the footing. Thus, the downward wave pressure has the intensities of $(p_c)_{x=0} = 22.7$ kN/m$^2$ at the wall and $p_c = 16.9$ kN/m$^2$ at the front ($x = b = 4.0$ m).

While the intensity of the uplift pressure under the Goda formula is $p_u = 27.9$ kN/m$^2$, the uplift pressure at the tip of the footing has the reduced intensity of $p_{ue} = 18.7$ kN/m$^2$.

## 4.4.3   *Wave Pressure on Horizontally-Composite Breakwaters and Other Special Breakwaters*

Quite a number of Japanese breakwaters have the upright sections covered with large mounds of wave-dissipating concrete blocks such as Tetrapods for the purpose of reducing wave forces acting on upright sections, wave overtopping, and/or wave reflection. They are sometimes called the *horizontally-composite breakwaters*, and the term is used in this book.

The Goda wave pressure formulas have been modified to be capable of estimating the wave pressure acting on the horizontally-composite breakwaters and other breakwaters of special configurations. The *pressure correction factors* $\lambda_1$, $\lambda_2$, and $\lambda_3$ that have been introduced in Eqs. (4.2), (4.3), and (4.9) have been evaluated with the results of various model tests. For the horizontally-composite breakwaters, Takahashi *et al.*[31] have proposed the following empirical formulas:

$$\lambda_1 = \begin{cases} 1.0 & : H_{\max}/h \leq 0.3, \\ 1.2 - 2H_{\max}/3h & : 0.3 < H_{\max}/h \leq 0.6, \\ 0.8 & : 0.6 < H_{\max}/h, \end{cases} \tag{4.42}$$

$$\lambda_2 = 0, \quad \lambda_3 = \lambda_1. \tag{4.43}$$

The rationale for the above formulation is as follows. The wave pressures by breaking waves in the Goda formula are represented by the term involving the coefficient $\alpha_2$. Because a mound of wave-dissipating blocks in front of the upright section significantly reduces the breaking wave pressure especially of impulsive nature, the correction factor for this term is set as $\lambda_2 = 0$. The wave pressures of pulsating nature is represented by the term involving the coefficient $\alpha_1$, and the pressure reduction effect of the block mound on this term is enhanced when the ratio of wave height to water depth is large. Thus, the correction factor is set at $\lambda_1 = 0.8$ when the height-to-depth ratio exceeds 0.6, while the reduction effect is regarded null ($\lambda_1 = 1.0$) when the height-to-depth ratio is less than 0.3. In between, the linear variation with respect to the height-to-depth ratio is assumed.

Reduction of wave pressures by block mounds is effective only under the condition that a mound of wave-dissipating blocks maintains its crest at the condition of sufficient width and elevation. If the crest elevation of the block mound is lower than the design water level and the upper part of the upright section is exposed to wave attack, for example, the wave force acting on the upright section may become larger than the force in the case of no block mound. Even though the design and construction was so made to have sufficient covering by blocks, there is a possibility that the block mound may settle down by scouring of the seabed in its front or by washing out of sand beneath the rubble foundation. If the amount of settling of the whole mound is large, there could be a danger of sliding of and/or structural damage to the breakwater caisson by generation of impulsive breaking wave forces at the occasion of storm wave event.

Another type of composite breakwaters which have appeared since the 1980s in Japan is the sloping-top caisson breakwater. The front part of the superstructure made of plain concrete is built as a slope with the inclination of 45° or so; sometimes the caisson itself is designed with a sloped front near the top.[32] Because the wave pressure on the sloped part is exerted normal to the slope, it has the downward-acting vertical component in addition to the horizontal component. The resulting downward wave force counteracts the uplift acting on the bottom of the upright section, thus increasing the stability against sliding and overturning. Thus, the required width of the upright section can be reduced to some extent.

A demerit of the sloping-top caisson breakwater is a high degree of wave run-up compared with the standard breakwater of rectangular cross-section. In order to reduce the wave transmission by overtopping, the crest of the sloping-top breakwater is often designed at the elevation of $1.0H_{1/3}$ above the design water level, while the conventional upright breakwater is designed with the crest elevation of $0.6H_{1/3}$ as a standard. The method of calculating the horizontal and vertical wave force components acting on the sloping-top caisson breakwater can be found in Ref. 33.

### 4.4.4  *Comments on Design of Concrete Caissons*

The upright section of a vertical breakwater can be composed of layers of solid concrete blocks, of layers of cellular concrete blocks with some filler, or of concrete caissons. Experience in Japanese breakwater construction has demonstrated the reliability of reinforced concrete caissons since the early 1910s.

As seen in the examples in Fig. 4.1, a concrete caisson is divided into rows and columns of inner cells with partition walls. The width of the caisson is determined by the stability condition against wave action. The length of the caisson is mostly determined by the maximum allowable size in the caisson manufacturing yard. As a precaution against uneven settling of the rubble mound and the ground, as well as possible scouring, the length is usually selected in the range of 0.5 to 2.0 times the width of the caisson. The height of the caisson is determined by considerations of the capacity of the caisson yard, an analysis of the total cost, efficiency of divers' work in leveling the top of the rubble mound foundation, and other factors. Concrete caissons must have sufficient buoyant stability, unless they are to be carried and set in position with the aid of a huge floating crane.

The maximum dimension of the inner cells is usually designed to be less than 5 m. This determination is done in conjunction with selection of the thickness of the outer walls (between 40 to 50 cm) and the thickness of the partition walls (between 20 to 25 cm). The thickness of the bottom slab is usually 50 to 70 cm. The outer walls must be designed against the pressure exerted by the filler material and internal hydrostatic pressure. In addition, the front wall must be able to withstand the stress caused by wave pressure. If the dimensions of the inner cells are designed to be much greater than 5 m, the downward wave pressure by green water on the crown of the breakwater may need to be examined. Further details of caisson design can be found in *Technical Standards for Port and Harbour Facilities in Japan.*[33]

Few breakwater caissons in Japan have shown indications of deterioration due to corrosion. The standard design strength of the concrete for breakwater caissons in Japan is 24 MPa, and for many years the thickness of the covering for reinforcement has been specified to be not less than 7 cm on the seaward side and 5 cm on the landward side. It appears that continuous submergence of the major portion of a caisson acts to prevent deterioration of the reinforced concrete.

## 4.5 Design of Rubble Mound Foundation of Composite Breakwaters

### 4.5.1 *Dimensions of Rubble Mound*

It is best to set the height of the rubble mound as low as possible to prevent the generation of large wave pressure. But the function of a rubble mound — to spread the vertical load due to the weight of the upright section and the wave force over a wide area of the seabed — necessitates a minimum thickness, which is required to be not less than 1.5 m in Japan. Furthermore, the top of the rubble mound should not be too deep, in order to facilitate underwater operations of divers in leveling the surface of the mound for even setting of the upright section. A cost analysis on breakwater construction will yield the optimum height of the rubble mound under the above constraints.

The berm width is a factor to be selected empirically. If the seabed is soft, the dimensions of the rubble mound should be determined by safety considerations against circular slip of the ground. The berm in front of an upright section functions to provide protection against possible scouring of the seabed. A wide berm is desirable in this respect, but the cost and danger of inducing impulsive breaking wave pressure precludes the design of too great a berm width. The practice in Japan is for a minimum of 5 m under normal conditions and about 10 m in areas attacked by large storm waves. The berm to the rear of an upright section has the function of safely transmitting the vertical load to the seabed. It also provides an allowance of some distance if sliding of an upright section should occur. The gradient of the slope of the rubble mound is usually set at $1:2$ to $1:3$ for the seaward side and $1:1.5$ to $1:2$ for the harbor side.

### 4.5.2  Foot-Protection Blocks

In breakwater construction in Japan, it is customary to provide a few rows of *foot-protection concrete blocks* at the front and rear of the upright section, as seen in Fig. 4.2. The foot-protection usually consists of rectangular blocks weighing from 100 to 400 kN (10 to 40 tons), depending on the design wave height. Foot-protection blocks are indispensable, especially when storm waves attack a vertical breakwater at an oblique angle. The stability of a foot-protection block is governed by its height $t$. Kimura *et al.*[34] have proposed the following formula based on a series of model tests:

$$t = AH_{1/3} \left( \frac{h'}{h} \right)^{-0.787} \quad : \quad 0.4 \leq \frac{h'}{h} \leq 1.0, \qquad (4.44)$$

where $h'$ is the water depth at which concrete blocks are to be placed and $h$ is the water depth. The constant $A$ is assigned the value of 0.21 at the breakwater head and 0.18 for the trunk section.

The remainder of the berm and slope of the rubble mound foundation must be protected with armor units of sufficient mass to withstand wave action. The question of the minimum mass of armor units baffled engineers for many years, forcing them to rely on experience. However, Tanimoto *et al.*[35,36] succeeded in formulating a calculation method for the mass required of armor units, based on irregular wave tests and theoretical analysis. According to their study, the minimum mass of armor units for a rubble mound foundation can be calculated by a formula of the Hudson type:

$$M = \frac{\rho_r}{N_s^3 (S_r - 1)^3} H_{1/3}^3, \qquad (4.45)$$

where $M$ is the mass of the armor unit, $\rho_r$ the density of the armor unit, $S_r$ the ratio of $\rho_r$ to the density of seawater, $H_{1/3}$ the design significant height, and $N_s$ the *stability number*. The value of the stability number depends on the wave conditions and mound dimensions, as well as on the shape of the armor units. The nominal diameter of armor units which is defined as $D_n = (M/\rho_r)^{1/3}$ is calculated by

$$D_n = \frac{H_{1/3}}{N_s \Delta} \; : \; \Delta = S_r - 1. \tag{4.46}$$

For normally incident waves, irregular wave tests have shown that the most unstable location along a breakwater is the corner between the slope and the horizontal section of the berm. Through analysis of the fluid dynamic forces of drag and lift acting on a single armor unit, together with the results of laboratory experiments, Tanimoto *et al.* proposed the empirical formula for the stability number $N_s$ for armor stones. The formula has further been expanded to include the effect of oblique wave attack by Takahashi *et al.*[37] and Kimura *et al.*[38] as listed below.

$$N_s = \max\left\{1.8, \left(1.3\frac{1-\kappa}{\kappa^{1/3}}\frac{h'}{H_{1/3}} + 1.8\exp\left[-1.5\frac{(1-\kappa)^2}{\kappa^{1/3}}\frac{h'}{H_{1/3}}\right]\right)\right\} :$$

$$B_M/L' < 0.25, \tag{4.47}$$

in which

$$\kappa = \frac{4\pi h'/L}{\sinh(4\pi/L)}\sin^2\left(\frac{2\pi B_M}{L'}\right), \tag{4.48}$$

where max $\{a, b\}$ denotes the larger one of $a$ or $b$, ande $h'$ is the water depth at which armor units are to be placed, $L'$ the wavelength at the depth $h'$, and $B_M$ the berm width.

### Example 4.5

Calculate the minimum mass of the rubble stone required to armor the rubble mound foundation of the breakwater presented in Fig. 4.8, assuming the waves are incident normal to the structure.

### Solution

(i) Wave and mound dimensions

$$H_{1/3} = 5.8 \text{ m}, \quad T_{1/3} = 11.4 \text{ s},$$

$$h' = 7.1 \text{ m}, \quad L' = 91.6 \text{ m} \quad \text{and} \quad B_M = 8.0 \text{ m}.$$

(ii) Parameter $\kappa$ and stability number $N_s$

$$\kappa = \frac{4\pi \times 7.1/91.6}{\sinh(4\pi \times 7.1/91.6)} \times \sin^2\left(\frac{2\pi \times 8.0}{91.6}\right) = 0.233\,,$$

$$N_s = \max\left\{1.8,\ \left(1.3 \times \frac{1 - 0.233}{0.233^{1/3}} \times \frac{7.1}{5.8} + 1.8\right.\right.$$

$$\left.\left.\times \exp\left[-1.5 \times \frac{(1 - 0.233)^2}{0.233^{1/3}}\frac{7.1}{5.8}\right]\right)\right\},$$

$$= \max\{1.8, 2.29\} = 2.29\,.$$

(iii) Mass of armor stone

Assuming $\rho_r = 2650$ kg/m$^3$ and $S_r = 2.57$, we have

$$M = \frac{2650}{2.29^3 \times (2.57 - 1)^3} \times 5.8^3 = 11.1\ \text{Mg(ton)}\,.$$

In actual design, heavy concrete blocks may be required as the armor units of this breakwater.

For obliquely incident waves, Kimura et al.[38] have modified the parameter $\kappa$ as in the following, based on calculations of the orbital velocity of water particles in a three-dimensional standing wave system:

$$\kappa = \frac{4\pi h'/L'}{\sinh(4\pi h'/L')}\kappa_2\,, \tag{4.49}$$

$$\kappa_2 = \max\left\{\alpha_S \sin^2\beta \cos^2\left(\frac{2\pi x}{L'}\cos\beta\right),\ \cos^2\beta \sin^2\left(\frac{2\pi x}{L'}\cos\beta\right)\right\}:$$

$$0 \le x \le B_M\,, \tag{4.50}$$

in which the distance $x$ from the foot of the upright section should be varied between 0 and $B_M$ to give the maximum value of $\kappa_2$. For normal wave incidence ($\beta = 0$), Eqs. (4.49) and (4.50) reduce to Eq. (4.48) at $x = B_M$. The factor $\alpha_S$ has been introduced by Kimura et al.[38] to account for the effect of slope, and the value of $\alpha_S = 0.45$ is given based on the measured data. According to irregular wave tests for oblique incidence by Kimura et al., waves with the incidence angle $\beta = 60°$ caused greater damage than waves with $\beta = 0°$ and $45°$: damage occurred at $x = 0$ when $\beta = 60°$.

Around the corners of caisson at a breakwater head, strong oscillatory flow is generated at the sea bottom and on top of the rubble mound. The armor units around a breakwater head should be provided with the weight much greater than those along the trunk section. Kimura *et al.*[38] have proposed to calculate the $\kappa$ value for Eq. (4.33) as follows:

$$\kappa = \frac{4\pi h'/L'}{\sinh(4\pi h'/L')} \alpha_S \tau^2 , \qquad (4.51)$$

where $\tau$ is the correction factor for local rapid flow around the corners. Kimura *et al.*[34,38] recommend the following value for $\tau$, based on their test results:

$$\tau = \begin{cases} 1.4 : \beta = 0°, 45° , \\ 2.5 : \beta = 60° . \end{cases} \qquad (4.52)$$

When rubble stone of sufficient size for armoring the foundation mound is difficult to obtain at construction sites, concrete blocks of deformed shapes are often employed as the armor units. The stability of concrete armor units depends not only on their size but also on their shape and placement method. Thus, the selection of the type and size of concrete armor units should be made by referring to the irregular wave model tests and previous construction experiences at other sites. Nevertheless, Fuji-ike *et al.*[39] have presented the following formula for the stability number $N_s$ of concrete armor units:

$$N_s = N_{so} \times \left\{ 1.0, \left( 0.525 \frac{(1-\kappa)h'}{\kappa^{1/2} H_{1/3}} + \exp\left[ -0.9 \frac{(1-\kappa)^2 h'}{\kappa^{1/2} H_{1/3}} \right] \right) \right\} , \qquad (4.53)$$

where $N_{so}$ denotes the standard stability number, which can be derived from the $K_D$ value for concrete blocks employed for armoring the slope of mound structure as $N_{so} = (K_D \cot \theta)^{1/3}$. The parameter $\kappa$ in Eq. (4.53) is estimated by

$$\kappa = C_R \frac{4\pi h'/L'}{\sinh(4\pi h'/L')} \times \kappa_3 , \qquad (4.54)$$

$$\kappa_3 = \begin{cases} \sin^2(2\pi B_M/L') & : B_M/L' < 0.15 , \\ 1.309 - \sin^2(2\pi B_M/L') & : 0.15 \leq B_M/L' < 0.15 , \\ 0.309 & : 0.25 \leq B_M/L' , \end{cases} \qquad (4.55)$$

where the coefficient $C_R$ represents the *influence of the breakwater shape*. It is given the value of $C_R = 1.0$ for conventional composite breakwaters and $C_R = 0.4$ for horizontally-composite breakwaters.

In any case, the stability of the foot-protection blocks and armor units of a rubble mound foundation should be examined with a hydraulic scale model test employing irregular trains of laboratory waves. The previously mentioned studies by Tanimoto *et al.*[35,37] indicated that trains of regular waves should have a height of 1.37 times the significant height of irregular waves in order to produce the same amount of damage on the armor units of a rubble mound foundation.

### 4.5.3   *Protection Against Scouring of the Seabed in Front of a Breakwater*

A vertical breakwater reflects most of the wave energy incident to it, thus creating greater agitation at its front than that created by a mound breakwater. This agitation is thought to produce scouring of the bed in front of the breakwater. The disaster of the Mustapha breakwater in Alger Port in 1934 is a well-known example of seabed scouring, which is considered to have accelerated the collapse of the upright section (Pierre[40]). There is a possibility, however, that the collapse might have been caused by geotechnical failure of the foundation. One incidence of the geotechical failure of a Japanese breakwater produced a forward tilting of the upright section, which was similar in shape as the Mustapha breakwater.

It is interesting to note that in Japan there has been no case of the collapse of a breakwater due to seabed scouring, even though several hundred kilometers of vertical breakwaters have been built along the coast. Scouring has been taking place at a number of breakwaters and the tips of the rubble mounds are dislocated, but repairs are always made before the stability of the upright sections is threatened. The absence of breakwater collapse due to scouring may also owe to the practice of providing a quite broad berm and gentle slope for the rubble mound in front of the upright section.

Nonetheless, dislocation of the rubble mound foundation due to scouring of the seabed is a phenomenon against which precautions must be taken. For this purpose, various materials such as plastic filters and asphalt mats are spread beneath the area of the tip of the rubble mound and extended beyond it. However, no effective method has yet been found to stop scouring of the bed and dislocation of the tip of a rubble mound. Recent practice is toward the revival of gravel matting, i.e., a thin layer of quarry run extended beyond the tip of the armor layer. Gravel may be dispersed by strong wave agitation, but a mixture of gravel and original bed material will withstand wave action for a longer duration than an unprotected seabed. Presently use of non-woven geotextile sheets is favored in many countries, though they are not much employed in Japan yet.

# References

1. J. Larras, *Cours d'Hydraulique Maritime et de Travaux Maritimes* (Dunod, Paris, 1961), pp. 244–245.
2. International Organization for Standardization, "*Actions from waves and currents on coastal structures*, ISO 21650 (2007).
3. Y. Ito, "A treatise on historical development of breakwater design," *Tech. Note of Port and Harbour Res. Inst.* (69) (1969), 78p (*in Japanese*).
4. K. Tanimoto and Y. Goda, "Stability of deep water caisson breakwater against random waves," *Coastal Structures and Breakwaters* (The Inst. Civil Engrs., Thomas Telford, 1992), pp. 221–206.
5. S. Takahashi, "Design of vertical breakwaters," *Reference Document* (34) (Port and Harbour Res. Inst., 1996), 85p.
6. Th. Stevenson, *The Design and Construction of Harbours* (3rd Ed.) (Adam and Charles Blacks, 1886).
7. B. Gaillard, "Wave action in engineering structure," *Engineering News* 23 Feb. 1905.
8. I. Hiroi, "On a method of estimating the force of waves," *Memoirs of Engrg. Faculty, Imperial Univ. Tokyo*, **X**(1) (1919), p. 19.
9. G. Sainflou, "Essai sur les digues maritimes, verticales," *Annales Ponts et Chaussées* **98**(4) (1928).
10. R. R. Minikin, *Winds, Waves and Maritime Structures* (Griffin, London, 1950), pp. 38–39.
11. R. A. Bagnold, "Interim report on wave-pressure research," *J. Inst. Civil Engrs.* **12** (1939), pp. 202–226.
12. Y. Ito, M. Fujishima and T. Kitatani, "On the stability of breakwaters," *Rept. Port and Harbour Res. Inst.* **5**(14) (1966), 134p. (*in Japanese*) or *Coastal Engineering in Japan* **14** (1971), pp. 53–61.
13. Y. Goda and T. Fukumori, "Laboratory investigation of wave pressures exerted upon vertical and composite walls," *Rept. Port and Harbour Res. Inst.* **11**(2) (1972), pp. 3–45 (*in Japanese*) or *Coastal Engineering in Japan* **15** (1972), pp. 81–90.
14. Y. Goda and S. Kakizaki, "Study on finite amplitude standing waves and their pressures upon a vertical wall," *Rept. Port and Harbour Res. Inst.* **5**(10) (1966), 57p. (*in Japanese*) or *Coastal Engineering in Japan* **10** (1967), pp. 1–11.
15. Y. Goda, "A new method of wave pressure calculation for the design of composite breakwater," *Rept. Port and Harbour Res. Inst.* **12**(3) (1973), pp. 31–70 (*in Japanese*) or *Proc. 14th Int. Conf. Coastal Engrg.* (Copenhagen, 1974), pp. 1702–1720.
16. K. Tanimoto, K. Moto, S. Ishizuka and Y. Goda, "An investigation on design wave force formulae of composite-type breakwaters," *Proc. 23rd Japanese Conf. Coastal Engrg.* (1976), pp. 11–16 (*Japanese*).
17. S. Takahashi and K. Tanimoto, "Generation mechanism of impulsive pressure by breaking wave on a vertical wall," *Rept. Port and Harbour Res. Inst.* **22**(4) (1983), pp. 3–31 (*in Japanese*).

18. Y. Goda, "Dynamic response of upright breakwaters to impulsive breaking wave forces," *Coastal Engineering* **22**(1 and 2) (1994), pp. 135–158.

19. A. Lamberti and L. Martinelli, "Prototype measurements of the dynamic response of caisson breakwater," *Coastal Engineering 1998 (Proc. 26th Int. Conf.*, Copenhagen, ASCE) (1999), pp. 1972–1985.

20. S. Takahashi, K. Tanimoto and K. Shimosako, "Experimental study of impulsive pressures on composite breakwaters — Fundamental feature of impulsive pressure and the impulsive pressure coefficient," *Rept. Port and Harbour Res. Inst.* **31**(5) (1993), pp. 33–72.

21. M. Tsuda and T. Takayama, "Design procedures for a caisson wall against impulsive wave forces," *Annual J. Civil Engrg. Ocean, JSCE* **22** (2006), pp. 667–672 (*in Japanese*).

22. S. Takahashi, K. Tanimoto and K. Shimosako, "A proposal of impulsive pressure coefficient for the design of composite breakwaters," *Proc. Int. Conf. Hydro-Technical Engrg. for Port and Harbor Constr. (Hydro-Port '94)* (Yokosuka, Japan 1994), pp. 489–504.

23. K. Tanimoto, S. Takahashi and T. Kitatani, "Experimental study of impact breaking wave forces on a vertical-wall caisson of composite breakwater," *Rept. Port and Harbor Res. Inst.* **20**(2) (1981), pp. 3–39 (*in Japanese*).

24. K. Shimosako and N. Osaki, "Study on the application of the calculation methods of wave forces acting on the various type of composite breakwaters," *Tech Note, Prot and Airport Res. Inst.* No. 1107 (2005), pp. 1–14 (*in Japanese*).

25. K. Shimosako and S. Takahashi, "Calculation of expected sliding distance of composite breakwaters," *Proc. Coastal Engrg., JSCE* **41** (1994), pp. 756–760 (*in Japanese*).

26. K. Tanimoto, K. Furukawa and H. Nakamura, "Fluid resistance of upright section of composite breakwater against sliding and new model of sliding distance estimation," *Proc. Coastal Engrg., JSCE* **43** (1996), pp. 846–850 (*in Japanese*).

27. K. Shimosako and S. Takahashi, "Reliability design method of composite breakwaters using expected sliding distance," *Rept. Port and Harbor Res. Inst.* **37**(3) (1998), pp. 3–30 (*in Japanese*).

28. K. Tanimoto, K. Kimura and K. Miyazaki, "Study on stability of deep water breakwaters against waves (1st report) — Wave forces acting on upright section of trapezoidal shape and its stability against sliding," *Rept. Port and Harbor Res. Inst.* **27**(1) (1988), pp. 3–29 (*in Japanese*).

29. K. Esaki, T. Takayama, T.-M. Kim and M. Arai, "Experimental investigation of uplift and compressive pressures on a footing of a hybrid caisson," *Annual J. Civil Engrg. in the Ocean, JSCE* **20** (2004), pp. 73–78 (*in Japanese*).

30. K. Esaki, T. Takayama and T. Yasuda, "Verification of wave pressure acting on a breakwater with footing," *Annual J. Civil Engrg. in the Ocean, JSCE* **22** (2006), pp. 319–324 (*in Japanese*).

31. S. Takahashi, K. Tanimoto and K. Shimosako, "Wave and block forces on a caisson covered with wave dissipating blocks," *Rept. Port and Harbor Res. Inst.* **29**(1) (1990), pp. 54–75 (*in Japanese*).

32. Ports and Harbours Bureau (Ministry of Land, Infrastructure, Transport and Tourism), National Institute for Land and Infrastructure Management, and Port and Airport Research Institute, *Technical Standards and Commentaries for Port and Harbour Facilities in Japan*, Overseas Coastal Area Development Institute of Japan (2009), Part III, Chap. 4, Sec. 3.7.

33. *Loc. cit.* Ref. 39, Part II, Chap. 2, Sec. 4.7.2(6).

34. K. Kimura, Y. Mizuno, K. Sudo, N. Kuwabara and M. Hayashi, "Characteristics of damage at the foundation mound of composite breakwaters around their heads and the estimation method of stable weight of armor units," *Proc. Coastal Engrg., JSCE* **43** (1996), pp. 806–810 (*in Japanese*).

35. K. Tanimoto, T. Yagyu, T. Muranaga, K. Shibata and Y. Goda, "Stability of armour units for foundation mounds of composite breakwaters by irregular wave tests," *Rept. Port and Harbour Res. Inst.* **21**(3) (1982), pp. 3–42 (*in Japanese*).

36. K. Tanimoto, T. Yagyu and Y. Goda, "Irregular wave tests for composite breakwater foundations," *Proc. 18th Int. Conf. Coastal Engrg.* (Cape Town, 1982), pp. 2144–2163.

37. S. Takahashi, K. Kimura and K. Tanimoto, "Stability of armour units of composite breakwater mound against oblique waves," *Rept. Port and Harbour Res. Inst.* **29**(2) (1990), pp. 3–36 (*in Japanese*).

38. K. Kimura, S. Takahashi and K. Tanimoto, "Stability of rubble mound foundations of composite breakwaters under oblique wave attack," *Proc. 24th Int. Conf. Coastal Engrg.* (Kobe, ASCE, 1994), pp. 1227–1240.

39. T. Fuji-ike, K. Kimura T. Hayashi and Y. Doi, "Armor block stability of rubble mound foundation for horizontally composite breakwaters," *Proc. Coastal Engrg., JSCE* **46** (1999), pp. 881–885 (*in Japanese*).

40. R. Pierre, "La jetée de Mustapha au Port d Alger," *Annales des Ponts et Chaussées* Avr.-Mai, 1935.

# Chapter 5

# Design of Coastal Dikes and Seawalls

## 5.1 Random Wave Run-Up on Coastal Dikes and Seawalls

### 5.1.1 *Run-Up Height by Standing Waves at Vertical Wall*

When waves hit a structure, water rushes onto the front slope or along the wall of the structure to the elevation higher than the incident wave crest. It is the phenomenon of *wave run-up*. The vertical distance between the highest point of water rise and the still water level is called the *run-up height*.

When waves are not breaking in front of a vertical wall located in relatively deep water, the run-up height is equal to the wave crest height of standing waves and it can be estimated with the finite amplitude standing wave theory. Previously the author[1] prepared a design diagram for estimation of the crest height $\eta_{\max}$ of the fourth order standing waves for the range of the depth water depth of $h/L_A = 0.05 \sim 0.30$, where $h$ is the water depth and $L_A$ is the wavelength calculated by the small amplitude wave theory. A set of curves depicting the relationship between the relative crest height and the wave steepness for various values of the relative water depth are now empirically fitted with the following formula:

$$\eta_{\max}/H_I = 2 - \exp[-A(H_I/L_A)^b] \; : \; A = -0.153 - 2.153\ln(h/L_A),$$

$$b = 1 - 0.06[\ln(h/L_A)]^2, \qquad (5.1)$$

where $H_I$ is the incident wave height.

The relative crest height estimated by Eq. (5.1) is slightly larger than that obtained from the design diagram with the bias of 0.006 and the standard deviation of 0.031 in the absolute value of $\eta_{\max}/H_I$. Figure 5.1 is a graphical presentation of the relative crest height with Eq. (5.1).

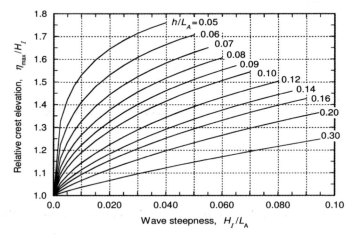

Fig. 5.1  Design diagram for relative crest height of standing waves.

The run-up height of random waves at a vertical wall has not been investigated yet, but it may be approximately evaluated by assuming that individual wave crest heights are equal to those of regular waves and the Rayleigh distribution is applicable for individual wave heights. Because the relative crest height is larger for waves of large height than waves of small height as indicated in Fig. 5.1 (nonlinear effect), the distribution of individual crest heights will be wider than the Rayleigh and the difference will be enhanced in shallow water area.

When the water depth in front of a vertical wall is not large enough with respect to the incident wave height, breaking waves act on the vertical wall. An uprush of green water is often observed along the wall together with high flying spray. The run-up height by breaking waves at a vertical wall is affected by many factors, evading the effort for empirical formulation. Nevertheless, Kimura *et al.*[2] measured the individual run-up heights of green water by random waves at a vertical wall in the relative depth of $0 < h/H_0 < 1$, though the model tests were made on a limited number of cases. In some case, the mean of the highest one-third run-up height was as high as 5.1 times the significant height of incident waves.

### 5.1.2  *Run-Up Height on Smooth Slopes and Coastal Dikes*

(A) *Distribution of run-up heights on smooth slopes*

The distribution of run-up heights of random waves on smooth slopes has been investigated by Mase,[3] who installed four slopes with the gradients of

Table 5.1   Constants $a$ and $b$ for estimation of random wave run-up on smooth slopes after Mase.[3]

| Parameter | $R_{\max}$ | $R_{2\%}$ | $R_{1/10}$ | $R_{1/3}$ | $\bar{R}$ |
|-----------|------------|-----------|------------|-----------|-----------|
| $a$ | 2.32 | 1.86 | 1.70 | 1.38 | 0.88 |
| $b$ | 0.77 | 0.71 | 0.71 | 0.70 | 0.69 |

1 on 5 to 1 on 30 in a wave flume with the uniform depth of $0.43 \sim 0.45$ m. He recorded the run-up of all waves consecutively and made statistical analysis of the characteristic run-up heights such as the maximum $R_{\max}$, the 2% exceedance $R_{2\%}$, the highest one-tenth $R_{1/10}$, the highest one-third $R_{1/3}$, and the mean $\bar{R}$. The characteristic run-up heights $R_X$ relative to the deepwater wave height $H_0$ were formulated as a function of the *Iribarren number* $I_{r,s}$ as in the following:

$$\frac{R_X}{H_0} = a\, I_{r,s}^b \; : \; 1/30 \leq \tan\alpha_s \leq 1/5, \; 0.007 \leq H_0/L_0, \qquad (5.2)$$

where the Iribarren number is defined as $I_{r,s} = \tan\alpha_s/\sqrt{H_0/L_0}$, $\alpha_s$ denotes the angle between the slope and the horizontal, and the deepwater wavelength $L_0$ is calculated with the significant wave period. Mase has given the values of the constant $a$ and $b$ as listed in Table 5.1 as best fitting ones to the experimental data.

### Example 5.1

Calculate the characteristic run-up heights when the deepwater waves with $H_0 = 4.5$ m and $T_{1/3} = 8.5$ s are normally incident to a smooth slope with the gradient of 1 on 8.

### Solution

The deepwater wave length is $L_0 = 112.7$ m and the Iribarren number is calculated as $I_{r,s} = (1/8)\sqrt{4.5/112.7} = 0.626$. Thus various characteristic run-up heights are estimated as below.

$$R_{\max} = 4.5 \times 2.32 \times 0.626^{0.77} = 7.28 \text{ m} = 1.63\,R_{1/3}\,,$$
$$R_{2\%} = 4.5 \times 1.86 \times 0.626^{0.71} = 6.00 \text{ m} = 1.34\,R_{1/3}\,,$$
$$R_{1/10} = 4.5 \times 1.70 \times 0.626^{0.71} = 5.49 \text{ m} = 1.23\,R_{1/3}\,,$$
$$R_{1/3} = 4.5 \times 1.38 \times 0.626^{0.70} = 4.47 \text{ m}\,,$$
$$\bar{R} = 4.5 \times 0.88 \times 0.626^{0.69} = 2.87 \text{ m} = 0.64\,R_{1/3}\,.$$

The above example suggests the distribution of run-up heights being slightly narrower than the Rayleigh as indicated in the rightmost figures of the run-up height relative to $R_{1/3}$. Such distribution seems to have originated from the test conditions of very gentle slopes installed in relatively deep water. For slopes located in shallow water, Mase *et al.*[4] have made laboratory tests for two slopes with the gradient of 1 on 0.5 and 1 on 3 which are installed on beaches of 1/20. In these tests, the run-up heights almost followed the Rayleigh distribution.

## (B) *Run-up heights on coastal dikes*

Coastal dikes, seawalls, and revetments are built along the natural coast and/or the waterlines of reclaimed land to protect the land area from high waves during storm tide. Definitions of these structures are not so clear, but *revetments* mainly refer to vertical structures standing in water of finite depth. The term of revetments is also applied to the surface layer armoring the fore slope of coastal structures. *Coastal dikes* are inclined structures with slopes gentler than 1 on 4 according to ISO 21650,[5] while *seawalls* are structures with slopes steeper than 1 on 4 inclusive of vertical structures. Coastal structures usually have broad crests and rear slopes, but seawalls may not have rear slopes.

Coastal dikes in North European countries are mostly earth structures protected with some armor layers, which include grass planting. Thus, they are so designed not to allow wave overtopping and their crests are determined by referring to the 2% run-up height $R_{2\%}$ under the design wave and tide conditions. The practice of employing the criterion of 2% run-up height is said to have been introduced when the design was made in 1936/1937 for the dikes for polders within the IJsselmeer, which was created by closure of the Southern Sea in the Netherlands in 1932.[6] Since then a number of studies have been carried out on the 2% run-up height on slopes. The Technical Advisory Committee on Flood Defense of the Netherlands (TAW) has published a review report on wave run-up on coastal dikes. A summary of its findings can be found in the new *Rock Manual.*[7] Reference 6 which may be called the *EurOtop Manual*[6] also lists various information on wave run-up on slopes and structures.

Among various results of research works, the following formula by Van Gent[8] is representative of estimating the 2% run-up height:

$$\frac{R_{2\%}}{\gamma_f \gamma_\beta H_{1/3}} = \begin{cases} 1.35 I_{r,\,s,\,-1} & : I_{r,\,s,\,-1} \leq 1.74, \\ 4.7 - 4.09/I_{r,\,s,\,-1} & : I_{r,\,s,\,-1} > 1.74, \end{cases} \qquad (5.3)$$

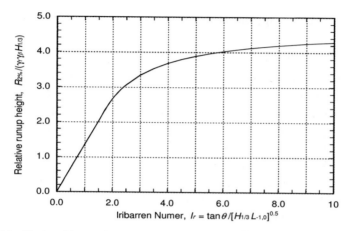

Fig. 5.2   Design diagram for random wave run-up height $R_{2\%}$ on a uniform slope.

where $\gamma_f$ and $\gamma_\beta$ denote the influence factors for roughness on a slope and oblique wave attack, respectively, and the Iribarren number $I_{r,s,-1}$ is defined by using the significant wave height $H_{1/3}$ and the deepwater wavelength $L_{0,-1}$ corresponding to the spectral significant wave period, $T_{m-1,0}$, i.e., $I_{r,s,-1} = \tan\alpha_s / \sqrt{H_{1/3}/L_{0,-1}}$ with $\alpha_s$ denoting the angle between the slope and the horizontal plane. The significant wave height is to be evaluated in front of a coastal dike by appropriately taking into account the wave attenuation by random breaking. The spectral significant wave period tends to increase in very shallow water by the process of random wave breaking, and van Gent recommends confirming any change in wave periods through numerical computation. Figure 5.2 is a graphical presentation of Eq. (5.3).

The influence factor $\gamma_f$ has been proposed to have the value of 1.0 for the armor layer of concrete, asphalt, and grass, 0.7 for single layer armor stone on impermeable base, and 0.55 for two layer armor stone on impermeable base.[7] The influence factor for oblique wave attack has been presented as in the following[7]:

$$\gamma_\beta = 1 - 0.0022\,|\beta(^\circ)| \;:\; 0^\circ \le |\beta| \le 80^\circ. \tag{5.4}$$

The run-up height estimation formula of Eq. (5.3) is applicable for the range of $1 < I_{r,s,-1} < 10$ and $1/6 \le \tan\alpha_s \le 1/2.5$. The estimated run-up height $R_{2\%}$ has the standard deviation of 0.37. For the case when the front slope of a coastal dike has different gradients in the lower and upper parts or a berm in the middle, some adjustment of the slope angle $\alpha_s$ is necessary; refer to the new *Rock Manual*[7] and other references.

## 5.2   Wave Overtopping Rate of Coastal Dikes and Seawalls

### 5.2.1   *Overtopping Rate by Random Sea Waves*

The main body of the coastal dike and seawall must be strong enough to withstand the attack of storm waves, and the crest must be high enough to prevent the intrusion of sea water onto the land by overtopping. Wave over-topping is primarily governed by the absolute heights of individual waves relative to the crest elevation of the seawall. Overtopping is not a continuous process but an intermittent occurrence at times of attack of individual high waves among the storm waves. The degree of wave overtopping is measured by the amount of overtopped water onto the land area, either as the amount (volume) per wave per unit length of seawall or as the mean rate of overtopping volume per unit length during the occurrence of storm waves.

The rate of overtopping averaged over the duration of the storm waves is hereby denoted with the symbol $q$. The overtopping rate is calculated by means of the amount of overtopping by individual waves $Q(H_i, T_i)$ as

$$q = \frac{1}{t_0} \sum_{i=1}^{N_0} Q(H_i, T_i) \,, \tag{5.5}$$

where

$$t_0 = \sum_{i=1}^{N_0} T_i \; : \text{duration of storm waves} \,, \tag{5.6}$$

$N_0$ : total number of waves,

$H_i, T_i$ : height and period of the $i$th individual wave attacking the seawall.

The mean rate of wave overtopping must be obtained through laboratory tests with irregular waves or through field measurements at seawalls with appropriate installations. But when a set of laboratory data on the rate of overtopping $q_0$ by regular waves with various combination of heights and periods is available, an approximate value of the mean rate of random wave overtopping can be estimated with the following formula:

$$q = \frac{1}{t_0} \sum_{t=1}^{N_0} T_i q_0(H_i, T_i) \,. \tag{5.7}$$

The above estimate is not expected to be too accurate, because the estimation neglects the random process of wave breaking, the presence of surf beat inherent to random waves, and the effects of interference by the preceding waves. Nevertheless, Eq. (5.7) provides engineers with a practical method of obtaining an estimate of random wave overtopping. A further simplification of Eq. (5.7) can be made by assuming that all wave periods are equal to some representative wave period, such as the significant period $T_{1/3}$. Thus,

$$q \simeq q_{EXP} = \int_0^\infty q_0(H|T_{1/3})p(H)dH , \qquad (5.8)$$

where

$q_0(H|T_{1/3})$ : overtopping rate by regular waves with height $H$

and period $T_{1/3}$ ,

$p(H)$ : probability density function of wave height .

Although Eq. (5.8) does not include the correlation between individual wave heights and periods, the correlation exists only among waves of smaller heights, being nil among waves with larger heights (refer to Sec. 10.3.2). Therefore the assumption is acceptable for the design of seawalls, because seawalls are designed to allow only occasional overtopping by a small number of high waves in a wave group. As for the probability density function $p(H)$, the Rayleigh distribution described in Sec. 2.2.1 can be employed if the laboratory data have been given in terms of the equivalent deepwater wave height.

The estimate of the mean overtopping rate, $q_{EXP}$ of Eq. (5.8), has been termed the *expected rate of wave overtopping* by the author.[9,10] A comparison was made between the estimated overtopping rate by Eq. (5.8), based on regular wave data, and the results of direct measurements with irregular waves in the laboratory for vertical walls and block mounds backed up by upright retaining walls. It was found that Eq. (5.8) gave reasonably good approximations to the measured values, except for seawalls close to the shoreline, where surf beat dominates.

An important feature of wave overtopping of prototype seawalls is its random nature. The analysis of wave overtopping requires incorporation of the probability distribution of individual wave heights and periods, especially of the former. If regular wave data for overtopping are directly applied to the design of a seawall by interpreting the regular wave height as equivalent to the significant height, the error introduced in the estimate of the overtopping rate may be quite large. For a seawall with a relatively high crest elevation, the overtopping rate will be underestimated in regular wave data, because the estimation ignores the existence of individual waves higher than the significant wave. Such an estimate of overtopping rate will lead to an unsafe design of the seawall.

### 5.2.2    Mean Rate of Wave Overtopping at Vertical and Block-Mound Seawalls

One of the typical shapes of a seawall is a vertical wall jutting directly from the seabed. There are several laboratory data sets available on wave overtopping of such vertical walls. Figures 5.3 and 5.4 are design diagrams compiled by Goda *et al.*[11] in 1975 for the estimation of the mean rate of wave overtopping at vertical seawalls. The *mean wave overtopping rate* is defined as the total quantity of overtopped water volume divided by the time duration over a few hundred waves. They were prepared on the basis of irregular wave tests and calculation of wave deformation in the surf zone.

Figure 5.3 is for a sea bottom slope of 1/10, and Fig. 5.4 is for a slope of 1/30. The mean rate of wave overtopping is non-dimensionalized with the equivalent deepwater wave height $H_0'$, which is discussed in Sec. 3.4, in the form of $q/\sqrt{2g(H_0')^3}$. The curves of dimensionless wave overtopping rate for several constant values of the relative crest elevation $h_c/H_0'$ are shown against the depth-to-height ratio $h/H_0'$. The symbol $h$ is the water depth at the toe of the seawall, $h_c$ the crest elevation of the seawall above the still water level, i.e., freeboard, and $g$ the acceleration of gravity ($= 9.8$ m/s$^2$). As seen in the insets of the figures, a simple wall with no recurved parapet and no foot-protection rubble mound is being considered. The diagrams cover a wide rage of depth-to-height ratio $h/H_0'$, including the land area with negative depth.

The diagrams of Figs. 5.3 and 5.4 have been prepared on the basis of irregular wave overtopping rate in a wave flume, which were carried out in the depth-to-height ratio range of $h/H_0 = -0.6 \sim 1.7$. The empirical curves in the offshore range of $h/H_0 > 2$ were prepared on the basis of calculation of the expected wave overtopping rate by means of the then-available empirical formulas on wave run-up height in consideration of random wave shoaling and breaking. However, several recent studies cited in Refs. 6 and 7 indicate that the overtopping rate for a given relative crest height is not affected by the depth-to-height ratio in the range of $h/H_0 > 4$. There is a possibility that the diagrams of Figs. 5.3 and 5.4 may yield underestimation of the wave overtopping rate in such range

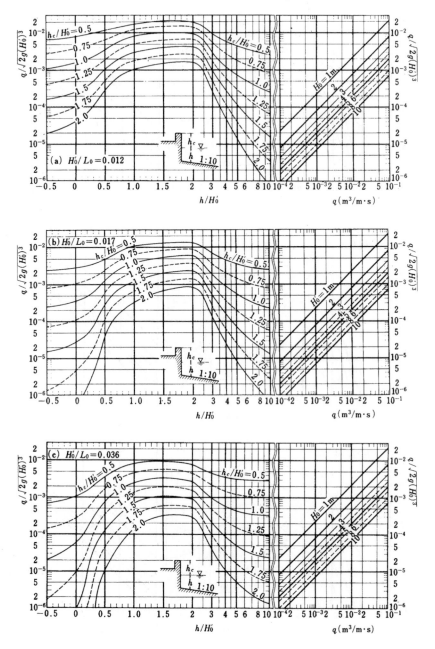

Fig. 5.3   Design diagrams of wave overtopping rate of vertical seawalls on a sea bottom slope of 1/10.[11]

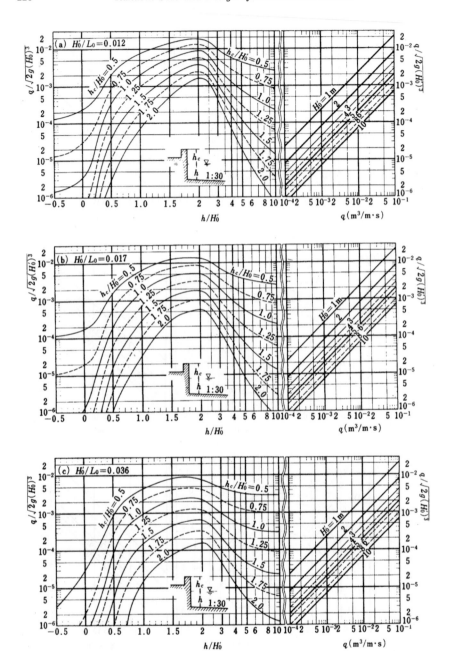

Fig. 5.4   Design diagrams of wave overtopping rate of vertical seawalls on a sea bottom slope of 1/30.[11]

of $h/H_0$. It is suggested to employ the new formulas to be presented in Sec. 5.2.4 when estimating the wave overtopping rate in relatively deep water areas.

### Example 5.2

A vertical seawall with a crest elevation $+ 6.0$ m above the datum level is built in water of depth $- 5.0$ m on a bottom with slope $1/30$. Estimate the mean rate of wave overtopping when waves with the equivalent deepwater height of $H_0' = 4.1$ m and $T_{1/3} = 8.5$ s attack the seawall at the tide level of $+ 1.5$ m.

### Solution

Since the deepwater wavelength is $L_0 = 113$ m and the wave steepness is $H_0'/L_0 = 0.040$, the depth-to-height ratio and crest elevation become

$$h/H_0' = (5.0 + 1.5)/4.5 = 1.44\,,$$

$$h_c/H_0' = (6.0 - 1.5)/4.5 = 1.0\,.$$

By referring to Fig. 5.4(c) we obtain

$$q/\sqrt{2g(H_0')^3} = 2 \times 10^{-3}\,. \quad \text{Thus} \quad q \simeq 0.08 \text{ m}^3/\text{m} \cdot \text{s}\,.$$

If either the wave steepness or the bottom slope differs from those in Figs. 5.3 and 5.4, interpolation or extrapolation becomes necessary. If the bottom slope is milder than $1/30$, the wave overtopping rate in water shallower than $2H_0'$ becomes less than that given by Fig. 5.4 in general. The rate of reduction in overtopping rate increases as the relative crest elevation $h_c/H_0'$ increases.

Seawalls made of sloping mounds of rubble stones and concrete blocks of the wave-dissipating type are more popular than vertical seawall. In Japan, block mound seawalls with the slope of 1 on 1.5 or so backed by a vertical retaining wall are quite common, especially along coasts facing rough seas. Figure 5.5 is a sketch of such a seawall. Such type of seawalls will be called *block-mound seawalls* in the present chapter. The wave overtopping rate of block-mound seawalls is governed not only by the characteristics of the incident waves, water depth and crest elevation, but also by the size and shape of the mound. Therefore, the compilation of generalized design diagrams for the overtopping rate of block-mound seawalls is more difficult than for the case of vertical seawalls.

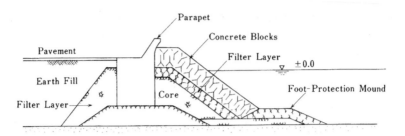

Fig. 5.5   Idealized typical section of a seawall built with concrete block mound of the wave-dissipating type.

Nevertheless, Goda *et al.*[11] have proposed the diagrams shown in Figs. 5.6 and 5.7.   The symbol $h_c$ in the figures refers to the crest elevation of the vertical parapet. The seawalls under consideration have two layers of Tetrapods resting on mounds of crushed stones placed in front of the vertical wall. The lower layer of Tetrapods at the crest of the block mound consists of two rows of Tetrapods. The top of the vertical parapet is set at a height of about $0.1\ H_0'$ above the crest of the block mound, and no recurvature is given to the parapet. Furthermore, no foot-protection mound beneath the block mound is provided. The slope of the mound has a gradient of $1:1.5$. As in the case of Figs. 5.3 and 5.4, laboratory tests were made with irregular trains of waves.

### Example 5.3

How much of a reduction in wave overtopping rate can be expected by changing the seawall from the vertical wall type to the block mound type for the same conditions as in Example 5.2?

### Solution

By reading the value of the ordinate corresponding to the input data of $h/H_0' = 1.44$ and $h_c/H_0' = 1.0$ in Fig. 5.7(c), we obtain

$$q/\sqrt{2g(H_0')^3} = 3 \times 10^{-4}\,. \quad \text{Thus} \quad q \simeq 0.013 \mathrm{m}^3/\mathrm{m}\cdot\mathrm{s}\,.$$

Thus, the overtopping rate is reduced to 1/6 that of the vertical seawall.

In the study of wave overtopping, scatter in the (laboratory) data is inevitable. The scatter becomes greatest when the crest of the seawall is high and the amount of wave overtopping is small. The diagrams presented as

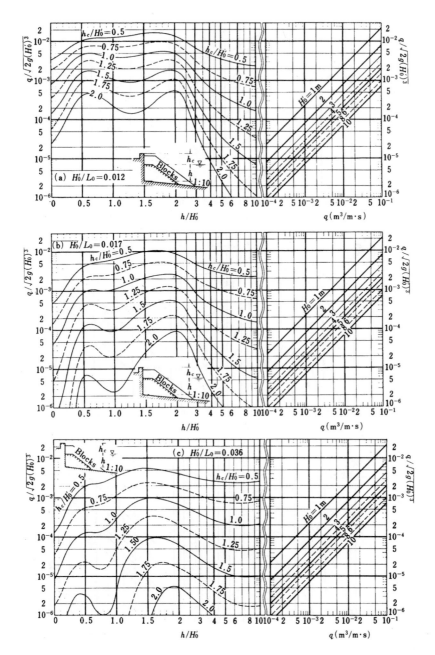

Fig. 5.6  Design diagrams of wave overtopping rate of block-mound seawalls on a sea bottom slope of 1/10.[11]

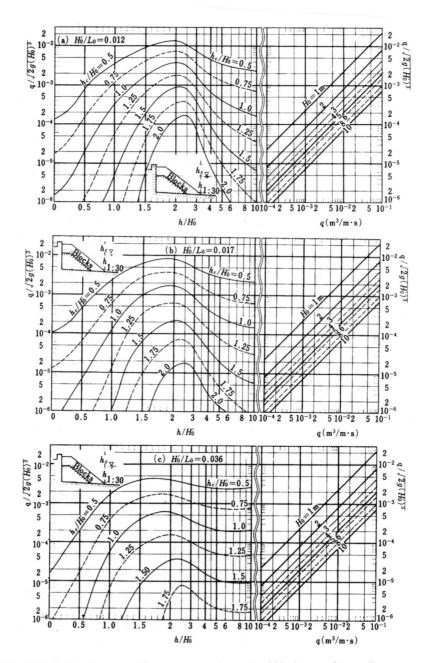

Fig. 5.7   Design diagrams of wave overtopping rate of block-mound seawalls on a sea bottom slope of 1/30.[11]

Table 5.2  Probable range of variation of estimated rate of wave overtopping to the true value.

| $q/\sqrt{2g(H_0')^3}$ | Vertical Seawall | Block-Mound Seawall |
|---|---|---|
| $10^{-2}$ | $0.7 \sim 1.5$ times | $0.5 \sim 2$ times |
| $10^{-3}$ | $0.4 \sim 2$ times | $0.2 \sim 3$ times |
| $10^{-4}$ | $0.2 \sim 3$ times | $0.1 \sim 5$ times |
| $10^{-5}$ | $0.1 \sim 5$ times | $0.05 \sim 10$ times |

Figs. 5.3, 5.4, 5.6 and 5.7 contain a certain range of variation due to this scatter in the original data. Comparing the limited available data of field measurements, the range of variation in the estimation of wave overtopping rate is assessed as listed in Table 5.2. The range of variation is greater for block-mound seawalls than for vertical seawalls. Although the range of variation is relatively large for the estimation of the overtopping rate of a seawall with specific dimensions, the variation is small when the crest elevation is sought for a given rate of overtopping; the range of the variation of estimated crest elevation is thought to be less than $\pm 20\%$. The difference in the ranges of variation stems from the nature of the equicontour lines of the overtopping rate in Figs. 5.3, 5.4, 5.6 and 5.7. Because the slope of the equicontour lines is quite steep in the zone of small values of nondimensional overtopping rate, a slight change in wave conditions and other factors produces a significant change in the rate of overtopping.

### 5.2.3  Mean Rate of Wave Overtopping at Coastal Dikes of Plane Slope

As for the mean rate of wave overtopping at plane slopes, Tamada *et al.*[12] have carried out a series of irregular wave tests for slope gradients of $\tan \alpha_s = 1/3, 1/5$, and $1/7$, and prepared three sets of design diagrams for respective slopes. Figures 5.8 to 5.10 have been redrawn by the author by using their original data, which Tamada *et al.* kindly made available for the author. The symbols are their original nondimensional data and the curves have been drawn by fitting to the data as smooth as possible.

Each set of design diagrams is composed of four graphs, which correspond to the combination of two beach slopes of $s = 1/10$ and $1/30$ and two wave steepness of $H_0'/L_0 = 0.017$ and $0.036$. The depth-to-height ratio covers the range of $h/H' = -0.25$ to $0.71$ as the sloped dikes are usually located in relatively shallow water areas. The format of Figs. 5.8 to 5.10 is the same as that of Figs. 5.3 and 5.4.

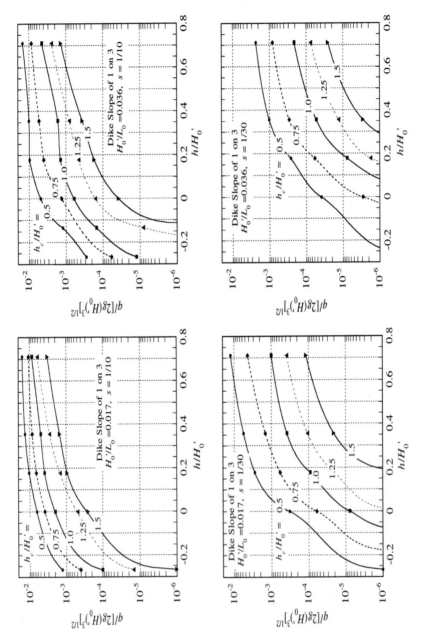

Fig. 5.8   Design diagrams of wave overtopping rate of plane slope with gradient of 1 on 3.[12]

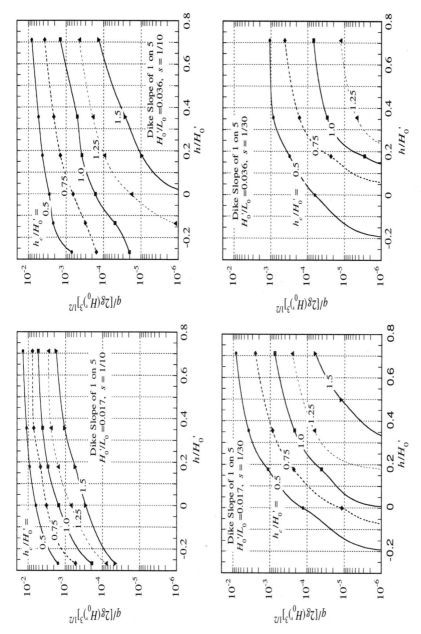

Fig. 5.9   Design diagrams of wave overtopping rate of plane slope with gradient of 1 on 5.[12]

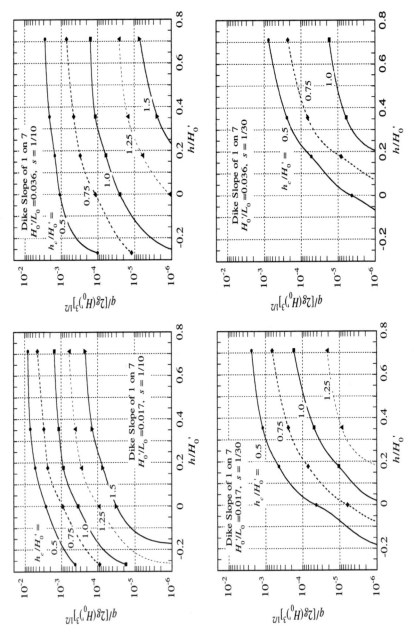

Fig. 5.10 Design diagrams of wave overtopping rate of plane slope with gradient of 1 on 7.[12]

The wave overtopping rate of plane slopes decreases as the crest is raised high in the manner similar as the case of vertical seawall, and it also decreases as the water depth in front of the slope becomes shallow within the range of depth-to-height ratio shown in the diagrams.

The plane slope with the gradient of 1 on 3 tends to yield the overtopping rate much larger than the case of vertical wall, probably because of higher wave run-up height. The plane slope with the gradient of 1 on 5 also tends to yield the overtopping rate larger than the case of vertical wall, but the difference becomes small for wind waves with the steepness of $H_0'/L_0 = 0.036$. In the case of the plane slope with the gradient of 1 on 7, the wave overtopping rate is less than the case of vertical wall. The decrease of wave topping rate with the decrease in the slope gradient is related with the phenomenon of lower wave run-up on a milder slope, as implied by Eq. (5.3); the run-up height is proportional to the slope gradient through the Iribarren number.

### 5.2.4 Unified Formulas for Wave Overtopping Rate of Vertical and Inclined Seawalls

Because of keen interests in the rational design of coastal dikes and seawalls, there have been many studies on the problem of wave overtopping, which have yielded various formulas for estimating wave overtopping rates. The *EurOtop Manual*[6] and the new *Rock Manual*[7] list these formulas and discuss their applications.

While the design diagrams of Figs. 5.3 to 5.10 employ the equivalent deepwater wave height $H_0'$ as the design parameter, most of the formulas developed in Europe employ the significant wave height at the toe of structure, i.e., $H_{1/3,\text{toe}}$. Because use of $H_0'$ as the design parameter is not well established outside Japan, the author has reanalyzed the original laboratory data employed for preparation of Figs. 5.3 to 5.7 and Figs. 5.8 to 5.10 for deriving the wave overtopping formulas using the local significant wave height $H_{1/3,\text{toe}}$. By making use of the large database of the CLASH project (De Rouck *et al.*[13,14] and van der Meer[15]), the author has come to propose the following formula for estimating the wave overtopping rate of coastal structures made of smooth planes including vertical and inclined seawalls[16]:

$$\frac{q}{\sqrt{g\,(H_{1/3})_{\text{toe}}^3}} \equiv q* = \exp\left\{-\left[A + B\frac{h_c}{(H_{1/3})_{\text{toe}}}\right]\right\}. \tag{5.9}$$

The symbol $q_*$ denotes the dimensionless wave overtopping rate. The symbols $A$ and $B$ are the intercept constant and the gradient constant, respectively. They are formulated as the functions of the depth-to-height ratio at the toe $h/H_{1/3,\text{toe}}$ and the seabed slope $s$ as in the following:

$$A = A_0 \, \tanh\{(0.956+4.44\,s) \times [h/(H_{1/3})_{\text{toe}}+1.242-2.032\,s^{0.25}]\}\,, \quad (5.10)$$

$$B = B_0 \, \tanh\{(0.822 - 2.22\,s) \times [(h/(H_{1/3})_{\text{toe}} + 0.578 + 2.22\,s]\}\,. \quad (5.11)$$

Equations (5.10) and (5.11) are applicable in the range of $0 \le h/(H_{1/3})_{\text{toe}} \le 6$. The constants $A_0$ and $B_0$ have been formulated as the function of the slope angle $\alpha_s$ as in the following:

$$A_0 = 3.4 - 0.734 \cot \alpha_s + 0.239 \cot^2 \alpha_s - 0.0162 \cot^3 \alpha_s\,, \quad (5.12)$$

$$B_0 = 2.3 - 0.50 \cot \alpha_s + 0.15 \cot^2 \alpha_s - 0.011 \cot^3 \alpha_s\,. \quad (5.13)$$

Equations (5.12) and (5.13) are applicable in the range of $0 \le \cot \alpha_s \le 7$. The both constants $A_0$ and $B_0$ take the minimum values around the slope of $\cot \alpha_s \simeq 2$, the condition at which the overtopping rate becomes largest.

Applications of Eqs. (5.12) and (5.13) include the case of vertical seawalls which are characterized with $\alpha_s = 90°$ and $\cot \alpha_s = 0$. Thus, the set of Eqs. (5.9) to (5.13) is universally applicable for both vertical and inclined seawalls regardless of the slope angle. To estimate the overtopping rate, one has to calculate the significant wave height at the location where a structure is going to be built. It can be done by using the simplified formula of Eq. (3.47) in Sec. 3.6.3(B) with the input of the equivalent deepwater wave height $H_0'$.

The accuracy of the new formula is shown in Fig. 5.11 in the form of the ratio of the measured to the estimated mean wave overtopping rates for the case of vertical walls. The legend such as DS-028 refers to the dataset number in the CLASH dataset, and the dataset DS-802 is that of Goda et al.[11] which include 34 data of seawalls located at the shoreline. In total, 715 data of laboratory measurements of mean wave overtopping rate are plotted there. The wave overtopping ratio $q*_{\text{meas}}/q*_{\text{est}}$ larger than 1.0 means an underestimation, while the ratio smaller than 1.0 means an overestimation. The scatted data of Fig. 5.11 suggest a slight tendency of underestimation when the geometric mean of the overtopping ratio is calculated.

The accuracy of predicting the overtopping rate decreases as the dimensionless overtopping rate becomes small. The decrease partly originates

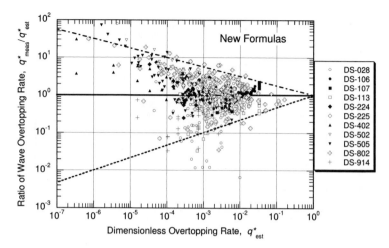

Fig. 5.11  Ratio of the measured to the estimated mean overtopping rates versus the dimensionless overtopping rate of vertical walls.[16]

from the nature of the random wave overtopping phenomenon. As the relative crest height becomes higher, wave overtopping is made by a smaller number of large waves, the occurrence probability of which is varied from a train of test waves to another. The two dashed lines for the upper and lower boundaries in Fig. 5.11 have been subjectively drawn by the author as the guidelines for the reliability of the estimated overtopping rate. They are expressed as below:

$$\text{Range} : \frac{q*_{\text{meas}}}{q*_{\text{est}}} = 1.0 \times q*_{\text{est}}^{1/3} \quad \leftrightarrow \quad 1.0 \times q*_{\text{est}}^{-1/4} . \tag{5.14}$$

The accuracy of the new formulas for the case of inclined seawalls with plane slopes is shown in Fig. 5.12. In total, 1254 data from the CLASH database are plotted there, having been classified with the gradient of slopes. The laboratory data of 198 cases by Tamada *et al.*[12] are not included in Fig. 5.12. The dispersion of the wave overtopping ratio is wider than the case of vertical walls. The guidelines for the upper and lower boundaries shown with dashed lines in Fig. 5.12 are expressed as below.

$$\text{Range} : \frac{q*_{\text{meas}}}{q*_{\text{est}}} = 1.0 \times q*_{\text{est}}^{2/5} \quad \leftrightarrow \quad 1.0 \times q*_{\text{est}}^{-1/3} . \tag{5.15}$$

Against the dataset of Tamada *et al.*,[12] the new formulas has a tendency of slight underestimation but with a narrower range of scatter than that in Fig. 5.12.

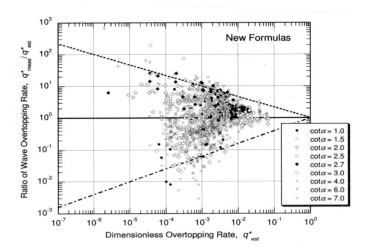

Fig. 5.12   Ratio of the measured to the estimated mean overtopping rates versus the dimensionless overtopping rate of inclined seawalls.[14]

The range of the spread of the wave overtopping ratio by the new set of the estimation formulas is wider than the range listed in Table 5.2 for the design diagrams shown in Figs. 5.3 to 5.7. It is because the CLASH database contains wide variations of the test conditions and the data acquisition and analysis methods, even though only the data with good reliability were employed in the formulation. It is suggested to refer to Figs. 5.3 to 5.7 when the design conditions are fit for applications, except for the range of $h/H_0' > 4$.

### Example 5.4

A seawall is planned at the depth of $h = 5.0$ m on a planar beach with the bed slope of $s = 1/50$ against the design wave with $H_0' = 4.0$ m and $T_{1/3} = 10.0$ s, which are normally incident to the coast. Estimate the mean rate of wave overtopping for both the cases of vertical seawall with $\cot \alpha_s = 0$ and inclined seawall with $\cot \alpha_s = 2.0$ when the crest elevation is set at $h_c = 5.0$ m.

**Solution** The design wave has the deepwater wave length of $L_0 = 156$ m and the steepness of $H_0'/L_0 = 0.0256$. Use of Eq. (3.47) yields the wave height at the toe as $(H_{1/3})_{\text{toe}} = 3.305$ m.

For the case of vertical seawall, the constants have the values of $A_0 = 3.4$ and $B_0 = 2.3$. By inserting $h/(H_{1/3})_{\text{toe}} = 5.0/3.305 = 1.380$ and $s = 0.02$ into Eqs. (5.10) and (5.11), the intercept and gradient constants are evaluated as $A = 3.296$ and $B = 2.140$. Then the mean overtopping rate is calculated as below.

$$q* = \exp[-(3.296 + 2.140 \times 5.0/3.305)] = 1.455 \times 10^{-3},$$

$$q_{est} = q * \times \sqrt{9.81 \times 3.305^3} = 0.0274\,\text{m}^3/\text{m/s}.$$

The range of variation of the overtopping rate is estimated with Eq. (5.14) as $q*_{meas}/q*_{est} = 0.113 \leftrightarrow 5.12$. By regarding $q*_{meas}$ as representative of the true overtopping rate, the overtopping rate of the vertical seawall is estimated with the following range of variation:

$$q_{est} = 0.0031 \leftarrow 0.027 \rightarrow 0.14\,\text{m}^3/\text{m/s}.$$

For the case of inclined seawall with $\cot \alpha_s = 2.0$, the constants have the values of $A_0 = 2.758$ and $B_0 = 1.812$ according to Eqs. (5.12) and (5.13). Thus, the intercept and gradient constants are evaluated as $A = 2.674$ and $B = 1.686$, and the mean overtopping rate is calculated as below.

$$q* = \exp[-(2.674 + 1.680 \times 5.0/3.305)] = 5.39 \times 10^{-3},$$

$$q_{est} = q * \times \sqrt{9.81 \times 3.305^3} = 0.101\,\text{m}^3/\text{s/m}.$$

The range of variation is estimated with Eq. 5.15 as $q*_{meas}/q*_{est} = 0.123 \leftrightarrow 5.70$. Thus, the overtopping rate of the inclined seawall is estimated with the following range of variation:

$$q_{est} = 0.0125 \leftarrow 0.101 \rightarrow 0.58\,\text{m}^3/\text{s/m}.$$

The inclined seawall with $\cot \alpha_s = 2.0$ is expected to experience the mean wave overtopping rate about four times that of the vertical wall for this calculation case.

## 5.3 Influence of Various Factors on Wave Overtopping Rate

### (A) Geometry of coastal dikes and seawalls

Wave overtopping is affected by many factors. Even a small modification of the geometry of a seawall may change the amount of wave overtopping. For example, wave overtopping can be reduced to zero by redesigning the parapet into a curved shape, if the seabed conditions and the frontal shape of the seawall are suitable, and if no wind is blowing from the sea. The wave overtopping rate of a block-mound seawall can be reduced to some extent by replacing the entire volume of rubble stones in the underlayer and the core of the mound with concrete wave-dissipating blocks such as Tetrapods. The crown width of the block mound also affects the wave overtopping rate. The shape of the foot-protection mound is another factor, but its influence is too complex for allowing anyone to draw usable guidelines. In comparison with a vertical seawall, a sloped seawall with a smooth surface, which is one of the typical coastal dikes, usually has a larger rate of wave overtopping as demonstrated in Example 5.4.

Thus, the wave overtopping rate of a seawall to be designed for a specific situation is best determined through hydraulic model tests. Figures 5.3 to 5.10 and Eqs. (5.9) to (5.17) are to be used for estimation of the order of magnitude of the overtopping rate when a conceptual design of a coastal dike and/or seawall is made. The basic and/or detailed design should be made with more reliable information on the wave overtopping rate based on hydraulic model tests using random waves. In case that an irregular wave flume is not available, regular wave tests may suffice for the purpose, provided that the expected rate of overtopping is estimated from the test results by means of Eq. (5.8). Use of regular wave tests may even be recommended for detailed analysis of the wave overtopping phenomenon as has been done by Daemrich *et al.*[17]

### (B) *Oblique wave incidence*

Waves of oblique incidence produce lower wave run-up compared with waves of normal incidence as indicated by Eq. (5.4). Oblique wave incidence also reduces the wave overtopping rate according to various laboratory test results. The *EurOtop Manual*[6] and the new *Rock Manual*[7] cite the influence factor of the following form, which should be multiplied to the incident wave height $(H_{1/3})_{\text{toe}}$ in an estimation formula of exponential type similar as that of Eq. (5.9):

$$\gamma_\beta = 1 - 0.0033\,|\beta(^\circ)| \ : \ 0^\circ \le |\beta| \le 80^\circ, \tag{5.16}$$

where $\beta$ is the wave incident angle to the structure.

Hiraishi *et al.*[18] have also proposed the following influence factor based on unidirectional random wave tests with the mean dimensionless overtopping rate in the order of $10^{-3}$ in the range of the depth-to-height ratio of $h/H_0' = 3.5 \sim 4.6$:

$$\gamma_\beta = \max\{0.75, \ [1 - \sin^2 \beta]\}. \tag{5.17}$$

The influence factor $\gamma_\beta$ decreases as the incident angle increases up to 30° and is given a constant value of 0.75 for $\gamma_\beta \geq 30°$. Hiraishi *et al.* also reports that multidirectionality of random waves reduces the wave overtopping rate by 30% for the case of normal incidence. The effect of multidirectionality is lessened, however, as the incident angle increases. At the incident angle of 30°, multidirectional waves produce the overtopping rate almost same as that of unidirectional random waves.

The effect of wave multidirectionality on overtopping rate has further been studied by Tomita *et al.*[19] with a 3-D scale model of a seawall protected with a mound of wave-dissipating blocks and by Ohno *et al.*[20] with a 2-D vertical wall in water of uniform depth. Both studies confirmed a certain amount of overtopping rate decrease by multidirectional waves, but the decrease of overtopping rate by oblique incidence was not so clear in the case of multidirectional waves. Around a corner of a L-shaped seawall layout, there were such cases that obliquely incident multidirectional waves produced a larger amount of wave overtopping than the multidirectional waves of normal incidence.

As discussed in Sec. 5.2.4, an estimate of wave overtopping rate of coastal structures is associated with a wide range of uncertainty. Because the effect of oblique wave incidence on overtopping rate is not so large, it is better to disregard the effect at the planning stage of a coastal structure. For an important project such as a large-scale reclamation, a 3-D hydraulic model tests using multidirectional waves should be carried out for obtaining reliable information on the spatial and temporal distribution of the amount of overtopping water, which is required for design of water drainage system behind the seawall.

### (C) *Wind actions*

Strong onshore winds accelerate wave overtopping and blow spray further inland. The wind effect is pronounced when the wave overtopping rate is small. The first quantitative field measurement on the effect of wind actions on the onshore transport of overtopped water was carried out by Fukuda *et al.*[21] in 1972. They presented a diagram of spatial distribution of overtopped water blown by winds. Yamashiro *et al.*[22] have reanalyzed their data, made additional field and laboratory measurements, and proposed an empirical formula for evaluation of wind effect on overtopping rate. Yamashiro *et al.* recommend to reduce the wind speed in a wind-wave flume to one-third of the prototype when executing wave overtopping tests in a laboratory.

Pullen *et al.*[23,24] measured the spatial distribution of water overtopped the seawall and blown by winds at Samphire Hoe, which is a land area reclaimed with chalk marl excavated from the Channel Tunnel. They have proposed the following formula for spatial distribution of overtopping discharge:

$$q(x)dx = q_{\text{all}} \times \frac{k}{L_0} \exp\left[-\frac{kx}{L_0}\right] dx \; : \; k = 29 \exp[-0.03V_w], \qquad (5.18)$$

where $q(x)\,dx$ denotes the mean overtopping discharge measured at the area between $x$ and $d + dx$ with $x$ being the distance from the crest of seawall, $q_{\text{all}}$ stands for the total rate of wave overtopping over the seawall crest, $L_0$ is the deepwater wavelength, and $V_w$ designates the wind velocity in the units of m/s.

Equation (5.18) predicts that 90% of overtopped water will fall within the area up to the distance $0.08L_0$ under no wind and $0.19L_0$ at $V_w = 30$ m/s. For waves with the period of $T = 12$ s and the wavelength of $L_0 = 225$ m, the distance is equivalent to 18 m and 43 m for $V_w = 0$ and $30\,\text{m/s}$, respectively. It has been observed at several reclaimed lands in Japan that a mass of overtopped water jumps over 30 m behind the seawall when overtopping is heavy. Thus, Eq. (5.18) seems to be applicable when the wave overtopping is not so severe.

Pullen *et al.* has also presented the following empirical formula for the increase of overtopping rate by wind actions:

$$q_{\text{wind}} = q_{\text{no-wind}}$$

$$\times \min\left\{4, 0, \left[1.0 + 3.0\left(\frac{-\log q - 2}{3}\right)^2\right]\right\} \; : \; q < 10^{-2}\text{m}^3/\text{s/m}.$$

$$(5.19)$$

Equation (5.19) assumes that the wind effect appears only when the wave overtopping is not severe with the rate being less than 0.01 m³/s/m and the maximum amplification is four times when the wave overtopping is less than $10^{-5}$ m³/s/m.

Both Eqs. (5.18) and (5.19) are based on a limited cases of field measurements and their applicability may not be universal. Nevertheless, they will provide a kind of guidelines on the effect of wind actions on wave overtopping.

(D) *Scale effects in hydraulic model tests*

Hydraulic model tests are indispensable tools for obtaining reliable information on wave overtopping characteristics of a coastal dike and/or seawall under design. Design diagrams and formulas serve only for the order-of-magnitude estimation of wave overtopping rate. A problem inherent in wave overtopping tests is the scale effect that model tests tend to underestimate the overtopping rate when the amount of overtopped water is small.

Sakakiyama and Kajima,[25] for example, carried out field measurements of wave overtopping at the reclamation seawall covered with Tetrapods at Fukui Port and obtained the mean rate of $10^{-5} \sim 10^{-3}$ m$^3$/s/m for waves with $H_{1/3} = 2.1 \sim 6.1$ m. When they tried to reproduce the wave overtopping in a 3-D model including the surrounding bathymetry, the model yielded the overtopping rate equal to or slightly larger than the prototype for waves with the height greater than 5 m. However, the dimensionless wave overtopping rate of the model for waves with the height of 2.8 m was two orders smaller than the prototype. They have proposed the threshold Reynolds number of $(\mathbf{R_e})_{\text{crit}} = 10^5$ below which the scale effect with underestimation will appear; the Reynolds number is defined as $\mathbf{R_e} = D_n U / \nu$, where the nominal diameter of concrete block is given as $D_n = (M/\rho_r)^{1/3}$ and the representative velocity is given by $U = (gH_{1/3})^{1/2}$.

Kortenhaus *et al.*[26] also reports the results of comparison between the field overtopping measurements at the mound breakwater of Zeerbrugge Port (De Rouck *et al.*[27]) and the hydraulic model tests with the linear scale of 1/30. Model tests produced the overtopping rate being generally less than the prototype and no overtopping for the high crest cases of $h_c/H_{m0} > 2$, even though wave overtopping actually occurred at such cases in the field. They express their suspect on the presence of scale effects. De Rouck *et al.*[13] also discusses the scale effect problem on wave overtopping at a mound breakwater.

The above scale effects observed in wave-dissipating seawalls and mound breakwaters seem to have originated from two sources, i.e., the roughness resistance due to viscosity and the surface tension of water. The problem of scale effects with smooth-surface slopes and vertical walls has not been investigated yet and no information is available. Nevertheless, when a hydraulic model test of a coastal dike and/or seawall is contemplated, due attention should be paid for the possibility of scale effects on wave overtopping phenomenon.

## 5.4 Tolerable Rate of Wave Overtopping and Determination of Crest Elevation

### 5.4.1 *Design Principles for the Determination of Crest Elevation*

Planning of a coastal dike/seawall requires comparative designs of several structural types of coastal dikes or seawalls. For each design, the crest elevation must be determined according to the wave run-up or overtopping characteristics. The crest elevation should be specified in order to give a height above the design storm tide sufficient to prevent wave overtopping during the attack of design storm waves. The determination of the design storm tide and design waves is a difficult and intricate problem in itself. The statistical method to derive design values for a given return period is discussed in Chapter 13.

The crest elevation of a seawall is determined by one of the following two principles. One is to take the wave run-up height as the reference and to set the crest of the seawall higher than the run-up height so that no wave overtopping will occur. The other is to take the wave overtopping amount as the reference and to set the crest of the seawall at such a height as to keep the overtopping below some maximum tolerable quantity.

An unknown factor entering into the first method is the problem of wave irregularity; that is, which characteristic wave height, e.g., $H_{1/3}$, $H_{1/10}$, or $H_{\max}$, should be used in estimating the run-up height? Even when the information of random wave run-up height is available, the same question is asked: which characteristic run-up height, $\bar{R}$, $R_{1/3}$, or $R_{2\%'}$, for example, should be considered in seawall design? For a seawall to be built on land with a ground elevation higher than the design storm tide, its crest elevation may be set higher than the maximum run-up height of random waves without much difficulty, because the maximum run-up is not too large. For a seawall to be built in the sea, however, the same procedure cannot be adopted because the maximum run-up height is very large. Even for the planning of a seawall to be built on land, the same problem may be encountered if the beach slope is very steep.

As an example of difficulty in designing seawalls with the run-up criteria, a case of a calamity is cited here. When Typhoon No. 6626 attacked the Yoshihara Coast along the Bay of Suruga in Japan in 1966, thirteen people lost their lives behind the coastal dike. Although the coast was protected by a dike with an elevation of 13 m above the mean sea level, storm waves overran the dike and flooded the area behind it. The situation was later examined through a hydraulic model test (Tominaga *et al.*[28]), and the waves which had caused such calamity were estimated to have had $H_{1/3} \simeq 11$ m, $H_{\max} \simeq 20$ m, and $T_{1/3} \simeq 20$ s. The maximum run-up height for this wave condition would have been greater than 20 m above the mean sea level, though this was not confirmed in the test.

Another problem inherent to the principle of runup-based seawall design is the possible danger of incurring excessive damage in the hinterland upon the attack of the design storm. People may become overconfident because little wave overtopping occurs under normal conditions, and they utilize the land directly behind the seawall without providing a drainage system capable of dealing with possible overtopping water. However, the magnitudes of natural phenomena such as storm surge and storm waves are most difficult to assess as they occur very infrequently. There always remains some probability that an extraordinary storm may hit the design site with a magnitude in excess of the design level. Therefore, the concept that a seawall can completely stop the overtopping of storm waves at any time may lead to a coastal disaster in the event of the attack of an extraordinary storm.

The problem with applying the principle of overtopping-based seawall design is how to determine the tolerance limit of wave overtopping amount. Nevertheless, at least in this method the designer is aware of the existence of an overtopped water mass behind the seawall and is prepared to deal with it. Thus, the extent of damage even in the event of an extraordinary storm will be much less than in the case of the runup-based seawall design. With much information is available on the wave overtopping rate nowadays, coastal dikes and seawalls are to be designed with the basis of the tolerable rate of wave overtopping. The performance of a coastal dike/seawall under design should be confirmed by hydraulic model tests using random waves.

### 5.4.2 *Tolerable Rate of Wave Overtopping*

The amount of wave overtopping tolerable with regard to the structural integrity of a seawall is greater than the tolerable amount for the protection of the land behind the seawall. In this section, the term of seawall is used in a broader sense inclusive of coastal dikes. Even a seawall constructed with utmost care cannot escape the danger of structural failure if it is exposed to heavy wave overtopping for many hours. The mode of failure may be loss of earth-fill from the core of the seawall by leakage, cracks and breakage

Table 5.3   Various proposals for tolerable limits of wave overtopping rate.

| Type | Surface Armoring and Others | Overtopping rate $(m^3/s/m)$ | References |
|------|------------------------------|------------------------------|------------|
| Vertical seawall | Pavement on ground | 0.2 | Goda[10] |
| | No pavement on ground | 0.05 | ditto |
| Inclined seawall | Concrete on front slope, crown and back slope | 0.05 | Goda[10] |
| | Concrete on front slope and crown, with soil on back slope | 0.02 | Ditto |
| | Concrete on front slope, with soil on crown and back slope | <0.005 | Ditto |
| Pavement | Interlocking blocks | 0.01 | Endo et al.[29] |
| Vehicles | Stop or low speed | 0.01 ∼ 0.05 | Allsop et al.[30] |
| | Driving or damage | $1.1 \times 10^{-5}$ | Kimura et al.[31] |
| Pedestrians | Watchfull of waves | $10^{-4}$ | Allsop et al.[30] |
| | Careless ones | $3 \times 10^{-5}$ | Ditto |
| Buildings | Structural damage | $3 \times 10^{-5}$ | Allsop et al.[30] |
| | No damage | $10^{-6}$ | Ditto |

of the armoring surfaces of the crown and back slope, or total collapse. The structurally tolerable limit of wave overtopping depends on the type of seawall structure.

The author analyzed about 30 cases of vertical and inclined seawalls damaged in the aftermath of typhoons, estimating the wave overtopping rate for each seawall.[9,10] This yielded an estimate of the maximum tolerable limit of wave overtopping rate as listed in Table 5.3, which includes many more tolerable limits proposed in several recent studies for various types of usage behind seawalls.

The tolerable limits of wave overtopping rate for vertical and inclined seawalls in Table 5.3, proposed by the author,[10] are based on the damage caused by storm waves with the height of a few meters and a duration of a few hours only upon structures located within bay areas. It is believed that the tolerable limit should be lowered for seawalls facing the ocean and exposed to the attack of large waves, or for seawalls subject to many hours of storm wave actions. For example, Fukuda et al.[21] reported that an inclined seawall with concrete surfacing at the front slope, crown, and back slope at Himekawa Port collapsed at the estimated overtopping rate of $q_{EXP} = 0.015\,\mathrm{m^3/s/m}$. The seawall was struck by storm waves of $H_0' = 6$ m generated in the Sea of Japan by a slowly moving extratropical cyclone.[21]

They also reported another example of a sloped seawall on the Niigata Coast which lost part of its sand fill due to wave suction, and some of the paving concrete blocks at the crown slumped. The wave overtopping was estimated as having the rate of only $q_{EXP} = 0.002$ m$^3$/s/m.

The tolerable limits of wave overtopping rate vary over a wide range depending on the viewpoint of structural safety and utilization of the area behind a seawall. The limits also reflect the subjective judgment of researchers who have proposed them. For the protection of a relatively densely populated coastal area, the overtopping rate of $q = 0.01$ m$^3$/s/m is currently adopted as a guideline in port areas in Japan. This guideline evolved from the previous coastal protection policy based on run-up height.

If the safe passage of vehicles is to be secured at all times along a coastal highway protected by a continuous seawall, the tolerable limit seems to be on the order of $10^{-5}$ m$^3$/s/m.[31] The overtopping rate is not the field measurement value but the one estimated using Fig. 5.6. The reliable measurement of such a small wave overtopping rate is rather difficult in laboratory tests due to limitations in the accuracy of the measurement as well as due to the inherent variability of the wave overtopping phenomenon. If a seawall is designed to have such a small overtopping amount, the crest elevation should be determined with due consideration for the uncertainty in estimation of wave overtopping rate.

In any case, the tolerable limit for the wave overtopping rate needs to be set by consideration of not only technical aspects but also many other factors. It is regarded as a constraint attached to the design of seawalls.

### 5.4.3 *Examples of Determining Crest Elevation of Seawalls*

Examples will be given of the decision process for the crest elevation as determined by the criterion of tolerable overtopping rate. By arbitrarily setting the tolerable limit at $q = 0.01$ m$^3$/s/m, the required crest elevation for a vertical seawall is estimated from Figs. 5.3 and 5.4, and the result is shown in Fig. 5.13 for wind waves with steepness of $H_0'/L_0 \simeq 0.036$. The crest elevation of a block-mound seawall is shown in Fig. 5.14. The crest elevation is displayed as a ratio formed with the equivalent deepwater wave height. The solid lines are for a sea bottom slope of 1/30 and the dash lines for a slope of 1/10. It will be observed in these figures that the relative crest elevation must be increased for larger waves in order to maintain the overtopping rate below a fixed value. Furthermore, it is seen that the required crest elevation for wind waves is highest in an area where the effective

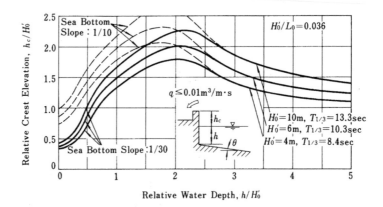

Fig. 5.13　Crest elevation of vertical seawall for the condition of overtopping rate not greater than 0.01 m$^3$/m/s.[11]

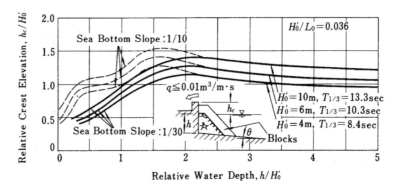

Fig. 5.14　Crest elevation of block-mound seawall for the condition of overtopping rate not greater than 0.01 m$^3$/m/s.[11]

water depth is 1.5 to 2.5 times the equivalent deepwater wave height and that the effect of bottom slope is greatest in an area of shallow water. Similar diagrams for the required crest elevation for different values of the tolerable wave overtopping rate or for incident waves of lesser steepness such as swell can be prepared from Figs. 5.3, 5.4, 5.6 and 5.7.

By comparing the curves in Figs. 5.13 and 5.14, it can be concluded that the required crest elevation of a seawall protected by a concrete block mound of the wave-dissipating type is about 60% to 70% of that of a vertical seawall for wind waves. Similar comparisons have been made for other wave steepnesses, and the results are shown in Figs. 5.15 and 5.16. Although the tolerable overtopping rate (dimensionless) varies in the range of

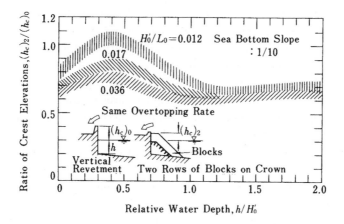

Fig. 5.15 Ratio of crest elevations of block-mounds to vertical seawalls for the same overtopping rate (sea bottom slope of 1/10).[11]

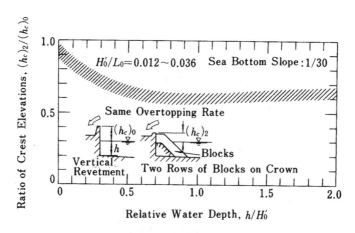

Fig. 5.16 Ratio of crest elevations of block-mounds to vertical seawalls for the same overtopping rate (sea bottom slope of 1/30).[11]

$q/\sqrt{2g(H_0')^3} = 5 \times 10^{-5} \sim 2 \times 10^{-3}$, no appreciable difference is observed in the ratio of the required crest elevation of a block-mound seawall to that of a vertical seawall.

It should be noted in Fig. 5.15 that the placement of a mound of concrete wave-dissipating blocks in front of a vertical wall causes a rise in the required crest elevation when the seawall is designed against swell of low steepness incident to a coast with a bottom slope of 1/10. This may be understood as follows: the effect of the rough, porous surface of the block mound in reducing wave run-up height has been superseded by the effect of the slope in enhancing wave run-up in this particular situation, especially because the block mound is small compared to the wavelength of the incident waves. As previously discussed, the amount of wave overtopping is quite sensitive to the geometry and material making up the seawall, and the above result cannot be considered as general. But it will be necessary to check the design section of a seawall through the aid of hydraulic model tests with irregular waves, when the seawall is to be located at a water depth of about 0.4 times the equivalent deepwater wave height on a coast of steep slope.

The effect of the crown width of a block-mound seawall upon the overtopping rate has been tested with irregular model waves for a few cases. Figure 5.17 illustrates the results of such tests, conducted at a

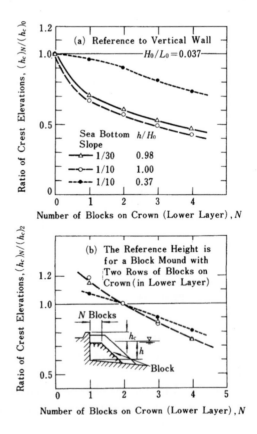

Fig. 5.17   Effect of crown width of block mound on design crest elevation.[32]

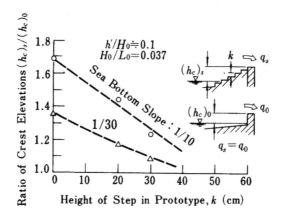

Fig. 5.18   Ratio of crest elevations of sloped seawall to vertical seawall.[32]

depth-to-height ratio of $h/H_0' = 0.98$ on a slope of $1/30$, and at $h/H_0' = 0.37$ and $1.00$ on a slope of $1/10$. The crown width of a block-mound seawall was varied by using one to four rows of blocks in the lower layer of a two-layer arrangement. The upper figure shows the ratio of required elevations of the block-mound to the vertical seawall. The lower figure shows the ratio of the required crest elevations of the block-mound seawall to that of a seawall having two rows of concrete blocks in the lower layer. The range of wave overtopping rate (dimensionless) considered is $q/\sqrt{2g(H_0')^3} = 2 \times 10^{-4} \sim 2 \times 10^{-3}$. As seen in Fig. 5.17, the required crest elevation decreases with widening of the crown of the block mound.

Figure 5.18 compares seawalls having a front slope of $1:2$ with vertical seawalls (Goda and Kishira[32]). The model seawalls were set near the shoreline and the wave overtopping rate was measured for several crest elevations. From the resultant data on the overtopping rate, the design crest elevations of both types of seawall for a given tolerable overtopping rate were estimated and compared. As seen in Fig. 5.18, an inclined seawall requires a higher crest elevation than a vertical seawall for a given tolerable overtopping rate. The replacement of a smooth slope with a stepped one can reduce the required crest elevation to the extent that the crest elevation will be only 10% to 20% greater than that of a vertical seawall, if the height of each step is about 30 cm for an incident significant wave height of about 5 m.

In the practical design of a seawall, the crest elevation as calculated from the above considerations of wave run-up and overtopping is usually given a small amount of additional rise to allow for unknown factors. If some ground settlement or subsidence is expected owing to the foundation characteristics, the estimated amount of settlement must be taken into account in the determination of the final crest elevation.

## 5.5   Additional Design Problems Related to Seawalls

(A) *Temporal fluctuation of wave overtopping quantity*

The quantity of overtopped water by random waves varies greatly from wave to wave. The data on the overtopping rate presented in Figs. 5.3 to 5.9 and estimated using Eqs. (5.9) to (5.17) pertain to an average over several hundred waves; i.e., an average over a duration of 30 minutes to one hour. In a shorter time interval, a much larger amount of wave overtopping than the overall mean would occur.

Inoue *et al.*[33] have measured the overtopping quantities by individual waves in irregular trains for a vertical wall on a steep slope of 1/10. They reported that the largest quantity per wave varies between five to ten times the average quantity and may reach to twenty times in some cases. Similar observation was also made by Sekimoto *et al.*,[34] who carried out a series of scale model tests of deepwater seawalls.

A unique field measurement of wave run-up flow over a wide crest berm having a slope of 7% was carried out by Izumiya *et al.*[35] at the block-mound seawall beneath the coastal viaduct along the Oyashirazu Coast in Toyama Prefecture, Japan, in 2003 to 2004. The seawall has the crest elevation of 3.1 m above the mean water level and the frontal water depth of about 4.9 m, and the significant wave height during the measurement period was $1.0 \sim 3.5$ m. By measuring the flow speed of water running over the wide crest berm using the radio-wave current meter of non-contact type and its thickness using the acoustic wave gauge of air-emission type, they assessed the overtopping quantities of individual waves. The distribution of individual overtopping quantity, when the data of zero quantity is excluded, approximately followed the exponential type for the range of $q_{\mathrm{indiv.}} = 0 \sim 0.1$ m$^3$/s/m. The 1% and 2% exceedance values were about six and five times the average overtopping rate, respectively.

The temporal variation of wave overtopping rate is related with the phenomenon of wave grouping, i.e., appearance of high waves in succession (refer to Sec. 10.2). Kimura *et al.*[36] have theoretically discussed the effect of wave grouping on the short-term fluctuation of wave overtopping quantity.

The temporal fluctuation of wave overtopping quantity such as the maximum quantity directly affects the safety of facilities behind the seawall. The drainage system for overtopped water must be designed by taking into account of temporal and spatial variation of overtopping water. Though no specific guideline is available persently, engineers should pay due caution for this aspect.

### (B) *Facilities for draining overtopped water*

In the design of seawalls, the drainage system for the overtopped water should be well planned, because the overtopping amount by storm waves is quite large. In the preceding section, the tolerable rate of wave overtopping was arbitrarily set at $q = 0.01$ m$^3$/s/m as an example of the determination of the crest elevation. This amount is equivalent to a discharge of 10 m$^3$/s for every 1000 m of seawall extension.

Tonomo et al.[37] have made a multidirectional random wave model test for a vertical seawall along a reclaimed land of 800 m long in water of 20 m deep, reproduced with a linear scale of 1/100. The seawall was given the crest elevation of $h_c = 11.0$ m against the waves with $H_{1/3} = 9.0$ m and $T_{1/3} = 15.0$ s. A drainage channel of 30 m wide was provided behind the seawall with a gradient of 1/500 toward the outfall. Measurements were made for the temporal and spatial overtopping quantity along the seawall as well as the flow depth in the drainage channel. When unidirectional random waves were normally incident to the seawall, the wave overtopping rate varied between $0.05 \sim 0.20$ m$^3$/s/m depending on the location, and the maximum flow depth was about 4 m. When the incident angle was 30°, the overtopping rate was $0.02 \sim 0.12$ m$^3$/s/m and the maximum flow depth was about 3 m. With introduction of multidirectional waves, the flow depth decreased by about 50%.

The model test is a good example for the necessity of a large discharge channel behind a seawall. The channel needs to have a sufficient width and its floor must be well protected against the impact of the falling water mass immediately behind the seawall. The overtopped water would then flow down the channel to the sea at appropriate outlets. The required channel width may be a few tens of meters if the overtopping rate is on the order of 0.01 m$^3$/s/m. If the ground elevation behind the seawall is lower than the storm tide level or if the hinter area is too extensively utilized to allow provision of a wide outlet, then the rate of wave overtopping should be reduced to the order of $10^{-3}$ m$^3$/s/m or less.

(C) *Countermeasures against leakage of fill material and scouring of seabed*

Structural details of seawall designs may be found elsewhere, e.g., Ref. 38. In Japan, the technical standards for design and construction of coastal protection facilities were officially established in 1958 and have undergone subsequent revisions. From the perspective of vulnerability to wave attack, precautions should be taken against scouring of the seabed in front of a seawall and leakage of filling material in the core of an inclined seawall or in the rear of a vertical seawall. Unfortunately, both factors are difficult to control, especially on a sandy coast. Nevertheless, frequent inspections and repairs should be made if necessary, so as to maintain the structural strength of a seawall at a good level after construction. The design of the seawall itself should provide maximum possible protection against both scouring and leakage.

When a vertical seawall is constructed of concrete caissons or L-shaped concrete blocks, gaps between the caissons or L-shaped blocks create a zone of strong oscillatory water flow. The filler material in the rear of the front wall may be washed away by the water flow through such gaps. Plastic sheets and plates are sometimes applied to the rear of the caissons or L-shaped blocks in order to shield the gap, but they are easily torn off or become dislocated by strong wave action in coastal areas subject to high waves, before they perform their intended function. In such cases, two or three layers of rubble and gravel filters with varying grain diameters should be provided just behind the front wall, so that the strength of the oscillatory flow is weakened gradually and no appreciable washing of filler material takes place.

## References

1. Y. Goda and S. Kakizaki, "Study on finite amplitude standing waves and their pressures upon a vertical wall," *Rept. Port and Harbour Res. Inst.* **5**(10) (1966), 57p. (*in Japanese*) or *Coastal Engineering in Japan* **10** (1967), pp. 1–11.
2. K. Kimura, K. Yasuda, Y. Yamamoto, N. Umezawa, T. Shimizu and T. Sato, "Traffic hindrance on coastal road by wave overtopping and its countermeasures," *Proc. Coastal Engrg., JSCE* **48** (2001), pp. 756–760 (*in Japanese*).
3. H. Mase, "Random wave runup height on gentle slope," *J. Waterway, Port, Coastal, and Ocean Engrg.* **115**(5) (1989), pp. 649–661.

4. H. Mase, A. Miyahira, H. Sakurai and M. Inoue, "Radom wave runup on seawall with shallow foreshore — Relation between prediction and representative runup heights," *J. Hydraulic, Coastal and Environmental Engrg. JSCE* No. 726 (II-62) (2003), pp. 99–107 (*in Japanese*).

5. International Standardization Organization, "*Actions from Waves and Currents on Coastal Structures,*" ISO 21650 (2007), Sec. F.1.1.

6. EA (UK); ENW (NL); KFKI (DE), "*EurOtop Wave Overtopping of Sea Defenses and Related Structures — Assessment Manual*" (June 2007), Section 5.2.1.

7. CIRIA; CUR; CETMEF, "*The Rock Manual. The use of rock in hydraulic engineering* (2nd edition)" C683 (CIRIA, London) (2007), Sec. 5.1.2.

8. M. R. A. Van Gent, "Wave runup on dikes with shallow foreshores," *J. Waterway, Port, Coastal, and Ocean Eng.*", **127**(5) (1989), pp. 254–262.

9. S. Tsuruta and Y. Goda, "Expected discharge of irregular wave overtopping," *Proc. 11th Int. Conf. Coastal Engrg.* (London, ASCE) (1968), pp. 833–852.

10. Y. Goda, "Estimation of the rate of irregular wave overtopping of seawalls," *Rept. Port and Harbour Res. Inst.* **9**(4) (1970), pp. 3–41 (*in Japanese*).

11. Y. Goda, Y. Kishira and Y. Kamiyama, "Laboratory investigation on the overtopping rate of seawalls by irregular waves," *Rept. Port and Harbour Res. Inst.* **14**(4) (1975), pp. 3-44 (*in Japanese*).

12. T. Tamada, M. Inoue and T. Tezuka, "Experimental studies on diagrams of wave overtopping rate on gentle slope-type seawalls and their reduction effect on wave topping," *Proc. Coastal Engrg., JSCE* **49** (2002), pp. 641–645 (*in Japanese*).

13. J. De Rouck, J. Geeraerts, P. Troch, A., Kortenhasu, T. Pullen and L. Franco, "New results on scale effects for wave overtopping at coastal structures," *Coastlines, Structures and Breakwaters 2005 (Proc. Conf.,* Inst. Civil Engrs., Thomas Telford) (2006), pp. 29–43.

14. J. De Rouck, H. Verhaeghe and J. Geeraerts, "Crest level assessment of coastal structures — General overview," *Coastal Engineering* **56**(2) (2009), pp. 99–107.

15. J. W. van der Meer, H. Verhagen and G. J. Steendam, "The new wave overtopping database for coastal structures," *Coastal Engineering* **56**(2) (2009), pp. 108–120.

16. Y. Goda, "Derivation of unified wave overtopping formulas for seawalls with smooth, impermeable surfaces based on selected CLASH datasets," *Coastal Engineering,* **56** (2009), pp. 385–399.

17. K.-F. Daemrich, J. Meyering, N. Ohle and C. Zimmerrmann, "Irregular wave overtopping at vertical walls — Learning from regular wave tests," *Coastal Engineering 2006 (Proc. 30th Int. Conf.,* San Diego, World Scientific) (2007), pp. 4740–4752.

18. T. Hiraishi, N. Mochizuki, S. Sato, H. Maruyama, T. Kanazawa and T. Masumoto, "Effect of wave directionality on overtopping at seawall," *Rept. Port and Harbour Res. Inst.* **35**(1) (1996), pp. 39–64 (*in Japanese*).

19. T. Tomita, T. Kawai and T. Hiraishi, "Wave overtopping on wave-dissipating seawall with corner part in obliquely unidirectional and multidirectional waves," *Tech. Note, Port and Harbour Res. Inst.* No. 968 (2000), pp. 1–22 (*in Japanese*).

20. K. Ohno, Y. Matsumi, R. Takeda, M. Tsukamoto and A. Kimura, "Characteristics of spatial distribution of wave overtopping rate along sea wall in multidirectional waves," *Annual J. Coastal Engrg., JSCE* **51** (2004), pp. 631–635 (*in Japanese*).

21. N. Fukuda, T. Uno and I. Irie, "Field measurements of wave overtopping of seawalls (2nd Report)," *Proc. 20th Japanese Conf. Coastal Engrg.* (1973), pp. 113–118 (*in Japanese*).

22. M. Yamashiro, A. Yoshida, H. Hashimoto and I. Irie, "Conversion ratio of wind velocity from prototype to experimental model on wave-overtopping," *Coastal Engineering 2006 (Proc. 30th Int. Conf.,* San Diego, World Scientific) (2007), pp. 4753–4765.

23. T. Pullen, W. Allsop and T. Bruce, "Wave overtopping at vertical sealls: Field and laboratory measurements of spatial distributions," *Coastal Engineering 2006 (Proc. 30th Int. Conf.,* San Diego, World Scientific) (2007), pp. 4702–4713.

24. T. Pullen, W. Allsop, T. Bruce and J. Pearson, "Field and laboratory measurements of mean overtopping discharges and spatial distributions at vertical seawalls," *Coastal Engineering* **56**(2) (2009), pp. 121–140.

25. T. Sakakiyama and R. Kajima, "Comparison of field measurements and hydraulic model tests of wave overtopping at a wave-dissipating seawall," *Proc. Coastal Engrg., JSCE* **44** (1997), pp. 736–740 (*in Japanese*).

26. A. Kortenhaus, H. Oumeraci, J. Geeraerts, J. De Rouck, J. R. Medina and J. A. González-Escrivá, "Laboratory effects and further uncertainties associated with wave overtopping measurements," *Coastal Engineering 2004 (Proc. 29th Int. Conf.,* Lisbon, World Scientific) (2005), pp. 4456–4468.

27. P. Troch, J. Geeraerts, B. Van de Walle, J. De Rouck, L. Van Damme, W. Allsop and L. Franco, "Full-scale wave overtopping measurements on the Zeerbrugge rubble mound breakwater," *Coastal Engineering* **51** (2004), pp. 609–628.

28. M. Tominaga, H. Hashimoto and T. Nakamura, "On the disaster at Yoshihara Coast caused by Typhoon No. 6626," *Proc. 14th Japanese Conf. Coastal Engrg.* (1967), pp. 206–213 (*in Japanese*).

29. T. Endo, K. Kimura, S. Kikuchi and K. Sudo, "Durability of pavement behind a seawall against wave actions and tolerable wave overtopping rate," *Proc. Coastal Engrg., JSCE* **42** (1995), pp. 1276–1280 (*in Japanese*).

30. N. W. H. Allsop, L. Franco, G. Belloti, T. Bruce and J. Geeraerts, "Hazards to people and property from wave overtopping at coastal structures," *Coastlines, Structures and Breakwaters 2005 (Proc. Conf.,* Inst. Civil Engrs., Thomas Telford) (2006), pp. 153–165.

31. K. Kimura, M. Hamaguchi, M. Okada and T. Shimizu, "Modeling of wave spray on block mound seawall for coastal highway," *Proc. Coastal Engrg., JSCE* **50** (2003), pp. 796–800 (*in Japanese*).

32. Y. Goda and Y. Kishira, "Experiments on irregular wave overtopping characteristics of seawalls of low crest types," *Tech. Note of Port and Harbour Res. Inst.* (242) (1976), 28p. (*in Japanese*).

33. M. Inoue, H. Shimada and K. Tonomo, "Wave overtopping characteristics under irregular waves," *Proc. Coastal Engrg., JSCE* **36** (1989), pp. 618–622 (*in Japanese*).

34. T. Sekimoto, H. Kunishi, T. Shimizu, O. Kyotani and R. Kajima, "Characteristics of short-term wave overtopping at the seawall around an artificial island," *Proc. Coastal Engrg., JSCE* **39** (1992), pp. 581–585 (*in Japanese*).

35. T. Izumiya, R. Hamada and K. Ishibashi, "Probabilistic characteristics of wave overtopping rate of block-mound seawall," *Annual J. Coastal Engrg., JSCE* **53** (2006), pp. 716–720 (*in Japanese*).

36. A. Kimura, A. Seyama and T. Yamada, "Statistical properties of the short-term overtopping discharge," *Proc. 28th Japanese Conf. Coastal Engrg.* (1981), pp. 335–338 (*in Japanese*).

37. K. Tonomo, M. Inoue, T. Memita and T. Tamada, "Sudy on discharge capacity of drainage of wave overtopping," *Proc. Coastal Engrg., JSCE* **49** (2002), pp. 651–655 (*in Japanese*).

38. U.S. Army Coastal Engineering Research Center, *Shore Protection Manual* (1977), pp. 6-1–6-15.

# Chapter 6

# Probabilistic Design of Breakwaters

## 6.1 Uncertainty of Design Values

### 6.1.1 *Overview*

In the conventional design methods, all values of loads and resistances are assumed to have been determined with certainty. Stability of a structure as a whole and/or the strength of structural members are examined by using the deterministic design values. The preliminary design of upright sections of composite breakwaters discussed in Sec. 4.3 is such an example. However, most of design values are not deterministic but carries certain degrees of uncertainty. The maximum wave height of a random wave train, for example, differs from one train to another even when the duration of random wave group is the same. We can only describe the probability density function of the maximum wave height for a given number of waves. The design significant wave height also contains a number of uncertain factors such as the appropriateness of design storm event, reliability of wave hindcasting event, and fitting error in extreme wave analysis.

The conventional design methods have covered up the problem of the uncertainty in design values with the concepts of the *safety factor* and the *allowable stress*. The stability of vertical/composite breakwater against sliding, for example, was judged as good enough when the resistance of upright section against sliding is equal to or greater than 1.2 times the wave force. The figure of 1.2 is an example of the safety factor, which has been accepted in view of the past experience. The allowable stress of materials for structural members was determined by discounting the breakage strength in laboratory tests by several tens of percents in consideration of statistical variability in material strength and/or by empirical rules. Such the conventional method is called the *deterministic design method*.

Another approach is to introduce the uncertainty of design values and its influence in the design process in a manner as concrete as possible. It is called the *probabilistic design method*. When the reliability in terms of the probability of failure is dealt with, it is called the *reliability-based design*. When the performance of a structure such as deformation under loading is to be analyzed, it is called the *performance-based design*.

The first step of the probabilistic design method is quantification of the uncertainty of design values. ISO 2394 *General Principles on Reliability of Structures*[1] defines the model uncertainty as related to the accuracy of a model, e.g., physical or statistical uncertainties, and the statistical uncertainty as related to the accuracy of the distribution and estimation of parameters. ISO 21650 *Actions from Waves and Currents on Coastal Structures*[2] cites major sources contributing to the uncertainties, among which those of hydraulic parameters related to the design of maritime structures can be itemized as below.

A. Statistical uncertainty inherent to the natural process:

   A-1. Variability of the heights and periods of individual waves (wave heights approximately follow the Rayleigh distribution),

   A-2. Temporal variations of significant wave height and period, and wave direction (their statistics are expressed as wave climate at a specific site),

   A-3. Variability of extreme wave heights and periods of individual storm events (extreme wave heights can be analyzed statistically),

   A-4. Tide level at arbitrary date and hours.

B. Errors and statistical variability associated with wave measurements:

   B-1. Reliability of visual measurements,

   B-2. Errors in instrumental measurements and data analysis,

   B-3. Variability of significant wave height and other characteristic heights and periods owing to the limitation in recording time of around 20 minutes (standard deviation of about 6% for the significant wave height).

C. Uncertainty in extreme wave analysis:

   C-1. Error in parameter estimation owing to limited record length of storm wave data,

   C-2. Statistical variability of return wave height for a selected extreme distribution (confidence interval),

C-3. Influence of selecting an inappropriate distribution function owing to the lack of the information on population distribution.

D. Accuracy of models used for hindcasting:

D-1. Accuracy of barometric and wind fields of storm events estimated from weather maps,

D-2. Accuracy of wave hindcasting models,

D-3. Accuracy of storm surge models.

E. Accuracy of models used for wave transformation:

E-1. Accuracy of wave transformation models in shallow water for shoaling, refraction, diffraction, random breaking, and others,

E-2. Accuracy of wave and current interaction models.

F. Reliability of models and experiments for predicting actions and responses:

F-1. Bias and standard deviation of predicted actions and responses (wave pressure, mass of armor units, wave overtopping rate, etc.),

F-2. Reliability of hydraulic model tests with random waves (reduction of variability by repeated tests),

F-3. Reliability of structural tests on deformation and breaking strength.

G. Variability and reliability of parameters related to resistance:

G-1. Variability of mass density, material strength and others of structures,

G-2. Variability of friction coefficient between concrete and rubble stone,

G-3. Reliability of roughness and permeability of slope concerning wave run-up and overtopping.

The design parameters on geotechnical aspects such as subsidence and slip failure are not discussed here. Readers are requested to examine their uncertainty by obtaining the necessary information from appropriate references.

### 6.1.2 *Examples of Uncertainty of Design Parameters for Breakwater Design*

For examination of the uncertainty of design values, one needs to know how they vary, i.e., their probability distributions. However, a common approach is to assume a normal distribution of a design value which can be designated with two parameters of the mean $\mu$ and the standard deviation $\sigma$. The difference between the mean and the nominal or characteristic value[a] of the action or the material property is called the bias. In the present chapter, the ratio of the mean of a design value to the characteristic value denoted by $X$ is introduced under the name of *characteristic ratio* $(= \mu/X)$. It has the value of 1 when the mean is equal to the characteristic value and greater than 1 when the mean exceeds the characteristic value. The standard deviation is used to define the coefficient of variation $V$ that is calculated as its ratio to the mean, i.e., $V = \sigma/\mu$. The characteristic ratio indicates how much the estimated design value differs from the characteristic value as a whole and the coefficient of variation expresses the degree of spread around the mean.

The data of the characteristic ratio $\mu/X$ and the coefficient of variation $V$ must be collected through various tests, measurements, and investigations. It is desirable to construct the database on the uncertainty of various design parameters so that future workers could add more data to the database and improve the accuracy of $\mu/X$ and $V$. Though several authors have presented the data of $\mu/X$ and $V$ for a number of design parameters, most of their reports do not have the lists of original data unfortunately. Nevertheless, Table 6.1 lists the proposed data of $\mu/X$ and $V$ for design parameters related to breakwater design.

The variability of the first item of design parameters in Table 6.1, *offshore significant wave height*, was tentatively assumed by Takayama and Ikeda[3] for the sake of convenience. It should be assessed with due consideration of extreme wave statistics to be discussed in Sec. 6.1.3.

The offshore significant wave height needs to be converted to the maximum and significant wave heights at the site of breakwater construction through computation with wave transformation models. This is the second item, *in-situ wave height*, in Table 6.1. Takayama and Ikeda[3] made comparison of the significant wave heights measured in a few model tests of 3-D multidirectional random wave transformation and those computed with the energy balance equation, and obtained the characteristic ratio of 0.92 and the coefficient of variation of 0.04 for a set of 66 data. In their

---

[a]ISO 2394 defines the characteristic value of an action as a principal representative value of a design parameter. It is chosen either on a statistical basis, so that it can be considered to have a specified probability of not being exceeded towards unfavorable values during a reference period, or on acquired experience, or on physical constraints. ISO 2394 also defines the characteristic value of a material property as an *a priori* specified fractile of the statistical distribution of the material property in the supply produced within scope of the relevant material standard.

Table 6.1    Various proposals of the characteristic ratio and the coefficient of variation of design parameters for breakwaters.

| Design Parameter | Characteristic Value | Characteristic Ratio $\mu/X$ | Coef. of Variation $V$ | References |
|---|---|---|---|---|
| 1. Offshore significant wave height | – | 1.00 | 0.10 | Takayama and Ikeda[3] |
| 2. In-site wave height | | | | |
|   Energy balance equation | – | 0.92 | 0.04 | Takayama and Ikeda[3] |
|   Bottom slope milder than 1/30 | – | 0.97 | 0.04 | Nagao[4,5] |
|   Bottom slope equal to or steeper than 1/30 | – | 1.06 | 0.08 | Nagao[4,5] |
|   Breaking deformation | – | 0.87 | 0.10 | Nagao[4,5] |
|   Overall change | – | 1.00 | 0.10 | Shimosako and Takahashi[6] |
| 3. Wave period | | | | |
|   Significant period | – | 1.00 | 0.10 | Shimosako and Takahashi[6] |
|   Individual period | – | 1.00 | 0.10 | Ditto |
| 4. Estimated wave force | | | | |
|   Caisson breakwater | – | 0.91 | 0.19 | Takayama and Ikeda[3] |
|   Horizontally-composite breakwater | – | 0.84 | 0.12 | Nagao[4,5] |
| 5. Stability number of armor units | | | | |
|   Rubble stone by van der Meer formula | – | 1.00 | $0.065 \sim 0.08$ | ISO 21650[7] |
|   Concrete cube | – | 1.00 | 0.10 | Ditto |
| 6. Friction coefficient | 0.60 | 1.06 | 0.15 | Takayama and Ikeda[3] |
| 7. Specific weight of material | | | | |
|   Reinforced concrete | 24.0 kN/m$^3$ | 0.98 | 0.02 | Nagao[4,5] |
|   Plain concrete | 22.6 kN/m$^3$ | 1.02 | 0.02 | Ditto |
|   Saturated fill sand | 18.0, 19.0, or 20.0 kN/m$^3$ | 1.02 | 0.04 | Ditto |

comparison, the ratio of the measured to the computed heights was employed to define the ratio $\mu/X$ and $V$. Nagao[4,5] has presented the values of $\mu/X$ and $V$ for three categories of wave transformations for wave force estimation: waves over sea bottoms of mild and steep slopes, and breaking deformation. However he did not mention about the calculation models or the number of data for analysis. Shimosako and Takahashi[6] has proposed a set of the values of $\mu/X$ and $V$ for the overall change of wave height between the offshore and the construction site, but it is a subjective one for convenience of carrying out the calculation of the expected sliding distance of breakwaters.

The third item of *wave period* is also subject to variability. The proposals by Shimosako and Takahashi[6] are subjective ones without any supporting data.

The variability of *estimated wave force* listed as the fourth item in Table 6.1 has been investigated by Takayama and Ikeda[3] for caisson breakwaters. They examined the reliability of the Goda formulas by using a set of 66 data from the original pressure measurement data of regular waves by Goda and Fukumori.[7] The characteristic ratio of 0.91 thus obtained indicates the tendency of overestimation by the Goda formulas. Takayama and Ikeda also examined the variability of estimated wave force acting on horizontally-composite breakwater by using a few sets of laboratory data. Nagao[4,5] slightly modified the values of $\mu/X$ and $V$ as listed in Table 6.1, but he did not present the details of calibration data.

The information of the variability of the fifth item, *stability number of armor units*, has been taken from ISO 21650.[8]

The basic data for the variability of the sixth item, *friction coefficient*, are the prototype test results with the contact pressure in the range of $25 \sim 100$ kN/m$^2$. Takayama and Ikeda[3] analyzed 42 data from two sources and obtained the values listed in Table 6.1.

The variability of the seventh item, *specific weight of materials*, is due to Nagao[4,5] based on the investigation by Minami and Kasugai,[9] who collected more than ten thousands data at port construction sites through questionnaires. The specific weight of fill sand was grouped according to the design value of 18.0, 19.0, or 20.0 kN/m$^3$. As seen in Table 6.1, the coefficient of variation of the specific weight of concrete and fill sand is rather small.

### 6.1.3   Uncertainty of Offshore Significant Wave Height

The offshore significant wave height has many sources of uncertainties as listed in A to D in Sec. 6.1.1. Takayama and Ikeda[3] have assumed the characteristic ratio of 1.00 and the coefficient of variation of 0.10 in their analysis of sliding probability of composite breakwaters and many researchers adopted these values. However, these values are subjective ones and should be reexamined in light of the original wave data whenever a probabilistic design is made of a breakwater.

The estimated value of offshore significant wave height to be employed in design could be regarded as representing the true height, unless there is some evidence of bias in wave hindcasting, extreme analysis and other processes. Thus, the characteristic ratio can be set at 1.00. On the other hand, the coefficient of variation should be set in considerations of various factors as discussed hereinafter.

The first factor is the probabilistic nature of the largest significant wave height for a given duration of time, say $L$ years; it is called the *L-year maximum wave height*. Even under the condition that the population distribution of extreme significant wave height is known, a series of storm events for $L$ years yields a $L$-year maximum height which differs from one realization from another, because the $L$-year maximum height has its own probability density function depending on the population function. Its variability can be assessed with its standard deviation, which is presented in Sec. 13.4.3, and can be incorporated into the coefficient of variation of the offshore significant wave height. When the expected sliding distance method (see Sec. 6.3.2) is carried out, however, this type of variability is materialized in the Monte Carlo simulation and thus unnecessary to be considered.

The second factor is the uncertainty associated with the parameter estimation of an extreme distribution function owing to limitation in the sample size or the number of storm wave heights employed in the analysis. This is a matter of statistical variability of the parameter values and the resultant return wave height. As will be discussed in Sec. 13.3, the coefficient of variation of the return wave height can be estimated with the information of the distribution function and the sample size. Uncertainty of the population distribution other than the distribution fitted best to the sample increases the range of variability. The coefficient of variation may range from several percent to over ten percent depending on the sample size.

The third factor is the accuracy of wave measurements and hindcasting. Even if instrumental wave records themselves are reliable, the significant wave height analyzed from the records has a statistical variability of several percent as will be discussed in Sec. 10.6. Most of hindcasted wave heights have the coefficient of variation in excess of ten percent mainly because of the inaccuracy in wind field hindcasting.

The variability in the above three factors should be added together in the form of the sum of variance. For example, if the coefficients of variation of the first to third factors are estimated as 10%, 5%, and 6%, respectively, then the overall coefficient of variation becomes $V = \sqrt{10^2 + 5^2 + 6^2} = 12.7\%$. The value of $V = 0.10$ proposed by Takayama and Ikeda[3] seems to be an underestimation for many cases, but it will be an overestimation for the cases when the expected sliding distance method is employed.

One limiting factor to the variability of wave height is the wave breaking process in nearshore waters. The wave height within the surf zone is governed by the relative water depth and the seabed slope. It has its own coefficient of variation from 14% for a steep slope of $1/10$ to 6% for a mild slope of $1/100$ as discussed in Sec. 3.6.1. The depth-limited breaking wave height is independent of the deepwater wave height, though it is slightly affected by the wave period through the relative water depth. Thus, all other sources of uncertainties related to wave heights are superseded by the variability of breaking wave height in nearshore waters.

## 6.2    Reliability-Based Design of Breakwaters

### 6.2.1    *Classification of Reliability-Based Design Method*

In the reliability-based design method, possible modes of failure of a structure are examined whether a failure may occur during the design working life. If a failure is probable, its occurrence probability is evaluated and the design is so made to control the failure probability below a preset level.

The reliability-based design method is often classified in three levels depending on the minuteness of assessing the probability of failure. *Level III method* examines the uncertainties of all design parameters related to failure, expresses the probability distribution functions of individual design variables in explicit forms, and computes the failure probability by multiple integration of probability distributions or simulation techniques. It is most reliable method, but assignment of distribution functions to all design variables is very difficult even if it is feasible. The expected sliding distance method to be introduced in Sec. 6.3.2 is a kind of Level III method, because it introduces the extreme wave height distribution at the site, the Rayleigh distribution of individual wave heights, and other distribution functions in the Monte Carlo simulations.

*Level II method* simplifies the probability distributions of design parameters by approximating them with the normal distributions; only the means and the standard deviations of individual parameters are called for. The probability of failure is also approximated by a normal distribution, and the reliability index that corresponds to the non-exceedance probability of the standard normal distribution is calculated. The safety of a structure is judged with the value of the reliability index thus calculated. For this purpose, a limit state function is defined for each mode of failure with the relevant design variables. The function is expanded in a Taylor series, and its first derivative terms are usually employed in the calculation of the reliability index; the method is called *FORM* (First-Order Reliability Method).

*Level I method* does not assess the failure probability directly but introduces the partial factors to be multiplied to the relevant design variables of the limit state function. The values of the partial factors are usually calculated for a certain acceptable probability of failure, which needs to be set through calibration with a number of existing structures.

Both the methods of Level II and Level I for caisson breakwater design are introduced in the subsequent sections.

### 6.2.2 *Evaluation of External Safety by Level II Method*

(A) *Limit state function and computation formula for reliability index*

The first step in the reliability design method is to define a *limit state function*, which is also called the performance function, for each mode of failure by taking all design variables related to that mode. With the notation $Z$ for the limit state function, the failure is judged not to occur when $Z > 0$, but the failure will occur when $Z \leq 0$. When $n$ variables of $X_1, X_2, \cdots, X_n$ are involved in a particular failure mode, the limit state function is defined as

$$Z = g(X_1, X_2, \cdots, X_n) \begin{cases} < 0 : \text{undesirable state (unsafe set)}, \\ = 0 : \text{limit state}, \\ > 0 : \text{desirable state (safe set)}. \end{cases} \qquad (6.1)$$

For the case of the stability of armor stone to be analyzed with the Hudson formula, Burcharth[10,11] expresses the limit state function as follows:

$$g = A \, \Delta \, D_n \, (K_D \cot \alpha)^{1/3} - H_s, \qquad (6.2)$$

where $A$ represents a stochastic variable with the mean of 1.0, $\Delta$ is the relative density of stone in water, $K_D$ is a damage factor, $\alpha$ is the slope angle, and $H_s$ is the design significant wave height.

For the case of sliding of the upright section of a composite breakwater, the limit state function can be defined by rewriting Eq. (4.36) as follows:

$$g = f(W - U) - P, \qquad (6.3)$$

where $f$ denotes the friction coefficient between the upright section and the rubble stone foundation, $W$ is the weight of upright section in still water, $U$ is the total uplift, and $P$ is the horizontal wave force exerted upon the upright section. Because these design variables have certain ranges of variation as indicated in Table 6.1, the safety analysis must be performed by duly taking the variability into consideration.

As mentioned before, Level II method assumes that the value of limit state function follows the normal distribution with the mean $\mu_Z$ and the standard deviation $\sigma_Z$. Then the probability $P_f$ that the structure will not belong to a safe set, i.e., $Z \leq 0$, is given by the non-exceedance probability of the standard normal distribution as below.

$$P_f = \mathrm{P_r}(Z \leq 0) = 1 - \Phi(\mu_Z/\sigma_Z) : \Phi(x) = \frac{1}{\sqrt{2\pi}} \int_{-\infty}^{x} \exp[-t^2/2]\,dt.$$
$$(6.4)$$

The ratio of the standard deviation to the mean is called the *reliability index* and is given the symbol of $\beta$, i.e.,

$$\beta = \frac{\mu_Z}{\sigma_Z}. \qquad (6.5)$$

Once the reliability index is calculated, the probability of failure is immediately obtained from a table of the normal distribution. The values of $\beta = 0, 1.0, 2.0$, and $3.0$, for examples, indicates the failure probability of $P_f = 0.50, 0.159, 0.0228$, and $0.0013$, respectively.

The general expression of the limit state equation in Eq. (6.1) is now expanded into a Taylor series at an arbitrary point $x* = (x_1^*, x_2^*, \cdots, x_n^*)$ up to the first derivatives as in the following:

$$Z \simeq g(x_1^*, x_2^*, \cdots, x_n^*) + \sum_{i=1}^{n}(X_i - x_i^*)\frac{\partial g}{\partial X_i}\Big|_{x^*}. \qquad (6.6)$$

When the design variables can be assumed as mutually independent, a Taylor expansion is made at a linearization point with the mean values of the design variable; the point is called the *mean point*. The value of the limit state function at the mean point is obtained by substituting $X_i = \mu_{X_i}$ into Eq. (6.1) as

$$\mu_Z = g(\mu_{x_1}, \mu_{x_2}, \cdots, \mu_{x_n}). \qquad (6.7)$$

The variance of the limit state function can be calculated as follows:

$$\sigma_Z^2 = \sum_{i=1}^{n} \left( \sigma_{X_i} \frac{\partial g}{\partial X_i} \bigg|_{\mu_{x_i}} \right)^2 . \tag{6.8}$$

The point with the mean values of design variables is not necessarily the point being most unsafe. We need to search for the point that is located on the surface $Z = 0$ in the multidimensional space of $X = (X_1, X_2, \cdots, X_n)$. This is the point at which the safety of structure should be examined, and it is called the *design point*. The search is made by solving the following equations through iterations:

$$g(x_1^*, x_2^*, \cdots, x_n^*) = 0 , \tag{6.9}$$

$$x_i^* = \mu_{X_i} - \alpha_i \frac{\mu_Z}{\sigma_Z} \sigma_{X_i} \; : \; i = 1, 2, \ldots, n , \tag{6.10}$$

$$\alpha_i = \frac{\sigma_{X_i} \dfrac{\partial g}{\partial X_i} \bigg|_{x^*}}{\sigma_Z} \; : \; i = 1, 2, \ldots, n , \tag{6.11}$$

$$\mu_Z = \sum_{i=1}^{n} (\mu_{X_i} - x_i^*) \frac{\partial g}{\partial X_i} \bigg|_{x^*} , \tag{6.12}$$

$$\sigma_Z = \sqrt{\sum_{i=1}^{n} \left( \sigma_{X_i} \frac{\partial g}{\partial X_i} \bigg|_{x^*} \right)^2} . \tag{6.13}$$

The parameter $\alpha_i$ defined by Eq. (6.11) is called the *sensitivity factor*. Iterations are made first by introducing the $\mu_Z$ and $\sigma_Z$ calculated by Eqs. (6.7) and (6.8). The initial estimate of the sensitivity factor is obtained with Eq. (6.11), and modification to respective design values $x_i^*$ is made with Eq. (6.10). The condition of Eq. (6.9) is not satisfied initially, but non-zero value is rapidly converged to zero as iterations are progressed. The converged values of $x_i^*$ constitute the design point.

(B) *Calculation of reliability index for sliding of breakwater — Case 1*

To illustrate calculation of the reliability index, safety of the composite breakwater shown in Example 4.1 with the preliminary stability analysis in Example 4.3 is examined. The limit state equation of Eq. (6.3) is employed with the design variables of the friction coefficient $f$, the weight of upright

Table 6.2    Design variables for reliability analysis of the breakwater in Example 4.1 — Case 1.

| No. | Design Variable | Symbol | Mean $\mu_X$ | Standard Deviation $\sigma_X$ | Character-istic Ratio $\mu/X$ | Coefficient of Variation $V$ |
|------|-----------------|--------|------|----------|------|------|
| $X_1$ | Friction coefficient | $f$ | 0.636 | 0.095 | 1.06 | 0.15 |
| $X_2$ | Weight of upright section (kN/m) | $W$ | 2331 | 45.7 | 1.02 | 0.02 |
| $X_3$ | Uplift (kN/m) | $U$ | 377 | 90.5 | 0.77 | 0.24 |
| $X_4$ | Horizontal wave force (kN/m) | $P$ | 714 | 171.4 | 0.77 | 0.24 |

section in still water $W$, the uplift $U$, and the horizontal wave force $P$, which are assigned the notations of $X_1$ to $X_4$, respectively. The figures of the design variables in Examples 4.1 and 4.3 are to be modified by using the characteristic ratios listed in Table 6.1 and the coefficients of variation of these variables also needed to be evaluated. As for the variables $X_1 = f$ and $X_2 = W$, the figures listed in Table 6.1 are applied.

As for the uplift $X_3 = U$ and the wave force $X_4 = P$, the characteristic ratio is calculated as the product of those for offshore significant wave height, in-site wave height on mild bottom slope with breaking deformation, and wave force estimation. It becomes as follows:

$$\mu_3/X_3 = \mu_4/X_4 = 1.00 \times 0.97 \times 0.87 \times 0.91 = 0.768\,.$$

The coefficient of variation is estimated as the square root of the sum of their variances by assuming the variables $X_1$ to $X_4$ are uncorrelated each other. The result becomes

$$V_3 = V_4 = \sqrt{0.10^2 + 0.04^2 + 0.10^2 + 0.19^2} = 0.240\,.$$

With these characteristic ratios and coefficients, the means and standard deviations of the design variables become as listed in Table 6.2.

Calculation with Eqs. (6.11) to (6.13) requires the derivation of the partial derivatives $\partial g/\partial X_i$, the results of which are shown below.

$$\frac{\partial g}{\partial X_1} = X_2 - X_1, \quad \frac{\partial g}{\partial X_2} = X_1, \quad \frac{\partial g}{\partial X_3} = -X_1, \quad \frac{\partial g}{\partial X_4} = -1\,. \quad (6.14)$$

The initial estimates of the partial derivatives are calculated by using the mean values in Table 6.2 as in the following:

$$\frac{\partial g}{\partial X_1} = 1,954, \quad \frac{\partial g}{\partial X_2} = 0.636, \quad \frac{\partial g}{\partial X_3} = -0.636, \quad \frac{\partial g}{\partial X_4} = -1\,.$$

The mean and the standard deviation of the limit state function at the mean point are calculated by Eqs. (6.7) and (6.8) by using the means of design variables listed in Table 6.2, together with the resultant value of the reliability index as follows:

$$\mu_Z = 528.7\text{kN/m}, \quad \sigma_Z = 260.8\text{kN/m}, \quad \beta = 2.028. \tag{6.15}$$

The value of $\beta = 2.028$ corresponds to the probability of failure of 0.0213 according to the standard normal distribution table. The breakwater under examination has the overall safety factor of 1.16 against sliding as calculated in Example 4.2, the value of which is less than the acceptable threshold of 1.20. A relatively large failure probability indicates a small margin of safety.

As discussed in Clause (A), the reliability index should be evaluated at the design point by solving Eqs. (6.9) to (6.13). By using the initial estimate of Eq. (6.15) and repeating iterations a few times, the following converged values are obtained:

$$\left.\begin{array}{l} X_1 = f = 0.497, \ X_2 = 2,322.6\text{kN/m}, \ X_3 = U = 409.8\text{kN/m}, \\[2mm] X_4 = P = 950.6\text{kN/m}, \\[2mm] \alpha_1 = 0.7131, \ \alpha_2 = 0.0891, \ \alpha_3 = -0.1765, \ \alpha_4 = -0.6726, \\[2mm] \mu_Z = 523.0\text{kN/m}, \ \sigma_Z = 254.8\text{kN/m}, \ \beta = 2.052, \ P_f = 0.0201. \end{array}\right\} \tag{6.16}$$

### (C) Interpretation of failure probability in Level II method

We have calculated the probability of sliding failure at the mean and the design points. At this stage it would be helpful for understanding of the concept of the reliability-based design to ponder about the meaning of failure probability. Let us imagine that we design and construct one thousand caisson breakwaters with the dimensions same as those of Example 4.1 and let those breakwaters be subject to the design wave conditions. Because the friction coefficient, the horizontal wave force, the uplift, and the weight of upright section are subject to the statistical and model uncertainties with the given coefficient of variation, actions to each breakwater and its resistance differ from one breakwater to another. The mean values of $X_1$ to $X_4$ in Table 6.2 are the ensemble averages of 1,000 breakwaters. The reliability index of $\beta = 2.028$ given by Eq. (6.15) is that of a breakwater just happened to be located at the mean point in the multidimensional space with the mean values of $X_1$ to $X_4$. Other 999 breakwaters occupy various positions in the space $(X_1, X_2, X_3, X_4)$ and have different values of the limit state function $Z$. Among 1,000 breakwaters, about 21 cases would have the zero or negative $Z$ values and this yields the failure probability of $P_f \simeq 21/1000 = 0.021$.

The results of Eq. (6.16) indicates that the least safe situation occurs when the design parameters of $X_1$ to $X_4$ happen to have the values of Eq. (6.16), which are the coordinate values of the design point. According to this FORM analysis, there would be about 20 cases of breakwaters that have the zero or negative $Z$ values and this yields the failure probability of $P_f \simeq 20/1000 = 0.020$.

Theoretically speaking, the design point analysis should yield the failure probability higher than that derived by the mean point analysis. The reason why the opposite result was obtained in the above analysis is not clear. Use of only the first order derivatives in the Taylor expansion of Eq. (6.6) might be the cause for the contradictory result. Use of the first and second order derivatives, which is called *SORM* (Second-Order Reliability Method), may solve the contradiction.

(D) *Calculation of reliability index for sliding of breakwater — Case* 2

Among the four design variables of the limit state function of Eq. (6.3), the friction coefficient $f$ and the weight $W$ have the coefficient of variation independent of other parameters. However, the uplift $U$ and the horizontal wave force $P$ are affected by other design parameters such as the wave height and period. For making more detailed examination of the reliability index, it is necessary to employ another form of the limit state function which includes more variables. One approach is to introduce the wave height as a direct design variable, because both the uplift and wave force are approximately proportional to the wave height.

A new limit state function is set for sliding of the upright section of a composite breakwater as in the following:

$$Z = f W - f C_U A_U H_D - C_P A_P H, \qquad (6.17)$$

where $C_U$ and $C_P$ are the coefficients representing the accuracy of estimating the uplift and horizontal wave force, respectively, and $A_U$ and $A_P$ are the linear factors for estimating the uplift and horizontal wave force as functions of the wave height $H_D$. The linear factor $A_U$ is obtained from Eqs. (4.9) and (4.14) as $A_U = (1/4)(1 + \cos\beta)\alpha_1\alpha_3\rho_w g B$, while $A_P$ is proportional to the pressure coefficient $\alpha_1$. The pressure coefficients $\alpha_1$ and $\alpha_3$ in Eqs. (4.6) and (4.8) are the functions of the relative water depth $h/L$

Table 6.3    Design variables for reliability analysis of the breakwater in Example 4.1 — Case 2.

| No. | Design Variable | Symbol | Mean $\mu_X$ | Standard Deviation $\sigma_X$ | Character-istic Ratio $\mu/X$ | Coefficient of Variation $V$ |
|---|---|---|---|---|---|---|
| $X_1$ | Friction coefficient | $f$ | 0.636 | 0.095 | 1.06 | 0.15 |
| $X_2$ | Weight of upright section (kN/m) | $W$ | 2331 | 45.7 | 1.02 | 0.02 |
| $X_3$ | Coefficient for uplift | $C_U$ | 0.95 | 0.095 | 0.95 | 0.10 |
| $X_4$ | Coefficient for wave force | $C_P$ | 0.91 | 0.137 | 0.91 | 0.15 |
| $X_5$ | Factor for uplift (kN/m$^2$) | $A_U$ | 61.1 | 3.7 | 1.00 | 0.06 |
| $X_6$ | Factor for wave force (kN/m$^2$) | $A_P$ | 115.9 | 4.6 | 1.00 | 0.04 |
| $X_7$ | Wave height (m) | $H$ | 7.2 | 1.01 | 0.90 | 0.14 |
| $X_8$ | Wave period (s) | $T$ | 11.4 | 1.14 | 1.00 | 0.10 |

and their values vary with the wave period. Thus, the factors $A_U$ and $A_P$ contain the wave period as the internal design variable.

The limit state function of Eq. (6.17) has seven design variables, the means and standard deviations of which are given as listed in Table 6.3 together with the information on wave period.

The characteristic ratio and the coefficient of variation of $C_P$ are set as equal to those of the horizontal wave force in Table 6.1. Those of $C_U$ for the uplift are slightly modified by the author's subjective judgment. The means of the uplift and force factors $A_U$ and $A_P$ are calculated by dividing the values of the uplift and horizontal wave force in Example 4.1 by the design wave height of $H_{\max} = 8.0$ m. Their standard deviations are evaluated from the range of variations of the values of $A_U$ and $A_P$ when the wave period is varied by the amount of one standard deviation. The characteristic ratio and the coefficient of variation of the wave period are taken from Table 6.1. The wave height in front of the breakwater is tentatively assumed to have been biased with the characteristic ratio of 0.9 in consideration of the characteristic ratio for wave breaking deformation listed in Table 6.1. The standard deviation of wave height is calculated by considering the variances in offshore wave height estimation and wave breaking deformation, because the estimation error in wave transformation is superseded by the variability in wave breaking deformation.

Evaluation of the reliability index with the limit state function of Eq. (6.17) is first made by calculating its partial derivatives with respect to the design variables. The results become as below.

$$
\left.\begin{array}{l}
\dfrac{\partial g}{\partial X_1} = X_2 - X_3 X_5 X_7, \quad \dfrac{\partial g}{\partial X_2} = X_1, \quad \dfrac{\partial g}{\partial X_3} = -X_1 X_5 X_7, \\[2mm]
\dfrac{\partial g}{\partial X_4} = -X_6 X_7, \quad \dfrac{\partial g}{\partial X_5} = -X_1 X_3 X_7, \quad \dfrac{\partial g}{\partial X_6} = -X_4 X_7, \\[2mm]
\dfrac{\partial g}{\partial X_3} = -X_1 X_3 X_5 - X_4 X_6.
\end{array}\right\} \tag{6.18}
$$

Substitution of the mean values in Table 6.3 into Eq. (6.18) yields the following values of the initial estimates of the derivatives as follows:

$$
\dfrac{\partial g}{\partial X_1} = 1{,}913.1, \quad \dfrac{\partial g}{\partial X_2} = 0.636, \quad \dfrac{\partial g}{\partial X_3} = -279.8, \quad \dfrac{\partial g}{\partial X_4} = -834.5,
$$

$$
\dfrac{\partial g}{\partial X_5} = -4.350, \quad \dfrac{\partial g}{\partial X_6} = -6.552, \quad \dfrac{\partial g}{\partial X_3} = -142.4.
$$

The mean and the standard deviation of the limit state function at the mean point are calculated by Eqs. (6.7) and (6.8) by using the means of design variables listed in Table 6.3, together with the resultant value of the reliability index as follows:

$$
\mu_Z = 457.3 \text{kN/m}, \quad \sigma_Z = 263.6 \text{kN/m}, \quad \beta = 1.735. \tag{6.19}
$$

The value of $\beta = 1.735$ corresponds to the probability of failure of 0.0413 according to the table of normal distribution. This value of reliability index is smaller than the case using the limit state function of Eq. (6.3). Thus, the calculated value of the reliability index is not unique but depends on the functional form defining the limit state. Superiority among several comparative designs of breakwaters can be made based on the reliability index from the viewpoint of safety, but comparison is only meaningful when the same limit state function is employed.

Examination of the reliability index is next carried out at the design point $x^*$. The procedure is same as the case 1, but use is made of the initial mean and standard deviation in Eq. (6.19) and the partial derivatives are calculated at each iteration step by using Eq. (6.18). Results of iterations are listed in Table 6.4, in which three additional cases with different settings of variability: the case with no variation in wave period, and the cases with the coefficient of variation of wave height at $V_H = 0.10$ and 0.20.

The results of Table 6.4 indicate the following features of the reliability index:

(1) A switch of the limit state function from Eq. (6.3) to Eq. (6.17) causes to decrease the reliability index and to increase the probability of failure,

Table 6.4  Calculation of sensitivity factor and reliability index of the breakwater in Example 4.1 — Case 2.

| Item | Coefficient of Variation of Wave Height | | | | |
|---|---|---|---|---|---|
| | With Period Variation $V_H = 0.14$ | | Period Fixed | Coef. Var. $V_H = 0.10$ | Coef. Var. $V_H = 0.20$ |
| | Initial | Converged | Converged | Converged | Converged |
| 1. Design point $x_i^*$: | | | | | |
| Friction coefficient $f$ | 0.5224 | 0.5316 | 0.5286 | 0.5099 | 0.5585 |
| Weight of upright section $W$ (kN/m) | 2322.3 | 2324.0 | 2323.9 | 2323.0 | 2325.5 |
| Coefficient for uplift estimation $C_U$ | 0.9666 | 0.9652 | 0.9654 | 0.9666 | 0.9627 |
| Coefficient for force estimation $C_P$ | 1.0131 | 1.0228 | 1.0257 | 1.0387 | 1.0000 |
| Uplift factor $A_U$ (kN/m²) | 61.49 | 61.46 | 61.10 | 61.49 | 61.40 |
| Force factor $A_P$ (kN/m²) | 116.81 | 117.01 | 115.90 | 117.18 | 116.77 |
| Wave height $H$ (m) | 8.156 | 8.170 | 8.189 | 7.792 | 8.671 |
| 2. Sensitivity factor $\alpha_i$ | | | | | |
| Friction coefficient $f$ | 0.6894 | 0.6425 | 0.6533 | 0.7124 | 0.5480 |
| Weight of upright section $W$ (kN/m) | 0.1103 | 0.0893 | 0.0902 | 0.0940 | 0.0812 |
| Coefficient for uplift estimation $C_U$ | -0.1008 | -0.0932 | -0.0939 | -0.0936 | -0.0899 |
| Coefficient for force estimation $C_P$ | -0.4337 | -0.4816 | -0.4858 | -0.5044 | -0.4414 |
| Uplift factor $A_U$ | -0.0611 | -0.0570 | 0 | -0.0573 | -0.0549 |
| Force factor $A_P$ | -0.1143 | -0.1413 | 0 | -0.1501 | -0.1269 |
| Wave height $H$ (m) | -0.5455 | -0.6029 | -0.5606 | -0.4403 | -0.7702 |
| 3. Reliability index $\beta$ | 1.735 | 1.710 | 1.731 | 1.863 | 1.488 |
| 4. Failure of probability $P_f$ | 0.0413 | 0.0436 | 0.0417 | 0.0312 | 0.0683 |

even though the coefficient of variation of the wave force estimate is decreased. Thus, assessment of safety is dependent on the definition of the limit state function.

(2) Judging from the magnitude of the sensitivity factors, the friction coefficient exercises the largest influence on the reliability index, being followed by the wave height and then by the estimation of horizontal wave force.

(3) A degree of variation of the wave height strongly affects the safety against sliding. When the coefficient of variation of the wave height is 0.10, the failure probability is 3.1%. However it increases to 6.8% when the coefficient of variation of the wave height is 0.20.

(4) Introduction of the variability of wave period causes to decrease the reliability index by 0.02 and increases the failure probability from 4.2% to 4.4%.

(5) In Case 2, the reliability index at the design point is lower than that at the mean point as expected.

### 6.2.3  *Design of Breakwaters with Partial Factor System*

The reliability-based design method at Level II is capable of assessing the risk of failure in a quantitative manner. There remains some ambiguity however such that the characteristic ratio and the coefficient of variation of various design variables are subject to the judgment of an analyst to a certain degree and the consensus has not been established yet on the acceptable level of failure probability of maritime structures. Because of such ambiguity and/or other reasons, Level II method has not been practiced so often for design of breakwaters and maritime structures. The 2007 version of the *Technical Standards for Port and Harbour Facilities in Japan*,[12] for example, is promoting use of Level I method instead.

Level I method employs the partial factors to be multiplied to the design variables in the limit state function, and design is so made to keep the function at a non-negative value. For the case of breakwater design against sliding, the limit state function is modified as below.

$$Z = \gamma_f \left( \sum_i \gamma_{W_i} W_i - \gamma_U U \right) - \gamma_P P \geq 0 \,, \qquad (6.20)$$

where $\gamma_f, \gamma_{W_i}, \gamma_U$, and $\gamma_P$ denotes the partial factors for the friction coefficient, specific weight of materials, uplift, and horizontal wave force, respectively. The subscript $i$ to the weight $W_i$ is applied to different materials such as reinforced/plain concrete and fill sand. Equation (6.20) differs from the load and resistance format described in ISO 2394[1] in which the partial factor for resistance is employed as a denominator instead of multiplier.

The above partial factors for the sliding failure mode have been set in Japan for the target reliability index of $\beta_T = 2.40$ for caisson breakwaters and horizontally-composite breakwaters as listed in Table 6.5. The partial factors for the failure modes of overturning and bearing capacity are also available in Ref. 12 together for other gravity type special breakwaters such as sloping-top breakwaters, wave-absorbing caisson breakwaters, and upright wave-absorbing breakwaters.

Assessment of the partial factors and others has been made by Yoshioka and Nagao,[13] who analyzed 76 prototype breakwaters in Japanese ports (38 caisson breakwaters and 38 horizontally-composite ones). By employing the characteristic ratio and the coefficient of variation listed in Table 6.1 and others, they calculated the reliability indices of these breakwaters against the individual failure modes of sliding, overturning and bearing capacity at the design wave and tide conditions by Level II method. Among the three failure modes, sliding was confirmed as being the governing mode. The system reliability index varied between 1.6 and 3.0 with the mean of 2.38 and the standard deviation of 0.30. The mean value of $\beta = 2.38$ corresponds to the failure probability of $8.7 \times 10^{-3}$.

Sliding events of actual breakwaters have been investigated by Kawai *et al.*[14] from the damage reports over five years of 1989 to 1993 of about 16,000 upright and horizontally-composite caissons of Japanese breakwaters. During this period, 32 caissons were slid by storm waves. The sliding failure rate defined as the ratio of the number of slid caissons to the total number of caissons was $4.1 \times 10^{-4}$ per year, which corresponds to the failure probability of $2.0 \times 10^{-2}$ over 50 years. Thus the above failure probability calculated by Yoshioka and Nagao is in the same order of magnitude with the actual sliding failure rate.

The partial factors are given by the following formula, once the target value of the reliability index $\beta_T$ is set:

$$\gamma_X = (1 - \alpha_X \beta_T V_X)\frac{\mu_X}{X_k}, \qquad (6.21)$$

where the subscript $X$ denotes the variable being related to the design parameter $X$, $\alpha$ is the sensitivity factor, $V$ is the coefficient of variation, $\mu$ is the mean value, and $X_k$ is the true or characteristic value.

Yoshioka and Nagao set the target reliability index at $\beta_X = 2.40$ which is at a slightly safer side than the mean. Then, they redesigned the

Table 6.5 Patial factor, sensitivity factor, characteristic ratio, and coefficient of variation against sliding failure mode of caisson and horizontally composite-breakwaters.

| Design Parameter | Partial Factors | Description | Caisson Breakwater | | | | Horizontally-Composite Breakwater | | | |
|---|---|---|---|---|---|---|---|---|---|---|
| | | | $\gamma$ | $\alpha$ | $\mu/X$ | $V$ | $\gamma$ | $\alpha$ | $\mu/X$ | $V$ |
| Friction coefficient | $\gamma_f$ | — | 0.79 | 0.689 | 1.060 | 0.150 | 0.77 | 0.750 | 1.060 | 0.150 |
| Horizontal wave force and uplift | $\gamma_P, \gamma_U$ | Slope < 1/30 | 1.04 | −0.704 | 0.740 | 0.239 | 0.91 | −0.636 | 0.702 | 0.191 |
| | $\gamma_P, \gamma_U$ | Slope ≥ 1/30 | 1.17 | −0.704 | 0.825 | 0.251 | 1.01 | −0.636 | 0.772 | 0.205 |
| Specific weight | $\gamma_{W_{RC}}$ | R. concrete | 0.98 | 0.030 | 0.980 | 0.020 | 0.98 | 0.030 | 0.980 | 0.020 |
| | $\gamma_{W_{NC}}$ | Plain concrete | 1.02 | 0.025 | 1.020 | 0.020 | 1.02 | 0.031 | 1.020 | 0.020 |
| | $\gamma_{W_{SAND}}$ | Fill sand | 1.01 | 0.150 | 1.020 | 0.040 | 1.01 | 0.150 | 1.020 | 0.040 |

cross-sections of 76 breakwaters so that they satisfy the condition of $\beta_X = 2.40$. As exemplified in Table 6.4, Level II analysis yields the converged values of the sensitivity factors of design parameters in the process of calculating the reliability index. The variation of the sensitivity factors thus obtained for respective design parameter was relatively small. Yoshioka and Nagao took the mean value as representative of each sensitivity factor, which is listed in Table 6.5. The terms of $\mu/X$ and $V$ listed in Table 6.5 are those employed in the analysis by Yoshioka and Nagao.

### Example 6.1

Design the caisson breakwater shown in Fig. 4.8 with Level I method using the partial factors listed in Table 6.1. The weight of the upright section is assumed to have the weight of $W = 2285$ kN/m as in Example 4.3.

### Solution

The friction coefficient is taken at $f = 0.60$, while the horizontal wave force and the uplift have been calculated as $P = 927$ kN/m and $U = 489$ kN/m. By assuming the partial factor of the weight of the upright section as $\gamma_W = 1$, the value of the limit state function of Eq. (6.20) is calculated as follows:

$$Z = 0.79 \times 0.6 \times (1.0 \times 2285 - 1.04 \times 489) - 1.04 \times 927 = -122 \text{ kN/m}.$$

Because $Z$ is negative, the breakwater is judged unsafe. This is expected because the reliability index analyzed in Sec. 6.2.2 was $\beta = 2.05$ which was smaller than the target value of $\beta_T = 2.40$. A remedy for the design is to introduce an asphalt mat to be laid down underneath the caisson bottom. The friction coefficient between asphalt mat and rubble stone varies between 0.7 and 0.8, as specified in Ref. 15. By using the safe-side value of $f = 0.70$, the limit state function is calculated to have the value of 18.3 kN/m, which is judged as acceptable.

The values of the partial and the sensitivity factors vary, depending on respective design parameters and breakwater types. Compared with caisson breakwaters, horizontally-composite breakwaters indicate smaller values of the partial factors for the horizontal wave force and uplift. It is related to a smaller characteristic ratio for wave force in Table 6.1 and seems to have originated from a tendency of overestimating wave force with the pressure correction factor $\lambda_1$ of Eq. (4.42) in Sec. 4.4.3.

The values of the partial factors listed in Table 6.5 are recommended ones for breakwaters to be built in Japan with the design working life of 50 years or so. For a breakwater to be served for a short working life, say 10 years, the partial factor can be lowered by using Eq. (6.21). If the probability of failure during 10 years is tentatively set at $P_{f,10} = 0.01$, a temporary breakwater designed with this value of probability will have the failure probability of $P_{f,50} = 0.049$ for 50 years. The corresponding value of the reliability index is obtained as $\beta_T = 1.65$ from the standard normal distribution table. Then use of Eq. (6.21) yields the partial factors of $\gamma_f = 0.88$ and $\gamma_P = \gamma_U = 0.95$ for caisson breakwaters on the sea bottom of mild slope. Because of a larger $\gamma_f$ and a smaller $\gamma_P$ and $\gamma_U$, the breakwater with the 10-year working life will be designed with a smaller cross section.

## 6.3 Performance-Based Design of Breakwaters

### 6.3.1 *Outline of Performance-Based Design Method*

In comparison to the reliability-based design method which tries to control the probability of failure of a structure below a preset threshold level as described in Sec. 6.2, the performance-based design sets the criteria for the performance of a structure to be maintained during its design working life and the design is so made to satisfy the criteria. The *Technical Standards for Port and Harbour Facilities in Japan*,[15] for example, define the following four requirements of performance:

(1) *Serviceability*: This is the ability that the facilities should be serviceable without any inconvenience. The structural response of the facilities against possible actions during service should be such that no damage will occur or the damage should be a minor one to the extent that the performance of the facilities will be quickly recovered with minimal repair works.

(2) *Restorability*: This is the ability that the facilities can be in service continuously with technically feasible and economically reasonable repair works. The structural response of the facilities against possible actions during service should be such that the damage will remain within the range in which the required performance will be recovered, in a short period by a small amount of repair works.

(3) *Safety*: This refers to the ability such that the safety will be secured of the people using the facilities and others. The structural response of the facilities against possible actions during service should be such that the damage of a certain level may occur, but the damage would remain below the level which is not be critical for the structural integrity and does not cause significant influence for securing the safety of people.

(4) *Usability*: This is the ability that the facilities should have from the viewpoint of its service and convenience in use. Facilities in ports and harbors are to be arranged at appropriate locations, their structural dimensions (facilities' lengths, widths, water depths, crown elevations, clearance limits, etc.) should satisfy the predetermined values, the required tranquility of harbor basins etc. is to be maintained, and auxiliary facilities are to be provided wherever necessary.

The last item of usability needs to be examined at the stage of the layout plan of port and harbor facilities and/or their basic designs. The other three items of serviceability, restorability and safety deal with the response of facilities when actions from storm waves and others are exercised on them and the degree of damage that may occur by these actions; the tolerable level of damage is raised from the serviceability to safety requirement. The serviceability requirement may corresponds to the serviceability in ISO 2394,[1] which is defined as the ability of a structure or structure element to perform adequately for normal use under all expected actions. ISO 2394 also defines the serviceability limit state and the ultimate limit state, the latter being defined as a state associated with collapse, or with other similar forms of structural failure. The safety performance means that the facilities should remain below the ultimate limit state.

Judgment whether the facilities satisfy the respective performance requirements must be made with quantitative estimate of deformation of a structure or structural elements under external actions and/or corrosion by environmental influence. Prediction is required of not only the danger of collapse but also the degree of deformation. Performance-based design method aims at predicting the magnitude of deformation during the design working life of a structure under design.

Prediction methods are diverse depending on the type of deformation. In the case of coastal dikes and seawalls, for example, the wave overtopping rate is a measure of structural performance and the inundation level behind the facilities can be evaluated with the overtopping rate, provided that the facilities themselves including parapet walls would not collapse during the storm event. In the case of mound breakwaters, design formulas for selection of armor units have the damage level as the design parameter as exemplified in the van der Meer formula.[16] Thus, the restorability and safety performance can be evaluated for various levels of storm conditions.

In the case of vertical/composite breakwaters, sliding distance, tilting angle, and settlement of the upright section are the measure of deformation provided that the front wall and other structural members are not damaged. Among these deformations, sliding distance is the most direct measure of breakwater performance. Shimosako and Takahashi[17] presented the methodology for calculating the expected sliding distance of a composite breakwater during its design working life. They further extended the methodology as a tool of the performance-based design of composite breakwaters[18] in 1998. Since then, the expected sliding distance method has been employed in a large number of breakwater designs in Japan. The next section describes this design method.

### 6.3.2 *Performance-Based Design with Expected Sliding Distance Method*

(A) *Flow for calculating expected sliding distance*

The distance over which an upright section may slid by the action of a single wave can be calculated by the procedure described in Sec. 4.2.4. Under a severe storm wave event, a breakwater may slid by the actions of several large waves and the sliding by such waves would be added to yield the *accumulated sliding distance* by the storm event. During a design working life of the breakwater, it may or may not experience a few storm wave events exceeding the design conditions. If there occur several storm wave events that yield breakwater sliding, the breakwater displacements by individual storms are added to yield the *total sliding distance*. The amount of the total sliding distance wholly depends on the probabilistic nature of storm events. Nevertheless, we can simulate the occurrence of annual maximum storm events and individual waves in storm wave events under respective probability distributions through the technique of Monte Carlo (simulation) method. Each cycle of simulation for one design working life yields an estimate of total sliding distance, which differs from one cylcle to another. The ensemble average of these total sliding distances is an expected value, and it is called the *expected sliding distance* during the design working life of a breakwater.

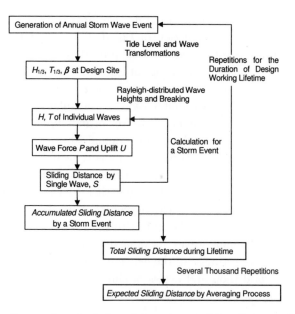

Fig. 6.1 Flow diagram for calculation of the expected sliding distance taken from Shimosako and Takahashi[18] with some modification.

Figure 6.1 shows the flow for calculating the expected sliding distance developed by Shimosako and Takahashi[18] with some modification. First, a cross section of composite breakwater is designed for examination at a site of proposed construction. The dimensions of breakwater are set and the weight of the upright section in still water is calculated.

Using a random-number generation algorithm, one event of annual maximum storm waves is simulated from the extreme distribution function of significant wave height at a designated site for every year during the design working life, say 50 years. The output at this stage is the significant wave height and period and the wave direction at the offshore. Wave transformations from the offshore to the design site including random breaking are computed to yield the data of design significant wave data in front of the breakwater. An annual storm wave event is assumed to continue for two hours for the sake of convenience, the number of waves during which is calculated with the mean wave period. The heights of individual waves are calculated according to the probability density function that is the Rayleigh in the offshore but may be transformed by the random wave breaking process in the surf zone. The periods of individual waves are assumed to be normally distributed or given as equal to $T_{1/3}$.

For each single wave, the horizontal wave force and the uplift are computed and the limit state function of Eq. (6.3) is calculated. If $Z < 0$, the sliding distance is estimated with the procedure in Sec. 4.2.4. Initially Shimosako and Takahashi[17] made use of Eq. (4.35) which has been derived under the approximation of triangular wave force pulse, but later they have come to use the wave force in combination of triangular pulse and pulsating force.[18] Summation of the sliding distances of individual waves yields the accumulated sliding distance during one annual storm wave event.

Simulation of annual storm wave events is repeated for the number of years during the design working life, and the accumulated sliding distance of each year when not zero is added together to yield the total sliding distance. This concludes one cycle of simulation trial. Because the total sliding distance thus obtained is subject to the variation of the occurrence probability of large annual waves, its value fluctuates greatly from one cycle to another. A stable estimate of the expected sliding requires an averaging process over several thousand cycles of simulation trials. Shimosako and Takahashi use the average over 5000 cycles for the final result.

## (B) *Selection of design parameters and their probabilistic description*

Similarly as Level II design method, the performance-based design with the expected sliding distance method needs to make appropriate selection of design parameters and to assign their probability density functions. Respective engineers can make their own selection, but examples from two previous studies[18,19] are cited here for their reference. Table 6.6 summarizes the design parameters and the distribution functions, which were employed in the two studies.

A few explanations are added here to some items in Table 6.6. Shimosako and Takahashi[18] assigned a uniform distribution for the principal wave direction in the range of $\pm 11.25°$ in consideration of customary assignment of the wave direction at one of 16 points bearings. They approximated the tidal level variation with a triangular distribution between the mean spring high and low waters by referring to the probability distributions of astronomical tide levels around Japanese coasts, which were computed by Kawai *et al.*[14] by using the four major tidal constituents. To this tidal level variation, Shimosako and Takahashi added a normal distribution of storm tide, the amplitude of which was pre-assigned. Goda and Takagi[19] did not consider storm tide level, because the study was not site-specific one. They set the design tide level at the mean spring high water and gave the tidal level variation down to the mean spring low water.

Table 6.6 Examples of design parameters employed in calculation of expected sliding distance with their distribution functions.

| Design Parameter | Shimosako and Takahashi[18] | | | Goda and Takagi[19] | | | Remarks |
|---|---|---|---|---|---|---|---|
| | Distribution | $\mu/X$ | $V$ | Distribution | $\mu/X$ | $V$ | |
| Offshore height $(H_{1/3})_0$ | Normal | 1.00 | 0.10 | Normal | 1.00 | 0.10 | Extremal distribution for $(H_{1/3})_0$ |
| Significant period | Normal | 1.00 | 0.10 | Fixed to $H_{1/3}$ | – | – | |
| Wave direction | Uniform | 1.00 | [±11.25°] | Not considered | – | – | |
| Tide level | Astronomical | – | – | Triangular | – | – | Input of tidal amplitudes |
| Storm tide | Normal | 1.00 | 0.10 | Not considered | – | – | |
| Wave transformation | Normal | 1.00 | 0.10 | Normal | 0.87 | 0.10 | |
| Individual height | Rayleigh | – | – | Rayleigh | – | – | |
| Wave actions | Normal | 1.00 | 0.10 | Normal | 0.91 | 0.10 | |
| Friction coefficient | Normal | 1.00 | 0.10 | Normal | 1.10 | 0.10 | $X_f = 0.60$ |

The period of the annual maximum significant wave was calculated from the condition of the deepwater wave steepness of $H_0/L_0 = 0.039$ in the study by Goda and Takagi, but a later study by Goda[20] gave the significant wave period on the basis of Eq. (3.6) in Sec. 3.1.2.

The statistical variation of the weight of upright section is small compared with other design parameters, as exemplified in its small value of the sensitivity factor in Table 6.5 in Sec. 6.2.3. The above two studies treated the weight of upright section as a deterministic variable.

The design parameter of friction coefficient is usually represented with a normal distribution in the probabilistic design, but its simple application in the Monte Carlo simulation may yields an excessively small or large value of friction coefficient. The original data of friction coefficient examined by Takayama and Ikeda[3] were bounded by the smallest value of 0.45 and the largest value of 0.91 with the mean of 0.636 and the standard deviation of 0.096. Kim and Takayama[21] have proposed to apply a doubly-truncated normal distribution to the friction coefficient for calculation of the expected sliding distance. They also employed a doubly-truncated normal distribution to the estimated wave force. With these modifications, they calculated the expected value of the accumulated sliding distance of the horizontally-composite breakwater in Susami Fishing Harbor, Wakayama Prefecture, Japan, which was damaged by the storm waves generated by the Typhoon Tokage (No. 0423). The calculated sliding distance agreed well with the actual damage to the breakwater (Kim *et al.*[22]).

Computed results of the expected sliding distance are affected by the selection of design parameters and their probability distribution to be incorporated in the simulation works. As seen in Table 6.6, the two previous simulation studies made use of different sets of design parameters, partly because both studies had the objectives of gaining general conclusions on the characteristics of the expected sliding distance. When a specific design of breakwater is to be made at a particular site, the characteristics of waves and tides at the site should be fully investigated and the most appropriate distribution functions should be incorporated in the simulation works.

Among various design parameters, the extreme distribution of deepwater significant wave height exercises the largest influence on the computed result of expected sliding distance. An extreme wave height distribution with a heavy tail (extending long toward the large value) yields a few extremely large heights occasionally and causes to increase the expected sliding distance. Selection of an extreme distribution function with a light tail brings forth a small value of the expected sliding distance. When the extreme distribution of storm wave heights has not been well established at the design site, the information of the extreme wave statistics at neighboring stations should be investigated and an appropriate extreme distribution function should be selected together with the parameters for scale, location, and shape. The values of these parameters should not be assigned subjectively but with due consideration for the spread parameter of the extreme distribution of wave heights; the spread parameter is defined as the ratio of the 50-year wave height to the 10-year wave height as a measure of the degree of extension of the distribution tail toward the right (see Sec. 13.1.5 for detailed discussion).

## (C) *Calculation of expected sliding distance and its interpretation*

Calculation with the flow in Fig. 6.1 requires only a small computation load on PC and is easily adapted for routine design works. The output is the value of the expected sliding distance such as 0.18 m during the design working life of 50 years for one design, 0.12 m for another design, and so forth. Because the development of the expected sliding design method is relatively new, consensus has not been reached on the acceptable amount of the expected sliding distance. Nevertheless, Takahashi and Shimosako[23] have proposed a matrix for the acceptable amount of the accumulated sliding distance for a designated storm wave event, which depends on the importance of breakwaters, as listed in Table 6.7. Calculation of the accumulated sliding distance is made by fixing the offshore waves at the designated event, but all other design parameters are given probabilistic variations. A large number of simulations are to be made to yield a reliable average value.

Table 6.7 reads such that a very important breakwater (A) may have the accumulated sliding distance up to 0.03 m when encountered with a storm event with the return period of 500 years and up to 0.10 m against an extraordinary 5000-year storm event. A less important breakwater (C) may be collapsed by the storm event with the return period of 500 years, but its accumulated sliding distance should be equal to or less than 0.30 m against the 50-year storm event.

Another proposal has been made by Shimosako and Tada[24] for the exceedance rate of the total sliding distance, which should be maintained at the level listed in Table 6.8. To apply this proposal, computation flow of the expected sliding distance needs to be modified so as to store the output of the total sliding distance during the design working life for individual simulation cycles in a computer memory and to make a frequency analysis of the total sliding distance.

Table 6.7 Acceptable amount of accumulated sliding distance by a storm event for breakwaters of different levels of importance, as proposed by Takahashi and Shimosako.[23]

| | Acceptable Sliding Distance | | | |
|---|---|---|---|---|
| Storm Wave Event | 0.03 m (Serviceability Limit) | 0.10 m (Restorability Limit) | 0.30 m (Ultimate Limit) | 1.0 m (Collapsed State) |
| 5-year event | B | C | – | – |
| 50-year event | – | B | C | – |
| 500-year event | A | – | B | C |
| 5000-year event | – | A | – | – |

Note: The letters A, B, and C refer to the importance of a breakwater under design such as 'very important,' 'important,' and 'less important,' respectively.

Table 6.8 Acceptable exceedance rate of total sliding distance for breakwaters of different levels of importance, as proposed by Shimosako and Tada.[24]

| Importance of Breakwater | Exceedance Rate of Total Sliding Distance | | |
|---|---|---|---|
| | 0.1 m | 0.3 m | 1.0 m |
| Low | ≤ 50% | ≤ 20% | ≤ 10% |
| Medium | ≤ 30% | ≤ 10% | ≤ 5% |
| High | ≤ 15% | ≤ 5% | ≤ 2.5% |

Both proposals of acceptable sliding distance listed in Tables 6.7 and 6.8 are based on observations of several cases of sliding failure damage of breakwaters in Japan, but no systematic calibrations on a large number of existing breakwaters have been undertaken yet. Such calibrations with classification of slid cases and non-slid cases remain as the future task of harbor engineers in Japan.

## (D) *Optimum design of breakwaters with the concept of expected sliding distance*

A vertical/composite breakwater can be designed by satisfying the acceptable value of the accumulated or total sliding distance listed in Table 6.7 or 6.8. Several trial designs will be required until the optimum cross section of the breakwater is found. A horizontally-composite breakwater will also be designed with the concept of expected sliding distance by using appropriate wave pressure formulas as done by Shimosako *et al.*[25] Examination of the breakwater damage at Susami Fishing Harbor by Kim and Takayama[21] with the expected sliding distance method is an example of application to a horizontally-composite breakwater.

The performance-based design method with the expected sliding method can also be expanded to the design optimization of breakwaters. One approach is the minimization of the total cost that is the sum of the costs of initial construction and expected repair works necessitated by possible damage during the design working life. The design wave is fixed and the sliding failure damage will be evaluated for a number of alternative breakwater designs. Several studies have been made in Japan on the methodology of optimum breakwater design using the expected sliding distance method.

Another approach was taken by Goda and Takagi,[19] who varied the design wave height over a wide range of return period in search of the optimum breakwater design having the minimum total cost and keeping the expected sliding distance below some acceptable amount. They gave a name of *rational return period* to the return period of the wave height of the optimum design, and calculated the ratio of the rational return period to the design working life. It was found that the ratio is affected strongly by the water depth relative to the design significant wave height $(H_s)_D$. If the water depth is less than about $1.5(H_s)_D$, for example, a composite breakwater can be safely designed with the waves corresponding to the return period shorter than the design working life of the breakwater. If the water depth is deeper than about $3.0(H_s)_D$, on the other hand, a breakwater needs to be designed against the waves corresponding to the return period being 2 to 8 times the design working life. Shortening of the design return period in relatively shallow waters is induced by the limitation in the maximum wave height by wave breaking. Because the maximum wave height is not limited by wave breaking in relatively deep waters, there is a possibility of large wave height and of sliding failure.

Goda[20] further explored this aspect by employing the acceptable exceedance level of the total sliding distance for a breakwater of medium importance, i.e., keeping the exceedance rate of the total sliding distance of 0.3 m below 10%. He called the required minimum caisson width as *optimum width* $B_{opt}$ and calculated the ratio of this width to the nominal design wave height $(H_{1/3})_{0D}$, which is the significant wave height corresponding to the return period of 50 years. He calculated the optimum caisson width ratio $B_{opt}/(H_{1/3})_{0D}$ for the wave heights corresponding to the return

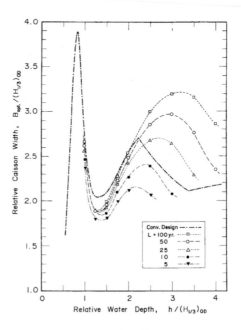

Fig. 6.2   Calculated examples of optimum caisson width satisfying the condition of acceptable total sliding distance.[20]

period ranging from 5 to 100 years. The calculated results of the optimum caisson width ratio are shown in Fig. 6.2. The rubble mound foundation is given the thickness of a larger one of 2.5 m or 20% of water depth.

The caisson width designed by conventional method with the overall safety factor of 1.20 for the design return period of 50 years is shown with the dash-dot line. The peak around the relative depth of $h/(H_{1/3})_{0D} \simeq 0.85$ represents the increase of caisson width against the impulsive breaking wave force, which is predicted by the impulsive pressure coefficient introduced in Sec. 4.2.3. The conventionally-designed caisson width shows the second peak at $h/(H_{1/3})_{0D} \simeq 2.2$, beyond which the maximum wave height takes a constant value of $\beta^*_{max} H'_0$. Beyond $h/(H_{1/3})_{0D} \simeq 3.3$, the caisson width gradually increases, because the failure mode of overturning becomes dominant over the sliding failure.

The performance-based design shows a strong influence of the design return period on the optimum caisson width. In the zone of relatively shallow waters with $h/(H_{1/3})_{0D} \leq 2.0$, the optimum caisson width is equal to or less than the conventionally-designed width even if the design return period is taken at $L = 100$ years. However, in the zone of relatively deep waters with $h/(H_{1/3})_{0D} \geq 2.5$, the performance-based design demands the caisson width larger than the conventional one when the design return period is longer than 25 years. The result of Fig. 6.2 suggests that a conventionally-designed composite breakwater in relatively deep waters may slid over a large distance when attacked by the design storm waves. Thus, breakwater design in deeper water areas must be done with every care, because wave heights there are no more bounded by the depth-limited breaking process.

### 6.3.3 Vertical Breakwater Design with Modified Level I Method

In breakwater designs, examination of the performance requirements of *serviceability*, *restorability*, and *safety* introduced in Sec. 6.3.1 is not possible unless estimation is made of the expected sliding distance against the design wave condition or some extraordinary waves exceeding the design condition. A system of partial factors under Level I method such as listed in Table 6.5 is easy for use by practitioners, but it cannot predict the amount of possible deformation of sliding and others. Furthermore, the partial factors recommended in the *Technical Standards for Port and Harbour Facilities in Japan* have been set without taking into account the characteristics of extreme wave distribution, the relative water depth, and other factors.

To remedy the above shortcomings, Yoshioka *et al.*[26] have collected the design data of 33 existing caisson breakwaters and made conventional design calculations with the overall sliding safety factor of 1.2. Then they carried out the simulation calculations of expected sliding distance for these breakwaters and assessed the probability that the total sliding distance exceeds 0.30 m for each breakwater. The probability ranged from 0.5% to 11.7% with the expected sliding distance ranging from 0.7 to 42.5 cm. The mean failure probability converted from the mean of the reliability index was 2.6%. Then Yoshioka *et al.* redesigned the breakwaters with the expected sliding distance method by taking the mean exceedance probability of 2.6% as the target probability. For these redesigned cross-sections of breakwaters, Yoshioka *et al.* evaluated the reliability index with Level II method. The resultant values of the reliability index were subjected to the linear multiple regression analysis in search of major factors which affect the reliability index in consideration of sliding distance.

Through this process, Yoshioka *et al.* have come to propose the following estimation formula:

$$\beta_{\text{SLT}} = 2.40 \times \max\left\{0.60, \ (2.7 + 0.2\alpha_1 - \alpha^* \cos^2\beta_{\text{dir}} - 1.6\kappa)\right\}, \quad (6.22)$$

where $\beta_{\mathrm{SLT}}$ denotes the target reliability index in consideration of sliding performance, $\alpha_1$ is the pressure coefficient defined by Eq. (4.6) in Sec. 4.2.2, $\alpha^*$ is the pressure coefficient defined by Eq. (4.19) in Sec. 4.2.3, $\beta_{\mathrm{dir}}$ is the wave incident angle, and $\kappa$ is a new parameter defined as in the following:

$$\kappa = \begin{cases} \gamma_{50} & : \gamma_{50} \leq 1.2 \,, \\ 1.2 & : \gamma_{50} > 1.2 \quad \text{and} \quad h/(H_{1/3})_{0D} \leq 3.0 \,, \\ \max\{1.0, (2.4 - \gamma_{50})\} & : h/(H_{1/3})_{0D} > 3.0 \,, \end{cases}$$

$$(6.23)$$

where $\gamma_{50}$ denotes the spread parameter as defined by $H_{50}/H_{10}$ in Sec. 13.1.5 for describing the degree of extension of the distribution tail toward the right, or heaviness of the tail of the distribution. The term $(H_{1/3})_{0D}$ represents the design significant wave height corresponding to the return period of 50 years. According to Eq. (6.23) the parameter $\kappa$ has the maximum value of 1.2 at which the target reliability index $\beta_{\mathrm{SLT}}$ takes a minimum value.

Although the influence of the spread parameter on the optimum breakwater design is not in accord with the result of a simulation study by Goda,[20] introduction of the target reliability index of Eq. (6.22) is an improvement over the partial factor system with the values in Table 6.5. Yoshioka *et al.* has called the method with Eqs. (6.22) and (6.23) as the *modified Level I method*. The values of the partial factors are calculated by Eq. (6.21) with the new target reliability index of $\beta_{\mathrm{SLT}}$. Yoshioka *et al.* state that the sensitivity factors, the coefficient of variation, and the characteristic ratio can be set as same as those in Table 6.5.

### Example 6.2

A 50-year return wave height of $(H_{1/3})_{0D} = 7.5$ m was obtained from an extreme wave height distribution with the spread parameter of $\gamma_{50} = 1.17$ at a design site. The significant period of this design wave is $T_{1/3} = 12$ s. Calculate the partial factors with the modified Level I method for designing a caisson breakwater at the depth of $h = 18$m on the bottom slope of 1/20. The incident wave angle is $\beta_{\mathrm{dir}} = 15°$. The rubble mound foundation has the thickness of 4 m $(h' = 14)$ m and the foot protection layer is 2 m thick $(d = 12$ m$)$.

### Solution

Under the given conditions, the pressure coefficients are calculated as $\alpha_1 = 0.838$ and $\alpha^* \cos^2 \beta_{\mathrm{dir}} = 0.118$. The parameter $\kappa$ is equal to

Table 6.9 Examples of modified values of partial factors by the modified Level I method.

| Design Parameter | Sensitivity Factor $\alpha$ | Characteristic Ratio $\mu/X$ | Coefficient of Variation $V$ | Partial Factor | |
|---|---|---|---|---|---|
| | | | | Recommended | Modified |
| Friction coefficient: $f$ | 0.689 | 1.060 | 0.150 | 0.79 | 0.83 |
| Wave force and uplift: $P, U$ | −0.704 | 0.740 | 0.239 | 1.04 | 1.00 |
| Specific weight (sample): $\gamma_W$ | 0.030 | 0.980 | 0.020 | 0.98 | 0.98 |

$\gamma_{50} = 1.17$ by Eq. (6.23). The target reliability index is calculated as below.

$$\beta_{\text{SLT}} = 2.40 \times \max\{0.60, \ (2.7 + 0.2 \times 0.838 - 0.118 - 1.6 \times 1.17)\}$$
$$= 2.40 \times \max\{0.60, 0.878\} = 2.11.$$

The partial factors calculated with Eq. (6.20) together with the data in Table 6.5 are obtained as listed in Table 6.9.

The partial factor for friction coefficient increases, while that for wave force and uplift decreases in comparison with the recommended values. Thus the caisson will be designed with a smaller width than the one with the partial factor values recommended in Table 6.6 of Level I method.

Adjustment of the target reliability index for respective design conditions leads to the use of the target failure probability being different from one design to another. It may sound unreasonable from the principle of reliability-based design method, but the breakwaters designed under the modified Level I method maintain the same level of safety from the viewpoint of the exceedance probability of the total sliding distance. Future design of vertical breakwaters would be proceeded by making a preliminary design with the modified Level I method, and then the design would be scrutinized with the expected sliding distance method based on the Monte Carlo simulation technique.

## References

1. International Organization for Standardization, *General principles on reliability for structures*, ISO 2394 (1998).

2. International Organization for Standardization, *Actions from waves and currents on coastal structures*, ISO 21650 (2007), Clause 6.1.

3. T. Takayama and N. Ikeda, "Estimation of sliding failure probability of present breakwaters for probabilistic design," *Rept. Port and Harbour Res. Inst.* **31**(5) (1992), pp. 3–32.

4. T. Nagao, "Reliability based design way for caisson type breakwaters," *J. Structural Engrg. JSCE* No. 699 (I-57) (2001), pp. 173–182 (*in Japanese*).

5. T. Nagao, "Reliability based design method for checking the external safety of caisson type breakwaters," *Res. Rept. National Inst. Land and Infrastructure Management* No. 4 (2002), 26p. (*in Japanese*).

6. K. Shimosako and S. Takahashi, "Reliability design method of composite breakwater using expected sliding distance," *Rept. Port and Harbour Res. Inst.* **37**(3) (1998), pp. 3–30 (*in Japanese*).

7. Y. Goda and T. Fukumori, "Laboratory investigation of wave pressures exerted upon vertical and composite walls," *Rept. Port and Harbour Res. Inst.* **11**(2) (1972), pp. 3–45 (*in Japanese*).

8. International Organization for Standardization, *Loc. cit.* Ref. 2, Annex D.

9. K. Minami and Y. Kasugai, "The application of the limit state design method to RC structures in port facilities," *Tech. Note Port and Harbour Res. Inst.* No. 716 (1991), 47p. (*in Japanese*).

10. H. F. Burcharth, "Reliability-based design of coastal structures," *Advances in Coastal and Ocean Engineering* **3** edited by P. L.-F. Liu, World Scientific (1997), pp. 145–214.

11. H. F. Burcharth, "Reliability based design of coastal structures," Chapter VI-6 of *Coastal Engineering Mannual* EM 1102-2-1100, U.S. Army Corps of Engineers (2002).

12. Ports and Harbours Bureau (Ministry of Land, Infrastructure, Transport and Tourism), National Institute for Land and Infrastructure Management, and Port and Airport Research Institute, *Technical Standards and Commentaries for Port and Harbour Facilities in Japan*, Overseas Coastal Area Development Institute of Japan (2009), Part III, Chap. 4, Sec. 3.1.4.

13. T. Yoshioka and T. Nagao, "Level-1 reliability-based design method for gravity-type breakwaters," *Res. Rept. National Inst. Land and Infrastructure Management* No. 20 (2005), 38p. (*in Japanese*).

14. H. Kawai, T. Takayama, Y. Suzuki and T. Hiraishi, "Failure probability of breakwater caisson for tidal level variation," *Rept. Port and Harbour Res. Inst.* **36**(4) (1997), pp. 3–41 (*in Japanese*).

15. *Loc. cit.* Ref. 12, Part I, Chap. 1, Sec. 1.4.

16. J. W. van der Meer, "Stability of breakwater armor layer - Design formulae," *Coastal Engineering* **11** (1987), pp. 219–239.

17. K. Shimosako and S. Takahashi, "Calculation of expected sliding distance of composite breakwaters," *Proc. Coastal Engrg., JSCE* **41** (1994), pp. 756–760 (*in Japanese*).

18. K. Shimosako and S. Takahashi, "Reliability design method of composite breakwaters using expected sliding distance," *Rept. Port and Harbor Res. Inst.* **37**(3) (1998), pp. 3–30 (*in Japanese*).

19. Y. Goda and H. Takagi, "A reliability design method of caisson breakwaters with optimal wave heights," *Coastal Engineering Journal* **42**(4) (2000), pp. 357–387.

20. Y. Goda, "Performance-based design method of caisson breakwaters with new approach to extreme wave statistics," *Coastal Engineering Journal* **43**(4) (2001), pp. 289–316.

21. T.-M. Kim and T. Takayama, "Computational improvement for expected sliding distance of a caisson-type breakwater by introduction of a doubly-truncated normal distribution," *Coastal Engineering Journal* **45**(3) (2003), pp. 387–419.

22. T.-M. Kim, T. Yasuda, H. Mase and T. Takayama, "Computational analysis of caisson sliding distance due to Typhoon Tokage," *Proc. 3rd Int. Conf. Asian and Pacific Coasts (APAC 2005)* (Jeju, Korea) (2005), pp. 565–576 (in CD-ROM).

23. S. Takahashi and K. Shimosako, "Performance design for maritime structure and its application to vertical breakwaters — Caisson sliding and deformation-based reliability design," *Proc. Advanced Design of Maritime Structures in the 21st Century* edited by Y. Goda and S. Takahashi, Port and Harbour Tech. Res. Inst., Yokosuka, Japan (2001), pp. 63–73.

24. K. Shimosako and T. Tada, "Study on the allowable sliding distance based on the performance design of composite breakwaters," *Proc. Coastal Engrg., JSCE* **50** (2003), pp. 788–770 (*in Japanese*).

25. K. Shimosako, T. Tomimoto, E. Nakagawa, N. Osaki and F. Nakano, "Performance design of horizontally composite breakwaters based on the sliding distance," *Annual J. Coastal Engrg, JSCE* **53** (2006), pp. 896–900 (*in Japanese*).

26. T. Yoshioka, T. Nagao and Y. Moriya, "The method on determination of partial factors for caisson-type composite breakwaters in regard to sliding deformation," *Annual J. Coastal Engrg, JSCE* **52** (2005), pp. 811–815 (*in Japanese*).

# Chapter 7

# Harbor Tranquility

## 7.1 Parameters Governing Harbor Tranquility

The fundamental function of a harbor is to provide safe anchorage for vessels and to facilitate smooth and unhindered transfer of passengers and cargo between vessels and land. Guaranteed harbor tranquility is not only essential for safe anchorage, but it is also important for efficient port operation. The problem of harbor tranquility essentially reduces to the questions of the motions of ships moored at anchorage or along a wharf and of the mooring forces. Many hydraulic laboratories in Europe have been testing harbor tranquility by measuring the motions of model ships moored at strategic places in a harbor and thus judging the goodness of harbor planning by the degree of ship motions, e.g., Russel.[1]

A number of hydraulic model tests of harbor tranquility, however, are still made by measuring the distribution of wave height in a given harbor layout and comparing various harbor layouts by means of either the absolute magnitude of wave height or its ratio to the incident wave height. But it should be kept in mind that the effect of wave action on a ship is different for wind waves with a period of 5 s and for swell with a period of 15 s, even if the wave height is the same. Also, a large ship may not feel wave agitation whereas a small boat may be violently swung by the same waves. Thus, harbor tranquility needs to be judged from the viewpoint of ship motions.

The motions of ships, especially of ships moored in water of finite depth, is a very complex hydrodynamic problem. The layout of the mooring lines, their elastic characteristics, and the response of fenders further complicate the problem. Although much progress has been made in the numerical simulation analysis of the problem, it will still require some time before engineers can have detailed information on the characteristic of ship motions. From the viewpoint of port operation, the relationship between ship motion and cargo handling works also enters into the judgment of harbor tranquility. For example, loading and unloading operations by means of mast cranes of a ship will not be hindered much by movements of a ship by a meter or so. But the handling of containers stacked inside the deep hull of a container ship will have to be halted if the ship moves to and fro more than half a meter. Wind is an additional factor affecting the safety of ships and the efficiency of cargo handling in a harbor, and thus the wind also enters into the problem of harbor tranquility.

In addition to physical factors, there are several other factors connected with the evaluation of harbor tranquility, such as the ease of ship navigation at the harbor entrance, the question of whether ship will remain within the harbor during severe storm conditions, and the limiting conditions of maritime operations. Also important are economic factors, such as the efficiency of cargo handling works, the cost of demurrage during the suspension of ship operation, and the construction cost of breakwaters and other protective facilities.

Figure 7.1 is an example of a flow chart which illustrates the complicated and diverse factors connected with the problem of harbor tranquility and the way to deal with them. The solution should begin with an understanding of the offshore waves and storm winds, and then proceed to the estimation of the waves and winds in the harbor by taking into account the sheltering effects of the breakwaters and other protective facilities. The data on the waves and winds in a harbor should be combined with the motions of the ships and cargo handling equipment to yield the limit for workable conditions in the harbor. Then the questions of ship navigation at the entrance and ship refuge at the times of heavy storms must be taken into account in the judgment of safety. Finally, the tranquility of a harbor under planning should be judged with due consideration to various economic factors. Planning and design of harbor facilities should proceed by such steps to guarantee the desired degree of harbor tranquility.

The above or a similar procedure for the analysis of harbor tranquility is rather difficult to carry out if all the factors are to be evaluated quantitatively (because of the lack of various necessary data). Nevertheless, efforts are being directed at clarifying the roles of these factors, and it is hoped that a systematic and comprehensive procedure such as shown in Fig. 7.1 will be followed in the near future. For the time being, a practical approach to the problem of harbor tranquility may still be the evaluation of the wave

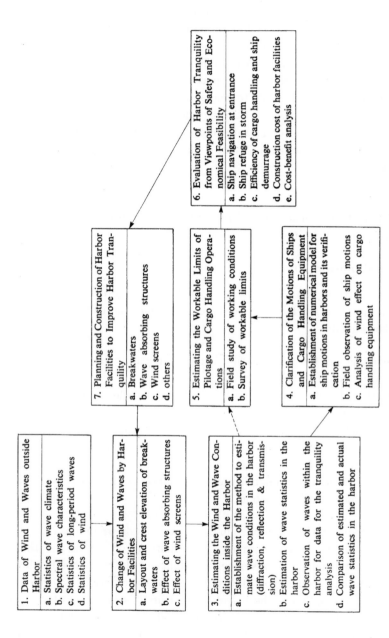

Fig. 7.1 Elements of harbor tranquility and a flow chart for tranquility analysis.

height in the harbor under planning, especially because engineering practice prefers continuity of design method based on conventional techniques
of hydraulic model tests, as well as on numerical analysis of harbor tranquility. But the description of the wave height in a harbor either as an
absolute value or as its ratio to the offshore height under certain storm
conditions alone represents only one part of the complex problem of harbor tranquility. It is necessary to reanalyze the wave height data in the
form of the days of exceedance of the significant wave height above certain
levels during a year at several strategic points in the harbor. By doing so,
the safety and economic feasibility of a planned harbor can be evaluated
somewhat quantitatively.

## 7.2   Estimation of the Probability of Wave Height Exceedance Within a Harbor

### 7.2.1   *Estimation Procedure*

When harbor tranquility is to be evaluated by means of the number of
days of a certain wave height exceedance, the following calculation steps
are recommended:

(i)   Preparation of the joint distribution of significant wave height, period
      and direction outside the harbor.
(ii)  Selection of strategic points within the harbor at which the wave height
      is to be estimated.
(iii) Estimation of the ratio of wave height inside to outside the harbor.
(iv)  Calculation of the absolute height of waves in the harbor at various
      levels of the offshore wave height.
(v)   Calculation of the probability of exceedance of the wave height at
      selected points in the harbor.

   The procedure may be compiled in a flow chart as shown in Fig. 7.2. The
flow chart combines various sources of information and synthesizes them
for the estimation of the exceedance probability of the wave height. The
procedure shown here assumes that the effect of wave period is negligible
for the sake of simplifying the calculation. But it is apparent that ship
motion is strongly dependent on the wave period. Thus, it is recommended
that in practice the analysis of the exceedance of the probability of wave
height be carried out separately for a few classes of wave period.

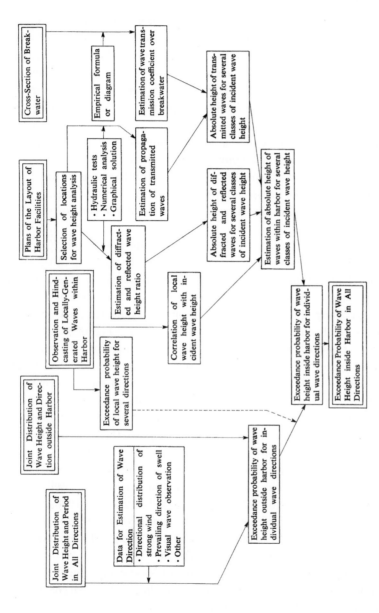

Fig. 7.2  Flow chart for the exceedance probability of wave height within a harbor.

### 7.2.2   *Joint Distribution of Significant Wave Height, Period and Direction Outside a Harbor*

The statistics of the wave climate outside a harbor under planning, obtained either by instrumental wave measurements (backed up by visual observations) or by wave hindcasting, are best analyzed in the form of exceedance probabilities of significant wave height for several directions of wave approach, and separately for a few classes of wave period. In the following, however, an explanation is given for an example without separation by wave period for simplicity.

The probability of exceedance is to be expressed as a percentage over the total number of wave data points inclusive of calm days. It is a standard technique to analyze wave climate statistics on the basis of wave data covering a period of more than five years, but it might be necessary to be content with the data for only three years if a reliable record of sufficient length is not available.

The wave direction is usually described by using a system of 16-point bearings. This system should also be employed in the analysis of the wave height exceedance probability. If the available wave data are scant, a compromise by using an 8-point bearing system may become necessary.

Although many stations are carrying out instrumental measurements of the wave heights and periods, only a limited number of stations are making simultaneous measurements of the wave direction. Therefore, it is quite a fortunate circumstance if the required statistics of wave height and direction can be obtained from the records of field measurements. In many instances, the wave direction needs to be estimated subjectively by the analyst by referring to the frequency diagrams of strong wind direction, the characteristics of the swell persisting at the site, the available data of visual observations of the wave direction, etc.

For a site where no field measurement data is available, a wave hindcasting project should yield an estimate of wind waves and swell several times a day for the duration of the given period. Since the 1990s, several databases of world wave forecasting have become accessible through meteorological institutions. These databases could be the good source of wave climate information inclusive of wave direction.

Table 7.1 is an example of the direction-wise exceedance probability of significant wave height at Akita Port located at the northeasern coast of the Japan Sea, based on the instrumental wave records from January through December, 1974. It was prepared by the First District Port Construction

Table 7.1 Example of exceedance probability of wave height in different directions.

−Akita Port (1974)−

| Wave Direction | Percentage (%) | Significant Wave Height $H_{1/3}$ | | | | | | |
|---|---|---|---|---|---|---|---|---|
| | | Over 0 m | Over 1 m | Over 2 m | Over 3 m | Over 4 m | Over 5 m | Over 6 m |
| SW | 1.4 | 1.4 | 0.4 | 0.1 | 0 | 0 | 0 | 0 |
| WSW | 29.0 | 29.0 | 9.2 | 2.5 | 1.2 | 0.5 | 0.3 | 0.1 |
| W | 68.2 | 68.2 | 21.6 | 6.2 | 2.8 | 1.2 | 0.7 | 0.2 |
| WNW | 1.4 | 1.4 | 0.4 | 0.1 | 0 | 0 | 0 | 0 |
| Total | 100.0 | 100.0 | 31.6 | 8.9 | 4.0 | 1.7 | 1.0 | 0.3 |
| Prevailing period $T_{1/3}$ | | 7.0 s | | | 9.0 s | | 11.0 s | |

Bureau of the Ministry of Transport, Japan.[a] The daily wave direction was analyzed off the screen of a special imaging radar. Although the wave observation itself has continued for many years, the exceedance probability was analyzed for only one year for the purpose of comparing the result of the estimation of the wave height within the harbor to the observed wave height inside the harbor for the same period.

The example shown here employs the class interval of 1 m for the wave height. A narrower class interval of 0.5 m or less would be more suitable for a coast with a mild wave climate, such as a site located within a bay or in another relatively confined body of water. The present example will not involve making a period-wise analysis, as discussed earlier. Even without the distinction of wave period, it is necessary to assign the predominant wave period with respect to wave direction, or to the respective class of wave height, by referring to the joint distribution of significant wave height, period and direction. The predominant wave period associated with certain directions or heights should be employed in the estimation of the wave height ratio in a harbor.

In a harbor of large dimensions, wind waves generated within the harbor may become a source of disturbance for small craft, such as coastal vessels and lighters.[b] In this situation, a statistical analysis of the local wind waves

---

[a]Through the reorganization of the ministerial government system of Japan in 2001, it is now the Hokuriku Regional Development Bureau, the Ministry of Land, Infrastructure, Transport and Tourism.

[b]In a harbor, accommodating large ocean-going vessels along piers and wharves, local wind waves do not in any way hinder a ship's mooring and cargo handling operations, because the sizes of such vessels are very large compared to the wavelength.

must be performed mainly on the basis of hindcast data inferred from the annual wind statistical data. If possible, the data of the local wind waves should be related to information of the offshore waves so that a certain class of offshore wave heights is associated with a certain height of the local wind waves.

### 7.2.3 *Selection of the Points for the Wave Height Estimation*

The estimation of wave height within a harbor is made at a limited number of points taken to be representative of the situation for port operation. When harbor tranquility is investigated through hydraulic scale models, the harbor is divided into areas; in each area several points are selected for measurement of the wave height. The exceedance probability of the wave height in a harbor is usually calculated with the wave height ratio averaged over individual areas. When the ratio of wave height inside to outside the harbor is estimated by numerical computations, to be described below, the wave height ratio is obtained at a number of points. In such cases, too, an area average of the wave height ratio facilitates the overall analysis of harbor tranquility.

If the graphical solution method to be described in Sec. 7.3 is used, the number of estimation points is best limited to a dozen or so in consideration of the amount of manual labor involved and the inherent low level of accuracy of the graphical solution itself. The points to be used for the wave height analysis should not be in the proximity of quay walls and recessed corners of structures, where interference by reflected waves is intense; they should be at the centers of channels, basins and other such locations.

### 7.2.4 *Estimation of Wave Height in a Harbor Incident Through an Entrance*

The majority of waves observed in a harbor are those incident through the harbor entrance. Their height is estimated through hydraulic model tests, numerical computations and/or graphical solution methods. The discussion on hydraulic model tests will be made in Chapter 8. As for numerical computations, several techniques have been proposed, based on different principles. For example, Barailler and Gaillard[2] solved the problem by means of Green functions, Tanimoto et al.[3] calculated the propagation of individual waves by solving the wave equation, and Abbott et al.[4] used the

Bernoulli equation for an efficient solution of wave propagation. All these methods require the harbor area to be represented on a grid with a mesh spacing of an eighth of the wavelength or so, thus limiting their applicability to a harbor of several wavelengths or less in size.

On the other hand, Takayama[5,6] has extended the solution of wave diffraction in 1977, so as to include the effects of wave reflection from various boundaries within a harbor as well as the effects of secondary diffraction by auxiliary breakwaters and groins. Although the technique is limited to water of constant depth, there are no limitations concerning the size of the harbor, and in addition it has the capability of dealing with random sea waves specified with directional spectra. Because of such merits, the numerical method by Takayama has been widely utilized in harbor planning in Japan. Since the late 1990s, the Boussinesq equation model has become a favored tool for analyzing wave height distribution within a harbor of complicated geometry, thanks to the rapid growth in computation power. The aspect of numerical wave analysis will be discussed in Chapter 12.

The wave height in a harbor is in essence estimated as its ratio to the height of the incident waves at the entrance or in the offshore. The estimation must be done with due consideration of the random nature of sea waves, especially of the directional spreading of wave energy. With installation of multidirectional wave generators in many hydraulic laboratories since the 1990s, it has become possible to carry out harbor model tests using directional random waves. If model tests are made with unidirectional irregular waves owing to the unavailability of multidirectional wave generators, the tests must be done for several wave directions so as to take into account the effect of directional spreading of wave energy. The test results are synthesized by the following formula:

$$K_{\text{eff}} = \left[ \frac{\sum K_j^2 D_j}{\sum D_j} \right]^{1/2}, \tag{7.1}$$

where $K_j$ denotes the ratio of wave height at a point or of an area within the harbor to the incident height for the $j$th direction of wave approach, $D_j$ represents the relative wave energy in the respective directions listed in Table 3.2 in Sec. 3.2.2, and $K_{\text{eff}}$ is the effective wave height ratio accounting for the directional wave characteristics. The operation expressed by Eq. (7.1) is simply to take a weighted mean of the wave energy for several directional components, on the basis of unidirectionally estimated ratios of wave height.

Table 7.2 Diffraction coefficient of random sea waves at the location under study.

| Wave Direction | Diffraction Coefficient | | |
| --- | --- | --- | --- |
| | $T = 7$ s | $T = 9$ s | $T = 11$ s |
| SW | 0.101 | 0.116 | 0.130 |
| WSW | 0.277 | 0.290 | 0.303 |
| W | 0.508 | 0.587 | 0.595 |
| WNW | 0.849 | 0.852 | 0.855 |

When hydraulic model tests are carried out with regular waves, the resultant wave height ratio must be modified so as to yield the response of the harbor to random directional incident waves. Wave height ratios are to be obtained for several representative values of the wave period and direction, just as in the case of the estimation of random wave refraction discussed in Sec. 3.2.

In the example of Akita Port, for which the wave statistics are listed in Table 7.1, the wave measurements[c] continued for one year at a location 800 m inward from the tip of the south breakwater (with its axis approximately in the direction of NW) and with an offset of 620 m to the breakwater. The ratio of the wave height at that location to the incident wave height was obtained in calculation of the diffraction of random sea waves by a semi-infinite breakwater. The results are listed in Table 7.2. In this example, the differences in diffraction coefficients between the three wave periods are rather small, indicating relatively little effect of the incident wave period in this case.

### 7.2.5 *Estimation of Waves Transmitted Over a Breakwater*

If some wave energy is transmitted to the interior of a harbor by wave over-topping or passing through breakwaters, the wave transmission coefficient is to be estimated by means of Fig. 3.56 in Sec. 3.9, or by other laboratory data. Then the absolute height of the transmitted waves (see Fig. 3.56 for the notations) is calculated for several classes of the incident wave height. In the example of Akita Port, the south breakwater had values of the pertinent parameters of $h_c = 5.0$ m, $d = 8.5$ m and $h = 12.0$ m. Thus, the transmitted wave height in terms of the incident height was estimated as listed in Table 7.3.

[c] A portion of these results was presented in Fig. 3.20 as the verification data for the random wave diffraction diagrams.

Table 7.3   Estimated height of transmitted waves.

| Incident height $H_I$ (m) | 1.0 | 2.0 | 3.0 | 4.0 | 5.0 | 6.0 |
|---|---|---|---|---|---|---|
| Transmitted height $H_T$ (m) | 0.03 | 0.06 | 0.09 | 0.24 | 0.45 | 0.81 |

As discussed in Sec. 3.9.3, little is known about the pattern of propagation of waves transmitted over a breakwater. Nevertheless, some estimate of wave propagation may be necessary in the analysis of harbor tranquility. Therefore, unless a hydraulic model test is done with a scale large enough to be capable of simulating wave transmission by overtopping (which is not the case in most harbor tranquility tests), a practical method is to treat the extent of the length of overtopped breakwater as an imaginary opening in a breakwater and to estimate the propagation of transmitted waves as waves diffracted from the opening. A further simplification which may be used, depending on the harbor layout, is the assumption of nondispersive propagation of transmitted waves to the points of interest for a certain range of wave direction and no propagation from outside that range, that is, to assume a coefficient of propagation of unity for the former and zero for the latter.

### 7.2.6   *Estimation of the Exceedance Probability of Wave Height Within a Harbor*

The first step is to calculate the absolute height of waves at respective locations within a harbor for various classes of incident wave height, separately for several directions. The absolute height of the waves incident through the entrance is calculated with the wave height ratio data obtained as discussed in Sec. 7.2.4, and the absolute height is combined with the information on transmitted wave height over the breakwater. The combined height of waves is estimated as the square root of the sum of the squares of both heights on the basis of the addition of wave energy as expressed by Eq. (3.51) in Sec. 3.7.3. The calculation for the data of Akita Port is listed in Table 7.4. In a case where the locally generated wind waves within the harbor may present some hindrance to port operation, their contribution must be added in the above calculation at the respective class of incident height using the same principle.

From the results of the estimated wave height within the harbor for the relevant wave directions, the exceedance probability curves of wave height within the harbor are drawn by combining the exceedance probability of

Table 7.4   Superposition of diffracted and transmitted wave height.

|  |  | $H_I$ (m) |  |  |  |  |  |
|---|---|---|---|---|---|---|---|
| Wave Direction |  | 1.0 | 2.0 | 3.0 | 4.0 | 5.0 | 6.0 |
| SW | $H_d$ | 0.10 | 0.20 | 0.35 | 0.46 | 0.65 | 0.78 |
|  | $H_T$ | 0.03 | 0.06 | 0.09 | 0.24 | 0.45 | 0.81 |
|  | $H_S$ | 0.10 | 0.21 | 0.36 | 0.52 | 0.80 | 1.12 |
| WSW | $H_d$ | 0.28 | 0.55 | 0.87 | 1.16 | 1.52 | 1.82 |
|  | $H_T$ | 0.03 | 0.06 | 0.09 | 0.24 | 0.45 | 0.81 |
|  | $H_S$ | 0.28 | 0.55 | 0.87 | 1.18 | 1.59 | 1.99 |
| W | $H_d$ | 0.51 | 1.02 | 1.76 | 2.35 | 2.98 | 3.57 |
|  | $H_T$ | 0.03 | 0.06 | 0.09 | 0.24 | 0.45 | 0.81 |
|  | $H_S$ | 0.51 | 1.02 | 1.76 | 2.36 | 3.01 | 3.66 |
| WNW | $H_d$ | 0.85 | 1.70 | 2.56 | 3.41 | 4.28 | 5.13 |
|  | $H_T$ | 0 | 0 | 0 | 0 | 0 | 0 |
|  | $H_S$ | 0.85 | 1.70 | 2.56 | 3.41 | 4.28 | 5.13 |

Note: Transmitted waves were assumed to arrive at the site without dissipation for the direction of SW to W, but not to arrive for the direction of WNW.

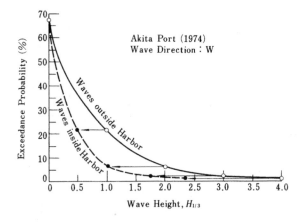

Fig. 7.3   Preparation of exceedance probability curve of wave height inside a harbor.

the offshore wave height. This can be done as shown in Fig. 7.3. First, the exceedance curve for the offshore waves is drawn for one wave direction, then the abscissas of the respective wave height classes are shifted toward the left according to the values of the wave height such as listed in Table 7.4,

Table 7.5 Result of estimation of the exceedance probability of wave height inside the harbor (percent).

| Direction of Offshore Waves | Interior Wave Height $H_s$ | | | | | |
|---|---|---|---|---|---|---|
| | Over 0 m | Over 0.5 m | Over 1.0 m | Over 1.5 m | Over 2.0 m | Over 2.5 m |
| SW | 1.4 | 0 | 0 | 0 | 0 | 0 |
| WSW | 29.0 | 3.0 | 0.8 | 0.4 | 0.1 | 0 |
| W | 68.2 | 21.5 | 6.5 | 3.0 | 2.0 | 1.2 |
| WNW | 1.4 | 0.4 | 0.1 | 0 | 0 | 0 |
| Total (all directions) | 100.0 | 24.9 | 7.4 | 3.4 | 2.1 | 1.2 |

and finally the ordinates at shifted locations are connected with a smooth curve.

The probability of exceedance at predetermined values of the wave height such as 0.25 m, 0.5 m and 1.0 m is then read from the exceedance curves obtained for each wave direction. The probabilities in percentages are listed in a form such as shown in Table 7.5. By repeating this procedure for every wave direction and then taking the sum of probabilities of exceedance at each class of wave height, the overall exceedance probability of wave height within the harbor can be obtained as listed in the bottom row of Table 7.5. Figure 7.4 plots this result, with the open circles denoting results for offshore waves and filled circles denoting results of estimation of wave height inside the harbor. The diagonal crosses give the results of a wave observation of one year's duration with a rate of return of 84%. Although the calculation of the wave height in the harbor yielded an estimate of exceedance probability slightly higher than the observation in the range of wave height in excess of 1.5 m, the reliability of the wave height estimation seems to be rather satisfactory considering the accuracy of each step involved.

By following the procedure described above, the exceedance probability of wave height within a harbor under planning can be evaluated quantitatively. The exceedance probability is best converted to the number of days of wave height exceedance per year by multiplying the probability by 365 days so that the results are more easily understood. The interpretation of the exceedance probability data available for a harbor for the evaluation of harbor tranquility is a task which will be improved and refined through further applications. There are no established criteria for the allowable exceedance probability of wave height in a harbor, for example, how many

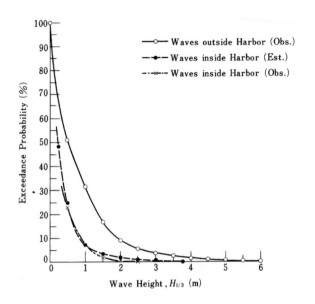

Fig. 7.4   Exceedance probability of wave height within Akita Port.

days can be tolerated with the wave height exceeding 0.5 m in a mooring basin or along a wharf, because judgment can only be made with various other operational and economic factors taken into consideration. Nevertheless, the description of the degree of wave agitation in a harbor with the exceedance probability such as shown in Table 7.5 and Fig. 7.4 provides a common basis for the quantitative evaluation of the workability of port facilities.

Many more investigations need to be done on the wave height in a harbor, the motions of a ship at berth, the working limits of port operation, etc., in order to make a more quantitative and rational assessment of the serviceability of a particular harbor.

### 7.2.7   *Examination of Storm Wave Height in a Harbor*

The discussion of the wave height exceedance probability in a harbor in the previous sections was mainly concerned with the workability of port operation. The waves in question were mostly waves of medium height, which comprise the major portion of wave climate statistics. As stated in the beginning of the present chapter, harbors also have the important function of providing safe anchorage for vessels in times of storms. The breakwaters must be capable of providing sufficient shelter for vessels from attacking storm waves and for preventing large waves from reaching port facilities. Thus, the waves of concern are those rare ones that occur only once in several years or once in several tens of years. The wave conditions in a harbor thus need to be investigated for rare storms too. The data for the statistics of such storm waves usually come from a source different from that of the wave climate statistics, which are based on wave measurements over several years at most. Wave hindcasting for major storms which occurred over a few score of years is the major source of data. Preferably the hindcasting technique should be calibrated with some instrumentally recorded storm wave data. The hindcast wave data are from the same data source used to determine the design wave for breakwaters and other structures. But the waves of extreme height must be chosen corresponding to several separate directions of wave approach for the examination of harbor tranquility at times of storm attacks.

The method of utilization of the estimated storm wave height information is not well established, as no definite criteria exist for the maximum allowable height of storm waves in a harbor. Generally speaking, if the significant wave height in a mooring basin can be kept below 1 m even for the design storm, the harbor may be considered calm. In any case, the description of the storm wave condition in a harbor in terms of the absolute value of the significant wave height will facilitate the judgment of the safety of the harbor in a quantitative manner.

## 7.3 Graphical Solution of the Distribution of Wave Height in a Harbor

Diffraction diagrams provide an indispensable means for estimation of the wave height behind breakwaters and other barriers. If all the waterfronts of a harbor are composed of natural beaches and/or wave-absorbing structures, direct application of diffraction diagrams to the harbor can yield a good estimate of the wave height distribution. In most cases, however, there exist some sort of reflective structures, such as vertical quay walls within a harbor, which cause additional wave agitation. Some modification in the usage of diffraction diagrams is necessary for a harbor with reflective waterfront lines.

For this purpose the author devised a so-called *mirror-image method*,[7] inspired by the suggestion of Carr quoted in the textbook edited by Ippen.[8]

The method is a graphical technique of applying diffraction diagrams (for random sea waves) to a harbor, by transferring the geometry of the harbor layout in the plane of a mirror image along the boundary of wave reflection, and by treating the reflected waves as waves progressing in the mirror-image plane. Takayama[5,6] adopted the concept of the mirror-image method when he developed a numerical model for analyzing the distribution of wave heights within a harbor. Even though various advanced numerical models such as the Boussinesq equation are employed nowadays, a concept of evaluating the effects of reflected waves will be instructive in understanding the nature of wave disturbances within a harbor. From this viewpoint, an explanation will be given in this section with an illustrative example.

Let us consider a harbor sketched in Fig. 7.5. Wharves G and H are built with vertical walls, and the seawalls indicated with diagonal crosses are provided with wave-absorbing structures. Points A, B and C are the locations at which the wave height is to be estimated. The first step in the graphical solution is to draw the diffraction diagram corresponding to the geometry of the harbor entrance and the conditions of incident waves. In this example, swell with $H'_0 = 4.3$ m and $T_{1/3} = 12$ s will be assumed to be incident to the harbor. The opening of the entrance is measured as $B = 310$ m along the line between the tips of the two breakwaters, and this line makes an angle of 41° with the direction of the approaching swell.

By assuming the tide level of $+ 1.0$ m above the datum, the water depth at the entrance becomes $h = 11$ m, and the wavelength at this depth is read from Table A.3 in the Appendix as $L = 118$ m. Thus, the relative

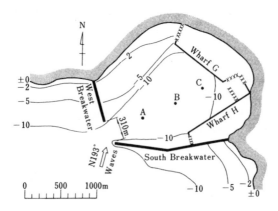

Fig. 7.5  Harbor geometry for application of graphical solution of wave height distribution.

opening width becomes $B/L = 2.63$. The value of the directional spreading
parameter is given as $(s_{max})_0 = 25$, because the swell is considered to have
a short decay distance with $H'_0/L_0 = 0.019$. At the harbor entrance, the
parameter is thought to have increased to the value of $s_{max} = 100$ due
to the effect of wave refraction corresponding to the relative water depth
of $h/L_0 = 0.049$ there, according to Fig. 2.14 in Sec. 2.3.2. Then the
deviation angle of the axis of the diffracted waves is estimated as $\Delta\Theta \simeq 8°$
with the data listed in Table 3.4 in Sec. 3.3.3. The apparent direction of the
diffracted waves thus makes an angle of 49° with the line connecting the
tips of the two breakwaters. The apparent opening width of the entrance
from this direction is $B' = 234$ m, and therefore $B'/L = 1.98$. The random
wave diffraction diagram which has the conditions closest to the present
situation is Fig. 3.17 (2) in Sec. 3.3.2. By transferring this diagram to the
plan shape of the harbor under consideration, we obtain the contour lines
of the wave height ratio as sketched in Fig. 7.6. The diffraction diagram
should be extended to an area larger than the harbor area (Zone I) itself.

If there were no wave reflection from the wharves, the wave height ratios
at Points A, B and C could be read immediately from Fig. 7.6 with the aid
of contour lines. But it is necessary to estimate the effect of wave reflection

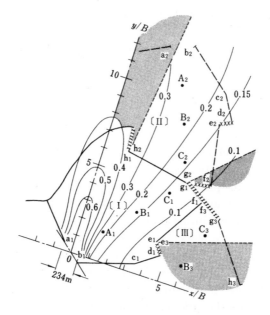

Fig. 7.6   Mirror-imaged diffraction diagram applied to a particular harbor geometry.

from Wharves G and H. In order to analyze the wave reflection from Wharf G, the plan shape of the harbor is transferred in the plane of the mirror image (Zone II) with the wharf line $\overline{g_1 h_1}$ as the reflective boundary. This operation can be performed easily, first by tracing the plan shape of the harbor on transparent paper, then by turning it over with the line $\overline{g_1 h_1}$ fixed in place and finally by tracing the outline of the harbor in Zone II on the other side of the transparent paper. On the mirror image of harbor in Zone II, Points $A_2$, $B_2$ and $C_2$, which are the mirror images of A, B and C, are plotted. In the same manner the mirror image of the harbor for the wave reflection from Wharf H is drawn in Zone III. Among the mirror images of the harbor in Zones II and III, the shaded portions represent areas to which reflected waves will not reach directly as they are in shadow zones as viewed from the harbor entrance. For example, Point $B_3$ in Zone III does not receive waves reflected from Wharf H.

In addition to primary reflection, there is secondary wave reflection from the back face of the south breakwater (if it is of vertical type) in the area $\overline{b_2 c_2 d_2}$ in Zone II and from Wharf H in the area $\overline{e_2 f_2}$ in Zone II. The graphical solution of the wave height distribution must include such secondary wave reflections, which are obtained by drawing the mirror images of the harbor with secondary reflective mirror images taken separately along lines $\overline{b_2 c_2}$, $\overline{c_2 d_2}$ and $\overline{e_2 f_2}$. However, these secondary wave reflections are omitted in the present explanation for the sake of brevity.

Next the values of the diffraction coefficient at the selected points in the respective zones are read and compiled as shown in Table 7.6. The resultant wave height at each point is estimated as the square root of the sum of the squares of the incident and all reflected wave heights, based on the principle of the summation of wave energy. If Wharf G does not have a vertical wall but consists of some sort of energy-dissipating structure, the wave height ratio entered in the column for Zone II should be reduced by a rate equal to the reflection coefficient. Furthermore, in a harbor of large dimensions principally planned against wind waves, the reflection coefficient of a vertical quay wall may be reduced to 80% or so in consideration of the wave decay experienced during propagation in the harbor, as discussed in Sec. 3.7.2.

A reminder should be given that the application of diffraction diagrams for regular waves produces a wave height ratio quite different from actual situations. In the present example, regular wave diffraction would produce wave height ratios on the order of one half of those listed in Table 7.6, thus resulting in an underestimation of wave agitation in the harbor.

Table 7.6 Superposition of diffracted and reflected wave heights.

| Point | Diffraction Coefficient | | | Ratio of Superposed to Offshore Wave Height |
|-------|--------|---------|----------|---------------------|
|       | Zone I | Zone II | Zone III | |
| A | 0.27 | 0.27 | – | 0.38 |
| B | 0.17 | 0.22 | – | 0.28 |
| C | 0.14 | 0.16 | 0.07 | 0.22 |

This mirror-image method of graphical solution cannot be applied to a harbor of complicated shape, as might be understood from the above illustrative example. Thus, the shape of a harbor needs to be approximated by a simpler form before the application of the graphical method. There is also the need of a somewhat subjective modification of the diffraction coefficient around the vicinity of the boundaries of geometric shadows viewed from the harbor entrance, such as the areas along the dashed lines extending from the points $e_1$, $g_1$ and $h_1$ in Fig. 7.6, because reflected waves penetrate into these shadow zones just as in the case of the diffraction of waves in the sheltered area behind a semi-infinite breakwater. It would not be appropriate to have an abrupt change in wave height across the boundary of a geometric shadow of reflected waves. There should be a smooth variation in wave height similar to that of the diffraction coefficient for a semi-infinite breakwater.

If a reflective structure in a harbor has an extent equivalent to only a few wavelengths or less, waves reflected by it will disperse in the harbor in a manner similar to the diffraction of waves through a narrow opening. In such a case, the propagation of the reflected waves is best estimated with the technique of fictitious wave diffraction, which treats the reflective boundary as an imaginary opening of a breakwater, as discussed in Sec. 3.7.2. A diffraction diagram corresponding to regular waves would be more suitable for this situation, because the waves incident to the reflected boundary have a narrow range of directional spreading, limited by the opening width of the entrance viewed from the location of the reflective boundary.

In summary, the graphical solution of the mirror-image method requires sound judgment of the analyst with appropriate modifications of the technique, depending on the harbor geometry and wave conditions.

## 7.4 Some Principles for Improvement of Harbor Tranquility

Although the tranquility of a harbor cannot be completely characterized by means of the wave height alone, as discussed at the beginning of the present chapter, in harbor planning, tranquility can only be achieved by suppressing the heights of waves within the harbor. This is the essence of harbor planning, which many experienced engineers have endeavored to realize. There are many good discussions in the textbooks of harbor engineering, but the author would also like to contribute to the subject by listing several principles for the improvement of harbor tranquility.

(i) *Broad interior for a harbor.* Some harbor is made of a long, narrow water area, probably due to the local topographic conditions or by reason of historical development. Such a harbor surely suffers from the problem of too little dispersion of the intruding waves and of multi-reflection of the waves inside. The initial planning of a harbor should provide a broad water area with sufficient room for future expansion.

(ii) *No wave reflection at the spot of first wave arrival.* The portion of the waterfront from where the outer sea can be viewed through the harbor entrance should be left as a natural beach or be provided with wave-absorbing revetments. The essential measure for reducing wave agitation in a harbor is to minimize the intrusion of waves from the entrance and to dissipate the energy of the intruding waves at the place where they first reach the waterfront. That is to say, a harbor planner should try to control wave agitation at the first encounter of the waves with the harbor facilities. If a wharf or the revetment of a vertical bulkhead is built at a location where the intruding waves arrive directly from the entrance, this implies the worst kind of harbor layout because waves reflected from the bulkhead will cause considerable agitation within the harbor.

Figure 7.7 is a sketch of the layout of a harbor discussed by Ozaki.[9] It was reported that the water surface at Quay AB could not be maintained calm because of the waves reflected by the newly-built Wharf DC, and engineers had difficulty in finding effective countermeasures.

The locations of the first arrival of intruding waves from the entrance will vary depending on the direction of the incident waves. Thus, it is best to assume that the waves can arrive from any direction and that a location from which the outer sea can be viewed through the harbor entrance may become an area of the first arrival of intruding waves. Such locations should not contain reflective structures.

Figure 7.8 gives three examples of such locations. The portions covered with diagonal crosses are to be kept as natural beaches or provided with revetments of the wave-absorbing type.

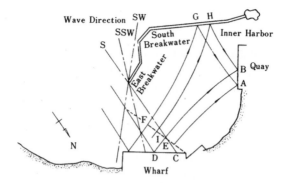

Fig. 7.7   Example of harbor agitation by reflected waves (after Ozaki[9]).

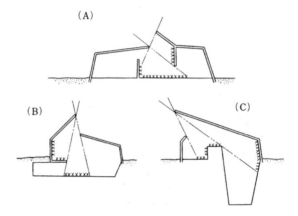

Fig. 7.8   Illustrations of areas to be provided with wave-absorbing structures.

(iii) *Small-craft basin at a recess of a harbor.* Small-craft basins should be located in an area which cannot be viewed directly from the harbor entrances. Small craft and many working vessels are easily set in motion by waves of small height. Thus, a basin for small craft must be given maximum protection against incoming waves. Secondary or inner breakwaters are often built at the entrances of such basins to assure tranquility there. It is important to overlap the outer and inner breakwaters against the direction of wave propagation so that no direct penetration of the incident waves will reach the small-craft basin. This is the fundamental principle in the planning of a marina. Also important are the provisions for a wave-spending beach, slipway, or wave-absorbing revetment at areas where waves diffracted by secondary breakwaters first reach. This is basically the same principle as (ii).

(iv) *Reservation of wave dissipation area along the warterfront.* A portion of the waterfront of a harbor should be reserved as an area of wave energy dissipation. If one attempts to apply the mirror-image method to a harbor with a rectangular shape for which all the waterfront lines are made up of vertical walls, mirror images of the harbor geometry appear one after another in an endless fashion. This implies multirepetition of wave reflection. It presents the worst possible situation for harbor tranquility. The need to increase the cargo handling capacity of a port sometimes leads to the construction of a new wharf or quay at the location where a sandy beach or rocky reef existed previously. But this should be done with caution, because the removal of a wave-spending beach often results in a worsening of harbor tranquility by addition of reflected waves. The total cargo handling capacity of the port may well go down by the construction of an additional quay in such a strategic place.

For example, the beach between the west breakwater and Wharf G in Fig. 7.5 should not be converted into a wharf, because the beach is functioning as an important wave-spending beach. In the example shown in Fig. 7.9, the block mound revetment at A and the sandy beach at B should be kept as they are. It is not recommended to convert them into wharves of vertical walls. In fact, a port with a layout similar to that in Fig. 7.9 had problems with worsened tranquility after the area in front of the beach at B was reclaimed by construction of a vertical revetment at its waterfront.

(v) *Wave reflection from the back face of breakwater.* Precautions should be taken against the reflection of waves from the back faces of vertical breakwaters. For some harbor layouts, waves incident through the entrance

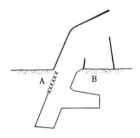

Fig. 7.9   A harbor layout.

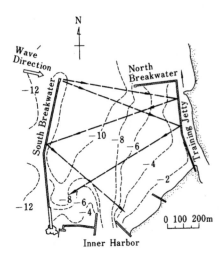

Fig. 7.10  Example of wave reflection by back face of a vertical breakwater (after Ozaki[9]).

or waves reflected from a wharf hit the back faces of breakwaters. If a breakwater is of the sloping mound type made of rubble stones or concrete blocks, wave reflection by the back face will be weak and the tranquility of the harbor will not be much disturbed. In the case of vertical breakwaters, their back faces are formed of vertical walls and they reflect incident waves almost completely. In the example of Fig. 7.7, the back face of the south breakwater at section GH is reflecting the waves first reflected by Wharf CD, and this is causing additional disturbance at Quay AB.

Figure 7.10 is another example given by Ozaki.[9] At the time he discussed the situation, the extension of the south breakwater was insufficient for suppressing the intrusion of waves into the harbor. The intruding waves were first reflected by a training jetty of the composite type. These reflected waves and re-reflected waves from the back face of the south breakwater were causing a disturbance in the water area around the entrance of the inner harbor.

Although the alignment of a breakwater is usually investigated and designed from the viewpoint of its effectiveness in diminishing the energy of waves intruding into the harbor, it is also necessary to examine the possibility of wave reflection from the back face of the breakwater. If the harbor layout is such that wave reflection is unavoidable, some measure such as the redesign of the breakwater into a non-reflective structure will become necessary.

(vi) *Caution on quay walls of the wave-absorbing type.* In recent years, special types of quay walls and vertical revetments capable of dissipating wave energy have been built in a number of harbors for the purpose of maintaining tranquility while providing additional facilities for port operation. They include concrete caissons with perforated or slit front walls, and several intricate forms of concrete blocks to be layed up to a required height. They all have hollow spaces which contain a mass of water connected to the outside water through many holes, slits, or irregular openings. In resonance with the undulation of the waves in front of structures, an alternating jet flow runs through there openings and wave energy is dissipated in the process by turbulence in the form of wakes and eddies.

By installation of such structures, it is possible to decrease the reflection coefficient to 30% or less as indicated in Fig. 3.44 in Sec. 3.7.1, if the size of the hollow spaces and the opening ratio of the front wall are appropriately designed for the conditions of the incoming waves (mainly the wave period, or wavelength). However, the reflection coefficient may remain near 100% if the dimensions of the structure do not fit the wave conditions. Therefore, execution of hydraulic model tests is recommended before the adoption of structures of the energy dissipating type. Another check point is that the upper deck of such a structure should have a sufficient clearance height above the design water level so as not to hamper the free movement of the water mass in the hollow space, which is essential for the dissipation of wave energy. This aspect should be carefully examined in the case of harbors designed for small boats on a coast with a low tidal range, as users of such boats prefer the lowest possible elevation of the apron for easy landing and cargo handling.

## 7.5   Motions of Ships at Mooring

### 7.5.1   *Modes and Equations of Ship Motions*

As discussed in Sec. 7.1, tranquility of a harbor is evaluated with the safety of ships and the efficiency of cargo handling operations under constraints of construction and maintenance costs. In such evaluations, the amplitudes of ship motions at mooring provide the most important information. In contrary to fixed structures such as breakwaters and seawalls, ships floating on water have six degrees of freedom of motions and thus no straightforward formula are available for calculation of ship motions and mooring forces. If we try to give no allowance for ship motions under wave actions for example, we must employ extremely strong mooring chains, and the resultant mooring force will be great. If we provide a loose mooring just sufficient to prevent drifting of a ship on the other hand, the amplitudes of ship motions become large but the mooring forces remain small. Because ship motions at mooring are dependent on the mooring system, design of a mooring facility is often made through repetition of the analysis of ship motions by varying mooring characteristics until the ship motions are kept below an acceptable level.

To solve the problem of ship motions at mooring, we must understand the characteristics of the motions of a body floating on water, which has three degrees of freedom of reciprocating motions and three degrees of freedom of rotational motions. The six modes of motions are named individually as follows:

 (i) *surge* or *surging*: horizontal, longitudinal motion of floating body,
 (ii) *sway* or *swaying*: horizontal, lateral motion of floating body,
 (iii) *heave* or *heaving*: vertical reciprocating motion of floating body,
 (iv) *pitch* or *pitching*: rotational motion around the lateral axis of floating body,
 (v) *roll* or *rolling*: rotational motion around the longitudinal axis of floating body,
 (vi) *yaw* or *yawing*: rotational motion around the vertical axis through the center of gravity of floating body.

The six modes of ship motions are sketched in Fig. 7.11.

The six modes of motions do not occur independently but are excited in a coupled manner. For example, rolling of a floating body occurs simultaneously with swaying motion, and it is further coupled with a yawing motion in many instances. Heaving of floating body excites a pitching motion, and it is almost impossible to restrict either one of the motions without interference with the other motion. However, surging of a long and symmetric body such as a ship can occur independently without coupling with other modes of motions.

The motions of a floating body under actions of winds and waves are analyzed by solving the equations of motions of six modes. Because the six modes of motions are coupled together, the equations of motions become the simultaneous differential equations of the second order with six unknowns of $x_j$ $(j = 1 \sim 6)$. That is

$$\sum_{j=1}^{6}\{(M_{kj}+m_{kj})\ddot{x}_j + N_{kj}\dot{x}_j + C_{kj}\dot{x}_j|\dot{x}_j| + B_{kj}x_j + R_{kj}(x_j)\} = X_k(t)\,, \quad (7.2)$$

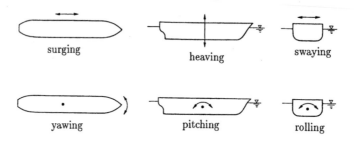

Fig. 7.11   Six modes of ship motions.

where $k = 1 \sim 6$ corresponds to the six modes of motions respectively, $j$ refers to the mode of motion coupled with the mode $k$, and $x_j$ is the displacement or rotation of the coupled motion in the mode $j$. The term $M_{kj}$ is called the *inertia matrix,*, which represents the mass or the moment of inertia of the floating body in the direction of $k$ when the body makes a motion in the mode $j$. The term $m_{kj}$ is called the *added mass*, which represents the component of fluid resistance proportional to the acceleration in the direction of $k$ when the floating body generates waves by moving in the mode $j$: for a rotational motion, the added moment (instead of mass) of inertia is defined and the same is applied for other terms. The term $N_{kj}$ is called the *wave damping coefficient,*, which is the coefficient of the component of fluid resistance proportional to the velocity. The term $C_{kj}$ represents the *coefficient of nonlinear damping force* such as the drag force. The term $B_{kj}$ is the *coefficient of restoration force due to buoyancy* which is proportional to the displacement (rotation) of the floating body. The term $R_{kj}$ stands for the *reaction force* of the mooring system system, and a functional representation such as $R_{kj}(x_j)$ is employed because the reaction force is nonlinear with the displacement $x_j$. The term $X_k(t)$ in the right side of Eq. (7.2) represents the external forces such as wave, wind and current loads acting in the direction $k$, which are evaluated as the floating body being set at the stationary state.

The motion of a floating body is determined as the result of equating the external force $X_k(t)$ with the inertia and fluid resistance generated by the motion of floating body as represented on the left side of Eq. (7.2). Because a freely floating body has a null reaction force of mooring system, its motion is governed by the external forces alone. If the motions of a floating body could be suppressed completely ($x_j = 0$), then the external forces $X_k(t)$ become the mooring force $R_k$. However, such a complete stop of motion is impossible, because conventional mooring arrangements do not have the capacity to withstand the external forces.

A unique component of the external forces inherent to a floating body is the *wave drift force*. It refers to a constant component of wave force which causes the floating body to move slowly in the direction of wave propagation. For a floating body of slender shape subject to beam seas (waves acting normal to the side), the wave drift force has been derived by Maruo[10] and Newman[11] as in the following:

$$F_D = \frac{1}{16}\rho g H_I^2 B \left(1 + K_R^2 - K_T^2\right) \left(1 + \frac{4\pi h/L}{\sinh 4\pi h/L}\right), \qquad (7.3)$$

where $\rho$ is the density of sea water, $g$ the acceleration of gravity, $H_I$ the incident wave height, $B$ the projected width of a floating body, $K_R$ and $K_T$ the coefficients of reflection and transmission of a floating body respectively, $h$ the water depth, and $L$ the wavelength. For a floating body subject to head seas (waves approaching from the bow) or a floating body of square or complex shape, the computation of wave drift force becomes complicated. Nevertheless, the characteristics of wave drift force being proportional to the square of wave height remain unchanged. The constant wave drift force provides the base value of mooring force for a ship at an offshore multiple-buoy berth or a floating breakwater.

In random sea waves, the mean wave drift force is evaluated with the root-mean-square height $H_{rms}$. However, the wave drift force slowly fluctuates in time in response to the gradual variation of wave height in a train of waves, which is called the phenomenon of wave grouping as to be discussed in Sec. 10.2. Such slow variations of wave drift force cause the long-period oscillations of moored vessels.

### 7.5.2 Ship Mooring and Natural Frequency of Ship Mooring System

The method of mooring a ship varies depending on the place of mooring. At the anchorage in stormy conditions, a ship relies upon her anchors and the thrust of propellers to keep her position firmly. At a single buoy berth, a heavy-duty hawser moors a tanker to a buoy. In the analysis of vessel movements at such a single-point mooring, the motion of buoy must be solved simultaneously with that of a vessel. At dolphin berths in the offshore and conventional wharves in a harbor, mooring lines and fenders

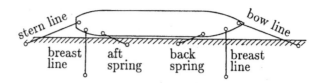

Fig. 7.12   Mooring lines for a ship at berth.

constitute a mooring facility. A combination of ship, mooring lines, and fenders consitute a *mooring system.*

*Mooring lines* are made of steel wires and/or synthetic fiber lines (nylon ropes and polyester ropes), which are equipped on board a ship. The number and size of ropes are specified according to the equipment number which is calculated with the rules of ship classification societies. These ropes are intended to hold a ship alongside a berth for smooth cargo handling operations under normal weather conditions. They are not strong enough to hold a ship at stormy conditions, and they are sometimes broken at times of strong winds and/or high waves. Some harbor authorities have heavy-duty mooring lines in store to assist vessels at mooring in their harbors at stormy conditions, but their effectiveness is limited by the structural strength of bollards on board the ships.

When a ship is moored at a berth, mooring lines are given the names depending on the points of mooring as shown in Fig. 7.12. The bow (head) lines and stern lines are mainly employed to harness surging motions, while breast lines are to constrain swaying and yawing of a ship. Spring lines (aft and back) are used to fix a vessel at normal weather conditions, and are recommended to let released at stormy conditions.

Of the two types of mooring lines, steel wires exhibit a nearly linear relationship between the tension and elongation, and the elongation at breakage is relatively small. On the other hand, synthetic fiber ropes exhibit nonlinear behavior of elongation with the tension, with a large elongation at breakage. In scale model tests of ship mooring, various efforts are made to simulate such a nonlinear relationship between tension and elongation of mooring lines in the model: use of a series of multiple springs with different stiffness is one of such techniques.

*Fenders* are primarily installed to absorb the berthing impact and to prevent damage on the vessel's hull and wharf structures. At times of mooring at stormy conditions, fenders serve as the important element to constrain swaying and yawing of a vessel. However, the load and

Table 7.7 Restoring forces of floating body at mooring.

| Mode | Buoyancy | Mooring lines | Fender |
|------|----------|---------------|--------|
| Surging | × | ○ | × |
| Swaying | × | ○ | ○ |
| Heaving | ○ | × | × |
| Rolling | ○ | △ | × |
| Pitching | ○ | × | × |
| Yawing | × | ○ | △ |

Note: ○ active △ marginal × nonexistent.

displacement relationship of a fender is a special one: it works on the compression side only, and the load and displacement curve is highly nonlinear.

The natural frequency of a mooring system is governed by the mass (moment of inertia) of a vessel and the restoring forces. Table 7.7 lists the elements of restoring forces for the six modes of ship motions. For heaving and pitching motions, mooring lines and fenders can exercise no effect, and the restoring forces due to buoyancy of a floating body are the sole forces for restoration. Thus, the natural frequencies of heave and pitch are in the range of several seconds to less than 20 seconds. Rolling motions can be suppressed to a certain extent by the selection of mooring lines and their points of attachment. However, the natural frequency of rolling remains almost the same as of a freely floating body. For surging, swaying and yawing motions, a floating body has no restoring force due to buoyancy and it is brought back to the equilibrium position mainly through the relatively weak tension of mooring lines. The natural periods of these motions thus become long in the range of several tens of seconds to a few minutes. As the winds and waves have the components of long-period fluctuations, moored vessel can be brought into resonance with them, resulting in a slow motion of large amplitudes. Tightening of mooring lines shortens the natural periods and is effective in suppressing resonant motions. However, the frequent tension adjustments of mooring lines must be executed with much care because tightening of mooring lines enhances the danger of line breakage.

### 7.5.3 *Time-Domain Analysis of Moored Ship*

The motions of a moored ship is analyzed by solving Eq. (7.2). In case of a ship navigating in the sea, no mooring force exists. The equations can be transformed into a set of linear differential equations by introducing an equivalent linearization of the term $C_{kj}$ or by simply neglecting it. Once the terms $M_{kj}$, $m_{kj}$, $N_{kj}$, $B_{kj}$, etc. are evaluated by the strip theory or by some other numerical schemes, the simultaneous equations can be solved with relative ease. Ship motions in directional seas can be analyzed by linearly superposing the ship responses to wave spectral components. This is a frequency-wise solution and is widely utilized in naval hydrodynamics.

In case of a ship moored at berth however, the nonlinear restoring forces of mooring lines and fenders are the important parameters and the wave drift force which is proportional to the square of wave height fluctuates with respect to time. Thus, the frequency-wise linear superposition technique cannot be employed in the analysis. Instead, a time-domain analysis is used for the motion of a moored vessel, by giving the input of time-varying external forces $X_k(t)$ and by solving Eq. (7.2) with a time step of small increment. The time series of external forces are prepared for each mode of the six degrees of freedom of vessel motions by numerical simulation techniques based on the spectra of forces and moments exerted upon a vessel by winds and waves.

Numerical simulations of moored-ship motions have been carried out by many researchers for clarification of the safe mooring problems in harbors, and they were often combined with field measurements of ship motions. Ueda[12] has made a series of analysis with applications to field data in the 1980s, which led to a later proposal of allowable ship movements by Ueda and Shiraishi[13] in 1988.

An important feature of ship motions in harbors is their long periodicity in the range of several tens of seconds to several minutes. The external force $X_k(t)$ must include the long period components such as the time-varying wave drift force as done by Hsu and Blenkarn[14] and Pinkster.[15] Wind forces are another important factor contributing to the long-period ship motions. A frequency spectrum of wind speed such as Davenport's spectrum[16] needs to be introduced in the generation of external forces. The linear and nonlinear damping coefficients $N_{kj}$ and $C_{kj}$ due to viscosity in Eq. (7.2) take the values smaller for long period motions than those for short period motions (Fujihata *et al.*[17]). The above is just an overview of the technique of the time-domain analysis of moored ship motions in a harbor. Interested readers should seek for the assistance of experts in this field if they have to solve this kind of problems.

### 7.5.4 *Some Remarks on Ship Mooring*

It is desirable to understand the characteristics of the movements of moored ships, even before a numerical simulation and/or hydraulic model test are commissioned for examination of harbor tranquility. The following is several remarks for port planners from the viewpoint of ship motions at mooring.

(i) *Waves of long periodicity cause large ship motions.* In conventional analysis of harbor tranquility, the measure of tranquility is the wave height at respective locations in a harbor as exemplified in Sec. 7.2. While the ship motions increase in proportion to the wave height, they are also affected by the wave period. Generally speaking, the ship motions become larger for long-period waves than for short-period waves, although the wave period effect varies depending on the ship size and mooring arrangement. A port which is opened to the ocean and exposed to swell should lower the threshold wave height for safe working conditions, as compared with a port which is located in an embayment sheltered from swell.

(ii) *Ships are most susceptible to beam waves and winds.* The wind and wave forces are linearly proportional to the exposed area of a ship's body, which is the largest to the direction from the side. A berth should be located at the position where a ship moored there would not be subjected to winds and waves from abeam directions. If siting of a berth at a location of beam waves is inevitable by the constraints in port planning, the threshold wave height for safe working conditions at that berth should be lowered from that of other berths. Ueda and Shiraishi[18] recommend to include a deviation of 10° to 15° in the wave direction when carrying out numerical simulation of moored ship motions.

(iii) *Avoid a location of strong offshore seasonal winds for a berth.* Seasonal winds often have the prevailing directions at respective ports. A ship at a berth should not be exposed to strong offshore winds, because the breast lines are not strong enough to withstand the full force of offshore winds: a berth should be located at the other side of slip at which the onshore winds would act on a moored ship. If setting of a berth at a location of offshore wind is inevitable, tall sheds and other buildings are to be erected behind a quay to serve as a wind screen for reduction of wind forces.

(iv) *Winds are to be specified by means of spectrum.* In design of fixed structures, the wind load is often treated as a steady force. In the analysis of the motion of a floating body moored on water, however, temporal fluctuations of wind speed need to be taken into account, because the moored body would respond to the long-period component of wind force and its movement may become large. The wind speed fluctuations are simulated with the wind spectrum such as Davenport's one.[16] In a preliminary analysis, the maximum instantaneous wind speed in consideration of the gust factor of winds may be used in estimating the mooring force. However, the time-domain numerical analysis of Eq. (7.2) must be carried out with fluctuating wind forces for obtaining reliable information of ship motions and mooring forces.

(v) *The ship motions become the largest at the ballast condition.* The wind force is the largest for a ship in ballast, because of the largest area of exposure. As to the wave-induced ship motions, a ship in ballast also experiences the largest movements, contrary to a layman's expectation. It is so because the inertial force matrix and damping coefficients in the left side of Eq. (7.2) decrease much faster than the external force of waves in the right side of Eq. (7.2) does, as the ship draft decreases.

(vi) *Motions of freely floating ships are the first approximation to those of moored ships.* Currently available mooring arrangements are not strong enough to suppress the ship movements at medium to storm weather conditions. It is difficult to reduce the ship movements below those of a freely floating ship. If resonance occurs, surging and swaying motions become quite large. The motions of a freely floating ship provide the first approximation to the motions of a moored ship and mooring forces, though more accurate information must be sought for by solving Eq. (7.2).

## 7.6 Allowable Ship Movements and Mitigation of Mooring Troubles

### 7.6.1 *Allowable Movements of Moored Ships*

A ship moored at a berth is always moved by winds, waves and currents, and so a complete suppression of ship motions is impossible. The allowable movements of a moored ship at extreme sea states are determined by the breakage strength of mooring lines, while at normal conditions they are dictated as the amounts below which cargo handling works can be carried out safely and efficiently. The latter amounts depend on the size and type of ships, the types of cargo handling equipment and the modes of ship motions. Thus, the allowable ship movements at mooring must be established through detailed surveys of actual cargo handling works at respective ports.

Bruun[19] made a first proposal for criteria for allowable ship movements in 1980. Then, Ueda and Shiraishi[13,20] proposed a slightly different set of allowable ship movements in 1988. They collected about 110 incidents of suspension of cargo handling works in Japanese ports, estimated the motion amplitudes of individual ships through numerical simulation, derived the threshold values of allowable ship movements, and made some corrections to them in consultation with the opinions of stevedores. These and other survey results by Jensen *et al.*[21] have been compiled by the Working Group No. 24 of the Permanent International Association of Navigation Congresses (PIANC), which published a report on criteria for movements of moored ships[22] in 1995. Furthermore, Sato *et al.*[23] have presented the survey results on allowable movements of moored container ships and ferries in 2003.

These various proposals are compiled in Table 7.8, being classified by vessel types. The criteria by Ueda and Shiraishi are given priority to other proposals for the same category. The amount of movements is the amplitude from the mean position as indicated with a double sign. However, the allowable amplitude of sway movement is positive (offshore motion) only, because the onshore movement represents small amount of fender compressions.

The entry of 50% efficiency on the row of container ships refers to the case such that the working rate of container handling is allowed to one half of the case of smooth operation without any hindrance, while the entry of 100% refers to the latter case. The allowable movements of gas carriers proposed by Bruun[19] seem too strict compared with those by PIANC.[22] Bruun's criteria reflected the situations in the late 1970s when he collected relevant data. There have been built many offshore berths for LNG carriers since then, and it would be appropriate to reexamine his criteria with a survey on recent data.

### 7.6.2 *Mitigation of Mooring Troubles*

(A) *Mitigation for small crafts*

While large vessels can take refuge at open anchorage during storms, small fishing boats and leisure crafts must seek refuge within a well protected

Table 7.8    Allowable amplitudes of ship movements at berth

| Ship Type | Cargo Handling Equipment Etc. | Surge (m) | Sway (m) | Heave (m) | Yaw (°) | Pitch (°) | Roll (°) | References |
|---|---|---|---|---|---|---|---|---|
| Coasters | Ship's gears | ±0.5 | +1.2 | ±0.3 | ±1.0 | ±0.5 | ±0.5 | PIANC[22] |
| Ferries | Bow ramp | ±0.4 | +0.5 | ±0.4 | ±1.0 | ±0.5 | ±0.5 | Sato et al.[23] |
|  | Side ramp | ±0.3 | +0.6 | ±0.3 | ±1.0 | ±0.5 | ±0.5 | PIANC[22] |
| General cargo | Ship's gear and quay cranes | ±1.0 | ±0.75 | ±0.5 | ±2.5 | ±1.0 | ±1.5 | Ueda and Shiraishi[13] |
| Container ships | 50% efficiency | ±1.0 | ±1.0 | ±0.6 | ±3.0 | ±1.0 | ±1.0 | Sato et al.[23] |
|  | 100% efficiency | ±0.5 | ±0.5 | ±0.4 | ±1.5 | ±0.5 | ±0.5 | Ditto |
| Ore carriers | Bucket cranes | ±1.0 | ±1.0 | ±0.5 | ±3.0 | ±1.0 | ±1.0 | Ueda and Shiraishi[13] |
| Grain carriers | Elevator | ±1.0 | ±0.5 | ±0.5 | ±1.0 | ±1.0 | ±1.0 | Ditto |
| Bulk carriers | Elevator/bucket | ±1.0 | ±0.5 | ±0.5 | ±1.0 | ±1.0 | ±1.0 | PIANC[22] |
| Oil tankers |  |  |  |  |  |  |  |  |
| Coastal | Loading arms | ±1.0 | +0.75 | ±0.5 | ±4.0 | ±2.0 | ±2.0 | Ueda and Shiraishi[13] |
| Ocean-going | Loading arms | ±1.5 | +0.75 | ±0.5 | ±3.0 | ±1.5 | ±1.5 | Ditto |
| VLCC | Loading arms | ±1.5 | +3.0 | – | – | – | – | PIANC[22] |
| VLCC | Single buoy mooring | ±2.3 | +1.0 | ±0.5 | ±4.0 | – | ±3.0 | Bruun[19] |
| Gas carriers | Loading arms | ±1.0 | +2.0 | – | ±1.0 | ±1.0 | ±1.0 | PIANC[22] |
| LNG carriers | Loading arms | ±0.1 | +0.1 | minimal | minimal | – | minimal | Bruun[19] |

harbor. The harbor needs to be well sheltered by breakwaters against waves. The breakwaters should also have the crests sufficiently high to prevent damage to boats by the falling mass of overtopped water. Allsop *et al.*[24] have cited the wave overtopping rate of $0.01\text{m}^3/\text{s}/\text{m}$ as a tolerable threshold for marinas for small boats. In Japan, breakwaters of harbors for small vessels are recommended to have their crests at the elevation of $1.25H_{1/3}$ above the mean monthly-highest water level.

As for tolerable movements of fishing boats moored in harbors, Sato and Saeki[25] reported the results of surveys at 24 fishing harbors in Hokkaido, Japan. According to their survey, the vessel movements tolerable for fish unloading operations are $\pm0.5$ m for surging, 0.5 m for swaying, and $\pm0.5$ m for heaving.

Mooring troubles would appear in a harbor for small vessels if the whole length of waterfront is utilized as the quay, because waves once entered in the harbor will be reflected many times within the harbor and vessels will be moved roughly by high waves. Construction of reflective quay walls at the location facing to the harbor entrance should be prohibited. As discussed in Sec. 7.4, spending beaches or facilities capable of wave absorption should be arranged at strategic locations. Alteration of vertical quay walls to a wave-absorbing type such as perforated-wall caissons may improve the harbor tranquility. When mooring troubles are mainly caused by waves of swell type at a harbor facing the open sea, waves have fairly long wavelength. Conventional wave-absorbing caissons have relatively narrow wave chambers, which are not effective for absorbing the energy of waves with large wavelengths. Hydraulic model tests are recommended for search of best solutions of harbor layout and structural geometry of wave-absorbing facilities.

## (B) *Mitigation of mooring troubles due to swell*

When a severe storm is forecasted, a harbor master usually orders the captains of large vessels to move out of the harbor and take refuge at open anchorage. It is because the mooring lines of ships at berths will be broken by strong wind forces and ships will collide with the berth facilities even though the harbor may be well protected against storm waves.

Even at calm weathers, there may appear some incidences of large ship movements at berth and severance of mooring lines. Port of Cape Town, for example, had a large number of ships' call during World War II as a result of closure of the Suez Canal. There were many reports of mooring troubles in this period, however, which were thought to have been caused by the oscillations of the harbor basin with the period of one minutes or so. Port of Long Beach also reported several incidents of mooring troubles with large ship movements at berth.

In the 1950s and 1960s, mooring troubles was attributed to resonant oscillations of a harbor basin and many studies were carried out along this line, including the author's one.[26,27] Most of harbor oscillations, however, have the resonance period of several minutes to longer than ten minutes. Because the oscillation period of the mooring system in surging, swaying, and yawing modes is several tens of seconds to a few minutes as discussed in Sec. 7.5.2, resonant harbor oscillations are not considered as the culprit of mooring troubles nowadays. Instead, swell of long period and/or the infragravity waves are thought to excite large movements of moored ships.

As explained in Sec. 7.5.4, the movements of moored ships increases as the wave period becomes long. Ports facing the open seas from where swell often come are susceptible to mooring troubles. Mitigation measures are firstly improvement of mooring system such as provision of storm lines at the berth side, automatic tension control winches at the berth, and so on. Secondly, it is important to enhance the level of sheltering against swell by extension of breakwaters for restricting intrusion of swell into harbor basins. Thirdly measures are to be taken for enhancement of wave absorption capacity within harbor basins by installation of berthing facilities of wave-absorbing type and/or removal of wave-reflecting facilities at strategic location.

(C) *Mitigation of mooring troubles due to long-period waves*

As discussed in Sec. 2.5, the waves with the period ranging from around 20 seconds to several minutes may accompany wind waves and swell. Long-period waves with the height less than 10 cm can excite large ship movements, because the natural periods of the surge and sway of moored ships often fall in this period range. The natural periods of oscillations of moored ships could be shortened by changing mooring lines of synthetic fiber lines to steel wires with synthetic tails if applicable, and thus minimizing the possibility of exciting resonance phenomena. In port planning, field investigations should be undertaken for the magnitude of long-period waves at the site, and if their energy is not of negligible amount, the harbor layout should be so made to reduce the intrusion of long-period waves into the harbor; hydraulic model tests are recommended for search of the optimum harbor layout.

Dissipation of the energy of long-period waves within a harbor basin is difficult because of their long periodicity. Energy dissipation facilities against long-peirod waves will require a large width of 30 to 60 m, according to the results of model studies by Hiraishi and Nagase[28] and Ikeno et al.[29]

Another countermeasure against mooring troubles is the in situ forecast of long-period waves. Correlation between the ordinary waves with the period less than 20 seconds and the long-period waves at the site is sought from the records of wave measurements. Then the forecast of long-period waves is made on the basis of daily forecast of ordinary waves. A few ports in Japan have introduced and been operating such systems.

## References

1. R. C. H. Russel, "Modern techniques for protecting the movement of ships inside harbours," *Analytical Treatment of Problems in the Berthing and Mooring of Ships* (NATO & HRS, 1973), pp. 267–275.
2. L. Barailler and P. Gaillard, "Evolution récent des modèles mathématiques d'agitation due á la houle, Calcul de la diffraction en profondeur non uniforme," *La Houille Blanche* **22**(8) (1976), pp. 861–869.
3. K. Tanimoto, K. Kobune and K. Komatsu, "Numerical analysis of wave propagation in harbours of arbitrary shape," *Rept. Port and Harbour Res. Inst.* **14**(3) (1975), pp. 35–58 (*in Japanese*).
4. M. B. Abbott, H. M. Petersen and O. Skovgaard, "Computation of short waves in shallow water," *Proc. 16th Int. Conf. Coastal Engrg.* (Hamburg, ASCE) (1978), pp. 414–433.
5. T. Takayama and Y. Kamiyama, "Diffraction of sea waves by rigid or cushion type breakwaters," *Rept. Port and Harbour Res. Inst.* **16**(3) (1977), pp. 3–37.
6. T. Takayama, "Wave diffraction and wave height distribution inside a harbor," *Tech. Note Port and Harbour Res. Inst.* No. 367 (1981), 140p (*in Japanese*).
7. S. Sato and Y. Goda, *Coastal and Harbour Engineering* (Shokoku-sha, Tokyo, 1972), pp. 72–77 (*in Japanese*).
8. Ed. A. T. Ippen, *Estuary and Coastline Hydrodynamics* (McGraw-Hill, 1966), pp. 319–323.
9. A. Ozaki, "Wave energy dissipating structures in harbors," *Lecture Series on Hydraulic Engineering* **65–17**, Hydraulics Committee, Japan Soc. Civil Engrs, (1965), 25p (*in Japanese*).
10. H. Maruo, "The drift of a body floating on waves," *J. Ship Res.* **4**(3) (1960), pp. 1–10.

11. H. Newman, "The drift force and moment on ships in waves," *J. Ship Res.* **11**(1) (1967).

12. S. Ueda, "Analytical method of ship motions moored to quay walls and the applications," *Tech. Note Port and Habour Res. Inst.* No. 504 (1984), 372p (*in Japanese*).

13. S. Ueda and S. Shiraishi, "The allowable ship motions for cargo handling at wharves," *Rept. Port and Harbour Res. Inst.* **27**(4) (1988), pp. 3–61.

14. F. H. Hsu and K. A. Blenkarn, "Analysis of peak mooring force caused by slow vessel drift oscillation in random seas," *Prepr. 2nd Offshore Tech. Conf.* (1970), OTC1159.

15. J. A. Pinkster, "Low frequency phenomena associated with vessels moored at sea," *Soc. Petroleum Engrs, AIME, SPE* No. 4837 (1974).

16. A. G. Davenport, "Gust loading factors," *Proc. ASCE* **98** (ST3) (1967), pp. 11–34.

17. S. Fujihata, S. Hata, S. Nakayama, Y. Moriya, T. Sekimoto, M. Ikeno and K. Sasa, "Numerical simulation on moored vessel motions in harbors and estimation of viscous damping coefficients," *Proc. Coastal Engrg., JSCE* **46** (1999), pp. 856–860 (*in Japanese*).

18. S. Ueda and S. Shiraishi, "On the effects of wave direction and mooring lines on the motions of moored ships," *Proc. 31st Japanese Conf. Coastal Engrg.* (1984), pp. 451–455 (*in Japanese*).

19. P. Bruun, "Breakwater or mooring system?" *The Dock & Harbour Authority* (Dec. 1980), pp. 260–262.

20. S. Ueda, "Motions of moored ships and their effect on wharf operations efficiency," *Rept. Port and Harbour Res. Inst.* **26**(5) (1987), pp. 319–373.

21. O. Jensen, J. G. Viggósson, J. Thomsen, S. Bjørdal and J. Lundgren, "Criteria for ship movements in harbours," *Proc. 23rd Int. Conf. Coastal Engrg.* (Venice, ASCE) (1992), pp. 3074–3087.

22. Permanent Int. Association of Navigation Congresses (PIANC), "Criteria for movements of moored ships in harbours: A practical guide," *Rept. Working Group No. 24, Permanent Technical Committee II, Supplement to Bulletin No. 88* (1995), 35p.

23. H. Sato, S. Shiraishi and H. Yoneyama, "Investigations on allowable movements of container ships and ferries for cargo handling works," *Tech. Note Port and Airport Res. Inst.* **1055** (2003), 43p (*in Japanese*).

24. N. W. H. Allsop, L. Franco, G. Belloti, T. Bruce and J. Geeraerts, "Hazards to people and people and property from wave overtopping at coastal structures," *Coastlines, Structures and Breakwaters 2005* (*Proc. Conf.*, Inst. Civil Engrs., Thomas Telford) (2006), pp. 153–165.

25. N. Sato and H. Saeki, "A investigation of moored ship motions at fishing ports in Hokkaido," *Annual J. Coastal Engrg. in the Ocean, JSCE* **19** (2003), pp. 637–642 (*in Japanese*).

26. T. Ippen and Y. Goda, "Wave induced oscillations in harbors; the solution for a rectangular harbor connected to the open sea," *M.I.T. Hydrody. Lab. Rept.* **59** (1963), 90p.

27. Ed. A. T. Ippen, *Loc. cit.* Ref. 8, pp. 310–315.

28. T. Hiraishi and K. Nagase, "Experiment on optimization of sea wall for long period waves," *Annual J. Coastal Engrg., JSCE* **51** (2004), pp. 721–725 (*in Japanese*).

29. K. Ikeno, T. Kumagai, Y. Moriya, K. Ooshima and T. Sekimoto, "Development of upright wave-dissipating structure for long waves," *Annual J. Coastal Engrg., JSCE* **51** (2004), pp. 731–735 (*in Japanese*).

# Chapter 8

# Hydraulic Model Tests with Random Waves

## 8.1 Similarity Laws and Scale Effects

### 8.1.1 *Selection of Model Scales with the Froude Similarity Law*

As discussed in the previous chapters, many problems in the planning and design of harbor facilities and coastal structures must be solved by means of hydraulic model tests. Although most tests before the 1970s had been performed with regular trains of waves, the advantages of hydraulic model tests with random waves are now well acknowledged and facilities for such tests are becoming available in laboratories around the world. Thus, the present chapter mainly discusses hydraulic model tests with random waves.

The fundamental condition to be satisfied in a model test is that the model must behave in a manner similar to the prototype. Similitude of the model to the prototype is required in the three general categories of geometric shape, kinematics of the various motions, and dynamic forces acting in the model and the prototype. *Geometric similarity* means that all lengths in the prototype are scaled down in the model by a certain ratio or ratios. *Kinematic similarity* requires proportionality of the velocities and accelerations of the various bodies and the fluid between the model and the prototype. *Dynamic similarity* demands that all dynamic forces in the prototype must be reproduced in the model with the same scale ratio. Complete dynamic similarity is seldom possible, however, because of the varied nature of the many forces acting in the prototype. Depending on the predominant forces for which dynamic similarity is to be maintained, several model laws of similitude can be derived.

In hydraulic model tests concerning sea waves, the viscosity and surface tension of water usually do not play significant roles, leaving

Table 8.1    Model scales according to the Froude law.

| Item | Scale | Example | Prototype | Model |
|---|---|---|---|---|
| Horizontal length and wavelength | $l_r$ | 1/25 | 50 m | 2.0 m |
| Water depth | $h_r = l_r$ | 1/25 | 15 m | 60 cm |
| Wave height | $H_r = l_r$ | 1/25 | 6 m | 24 cm |
| Wave period and time | $T_r = l_r^{1/2}$ | 1/5 | 10 s | 2.0 s |
| Wave pressure | $P_r = l_r$ | 1/25 | 90 kPa | 3.6 kPa |
| Force per unit length | $P_r = l_r^2$ | 1/625 | 1500 kN/m | 24 N/cm |
| Weight per unit length* | $w_r = l_r^2$ | 1/625 | 2800 kN/m | 44.8 N/cm |
| Weight of armor unit* | $W_r = l_r^3$ | 1/15,625 | 300 kN | 19.2 N |
| Overtopping amount per wave per unit length | $Q_r = l_r^2$ | 1/625 | 0.6 m³/m | 9.6 cm³/cm |
| Overtopping rate per unit length | $q_r = l_r^{3/2}$ | 1/125 | 0.06 m³/m·s | 4.8 cm³/cm·s |

*With the same specific weight of material.

inertia and gravitational forces as the governing forces. The scaling law for such a situation is the *Froude law*, which dictates that the scales for the time and velocity should be equal to the square root of the length scale. Table 8.1 gives examples of *model scales* obtained from the Froude law, with explanatory figures.

If a model is made with the same geometric scale in the horizontal and vertical directions, it is called an *undistorted model*. If the geometric scales are different in the horizontal and vertical directions, the model is called a *distorted model*. Hydraulic model tests on tidal currents are prominent examples of distorted models, with the horizontal scale much smaller than the vertical scale, because these models usually have to cover a large water area. On the other hand, most model tests on sea waves are done with undistorted models, because the horizontal and vertical motions of water particles by wave action must be reproduced with the same scale. One possible exception may be the use of a distorted model for the analysis of tranquility of a large harbor against short wind waves, in which case only the effects of wave diffraction and reflection enter, because wave refraction is negligible.

It is desirable to make use of the largest model possible to obtain results of highest possible accuracy. The disadvantages of large models are the increased costs of construction and operation, as well as elongation of the required test time. The model scale must be selected based on considerations of the size of the prototype structure (or harbor) and the size of the available test facilities. No definite criteria can be cited on the acceptable scales of hydraulic models. Many laboratories, however, have been making model tests on harbor tranquility with scales of 1/50 to 1/150. Tests on breakwater stability and wave overtopping of seawalls are often carried out with scales of 1/10 to 1/50.

Despite the large costs involved, there are quite a number of requests for making large-scale tests mainly because of avoiding the scale effect problems. In response to such requests, wave flumes capable of generating large waves with the height exceeding 2 m have been built and operated in several countries. The largest wave in laboratory is 3.5 m high in the Large Hydro-Geo Flume at the Port and Airport Research Institute in Japan.

### 8.1.2 *Possible Scale Effects in Model Tests*

Selection of the model scale needs to be made with due considerations for possible *scale effects*, which means that a phenomenon in the prototype is not well reproduced in a model as predicted by the similarity law because of the size limitation. The phenomena in which scale effects may appear are listed below.

(1) Wave attenuation during propagation by surface tension and other factors,
(2) Wave breaking limit in shallow water area,
(3) Wave run-up on a slope with rough surface,
(4) Wave overtopping quantity (rate) when the relative crest height is close to the run-up limit,
(5) Wind effects on wave overtopping,
(6) Stability of armor units,
(7) Phenomenon involving air compression such as the uplift to the ceiling slab of open-type wharf or perforated-wall caisson, and
(8) Others.

The first phenomenon of *wave attenuation during propagation* often occurs when the wave period is less than 0.5 s (Goda[1]). It is because the water surface of a test flume or basin is often covered by a thin film made of dusty, minute foreign materials. This thin film exerts retardation of wave motion through the surface tension and gradually diminishes the wave amplitude. For random waves, the limiting wave period is even greater because the spectral peak must be well above the threshold of 0.5 s. The significant period of model waves is preferably longer than 1.0 s and never shorter than 0.8 s.

The second item of *wave breaking limit* may be affected by the surface tension of water. When the height of test waves is not large enough, plunging of wave crests is not well reproduced and plunging breakers take a form somewhat similar to partial spilling breaker. According to a comparative test of breaking wave heights with different scales by Goda and Morinobu,[2] model regular waves with the height equal to or larger than 7 cm are not subject to scale effects; waves of low height break with relatively smaller height. In random wave tests, a threshold significant height of 10 cm would be recommended to avoid scale effects on wave breaking phenomena.

The third item of *wave run-up* on a slope with rough surface is the phenomenon that a small-scale model holds wave run-up heights at a relatively low elevation, because the effects of the viscosity and surface tension of water are exercised on wave run-up much stronger for small waves than for large waves. Burcharth and Lykke Andersen[3] have demonstrated the scale effects on wave run-up using a few simplified physical models. We can sometime observe that uprushing spray of waves hitting a vertical breakwater rises higher than 100 m in the field, which is ten to twenty times the wave height. It would be difficult to reproduce such uprushing spray in small models.

The fourth item of *wave overtopping* is also related with the influence of the viscosity and surface tension of water. As discussed in Sec. 5.3 (D), the wave overtopping rate of seawalls by model tests becomes smaller than that converted from the prototype when the absolute volume of wave overtopping is small. The scale effects have been observed on seawalls and breakwaters having frontal mounds of concrete blocks. No report on scale effects has been made so far for vertical seawalls or inclined seawalls with smooth surface. Sakakiyama and Kajima[4] have proposed the threshold Reynolds number of $(\mathbf{R_e})_{\text{crit}} = D_n U/\nu = 10^5$ below which the scale effects on wave overtopping rate appear, where $D_n$ is the nominal block diameter defined as equal to $(M/\rho_r)^{1/3}$ and $U$ is the representative velocity defined as equal to $(g\,H_{1/3})^{1/2}$. To satisfy this threshold condition, a model needs a block diameter of $D_n = 0.08$ m and a significant wave with $H_{1/3} = 0.15$ m, for example.

The fifth item of *wind effects on wave overtopping* has not been clarified yet. As discussed in Sec. 5.3 (C), wave overtopping quantity tends to increase by wind actions when wave overtopping quantity is small. Although an empirical formula to account for the wind effects has been proposed, its range of application seems to be limited. A test has been done to blow winds by a fan installed above a vertical seawall and to measure the increase of wave spray quantity over that without winds (De Waal *et al.*[5]), but only qualitative results have been obtained.

As for the sixth item of *armor unit stability*, Sakakiyama *et al.*[6] cite a threshold Reynolds number of $(\mathbf{R_e})_{\text{crit}} = (3 \sim 5) \times 10^5$, by referring to the studies by Thomsen *et al.*[7] and Shimada *et al.*[8]

The seventh item concerns with the *compressibility of air* in the model and the prototype. This involves the problems of the uplift exerted below the deck of an open-type pier, the air and water pressures inside a wave chamber of wave-absorbing caisson when the wave crest rises above the upper end of perforation and the air is trapped inside the chamber (Tanimoto *et al.*[9]), the air pressure inside an oscillating water-column wave power caisson (Ojima *et al.*[10]), and others. Because the air property is the same in the model and the property, a model test with a reduced scale is carried out with the air much harder to compress than the prototype. Thus the measured air pressure becomes higher than the prototype value. The pressure intensity converted by the Froude law must be corrected with an appropriate theory of adiabatic air compression.

## 8.2   Necessity of Hydraulic Model Test with Random Waves

Nowadays, many problems are being solved numerically with the advancement of computer capacity and numerical models. Some problems such as storm surge generation and tsunami propagation can only be solved by means of numerical computation. Some problems such as harbor tranquility are competing with hydraulic model tests in terms of the costs, time, and accuracy involved. Nevertheless, many problems need to be solved through hydraulic model tests to obtain reliable answers. The following is a list of problems waiting for the solutions by hydraulic model tests:

A. Wave transformation and deformation (3-D tests):

A-1. Spatial distribution of wave height in a harbor with partially reflective facilities inside,

A-2. Wave deformation on complicated bathymetry involving wave breaking,

A-3. Wave transformation in presence of currents,

A-4. Hydrodynamic features around natural reefs,

A-5. Nearshore currents and wave setup induced by low crested structures.

B. Optimum geometry of breakwaters and other facilities:

  B-1. Optimum cross-section of wave-absorbing type and special type of breakwaters (2-D tests),

  B-2. Optimum cross-section of coastal dikes and seawalls (2-D tests),

  B-3. Optimum layout of low crested structures for beach protection (3-D tests).

C. Stability of breakwaters (2-D tests):

  C-1. Wave forces exerted on composite breakwaters and their sliding distance,

  C-2. Wave forces exerted on and stability of wave-absorbing type and special type of breakwaters,

  C-3. Stability of armor stone and concrete blocks against wave actions,

  C-4. Stability of foot protection blocks and armor units of rubble foundation mounds of composite breakwaters.

D. Performance and stability of breakwaters and seawalls (3-D tests):

  D-1. Stability of the main body and rubble mound foundation of a breakwater head,

  D-2. Stability of a concrete block mound of horizontally-composite breakwater at its transition part to a conventional section,

  D-3. Spatial distribution of wave overtopping quantity along the seawall of a large-scale reclamation site and confirmation of the capacity of draining facilities.

E. Performance and stability of floating facilities:

  E-1. Performance of floating breakwaters and safety of mooring lines,

  E-2. Stability of floating pontoons and mooring forces,

  E-3. Motions of ships moored at buoys and mooring forces,

  E-4. Motions of ships moored at fixed berths and mooring forces.

Among the above problems, the item A-1 used to be the major topics of hydraulic model tests. With the development of numerical computation techniques, requests for this type of tests have become less than before. However, a port which is suffering the ship mooring problems by swell and/or long-period waves needs appropriate assessment of wave-absorbing facilities to be installed within a harbor. Wave absorbing capacity of such facilities should be investigated with a large model, and their reduced models are to be placed in a 3-D harbor model with the wave reflection performance appropriately reproduced.

The items A-2, A-4, and A-5 involve the process of random wave breaking, which has not been well reproduced by many numerical models yet. Physical model tests are required to solve these problems.

Recent development in numerical wave flumes has begun to solve some problems under the item B on optimum geometry of breakwaters and other facilities. Selection of the optimum solution among several alternatives may be made with computation in numerical wave flumes, but the confirmation of its performance needs to be made in a physical model test.

For the problems of stability and performance of structures under the items C and D, hydraulic model tests can yield the solution more reliable than numerical computation. Wave breaking process, especially on 3-D topography, is one reason of the superiority of physical tests to numerical computation. The action of waves on and around the breakwater head is another topic which should be studied through hydraulic model tests.

The problems concerning floating structures including ships under the item E may be solved by numerical computations, but hydraulic model tests can also provide appropriate answers. Wave mitigation performance of a floating breakwater depends on its cross-sectional shape, which is usually investigated with hydraulic model tests. Good reproduction of the load and elongation characteristics of mooring lines is the key factor in model tests.

Problems related with low crested structures listed in the items A-5 and B-3 concern with the conservation of beach against erosion by waves and currents. Although direct simulation of beach morphology with physical models is not feasible because of the limitation in grain size reduction of sand, the information on waves, currents, and water levels around low crested structures can be obtained through model tests and it provides the basic data for judgment of beach morphological changes.

## 8.3 Generation of Random Waves in Test Basins

### 8.3.1 *Random Wave Generator*

The working principle of a random wave generator is completely different from that of a conventional regular wave generator. Figure 8.1 illustrates the two kinds of wave generators. The upper figure (a) shows the typical composition of a conventional wave generator. The motor is usually of the three-phase alternating current type, which rotates at a fixed speed. The rotational speed is set to a predetermined value by a nonstep speed-variator, and then the speed is reduced by 1/10 to 1/20 with gears. The rotational speed can thus be adjusted to the required wave period.

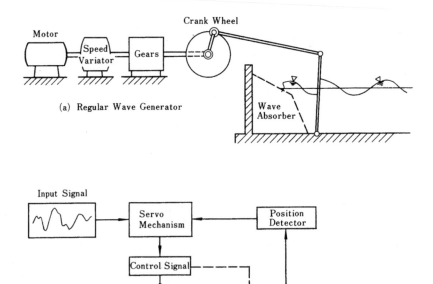

Fig. 8.1   Outline of regular and random wave generators.

The rotation of the output axis of the gears is transmitted to a crank wheel which rotates at a constant speed and produces a reciprocating motion of the wave paddle through a crank arm and linkage system. The figure shows the mechanism for a flap-type wave paddle with a rotational hinge at the bottom, but any type of mechanism such as piston type or a double-hinged flap type can be designed as required. The stroke of the crank arm or the amplitude of the paddle motion is usually designed to vary continuously from zero to some maximum. There are also sophisticated machines which enable changes of paddle amplitude during operation of the wave generator.

The lower figure (b) shows the basic concept of a random wave generator, which is characterized by a servo-mechanism system. The motor may be an electric servo-motor, a special direct-current motor with low rotor inertia, a hydraulic pump, or a hydraulic pulse motor. The output of the motor is converted to reciprocating motion of the driving unit of the wave paddle through a long screw bar by means of coupling by a peripherical ball bearing or through a hydraulic piston. The position of the wave paddle is continuously monitored by a sensing device, and the information is fed back to the servo-mechanism, which compares it with the input signal and sends a control signal to the direct current motor or the control valve of the hydraulic pump. Thus, a random wave generator is essentially a signal-following driving machine, and it generates a train of regular waves when the input signal is a sinusoid.

Generators for random waves first appeared in the 1960s at ship testing basins. In the 1970s, they became popular at hydraulic facilities of civil engineering laboratories. It may be said that the production of such generators was stimulated by rapid progress in sophisticated equipment for testing of severe vibration and durability of the components of missiles and rockets, as well as by developments in numerically controlled machines and robots in industry. The level has been reached where a random wave generator is not a machine of special design, but is assembled with a number of standard industrial components.

Laboratory wave generators have evolved into multidirectional random wave generators nowadays. Such a generator is composed of a large number of segmented wave paddles, each with the width 30 to 80 cm, which are independently controlled by servo systems. The computer-generated control signals are fed to individual wave paddles so that the water surface in front of the wave paddles will follow the temporal and spatial fluctuations of random wave profiles which are specified by a target directional spectrum. When individual wave paddles are driven with the same amplitude and frequency but with a constant phase lag between adjacent paddles, a regular train of oblique waves is generated.

In the 1950s, there were mechanical wave generators called the snake-type or serpent-type generator, which was capable of generating oblique waves in laboratory basins. The multidirectional random wave generators of today are the modernization of the snake-type wave generator of the past, having been given the greater capability of generating diverse profiles of random waves. Salter[11] manufactured the first wave generator of this type at Edinburgh University in the mid-1970s. Since then, many institutions throughout the world have installed the multidirectional wave generators in their hydraulic laboratories. The number of institutions was 32 in 1992 according to a survey by Funke and Mansard[12] and it is increasing ever since.[a]

### 8.3.2 *Preparation of Input Signal to the Generator*

(A) *Target spectrum of wave paddle motions*

The generator itself is manufactured in the form sketched in Fig. 8.1(b). However, the most important aspect of random wave generation is the preparation of the input signal to the generator for obtaining random waves of prescribed characteristics. This is basically done according to the following steps.

First, the target frequency spectrum of the random waves is chosen. If only the significant height and period of the test waves are given, the modified Bretschneider–Mitsuyasu type spectrum, Eq. (2.11) in Sec. 2.3.1, provides a convenient formula for the spectrum of the model waves. If the frequency spectrum of the random waves at the site is known, it may be used as the target. Depending on the nature of the tests, many other spectra might be used, such as a narrow band spectrum of long-traveled swell, a sharply peaked spectrum of the JONSWAP type described by Eq. (2.12) in Sec. 2.3.1, which represents wind waves generated by strong winds, or a multimodal spectrum for the superposition of wind waves and swell.

The target spectrum of the water waves must be converted to the target spectrum of the wave paddle motion by means of the transfer function for wave generation. This function represents the relationship between the height of the generated waves and the amplitude of the wave paddle for waves of constant period. According to the solution of the velocity potential given by Biésel and Suquet,[14] the *transfer function* for wave gneration is given by

$$\text{Piston type}: F_1(f, h) = \frac{H}{2e} = \frac{4\sinh^2(2\pi h/L)}{4\pi h/L + \sinh(4\pi h/L)}, \tag{8.1}$$

$$\text{Flap type}: F_2(f, h) = \frac{H}{2e} = \frac{4\sinh(2\pi h/L)}{2\pi h/L}$$

$$\times \frac{1 - \cosh(2\pi h/L) + (2\pi h/L)\sinh(2\pi h/L)}{4\pi h/L + \sinh(4\pi h/L)}, \tag{8.2}$$

---

[a]A further survey by Mansard *et al.*[13] in 1997 lists 42 facilities at 40 institutions, but there are many more facilities not covered by the survey.

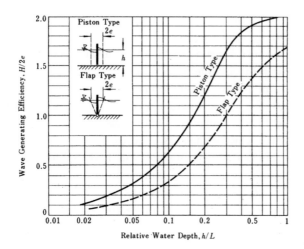

Fig. 8.2 Transfer function for wave generation by paddle motion of piston type and flap type.

where $H$ denotes the height of the generated waves, $e$ is the amplitude of wave paddle at the mean water level, and $f$ stands for the wave frequency $(= 1/T)$ which implicitly enters in Eqs. (8.1) or (8.2) through the wavelength $L$.

The transfer functions given by Eqs. (8.1) or (8.2) are plotted in Fig. 8.2 against the relative water depth $h/L$. As seen in the figure, short-period waves (high frequency) have a high efficiency of wave generation, whereas long-period waves (low frequency) have low efficiency. Because the transfer function for wave generation varies with the wave frequency, the target spectrum of the wave paddle motion $S_G(f)$ is calculated for the target wave spectrum $S_w(f)$ as follows:

$$S_G(f) = S_w(f)/F_j^2(f/h) : j = 1 \text{ or } 2. \qquad (8.3)$$

The target paddle spectrum takes a shape enhanced at the low frequency side and damped at the high frequency side in comparison with the target wave spectrum.

## (B) *Preparation of paddle motion signals*

The next step is a simulation of a random wave profile for the motion of wave paddle which satisfies the target paddle spectrum. At the early stage of random wave tests, the analog simulation technique was used. It employed an analog white noise generator and a set of 10 to 15 bandpass filters, which tailored the white noise signal in such a way that a desired spectral shape can be obtained by adjusting their outputs. The output signal was fed directly into the control system of the random wave generator. At present, however, the digital technique is prevailing.

The digital simulation technique involves numerical generation of random signals with the aid of a personal computer. Three methods are currently employed for this purpose. One makes use of the impulse response function to digital white signals as done by Freyer *et al.*,[15] another uses the inverse fast Fourier transform algorithm, and the third method synthesizes random signals as the sum of finite number of sinusoidal waves (refer to Sec. 11.5). Each has its own advantages and disadvantages; however, the differences are not essential but operational. The digital signals thus generated are fed into a digital–analog converter and the resultant analog signals are sent to the control system of the random wave generator. All the digital methods involves generation of quasi-random numbers with the input of a *seed number* as described in Sec. 11.5.4. By changing the input seed number, different profiles of random waves are generated. When the same seed number is input, the same profile of random waves is produced in a test flume.

After the random signals for driving the wave paddle have been prepared, a test run of wave generation is carried out and the records of the generated waves are analyzed for the wave spectrum as well as for various representative wave heights and periods. If the spectrum, wave height, or period is found to have significant deviations from the expected wave characteristics, the target paddle spectrum is adjusted and the procedure is repeated until satisfactory results are obtained. In this process, attention should be paid to the statistical variability of the actual data, which has an inherent characteristic of random water waves (refer to Sec. 10.6). Just as in the problem of quality control in mass production in a factory, we cannot decide whether we should adjust the input signal to the generator on the basis of one set of data which shows some deviation from the target value. We need several measurements or a sufficiently long record of the wave profile before we can make a decision on the adjustment of the input signal.

### (C) *Phase-delayed signals to wave generators*

When the random control signals are fed to a wave generator, all components of wave spectrum are generated simultaneously. Low frequency component waves travel faster in a test flume than high frequency components, and waves arriving at a measurement site in the beginning contain only low frequency waves. Most of wave tests can be carried out by discarding initial parts of wave records before the major part of spectral components reach to the test site. It may be desirable however in some cases that the full spectral component of waves would arrive at the site of a model structure simultaneously.

It can be realized by means of phase-delayed generation of the wave spectrum; high frequency components are started early and lower frequency components are gradually added to the random signal for driving the wave paddle. The degree of phase delay must be calculated from knowledge of the propagation time of the individual component waves over the distance between the wave paddle and the model. By preparing the random input signal to the generator in this way, the test structure will receive a train of irregular waves having full spectral components from the beginning of the arrival of waves.

### 8.3.3 *Input Signals to a Multidirectional Random Wave Generator*

The principle of generating multidirectional random waves is the superposition of a large number of regular trains of oblique waves with different frequencies and directions of propagation. For a given target directional spectrum, the amplitudes and initial random phases of individual component waves are calculated for respective frequencies and directions. For each component wave, the amplitude of wave paddle motion is obtained by means of the transfer function of Eq. (8.1) or (8.2), and the phase lag of each paddle is evaluated with the frequency and direction of the component wave and the distance of the paddle from the tip of the wave generator. The phase angle of a particular wave paddle for each component wave is given by the sum of initial random phase and the phase lag of that paddle. The data of wave paddle amplitudes and phase angles of all component waves are stored for each wave paddle, and the control signal for each wave paddle motion is constructed by summing up the various sinusoidal motions prescribed for all component waves.

Surface wave profiles satisfying a target directional spectrum were initially computed by the *double summation method*: i.e., representation of wave profiles with the product of a series of various frequency components and that of various directional components. However, the problem of phase locking among the directional components at the same frequency was pointed out by Jefferys[16] and Miles and Funke.[17] Thus, the *single summation method* which assigns a single direction to each frequency component is currently employed in many laboratory basins as a standard procedure in generating multidirectional random waves. The technique for simulation of random wave profiles is discussed in Sec. 11.5.

### 8.3.4   *Non-Reflective Wave Generator*

Many random wave generators are now equipped with non-reflecting capacity, i.e., to absorb the waves reflected back from the model structures installed in test flumes/basins. Old type generators re-reflected these reflected waves toward the model structures, and reflected waves traversed back and forth many times, changing wave conditions significantly. If the first reflected waves are detected at the generator side and the generator can be driven in a manner to absorb the reflected waves, multi-reflection of test waves can be avoided. Such function of wave generator also serves in minimizing the waiting time of tests by quickly suppressing waves remaining in a test flume after a test is stopped.

Milgram[18] was the first in testing the idea. He installed a capacitance wave gauge in front of the wave paddle for detection of reflected waves. On the other hand, Salter[11] utilized the fluid reaction force to the wave paddle for detection of reflected waves and controlled the generator in a mode of counteracting the fluid force by reflected waves. Since then the both methods are employed in the wave generating system with the choice being dependent on the manufacturers.

Another method for detecting and controlling reflected waves has been proposed by Frigaard and Christensen,[19] who set two wave gauges in the middle of a wave flume and dissolve the incident and reflected waves by means of digital filters. Matsumoto *et al.*[20] applied this technique in wave flume tests with good results of reflected wave suppression.

### 8.3.5   *Other Topics on Wave Generation in Test Flumes*

(A) *Single breaking wave at a designated location*

In the field of marine engineering, a technique of generating and applying a *single breaking wave* to a test structure positioned at a specified location is sometimes used. In this technique, a spectrum of waves are generated by adjusting the start time of each component wave such a way that a high frequency wave starts earlier and a low frequency wave moves out later so that all component waves will arrive at the designated position at the same time. Adjustment of start time is so made by taking into account the travel time by group velocity of each component wave from the wave maker to the test position. The technique is applied not only in a wave flume but also in a multidirectional wave basin. The input signal to the multidirectional wave generator is prepared with the single summation method with incorporation of appropriate time lags for respective component waves. A few trials and refinement of time lags for component waves will be necessary before the desired result of single wave breaking is attained.

(B) *Traveling secondary wave crests associated with regular waves of large amplitudes*

When we generate regular waves of large heights in shallow water, there sometime appears a *secondary wave crest* between the main wave crests. The secondary crest propagates with the speed slightly slower than the main crests, and the position of the secondary crest gradually moves from the preceding crest toward the following crest. As waves continue propagate, the secondary crest is absorbed by the following crest and then reappear behind it. The wave height takes a large value around the location where the secondary wave crest appear at the middle of two main crests, while the wave height becomes small at the location where the secondary crest is absorbed by the following main crest. Thus, the constancy of wave height is not maintained within a test flume.

This is one of the phenomena involving nonlinear interactions between wave components. Large waves in shallow water have the phase-bounded harmonic components with twice, thrice and higher frequencies of appreciable amplitudes as exemplified by the higher order Stokes waves. When waves are generated by sinusoidal motions of a wave paddle, no harmonic component waves are produced. In order to satisfy the free surface condition, the harmonic component waves with the phases opposite to those of the phase-bounded ones are automatically generated. These harmonic waves of opposite phases are free waves and propagate with the speed governed by the dispersion relationships. These free harmonic waves and the phase-bounded ones interfere each other and produce the nonlinear bound components with the fundamental and thrice frequencies. The amplitudes of the fundamental, second, and third harmonics vary from place to place as waves propagate. For detailed discussion of this phenomenon, see Ref. 21.

The remedy to this phenomenon is to provide the wave generator with non-sinusoidal signals, which nearly fit to the horizontal orbital velocity of water particles prescribed by the Stokes waves or cnoidal waves that correspond to the wave generating conditions.

(C) *Control of long-period waves in wave flumes*

Random waves generated in a test flume are accompanied with long-period waves bounded to the group of waves. *Bounded long-period waves* are the result of nonlinear wave interactions to be discussed in Sec. 10.5.3, and they may have appreciable amplitudes in shallow water conditions. When the input signal for linear random waves is given to the wave generator, there are simultaneously produced the group-bounded long-period waves and the free long-period waves that compensate the former. As a result, two long-period waves, one bounded to the original wave group and the other propagating freely, move in a test flume with different propagation speed.

In order to reproduce the long-period waves correctly, the wave generation signal should have the second order components that suppress the generation of free long-period waves. The problem was first discussed by Ottesen Hansen *et al.*[22] and Bowers[23] in 1980, and discussions were followed by Sand[24] and Barthel *et al.*[25] In order to make good control of long-period wave generation, the wave paddle needs to have a sufficiently long stroke, because the wave generation efficiency of wave paddle decreases as the relative depth becomes small as indicated by the transfer function of Eq. (8.1) and shown in Fig. 8.2. In case of piston-type generator, the efficiency is inversely proportional to the wave period as indicated by the following:

$$\frac{H}{2e} \simeq kh = \frac{2\pi\sqrt{h/g}}{T} . \tag{8.4}$$

(D) *Tsunami generation in wave flume*

When a wave paddle is driven for one cycle of sinusoidal wave, a solitary wave is generated in a wave flume. The solitary wave can propagate in water of uniform depth without changing its profile. When it moves onto a slope of very gentle inclination, it may split into a series of solitons which break one by one at shallow depth. Breaking of solitons at the front of tsunami was observed at the coast of Akita at the time of the tsunami generated by the Nihonkai-Chubu (Central Japan Sea) Earthquake in 1983. Tsuruya *et al.*[26] reproduced this phenomenon in a wave flume of 163 m long and reported the characteristics of soliton breaking. They drove the wave paddle in one cycle of sinusoid beginning with the positive half-cycle. With the stroke of about 40 cm and the period of 40 s, soliton waves of about 5 cm high were generated.

Recent technique of tsunami generation is to set the wave paddle at the rearmost position and push it forward to the foremost position in one stroke. As indicated in Eq. (8.4), the wave paddle needs to have a long stroke to be able to generate tsunami of appreciable height.

## 8.4   Model Tests Using Multidirectional Random Waves

Salter[11] invented the first multidirectional random wave generator to study the effect of wave directionality on a special wave power extraction device of the floating type called the Salter Duck. Offshore structures, especially of the floating type, have been the topics of intensive studies using multidirectional random wave generators. The movements of an oil tanker berthed at a single-point mooring buoy, for example, are greatly enhanced in the multidirectional random wave field in comparison with those in unidirectional irregular waves. On the other hand, an offshore platform experiences a reduction in the in-line wave load and an increase in the transverse wave load by introduction of multidirectional wave systems in model tests. Funke and Mansard[12] cite a number of cases in which the wave directionality becomes important in hydraulic model tests. Hiraishi[27] also presents several cases of model tests with multidirectional random waves and discusses the directionality effects.

A constraint in the use of multidirectional random waves in model tests is the relatively narrow test area in a wave basin, where the wave condition can be regarded homogeneous. Oblique waves generated from a finite width of a generator cannot propagate into the full area of basin; they are attenuated in the zone of geometric shadow by the diffraction effect, and the wave height distribution along the wave crest becomes non-uniform. Because multidirectional random waves are made up of many oblique wave components of various directions, the effective width of the homogeneous wave area decreases as the distance from the generator increases. Some laboratories are trying to eliminate the restriction of the limited test area by providing two sets of multidirectional random wave generators along the two sides of a rectangular basin and by employing sophisticated algorithms to control the two sets of wave generators in a coordinated manner; e.g., Ito *et al.*[28]

Another technical problem in a multidirectional wave basin is the absorption of reflected waves from a structure in a basin. In the case of wave flume tests, the technique for absorption of reflected waves by wave paddles has well been established and incorporated in the design of wave generators. In a multidirectional wave basin, reflected waves come back to the wave paddles with oblique angles of approach, the values of which are difficult to detect. Nevertheless, the wave absorption problem has been investigated by many researchers and given some practical solutions. Wave absorption is usually made by assuming the normal incidence of reflected waves, though attempts are being made to realize full absorption of oblique reflected waves. A review on this wave absorption problem is found in Schäffer and Klopman,[29] who cite more than ten related papers.

A model test of harbor tranquility in a multidirectional wave basin is feasible if the harbor entrance is single and fits within the zone of effective test area. A three-dimensional stability test of a breakwater head seems to be suited for a multidirectional wave basin, as Matsumi *et al.*[30] have observed larger damage on a breakwater head by multidirectional waves than by unidirectional irregular waves. Another subject to be tested in a multidirectional wave basin is the spatial variation of wave run-up and/or overtopping along a finite extension of seawall. It is known that wave heights along a barrier of finite extension vary by the effect of wave diffraction from the tips of the barrier. Because the diffraction effect is a function of wave frequency and direction, a test with multidirectional random waves is expected to yield a smaller variation of wave heights, and thus more uniform run-up and/or overtopping than a test with unidirectional irregular waves. A few cases concerning this problems have been discussed in Sec. 5.3 (B).

## 8.5   Some Remarks on Execution of Random Wave Tests

### 8.5.1   *Number of Test Runs and Their Durations*

A record of wave measurements in the sea is just one sample of many possible realizations of the true sea state. Likewise, random waves in test flumes/basins should be regarded as statistical samples of the phenomena we are looking at. Even though we cannot make field wave measurements twice with the same condition, we can repeat random wave tests many times so as to minimize the range of statistical variability, which is discussed in detail in Sec. 10.6. For example, the significant wave height calculated from a record of 100 individual waves has a standard deviation of some 6%; i.e., a significant height of 6.0 m means that the probability of the true value being in the range between 5.6 m to 6.4 m is about 73%. The standard deviation of a measured value generally decreases in proportion to the inverse of the square root of the number of waves. Reduction of the range of uncertainty is possible only through increase of the number of waves included in test runs. It can be achieved by making a long test run and/or several test runs for the same test conditions.

A guideline for planning of test runs is to measure around 200 waves after the wave condition in a test flume/basin becomes stable and to have three test runs for a given test condition. The input seed number for quasi-random numbers is to be changed for the three runs so that the profiles of random waves become different every time. The significant wave height, for example, would be different among the three runs. Because the standard deviation of the significant wave height calculated for 200 waves is about 4%, one run among the three runs would yield the significant wave height being different from the target value by more than 4%.

In the stability tests of armor stone, we are interested in the damage ratio or the ratio of dislocated number of armor units to the total number. The damage ratio increases as a test continues long, and it should be expressed as the function of the number of waves acted on armor units. The stability test should be planned with the duration equivalent to three to six hours in the prototype.

In the wave overtopping tests of coastal dikes and seawalls, the overtopping quantity has a large range of variation when wave overtopping occurs infrequently and the quantity itself is little. In such a case, it is recommended to make more than five runs and to take the average so as to have a reliable result.

### 8.5.2  *Calibration of Test Waves*

Hydraulic model tests using random waves in a wave flume begin with calibration of test waves, which are to be measured at both the section of uniform depth and the location at which model structures are to be placed. The former station is for measurement of offshore waves, and the latter station is for measurement of waves incident to the model. In structural stability tests, model structures are tested against not only the design wave but also several trains of waves higher than the design condition, and such higher waves should also be calibrated before structural tests. Periods of test waves with larger heights would be better elongated according to the relationship between the significant wave height and period such as Eq. (3.6) in Sec. 3.1.1. When the test site is not deep enough, test waves may break before reaching the test site and the information on incident waves to the test site may be blurred. In such case, another measurement station is set between the offshore and the test stations.

Wave measurements in calibration are to be made for three or more runs of test waves at each level of wave height as discussed above. By saving the seed number and using the same number for each run, we can make tests of model structures with the waves same as those calibrated. During the tests on model structures, the resolution technique of incident and reflected waves (see the next subsection) should be employed and it should be confirmed that the resolved incident waves are almost the same as the calibrated incident waves.

Analysis of wave data for statistical wave characteristics such as significant wave height and period is discussed in Sec. 11.1.2, while analysis of frequency wave spectrum is discussed in Sec. 11.2. In multidirectional random wave model tests, a star array of four wave gauges or a combination of a wave gauge and bi-directional current meters are set at each measuring station for detection of directional spreading characteristics. The multi-sensor data are analyzed with the method described in Sec. 11.3.

### 8.5.3   *Resolution of Incident and Reflected Waves in a Test Flume*

A model structure reflects incident waves toward the wave paddle, and they are superposed on a train of incident waves. Because most of random wave generators are equipped with the wave absorption function nowadays as described in Sec. 8.3.4, the reflected waves themselves do not interfere with the test results. Nevertheless, it is customary to make resolution of incident and reflected waves in the test flume so that we can confirm the incident waves being same as those in calibration and calculate the reflection coefficient of a model structure. A set of two wave gauges is installed at a station in the middle of the test flume, and the wave records of two gauges are analyzed with the Fast Fourier Transform technique for resolution of incident and reflected waves. The method of resolution is discussed in Sec. 11.4.

A keypoint in this system is the distance between the two wave gauges, which is denoted by $\Delta l$ here. A set of two wave gauge with the distance $\Delta l$ can work for linear component waves having the wavelength of $(2.2 \sim 20)\Delta l$; the system cannot work for nonlinear wave components such as those discussed in Sec. 10.5.3. Therefore, the distance $\Delta l$ should be selected in such a way that the wavelengths of the component waves with the frequency of $(0.5 \sim 1.8)f_p$ will fall in the effective range of resolution, where $f_p$ denotes the frequency at the spectral peak. When tests are made with several different wave periods, it may become necessary to adjust the gauge distance to satisfy the above requirement.

### 8.5.4 *Statistical Variability of Damage Ratio of Armor Units*

Model tests involving stability of armor units are carried out for the objective of clarifying the damage ratio of armor units. Examples are the stability tests of armor units for mound breakwaters, those for horizontally-composite breakwaters, and those for rubble mound foundation of composite breakwaters. They may be 2-D tests in a wave flume or 3-D tests in a wave basin.

The *damage ratio* is counted as the ratio of the number of units dislocated from their original positions to the total number of units placed for the tests. When $r$ units are dislocated among $n$ units initially placed, the damage ratio is simply $p = r/n$. It is often regarded as a deterministic value, but it is a stochastic variable with a certain amount of uncertainty. This is a typical problem of statistics, which dictates the confidence interval of the measured ratio with probabilistic theory; the problem is same as the cases of opinion polls. A measure of confidence interval is the standard deviation, and any textbook on statistics gives the following formula for the standard deviation of a ratio $p$:

$$\sigma_p = \sqrt{\frac{p(1-p)}{n}}. \qquad (8.5)$$

Suppose we have placed 200 armor units in a test area, tested their stability with 3 different trains of waves, and observed dislocation of 7, 5, and 9 units by 3 wave trains, respectively. The mean number of dislocated units is 7 and the damage ratio is 3.5%. By counting the total number as $n = 3 \times 200 = 600$, the standard deviation of the above damage ratio is calculated as

$$\sigma_p = \sqrt{0.035 \times (1 - 0.035)/600} = 0.0075.$$

The 90% confidence interval is estimated as $\pm 1.64\sigma_p = \pm 0.0123$. Thus, what we can say is that the true damage ratio is somewhere from 2.3% to 4.7% with the probability of 90%. If the confidence interval is considered too wide, it is necessary to increase the total number of armor units for tests by expanding the test area and/or increasing the number of wave trains for tests.

When several kinds of armor units are tested for their performance, we tend to make comparison with the mean damage ratio alone. However, we must take into account the confidence interval of the damage ratio for performance comparison of the armor units. The test report should be written with the range of confidence interval of damage ratio, not with the mean ratio alone.

## References

1. Y. Goda, "Discussion of several factors for the improvement of the accuracy of wave tests in laboratory tanks," *Proc. 15th Japanese Conf. Coastal Engrg.* (1968), pp. 50–56 (*in Japanese*).
2. Y. Goda and K. Morinobu, "Breaking wave height on horizontal bed," *Proc. Pacific Coasts and Ports '97* (1997), pp. 953–958, or "Laboratory study on wave breaking limit on horizontal step," *Proc. Coastal Engrg., JSCE* **44** (1997), pp. 66–70 (*in Japanese*).
3. H. F. Burcharth and T. Lykke Andersen, "Scale effects related to small scale physical modelling of overtopping of rubble mound breakwaters," *Coastal Structures 2007 (Proc. 5th Int. Conf.*, Venice, World Scientific) (2009), pp. 1522–1541.
4. T. Sakakiyama and R. Kajima, "Comparison of field measurements and hydraulic model tests of wave overtopping at a wave-dissipating seawall," *Proc. Coastal Engrg., JSCE* **44** (1997), pp. 736–740 (*in Japanese*).
5. J. P. De Waal, P. Tönjes and J. W. van der Meer, "Wave overtopping of vertical structures including wind effect," *Coastal Engineering 1996 (Proc. 25th Int. Conf.*, Orland, Florida, ASCE) (1996), pp. 2216–2229.
6. T. Sakakiyama, R. Kajima and Y. Kubo, "Experimental study on assessment of wave overtopping at seawalls of a power plant on a reclaimed island," *Proc. Coastal Engrg., JSCE* **41** (1994), pp. 661–665 (*in Japanese*).
7. A. L. Thomsen, P. E. Wohlt and A. S. Harrison, "Rip-rap stability on earth embankment tested in large and small scale wave tank" *CERC Technical Memorandum*, No. 37 (1972).
8. M. Shimada, Y. Fujimoto, S. Saito, T. Sakakiyama, R. Kajima and H. Hirakuchi, "On scale effects concerning stability of wave-dissipating concrete blocks," *Proc. 33rd Japanese Coastal Engrg. Conf.* (JSCE, 1986), pp. 442–445 (*in Japanese*).

9. K. Tanimoto, S. Takahashi and T. Muranaga, "Uplift forces on ceiling slab of wave dissipating caisson with perforated front wall — Analytical model for compression of an enclosed air layer," *Rept. Port and Harbour Res. Inst.* **19**(2) (1980), pp. 3–31 (*in Japanese*).

10. R. Ojima, S. Suzumura and Y. Goda, "Theory and experiments on extractable wave power by an oscillating water-column type breakwater caisson," *Coastal Engineering in Japan* **27** (1984), pp. 315–326

11. S. H. Salter, "Absorbing wave-makers and wide tanks," *Proc. Symp. Directional Wave Spectra Applications*, ASCE (1981), pp. 185–202.

12. E. R. Funke and E. P. D. Mansard, "On the testing of models in multidirectional seas," *Proc. 23rd Int. Conf. Coastal Engrg.* (Venice, ASCE) (1992), pp. 3454–3465.

13. E. P. D. Mansard, B. Manoha and E. R. Funke, "A survey of multidirectional wave facilities," *Proc. IAHR Seminar on Multidirectional Waves and Their Interaction with Structures (27th IAHR Congress*, San Francisco, National Research Council of Canada) (1997), pp. 195–226.

14. F. Biésel and F. Suquet, "Les apparails générateurs de houle en laboratoire," *La Houille Blanche*, **6**(2, 4, et 5) (1951) (translated by St. Anthony Falls Hydr. Lab., Univ. Minnesota, Rept. No. 39).

15. D. K. Freyer, G. Gilbert and M. J. Wilkie, "A wave spectrum synthesizer," *J. Hydraulic Res.* **11**(3) (1973), pp. 193–204.

16. E. R. Jefferys, "Directional seas should be ergodic," *Applied Ocean Res.* **9**(4) (1987), pp. 186–191.

17. M. D. Miles and E. R. Funke, "A comparison of methods for synthesis of directional seas," *Proc. 6th Int. Offshore Mech. and Arctic Engrg.* (1987), pp. 247–255.

18. J. H. Milgram, "Active water-wave absorbers," *J. Fluid Mech.* **43**(4) (1970), pp. 845–859.

19. P. Frigaard and M. Christensen, "An absorbing wave-maker based on digital filters," *Proc. 24th Int. Conf. Coastal Engrg.* (Kobe, ASCE) (1994), pp. 168–180.

20. A. Matsumoto, M. Tayasu and S. Matsuda, "An improved time domain procedure for separating incident and reflected water waves," *Proc. Civil Engrg. in the Ocean, JSCE* **18** (2002), pp. 209–213 (*in Japanese*).

21. Y. Goda, "Recurring evolution of water waves through nonlinear interactions," *Proc. 3rd Int. Symp. Ocean Measurement and Analysis (WAVES '97*, ASCE) (1997), pp. 1–23

22. N.-E. Ottesen Hansen, S. E. Sand, H. Lundgren, T. Sørensen and H. Gravesen, "Correct reproduction of group-induced long waves," *Proc. 17th Int. Conf. Coastal Engrg.* (Sydney, ASCE) (1980), pp. 784–800.

23. E. C. Bowers, "Long period disturbances due to wave group," *Proc. 17th Int. Conf. Coastal Engrg.* (Sydney, ASCE) (1980), pp. 610–623.

24. S. E. Sand, "Long wave problems in laboratory models," *J. Waterway, Port, Coastal, and Ocean Engrg. Div.* (ASCE) **108** (WW4) (1982), pp. 492–503.

25. V. Barthel, E. P. D. Mansard, S. E. Sand and F. C. Vis, "Group bounded long waves in physical models," *Ocean Engrg.* **10**(4) (1983), pp. 261–294.

26. H. Tsuruya, S. Nakano and H. Ichinohe, "Experimental study of tsunami transformation and run-up in shallow water — Reproduction of the behavior of the 1983 Nihonkai-Chubu Earthquake Tsunami," *Proc. 31st Japanese Coastal Engrg. Conf.* (JSCE, 1984), pp. 237–241 (*in Japanese*).

27. T. Hiraishi, "Effect of wave directionality to wave action on coastal structures," *Proc. IAHR Seminar on Multidirectional Waves and Their Interaction with Structures* (*27th IAHR Congress*, San Francisco, National Research Council of Canada) (1997), pp. 399–412.

28. K. Ito, H. Katsui, M. Mochizuki and M. Isobe, "Non-reflected multidirectional wave maker theory and experiments of verification," *Proc. 25th Int. Conf. Coastal Engrg.* (Orlando, Florida, ASCE) (1996), pp. 443–456.

29. H. A. Schäffer and G. Klopman, "Review of multidirectional active wave absorption method," *J. Waterway, Port, Coastal and Ocean Engrg.* **126**(2) (2000), pp. 88–97.

30. Y. Matsumi, E. P. D. Mansard and J. Rutledge, "Influence of wave directionality on stability of breakwater heads," *Proc. 24th Int. Conf. Coastal Engrg.* (Kobe, ASCE) (1994), pp. 1397–1441.

# Part II

# Statistical Theories of
# Random Sea Waves

# Chapter 9

# Description of Random Sea Waves

## 9.1 Profiles of Progressive Waves and Dispersion Relationship

In Part I, we discussed engineering applications of random wave theories to the design of coastal and harbor structures. In Chapters 9 to 12, we are going to describe the theories of random sea waves themselves and techniques for the analysis of random wave data.

As presented in Sec. 2.3, random sea waves are assumed to be representable as a superposition of an infinite number of small amplitude waves having different frequencies and directions of propagation. The profile of an individual wave is expressed as

$$\eta = a \, \cos(kx \, \cos \theta + ky \, \sin \theta - 2\pi ft + \varepsilon) \,, \tag{9.1}$$

where $\eta$ denotes the elevation of the water surface above the mean water level, $a$ is the wave amplitude, $k = 2\pi/L$ is the wavenumber with $L$ being the wavelength, $\theta$ is the angle between the $x$-axis and the direction of wave propagation, $f$ is the wave frequency, $\varepsilon$ is the phase angle, and $x$, $y$, and $t$ are the space and time coordinates.

This component wave has the properties of a small amplitude wave as derived from velocity potential theory, and it is assumed to propagate freely without interacting with the other component waves. While this component wave propagates, an energy $E$ of the following amount is added to the body of water having unit surface area:

$$E = \frac{1}{2}\rho g a^2 \,. \tag{9.2}$$

This is actually the density of the wave energy. The symbol $\rho$ denotes the density of sea water and $g$ denotes the acceleration of gravity. There exists the following relationship between the wavenumber $k$ and the frequency $f$:

$$\omega^2 = 4\pi^2 f^2 = gk \tanh kh, \tag{9.3}$$

in which $\omega = 2\pi f$ denotes the angular frequency and $h$ is the water depth.

Equation (9.3) is called the *dispersion relationship*, which is obtained from the surface boundary condition at the water surface. This equation can be rewritten to give the following relationship between the wavelength $L$ and the wave period $T$:

$$L = \frac{g}{2\pi}T^2 \tanh \frac{2\pi h}{L}. \tag{9.4}$$

This is a transcendental equation, and the value of $L$ for the given $h$ and $T$ can be obtained numerically to a specified accuracy by an iterative computation. In any calculation related to the problems of sea waves, either Eq. (9.3) or (9.4) must be solved to obtain $L$. Here we shall give an iterative solution procedure.

First, Eq. (9.3) is transformed to

$$x \tanh x = D, \tag{9.5}$$

in which

$$\left. \begin{array}{l} D = \omega^2 h/g = 2\pi h/L_0, \quad L_0 = 2\pi g/\omega^2 = gT^2/2\pi, \\ x = kh = 2\pi h/L. \end{array} \right\} \tag{9.6}$$

Although Eq. (9.5) can be solved directly using Newton's method, it is a good procedure to first rewrite it as follows in order to remove the inflection point:

$$y(x) = x - D \coth x. \tag{9.7}$$

By this technique, an iterative solution can be obtained through the following equation:

$$x_2 = x_1 - \frac{y(x_1)}{y'(x_1)} = x_1 - \frac{x_1 - D \coth x_1}{1 + D(\coth^2 x_1 - 1)}. \tag{9.8}$$

The best estimate for the initial value is

$$x_1 = \begin{cases} D & : D \geq 1, \\ D^{1/2} & : D < 1. \end{cases} \tag{9.9}$$

The error in Eq. (9.8) rapidly decreases with increase in the iteration number. In fact, the absolute error $|1 - x_2/x_1|$ is less than 0.05% by the

```
PI2=2. *3.141592654
D=PI2*DEPTH/(9.8*T**2/PI2)
WAVEL=PI2 *DEPTH/WAVE(D)

     FUNCTION WAVE(D)

     IF(D-10.0) 2 , 2 , 1
  1  XX=D
     GO TO 6
  2  IF(D-1.0) 3 , 4 , 4
  3  X=SQRT(D)
     GO TO 5
  4  X=D
  5  COTHX=1./ TANH(X)
     XX=X-( X -D* COTHX)/
    +      (1.+D*( COTHX**2-1.))
     E=1.-XX/X
     X=XX
     IF(ABS(E)-0.0005) 6 , 5 , 5
  6  WAVE=XX
     RETURN
     END
```

**Water Depth : DEPTH (m)**
**Wave Period : T (sec)**
**Wavelength  : WAVEL (m)**

Fig. 9.1  FORTRAN program for the computation of wavelength.

third iteration (for $x_4$). Figure 9.1 provides the FORTRAN program for the computation of the wavelength by the above method. Tables A-1 to A-4 in the Appendix list values of the wavelength and wave celerity as a function of water depth and wave period as computed with this program.

For direct calculation of the wavelength, Hunt[1] has derived the following polynomial approximation for $x = kh = 2\pi h/L$ with the accuracy of 0.1% based on the Taylor expansion of Eq. (9.5):

$$x^2 = D*(D+1/(1+D*(0.6522+D*(0.4622+D^2*(0.0864+0.0675*D))))) ,$$
(9.10)

in which $D$ is defined by Eq. (9.6). Hunt has also shown another polynomial approximation with the accuracy of 0.01%. In a discussion to Hunt's paper, Venezian[2] has introduced the following simple approximation with the accuracy of 0.04% for a limited range of $0 \leq D < 1$:

$$x = \sqrt{D}/(1 - D/6) : 0 \leq D < 1 \ (h/L_0 < \text{about } 0.16) .$$
(9.11)

## 9.2    Description of Random Sea Waves by Means of Variance Spectrum

Profiles of random sea waves can be expressed by means of various mathematical representations, including one using complex vectors. The differences among these representations are purely a matter of convenience for the particular mathematical manipulations of interest. For general purposes, the following series expression is the most easily understood:

$$\eta = \eta(x, y, t)$$

$$= \sum_{n=1}^{\infty} a_n \, \cos(k_n x \, \cos \theta_n + k_n y \, \sin \theta_n - 2\pi f_n t + \varepsilon_n). \qquad (9.12)$$

This expression assigns a number to each combination of frequency and propagation direction of the infinite number of component waves and then sums up the components; it was first employed by Longuet-Higgins.[3]

The interpretation of random sea waves as a linear superposition of free progressive waves is an assumption, the correctness of which cannot be proven but, rather, must be supported through evidence of agreement between the properties of real sea waves and those derived from the mathematical model. Most properties of sea waves have been successfully explained with the model of Eq. (9.12), except for some nonlinear behavior to be discussed in Sec. 10.5. Thus, Eq. (9.12) can be considered to be a valid assumption.

The validity of Eq. (9.12) rests on four conditions. First, the frequencies $f_n$ must be densely distributed between zero and infinity in such a manner that any infinitesimal interval $df$ contains an infinite number of frequencies $f_n$. Second, the directions $\theta_n$ must be densely distributed between $-\pi$ and $\pi$ with an infinite number of $\theta_n$ contained in an infinitesimal interval $d\theta$. Third, the phase angles $\varepsilon_n$ must be randomly and uniformly distributed between 0 and $2\pi$. Fourth, though the amplitude of each wave is infinitesimal, the summation of its square should have a finite and unique value. By denoting this value as $S(f, \theta)$, it is expressed as

$$\sum_{f}^{f+df} \sum_{\theta}^{\theta+d\theta} \frac{1}{2} a_n^2 = S(f, \theta) df \, d\theta. \qquad (9.13)$$

The function $S(f, \theta)$ defined by Eq. (9.13) is called the directional wave spectrum density function, or the *directional wave spectrum*. It represents

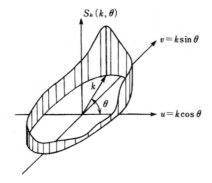

Fig. 9.2　Schematic representation of directional wave spectrum.

the manner in which wave energy[a] is distributed with respect to frequency $f$ and angle $\theta$.

The directional wave spectrum can also be expressed as a function of wavenumber $k$ and angle $\theta$. By considering the energy density contained in the interval from $k$ to $k + dk$ and from $\theta$ to $\theta + d\theta$, we can obtain the following expression for the directional wave spectrum in the wavenumber domain:

$$\sum_{k}^{k+dk} \sum_{\theta}^{\theta+d\theta} \frac{1}{2}a_n^2 = S_k(k, \theta)dk\, d\theta. \tag{9.14}$$

Figure 9.2 schematically shows $S_k(k, \theta)$ in the $k - \theta$ domain; the coordinate $(u, v)$ is the Cartesian representation of the polar coordinate $(k, \theta)$. The figure only exhibits the density function for a given wavenumber $k$; the general representation of $S_k(k, \theta)$ is an envelope surface in the range of $0 < k < \infty$. Often the directional wave spectrum is expressed in the form of a map of contour lines, the value on each line representing the absolute value of $S_k(k, \theta)$.

The directional wave spectrum in terms of the wave frequency, $S(f, \theta)$, and that in terms of the wavenumber, $S_k(k, \theta)$, are connected through the dispersion relationship, Eq. (9.3). The conversion between $S(f, \theta)$ and $S_k(k, \theta)$ is not a linear transformation but a one-to-one correspondence. In a detailed analysis of actual wave spectra, however, the conversion between $S_k(k, \theta)$ and $S(f, \theta)$ cannot be made in a one-to-one manner because of

---

[a]Although the energy of a wave is given by $\rho g a^2/2$ as in Eq. (9.2), it is customary to omit the factor $\rho g$ for simplicity.

the existence of nonlinear wave components. In the case of wind waves, the presence of drift currents deforms the dispersion relation and the conversion becomes somewhat complex.

Equation (9.12) provides a description of random sea waves, the profiles of which are changing from place to place and from time to time. When we consider the profile of random waves observed at a fixed point in the sea such as recorded by a wave gauge, the wave profile is expressed as

$$\eta = \eta(t) = \sum_{n=1}^{\infty} a_n \, \cos(2\pi f_n t + \varepsilon_n) \,. \tag{9.15}$$

The amplitudes $a_n$ and the phase angles $\varepsilon_n$ carry meanings slightly different from those associated with Eq. (9.12), in which $a_n$ and $\varepsilon_n$ represented the amplitudes and the phase angles of freely propagating independent waves. In the case of Eq. (9.15), however, the elementary amplitudes and phases are the result of mathematical manipulation of all waves propagating into different directions but having the same frequencies, such that they are added together, and the result is rewritten as a sum of sinusoidal functions. Thus, the component waves in Eq. (9.15) do not represent physical reality by themselves. Of course, we are always free to analyze an irregular time-varying function, such as shown in Fig. 2.2 in Sec. 2.1, in the form of a Fourier series without attaching any particular physical meaning.

In any case, Eq. (9.15) implies that the summation of the squares of wave amplitudes over an interval from $f$ to $f + df$ is finite and unique. The value of the sum is denoted by $S(f)$ and is given by

$$\sum_{f}^{f+df} \frac{1}{2} a_n^2 = S(f) df \,. \tag{9.16}$$

The function of $S(f)$ is called the *variance wave spectrum density* or, simply, the *frequency spectrum*.

## 9.3   Stochastic Process and Variance Spectrum

### (A) *Wave motions as a stochastic process*

The representation of random sea waves by Eqs. (9.12) and (9.15) in the foregoing section implies that the profile of sea waves $\eta$ can be expressed as a *stochastic process* satisfying the three conditions of stationarity, ergodicity, and a Gaussian process. Here, the terminology of stochastic process refers to the ensemble of variables such that the quantity in question varies randomly with time; its value at a specific time cannot be given deterministically, but each value occurs according to a certain probabilistic law. The independent variable can also be a space coordinate instead of time.

Let us assume that we could obtain an infinite number of wave records of $\eta_1(t)$, $\eta_2(t)$, ..., all of different profiles, but corresponding to the same sea state having the same significant wave height and period.[b] Then, the above wave profiles as a stochastic process are expressed as

$$\eta(t) = \{\eta_1(t), \eta_2(t), \ldots, \eta_j(t), \ldots\}. \tag{9.17}$$

The braces { } denote an ensemble, and Eq. (9.17) means that the sample records are considered to have come from the ensemble. Furthermore, we assume that the probability density of the wave profile $\eta$ is given not for one sample $\eta_j(t)$ but for the ensemble itself at a particular time $t$.

The first condition, *stationarity*, refers to the property that all statistical properties as ensemble means are time-invariant. For example, the stationarity condition applied to the arithmetic mean and the autocorrelation function gives

$$E[\eta(t)] = E[\eta(0)] \equiv \overline{\eta(t)} \quad -\infty < t < \infty, \tag{9.18}$$

$$E[\eta(t+\tau)\eta(t)] = E[\eta(\tau)\eta(0)] \equiv \Psi(\tau) \quad -\infty < t < \infty, \tag{9.19}$$

in which

$$E[\eta(t)] = \lim_{N\to\infty} \frac{1}{N} \sum_{j=1}^{N} \eta_j(t), \tag{9.20}$$

$$E[\eta(t+\tau)\eta(t)] = \lim_{N\to\infty} \frac{1}{N} \sum_{j=1}^{N} \eta_j(t+\tau)\eta_j(t). \tag{9.21}$$

We can define several other statistics. Nevertheless, when a process satisfies the conditions of Eqs. (9.18) and (9.19) alone and the variance $\mathrm{Var}[\eta(t)] = E[(\eta - E[\eta])^2]$ is finite, it is called a weakly stationary stochastic process, e.g., Koopmans.[4] If the probability density of $\eta$ is expressed by a Gaussian distribution, however, Eqs. (9.18) and (9.19) become sufficient conditions to make the process a stationary stochastic process in the strict sense, and all other statistics become stationary. When the independent

---

[b]This cannot be achieved in the real sea. But we can approach this situation in wind–water tunnel experiments by repeating the same test condition. The starting point of the time coordinate in such experiments should be taken after the same elapsed time of a sufficiently long duration after the turning-on of the wind blower.

variable is the space coordinate and the process is spatially stationary, it is called homogeneous.

The second condition, *ergodicity*, refers to the property that the time-averaged statistics for a particular sample $\eta_j(t)$ are equal to those of the ensemble average, that is,

$$E[\eta(t)] = \overline{\eta_j(t)} = \lim_{t_0 \to \infty} \frac{1}{t_0} \int_0^{t_0} \eta_j(t)dt \,, \tag{9.22}$$

$$E[\eta(t+\tau)\eta(t)] = \overline{\eta_j(t+\tau)\eta_j(t)} = \lim_{t_0 \to \infty} \frac{1}{t_0} \int_0^{t_0} \eta_j(t+\tau)\eta_j(t)dt \,. \tag{9.23}$$

A stochastic process which possesses ergodicity is always a stationary process, but the reverse is not necessarily true. However, if the probability density of the wave profile is described by a Gaussian distribution and the process does not contain periodic functions with certain periods (i.e., no components of regular waves), then the stationary process possesses the property of ergodicity.[5] For such a process, any statistic of the wave profile can be obtained by taking the time average of a sample; i.e., a wave record. The standard procedures for the spectral calculation of sea wave records to be discussed in Chapter 10 are based on the assumption of ergodicity.

The third condition, a *Gaussian process*, refers to the property that the probability density of a wave profile taking a value between $\eta$ and $\eta + d\eta$ at a certain time is given by a Gaussian (or normal) distribution as in the following:

$$p(\eta)d\eta = \frac{1}{(2\pi m_0)^{1/2}} \exp\left[-\frac{(\eta - E[\eta])^2}{2m_0}\right] d\eta \,, \tag{9.24}$$

in which

$$m_0 = E[(\eta - E[\eta])^2] = E[\eta^2] - E[\eta]^2 = \eta_{\text{rms}}^2 \,. \tag{9.25}$$

The quantity defined by Eq. (9.25) is the variance of the wave profile, which can be calculated from the time-average of $\eta^2(t)$ if the ergodicity condition is satisfied.

The applicability of the above three conditions for random sea waves has not been rigorously verified. First, it is impossible to collect a true ensemble of wave records for a prescribed sea state, because the sea state is always changing in time and space. This is easily understood when we think of the physical processes of wave generation, growth and decay. Therefore, stationarity and ergodicity can never be verified for sea waves. We may, in principle, collect an ensemble for a stochastic process by taking a wave record of long duration, but we cannot expect stationarity because the significant wave height and other statistics are known to vary with respect to time. However, we can expect constancy of a sea state for a short duration of several minutes to a few tens of minutes, and thus we can assume stationarity for such short wave records. As for ergodicity, we usually assume that it holds for sea waves as there is no reason or evidence to believe that it does not hold. Thus, we shall employ the theory of stationary stochastic processes for the analysis of random sea waves.

The assumption of a Gaussian distribution for the wave profile is known to be inapplicable, as demonstrated by a slight departure from the Gaussian distribution, to be discussed in Sec. 10.5.1. The discrepancy becomes quite noticeable for waves in shallow water. However, as the assumption of a Gaussian process is essential for decomposing random sea waves into an infinite number of component waves being superposed linearly, we intentionally make this assumption so that we can employ various statistical theories for the analysis of sea waves. It should be mentioned here that the theory of stationary stochastic processes is not limited to Gaussian processes (see Koopmans,[6] for example).

## (B) *Variance spectrum*

In the analysis of stationary stochastic process possessing ergodicity, the process is assumed to have a zero mean value, $E[\eta(t)] = 0$, in most cases. This condition is easily realized by subtracting the arithmetic mean from the original process. Thus, in the following, discussion is limited to processes satisfying

$$E[\eta(t)] = \overline{\eta(t)} \equiv 0. \tag{9.26}$$

For the case of a gradually varying mean water level such as encountered on wave records in waters with large tidal ranges, Eq. (9.26) can be satisfied by subtracting the time-varying mean water level from the record and by regarding the resultant wave profile as a stochastic process.

The autocorrelation function expressed by Eq. (9.19) is called the *autocovariance function* for a zero-mean process in the strict sense, but the terminology "autocorrelation function" is also often applied to a stochastic process with a zero mean value when the necessity to distinguish between them is slight. For a stationary stochastic process, there exists a function $S(f)$ as defined by the following equation for a given autocovariance function $\Psi(\tau)$ defined by Eq. (9.19):

$$\Psi(\tau) = \int_0^\infty S(f) \cos 2\pi f\tau \, df \, . \tag{9.27}$$

The function $S(f)$ can be expressed as follows, according to the theory of the Fourier transform, under the condition that the integral of $|\Psi(\tau)|^2$ over the full range of $\tau$ remains finite:

$$S(f) = 4 \int_0^\infty \Psi(\tau) \cos 2\pi f\tau d\tau \, . \tag{9.28}$$

The set of Eqs. (9.27) and (9.28) is called the *Wiener–Khintchine re-lations*, the generalized expression of which has symmetry of the following form defined in the ranges of $\tau$ and $f$ from $-\infty$ to $\infty$:

$$\left.\begin{aligned} \Psi(\tau) &= \int_{-\infty}^\infty S_0(f) e^{i2\pi f\tau} df \\ S_0(f) &= \int_{-\infty}^\infty \Psi(\tau) e^{-i2\pi f\tau} d\tau \end{aligned}\right\} \, . \tag{9.29}$$

The suffix '0' is attached to the frequency spectrum $S_0(f)$ in Eq. (9.29) in order to identify its definition in the range of $f = -\infty$ to $\infty$. By redefining Eq. (9.29) in the positive range $[0, \infty)$ for both $\tau$ and $f$, and by taking the Fourier cosine transform of Eq. (9.29), we can obtain Eqs. (9.27) and (9.28).

Let us examine the above relations by calculating the autocovariance function of the irregular wave profile expressed by Eq. (9.15).

$$\Psi(\tau) = \lim_{t_0 \to \infty} \frac{1}{t_0} \int_0^{t_0} \sum_{n=1}^\infty \sum_{m=1}^\infty a_n a_m \cos[(2\pi f_n(t+\tau) + \varepsilon_n] \cos[2\pi f_m t + \varepsilon_m] dt$$

$$= \lim_{t_0 \to \infty} \frac{1}{t_0} \int_0^{t_0} \sum_{n=1}^\infty \sum_{m=1}^\infty a_n a_m$$

$$\times [\cos(2\pi f_n t + \varepsilon_n) \cos(2\pi f_m t + \varepsilon_m) \cos 2\pi f_n \tau$$

$$- \sin(2\pi f_n t + \varepsilon_n) \cos(2\pi f_m t + \varepsilon_m) \sin 2\pi f_n \tau] dt$$

$$= \frac{1}{2} \sum_{n=1}^\infty a_n^2 \cos 2\pi f_n \tau \, . \tag{9.30}$$

In the above calculation, we have made use of the orthogonality property of trigonometric functions (integrals of the products of cosine functions with $n \neq m$ converge to zero as $t_0 \to \infty$ and integrals of all products of sine and cosine functions including $n = m$ converge to zero as $t_0 \to \infty$). By substituting Eq. (9.30) into (9.28), we obtain

$$S(f) = 2 \int_0^\infty \sum_{n=1}^\infty a_n^2 \cos 2\pi f_n \tau \cos 2\pi f \tau d\tau$$

$$= \sum_{n=1}^\infty a_n^2 \int_0^\infty [\cos 2\pi (f_n + f)\tau + \cos 2\pi (f_n - f)\tau] d\tau. \quad (9.31)$$

The frequency $f$ is regarded as a constant in the above manipulation. The integral on the right side of Eq. (9.31) gives two Dirac–delta functions, at $f_n = -f$ and $f_n = f$. Here we take only the delta function at $f_n = f$ as we are dealing with the problem in the half range of positive frequency, $0 \leq f < \infty$. Furthermore, the standard delta function is defined as an integral over the full range of $(-\infty, \infty)$, and therefore we should take one half the value of the delta function for Eq. (9.31). The result becomes

$$S(f) = \sum_{n=1}^\infty \frac{1}{2} a_n^2 \delta(f_n - f). \quad (9.32)$$

The delta function $\delta(f_n - f)$ is now approximated as $1/df$ for the frequency range from $f$ to $f + df$ and zero outside the range. Thus, we have

$$S(f) = \frac{1}{df} \sum_f^{f+df} \frac{1}{2} a_n^2. \quad (9.33)$$

This is in agreement with Eq. (9.16) which defines the frequency spectrum for irregular waves. Thus, the function $S(f)$ defined by Eq. (9.28) in terms of the autocovariance function is shown to be the variance spectral density function. The above relation forms the basis of a calculational procedure for making spectral estimates of random processes, known as the *Blackman–Tukey method*,[7] by means of the autocorrelation function.

It should be added here that by definition, the autocovariance function of Eq. (9.19) for $\tau = 0$ gives the variance of the wave profile $m_0$, and that the following relation between $m_0$ and the integral of the spectrum can be derived from Eq. (9.27) as

$$m_0 = \overline{\eta^2} = \Psi(0) = \int_0^\infty S(f) df. \quad (9.34)$$

This relation can also be derived from Eqs. (9.15) and (9.16).

# References

1.  J. N. Hunt, "Direct solution of wave dispersion equation," *Proc. ASCE* **105** (WW4) (1979), pp. 457–460.
2.  G. Venezian, Discussion to the Paper by J. N. Hunt in (1), *Proc. ASCE* **106** (WW4) (1980), pp. 501–502.
3.  M. S. Longuet-Higgins, "The statistical analysis of a random, moving surface," *Philos. Trans. R. Soc. London, Ser. A (966)* **249** (1957), pp. 321–387.
4.  L. H. Koopmans, *The Spectral Analysis of Time Series* (Academic Press, 1974), p. 38.
5.  Koopmans, *loc. cit.*, p. 54.
6.  Koopmans, *loc. cit.*, p. 258.
7.  R. B. Blackman and J. W. Tukey, *The Measurement of Power Spectra* (Dover Pub., Inc., 1958), 190p.

# Chapter 10

# Statistical Theory of Irregular Waves

## 10.1  Distribution of Wave Heights

### 10.1.1  *Envelope of Irregular Wave Profile*

In this chapter, we deal with the statistics of irregular wave trains on the basis of their mathematical expression by Eq. (9.15) in Sec. 9.1. The first item of the statistical description to be defined for an irregular wave profile is the wave amplitude, or height. The basic theory of the statistics of wave amplitudes was given by Rice.[1] Let us consider the case in which the energy of the wave spectrum is concentrated within a narrow range of frequency. Such a spectrum is said to be narrow-banded. A typical profile of waves with a *narrow-band spectrum* is schematically drawn in Fig. 10.1, for which it is seen that the individual waves have almost the same period but gradually varying amplitudes. In order to express such a feature of wave profiles more clearly, Eq. (9.15) is rewritten as follows:

$$\eta(t) = \sum_{n=1}^{\infty} a_n \cos(2\pi f_n t + \varepsilon_n) = Y_c(t) \cos 2\pi \bar{f} t - Y_s(t) \sin 2\pi \bar{f} t, \quad (10.1)$$

where

$$\left.\begin{aligned} Y_c(t) &= \sum_{n=1}^{\infty} a_n \cos(2\pi f_n t - 2\pi \bar{f} t + \varepsilon_n), \\ Y_s(t) &= \sum_{n=1}^{\infty} a_n \sin(2\pi f_n t - 2\pi \bar{f} t + \varepsilon_n). \end{aligned}\right\} \quad (10.2)$$

The frequency $\bar{f}$ can be any representative frequency in the spectral range where most of the wave energy is concentrated. For example, the following mean frequency defined with the first and zeroth spectral moments is one of the candidates:

$$\bar{f} = m_1/m_0, \quad (10.3)$$

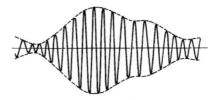

Fig. 10.1 Envelope of an irregular wave profile.

where

$$m_n = \int_0^\infty f^n S(f) df \, . \tag{10.4}$$

Furthermore, the amplitude $R$ and phase angle $\phi$ are defined in terms of $Y_c$ and $Y_s$ in Eq. (10.2) as

$$R = R(t) = \sqrt{Y_c^2(t) + Y_s^2(t)} \, , \tag{10.5}$$

$$\phi = \phi(t) = \tan^{-1}[Y_s(t)/Y_c(t)] \, . \tag{10.6}$$

These relations can be rewritten as

$$Y_c(t) = R \cos \phi, \quad Y_s(t) = R \sin \phi \, . \tag{10.7}$$

Then by using the quantities $R$ and $\phi$, the wave profile $\eta(t)$ can be expressed as follows:

$$\eta = R(t) \cos[2\pi \bar{f} t + \phi(t)] \, . \tag{10.8}$$

This equation shows that the amplitude $R$ and phase angle $\phi$ of the oscillations with the mean frequency $\bar{f}$ vary with elapsed time. The variation is quite gradual in the case of a narrow-band spectrum, and the amplitude $R = R(t)$ gives the amplitude of the wave envelope, which is indicated by the dashed lines in Fig. 10.1.

Next, the probability distribution of the envelope amplitude $R$ is considered. Examination of the nature of the amplitudes $Y_c$ and $Y_s$ reveals that both amplitudes are stationary Gaussian stochastic processes (by virtue of the central limit theorem). Both amplitudes have the variance of

$$E[Y_c^2] = E[Y_s^2] = E[\eta^2] = m_0 \, . \tag{10.9}$$

Also, we find $E[Y_cY_s] = 0$, and thus they are mutually independent. The probability that $Y_c$ and $Y_s$ simultaneously take values between $[Y_c, Y_c + dY_c]$ and $[Y_s, Y_s + dY_s]$, respectively, is given as the product of the probability density functions of two normal distributions. Thus,

$$p(Y_c, Y_s)dY_cdY_s = \frac{1}{2\pi m_0} \exp\left[-\frac{Y_c^2 + Y_s^2}{2m_0}\right] dY_cdY_s \, . \tag{10.10}$$

By changing variables to $R$ and $\phi$ by means of Eq. (10.7) and using the relation $dY_c dY_s = R\,dR\,d\phi$, we obtain

$$p(R, \phi)dR\,d\phi = \frac{R}{2\pi m_0} \exp\left[-\frac{R^2}{2m_0}\right] dR\,d\phi. \qquad (10.11)$$

However, $R$ and $\phi$ are uncorrelated independent variables. Therefore, the probability density function $p(R, \phi)$ can be expressed as the product of two functions $p(R)$ and $p(\phi)$. Moreover, since Eq. (10.11) does not contain a function of $\phi$, $p(\phi)$ must be constant in the range between $\phi = 0$ and $2\pi$. That is, $\phi$ is distributed uniformly between 0 and $2\pi$, with the probability density $1/2\pi$. Thus, the probability density function of $R$ is derived as

$$p(R)dR = \frac{R}{m_0} \exp\left[-\frac{R^2}{2m_0}\right] dR. \qquad (10.12)$$

### 10.1.2 The Rayleigh Distribution of Wave Heights

(A) *Theory for narrow-band spectral waves*

Equation (10.12) was derived by Rayleigh as the equation describing the distribution of the intensity of sound, which is the superposition of sound waves from an infinite number of sources. It is therefore called the *Rayleigh distribution*. From this theory of envelope amplitude, the probability density function of wave height is directly obtained as described below.

The assumption of a narrow-band spectrum leads to the very small probability that the maxima of the wave profile are located elsewhere other than as the wave crests, as will be described in Sec. 10.4. Therefore, the wave envelope $R = R(t)$ is regarded to represent the amplitudes of the individual waves themselves. Furthermore, the probabilities of wave crests and troughs are symmetric, because the theory presupposes linearity of the wave profile. Thus, the wave height $H$ can be regarded as twice the envelope amplitude $R$, and its probability density function is immediately obtained as

$$p(H)dH = \frac{H}{4m_0} \exp\left[-\frac{H^2}{8m_0}\right] dH. \qquad (10.13)$$

Longuet-Higgins[2] derived the relationships between several characteristic wave heights based on this probability density function. First, the mean height and the root-mean-square height are calculated to be

$$\overline{H} = \int_0^\infty H\,p(H)dH = (2\pi m_0)^{1/2}, \qquad (10.14)$$

$$H_{rms}^2 = \overline{H^2} = \int_0^\infty H^2 p(H)dH = 8m_0. \qquad (10.15)$$

With these results, Eq. (10.13) can be rewritten in a generalized form as

$$p(x)dx = 2a^2 x \exp[-a^2 x^2]dx, \qquad (10.16)$$

where

$$x = H/H_* \; : \; H_* \text{ being an arbitrary reference wave height}$$

$$a = \frac{H_*}{(8m_0)^{1/2}} = \begin{cases} 1/2\sqrt{2} & : \; H_* = m_0^{1/2} = \eta_{\text{rms}}, \\ \sqrt{\pi}/2 & : \; H_* = \overline{H}, \\ 1 & : \; H_* = H_{\text{rms}}, \\ 1.416 & : \; H_* = H_{1/3}. \end{cases} \qquad (10.17)$$

Equation (2.1) in Sec. 2.2.1 is the expression of Eq. (10.16) when the reference wave height is taken as $H_* = \overline{H}$.

Next, the mean of the highest $1/N$th waves is calculated. It is necessary to obtain the threshold wave height which has the probability of exceedance of $1/N$. By employing Eq. (10.16) for the purpose of a general discussion, the probability of exceedance[a] of the wave height ratio $x$ is obtained as

$$P(x) = P[\xi > x] = \int_x^\infty p(\xi)d\xi = \exp[-a^2 x^2]. \qquad (10.18)$$

Thus, the wave height ratio $x_N$, which has the exceedance probability of $P(x_N) = 1/N$, is calculated as

$$\exp[-a^2 x_N^2] = 1/N, \quad \text{or} \quad x_N = \frac{1}{a}(\ln N)^{1/2}. \qquad (10.19)$$

The mean of the highest $1/N$th waves, denoted as $x_{1/N}$, is then calculated as

$$x_{1/N} = \frac{\int_{x_N}^\infty xp(x)dx}{\int_{x_N}^\infty p(x)dx} = \frac{1}{1/N}\int_{x_N}^\infty xp(x)dx$$

$$= N\left\{ x_N \exp[-a^2 x_N^2] + \int_{x_N}^\infty \exp[-a^2 x^2]dx \right\}. \qquad (10.20)$$

Evaluation of the integral in the above equation yields the following result for $x_{1/N}$:

$$x_{1/N} = x_N + \frac{N}{a}\text{Erfc}[ax_N], \qquad (10.21)$$

where Erfc is the complementary error function,

$$\text{Erfc}[x] = \int_x^\infty e^{-t^2} dt. \qquad (10.22)$$

---

[a] $P(x)$ is equal to the difference between 1 and the value of the distribution function of $x$.

Table 10.1   Characteristic wave heights for the Rayleigh distribution.

| $N$ | $H_{1/N}/(m_0)^{1/2}$ | $H_{1/N}/\overline{H}$ | $H_{1/N}/H_{\mathrm{rms}}$ | Remarks |
|---|---|---|---|---|
| 100 | 6.673 | 2.662 | 2.359 | — |
| 50 | 6.241 | 2.490 | 2.207 | — |
| 20 | 5.616 | 2.241 | 1.986 | — |
| 10 | 5.090 | 2.031 | 1.800 | Highest 1/10th wave |
| 5 | 4.501 | 1.796 | 1.591 | — |
| 3 | 4.004 | 1.597 | 1.416 | Significant wave |
| 2 | 3.553 | 1.417 | 1.256 | — |
| 1 | 2.507 | 1 | 0.886 | Mean wave |

The relationships between several of the characteristic wave height have been numerically evaluated by Eq. (10.21), as listed in Table 10.1. From this table, we note that the constant $a$ in the generalized formula of the Rayleigh distribution, Eq. (10.16), takes the value of 1.416 when the reference wave height is $H_* = H_{1/3}$, i.e., significant wave height. The ratios between wave heights given by Eq. (2.3) in Sec. 2.2.2 and the relationship between the significant wave height and the zeroth spectral moment given by Eq. (2.33) in Sec. 2.4.1 were derived from such a numerical calculation.

The function Erfc$[x]$ of Eq. (10.22) can be asymptotically expanded as follows:

$$\mathrm{Erfc}[x] \sim \exp[-x^2] \sum_{n=0}^{\infty} (-1)^n \frac{(2n-1)!!}{2^{n+1}x^{2n+1}} \ : \ x \to \infty. \tag{10.23}$$

By making use of this expansion to second order and Eq. (10.19), the following approximate expression for $x_{1/N}$ is obtained:

$$x_{1/N} \simeq x_N + \frac{1}{2a(\ln N)^{1/2}} \left\{ 1 - \frac{1}{4 \ln N} \right\}. \tag{10.24}$$

For $N = 10$, this approximation gives an error of about 0.6% in excess of the exact value.

## (B) *Effect of spectral bandwidth*

The above theory of the Rayleigh distribution is intended to be applied to irregular wave profiles having narrow-band spectra. When Longuet-Higgins[2] presented the mutual relationship between characteristic wave heights in 1952, he did not specify the spectral narrowness in a quantitative manner. Shortly after, however, Cartwright and Longuet-Higgins[3] defined the following parameter and called it the *spectral width parameter* using the spectral moments of Eq. (10.4):

$$\varepsilon = [1 - m_2^2/(m_0 m_4)]^{1/2} \quad : \quad 0 < \varepsilon < 1.\tag{10.25}$$

They regarded the frequency spectrum as narrow-banded when $\varepsilon \simeq 0$ and as broad-banded when $\varepsilon \gg 0$.

Waves in the sea exhibit the frequency spectral form with the high frequency tail attenuating in proportion to $f^{-5}$ as represented by Eqs. (2.10) to (2.12) in Sec. 2.3.1. Thus, the fourth moment $m_4$ contained in the denominator in the right-hand side of Eq. (10.25) takes a very large value, making $\varepsilon$ close to 1 regardless of the shape of its peak. Although the spectral moments $m_n$ are defined by the integrtion up to the infinity in Eq. (10.4), actual calculation is made to the upper limit of the maximum frequency $f_{\max}$, which is determined by the record sampling interval $\Delta t$ as $f_{\max} = 1/2\Delta t$ (see Sec. 11.2.2). In other words, the calculated value of the spectral width parameter is a function of the maximum frequency in spectral analysis or the relative sampling interval $\Delta t/\overline{T}$, regardless of spectral shape as far as the waves in the sea are concerned. Examples of the spectral width parameter calculated with Eq. (10.25) for field measurement data are shown in Fig. 10.2 against the relative sampling interval (Goda and Kudaka[4]). Legends to the symbols refer to the location name and the dotted line represents an empirical relationship derived through numerical simulation of random wave profiles having a Wallops-type spectrum with $m = 4$.

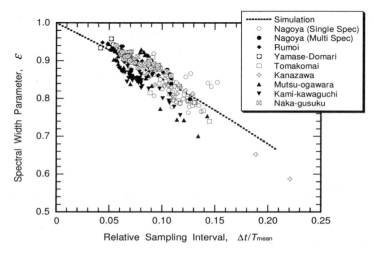

Fig. 10.2   Spectral width parameter $\varepsilon$ versus relative sampling interval $\Delta t/T_{\mathrm{mean}}$.[4]

As exhibited in Fig. 10.2, random sea waves possess broad-band spectra as far as the value of $\varepsilon$ is concerned, and application of the theory of Rayleigh distribution to their wave heights is not appropriate from the theoretical point of view. Nevertheless, the artifice of zero-upcrossing (or downcrossing) to define individual waves as described in Sec. 2.1.2 results in a wave height distribution quite close to the Rayleigh. It is said that the operations involved in the zero-upcrossing definition exercise the effect of making the spectrum narrow-banded, but the correctness of the above comment has not been scrutinized

The effect of spectral bandwidth on the distribution of wave heights has theoretically been examined by Tayfun[5] and Naess,[6] among others. Their theories, though not the same, are based on the inequality of the height of wave crest and the depth of the following wave trough; that is, a wave height $H$ is smaller than twice the envelope amplitude $R$ on the average. The tendency of the actual wave height distribution being narrower than the Rayleighan, as exemplified by Eq. (2.41) in Sec. 2.4.1, supports such theoretical derivations. Forristall[7] has demonstrated the applicability of Tayfun's theory with the field observation data for the case when the spectral peaks are fairly narrow. However, the theories do not succeed in predicting the deviation of wave heights from the Rayleigh distribution for wind waves with broad spectra such as the Bretschneider–Mitsuyasu and Wallops spectra, according to the comparison of theoretical results with numerical simulation data listed in Table 2.3 in Sec. 2.4.1.

A general expression of the wave height distribution may be the Weibull type of the following form:

$$P(x) = \int_x^\infty p(x)dx = \exp\left[-\left(\frac{x}{\alpha}\right)^\kappa\right], \tag{10.26}$$

$$p(x) = \frac{\kappa}{\alpha}\left(\frac{x}{\alpha}\right)^{\kappa-1}\exp\left[-\left(\frac{x}{\alpha}\right)^\kappa\right]. \tag{10.27}$$

The above Weibull distribution includes the Rayleigh distribution with the parameter values of $\kappa = 2, \alpha = \sqrt{2} = 2.828$ for $x = H/\eta_{\rm rms}$. The empirical expression of Eq. (2.41) by Forristall[8] has the parameter values of $\kappa = 2.126$ and $\alpha = 2.724$. Myrhaug and Kjeldsen[9] have also proposed the Weibull type wave height distribution with $\kappa = 2.39, \alpha = 1.05$ for $x = H/H_{\rm rms}$.

The mean of the highest $1/N$th waves, $x_{1/N}$, is calculated for the Weibull distribution as below. First, the threshold wave height having the exceedance probability of $1/N$ is obtained from Eq. (10.26) as

$$x_N = \alpha(\ln N)^{1/\kappa}. \tag{10.28}$$

The wave height $x_{1/N}$ is evaluated by numerical integration of the following equation:

$$x_{1/N} = x_N + N \int_{x_N}^{\infty} \exp\left[-\left(\frac{x}{\alpha}\right)^{\kappa}\right] dx. \qquad (10.29)$$

Forristall[8] carried out the calculation for the distribution with $\kappa = 2.126$ and $\alpha = 2.724$ and obtained the relationship of $H_{1/10} = 4.733\,\eta_{rms}$, $H_{1/3} = 3.774\,\eta_{rms}$, and $\overline{H} = 2.413\,\eta_{rms}$. These values fall between the wave height ratios of the Wallops type spectra with $m = 3$ and $5$ listed in Table 2.3. As discussed in Sec. 2.4.1, the effect of spectral bandwidth on wave height distribution is well described by means of the spectral shape parameter defined by Eq. (2.42).

## (C) *Effect of wave nonlinearity*

The distribution of individual waves is further affected by wave nonlinearity. Waves having strong nonlinearity in deep water may yield extremely large waves (more than twice the significant height) beyond the probability of occurrence predicted from the Rayleigh distribution, which are called *freak waves*. Mori *et al.*[10,11] have revealed that the occurrence probability of freak waves is related to the kurtosis of surface elevation and analyzed their exceedance probability.

As waves approach to the shore, wave profiles evolve into the form with sharp and high crests and shallow and flat troughs, and waves undergo the process of nonlinear shoaling. Because the heights of large waves are more enhanced than smaller waves, the distribution of individual wave heights becomes more stretched than the Rayleigh. The histogram of wave heights obtained by field measurements shown in Fig. 2.4 in Sec. 2.2.1 agreed quite well with the prediction by the Rayleigh, but the agreement owed to the nonlinear enhancement of wave heights in shallow water area. The subject of wave nonlinearity effect in shoaling water is further discussed in Sec. 10.5.

As waves further approach toward the shoreline, large waves begin to break and waves enter into the surf zone. The wave height distribution is gradually deformed through elimination of large heights due to the depth-controlled breaking process, which have been discussed in Sec. 3.6.2(B).

## 10.1.3   *Probability Distribution of Largest Wave Height*

The Rayleigh distribution does not have an upper bound. The probability density decreases exponentially as the independent variable $x$ becomes large, but never becomes zero. Thus, the largest wave height $H_{\max}$ is only statistically defined in such a manner that it is the largest value among a particular sample of wave heights arbitrarily chosen[b] from the population of wave heights. It cannot be a physically meaningful largest value of the wave height in a deterministic sense. The largest wave height is a statistical variable, which varies from one sample to another. We can only estimate the probability of $H_{\max}$. Derivations of the probability density function of $H_{\max}$ have been given by Longuet-Higgins[2] and Davenport,[12] the latter presenting a slightly simplified derivation for the largest value of wind gust loading.

Let us take $N_0$ wave heights from a population of wave heights which follows the Rayleigh distribution and denote the largest value among the group of $N_0$ wave heights (in nondimensional form) as $x_{\max}$. By denoting the probability density functions of $x_{\max}$ as $p^*(x_{\max})$, the probability that the largest value of $x$ takes a value in the range of $[x_{\max}, x_{\max} + dx_{\max}]$ is given by $p^*(x_{\max})dx_{\max}$ by definition. This probability is also calculated as the probability that only one wave among $N_0$ waves takes the height $x_{\max}$ in the range from $x_{\max}$ to $x_{\max} + dx_{\max}$ and the remainder of the $(N_0 - 1)$ waves have heights less than $x_{\max}$. Thus,

$$p^*(x_{\max})dx_{\max} = N_0[1 - P(x_{\max})]^{N_0-1}p(x_{\max})dx_{\max}$$
$$= d[1 - P(x_{\max})]^{N_0}, \qquad (10.30)$$

where $P(x)$ is the exceedance probability given by Eq. (10.18) and $p(x)$ is the probability density function given by Eq. (10.16).

When $N_0$ is very large, the right side of Eq. (10.26) can be approximated as follows:

$$\lim_{N_0 \to \infty} [1 - P(x_{\max})]^{N_0} = \lim_{N_0 \to \infty} \left[1 - \frac{\xi}{N_0}\right]^{N_0} = e^{-\xi}, \qquad (10.31)$$

where

$$\xi = N_0 P(x_{\max}) = N_0 \exp[-a^2 x_{\max}^2]. \qquad (10.32)$$

By substituting Eq. (10.31) into Eq. (10.30) and carrying out some manipulations, we obtain the probability density function of $x_{\max}$ as follows:

$$p^*(x_{\max})dx_{\max} = -e^{-\xi}d\xi = 2a^2 x_{\max} \xi e^{-\xi} dx_{\max}. \qquad (10.33)$$

---

[b]As pointed out by Longuet-Higgins,[2] ordinary records of continuous waves do not constitute a sample of wave heights in the above statistical sense, because successive wave heights are mutually correlated and not independent. However, the correlation is generally weak in case of wind waves, and thus we are allowed to apply the theory of the Rayleigh distribution to a sample of wave heights from continuous wave records.

The theoretical curves of $H_{\max}/H_{1/3}$ shown in Fig. 2.7 in Sec. 2.2.1 were calculated by means of Eq. (10.33) by employing the reference height $H_* = H_{1/3}$ and taking the value $a = 1.416$.

Once the probability density function is obtained, the mode, the mean and the variance can be calculated. First, the mode $(x_{\max})_{\text{mode}}$ is derived from the condition $dp^*/dx_{\max} = 0$ as

$$(x_{\max})_{\text{mode}} \simeq \frac{1}{a}(\ln N_0)^{1/2}\left\{1 + \frac{1}{4(\ln N_0)^2} + \cdots\right\}. \qquad (10.34)$$

Equation (2.4) in Sec. 2.2.2 for the expression of the most probable value of $H_{\max}$ has been obtained by regarding the second term on the right side of Eq. (10.34) as being negligible, and by using $a = 1.416$ for $H_* = H_{1/3}$. A more accurate expression for $(x_{\max})_{\text{mode}}$ than that given by Eq. (10.34) was presented by Longuet-Higgins.[2]

The arithmetic mean and the root-mean-square values are calculated by

$$E[x_{\max}] = \int_0^\infty x_{\max}\,p^*(x_{\max})dx_{\max} = \int_0^{N_0} x_{\max}\,e^{-\xi}d\xi, \qquad (10.35)$$

$$E[x_{\max}^2] = \int_0^\infty x_{\max}^2\,p^*(x_{\max})dx_{\max} = \int_0^{N_0} x_{\max}^2\,e^{-\xi}d\xi. \qquad (10.36)$$

Next, an explicit form of $x_{\max}$ in terms of $\xi$ is obtained by rewriting Eq. (10.32) as

$$x_{\max} = \frac{1}{a}(\ln N_0 - \ln\xi)^{1/2}$$
$$\simeq \frac{1}{a}(\ln N_0)^{1/2} - \frac{\ln\xi}{2a(\ln N_0)^{1/2}} - \frac{(\ln\xi^2)}{8a(\ln N_0)^{3/2}} + \cdots. \qquad (10.37)$$

This equation is substituted into Eq. (10.35) and the integration is carried out with respect to $\xi$. The result is[c]

$$E[x_{\max}] = (x_{\max})_{\text{mean}} \simeq \frac{1}{a}(\ln N_0)^{1/2} + \frac{\gamma}{2a(\ln N_0)^{1/2}} - \frac{\pi^2 + 6\gamma^2}{48a(\ln N_0)^{3/2}} + \cdots,$$
$$(10.38)$$

where

$$\gamma = -\int_0^\infty (\ln\xi)e^{-\xi}d\xi = 0.5772\ldots \quad \text{(Euler's constant)}.$$

---

[c]In this integration, the following formula by Cramer[13] is employed:

$$\int_0^\infty (\ln\xi)^2 e^{-\xi}d\xi = \frac{\pi^2}{6} + \gamma^2.$$

Equation (10.36) is also integrated with respect to $\xi$ by using Eq. (10.32). The results is

$$E[x_{max}^2] \simeq \frac{1}{a^2} \ln N_0 + \frac{1}{a^2}\gamma . \tag{10.39}$$

Therefore, the standard deviation of $x_{max}$ is calculated as follows:

$$\sigma(x_{max}) = \left\{ E[x_{max}^2] - E[x_{max}]^2 \right\}^{1/2} \simeq \frac{\pi}{2\sqrt{6}a(\ln N_0)^{1/2}} . \tag{10.40}$$

Furthermore, the probability $\mu$ that $x_{max}$ exceeds a specified value is calculated as

$$\mu = 1 - \int_0^{x_{max}} p^*(\zeta)d\zeta = 1 - [1 - P(x_{max})]^{N_0} \simeq 1 - \exp[-N_0 P(x_{max})] . \tag{10.41}$$

From this result, the largest wave height $(x_{max})_\mu$, such that the probability of being exceeded is $\mu$, is easily obtained as

$$(x_{max})_\mu \simeq \frac{1}{a} \left\{ \ln \left[ \frac{N_0}{\ln 1/(1-\mu)} \right] \right\}^{1/2} . \tag{10.42}$$

Equation (2.6) in Sec. 2.2.2 is based on the above result.

Comparison of Eqs. (10.34) and (10.38) with Eqs. (10.19) and (10.24) yields the following inequality for large values of $N_0$:

$$x_{N_0} \simeq (x_{max})_{mode} < (x_{max})_{mean} < x_{1/N_0} . \tag{10.43}$$

As approximations, we have

$$(x_{max})_{mean} \simeq x_{1.8N_0} , \quad x_{1/N_0} \simeq x_{2.6N_0} . \tag{10.44}$$

For the Weibull distribution of Eq. (10.26) which is narrower than the Rayleigh, Forristall[8] has derived the expected value of the largest wave height as follows:

$$E[x_{max}] \simeq \alpha(\ln \kappa)^{1/\kappa} \left[ 1 + \frac{\gamma}{\kappa \ln N_0} \right] , \tag{10.45}$$

where $\gamma$ is Euler's constant. A similar equation has been derived by Myrhaug and Kjeldsen.[9]

## 10.2 Wave Grouping

### 10.2.1 *Wave Grouping and Its Quantitative Description*

Although sea waves may look random, inspection of wave records indicates that high waves fall into groups rather than appear individually. Figure 10.3

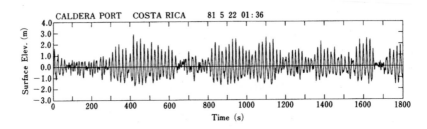

Fig. 10.3   Example of a wave record with conspicuous wave grouping.[14]

is an example of a wave profile exhibiting grouping observed at Caldera Port, Costa Rica, in Central America reported by Goda.[14] The record is of the swell that traveled over a distance of some 9000 km from the generating area in the Southwest Pacific Basin.

The grouping of high waves may influence i) effective number of consecutive waves necessary to produce resonance in structures and to capsize ships, ii) stability of armor stones and blocks of sloping breakwaters, and iii) fluctuation of wave overtopping quantity of seawalls. Furthermore, a well-developed wave grouping is often associated with the presence of long-period waves, which are called surf beats or infragravity waves as discussed in Sec. 2.5. A detailed analysis of these and other problems incorporating the effect of wave grouping remains for the future.

The length of wave grouping can be quantitatively described by counting the number of waves exceeding a specified value of the wave height $H_c$ without falling below that height. A succession of such high waves is called a *run* of high wave heights, and the number of waves is termed the run length. As sketched in Fig. 10.4, the run length will be denoted by $j_1$. We can also define a repetition length of high waves such that a run begins at the first

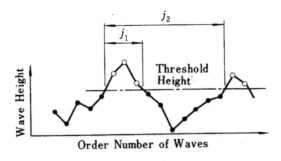

Fig. 10.4   Definition sketch for the run length.

Table 10.2   Frequency distribution of observed lengths of runs of wave heights.[16]

| Run Length | Ordinary Runs of Wave Heights | | Conditional Runs Inclusive of $H_{max}$ | |
|---|---|---|---|---|
| $j_1$ | $H > H_{1/3}$ | $H > H_{med}$ | $H > H_{1/3}$ | $H > H_{med}$ |
| 1 | 1,327 | 1,560 | 43 | 5 |
| 2 | 374 | 944 | 62 | 17 |
| 3 | 122 | 590 | 39 | 30 |
| 4 | 37 | 327 | 19 | 35 |
| 5 | 9 | 220 | 5 | 24 |
| 6 | 2 | 112 | 2 | 25 |
| 7 | 1 | 90 | 1 | 13 |
| 8 | | 46 | | 5 |
| 9 | | 30 | | 5 |
| 10 | | 16 | | 3 |
| 11 | | 13 | | 3 |
| over 12 | | 12 | | 6 |
| Total | 1,872 | 3,960 | 171 | 171 |
| Mean | 1.42 | 2.54 | 2.36 | 5.12 |
| Standard deviation | 0.77 | 1.99 | 1.18 | 3.08 |

Note: Total of 20,051 waves from 171 wave records.

exceedance of wave height over the threshold value, continuing through a sequence of waves exceeding and then falling below the threshold value, and ending at the first re-exceedance of threshold were height (Goda[15]). The repetition length of high waves is analogous to the definition of the zero-upcrossing period of the wave profile. We may call such a repetition of high waves as a *total run* and denote its length by $j_2$.

Measurements of the run length of wave heights as well as of the total run length have been made on many wave records. Table 10.2 gives some example results, based on 171 records of wind waves and young swell by Goda.[16] The threshold heights were set at $H_c = H_{1/3}$ and $H_c = H_{med}$ (median height $\simeq 0.94\ \overline{H}$). The table shows the distribution of individual length $j_1$ of the runs of wave height exceeding the threshold height $H_c$. The most frequently appearing run has the length of only one wave, and the frequency decreases monotonically as the run length increases. Table 10.2 also gives the length of the run that includes the highest wave in each wave record, here called the *conditional run*. There are 171 such conditional runs from 171 wave records, and they exhibit much longer run lengths than ordinary runs of wave heights. This indicates that the highest wave rarely appears as an isolated wave, but is accompanied by several other high waves.

There is another method of describing wave grouping, by means of the wave energy envelope, which is called the *smoothed instantaneous wave energy history* (abbreviated as SIWEH) by Funke and Mansard.[17] They introduced the concept of wave-groupiness and introduced the groupiness factor *GF*, which is a measure of the degree of the variation of wave heights in a given wave record. The interested reader is referred to their paper for details.

### 10.2.2  *Probability Distribution of Run Length for Uncorrelated Waves*

If successive wave heights are uncorrelated, the probability distribution of the run length is derived by a simple calculation of probability.[15,16] Let the occurrence probability that $H > H_c$ be denoted by $p_0$ and its nonoccurrence probability by $q_0 = 1 - p_0$. A run of length $j_1$ is such a process in that the first wave height exceeds the threshold height $H_c$, the succeeding $(j_1 - 1)$ wave heights also exceed $H_c$, and the $(j_1 + 1)$th wave height falls below $H_c$. Thus, the probability of a run with length $j_1$ is expressed as

$$P(j_1) = p_0^{j_1 - 1} q_0 . \tag{10.46}$$

Since $p_0 < 1$ by the definition of probability, a run with $j_1 = 1$ has the largest probability of occurrence. The mean and the standard deviation of the run length are calculated as follows:

$$\bar{j}_1 = \sum_{j_1=1}^{\infty} j_1 P(j_1) = \frac{q_0}{p_0} \sum_{j_1=1}^{\infty} j_1 p_0^{j_1} = \frac{1}{q_0} , \tag{10.47}$$

$$\sigma(j_1) = \left[ \sum_{j_1=1}^{\infty} j_1^2 P(j_1) - \bar{j}_1^2 \right]^{1/2} = \frac{\sqrt{p_0}}{q_0} . \tag{10.48}$$

The probability of a total run with the length $j_2$ can be derived by mathematical induction to give

$$P(j_2) = \frac{p_0 q_0}{p_0 - q_0} (p_0^{j_2 - 1} - q_0^{j_2 - 1}) . \tag{10.49}$$

The mean and the standard deviation of the total run length are calculated as

$$\overline{j_2} = \frac{1}{p_0} + \frac{1}{q_0}, \qquad (10.50)$$

$$\sigma(j_2) = \sqrt{\frac{p_0}{q_0^2} + \frac{q_0}{p_0^2}}. \qquad (10.51)$$

When the threshold wave height is taken as the median height $H_{\mathrm{med}}$, the probabilities of $p_0$ and $q_0$ are $p_0 = q_0 = 1/2$ by definition. In the case of $H_{\mathrm{c}} = H_{1/3}$, the probabilities are calculated as $p_0 = 0.1348$ and $q_0 = 0.8652$ by substituting the values $x = H_{\mathrm{c}}/m_0^{1/2} = 4.004$ and $a = 1/\sqrt{8}$ into Eq. (10.18). Substitution of these values of probability into Eq. (10.46) results in distributions of run lengths shorter than those recorded in the sea. The difference is due to the presence of some correlation between successive wave heights. For example, Rye[18] has found a correlation coefficient of $+0.24$ for successive heights of wind waves, while records of long-traveled swells such as those shown in Fig. 10.3 have the correlation coefficient in the range of 0.5 to 0.8.[14]

## 10.2.3 *Correlation Coefficient Between Successive Wave Heights*

The reliable theory of run length for sea waves requires incorporation of the correlation coefficient between successive wave heights. The theory was first given by Kimura,[19] and then re-presented by Battjes and van Vledder[20] and Longuet-Higgins,[21] both of whom expressed the relationship between the correlation coefficient by Kimura and the frequency spectrum in a more explicit manner. All the derivations are made on the calculation of the envelope amplitude sketched in Fig. 10.1, and they have the origin in the classical paper by Rice[1] for the statistical theory of random noise.

First, the envelope amplitudes at the time $t$ and $t + dt$ are denoted as $R_1$ and $R_2$ for the irregular wave profile expressed by Eq. (10.8), and their relationship is examined. For this purpose, the wave profiles at the above two time steps are expressed in a manner of Eq. (10.1) as follows:

$$\left.\begin{array}{l} \eta(t) = Y_{c1} \cos 2\pi \bar{f} t - Y_{s1} \sin 2\pi \bar{f} t, \\[2mm] \eta(t + dt) = Y_{c2} \cos 2\pi \bar{f}(t + \tau) - Y_{s2} \sin 2\pi \bar{f}(t + \tau), \end{array}\right\} \qquad (10.52)$$

where

$$
\left.\begin{aligned}
Y_{c1} &= \sum_{n=1}^{\infty} a_n \cos[2\pi(f_n - \bar{f})t + \epsilon_n] \,, \\
Y_{s1} &= \sum_{n=1}^{\infty} a_n \sin[2\pi(f_n - \bar{f})t + \epsilon_n] \,, \\
Y_{c2} &= \sum_{n=1}^{\infty} a_n \cos[2\pi(f_n - \bar{f})(t + \tau) + \epsilon_n] \,, \\
Y_{s2} &= \sum_{n=1}^{\infty} a_n \sin[2\pi(f_n - \bar{f})(t + \tau) + \epsilon_n] \,.
\end{aligned}\right\}
\tag{10.53}
$$

The amplitudes $Y_{c1}$, $Y_{s1}$, $Y_{c2}$ and $Y_{s2}$ defined above are the stationary stochastic processes satisfying the Gaussian distribution. Their variances and covariances are calculated as below.

$$
\left.\begin{aligned}
E[Y_{c1}^2] &= E[Y_{s1}^2] = E[Y_{c2}^2] = E[Y_{s2}^2] = m_0 \,, \\
E[Y_{c1}Y_{s1}] &= E[Y_{s1}Y_{c1}] = E[Y_{c2}Y_{s2}] = E[Y_{s2}Y_{c2}] = 0 \,, \\
E[Y_{c1}Y_{c2}] &= E[Y_{c2}Y_{c1}] = E[Y_{s1}Y_{s2}] = E[Y_{s2}Y_{s1}] \\
&= \int_0^{\infty} S(f) \cos 2\pi(f - \bar{f})\tau \, df = \mu_{13} \,, \\
E[Y_{c1}Y_{s2}] &= E[Y_{s2}Y_{c1}] = -E[Y_{s1}Y_{c2}] = -E[Y_{c2}Y_{s1}] \\
&= \int_0^{\infty} S(f) \sin 2\pi(f - \bar{f})\tau \, df = \mu_{14} \,.
\end{aligned}\right\}
\tag{10.54}
$$

The four stationary stochastic processes $Y_{c1}$, $Y_{s1}$, $Y_{c2}$ and $Y_{s2}$ are assigned the numbers 1 to 4 in this order. The covariance matrix is formed as follows:

$$
M = \begin{pmatrix}
m_0 & 0 & \mu_{13} & \mu_{14} \\
0 & m_0 & -\mu_{14} & \mu_{13} \\
\mu_{13} & -\mu_{14} & m_0 & 0 \\
\mu_{14} & \mu_{13} & 0 & m_0
\end{pmatrix} .
\tag{10.55}
$$

The determinant of the above matrix is

$$
|M| = (m_0^2 - \mu_{13}^2 - \mu_{14}^2)^2 = m_0^4(1 - \kappa^2)^2 \,,
\tag{10.56}
$$

where

$$\kappa^2 = \frac{1}{m_0^2}(\mu_{13}^2 + \mu_{14}^2)$$

$$= \left| \frac{1}{m_0} \int_0^\infty S(f) \cos 2\pi (f - \bar{f}) \tau \, df \right|^2 + \left| \frac{1}{m_0} \int_0^\infty S(f) \sin 2\pi (f - \bar{f}) \tau \, df \right|^2$$

$$= \left| \frac{1}{m_0} \int_0^\infty S(f) \cos 2\pi f \tau \, df \right|^2 + \left| \frac{1}{m_0} \int_0^\infty S(f) \sin 2\pi f \tau \, df \right|^2 . \tag{10.57}$$

The spectral shape parameter defined by Eq. (2.42) in Sec. 2.4.1 is based on the above equation with the time lag $\tau$ being taken as the mean wave period $\overline{T}$.

The joint probability density function of the normal variates having the covariance matrix of Eq. (10.55) can be written down by using the cofactor of the determinant $|M|$ as follows:

$$p(Y_{c1}, Y_{s1}, Y_{c2}, Y_{s2}) = \frac{1}{4\pi^2 m_0^2 (1 - \kappa^2)} \exp\left\{ -\frac{1}{2m_0^2(1 - \kappa^2)} \right.$$

$$\times \left[ m_0(Y_{c1}^2 + Y_{s1}^2 + Y_{c2}^2 + Y_{s2}^2) - \mu_{13}(Y_{c1}Y_{c2} + Y_{s1}Y_{s2}) \right.$$

$$\left. \left. - \mu_{14}(Y_{c1}Y_{s2} + Y_{s1}Y_{c2}) \right] \right\} . \tag{10.58}$$

Then, the following transformation of variates is introduced:

$$\left. \begin{array}{ll} Y_{c1} = R_1 \cos\phi_1, & Y_{s1} = R_1 \sin\phi_1, \\ Y_{c2} = R_2 \cos\phi_2, & Y_{s2} = R_2 \sin\phi_2. \end{array} \right\} \tag{10.59}$$

The joint probability density function of the four variables expressed by Eq. (10.53) is now rewritten for the variates $R_1$, $R_2$, $\phi_1$ and $\phi_2$ as

$$p(R_1, R_2, \phi_1, \phi_2) = \frac{R_1 R_2}{4\pi^2 m_0^2 (1 - \kappa^2)} \exp\left\{ -\frac{1}{2m_0^2(1 - \kappa^2)} \left[ m_0(R_1^2 + R_2^2) \right. \right.$$

$$\left. \left. - 2\mu_{13}R_1 R_2 \cos(\phi_2 - \phi_1) - 2\mu_{14}R_1 R_2 \sin(\phi_2 - \phi_1) \right] \right\} . \tag{10.60}$$

Because the phase angles $\phi_1$ and $\phi_2$ are independent of the envelope amplitudes $R_1$ and $R_2$, Eq. (10.60) can be integrated with respect to $\phi_1$ and $\phi_2$ for the range of $[0, 2\pi]$, respectively. For execution of integration, a new variable $\alpha = \phi_2 - \phi_1 - \tan^{-1}(\mu_{13}/\mu_{14})$ is introduced. By utilizing the fact that the exponential function on the right side of Eq. (10.60) is a periodic function of $\phi_2$, the terms containing $\phi_1$ and $\phi_2$ are integrated as below.[22]

$$\frac{1}{4\pi^2} \int_0^{2\pi} d\phi_1 \int_0^{2\pi} \exp\left\{ \frac{R_1 R_2}{m_0^2 (1-\kappa^2)} [\mu_{13}\cos(\phi_2-\phi_1) + \mu_{14}\sin(\phi_2-\phi_1)] \right\} d\phi_2$$

$$= \frac{1}{4\pi^2} \int_0^{2\pi} d\phi_1 \int_0^{2\pi} \exp\left\{ \frac{R_1 R_2}{m_0^2 (1-\kappa^2)} (\mu_{13}^2 + \mu_{14}^2)^{1/2} \cos\alpha \right\} d\alpha$$

$$= I_0 \left[ \frac{\kappa R_1 R_2}{(1-\kappa^2)m_0} \right]. \tag{10.61}$$

Full integration of Eq. (10.60) yields the following joint probability density function of $R_1$ and $R_2$:

$$p(R_1, R_2) = \frac{R_1 R_2}{m_0^2(1-\kappa^2)} \exp\left[ -\frac{R_1^2 + R_2^2}{2m_0(1-\kappa^2)} \right] I_0 \left[ \frac{\kappa R_1 R_2}{(1-\kappa^2)m_0} \right], \tag{10.62}$$

where $I_0$ denotes the modified Bessel function of the first kind of zeroth order.

With the joint probability density function thus obtained, the correlation coefficient between $R_1$ and $R_2$ is defined by the following:

$$r(R_1, R_2) = \frac{M_{11}}{(M_{20}M_{02})^{1/2}}, \tag{10.63}$$

where

$$M_{mn} = \int_0^\infty \int_0^\infty (R_1 - \bar{R}_1)^m (R_2 - \bar{R}_2)^n p(R_1, R_2) dR_1 dR_2. \tag{10.64}$$

Evaluation of Eq. (10.63) yields the following result for the correlation coefficient between $R_1$ and $R_2$:

$$r(R_1, R_2) = \frac{E(\kappa) - (1-\kappa^2)K(\kappa)/2 - \pi/4}{1 - \pi/4}, \tag{10.65}$$

where $K$ and $E$ are the complete elliptic integrals of the first and second kinds, respectively.

For the case that the frequency spectrum is relatively narrow-banded and the wave height $H$ can be assumed to be equal to twice the envelope amplitude $R$, the correlation coefficient between the successive wave heights is given by Eq. (10.65). The parameter $\kappa$ is evaluated from the wave spectrum by mean of Eq. (10.57), by setting the time lag $\tau$ as equal to the mean wave period $\overline{T}$. The joint probability density function of the successive wave heights $H_1$ and $H_2$ is also obtained by rewriting Eq. (10.62)

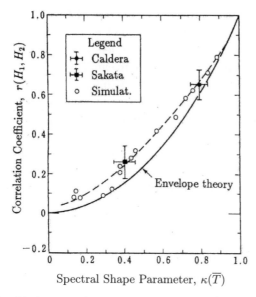

Fig. 10.5   Relationship between the correlation coefficient $r(H_1, H_2)$ and the spectral shape parameter $\kappa(\overline{T})$.[23]

as follows:

$$p(H_1, H_2) = \frac{4H_1 H_2}{(1 - \kappa^2)H_{\text{rms}}^4} \exp\left[-\frac{H_1^2 + H_2^2}{(1 - \kappa^2)H_{\text{rms}}^2}\right] I_0\left[\frac{2\kappa H_1 H_2}{(1 - \kappa^2)H_{\text{rms}}^2}\right].$$

$$(10.66)$$

The correlation coefficient between successive wave heights observed on field data and numerically simulated data is compared with prediction by the envelope theory of the above, as shown in Fig. 10.5 by Goda.[23] The abscissa represents the value of $\kappa(\overline{T})$, which has been called the spectral shape parameter in Sec. 2.4.1, and the ordinate is the correlation coefficient $r(H_1, H_2)$. Field data come from two sources: one from Caldera Port in Costa Rica (marked with the filled circle) and the other from Sakata Port in Japan (marked with the filled box). The former contains 51 records of long-traveled swell, including one record shown in Fig. 10.3, while the latter contains 68 records of wind waves. The lengths of the vertical and horizontal lines from the filled circle or box represent the magnitude of the standard deviations of respective data sources.

The numerical simulation of wave profiles have been made for 15 different spectral shapes so as to yield a wide range of $\kappa$ value: for each spectral shape one hundred different profiles were generated and analyzed. In the calculation of $\kappa$ by Eq. (10.57) for the data of Caldera Port, the spectral density in the range of the frequency below 0.5 times the peak frequency and that above 1.8 times the peak frequency were set to zero. Such a removal of spectral density was so made to eliminate a possible contamination of the frequency spectrum by nonlinear interaction terms to be discussed in Sec. 10.5. The data of Sakata Port were judged not to contain an appreciable amount of nonlinear interaction terms because of the large water depth (50 m), and thus no adjustment was made for the wave spectra.

Figure 10.5 indicates that the wave height correlation coefficient measured on wave profiles is slightly larger than the theoretical prediction. The difference is owing to the assumption of the wave height being twice the envelope amplitude. As discussed in Sec. 10.1.2 (B) on the effect of spectral bandwidth on wave height distributions, the wave height is slightly smaller than twice the envelope amplitude on the average. In other words, the temporal variation of wave heights is slightly smaller in comparison with that of envelope amplitudes, and thus produces a greater value of correlation coefficient. Despite the above discussion, the degree of the difference between the observed value of correlation coefficient and the theoretical prediction is rather small. The theory of correlation between wave envelope amplitudes introduced in the above can be said to be well applicable for waves in the sea.

### 10.2.4 Theory of Run Length for Mutually Correlated Wave Heights

By employing the definition by Kimura,[19] a notation of $p_{11}$ is given to the probability that an arbitrary wave height $H_2$ in a wave train does not exceed the threshold height $H_c$, under the condition that the preceding wave height $H_1$ has not exceeded $H_c$. Another notation of $p_{22}$ is given to the probability that $H_2$ exceeds $H_c$ under the condition that $H_1$ has already exceeded $H_c$. These probabilities are evaluated by the envelope theory as

$$
\left.
\begin{aligned}
p_{11} &= \int_0^{H_c} \int_0^{H_c} p(H_1, H_2)\,dH_1\,dH_2 \Big/ \int_0^{H_c} p(H_1)\,dH_1 \,, \\
p_{22} &= \int_{H_c}^{\infty} \int_{H_c}^{\infty} p(H_1, H_2)\,dH_1\,dH_2 \Big/ \int_{H_c}^{\infty} p(H_1)\,dH_1 \,,
\end{aligned}
\right\}
\tag{10.67}
$$

where $p(H_1)$ denotes the marginal probability density function of wave heights and is approximated with Eq. (10.13) of the Rayleigh distribution.

The probability of a run $H > H_c$ with length $j_1$ is then derived in a manner similar to the derivation of the run length for uncorrelated wave heights. Thus,

$$P(j_1) = p_{22}^{j_1-1}(1 - p_{22}).$$ (10.68)

The mean and the standard deviation of the run length are calculated as

$$\bar{j}_1 = \frac{1}{1 - p_{22}},$$ (10.69)

$$\sigma(j_1) = \frac{\sqrt{p_{22}}}{1 - p_{22}}.$$ (10.70)

The probability of the total run is also derived as follows:

$$P(j_2) = \frac{(1 - p_{11})(1 - p_{22})}{p_{11} - p_{22}}(p_{11}^{j_2-1} - p_{22}^{j_2-1}).$$ (10.71)

The mean and the standard deviation of the total run are calculated as

$$\bar{j}_2 = \frac{1}{1 - p_{11}} + \frac{1}{1 - p_{22}},$$ (10.72)

$$\sigma(j_2) = \left[\frac{1}{(1 - p_{11})^2} + \frac{1}{(1 - p_{22})^2} - \frac{1}{(1 - p_{11})} - \frac{1}{(1 - p_{22})}\right]^{1/2}.$$ (10.73)

When Kimura[19] applied his theory to the actual data, he did not calculate the spectral shape parameter $\kappa$ from the frequency spectrum. Instead, he used the wave height correlation coefficient by the temporal wave height records and estimated $\kappa$ by inversely solving Eq. (10.65). The same technique was applied for the three continuous wave records at Caldera Port, each containing 734 waves, 947 waves and 2278 waves, respectively, for calculation of the probabilities of the run length with $H_c = H_{\mathrm{med}}$ and $H_c = H_{1/3}$.[14] The observed probabilities agree well with the theory as shown in Fig. 10.6.

The variation of the mean run length with respect to the envelope correlation parameter is examined in Fig. 10.7 for the run of $H > H_{\mathrm{med}}$. Figure 10.8 shows the mean length of the total run of $H > H_{1/3}$.[23] The data sources are same as those employed in Fig. 10.5, and so is the calculation method of $\kappa$ for Caldera Port. As seen in Figs. 10.7 and 10.8, the agreement between the data and theory is generally good, though there are some deviations. Thus, the spectral shape parameter $\kappa$ defined by Eq. (10.57) can be concluded as the parameter governing the phenomenon of the run of wave heights.

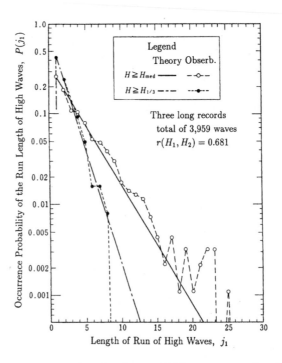

Fig. 10.6   Distribution of the lengths of runs of high waves exceeding the median and significant heights.[14]

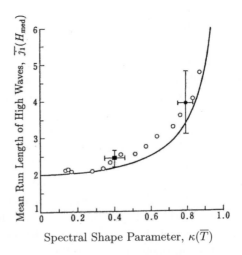

Fig. 10.7   Comparison of the theory and observations for the mean lengths of the runs of high waves of $H > H_{\text{med}}$.[23]

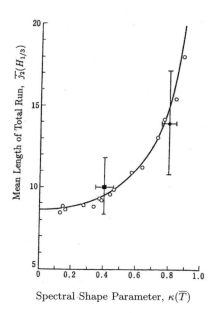

Fig. 10.8  Comparison of the theory and observations for the mean length of the total runs defined by $H_c = H_{1/3}$.[23]

The author[14] previously proposed the following parameter to describe the "peakedness" of the spectral peak:

$$Q_p = \frac{2}{m_0^2} \int_0^\infty f S^2(f) df . \qquad (10.74)$$

The above spectral peakedness parameter also has the capability to describe the statistics of run lengths. However, the spectral peakedness parameter $Q_p$ has a shortcoming that its value is sensitive to the resolution of spectral analysis.[14] Though the spectral shape parameter $\kappa$ is also affected by the spectral resolution,[23] its sensitivity is less as compared to $Q_p$. Thus, it is recommended to employ $\kappa$ for the analysis of the run of wave heights based on the information of wave spectrum.

## 10.3  Distribution of Wave Periods

### 10.3.1  *Mean Period of Zero-Upcrossing Waves*

The number of zero-upcrossing of irregular waves per unit time has been examined by Rice,[1] who presented a theoretical formula for the mean period of zero-upcrossing waves (Eq. (2.44) in Sec. 2.4.2). By using Eq. (9.15) for an irregular wave profile, the time derivative of the profile is expressed as

$$\dot\eta(t) = - \sum_{n=1}^{\infty} 2\pi f_n a_n \sin(2\pi f_n t + \epsilon_n)\,. \tag{10.75}$$

Let the event of zero-upcrossing be assumed of taking place in the time interval $t = t_0 \sim (t_0 + dt)$ with the time derivative in the range $\dot\eta \sim (\dot\eta + d\dot\eta)$. For the occurrence of such an event, the surface elevation at $t = t_0$, denoted by $\eta_0$, must be less than zero, but it must not be too small to have zero-upcrossing before the time $t = t_0 + dt$. The temporal variation of surface elevation can be approximated by a straight line if the time duration $dt$ is sufficiently short. Because the surface elevation at $t = t_0 + dt$ is approximated as $\eta = \eta_0 + \dot\eta dt$, which is greater than zero by the above assumption, $\eta_0$ must be greater than $-\dot\eta dt$. Thus, the permissible range of the surface elevation at the time $t$ is given as $-\dot\eta dt < \eta_0 < 0$. The occurrence probability of the above event is then expressed as

$$\int_{-\dot\eta dt}^{0} [p(\eta, \dot\eta d\dot\eta)]d\eta = p(0, \dot\eta)\dot\eta dt\, d\dot\eta\,, \tag{10.76}$$

in which $p(\eta, \dot\eta)$ denotes the joint probability density function of $\eta$ and $\dot\eta$. Because $\dot\eta$ can take any value between 0 and $\infty$, the probability that the surface elevation crosses the zero line upward in the interval $t = t_0 \sim (t_0 + dt)$ is given as

$$dP = dt \int_{0}^{\infty} \dot\eta p(0, \dot\eta)d\dot\eta\,. \tag{10.77}$$

The mean number of zero-upcrossings of the wave profile per unit time, denoted by $N_0^*$, is obtained by integrating Eq. (10.77) with respect to time. The result is

$$N_0^* = \int_{0}^{\infty} \dot\eta p(0, \dot\eta)d\dot\eta\,. \tag{10.78}$$

The wave profile $\eta$ and its time derivative $\dot\eta$ have ensemble means of zero and obey the normal distribution by virtue of the central limit theorem, as easily understood from their expressions, Eq. (10.75). Their variances are

$$E[\eta^2] = m_0, \quad E[\dot\eta^2] = (2\pi)^2 m_2\,. \tag{10.79}$$

The quantities $\eta$ and $\dot\eta$ are mutually independent because the covariance $E[\eta\,\dot\eta]$ is zero. Thus, $p(\eta, \dot\eta)$ is obtained as the product of two normal distribution functions as

$$p(\eta, \dot\eta) = \frac{1}{4\pi^2 (m_0 m_2)^{1/2}} \exp\left[ -\frac{1}{2}\left( \frac{\eta^2}{m_0} + \frac{\dot\eta^2}{4\pi^2 m_2} \right) \right]\,. \tag{10.80}$$

By substituting this equation into Eq. (10.78) and performing the integration, we obtain

$$N_0^* = (m_2/m_0)^{1/2} \,. \tag{10.81}$$

The mean period of the zero-upcrossing waves is the reciprocal of $N_0^*$. By denoting the mean period as $T_{02}$, we have the result

$$T_{02} = 1/N_0^* = (m_0/m_2)^{1/2} \,, \tag{10.82}$$

which was presented as Eq. (2.44) in Sec. 2.4.2.

## 10.3.2 Marginal Distribution of Wave Periods and Joint Distribution of Wave Heights and Periods

(A) *Joint probability density function of envelope amplitude and phase angle*

The next problem concerns the probability distribution of individual wave periods. Rice[1] gave an approximate formula for the wave period distribution. Longuet-Higgins[24] presented a theory of the joint distribution of the heights and periods of waves with narrow-band spectra and discussed its applicability to actual sea waves.[25]

The starting point of the theory is Eq. (10.8) for the amplitude $R$ and phase angle $\phi$ of the wave envelope, re-expressed in the following form:

$$\eta = R\cos\chi \,, \qquad \chi = 2\pi\bar{f}t + \phi = \bar{\omega}t + \phi \,, \tag{10.83}$$

where $\bar{f}$ is the mean frequency defined by Eq. (10.3), and $\chi$ may be called the total phase angle.

It is necessary to derive the joint probability of $R$, $\phi$, and their time derivatives for the analysis of wave period distributions. For this purpose, the following four variables are introduced:

$$\left.\begin{aligned}
\xi_1 &= R\cos\phi = \sum_{n=1}^{\infty} a_n \cos(2\pi f_n t - 2\pi\bar{f}t + \epsilon_n)\,, \\
\xi_2 &= R\sin\phi = \sum_{n=1}^{\infty} a_n \sin(2\pi f_n t - 2\pi\bar{f}t + \epsilon_n)\,, \\
\xi_3 &= \dot{\xi}_1 = -\sum_{n=1}^{\infty} 2\pi a_n(f_n - \bar{f})\sin(2\pi f_n t - 2\pi\bar{f}t + \epsilon_n)\,, \\
\xi_4 &= \dot{\xi}_2 = \sum_{n=1}^{\infty} 2\pi a_n(f_n - \bar{f})\cos(2\pi f_n t - 2\pi\bar{f}t + \epsilon_n)\,,
\end{aligned}\right\} \tag{10.84}$$

in which $\xi_1$ and $\xi_2$ are the same as $Y_c$ and $Y_s$ in Eq. (10.2), respectively. The variables $\xi_1$ to $\xi_4$ are normally distributed with zero mean by virtue of the central limit theory. Their variances are evaluated as follows by employing Eq. (9.16):

$$\left. \begin{aligned} E[\xi_1^2] &= E[\xi_2^2] = \sum_{n=1}^{\infty} \frac{1}{2} a_n^2 = m_0 \,, \\ E[\xi_3^2] &= E[\xi_4^2] = \sum_{n=1}^{\infty} \frac{1}{2} (2\pi a_n)^2 (f_n - \bar{f})^2 = \hat{\mu}_2 \,, \end{aligned} \right\} \tag{10.85}$$

where

$$\hat{\mu}_2 = (2\pi)^2 \mu_2 \,, \quad \mu_2 = \int_0^{\infty} (f - \bar{f})^2 S(f) df = m_2 - m_1^2/m_0 \,. \tag{10.86}$$

The moments $m_0, m_1$ and $m_2$ have been defined in Eq. (10.4). The covariance $E[\xi_i \xi_j]$ vanishes for $i \neq j$ if the mean frequency $\bar{f}$ is defined through the moments $m_0$ and $m_1$ as in Eq. (10.3), and thus $\xi_1 \sim \xi_4$ are mutually independent. Therefore, the joint probability density function for $(\xi_1, \xi_2, \xi_3, \xi_4)$ can be expressed as the product of individual probability density function of normal distributions. Thus,

$$p(\xi_1, \xi_2, \xi_3, \xi_4) = \frac{1}{4\pi^2 m_0 \hat{\mu}_2} \exp\left[ -\frac{\xi_1^2 + \xi_2^2}{2m_0} \right] \exp\left[ -\frac{\xi_3^2 + \xi_4^2}{2\hat{\mu}_2} \right] \,. \tag{10.87}$$

The variables $\xi_3$ and $\xi_4$ in Eq. (10.84) can also be expressed as

$$\left. \begin{aligned} \xi_3 &= \dot{R} \cos\phi - R\dot{\phi} \sin\phi \,, \\ \xi_4 &= \dot{R} \sin\phi + R\dot{\phi} \cos\phi \,. \end{aligned} \right\} \tag{10.88}$$

The Jacobian between $(\xi_1, \xi_2, \xi_3, \xi_4)$ and $(R, \phi, \dot{R}, \dot{\phi})$ is calculated to be

$$|J| = \frac{\partial(\xi_1, \xi_2, \xi_3, \xi_4)}{\partial(R, \phi, \dot{R}, \dot{\phi})} = R^2 \,. \tag{10.89}$$

By transforming variables, the following joint probability density function is derived from Eq. (10.87):

$$p(R, \phi, \dot{R}, \dot{\phi}) = \frac{R^2}{4\pi^2 m_0 \hat{\mu}_2} \exp\left[ -\frac{R^2}{2m_0} \right] \exp\left[ -\frac{\dot{R}^2 + R^2 \dot{\phi}^2}{2\hat{\mu}_2} \right] \,. \tag{10.90}$$

(B) *Marginal probability density function of wave period*

The wave period is related with the time derivative of the total phase angle $\chi$ as will be discussed later. The joint probability density function for $\phi$ and $\dot{\phi}$ is derived by integrating Eq. (10.90) with respect to $R$ from 0 to $\infty$ and $\dot{R}$ from $-\infty$ to $\infty$. The result is

$$p(\phi, \dot{\phi}) = \frac{(m_0/\hat{\mu}_2)^{1/2}}{4\pi[1 + (m_0/\hat{\mu}_2)\dot{\phi}^2]^{3/2}}. \tag{10.91}$$

By rewriting the above equation with the variables $\chi$ and $\dot{\chi}$ and using $\partial(\chi, \dot{\chi})/\partial(\phi, \dot{\phi}) = 1$, we obtain

$$p(\chi, \dot{\chi}) = \frac{(m_0/\hat{\mu}_2)^{1/2}}{4\pi[1 + (m_0/\hat{\mu}_2)(\dot{\chi} - \bar{\omega})^2]^{3/2}}. \tag{10.92}$$

The variable $\chi$ does not appear on the right side of the above equation, which means that $\chi$ is uniformly distributed in the range 0 to $2\pi$. The marginal distribution of $\dot{\chi}$ is then expressed by the following probability density function:

$$p(\dot{\chi}) = \frac{(m_0/\hat{\mu}_2)^{1/2}}{2[1 + (m_0/\hat{\mu}_2)(\dot{\chi} - \bar{\omega})^2]^{3/2}}. \tag{10.93}$$

Theoretically speaking, $\dot{\chi}$ can take any value between $-\infty$ and $\infty$. The probability of $\dot{\chi}$ taking a negative value is obtained by integrating Eq. (10.93) over the negative range as

$$\int_{-\infty}^{0} p(\dot{\chi})d\dot{\chi} = \frac{1}{2}\left\{1 - \frac{(m_0/\hat{\mu}_2)^{1/2}\bar{\omega}}{[1 + (m_0/\hat{\mu}_2)\bar{\omega}^2]^{1/2}}\right\}$$

$$= \frac{1}{2}\left\{1 - \frac{1}{(1 + \nu^2)^{1/2}}\right\} \simeq \frac{1}{4}\nu^2 - \frac{3}{16}\nu^4 + \cdots. \tag{10.94}$$

The parameter $\nu$ in the above was introduced by Longuet-Higgins[24] to represent the narrowness of the spectral bandwidth and is defined by

$$\nu = \frac{1}{\bar{\omega}}\left(\frac{\hat{\mu}_2}{m_0}\right)^{1/2} = \left[\frac{m_0 m_2}{m_1^2} - 1\right]^{1/2}. \tag{10.95}$$

Longuet-Higgins[25] mentions that $\nu$ is nearly equal to one half the value of another spectral width parameter $\epsilon$ of Eq. (10.25) when the parameter $\nu$ takes a sufficiently small value. In such a case, the probability of $\dot{\chi}$ being negative is negligibly small as indicated by Eq. (10.94), and the total phase angle $\chi$ can be regarded as an almost-always-increasing function.

Next, attention is given to the event that the wave profile $\eta = \eta(t)$ crosses the zero line upward. The event takes place when $\chi$ becomes

$(2n - 1/2)\pi$ while increasing, or when $\chi$ becomes $(2n + 1/2)\pi$ while decreasing. Thus, the probability that a zero-upcrossing event occurs in the time interval $[t, t + dt]$ is given in a manner similar to the derivation of Eq. (10.77):

$$H(\chi)dt = dt \int_0^\infty \dot{\chi}[p(\chi, \dot{\chi})]_{\chi=(2n-1/2)\pi} d\dot{\chi}$$

$$+ dt \int_{-\infty}^0 \dot{\chi}[p(\chi, \dot{\chi})]_{\chi=(2n+1/2)\pi} d\dot{\chi} . \tag{10.96}$$

The two terms on the right side of the above equation can be evaluated by using Eqs. (10.3), (10.86) and (10.92), and the result is

$$\left.\begin{array}{l}
\displaystyle\int_0^\infty \dot{\chi} p(\chi, \dot{\chi}) d\dot{\chi} = \frac{1}{2}\left(\frac{m_2}{m_0}\right)^{1/2}\left[1 + \frac{m_1}{(m_0 m_2)^{1/2}}\right] , \\[4mm]
\displaystyle\int_{-\infty}^0 \dot{\chi} p(\chi, \dot{\chi}) d\dot{\chi} = \frac{1}{2}\left(\frac{m_2}{m_0}\right)^{1/2}\left[1 - \frac{m_1}{(m_0 m_2)^{1/2}}\right] .
\end{array}\right\} \tag{10.97}$$

Thus, the probability $H(\chi)dt$ is obtained as

$$H(\chi)dt = (m_2/m_0)^{1/2}dt . \tag{10.98}$$

If we integrate the above probability over the duration of unit time, we get the mean number of zero-upcrossings per unit time, which is the same as Eq. (10.81).

The expressions inside the brackets on the right sides of Eq. (10.97) can be expanded as

$$\left.\begin{array}{l}
\displaystyle 1 + \frac{m_1}{(m_0 m_2)^{1/2}} = 1 + (1 + \nu^2)^{-1/2} \simeq 2 - \frac{1}{2}\nu^2 + \cdots , \\[4mm]
\displaystyle 1 - \frac{m_1}{(m_0 m_2)^{1/2}} = 1 - (1 + \nu^2)^{-1/2} \simeq \frac{1}{2}\nu^2 + \cdots .
\end{array}\right\} \tag{10.99}$$

Therefore, it is noted that the probability that $\chi$ takes the value $(2n+1/2)\pi$ while decreasing is very small if $\nu \ll 1$. This is another confirmation that the probability of $\dot{\chi}$ taking a negative value is very small, here demonstrated for the particular value $\chi = (2n + 1/2)\pi$.

The zero-upcrossing wave period can be defined as the difference between the time $\chi = [2(n + 1) - 1/2]\pi$ and the time $\chi = [2n - 1/2]\pi$ under the condition that $\nu \ll 1$. By assuming that $\ddot{\chi}$ is sufficiently small[d] and by approximating the variation of $\chi$ during the above time interval with

---

[d]Longuet-Higgins[24] has stated that $\ddot{\chi}$ is a quantity of the order of $\nu^2$.

a straight line, we obtain the following expression for the zero-upcrossing wave period:

$$T \simeq \frac{2\pi}{\dot{\chi}} \simeq T_{01}\left(1 - \frac{\dot{\phi}}{\bar{\omega}}\right), \tag{10.100}$$

where $T_{01}$ is one kind of mean wave period defined by

$$T_{01} = \frac{1}{\bar{f}} = \frac{2\pi}{\bar{\omega}} = \frac{m_0}{m_1}. \tag{10.101}$$

In order to derive the probability density function for the period $T$, we need to know the conditional probability density function of $\dot{\phi}$ when $\chi$ takes the value $(2n - 1/2)\pi$. This function can be derived by the same reasoning as employed in the derivation of Eq. (10.76), which describes the probability density function of $\dot{\eta}$ under the condition that the wave envelope $\eta = \eta(t)$ crosses the level $\eta = 0$ in the time interval $[t, t + dt]$. The result is

$$p(\dot{\phi}|\chi) = p(\dot{\chi}|\chi) = \frac{p(\chi, \dot{\chi})|\dot{\chi}|}{H(\chi)}. \tag{10.102}$$

The existence of a linear transformation between $\dot{\phi}$ and $\dot{\chi}$ as dictated by Eq. (10.83) was employed in the above derivation. It is possible to approximate the term $|\dot{\chi}|$ on the right side of the above equation as

$$\dot{\chi} = \bar{\omega}[1 + 0(\nu)]. \tag{10.103}$$

By employing only the first term in the above equation, Eq. (10.102) is evaluated as follows, after using Eqs. (10.82), (10.92), (10.95), (10.98), and (10.100):

$$p(\dot{\phi}|\chi) \simeq \frac{T_{02}\nu^2}{4\pi[\nu^2 + (1 - T/T_{01})^2]^{3/2}}. \tag{10.104}$$

The probability density function for the period $T$ is then obtained from the above result as

$$p(T) = \left|\frac{d\dot{\phi}}{dT}\right| p(\dot{\phi}|\chi) = \frac{\bar{\omega}T_{02}}{2\pi T_{01}} \frac{\nu^2}{2[\nu^2 + (1 - T/T_{01})^2]^{3/2}}. \tag{10.105}$$

The derivations are all based on the assumption that $\nu \ll 1$, which yields the quasi-equality $T_{02} \simeq T_{01}$. Thus, by introducing the new symbol $\bar{T} \simeq T_{02} \simeq T_{01} = 2\pi/\bar{\omega}$, the probability density function for the nondimensional wave period is finally obtained as

$$p(\tau) = \frac{\nu^2}{2[\nu^2 + (\tau - 1)^2]^{3/2}} \quad : \tau = T/\bar{T}. \tag{10.106}$$

(C) *Joint probability density function of wave height and period*

The joint distribution of wave heights and periods is next examined. By integrating the joint probability density function of Eq. (10.90) for $R$, $\phi$, $\dot{R}$ and $\dot{\phi}$ with respect to $\phi$ in the range 0 to $2\pi$ and $\dot{R}$ in the range $-\infty$ to $\infty$, we obtain the following joint probability density function of $R$ and $\dot{\phi}$:

$$p(R, \dot{\phi}) = \frac{R^2}{m_0 (2\pi \hat{\mu}_2)^{1/2}} \exp \left[ -\frac{R^2}{2m_0} \left( 1 + \frac{m_0}{\hat{\mu}_2} \dot{\phi}^2 \right) \right].$$ (10.107)

Now the variable $\dot{\phi}$ is replaced with $\tau$ by the following relation from Eq. (10.100):

$$\tau = T/\overline{T} \simeq 1 - \dot{\phi}/\bar{\omega}.$$ (10.108)

The envelope amplitude $R$ is also replaced with the wave height $H$, regarded as $H = 2R$, and the latter is then nondimensionalized in the form $x = H/H_*$ with the introduction of the constant $a$ in Eq. (10.17). The resultant joint probability density function for the nondimensional wave height $x$ and period $\tau$ becomes

$$p(x, \tau) = \frac{dR}{dx} \left| \frac{d\dot{\phi}}{d\tau} \right| p(R, \dot{\phi}) = \frac{2a^3 x^2}{\sqrt{\pi}\nu} \exp \left\{ -a^2 x^2 \left[ 1 + \frac{(\tau - 1)^2}{\nu^2} \right] \right\}.$$ (10.109)

The above function is characterized by symmetry about the $\tau$-axis at $\tau = 1$, and thus the correlation between $x$ and $\tau$ is zero. It is easily confirmed that the marginal probability density function expressed by Eq. (10.106) for the nondimensional wave period $\tau$ is obtained by integrating Eq. (10.105) with respect to $x$ over the range 0 to $\infty$, whereas the density function expressed by Eq. (10.16) for the nondimensional height $x$ is obtained by integrating Eq. (10.109) with respect to $\tau$ in the range $-\infty$ to $\infty$.

Finally, the distribution of wave periods in a certain range of wave height can be derived by the formula for the conditional probability density function as

$$p(\tau|x) = \frac{p(x, \tau)}{p(x)} = \frac{ax}{\sqrt{\pi}\nu} \exp \left[ -\frac{a^2 x^2}{\nu^2} (\tau - 1)^2 \right].$$ (10.110)

This probability density function is a normal distribution with a mean of unity. The standard deviation is immediately found from its functional form as

$$\sigma \left( \frac{T}{\overline{T}} \right)_x = \frac{\nu}{\sqrt{2}ax} = \left( \frac{2}{\pi} \right)^{1/2} \frac{\overline{H}}{H} \nu.$$ (10.111)

This equation indicates that the range of the distribution of the wave periods classified according to the wave height is inversely proportional to the wave height; that is, waves in the class of the smallest height exhibit the widest distribution in wave periods. Equation (10.111) also indicates that the period distribution is narrow for waves having small values of $\nu$. It should be mentioned that Eq. (10.111) is not applicable in the limit $H \to 0$ and the standard deviation of the period distribution of Eq. (10.106) as a whole cannot be evaluated because the integration for the variance diverges.

### (D) *Joint distribution of heights and periods of actual sea waves*

The above theory of period distribution by Longuet-Higgins[25] has been confirmed as being applicable to waves with very narrow-band spectra, based on the data of numerically simulated wave profiles.[26] In the case of actual sea waves, their spectra are rather broad-banded with large values of the parameter $\nu$. The standard frequency spectrum of Eq. (2.10) in Sec. 2.3.1 yields $\nu = 0.425$, whereas values of $\nu$ in observed wave spectra typically lie in the range of about 0.3 to 0.8. Apparently the theory cannot be applied to waves with such large values of $\nu$, as is understood from the process of its derivation. However, if the value of $\nu$ is determined from the inter-quartile range of the observed marginal distribution of wave periods,[25] the theory can become workable in explaining the properties of wave period distributions. Longuet-Higgins,[25] for example, has argued for the applicability of his theory to the data of a scatter diagram of $H$ and $T$ presented by Bretschneider[27] with the fitted value of $\nu = 0.234$.

Figure 10.9 shows the nondimensional period distribution for 13 records of surface waves (total number of 1686 waves), which had correlation coefficients between individual heights and periods in the range of $-0.25$ and $0.19$.[26] The abscissa is the nondimensional wave period normalized by the mean period of the individual records, and the ordinate represents the cumulative frequencies at respective period classes expressed in the form of a probability density. Although the mean value of $\nu$ calculated from the observed wave spectra of these 13 records is 0.51, the assignment of the value of $\nu = 0.26$, based on the period distribution, produces fairly good agreement between the theoretical distribution of Eq. (10.106) and the recorded one. Figure 10.10 gives the standard deviations of wave periods at various levels of the wave height classes for the same wave data as an index of the width of distribution range. Agreement between theory and observation is also acceptable.

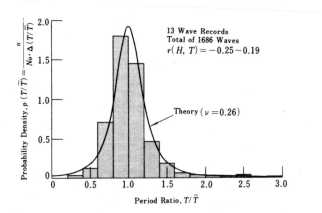

Fig. 10.9   Marginal distribution of wave periods with low correlation between individual heights and periods.[26]

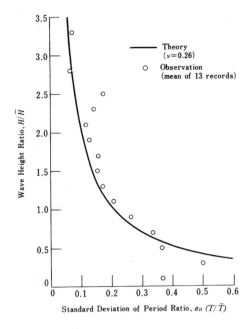

Fig. 10.10   Standard deviation of wave period with respect to the class of the height.[26]

The examples given in Figs. 10.9 and 10.10 represent the situation that the correlation between individual heights and periods is almost nil. As the correlation becomes stronger, there appears a tendency for waves of small height to have short periods and the joint distribution of wave heights

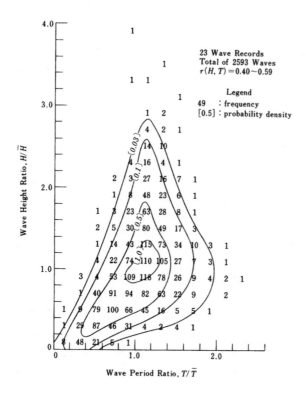

Fig. 10.11   Joint distribution of observed wave heights and periods.[26]

and periods becomes distorted from the symmetric shape indicated by
Eq. (10.109). Figure 10.11 shows a nondimensional scatter diagram of observed heights and periods of 2593 waves from 23 surface records, which had
correlation coefficients $r(H, T)$ in the range 0.40 to 0.59; the class bands
of nondimensional heights and periods are set as $\Delta H/\overline{H} = \Delta T/\overline{T} = 0.2$.
The scatter diagram shown here exhibits a tail toward the origin in the re-
gion of nondimensional wave height lower than about 1. The shape of the
tail becomes clearer as the correlation coefficient $r(H, T)$ further increases
(Goda[26]), though not shown here. At the same time, the range in the periods of waves having large heights gradually shifts toward the right; that
is, toward larger values of the nondimensional wave period. This tendency
causes an increase in period ratios such as $T_{1/3}/\overline{T}$.

It is interesting to note that in scatter diagrams such as Fig. 10.11, the
region of waves with large heights exhibits an almost symmetric distribution
with respect to the period. Symmetry of the distribution implies that the

wave heights and periods are uncorrelated in that region. For example, if the correlation coefficient is calculated for only those waves comprising the higher one-third of the record ($H/\overline{H} > 1.2$), which are used for calculating the significant wave height, the mean and the standard deviation of the correlation coefficients of the 23 records become 0.01 and 0.16, respectively. The feature of a lack of correlation among high waves does not disappear even if the overall correlation coefficient between wave height and period increases.

There is another theory of the joint distribution of the wave heights and periods, proposed by Cavanié et al.[28] It is a semi-empirical extension of the theory of the maxima of the wave profile with introduction of the spectral width parameter $\varepsilon$. By properly choosing the value of $\varepsilon$, the unsymmetrical feature of the joint height and period distribution can be demonstrated. But the selection of $\varepsilon$ is somewhat subjective and the noncorrelation among high waves cannot be maintained with increases in the overall correlation.

Longuet-Higgins[29] himself reformulated his theory of the joint distribution of wave heights and periods by expressing the zero-upcrossing wave periods as $T = T_{01}/(1 - \dot{\phi}/\overline{\omega})$, instead of the approximation by Eq. (10.100). His new theory can simulate the characteristics of observed waves such that short-period waves tend to have small heights. But the reformulated joint distribution function exhibits a slowly attenuating tail towards large wave periods. The mean wave period cannot be evaluated by the new theory, because the marginal probability density of wave periods behaves almost proportional to $\tau^{-2}$. Numerical evaluation of the mean wave period, which is made possible by introducing an upper limit in wave period, indicates that the mean period gradually increases as the spectral width parameter $\nu$ increases. The overall correlation coefficient between wave heights and periods by theory is slightly negative, which causes the significant period to be shorter than the mean period. These features do not agree with the observed characteristics of sea waves. Thus, the new theory by Longuet-Higgins[29] cannot quantitatively describe the observed joint distribution of wave heights and periods, although it can qualitatively express the general feature of the observed joint distribution by means of a relatively simple formula.

## 10.4   Maxima of Irregular Wave Profiles

If the profiles of the sea waves shown in Fig. 2.2 in Sec. 2.1.2 are closely inspected, one can notice several ups and downs in the wave profiles which are not of sufficient amplitude to cross the line of the mean water level. It is an interesting problem in statistics to investigate the probability distribution of such maxima and minima of irregular wave profiles. On this topic, Rice[1] has given a theory in conjunction with the statistical properties of random noise. Later, Cartwright and Longuet-Higgins[3] demonstrated that the theory is applicable to the distribution of the maxima of random sea wave profiles, and they introduced the spectral width parameter $\varepsilon$ defined by Eq. (10.25) in the derivation.

The maxima of an irregular wave profile expressed by $\eta = \eta(t)$ are defined as the points which satisfy the conditions $\dot{\eta} = 0$ and $\ddot{\eta} < 0$. By using Eq. (9.15) as the wave profile, the time derivatives are written as follows:

$$\left. \begin{aligned} \eta &= \sum_{n=1}^{\infty} a_n \cos(2\pi f_n t + \varepsilon_n), \\ \dot{\eta} &= -\sum_{n=1}^{\infty} 2\pi f_n a_n \sin(2\pi f_n t + \varepsilon_n), \\ \ddot{\eta} &= -\sum_{n=1}^{\infty} (2\pi f_n)^2 a_n \cos(2\pi f_n t + \varepsilon_n). \end{aligned} \right\} \tag{10.112}$$

Let us consider an event in which the wave profile $\eta$ exhibits a maximum in the interval $[\eta_0, \eta_0 + d\eta_0]$ and during the tine interval $[t_0, t_0 + dt_0]$. The probability of such an event is readily obtained as

$$dt \int_{-\infty}^{0} [p(\eta_0, 0, \ddot{\eta}) d\eta_0 |\ddot{\eta}|] d\ddot{\eta} \tag{10.113}$$

by replacing the variables in the derivation of Eq. (10.77) according to $\eta \to \dot{\eta}$ and $\dot{\eta} \to \ddot{\eta}$. The function $p(\eta, \dot{\eta}, \ddot{\eta})$ is the joint probability density function of $\eta$, $\dot{\eta}$ and $\ddot{\eta}$. The mean number of occurrences per unit time for which a maximum of $\eta$ takes a value between $\eta_0$ and $\eta_0 + d\eta_0$, which is expressed as $F(\eta_0) d\eta_0$, is given by

$$F(\eta_0) d\eta_0 = \int_{-\infty}^{0} [p(\eta_0, 0, \ddot{\eta}) |\ddot{\eta}| d\eta_0] d\ddot{\eta}. \tag{10.114}$$

On the other hand, the mean number of occurrences of maxima of $\eta$ per unit time is given by

$$N_1^* = \int_{-\infty}^{\infty} \left\{ \int_{-\infty}^{0} [p(\eta_0, 0, \ddot{\eta}) |\ddot{\eta}| d\eta_0] d\ddot{\eta} \right\}. \tag{10.115}$$

Thus, the probability density function of the maxima of an irregular wave profile is described as

$$p(\eta_{\max})d\eta_{\max} = F(\eta_{\max})d\eta_{\max}/N_1^* . \tag{10.116}$$

Actual calculation of $p(\eta_{\max})$ requires evaluation of $p(\eta, \dot{\eta}, \ddot{\eta})$. We know by the definition of Eq. (10.112) that the variables $\eta$, $\dot{\eta}$ and $\ddot{\eta}$ are normally distributed with the mean value of zero and that they have the following covariance matrix:

$$M = \begin{pmatrix} m_0 & 0 & -\hat{m}_2 \\ 0 & \hat{m}_2 & 0 \\ -\hat{m}_2 & 0 & \hat{m}_4 \end{pmatrix}, \tag{10.117}$$

where

$$\hat{m}_n = (2\pi)^n \int_0^\infty f^n S(f) df = (2\pi)^n m_n . \tag{10.118}$$

As a result of these properties, the function $p(\eta, \dot{\eta}, \ddot{\eta})$ can be shown to have the form[3]

$$p(\eta, \dot{\eta}, \ddot{\eta}) = \frac{1}{(2\pi)^{3/2}(\hat{m}_2\Delta)^{1/2}}$$

$$\times \exp\left\{-\frac{1}{2}\left[\frac{\ddot{\eta}^2}{\hat{m}_2} + \frac{1}{\Delta}(\hat{m}_4\eta^2 + 2\hat{m}_2\eta\dot{\eta} + m_0\ddot{\eta}^2)\right]\right\}, \tag{10.119}$$

where

$$\Delta = |M|/\hat{m}_2 = m_0\hat{m}_4 - \hat{m}_2^2 . \tag{10.120}$$

Substitution of Eq. (10.119) into Eq. (10.114) and evaluation of the integral yield the following result for $F(\eta_0)$:

$$F(\eta_0) = \frac{\Delta^{1/2}}{m_0(2\pi)^{3/2}(\hat{m}_2)^{1/2}} \exp\left[-\frac{x_0^2}{2}\right]$$

$$\times \left\{\exp\left[-\frac{x_0^2}{2\delta^2}\right] + \frac{x_0}{\delta}\int_{-x_0/\delta}^\infty \exp\left[-\frac{x^2}{2}\right] dx\right\}, \tag{10.121}$$

where

$$x_0 = \eta_0/m_0^{1/2}, \quad \delta = \Delta^{1/2}/\hat{m}_2 . \tag{10.122}$$

The number of maxima per unit time is then obtained from Eq. (10.115) as

$$N_1^* = \frac{1}{2\pi}\left(\frac{\hat{m}_4}{\hat{m}_2}\right)^{1/2} = \left(\frac{m_4}{m_2}\right)^{1/2} . \tag{10.123}$$

The above result can also be derived simply by the replacements $\eta \to \dot{\eta}$ and $\dot{\eta} \to \ddot{\eta}$, as in the derivation of Eq. (10.81) for the mean zero-upcrossing wave period, with the corresponding replacements $m_0 \to m_2$ and $m_2 \to m_4$.

The probability density function of the maxima of an irregular wave profile is obtained by substituting Eqs. (10.121) and (10.123) into Eq. (10.116). The result, expressed in terms of the nondimensional variable $x_* = \eta_{\max}/m_0^{1/2}$, is

$$p(x_*) = \frac{1}{(2\pi)^{1/2}} \left\{ \varepsilon \exp\left[-\frac{x_*^2}{2\varepsilon^2}\right] + (1-\varepsilon^2)^{1/2} x_* \exp\left[-\frac{x_*^2}{2}\right] \right.$$

$$\left. \times \int_{-\infty}^{x_*\sqrt{1-\varepsilon^2}/\varepsilon} \exp\left[-\frac{x^2}{2}\right] dx \right\}, \tag{10.124}$$

where

$$\varepsilon^2 = \frac{\delta^2}{1+\delta^2} = \frac{\Delta/\hat{m}_2^2}{1+\Delta/\hat{m}_2^2} = \frac{\Delta}{m_0\hat{m}_4} = \frac{m_0 m_4 - m_2^2}{m_0 m_4}. \tag{10.125}$$

The parameter $\varepsilon$ has been introduced in Eq. (10.25).

The probability density function of the maxima of a wave profile extends into the region $x_* < 0$, unless $\varepsilon = 0$. This explains the fact that some of the maxima in a wave profile appear in wave troughs. If we take the limiting condition $\varepsilon \to 0$, Eq. (10.124) reduces to Eq. (10.12) in Sec. 10.1.1 for the probability density function of the wave envelope amplitude. This is a natural consequence, because the wave profile does not have maxima other than at the wave crest for $\varepsilon = 0$, as shown in the profile of Fig. 10.1. The other limiting condition for $\varepsilon$ is $\varepsilon \to 1$, when $\delta \to \infty$. In this limit, Eq. (10.124) converges to the normal distribution expressed by $(1/\sqrt{2\pi})\exp[-x_*^2/2]$. A physical realization of such a situation is the case when ripples are riding on the surface of a swell of a long period. In this case, the individual ripple crests become the maxima of the overall wave profile, while the distribution of the heights of such crests is approximated by that of instantaneous swell profile, which follows the normal distribution.

The ratio of the number of negative maxima to total number of maxima of the wave profile can be found by integrating Eq. (10.124) in the range $(-\infty, 0)$. It can also be derived through simple reasoning as follows. First, the total number $N_1^*$ of maxima of the wave profile is divided into the numbers $N_1^+$ for $\eta_{\max} > 0$ and $N_1^-$ for $\eta_{\max} < 0$. Similarly, the numbers of positive and negative minima are introduced with the notation $N_2^+$ and

$N_2^-$, depending on whether $\eta_{\min} > 0$ or $\eta_{\min} < 0$. Thus,

$$\left.\begin{array}{l} \text{Number of maxima} : N_1^* = N_1^+ + N_1^- , \\[2mm] \text{Number of minima} : N_2^* = N_2^+ + N_2^- . \end{array}\right\} \qquad (10.126)$$

The profile of irregular waves under investigation is made up of linearly superposed wave components as described by Eq. (10.112), and the wave profile is statistically symmetric with respect to the zero line. Therefore, the number of minima is the same as the number of maxima given by Eq. (10.123), and the following equality exists with regard to $N_2^+$ and $N_2^-$.

$$\left.\begin{array}{l} N_2^+ = N_1^- = rN_1^* , \\[2mm] N_2^- = N_1^+ = (1-r)N_1^* , \end{array}\right\} \qquad (10.127)$$

where $r$ denotes the ratio of the number of negative maxima to the total number of maxima. By examining the relation between the number of zero-upcrossing points and the numbers of maxima and minima, we find that the number of maxima during the continuation of a wave crest ($\eta > 0$) from one zero-upcrossing to the succeeding zero-downcrossing point must be equal to the number of minima in the same period plus one. This leads to the following relation:

$$N_0^* = N_1^+ - N_2^+ = (1-r)N_1^* - rN_1^* = (1-2r)N_1^* . \qquad (10.128)$$

By rewriting the above relation, we obtain

$$r = \frac{1}{2}\left(1 - \frac{N_0^*}{N_1^*}\right) . \qquad (10.129)$$

By substituting Eqs. (10.81) and (10.123) into the above and utilizing Eq. (10.125), we can arrive at the following relationship between $r$ and $\varepsilon$:

$$r = \frac{1}{2}[1 - (1-\varepsilon^2)^{1/2}], \quad \text{or} \quad \varepsilon = [1 - (1-2r)^2]^{1/2} . \qquad (10.130)$$

The above equation indicates that the spectral width parameter $\varepsilon$ can be estimated by counting the number of negative maxima and calculating the ratio formed by it to the total number of maxima, without computing a wave spectrum. In the actual practice of wave data analysis, the following formula based on the numbers of maxima $N_1^*$ and zero-upcrossing points $N_0^*$ is often employed to estimate the value of $\varepsilon$:

$$\varepsilon = [1 - (N_0^*/N_1^*)^2]^{1/2} . \qquad (10.131)$$

For most sea wave records, it is found that the value of $\varepsilon$ estimated from $N_1^*$ and $N_0^*$ by Eq. (10.131) is slightly smaller than the value of $\varepsilon$ calculated from the moments of the wave spectrum, Eq. (10.125). The cause of this difference is attributed to the increase in the observed spectral density in the high frequency range due to the effects of wave nonlinearity and to noise in the record.

The problem of determining the largest value among the maxima of a wave profile was investigated by Cartwright and Longuet-Higgins.[3] They gave the following formula for the expected value of the highest maximum when an irregular wave profile has $N_1$ maximum points:

$$E[(x_*)_{\max}] \simeq \left[2\ln N_1 \sqrt{1-\varepsilon^2}\right]^{1/2} + \gamma \left[2\ln N_1 \sqrt{1-\varepsilon^2}\right]^{-1/2}, \quad (10.132)$$

where $\gamma$ is Euler's constant ($0.5772\ldots$). The above equation can be rewritten in terms of the number of zero-upcrossing points $N_0$, which is related to the number of maxima $N_1$ as $N_0 = N_1\sqrt{1-\varepsilon^2}$, by Eq. (10.131). The result is

$$E[(x_*)_{\max}] \simeq (2\ln N_0)^{1/2} + \gamma(2\ln N_0)^{-1/2}. \quad (10.133)$$

This expression has the same functional form as Eq. (10.38) for the largest wave height. In fact, Eq. (10.133) can be made to coincide with Eq. (10.38) by setting $H_{\max} = 2\eta_{\max}$.

## 10.5 Nonlinearity of Sea Waves

### 10.5.1 *Nonlinearity of Surface Elevation*

(A) *Maximum crest height*

The statistical theory of random sea waves is based on the assumption that random waves are made up of linear superposition of an infinite number of infinitesimal component waves. Because all component waves are expressed with trigonometric functions, wave crests and troughs are statistically symmetric and the surface elevation should follow the normal distribution. Real sea waves, however, exhibit the tendency that wave crests are high and peaked, while wave troughs are shallow and flat. Figure 10.12 is an example of the analysis of 171 wave records taken by surface measurement instruments (step-resistance gauges and ultrasonic wave sensors) as reported by Goda and Nagai.[30] For each wave record, the highest crest elevation $(\eta_c)_{\max}$ was detected and its ratio to the highest wave height $H_{\max}$ in the record was calculated; they may not belong to the same wave. Then the maximum crest height ratio $(\eta_c)_{\max}/H_{\max}$ was tabulated for several classes

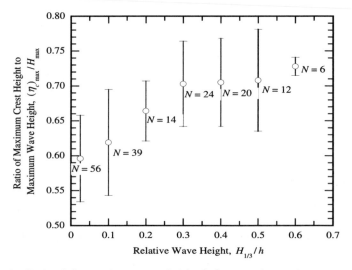

Fig. 10.12   Ratio of the maximum crest height $(\eta_c)_{max}$ to the maximum wave height $H_{max}$ in a wave record versus the ratio of significant wave height $H_{1/3}$ to water depth $h$.[30]

of the relative wave height $H_{1/3}/h$ and the arithmetic mean and standard deviation of each class were calculated. The open circles in Fig. 10.12 represent the mean value while the upper and lower vertical lines are given the length equivalent to the standard deviation. The legend $N$ indicates the number of wave records in each class of relative wave height.

As shown in Fig. 10.12, the maximum crest height is about 60% of the maximum wave height in deep water, and it rises up to about 70% in relatively shallow water area. When the significant wave height exceeds 60% of water depth, the maximum crest height is larger than 70% of the maximum wave height. The maximum wave crest height is one of the important design parameters for offshore platforms and has been investigated theoretically and experimentally for regular waves. Figure 10.2 presents an information on maximum crest heights of random waves in the sea.

## (B) *Skewness and kurtosis of surface elevation*

An example of the distribution of surface elevations in relatively shallow water is presented in Fig. 10.13, which is the result of wave data analysis with the significant wave height of 3.35 m in water of 11.5 m deep. Wave surface elevations are shown in the form of histograms, while the normal distribution fitted to the data is shown with a thick line. Compared with the normal distribution, the observed surface elevation has a distribution

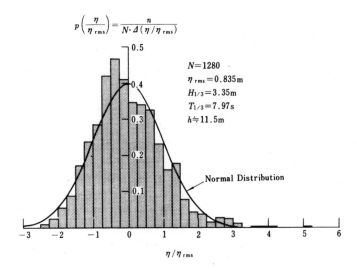

Fig. 10.13 Example of distribution of surface elevation in relatively shallow water.

shifted toward the left as a whole and a heavy tail at the right. The extent of the distortion of the statistical distribution from the normal is usually described by the skewness and kurtosis defined as

$$\text{Skewness}: \quad \sqrt{\beta_1} = \frac{1}{\eta_{\text{rms}}^3} \cdot \frac{1}{N} \sum_{i=1}^{N} (\eta_i - \overline{\eta})^3 , \qquad (10.134)$$

$$\text{Kurtosis}: \quad \beta_2 = \frac{1}{\eta_{\text{rms}}^4} \cdot \frac{1}{N} \sum_{i=1}^{N} (\eta_i - \overline{\eta})^4 . \qquad (10.135)$$

The normal distribution has the values $\sqrt{\beta_1} = 0$ and $\beta_2 = 3.0$, which is easily confirmed by calculation of Eqs. (10.134) and (10.135) with the probability density function of Eq. (9.24) in Sec. 9.3. As seen from the definition, the skewness is zero when the distribution is symmetric with respect to the mean value. When the mode of the distribution is located lower than the mean and the distribution has a heavy tail in the range greater than the mean, the skewness takes a positive value. The example shown in Fig. 10.13 has a skewness of $\sqrt{\beta_1} = 0.656$ and kurtosis of $\beta_2 = 4.09$, indicating clear deviation from the normal distribution.

Positive values of the skewness are commom in waves observed in the sea. For wind waves in deep water, the skewness is found to be proportional to the wave steepness. An empirical relation has been proposed by Huang and Long,[31] which is rewritten here as

$$\sqrt{\beta_1} \simeq 2\pi H/L.$$
(10.136)

The kurtosis gives an indication of the peakedness of the mode of the statistical distribution. If the peak is higher than that of corresponding normal distribution, the kurtosis becomes $\beta_2 > 3.0$, and vice versa. If $\beta_2 > 3.0$, the distribution shows heavy tails on both sides in compensation for the high peak. Mori and Yasuda[32] demonstrated in 1997 that the kurtosis is a key parameter describing the occurrence probability of large waves in a wave record: the larger the kurtosis, the greater the probability. As discussed in Sec. 10.1.2(C), Mori et al.[10,11] have clarified the relationship between the occurrence probability of freak waves and the kurtosis.

In relatively shallow water, both the skewness and kurtosis increase with increase in wave height. The author[33] has proposed the so-called *wave nonlinearity parameter* of the following:

$$(\Pi)_{1/3} = (H_{1/3}/L_A)\coth^3 k_A h,$$
(10.137)

where $H_{1/3}$ denotes the significant wave height, $L_A$ and $k_A = 2\pi/L_A$ are the wavelength and wave number calculated by the small amplitude (Airy) wave theory, respectively, and $h$ stands for the water depth.

In deep water the wave nonlinearity parameter becomes the wave steepness, i.e., $\Pi_{1/3} = H_{1/3}/L_A$, because $\tanh k_A h = 1$. In very shallow water, the wave nonlinearity parameter is an expression of the Ursell parameter, i.e., $\Pi_{1/3} \simeq 0.0040 H_{1/3} L_A^2/h^3$, because $\tanh k_A h \simeq 2\pi h/L_A$.

Figures 10.14 and 10.15 show the compiled results of the skewness and kurtosis of the field waves, respectively, by Goda.[34] The upper diagrams of the both figures correspond to the cases in which waves are regarded outside the surf zone, the outer edge of which is tentatively set at $h = 2.5H_0$. The data of skewness and kurtosis are plotted against the wave nonlinearity parameter. The lower diagrams of the both figures correspond to the cases in which waves are regarded inside the surf zone; the abscissa is the relative wave height of $H_0/h$. The swell data with low steepness have been measured by Hotta and Mizuguchi[35,36] and Ebersole and Hughes[37] (refer to Sec. 3.6.2 (B)). Other data with wave steepness greater than 0.01 are mainly the waves recorded by stationary wave observations in water of 10 m or deeper.[30]

When waves are not affected by the depth-controlled breaking outside the surf zone, the skewness and kurtosis monotonically increases with the wave nonlinearity parameter as seen in Figs. 10.14 and 10.15. Waves with the offshore steepness greater than around 0.01 may attain the skewness up

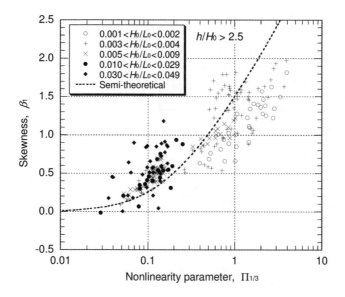

(a) Outside the surf zone ($h/H_0 > 2.5h$)

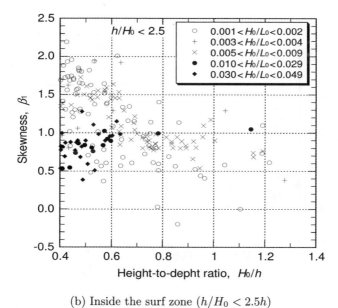

(b) Inside the surf zone ($h/H_0 < 2.5h$)

Fig. 10.14   Variation of the skewness of surface elevation outside and inside the surf zone.[34]

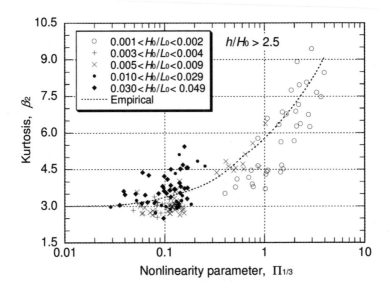

(a) Outside the surf zone $(h/H_0 > 2.5h)$

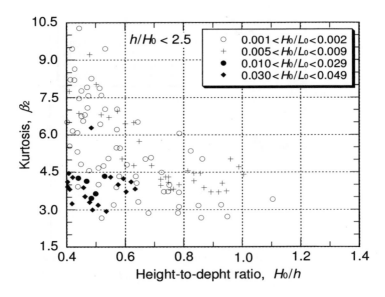

(b) Inside the surf zone $(h/H_0 < 2.5h)$

Fig. 10.15  Variation of the kurtosis of surface elevation outside and inside the surf zone.[34]

to about 1.2 and the kurtosis up to about 5.5, but no more. On the other hand, swell with the steepness less than 0.002 may have the skewness up to about 2.0 and the kurtosis up to about 10. As waves propagate into the surf zone, waves are gradually attenuated by breaking and the wave non-linearity is destroyed rapidly. With the increase of $H_0/h$ or the decrease of water depth, the skewness and kurtosis approach to the values of $\sqrt{\beta_1} = 0$, $\beta_2 = 3.0$, respectively, which are the case of the normal distribution.

The dotted line for the skewness in the upper diagram of Fig. 10.14 is based on an empirical relationship of the skewness of the profile of finite amplitude regular waves to the wave nonlinearity parameter. The skewness of irregular waves is evaluated by averaging the relationship with the weight of the occurrence probability of individual wave heights that follow the Rayleigh distribution. It is the technique similar as that employed by Longuet-Higgins[38] when he estimated the effect of wave nonlinearity on wave height distribution. The dotted line for the kurtosis in the upper diagram of Fig. 10.15 has been drawn to indicate the general trend without any theoretical background.

## (C) *Asymmetry of wave profiles*

The profiles of the crests and troughs of sea waves are asymmetric with respect to the mean water level (zero line). Figure 10.12 is an example of such asymmetry as represented with the crest height relative to the wave height. A quantitative assessment of the asymmetry with respect to the zero line is made with the skewness $\sqrt{\beta_1}$ of the wave profile defined by Eq. (10.134), the field data of which has been presented in Fig. 10.14.

Water waves also exhibit an asymmetry in the direction of wave propagation. It is common observation that the slope of a wave front becomes steep as waves approach to the surf zone and the crest point shifts toward the front from the center position; the plunging breaker is the limiting state of such an asymmetry of wave profiles. Wave forces exerted upon offshore structures are much affected by the degree of tilting of wave profiles. Myrhaug and Kjeldsen[39,40] have proposed an index of wave asymmetry in the direction of wave propagation by taking the ratio of the time interval between a zero-upcrossing point and the following crest point to the time interval between the crest point to the following zero-downcrossing point.

Apart from such an index for individual waves, the author[41] has proposed the following parameter called the *atiltness* to express the overall asymmetry of wave profiles of a wave record:

$$\beta_3 = \frac{1}{N-1} \sum_{i=1}^{N-1} (\dot{\eta}_i - \bar{\eta})^3 \Bigg/ \left[ \frac{1}{N-1} \sum_{i=1}^{N-1} (\dot{\eta}_i - \bar{\eta})^2 \right]^{3/2} . \qquad (10.138)$$

When the atiltness parameter takes a positive value, wave profiles show a forward tilting as a whole; when negative, a backward tilting. Wind waves which are growing by strong winds seem to indicate slightly positive values, but swell and ordinary sea waves in the offshore indicate $\beta_3 \simeq 0$ on the average without any correlation with the skewness $\sqrt{\beta_1}$. However, as waves approach to the beach, the atiltness parameter increases rapidly near the surf zone and exceeds 1.0 within the surf zone.

When the atiltness parameter has a large, positive value, the zero-upcrossing method yields slightly larger values of characteristic wave periods than the zero-downcrossing method (except for the mean period which remains the same), according to Goda.[41] The difference is due to the tendency that the waves of large heights defined by the former method have longer periods than the waves defined by the latter method. On the other hand, the characteristic wave heights remain statistically the same regardless of the method of wave definition. For the analysis of waves in and near the surf zone, it is safe to employ the zero-downcrossing method for definition of individual waves. For waves with $\beta_3 \simeq 0$, however, the zero-upcrossing method can be utilized as well, because both methods yield statistically the same results.

### 10.5.2    *Effects of Wave Nonlinearity on Characteristic Wave Heights and Periods*

(A) *Nonlinearity effect on wave heights*

The effect of wave nonlinearity on wave heights has theoretically been investigated by several researchers. Tayfun[42] computed the secondary interaction terms of irregular waves for examination of wave crest elevations and wave heights. He showed that the wave heights are not affected by the wave nonlinearity to the second order, while the crest elevations are heightened. The results concur with expectation by the finite amplitude theory of regular waves.

Longuet-Higgins,[38] on the other hand, clarified the relationship between the wave height and the potential energy of the third order Stokes waves in deep water. Then he computed the total potential energy of a train of irregular waves under the assumptions that individual waves can be replaced with theoretical finite amplitude waves and that the wave heights follow the Rayleigh distribution. From this potential energy, he evaluated the ratio $H_{\rm rms}/\eta_{\rm rms}$ and showed that the ratio increases several percent as the wave steepness becomes large.

The effects of wave nonlinearity are enhanced in shallow water. Figures 10.16 and 10.17 show the variations of the wave height ratios $H_{1/3}/H_{m0}$ and $H_{\rm rms}/H_{m0}$ outside and inside the surf zone. The data sources are common with those employed in calculation of the skewness and kurtosis shown in Figs. 10.14 and 10.15. The symbol $H_{m0}$ denotes the spectral significant wave height defined by Eq. (2.40) in Sec. 2.4.1, and it is equal to $4.004\eta_{\rm rms}$. The symbol $H_{\rm rms}$ denotes the root-mean-square value of individual wave heights defined by the zero-crossing method. As shown in Eq. (10.15) in Sec. 10.1.2(A), it is related with the zeroth spectral moment as $H_{\rm rms} = \sqrt{8m_0} = 2.828\eta_{\rm rms}$ when the wave heights follow the Rayleigh distribution.

Figure 10.16 shows the wave nonlinearity effect on the statistical significant wave height. If individual wave heights follow the Rayleigh distribution, then $H_{1/3}/H_{m0} = 1.0$. As discussed in Sec. 2.4.1, real sea waves have the distribution slightly narrower than the Rayleigh and thus the wave height ratio becomes $H_{1/3}/H_{m0} \simeq 0.95$ while wave nonlinearity is weak. With the increase in the wave nonlinearity parameter $\Pi_{1/3}$ beyond 0.2 or so, the wave height ratio $H_{1/3}/H_{m0}$ increases rapidly; swell with very low steepness may have the ratio in excess of 1.5.

The dashed lines in the upper diagrams of Figs. 10.16 and 10.17 represent semi-theoretical predictions based on the potential energy calculation of finite amplitude regular waves, which is made by referring to the method used by Longuet-Higgins.[38] The apparent increase in the height of regular waves in shallow water is calculated with the third-order Stokes wave theory and the second order cnoidal wave theory, and the results have been averaged over the range of wave heights with the weight of the occurrence probability of individual wave heights. The field data approximately follow the trend of the semi-empirical curve despite a large scatter, confirming that the observed nonlinearity effect is characteristic of water waves. As waves propagate further shoreward and enter in the surf zone, however, the wave nonlinearity is rapidly reduced and the wave height ratio $H_{1/3}/H_{m0}$ decrease toward the value of 0.95 as exhibited in the lower diagram of Fig. 10.6.

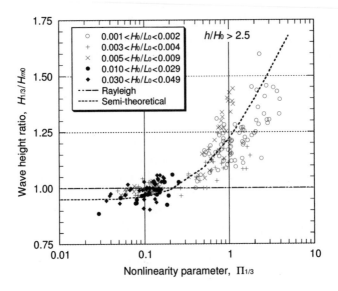

(a) Outside the surf zone $(h/H_0 > 2.5h)$

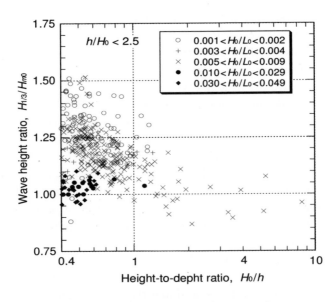

(b) Inside the surf zone $(h/H_0 < 2.5h)$

Fig. 10.16    Variation of the wave height ratio $H_{1/3}/H_{m0}$ outside and inside the surf zone.[34]

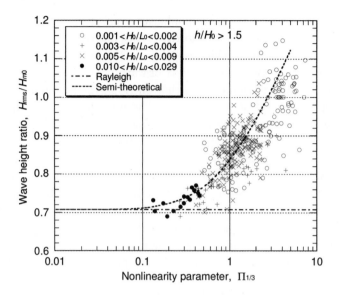

(a) Outside the surf zone ($h/H_0 > 2.5h$)

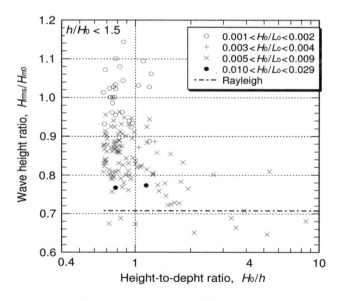

(b) Inside the surf zone ($h/H_0 < 2.5h$)

Fig. 10.17   Variation of the wave height ratio $H_{\mathrm{rms}}/H_{m0}$ outside and inside the surf zone.[34]

Figure 10.17 shows the wave nonlinearity effect on the statistical rms wave height. The wave height ratio $H_{\mathrm{rms}}/H_{m0}$ has the value of 0.706 under the Rayleigh distribution, which is shown as the horizontal dash-dots line in the left diagram. Because the effect of wave breaking on $H_{\mathrm{rms}}$ begins to appear in the water shallower than that on $H_{1/3}$, the boundary of the surf zone is tentatively set at $h/H_0 = 1.5$. The wave height ratio $H_{\mathrm{rms}}/H_{m0}$ increases with the wave nonlinearity parameter $\Pi_{1/3}$ outside the surf zone and decreases with the relative wave height $H_0/h$ inside the surf zone. The pattern of variation across the surf zone is the same as the case of $H_{1/3}$. For swell with very low steepness, the wave height ratio $H_{\mathrm{rms}}/H_{m0}$ may attain the value of 1.1. Such a high value may look odd, but it is a result of wave nonlinearity effect being assisted by the definition of wave heights by the two methods of statistical and spectral ones.

Another way of depicting the wave nonlinearity effect on the statistical significant wave height is to use the following parameter $\Pi_0$ instead of $\Pi_{1/3}$:

$$\Pi_0 = (H_0/L_A)\coth^3 k_A h\,, \tag{10.139}$$

where the deepwater significant wave height $H_0$ is used instead of $H_{1/3}$ in Eq. (10.137). Figure 10.18 is a re-plotting of the wave height ratio $H_{1/3}/H_{m0}$ against the nonlinearity parameter $\Pi_0$. The data shown in Fig. 10.16 are regrouped with a finer classification of wave steepness, and boundary lines between the classes of wave steepness are drawn subjectively to provide users with guidelines of wave height decrease inside the surf zone. Although waves with low steepness may have a large value of the wave height ratio $H_{1/3}/H_{m0}$, the wave height ratio remains below around 1.1 for waves with the steepness greater than about 0.01, because the depth-controlled breaking process restrains the wave nonlinearity before the latter exercises its full effect.

## (B) *Nonlinearity effect on wave periods*

The effect of wave nonlinearity does not appear on the periods of individual waves. However, the mean wave period $T_{02}$, estimated by the spectral moments $m_0$ and $m_2$ by means of Eq. (10.82), is grossly underestimated, because the second spectral moment $m_2$ is overestimated owing to the presence of the nonlinear spectral components to be discussed in Sec. 10.5.3 (B). The spectral mean wave period $T_{02}$ may become less than 70% of the mean period actually counted on a wave record when $\Pi_{1/3} > 0.3$.[33] Because of such wave nonlinearity effects, linearization of a wave spectrum by cutting

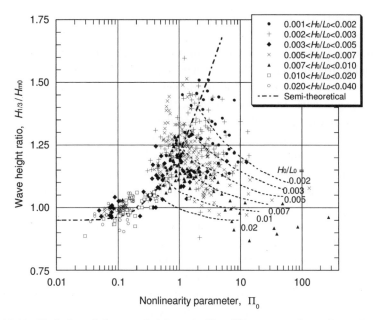

Fig. 10.18 Variation of the wave height ratio $H_{1/3}/H_{m0}$ across the surf zone in terms of the wave nonlinearity parameter $\Pi_0$.[34]

the low frequency range of $f < 0.5\ f_p$ and the high frequency range of $f > (1.5 \sim 1.8)f_p$ (where $f_p$ denotes the spectral peak frequency) is often undertaken, when statistical theories on wave heights and periods based on the frequency spectrum are tested against the field wave data.[29,43,44]

## 10.5.3 *Nonlinear Components of Wave Spectrum*

### (A) *Overview*

The deviation in the statistical distribution of surface elevation from the normal distribution indicates presence of nonlinear wave components. The theory of random sea waves is based on the linear superposition of an infinite number of infinitesimal amplitude component waves, as expressed by Eq. (9.12) or Eq. (9.15) in Sec. 9.2. The phase angle $\varepsilon_n$ is assumed to be uniformly and randomly distributed. The presence of nonlinear wave components means that some of the phase angles are not independent, but hold some fixed relation among each other, and that the component waves having such mutually dependent phase angles do not satisfy the dispersion relationship, Eq. (9.3).

In theoretical analysis of water waves, we have the nonlinear boundary condition at the surface. The Stokes waves, for example, satisfy the surface boundary condition by having the harmonic components of twice, thrice, and higher frequencies to the accuracy required. These harmonic component waves are bounded to the fundamental wave with the fixed phases and propagae with the same speed, without satisfying the dispersion relationship of Eq. (9.3). When two regular waves of different frequencies are superposed, the surface boundary conditions are not satisfied with the accompanying bounded harmonics alone but need to be supplemented with the secondary interaction waves having the sum and difference frequencies. With waves of large amplitudes, there appear tertiary interaction waves of appreciable amplitudes.

In case of the superposition of an infinite number of component waves, each pair of two component waves produce secondary interaction waves and each combination of three component waves produce tertiary interaction waves. These interaction waves constitute the so-called *nonlinear spectral components*, the computation of which was initiated by Tick,[45] in 1963 for irregular waves in water of arbitrary depth with a correction later made by Hamada.[46] Around the same time, Hasselmann[47] clarified the energy transfer among spectral components of wind-generated waves through nonlinear interactions in deep water. Masuda *et al.*[48] calculated the secondary and tertiary interaction terms for directional random waves in deep water, and resolved the linear and nonlinear spectral components of the observed spectra. The nonlinear energy transfer by Hasselmann has been introduced in the numerical computation program of wave forecasting as the indispensable mechanism of wave growth.

The presence of nonlinear spectral components means that some of frequency components of the observed spectra are not independent but phase-coupled. Mitsuyasu *et al.*[49] demonstrated that the apparent failure of the dispersion relationship for the high frequency range is due to the presence of nonlinear spectral components through minute analysis of wind-wave flume test data. Their measurements were supported by the theory by Masuda *et al.*

Detailed analysis of the secondary interaction among directional wave spectrum was made in 1979 by Sharma and Dean[50] for the evaluation of wave force exerted on offshore structures. Tuah and Hudspeth[51] examined the finite amplitude effect on secondary wave interactions. Hashimoto *et al.*[52] also calculated nonlinear interaction terms of directional random waves for reliable conversion of wave pressure records to surface profiles in routine works of stationary wave observations. Recently Kato *et al.*[53,54] have carried out numerical simulation of random wave profiles inclusive of secondary interaction components.

The secondary interaction terms of wave spectrum appear in the two frequency ranges, one higher and another lower than the peak frequency. The former has the frequency which is the sum of the frequencies of interacting free waves, while the latter has the frequency which is the difference of the frequencies of interacting free waves. The low frequency waves by secondary interaction are the bounded low-frequency waves, which have been discussed in Sec. 2.5. Sand[55,56] analyzed the group-bounded long waves as the result of secondary wave interactions. Kimura[57] and Okihiro *et al.*[58] also presented the computation scheme of the secondary wave interaction terms.

(B) *Resolution of nonlinear components of frequency spectrum*

For waves with small directional spreading such as long-traveled swell, secondary wave interaction may be calculated with frequency spectrum only. The theory by Tick[45] with a correction by Hamada[46] is employed for the calculation. The secondary interaction term of the frequency spectrum, denoted by $S^{(2)}(f)$, is given by

$$S^{(2)}(f_1) = \int_{-\infty}^{\infty} K(\omega, \omega_1) S^{(1)}(f_1 - f) S^{(1)}(f) df, \qquad (10.140)$$

in which

$$K(\omega, \omega_1) = \frac{1}{4} \left\{ \frac{gkk'}{\omega(\omega_1 - \omega)} + \frac{\omega(\omega_1 - \omega)}{g} - \frac{\omega_1^2}{g} \right.$$

$$\left. + \frac{\omega_1^2 \left[ \frac{g(\omega_1 - \omega)k^2 + g\omega k'^2}{\omega(\omega_1 - \omega)\omega_1} + \frac{2gkk'}{\omega(\omega_1 - \omega)} + \frac{\omega(\omega_1 - \omega)}{g} - \frac{\omega_1^2}{g} \right]}{g|k + k'| \tanh|k + k'|h - \omega_1^2} \right\}^2,$$

$$(10.141)$$

$$\omega^2 = gk \tanh kh, \quad (\omega_1 - \omega)^2 = gk' \tanh k'h. \qquad (10.142)$$

The original (linear) spectrum is denoted by $S^{(1)}(f)$ in the above. The recorded spectra of the long-traveled swell at Caldera Port in Costa Rica were resolved into linear components and their secondary interaction components by using the above formula and an iteration solution method as computed by Goda.[14] Figure 10.19 is an example of the resolution. The water depth is about 17 m. While the main peak in the frequency range of

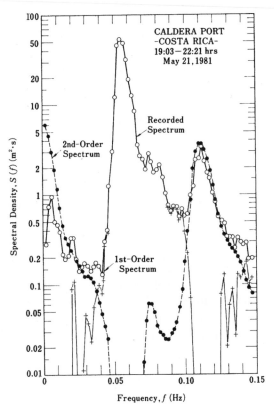

Fig. 10.19  Resolution of linear and nonlinear spectral components for a recorded spectrum.[14]

$f = 0.04 \sim 0.09$ Hz does not contain nonlinear terms, the auxiliary spectral peak around $f = 0.11$ Hz (twice the peak frequency) is seen to be entirely the product of secondary interactions. The secondary interaction terms at the low frequency zone, which corresponds to the difference frequency of interacting linear components, tend to be larger than the observation especially in the range of $f < 0.02$ Hz. Difference seems to have originated from neglect of directional spreading and possible presence of tertiary wave interactions. Although the nonlinear wave interaction theory in water of finite depth is limited to the second order at present, tertiary and higher-order interactions are taking place in nature. This is exemplified in the appearance of auxiliary spectral peaks at higher harmonics of the frequency of the primary peak in some of the wave spectra recorded in relatively shallow water.

(C) *Numerical simulation of surface waves with nonlinear interaction terms*

For numerical simulation of wave profiles with nonlinear interaction terms, Kato *et al.*[53] employed the following equation by referring to Tuah and Hudspeth[51] and Hashimoto *et al.*[52]:

$$\eta(\vec{x}, t) = \eta^{(1)}(\vec{x}, t) + \eta^{(2)}(\vec{x}, t) = \int_k \int_\sigma \{dA^{(1)} + dA^{(2)}\} \exp[i(\vec{k} \cdot \vec{x} - \sigma t)],$$
(10.143)

where $\vec{x}$ and $\vec{k}$ are the location and wave number vectors, respectively, and $\sigma$ is the angular frequency. The terms $dA^{(1)}$ and $dA^{(2)}$ denote the first-order and second-order complex amplitudes of the Fourier-Stieltjes components and they are respectively given by the following:

$$dA^{(1)} = \sqrt{S^{(1)}(\sigma, \theta) d\theta \, d\sigma},$$
(10.144)

$$dA^{(2)}(\vec{k}, \sigma) = \int_{k_1} \int_{\sigma_1} H(\vec{k}_1, \vec{k}_2, \sigma_1, \sigma_2) dA^{(1)}(\vec{k}_1, \sigma_1) \, dA^{(1)}(\vec{k}_2, \sigma_2),$$
(10.145)

where $S^{(1)}$ denotes the first order directional spectrum and $H(\vec{k}_1, \vec{k}_2, \sigma_1, \sigma_2)$ is the nonlinear second-order kernel function, which is given by

$$H(\vec{k}_1, \vec{k}_2, \sigma_1, \sigma_2) = \frac{1}{2g} \Bigg[ 2(\sigma_1 + \sigma_2) D(\vec{k}_1, \vec{k}_2, \sigma_1, \sigma_2)$$
$$- \frac{g^2 \vec{k}_1 \cdot \vec{k}_2}{\sigma_1 \sigma_2} + \sigma_1 \sigma_2 + \sigma_1^2 + \sigma_2^2 \Bigg],$$
(10.146)

where $g$ is the acceleration of gravity and $D(\vec{k}_1, \vec{k}_2, \sigma_1, \sigma_2)$ is evaluated by

$$D(\vec{k}_1, \vec{k}_2, \sigma_1, \sigma_2) = \frac{\begin{array}{c} 2(\sigma_1 + \sigma_2)[g^2 \vec{k}_1 \cdot \vec{k}_2 - (\sigma_1 \sigma_2)^2] \\ -\sigma_1 \sigma_2 (\sigma_1^3 + \sigma_2^3) + g^2 (k_1^2 \sigma_2 + k_2^2 \sigma_1) \end{array}}{2\sigma_1 \sigma_2 [(\sigma_1 + \sigma_2)^2 - g|\vec{k}_1 + \vec{k}_2| \tanh(|\vec{k}_1 + \vec{k}_2|h)]}.$$
(10.147)

The wave number vector $\vec{k}_i$ and the angular frequency $\sigma_i$ ($i = 1$ or 2) satisfies the dispersion relationship of Eq. (10.148) and have the relation of Eq. (10.149).

$$\sigma_i^2 = g|\vec{k}_i| \tanh(|\vec{k}_i|h),$$
(10.148)

$$\vec{k}_i(-\sigma_i) = -\vec{k}_i(\sigma_i).$$
(10.149)

The wave profiles of the secondary interaction terms to be computed by Eq. (10.143) can be separated into those of sum frequency and those of difference frequency as has been done by Kato *et al.*[53]

## 10.6    Sampling Variability of Sea Waves

(A) *General*

The irregularity of the profiles of sea waves suggests that the mean wave height, period and other statistical parameter must be treated not as deterministic quantities but as stochastic quantities. An example of the variation of a statistical quantity can be prepared by arbitrarily drawing a row of numbers from a table of random numbers as shown below and by taking the average of five consecutive numbers. The statistic of the average forms the row of a secondary random quantity as shown in the parentheses.

... 09025 07210 47765 77192 81842 62204 30413 47793 ...
... (3.2)  (2.0)  (5.8)  (5.2)  (4.6)  (2.8)  (2.2)  (6.0) ...

Such variability in statistical quantities is inherent to any phenomenon involving a stochastic process. A record of measured wave profiles is just a sample from the ensembel of sea wave records. Even if the sea state is homogeneous and stationary, another wave record taken at the next observation time or at a neighboring site will surely exhibit a different pattern of wave profiles. The statistical quantities calculated from the second record will not be the same as those from the first record. The reason we analyze wave records for various statistical quantities and calculate wave spectra is to gather clues to the sea conditions which existed at the time of our observation.

Statistically speaking, we are trying to estimate the properties of the ensemble from the data of the samples we have obtained. The quantities derived from a sample (i.e., record of wave profile) themselves are not our final objectives. If we are assured of the stationary state of the sea during the observation, we can take a number of successive wave records and improve the accuracy of the estimate by averaging the results of many records. But the stationary state of the sea is never realized as explained in Sec. 9.3(A), and we have to make our best efforts to estimate the true properties of the sea state from the wave records we have in our hands.

Any statistical quantity of the ensemble which has been estimated from a sample does not represent a true value but have a certain range of estimation error. For example, if the observed height of the significant wave is 6.24 m and the coefficient of variation of that observation is 6%, then what we can say is that there is a 50% probability that the true value of the significant wave height of that sea state lies in the range of 5.99 to 6.49 m and that there is a 90% probability that it lies in the range of 5.62

to 6.86 m. In order to evaluate of the range of estimation errors, it is necessary to know the distributions of the sample statistics. Many statistics of sea waves can be regarded to follow the normal distribution, with the exception of the statistics of $\eta_{max}$, $H_{max}$ and $T_{max}$. Thus, we can evaluate the range of estimation errors once we know the standard deviations of the various wave statistics.

If the data consist of randomly sampled ones, several established results of statistics can be used. Let us suppose that we take $n$ random samplings of a statistical variable $x$ and calculate the arithmetic mean $\bar{x}$, the root-mean-square value $x_{rms}$, the skewness $\sqrt{\beta_1}$, and the kurtosis $\beta_2$. These statistics have the following variances, e.g., Kendall and Stuart[59]:

$$\text{Mean}: \ \text{Var}[\bar{x}] = \mu_2/n = \sigma^2/n\,, \tag{10.150}$$

$$\text{Standard deviation}: \ \text{Var}[x_{rms}] = (\mu_4 - \mu_2^2)/4\mu_2 n\,, \tag{10.151}$$

$$\text{Skewness}: \ \text{Var}[\sqrt{\beta_1}] = 6/n\,, \tag{10.152}$$

$$\text{Kurtosis}: \ \text{Var}[\beta_2] = 24/n\,, \tag{10.153}$$

where

$$\text{Var}[X] = E[(X - E[X])^2] = E[X^2] - E[X]^2\,, \tag{10.154}$$

$$\mu_n = \int_{-\infty}^{\infty} (x - \bar{x})^n p(x)dx\,. \tag{10.155}$$

The quantity $p(x)$ denotes the probability density function of $x$, and $\sigma^2$ is the variance of the ensemble of $x$.

Among the above formulas, the first two can be applied to an ensemble with any statistical distribution, whereas the formulas for the skewness and kurtosis are only applicable to an ensemble possessing the normal distribution. Thus, the mean length of a run of wave height given by Eq. (10.69) in Sec. 10.2.4, for example, has the following variance:

$$\text{Var}[\bar{j}_1] = \frac{p_{22}}{N_R(1 - p_{22})^2}\,, \tag{10.156}$$

in which $N_R$ refers to the numbers of runs of the wave heights in a record, though it is not a random sampling in the strict sense. Because $N_R$ is on the order of ten for many wave records, relatively large variations are inevitably associated with the statistics of run length.

The correlation coefficient between two statistical variables has the following variance[60]:

$$\text{Var}[\rho] = (1 - \rho^2)^2/n\,, \tag{10.157}$$

where $n$ denotes the number of pairs of variables and $\rho$ is the correlation coefficient between the two ensembles. Equation (10.157) indicates that the variance of the correlation coefficient decreases as the correlation becomes higher.

(B) *Standard deviations of characteristic wave heights and periods*

Equations (10.150) to (10.153) cannot be applied directly to the data from wave records, because water surface elevations read off at constant time intervals are mutually correlated and do not constitute a data of random sampling. The correlation between instantaneous surface elevations is dictated by the frequency spectrum, and thus the variability of wave statistics is governed by the functional shape of wave spectrum. The coefficient of variation of the variance of surface elevation $\eta_{\mathrm{rms}}^2 = m_0$, for example, has theoretically been given by Tucker[61] and Cavanié.[62] The result can be rewritten in the following expression:

$$\mathrm{C.V.}\,[m_0] = \frac{\sigma(m_0)}{E[m_0]} = \frac{1}{\sqrt{N_0}\,m_0}\left[\int_0^\infty \overline{f}\,S^2(f)df\right]^{1/2}, \qquad (10.158)$$

where C.V. $[X]$ denotes the coefficient of variation of an arbitrary statistic $X$, $\sigma(X)$ is the standard deviation, $E[X]$ refers to the expected value, $N_0$ stands for the number of waves in a record, and $\overline{f}$ is the mean frequency. The coefficient of variation of the root-mean-square value of surface elevation $\eta_{\mathrm{rms}}$ is evaluated as one half of that of $m_0$ given by Eq. (10.158), unless the latter is not so large.

As for the coefficient of variation of the mean period $T_{02}$, estimated from the wave spectrum, Cavanié[62] has given a theoretical prediction, which is rewritten using the number of waves in a record as follows:

$$\mathrm{C.V.}\,[T_{02}] = \frac{1}{2\sqrt{N_0}}\left\{\overline{f}\int_0^\infty S^2(f)\left[\frac{f^4}{m_2^2} - \frac{2f^2}{m_0 m_2} + \frac{1}{m_0^2}\right]^{1/2}df\right\}.$$
$$(10.159)$$

Except for the above two, no theory is available for the variability of wave statistics in general. To fill the gap of information, Goda[63] has carried out a series of numerical simulations of wave profiles for various spectral shapes: these simulation data have been utilized for rewriting the JONSWAP and Wallops spectra in terms of the significant wave height and period as described in Sec. 2.3.1 and for preparing Tables 2.3 and 2.4 in Sec. 2.4. For each wave spectrum, 2000 samples of different wave profiles with 4096 data points (12 data per peak period) were generated with the inverse FFT method by means of the Monte Carlo simulation technique, and various wave statistics of surface elevation, characteristic wave heights and periods were analyzed by the zero-upcrossing method. The standard deviation and the coefficient of variation of each wave statistic were computed from this 2000 data set, respectively. Because the variability of most of the wave statistics is inversely proportional to the square root of the number of waves as expressed in Eqs. (10.158) and (10.159), the resultant standard deviation and coefficient of variation of a statistic $X$ have been formulated in the following equations by introducing a proportionally coefficient $\alpha$:

$$\sigma(X) = \frac{\alpha}{\sqrt{N_0}}, \quad \text{C.V.}[X] = \frac{\alpha}{\sqrt{N_0}}. \tag{10.160}$$

The proportionality coefficient $\alpha$ has been determined for each wave statistic as listed in Tables 10.3 and 10.4.

According to the above numerical simulations, the variations of the skewness $\sqrt{\beta_1}$ and the atiltness $\beta_3$ decrease as the spectral peak becomes sharp, while the variation of the kurtosis $\beta_2$ increases. The kurtosis indicates a slight departure from the power of $-1/2$ in its proportionality to $N_0$; its variability seems proportional to the power of $-1/3$ of $N_0$ when the exponent $m$ of the Wallops spectrum or the peak enhancement factor $\gamma$ of the JONSWAP spectrum is large.

The variability of the characteristic wave heights increases as the spectral peak becomes sharp. This tendency is expected by Tucker's theory: the coefficient of variation of $\eta_{\text{rms}}$ by numerical experiments almost agrees with the prediction by Eq. (10.146). Though the standard deviation of $H_{\text{max}}$ is not listed in Table 10.4, it can be evaluated by Eq. (10.40) in Sec. 10.1.3.

Table 10.3  Proportionality coefficient $\alpha$ for the standard deviation of parameters of surface elevation for various wave spectra[63]

$$\sigma(\beta_i) = \alpha/\sqrt{N_0}.$$

| Parameter | Wallops-type Spectrum | | | | JONSWAP-type Spectrum | | |
|---|---|---|---|---|---|---|---|
| | $m = 3$ | $m = 5$ | $m = 10$ | $m = 20$ | $\gamma = 3.3$ | $\gamma = 10$ | $\gamma = 20$ |
| Skewness $\sqrt{\beta_1}$ | 0.93 | 0.72 | 0.32 | 0.08 | 0.62 | 0.47 | 0.38 |
| Kurtosis $\beta_2$ | 2.29 | 2.57 | 3.05 | 3.49 | 2.77 | 3.27 | 3.68 |
| Atiltness $\beta_3$ | 0.88 | 0.77 | 0.40 | 0.09 | 0.75 | 0.68 | 0.60 |

Note: The values listed in this table are based on the results of numerical simulations of wave profiles with given wave spectra in the frequency range of $f = (0.5 \sim 6.0)f_p$.

Table 10.4   Proportionality coefficient $\alpha$ for characteristic wave heights and periods for various wave spectra[63]

$$\text{C.V.}[X] = \alpha/\sqrt{N_0}.$$

| Height and Period Ratio | Wallops-type Spectrum | | | | JONSWAP-type Spectrum | | |
|---|---|---|---|---|---|---|---|
| | $m = 3$ | $m = 5$ | $m = 10$ | $m = 20$ | $\gamma = 3.3$ | $\gamma = 10$ | $\gamma = 20$ |
| $H_{1/10}$ | 0.64 | 0.70 | 0.81 | 0.94 | 0.83 | 1.03 | 1.17 |
| $H_{1/3}$ | 0.57 | 0.60 | 0.69 | 0.80 | 0.72 | 0.91 | 1.04 |
| $\overline{H}$ | 0.61 | 0.64 | 0.70 | 0.81 | 0.77 | 0.96 | 1.08 |
| $\sigma(H)$ | 0.74 | 0.79 | 0.93 | 1.09 | 0.92 | 1.13 | 1.31 |
| $T_{1/10}$ | 0.64 | 0.48 | 0.31 | 0.22 | 0.35 | 0.24 | 0.19 |
| $T_{1/3}$ | 0.49 | 0.35 | 0.24 | 0.17 | 0.26 | 0.17 | 0.14 |
| $\overline{T}$ | 0.51 | 0.40 | 0.28 | 0.22 | 0.40 | 0.37 | 0.32 |
| $\sigma(T)$ | 0.66 | 0.66 | 0.74 | 0.88 | 0.66 | 0.97 | 1.32 |

Note: The values listed in this table are based on the results of numerical simulations of wave profiles with given wave spectra in the frequency range of $f = (0.5 \sim 6.0)f_p$.

However, the coefficient of variation of $H_{\max}$ observed in numerical experiments was smaller by up to 10% than the theoretical prediction, when the spectrum had a broad peak.[63] The difference seems to be related with the narrowness of the distribution of wave heights by simulations compared with the Rayleighan. Contrary to the variation of wave heights, the coefficient of variation of characteristic wave period decreases as the spectral peak becomes sharp.

Table 10.4 also reveals that the significant wave has smaller variations in both the height and period than the mean wave. In the case of $m = 5$ representing typical wind waves and for a record length of 100 waves, the coefficient of variation of $H_{1/3}$ is 6.0% against that of 6.4% for $\overline{H}$, while the coefficient of variation of $T_{1/3}$ is 3.5% against that of 4.0% for $\overline{T}$. The inferiority of the statistical stability of the mean wave to that of the significant wave seems to be caused by inclusion of many small waves in a wave record, which are more susceptible to sampling variability, for calculation of the mean wave height and period.

Variations in the characteristic wave heights and periods mean that wave height ratios such as $H_{1/3}/\overline{H}$ and the period ratios such as $T_{1/3}/\overline{T}$ are also subject to statistical variability. The fluctuations of the height and period ratios from wave records presented in Sec. 2.2.2 are examples of sampling variability. Although the full verification of the statistical variability of sea waves by the field data is not feasible because of the lack of stationarity, the data in Sec. 2.2.2 and the data of simultaneous wave measurements at three sites off Sakata Port (Fig. 3.38 in Sec. 3.6.3) provide the evidence of statistical variability: these data seems to indicate a slightly greater variability than those in Table 10.4 by numerical simulations.

Attention must be paid to the sampling variability of sea waves in the planning of wave observations, in the utilization of analyzed wave data, and in hydraulic model testing with random waves. For example, if wave recorders are set at two locations separated by several hundred meters in coastal water of nearly uniform topography, the two recorders may yield significant wave heights differing by 10%. It is not possible, however, to determine from one set of such data alone whether the difference is physically meaningful or was caused by sampling variability.

In general, the variance of a statistical variable defined by linear operations on normal variates is given by the sum of the variances of the individual variables. Thus, the difference between the characteristic wave heights and periods at two stations has a standard deviation of $\sqrt{2}$ times that of a single station. A comparison of hindcasted and recorded wave data should also be made with due allowance for he range of sampling variability of the recorded data.

Sampling variability of sea waves also appear in estimations of the spectral density. This subject is discussed in Sec. 11.2.1.

### References

1. S. O. Rice, "Mathematical analysis of random noise," 1944, reprinted in *Selected Papers on Noise and Stochastic Processes* (Dover Pub., 1954), pp. 132–294.
2. M. S. Longuet-Higgins, "On the statistical distributions of sea waves," *J. Marine Res.* **XI**(3) (1952), pp. 245–265.
3. D. E. Cartwright and M. S. Longuet-Higgins, "The statistical distribution of the maxima of random function," *Proc. R. Soc. London, Ser. A.* **237** (1956), pp. 212–232.
4. Y. Goda and M. Kudaka, " On the role of spectral width and shape parameters in control of individual wave height distribution," *Coastal Engineering Journal* **49**(3) (2007), pp. 311–335.
5. M. A. Tayfun, "Effects of spectrum band width on the distribution of wave heights and periods," *Ocean Engrg.* **10**(2) (1983), pp. 107–118.
6. A. Naess, "On the distribution of crest to trough wave heights," *Ocean Engrg.* **12**(3) (1985), pp. 221–234.
7. G. Z. Forristall, "The distribution of measured and simulated wave heights as a function of spectral shape," *J. Geophys. Res.* **89**(C6) (1984), pp. 10,547–10,552.

8. G. Z. Forristall, "On the statistical distribution of wave heights in a storm," *J. Geophys. Res.* **83**(C5) (1978), pp. 2353–2358.
9. D. Myrhaug and S. P. Kjeldsen, "Steepness and asymmetry of extreme waves and the highest waves in deep water," *Ocean Engrg.* **13**(6) (1986), pp. 549–568.
10. N. Mori and P. A. E. M. Janssen, "On kurtosis and occurrence probability of freak waves," *J. Physical Oceanogr.* **36**(7) (2006), pp. 1471–1483.
11. N. Mori, M. Onorato, P. A. E. M. Janssen, A. R. Osborne and M. Serino, "Exceedance probability for strongly nonlinear long crested waves," *J. Geophys. Res.*, doi:10.1029/2006JC004024 (2007).
12. A. G. Davenport, "Note on the distribution of the largest value of a random function with application to gust loading," *Proc. Inst. Civil Engrs.* **28** (1964), pp. 187–224.
13. H. Cramer, *Mathematical Methods of Statistics* (Princeton Univ. Press, 1946), p. 376.
14. Y. Goda, "Analysis of wave grouping and spectra of long-travelled swell," *Rept. Port and Harbour Res. Inst.* **22**(1) (1983), pp. 3–41.
15. Y. Goda, "Numerical experiments on wave statistics with spectral simulation," *Rept. Port and Harbour Res. Inst.* **9**(3) (1970), pp. 3–57.
16. Y. Goda, "On wave groups," *Proc. BOSS' 76* **1** (Trondheim, 1976), pp. 115–128.
17. E. R. Funke and E. P. D. Mansard, "On the synthesis of realistic sea states," *Proc. 17th Int. Conf. Coastal Engrg.* (Sydney, ASCE) (1980), pp. 3–57.
18. H. Rye, "Ocean wave groups," *Dept. Marine Tech., Norwegian Inst. Tech.* Rept. UR-82-18 (1982), 214p.
19. A. Kimura, "Statistical properties of random wave groups," *Proc. 17th Int. Conf. Coastal Engrg.* (Sydney, ASCE) (1980), pp. 2955–2973.
20. J. A. Battjes, and G. Ph. van Vledder, "Verification of Kimura's theory for wave group statistics," *Proc. 19th Int. Conf. Coastal Engrg.* (Houston, ASCE) (1984), pp. 642–648.
21. M. S. Longuet-Higgins, "Statistical properties of wave groups in a random sea state," *Phil. Trans. R. Soc. London, Ser. A* **312** (1984), pp. 219–250.
22. W. B. Davenport and W. L. Root, *Introduction to the Theory of Random Signals and Noise* (McGraw-Hill, 1958).
23. Y. Goda, "Numerical examination of several statistical parameters of sea waves," *Rept. Port and Harbour Res. Inst.* **24**(4) (1985), pp. 65–102 (*in Japanese*).
24. M. S. Longuet-Higgins, "The statistical analysis of a random, moving surface," *Phil. Trans. R. Soc. London, Ser. A* (*966*) **249** (1957), pp. 321–387.
25. M. S. Longuet-Higgins, "On the joint distribution of the periods and amplitudes of sea waves," *J. Geophys. Res.* **80**(18) (1975), pp. 2688–2694.
26. Y. Goda, "The observed joint distribution of periods and heights of sea waves," *Proc. 16th Int. Conf. Coastal Engrg.* (Hamburg, ASCE, 1978), pp. 227–246.
27. C. L. Bretschneider, "Wave variability and wave spectra for wind-generated gravity waves," *U. S. Army Corps of Engrg, Beach Erosion Board, Tech. Memo.* (113) (1959), 192p.

28. A. G. Cavanié, A. Arhan and R. Ezraty, "A statistical relationship between individual heights and periods of sea waves," *Proc. BOSS' 76* **II** (Trondheim, 1976), pp. 354–360.

29. M. S. Longuet-Higgins, "On the joint distribution of wave periods and amplitudes in a random wave field," *Proc. R. Soc. London, Ser. A* **389** (1983), pp. 241–258.

30. Y. Goda and K. Nagai, "Investigation of the statistical properties of sea waves with field and simulation data," *Rept. Port and Harbour Res. Inst.* **13**(1) (1974), pp. 3–37 (*in Japanese*).

31. N. E. Huang and S. R. Long, "An experimental study of the surface elevation probability distribution and statistics of wind-generated waves," *J. Fluid Mech.* **101** (1980), pp. 179–200.

32. N. Mori and T. Yasuda, "Weakly non-Gaussian model of wave height distribution for random wave train, *Proc. 16th Int. Conf. Offshore Mech. Arctic Engrg. (OMAE)* **II** (1997), pp. 99–104.

33. Y. Goda, "A unified nonlinearity parameter of water waves," *Rept. Port and Harbour Res. Inst.* **22**(3) (1983), pp. 3–30.

34. Y. Goda, "Reanalysis of regular and random breaking wave statistics," *Coastal Engineering Journal* **52**(1) (2010), pp. 71–106.

35. S. Hotta and M. Mizuguchi, "A field study of waves in the surf zone," *Coastal Engineering in Japan* **23** (1980), pp. 59–79.

36. S. Hotta and M. Mizuguchi, "Statistical properties of field waves in the surf zone," *Proc. 33rd Japanese Coastal Eng. Conf.* (JSCE, 1986), pp. 154–157 (*in Japanese*).

37. B. A. Ebersole and S. A. Hughes, "DUCK85 photopole experiment," *US Army Corps of Engrs., WES, Misc. Paper* (CERC-87-18) (1987) pp. 1–165.

38. M. S. Longuet-Higgins, "On the distribution of the heights of sea waves: Some effects of nonlinearity and finite band width," *J. Geophys. Res.* **85**(C3) (1980), pp. 1519–1523.

39. D. Myrhaug and S. P. Kjeldsen, "Parametric modelling of joint probability density functions for steepness and asymmetry in deep water waves," *Applied Ocean Res.* **6**(4) (1984), pp. 207–220.

40. D. Myrhaug and S. P. Kjeldsen, "Steepness and asymmetry of extreme waves and the highest waves in deep water," *Ocean Engrg.* **13**(6) (1986), pp. 187–224.

41. Y. Goda, "Effect of wave tilting on zero-crossing wave heights and periods," *Coastal Engineering in Japan* **29** (1986), pp. 79–90.

42. M. A. Tayfun, "Nonlinear effects on the distribution of crest-to-trough wave heights," *Ocean Engrg.* **10**(2) (1983), pp. 97–106.

43. K. G. Nolte and F. S. Hsu, "Statistics of larger waves in a sea state," *Proc. ASCE* **105**(WW4) (1979), pp. 389–404.

44. T. Honda and H. Mitsuyasu, "On the joint distribution of the heights and periods of ocean waves," *Proc. 25th Japanese Conf. Coastal Engrg.* (1978), pp. 75–79 (*in Japanese*).

45. L. J. Tick, "Nonlinear probability models of ocean waves," *Ocean Wave Spectra* (Prentice-Hall, 1963), pp. 163–169.

46. T. Hamada, "The secondary interactions of surface waves," *Rept. Port and Harbour Res. Inst.* (10) (1965), 28p.

47. K. Hasselmann, "On the non-linear energy transfer in a gravity wave spectrum," *J. Fluid Mech.* **12** (1962), pp. 481–500, *Ibid.,* **15** (1963), pp. 273–281 and pp. 385–398.

48. A. Masuda, Y. Kuo and H. Mitsuyasu, "On the dispersion relation of random gravity waves. Part 1. Theoretical framework," *J. Fluid Mech.* **92** (1979), pp. 717–730.

49. H. Mitsuyasu, Y. Kuo and A. Masuda, "On the dispersion relation of random gravity waves. Part 2. An experiment," *J. Fluid Mech.* **92** (1979), pp. 731–749.

50. J. N. Sharma and R. G. Dean, "Development and evaluation of a procedure for simulating a random directional second order sea surface and associated wave force," *Ocean Eng. Rept. No. 20, Dept. Civil Eng., Univ. of Delaware* (1979), 139p.

51. H. Tuah and R. T. Hudspeth, "Finite water depth effects on nonlinear waves," *J. Waterway, Port, Coastal, and Ocean Engrg.* **111** (1985), pp. 401–416.

52. N. Hashimoto, T. Nagai, K. Sugawara, M. Asai and K.-S. Park, "Conversion of wave pressure records to surface profiles in consideration of wave directionality and weakly nonlinear effect," *Rept. Port and Harbour Res. Inst.* **31**(3) (1993), pp. 27–51 (*in Japanese*).

53. H, Kato, N. Nobuoka and Y. Komatsuzaki, "Properties of the second order waves in numerical simulation for coastal area," *Annual J. Coastal Engrg, JSCE* **51** (2004), pp. 156–160 (*in Japanese*).

54. H, Kato and N. Nobuoka, "Properties of the second order waves in numerical simulation for coastal area (2)," *Annual J. Coastal Engrg, JSCE* **52** (2005), pp. 136–140 (*in Japanese*).

55. S. E. Sand, "Wave grouping described by bounded long waves," *Ocean Engrg.* **9**(6) (1982), pp. 567–580.

56. S. E. Sand, "Long waves in directional seas," *Coastal Engrg.* **6**(3) (1982), pp. 195–208.

57. A. Kimura, "Average two-dimensional low-frequency wave spectrum of wind waves," *Commun. Hydraulics, Dept. Civil Engrg., Delft Univ. Tech., Rept.* (84-3) (1984), 54p.

58. M. Okihiro, R. T. Guza and R. J. Seymour, "Bound infragravity waves," *J. Geophys. Res.* **97**(C7) (1992), pp. 11453–11469.

59. M. G. Kendall and A. T. Stuart, *The Advanced Theory of Statistics* **1** 3rd ed., (Griffin, 1969), p. 243.

60. Kendall and Stuart, *loc. cit.* p. 236.

61. M. J. Tucker, "The analysis of finite-length records of fluctuating signals," *Brit. J. Applied Phys.* **8** (Apr. 1957), pp. 137–142.

62. A. G. Cavanié, "Evaluation of the standard error in the estimation of mean and significant wave heights as well as mean period from records of finite length," *Proc. Int. Conf. Sea. Climatology* (Édition Technip, Paris) (1979), pp. 73–88.

63. Y. Goda, "Statistical variability of sea state parameters as a function of a wave spectrum," *Coastal Engineering in Japan* **31**(1) (1988), pp. 39–52.

# Chapter 11

# Techniques of Irregular Wave Analysis

## 11.1 Statistical Quantities of Wave Data

### 11.1.1 *Analysis of Analog Data*

Many wave recorders at present are equipped with electronic units which convert analog signals of the surface wave profile into digital signals and record them on a real-time basis. These signals are later processed by computers to yield information of the wave heights and periods as well as the spectra. Old types of wave recorders, however, processed information of the wave profile as an analog signal and registered it on a strip-chart. These records needed to be analyzed manually. Because such conventional wave recorders are still in some use, the analysis techniques for analog data are discussed in this subsection.

The first operation to be performed on a strip-chart record is to draw the line representing the mean water level. This is done by visual judgment. If a gradual variation in the mean water level is observable owing to the effects of tidal variations or other factors, a sloping straight line or even a curve can be fitted to the data. Then individual waves are defined by the zero-upcrossing (or zero-downcrossing) method described in Sec. 2.1.2. The zero line of the mean water level which was set initially must be observed strictly, as every crossing of the zero line must be counted as one wave. Then the heights and periods of the individual waves are read off the records, and the results are written on a sheet such as that in Table 2.1. The heights and periods of the highest wave, one-tenth highest wave, significant wave, and mean wave are calculated on the basis of such a table. If information on only the sea state is sought, the calculation may be limited to the significant wave height and period, because these quantities are the representative parameters among the various characteristic wave heights and periods.

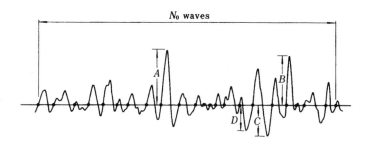

Fig. 11.1   Definition sketch for application of the Tucker method.

For a rapid analysis of a wave record, the method proposed by Tucker[1] may be employed. The method is based on the assumption of the Rayleigh distribution of the wave height, examined in Sec. 10.1.2. Procedures of the analysis are as follows. After drawing the zero line, readings are made for the largest positive amplitude $A$, the second largest positive amplitude $B$, the largest negative amplitude $C$, and the second largest negative amplitude $D$, as illustrated in Fig. 11.1. Also, the number of zero-upcrossing waves $N_0$ contained in the wave record is counted. Next, the heights $H_1$ and $H_2$ are defined as follows:

$$H_1 = A + C, \quad H_2 = B + D. \tag{11.1}$$

Then the root-mean-square surface elevation can be estimated as

$$\eta_{\text{rms}} = \frac{1}{2}(\eta_1 + \eta_2), \tag{11.2}$$

where

$$\eta_1 = \frac{H_1}{2\sqrt{2\ln N_0}\left[1 + \frac{0.289}{\ln N_0} - \frac{0.247}{(\ln N_0)^2}\right]}, \tag{11.3}$$

$$\eta_2 = \frac{H_2}{2\sqrt{2\ln N_0}\left[1 - \frac{0.211}{\ln N_0} - \frac{0.103}{(\ln N_0)^2}\right]}. \tag{11.4}$$

Once the value of $\eta_{\text{rms}} = m_0^{1/2}$ is estimated, characteristic wave heights such as $H_{1/10}$, $H_{1/3}$ and $\overline{H}$ can be estimated by means of the relations in Table 10.1. The mean wave period is simply obtained by dividing the record length by $N_0$. The periods $T_{1/10}$ and $T_{1/3}$ may be estimated with Eq. (2.9) of Sec. 2.2.3. The values of $H_{\text{max}}$ and $T_{\text{max}}$ are directly read from the wave record.

The reliablity of the Tucker method depends on the accuracy of the fit of the Rayleigh distribution to the wave height. If the highest wave tends to appear with a height greater than predicted by the Rayleigh distribution, $\eta_{rms}$ will be overestimated. On the other hand, if the upper portion of the wave height distribution tends to appear with less probability than given by the Rayleigh distribution, $\eta_{rms}$ will be underestimated. Even if the estimate of $\eta_{rms}$ is correct on the average, the statistical variability of these estimated wave heights is large because only four quantities are employed; the coefficient of variation is about 10% or greater.

If it is required to make a more detailed analysis of a strip-chart record, such as for the wave spectra, the analog record of the wave profiles should be digitized with the aid of a digitizing machine or by manual reading of the wave profiles at intervals of 1 s or so.

## 11.1.2  *Analysis of Digital Data*

The analysis of digital wave records by a computer is made in several steps. Figure 11.2 presents an example of the flow of the data analysis. Since the flow diagram is self-explanatory, only selected portions of the data processing techniques will be discussed.

### (A) *Data length and time interval of data sampling*

The standard duration of a sea wave recording is 20 min. If the mean wave period is 10 s, a wave record with the above duration contains 120 waves. This number of waves is considered sufficient to keep the sampling variability of the characteristic wave heights and periods below an acceptable level. At the same time, changes in the sea state may become appreciable if the recording length is much longer. The duration of 20 min is employed as a compromise between the requirements of wanting low sampling variability and having a stationary sea state. Some digital recordings employ the duration of 17 min 4 s, which yields 2048 data points for a sampling rate of twice per second. This is done to economize the computation time for the spectral analysis using the fast Fourier transform algorithm. In hydraulic model testing with irregular waves, however, wave records are best made for the duration of 200 waves or longer so as to decrease the sampling variation of the wave statistics. Even for field wave observations, the duration of 30 min or longer may be desirable if the sea state is considered rather stationary, as in the case of swell arriving from a remote source.

The sampling interval for the wave profile, on the other hand, should be set as short as practicable. The standard is less than a tenth of the

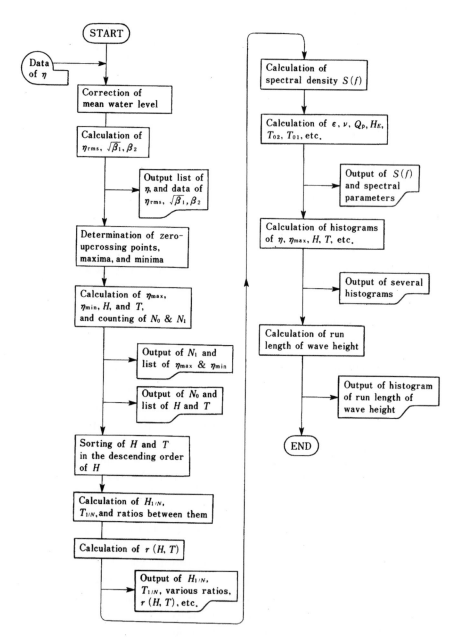

Fig. 11.2   Example of flow of wave data analysis.

significant wave period, and preferably a twentieth. A finer sampling is not considered beneficial, because the volume of the data processing increases excessively without producing a corresponding increase in the resultant information. A coarser sampling than the above standard introduces the problems of missing small waves, underestimation of the maxima and minima of the wave profiles, underestimation of wave heights, and so on.

## (B) *Correction for mean water level*

A simple procedure to determine the mean water level is to use the arithmetic mean of all the data points measured from any reference level. However, as most wave records contain the influence of the tide level variation, it is better to incorporate the following linear or parabolic correction in the analysis of wave data. The formula for the correction of linear change of the mean water level is derived from the least squares method as

$$\bar{\eta} = A_0 + A_1 n \ : \ n = 1, 2, \ldots, N \, , \tag{11.5}$$

where

$$A_0 = \frac{N_2 Y_0 - N_1 Y_1}{N_0 N_2 - N_1^2}, \quad A_1 = \frac{N_0 Y_1 - N_1 Y_0}{N_0 N_2 - N_1^2}, \tag{11.6}$$

$$N_r = \sum_{n=1}^{N} n^r, \quad Y_r = \sum_{n=1}^{N} n^r \eta_n, \tag{11.7}$$

in which $N$ denotes the total number of data points.

The formula for the correction of the mean water level for a parabolic change by the least squares method is derived as

$$\bar{\eta} = B_0 + B_1 n + B_2 n^2 \ : \ n = 1, 2, \ldots, N \, , \tag{11.8}$$

where

$$\left. \begin{array}{l} B_0 = \dfrac{1}{\Delta}[Y_0(N_2 N_4 - N_3^2) + Y_1(N_2 N_3 - N_1 N_4) + Y_2(N_1 N_3 - N_2^2)] \, , \\[2mm] B_1 = \dfrac{1}{\Delta}[Y_0(N_2 N_3 - N_1 N_4) + Y_1(N_0 N_4 - N_2^2) + Y_2(N_1 N_2 - N_0 N_3)] \, , \\[2mm] B_2 = \dfrac{1}{\Delta}[Y_0(N_1 N_3 - N_2^2) + Y_1(N_1 N_2 - N_0 N_3) + Y_2(N_0 N_2 - N_1^2)] \, , \\[2mm] \Delta = N_0 N_2 N_4 + 2 N_1 N_2 N_3 - N_2^3 - N_0 N_3^2 - N_1^2 N_4 \, . \end{array} \right\} \tag{11.9}$$

If a wave record indicates the presence of conspicuous surf beat or long-period oscillations in the mean water level, application of the numerical filter described in Sec. 11.5.4, is one method for their removal. In any case, correction of the mean water level is the first processing to be made on a wave record.

### (C) Analysis of zero-upcrossing points, maxima and minima

Zero-upcrossing of the wave profile is detected through the following criteria:

$$\eta_i \cdot \eta_{i+1} < 0 \quad \text{and} \quad \eta_{i+1} > 0, \tag{11.10}$$

where $\eta_i$ denotes the $i$th data point of the surface elevation after correction of the mean water level. The time of the zero-upcrossing is determined by linear interpolation between the sampling times of $\eta_i$ and $\eta_{i+1}$. The time difference from this point to the next zero-upcrossing point yields the zero-upcrossing wave period. If the zero-downcrossing method is employed, the second inequality in Eq. (11.10) is changed to $\eta_{i+1} < 0$.

The conditions defining a maximum in the wave profile are

$$\eta_{i-1} < \eta_i \quad \text{and} \quad \eta_i > \eta_{i+1}. \tag{11.11}$$

It is suggested that the time and the elevation of the maximum point be estimated by fitting a parabolic curve to the three points $\eta_{i-1}$, $\eta_i$ and $\eta_{i+1}$ in order to eliminate the problem of underestimating the true maximum between two discrete sampling points. The formula for a parabolic fitting is

$$\eta_{\max} = C - B^2/4A \quad \text{and} \quad t_{\max} = t_i - \Delta t B/2A, \tag{11.12}$$

where $\Delta t$ is the data sampling interval and

$$A = \frac{1}{2}(\eta_{i-1} - 2\eta_i + \eta_{i+1}), \quad B = \frac{1}{2}(\eta_{i+1} - \eta_{i-1}), \quad C = \eta_i. \tag{11.13}$$

In order to determine the zero-upcrossing wave height, the highest point on the surface elevation must be searched for in the time interval between two zero-upcrossing points. Once this point is found among the sampled points, it is designated as $\eta_i$, and then $\eta_{\max}$ is estimated by means of Eqs. (11.12) and (11.13) by use of the neighboring data points $\eta_{i-1}$ and $\eta_{i+1}$. The lowest surface elevation $\eta_{\min}$ is obtained by a similar process, and the wave height is calculated as the sum of the absolute values of $\eta_{\max}$ and $\eta_{\min}$. Unless the technique of parabolic fitting is employed, the data sampling interval must be quite narrow to avoid an artificially introduced decrease in wave height derived from digitized data.

(D) *Calculation of the correlation coefficient between wave height and period*

The correlation coefficient between individual wave heights and periods, $r(H, T)$, may not have immediate practical application, but it is a useful parameter in statistical studies of sea waves. For example, Goda[2,3] shows that the correlation coefficient is related to the distribution of wave periods, the ratios $T_{1/3}/\overline{T}$ and $\overline{T}/T_p$, and other quantities. Examination of the correlation coefficient among the higher one-third waves may also turn out to be useful in future studies.

In the analysis of wave grouping, the correlation coefficient between successive wave heights is a governing parameter as presented in the theory of Kimura.[4] Thus, it is recommended that this correlation coefficient be incorporated in standard programs of wave data analysis.

(E) *Calculation of spectral parameters*

After the frequency spectrum of the sea waves is analyzed as through the procedure described in the next section, the spectral width parameters $\varepsilon$ and $\nu$, the spectral shape parameter $\kappa$, and the spectral peakedness $Q_p$ can be calculated by Eqs. (10.25), (10.95), (10.57) and (10.74), respectively. These parameters are utilized in the comparison of statistical theories of sea waves with the data obtained. The spectral width parameter $\varepsilon$ can be estimated by Eq. (10.131) with the numbers of zero-upcrossings and maxima in the wave profiles, and the resultant value can be compared with the spectrally-estimated value.

From the estimate of the frequency spectral density, the mean wave periods $T_{02}$, $T_{01}$, and $T_{m-1,0}$ are obtained from Eqs. (10.82), (10.101), and (2.45), respectively. The various wave heights can be estimated from the relations in Table 10.1 by using the value $\eta_{\rm rms} = m_0^{1/2}$. These heights are denoted as $H_E$ in Fig. 11.2. As discussed in Sec. 10.5.3, attention should be paid to the presence of nonlinear spectral components in the high frequency range in the calculation of these spectrum-based parameters.

(F) *Frequency distribution of surface elevations, wave heights and periods*

The values of $\eta_i$, $\eta_{\max}$, $H$ and $T$ are conveniently tabulated in the form of frequency distribution. It is also recommended that the joint frequency distribution of $H$ and $T$ be prepared. Frequency distributions provide a convenient means for comparison of wave data with statistical theories, as described in Chapter 10. The distributions are displayed by using absolute values or by nondimensional forms normalized with the mean values. When a large number of wave data sets are analyzed statistically, the latter method is employed. The counting of frequencies in each class is facilitated by using the following algorithm; first, use a constant class interval $\Delta H$, convert the result of the operation $(H_i + \Delta H)/\Delta H$ into an integer, and then add the count of 1 to the class with the same order number as the above integer.

## 11.2  Frequency Spectral Analysis of Irregular Waves

### 11.2.1  *Theory of Spectral Analysis*

(A) *Fourier series representation of wave profiles*

Equation (9.16) is the fundamental equation for the analysis of the frequency spectrum of the irregular wave profile recorded at a fixed station, and it defines the spectral density function. This equation, however, contains an infinite number of amplitudes $a_n$ of the component waves, and it is not applicable to practical computations. In this section, we consider the situation of wave profiles given in the sequential form of $N$ data points $\eta(\Delta t), \eta(2\Delta t), \ldots, \eta(N\Delta t)$ sampled at a constant interval $\Delta t$, and we try to estimate the associated spectral density function. To simplify the explanation, $N$ is taken to be an even number.

By making a harmonic analysis of the time series of the wave profile $\eta(t)$, the profile can be expressed as the following finite Fourier series:

$$\eta(t_*) = \frac{A_0}{2} + \sum_{k=1}^{N/2-1} \left( A_k \cos \frac{2\pi k}{N} t_* + B_k \sin \frac{2\pi k}{N} t_* \right) + \frac{A_{N/2}}{2} \cos \pi t_* ,$$

$$(11.14)$$

where

$$t_* = t/\Delta t : t_* = 1, 2, \ldots, N .$$

Equation (11.14) gives the same surface elevations as the original wave profile $\eta(n\Delta t)$ at the points of $t_* = 1, 2, \ldots, N$, if the Fourier coefficients are determined as

$$A_k = \frac{2}{N} \sum_{t_*=1}^{N} \eta(t_*) \cos \frac{2\pi k}{N} t_* \ : \ 0 \le k \le N/2 \,, \tag{11.15}$$

$$B_k = \frac{2}{N} \sum_{t_*=1}^{N} \eta(t_*) \sin \frac{2\pi k}{N} t_* \ : \ 1 \le k \le N/2 - 1 \,. \tag{11.16}$$

(B) *Variance of Fourier coefficients*

Although a harmonic analysis provides definite values for the Fourier amplitudes of a deterministic process, it yields only a probabilistic answer for a stochastic process. Because sea waves form a random process, different values of the Fourier amplitudes $A_k$ and $B_k$ will be obtained whenever different sample records of the surface elevation $\eta(t)$ are analyzed. As described in Sec. 9.3, $\eta(t)$ can be regarded as a stationary stochastic process which follows the normal distribution. If the number of data points $N$ is sufficiently large, the amplitudes $A_k$ and $B_k$ also follow the normal distribution with a zero mean by virtue of the central limit theorem, as seen from the form of Eqs. (11.15) and (11.16). Their variances are calculated as follows.

First, $A_k^2$ is calculated for the case of $k \ne 0$ or $k \ne N/2$. The result is

$$A_k^2 = \frac{4}{N^2} \left\{ \sum_{t_*=1}^{N} \eta(t_*) \cos \frac{2\pi k}{N} t_* \right\}^2 = \frac{4}{N^2} \left\{ \sum_{t_*=1}^{N} \eta^2(t_*) \left( \cos \frac{2\pi k}{N} t_* \right)^2 \right.$$

$$\left. + 2 \sum_{\tau_*=1}^{N-1} \sum_{t_*=1}^{N-\tau_*} \eta(t_*) \eta(t_* + \tau_*) \cos \frac{2\pi k}{N} t_* \cos \frac{2\pi k}{N} (t_* + \tau_*) \right\} \,. \tag{11.17}$$

By taking the asymptotic condition of $N \to \infty$, the expected value of $A_k^2$ is calculated as

$$E[A_k^2] = \lim_{N \to \infty} \frac{2}{N^2} \left\{ \sum_{t_*=1}^{N} \eta^2(t_*) + 2 \sum_{\tau_*=1}^{N-1} \sum_{t_*=1}^{N-\tau_*} \eta(t_*) \eta(t_* + \tau_*) \cos \frac{2\pi k}{N} \tau_* \right\} \,. \tag{11.18}$$

To arrive at the above result, use was made of the fact that the expected value of $(\cos 2\pi k t_*/N)^2$ is $1/2$ and that of $(\cos 2\pi k t_*/N) \times (\sin 2\pi k t_*/N)$ is zero. By making use of the autocorrelation function defined by Eq. (9.19) in Sec. 9.3, Eq. (11.18) is rewritten as

$$E[A_k^2] = \lim_{N \to \infty} \frac{2}{N} \left\{ \Psi(0) + 2 \sum_{\tau_* = 1}^{N-1} \Psi(\tau_*) \cos \frac{2\pi k}{N} \tau_* \right\}. \qquad (11.19)$$

This equation is further rewritten as follows, by utilizing the fact that $\cos \theta$ is an even function of $\theta$:

$$E[A_k^2] = \lim_{N \to \infty} \frac{2}{N} \sum_{\tau_* = -\infty}^{\infty} \Psi(\tau_*) \cos \frac{2\pi k}{N} \tau_*$$

$$= \lim_{N \to \infty} \frac{2}{N \Delta t} \sum_{\tau = -\infty}^{\infty} \Psi(\tau)(\cos 2\pi f_k \tau) \Delta t, \qquad (11.20)$$

where

$$f_k = k/(N \Delta t) = k/t_0, \quad \tau = \tau_* \Delta t, \qquad (11.21)$$

in which $t_0$ denotes the duration of the wave record.

We recognize that the right side of Eq. (11.20) is the series representation of the right side of Eq. (9.29) for $S_0(f)$ in Sec. 9.3. Thus, the variance of $A_k$ is related to the spectral density function as

$$E[A_k^2] = \frac{2}{N \Delta t} S_0(f_k) = \frac{1}{N \Delta t} S(f_k) : 1 \le k \le \frac{N}{2} - 1, \qquad (11.22)$$

in which $S_0(f)$ is the two-sided spectral density function defined in the range $-\infty < f < \infty$ and is equal to $S(f)/2$. The variance of $B_k$ is also calculated in the same way to give

$$E[B_k^2] = \frac{1}{N \Delta t} S(f_k) : 1 \le k \le \frac{N}{2} - 1. \qquad (11.23)$$

We can also obtain the following results for $A_0$ and $A_{N/2}$:

$$E[A_0^2] = \frac{2}{N \Delta t} S(f_0), \quad E[A_{N/2}^2] = \frac{2}{N \Delta t} S(f_{N/2}). \qquad (11.24)$$

We have shown that the Fourier coefficients $A_k$ and $B_k$ are both stochastic variables which follow the normal distribution, and have a zero mean and variance $S(f)/(N \Delta t)$.[a] The correlation between $A_k$ and $B_k$ is examined below through their covariance:

$$E[A_k B_k] = \lim_{N \to \infty} \frac{4}{N^2} \left\{ \sum_{t_* = 1}^{N} \eta(t_*) \cos \frac{2\pi k}{N} t_* \right\} \left\{ \sum_{t_* = 1}^{N} \eta(t_*) \sin \frac{2\pi k}{N} t_* \right\}$$

$$= \lim_{N \to \infty} \frac{4}{N} \sum_{\tau_* = 1}^{N-1} \Psi(\tau_*) \sin \frac{2\pi k}{N} \tau_*. \qquad (11.25)$$

---

[a]The derivation given here is not mathematically rigorous. The reader is referred to Koopmans[5] or other references on statistics for details.

By recalling that $\Psi(\tau)$ can be expressed as the cosine transformation of $S(f)$ by means of Eq. (9.27), we conclude that the right side of Eq. (11.25) converges to zero. That is to say, $A_k$ and $B_k$ are statistically independent.

(C) *Estimation of spectral density*

With the above preparations, we can now proceed to the estimation of the spectral density. First, we introduce the variable $I_k$ defined by

$$
I_k = \begin{cases} N(A_k^2 + B_k^2) & : 1 \leq k \leq N/2 - 1, \\ NA_0^2 & : k = 0, \\ NA_{N/2}^2 & : k = N/2. \end{cases} \tag{11.26}
$$

Because $I_k$ is the sum of the squares of two independent stochastic variables of a normal distribution, $I_k$ is a stochastic variable which follows the chi-square distribution with two degrees of freedom. Only the $I_k$ for $k = 0$ and $N/2$ follow the chi-square distribution with one degree of freedom. The *chi-square distribution* is the distribution of stochastic variable defined by

$$
\chi_r^2 = \sum_{i=1}^{r} x_i^2/\sigma^2, \tag{11.27}
$$

where $x_i$ is a normally distributed variable with zero mean and variance of $\sigma^2$. The probability density function of $\chi_r$ is given by

$$
p(\chi_r^2) = \frac{1}{2^{r/2}\Gamma(r/2)}(\chi_r^2)^{r/2-1}e^{-\chi_r^2/2}, \tag{11.28}
$$

in which $\Gamma(\cdot)$ denotes the gamma function. The number of terms $r$ is called the degree of freedom of the chi-square distribution. By using $\chi_r^2$, the variable $I_k$ is rewritten as

$$
I_k = \begin{cases} \dfrac{1}{\Delta t}S(f_k)\chi_2^2 & : 1 \leq k \leq N/2 - 1, \\[2mm] \dfrac{2}{\Delta t}S(0)\chi_1^2 & : k = 0, \\[2mm] \dfrac{2}{\Delta t}S(f_{N/2})\chi_1^2 & : k = N/2. \end{cases} \tag{11.29}
$$

From the properties of the chi-square distribution, the expected value and variance of $I_k$ are obtained as

$$E[I_k] = \frac{2}{\Delta t} S(f_k) \,:\, 0 \le k \le N/2 \,, \tag{11.30}$$

$$\mathrm{Var}[I_k] = \frac{4}{(\Delta t)^2} S^2(f_k) \,:\, 0 \le k \le N/2 \,. \tag{11.31}$$

A plot of the variable $I_k$, defined by Eq. (11.29), against the frequency $f_k$ is called a *periodogram*. The variable $I_k$ itself is sometimes called a periodogram. Equation (11.30) indicates that the spectral density $S(f_k)$ at the frequency $f_k$ can be estimated through the expected value of the periodogram, but at the same time Eq. (11.31) indicates that the absolute value of $I_k$ varies greatly, with the magnitude of its standard deviation equal to its mean value. Furthermore, the variability does not decrease even if the data length $N$ is increased. Therefore, if we calculate the periodogram from the Fourier coefficients for a sample of wave profile $\eta(t)$, the resultant periodogram fluctuates greatly from one frequency to another, and the spectral density thus estimated for each frequency $f_k$ has low statistical reliability.

The great variability of the periodogram has been known from early times, and several techniques have been developed to suppress the fluctuation and to increase the reliability of the spectral estimate. One of the popular techniques is the *autocorrelation method* proposed by Blackman and Tukey[6] in 1958, which is based on Eq. (9.28) in Sec. 9.3. Another technique is a smoothing of the periodogram over a certain frequency band under the assumption that the spectral density function $S(f)$ varies only gradually with respect to the frequency. This is called the *smoothed periodogram method*[b] here. Smoothing is done by using one of several weight functions. In the following, a discussion is given for the case of simple averaging with uniform weight.

The formula for the simple averaging method to obtain the estimate of the spectral density $\hat{S}(f_k)$ is

$$\hat{S}(f_k) = \frac{1}{n} \sum_{j=k-[(n-1)/2]}^{k+[n/2]} I_j \frac{\Delta t}{2} \,, \tag{11.32}$$

in which $[n/2]$ denotes the largest integer not exceeding $n/2$.

---

[b]This method is usually called the *FFT method*, because the calculation of the periodogram is usually done with the fast Fourier transform technique.

If the condition

$$S(f_j) \simeq S(f_k) : k - [(n-1)/2] \leq j \leq k + [n/2] \qquad (11.33)$$

holds, then $\hat{S}(f_k)$ is the sum of $n$ chi-square variables with two degrees of freedom each, because the periodograms are mutually independent in the statistical sense. Therefore, $\hat{S}(f_k)$ follows the chi-square distribution with $2n$ degrees of freedom. That is,

$$\hat{S}(f_k) = S(f_k)\chi_{2n}^2/2n . \qquad (11.34)$$

Because the expected value of $\chi_{2n}^2$ is $2n$, the expected value $E[\hat{S}(f_k)]$ is the same as that of $S(f_k)$. Also, by increasing the data length $N$ and averaging number $n$ while keeping the condition $N \gg n$, the reliability of the spectral estimate can be increased as the variance of $\hat{S}(f_k)$ decreases inversely with increase in $n$. In fact, the coefficient of variation of the spectral estimate is obtained from the properties of the chi-square distribution as

$$\text{C.V.}[\hat{S}(f_k)] = \frac{\text{Var}[\hat{S}(f_k)]^{1/2}}{E[\hat{S}(f_k)]} = \frac{1}{\sqrt{n}} , \qquad (11.35)$$

in which C.V. and Var refer to the coefficient of variation and the variance, respectively.

## (D) *Confidence interval of spectral estimate*

A quantitative evaluation of the reliability of the spectral estimate is usually made by means of the confidence interval based on the chi-square distribution. That is, we estimate the interval in which the value of the true spectral density lies within a certain range corresponding to a preset level of probability. For example, suppose we have obtained a spectral estimate by averaging 20 periodograms and try to estimate the true spectral density with a 95% confidence interval. The number of degrees of freedom is $2n = 40$ for this case. The upper limit of the 95% confidence interval is obtained from the value of the chi-square variable that satisfies the condition of $P(\chi_{40}^2 \leq a) = 0.975$. From a probability table of the chi-square distribution, we obtain $a = 59.34$ or $a/2n = 1.48$. Similarly, the lower limit is obtained as $b/2n = 0.61$ for the condition of $P(\chi_{40}^2 \leq b) = 0.025$. Therefore, the true spectral density is estimated to lie in the interval from 0.61 to 1.48 times the observed spectral density. If a spectral estimate is obtained by averaging 40 periodograms, the lower and upper limits of the 95% confidence interval become 0.71 and 1.33 times the estimated values, respectively.

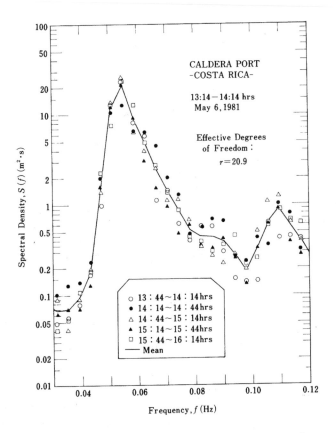

Fig. 11.3 Variations of independent spectral estimates.[7]

Owing to such variability in the estimated spectral density, the true spectrum of sea waves plotted against the frequency may take any shape within a confidence interval above and below the wave spectrum estimated from one sample of wave records. Figure 11.3 shows an example of the sampling variability of estimated spectra as reported by Goda.[7] A continuous record of swell for the duration of 2 hr 30 min was divided into five segments of 30 min length, and the wave spectrum was estimated for the five segments with an effective number of degrees of freedom of 20.9 (the parabolic smoothing function, Eq. (11.48) in the next section, was employed). The five spectral estimates are seen to be scattered around the mean, though the mean does not represent the true spectrum. The mean spectrum has a narrower band of reliability than individual estimates because of the increase in the number of degrees of freedom. A similar analysis of the sampling variability of spectral estimates of field and laboratory wind waves has been reported by Donelan and Pierson,[8] who demonstrated the accuracy of the spectral variability theory described above.

The reliability of the spectral estimate can be improved by increasing the number of periodograms to be averaged together and by increasing the number of degrees of freedom in the spectral estimate. However, this causes a decrease in spectral resolution. The term *"spectral resolution"* refers to the capability to distinguish two neighboring spectral peaks, and its measure is the minimum frequency distance between two independent spectral estimates. This distance is called the bandwidth. In the case of smoothing by simple averaging as in Eq. (11.32), the bandwidth is given by

$$f_B = f_{k+n} - f_k = \frac{n}{N\Delta t} = \frac{n}{t_0}. \tag{11.36}$$

The spectral resolution is high when the bandwidth $f_B$ is narrow. The denominator $t_0$ on the right side of Eq. (11.36) is the duration of the wave record. When $t_0$ is fixed, $f_B$ is linearly proportional to $n$, and therefore the resolution is inversely proportional to $n$. On the other hand, the reliability increases with $n$. Therefore, a simultaneous improvement in the reliability and resolution of the spectral estimate cannot be made for a wave record of fixed length. This situation is called the *Grenander uncertainty principle* (see Ref. 9). It should be mentioned here that the minimum possible bandwidth is $(f_B)_{\min} = 1/t_0$, which represents a component wave with period equal to the record length.

### 11.2.2 *Spectral Estimate with Smoothed Periodograms*

As discussed in the previous section, the autocorrelation method and the smoothed periodogram method are both employed for the estimation of the spectral density from a given record of a wave profile. But their theoretical basis is common, and their spectral resolutions and reliabilities are essentially the same. Although differences exist in their computation procedures, Rikiishi[10] has shown that the differences are essentially those of smoothing functions. Since the fast Fourier transform algorithm is available on almost all computers these days, a description and comments will be given on the practical procedures of making spectral estimates with the smoothed periodogram method.

## (A) *Record length and data sampling interval*

A wave record should be as long as possible, because the spectral resolution is governed by the record length as given by Eq. (11.36). The data sampling interval of $1/10$ to $1/20$ of the significant wave period is generally recommended for a wave analysis, as discussed in Sec. 11.1. Once the sampling interval $\Delta t$ is chosen, the highest frequency to which the spectrum can be estimated is automatically determined by

$$f_c = 1/2\Delta t. \tag{11.37}$$

This frequency is called the *folding frequency* or the Nyquist frequency, and the symbol $f_N$ is sometimes employed. The wave component with this frequency is sampled at the rate of two data points per wave. The word "folding" is used, because any wave energy contained in the frequency range $f > f_c$ is added to the wave energy in the range $0 < f < f_c$ in such a manner that the wave spectrum is folded with the axis at the frequency $f_c$. This phenomenon is called *aliasing*. In the case of sea waves, the spectral density attenuates in the high frequency range in proportion to $f^{-5}$, and thus aliasing does not present a serious problem in a spectral analysis if the data sampling interval is selected as discussed in Sec. 11.1.2. In the neighborhood of $f = f_c$, however, the spectral estimate may be increased slightly by the effect of aliasing.

## (B) *Correction for mean water level*

Before proceeding into spectral analysis, wave profile data must be so adjusted that the mean water level should become equal to zero. If the mean value of the surface elevation is not properly adjusted, or a possible gradual trend in the change of mean water level is not corrected, the spectral estimate near $f = 0$ appears large; this effect extends to other frequency ranges, too.

## (C) *Data window*

A wave profile data after making the correction for the mean water level is usually subjected to an operation of data modification before the harmonic analysis. The operation is expressed as

$$\eta(t_*) \to b(t_*)\eta(t_*) \,:\, t_* = 1, 2, \ldots, N. \tag{11.38}$$

The set of coefficients $b(t_*)$ is generally called the *data window*. In spectral analysis, a data window is usually designed to attenuate the front and rear portions of the wave record, and such a window is called a *taper*. Koopmans[11] cites the following two important tapers:

(a) Trapezoidal taper:

$$
b_1(t_*) = \begin{cases} t_*/l & : 0 \le t_* < l, \\ 1 & : l \le t_* \le N - l, \\ (N - t_*)/l & : N - l < t_* \le N. \end{cases} \tag{11.39}
$$

(b) Cosine taper:

$$
b_2(t_*) = \begin{cases} \dfrac{1}{2}[1 - \cos \pi t_*/l] & : 0 < t_* < l, \\ 1 & : l \le t_* \le N - 1, \\ \dfrac{1}{2}[1 - \cos \pi (N - t_*)/l] & : N - l < t_* \le N. \end{cases} \tag{11.40}
$$

The data window has the function of suppressing the amount of leakage of the sharply peaked spectral energy to the spectral estimates at neighboring frequencies. A window functions most effectively when the wave record contains single-frequency components of regular waves with appreciable energies. Kuwajima and Nagai[12] recommend the cosine taper with $l = 0.1N$ on the bais of numerical studies. A demerit of the data window is the decrease in the number of degrees of freedom, the rate of which is estimated by the following formula, according to Koopmans[11]:

$$
\frac{1}{\kappa_b} = \frac{\left[ \displaystyle\int_0^N b^2(t_*)dt_* \right]^2}{\displaystyle\int_0^N b^4(t_*)dt_*}. \tag{11.41}
$$

The rate of decrease is about the same for the trapezoidal and cosine tapers; for $l = 0.1N, 1/\kappa_b$ becomes about 0.9. In consideration of such a decrease in the number of degrees of freedom, Rikiishi and Mitsuyasu[13] discourage the use of a data window for records of sea waves with continuous spectra.

When a data window is applied to a wave record, the total energy level is decreased and all the values of the spectral estimate appear at a level lower than the true values. Therefore, the calculated values of periodograms must be corrected by multiplying them by the ratio $N/\Sigma b^2(t_*)$.

(D) *Computation of Fourier coefficients*

This is done with the fast Fourier transform (FFT) technique, which is one of the service software routines provided for modern computers. A limitation of the standard FFT algorithm is that the data length must be a power of 2, that is, $N = 2^m$. Thus, the recording of wave data is done to obtain this condition, or the data length is later adjusted by cutting the end of the record or adding a certain number of zero data. Addition of zero data is made to the tail of the record after the application of the data window, if it is to be applied. The addition of zero data decreases the total energy level as well as the number of degrees of freedom. The correction for the latter decrease is made by treating the addition of zero data as equivalent to the application of a data window of $b(t_*) = 0$ beyond the original data length in Eq. (11.43) below.

Kuwajima and Nagai[12] have proposed an extended algorithm for the FFT technique for a record of length $N = 2^m \times M$, with $M$ being an odd integer.

(E) *Calculation of periodogram*

After the Fourier coefficients $A_k$ and $B_k$ have been obtained, $I_k$ is calculated via Eq. (11.26), and corrections are made to the total energy level for the effects of the data window and adjustment of the number of data. That is,

$$I_k = \alpha(A_k^2 + B_k^2), \tag{11.42}$$

in which

$$\alpha = N_2/NU, \quad U = \left\{ \sum_{t_*=1}^{N_2} b^2(t_*) \right\} \Big/ N_2, \tag{11.43}$$

and $N_2$ denotes the number of data employed in the computation of the fast Fourier transform including zero data if they were added to the original wave record. In the calculation of Eq. (11.42), the coefficient $B_k$ is set to 0 for $k = 0$ and $N_2/2$.

(F) *Smoothing of periodograms*

Equation (11.32) represents the simple averaging method for the smoothing of periodograms. Koopmans[14] gives a general formula for the smoothing of periodograms for the estimation of spectral density as

$$\hat{S}(f_k) = \frac{\Delta t}{2} \sum_{j=k-[(n-1)/2]}^{k+[n/2]} K(f_k - f_j)\, I_j\,, \qquad (11.44)$$

in which

$$\sum_{j=k-[(n-1)/2]}^{k+[n/2]} K(f_k - f_j) = 1\,. \qquad (11.45)$$

The smoothing function $K(f)$ is also called the *weight function*, or simply the *filter*. The function $K(f)$ is equivalent to the spectral window employed in the autocorrelation method of spectral analysis.

Commonly used filters are as follows:

(a) Rectangular filter:

$$K_1(f_j) = \frac{1}{n}\ :\ -[(n-1)/2] \le j \le [n/2]\,, \qquad (11.46)$$

(b) Triangular filter:

$$K_2(f_j) = \frac{1}{\overline{K}_2}\left\{1 - \frac{|j|}{[(n-1)/2]}\right\}\ :\ -[(n-1)/2] \le j \le [n/2]\,, \quad (11.47)$$

(c) Parabolic filter:

$$K_3(f_j) = \frac{1}{\overline{K}_3}\left\{1 - \left(\frac{j}{[(n-1)/2]}\right)^2\right\}\ :\ -[(n-1)/2] \le j \le [n/2]\,,$$

$$\qquad (11.48)$$

in which $\overline{K}_2$ and $\overline{K}_3$ are the normalization constants to satisfy the condition of Eq. (11.45).

In practice, the number of periodograms to be used in the smoothing is taken as an even integer for the rectangular filter, and an odd integer for the triangular and parabolic filters. The central frequency of the smoothing is usually shifted by $n$ for the rectangular filter and by $(n+1)/2$ for the other filters with a half overlapping of periodograms in the spectral estimate.

The rectangular filter has $2n$ degrees of freedom, as described previously, and the bandwidth of the spectral resolution is given by Eq. (11.36). For other filters, the effective number of degrees of freedom is calculated with the following formula, according to Koopmans[14]:

$$r = \frac{2}{\sum_{j=-[(n-1)/2]}^{[n/2]} K^2(f_j)}\,. \qquad (11.49)$$

For a large $n$, the following approximations are available:

$$r \simeq 1.5n \; : \; \text{triangular filter}, \tag{11.50}$$

$$r \simeq \frac{5}{3}n \; : \; \text{parabolic filter}. \tag{11.51}$$

When the data window described in Clause (C) is employed, the effective number of degrees of freedom is evaluated by multiplying the value given in the above with $1/\kappa_b$ of Eq. (11.41). The bandwidth of the spectral resolution is evaluated similarly, and its equivalent value is given by

$$f_B \simeq \frac{3n}{4N\Delta t} \; : \; \text{triangular filter}, \tag{11.52}$$

$$f_B \simeq \frac{5n}{6N\Delta t} \; : \; \text{parabolic filter}. \tag{11.53}$$

The triangular and parabolic filters yield spectral estimates at a frequency interval of one half that of the rectangular filter, but it should be remembered that the adjacent two spectral estimates are not statistically independent for triangular and parabolic filters.

The number of periodograms to be smoothed is selected in consideration of the effective number of degrees of freedom and the bandwidth of the spectral resolution. Although one may think that the number of degrees of freedom can be increased by narrowing the sampling interval $\Delta t$ together with an increase of $N$ and $n$, the narrowing only extends the range of the spectral analysis to higher frequencies and the resolution around the spectral peak is rather worsened by an increase in $n$. The spectral resolution is improved only through the use of a long wave record as exhibited in Eq. (11.36). Although the minimum number of degrees of freedom to be maintained in a spectral analysis cannot be stated definitely, most studies have been done with the effective number greater than about 20. In the evaluation of the spectral peakedness parameter $Q_p$ defined by Eq. (10.69), Rye[15] recommend that the number of degrees of freedom be equal to about 16, because the evaluation of $Q_p$ requires quite fine resolution of the spectral peak.

The number of the periodograms to be smoothed is not necessarily a fixed value, and it can be varied over the range of frequency. Examination of low frequency spectral components is sometimes made with a small value of $n$ and a narrow bandwidth of the spectral resolution, but it should be remembered that low reliability of the spectral estimate will result in such an analysis.

## (G) *Final adjustment of energy level*

The adjustment of the total energy level for the effect of the data window may not be perfect when Eq. (11.43) alone is applied, because of some particular feature in the wave record, some residues arising from the treatment of overlapping of the periodograms in smoothing, especially for the lowest and highest frequencies, differences originating from the use of an uneven bandwidth for the spectral resolution, and other reasons. Thus, the zeroth moment of the estimated spectrum $m_0$ may be slightly different from the variance of the surface elevation $\eta_{\text{rms}}^2$, which was obtained in the initial step of the data analysis. A difference between $m_0$ and $\eta_{\text{rms}}^2$ is undesirable from consideration of the equality condition prescribed by Eq. (9.34). Therefore, the spectral estimates should be adjusted by multiplying all estimates by the ratio $\eta_{\text{rms}}^2/m_0$.

## 11.3 Directional Spectral Analysis of Random Sea Waves

As discussed in Chapter 3, wave transformations such as diffraction, refraction, and reflection are greatly influenced by the directional spectral characteristics of sea waves. Thus, we need to enrich our knowledge of directional wave spectra through the accumulation of a large amount of directional spectral data. However, measurements of directional wave spectra necessitate an effort several times greater than those of frequency spectra. While the latter can be obtained from a wave record at a single point, the former requires simultaneous recording of several wave components. With reference to the classification by Panicker,[16] the measuring techniques of directional wave spectra which have been tried so far are listed below.

(1) Direct measurement method

$$\left\{ \begin{array}{l} \text{Wave gauge array} \\ \text{Directional buoy} \\ \text{Two-axis current meter} \end{array} \right.$$

(2) Remote sensing method

$$\left\{ \begin{array}{l} \text{Optical technique} \left\{ \begin{array}{l} \text{Stereophotogrametry} \\ \text{Holography} \end{array} \right. \\ \text{Microwave technique} \end{array} \right.$$

Among the above, the direct measurement methods and stereophotogrametry are based on the same principle of analysis, namely digital analysis of the cross-spectra between various pairs of wave records. The hologram method[17] is an analog-type analysis of an aerial photograph by means of the diffraction pattern which is produced by exposing a negative film to a laser beam. Microwave techniques are more recent, and several approaches are being made. A synthetic aperture radar, mounted on a satellite or an airplane, is one promising method. King and Shemdin[18] reported a measurement of the distribution of waves in a hurricane, taken by means of air-borne synthetic aperture radar. A shore-based HF doppler radar has a also been employed for the measurement of directional waves, as reported by Vesecky *et al.*[19] Rapid progress is expected in the field of microwave techniques for directional wave measurements.

In the present section, however, the basic theories of direct measurement methods are described, following the chronological order of theoretical developments.

### 11.3.1 *Relation Between Direction Spectrum and Covariance Function*

The basic quantity used to estimate the directional spectrum is the covariance function[c] of wave profiles in the spatial and time domains. By generalizing the definition Eq. (9.19) in Sec. 9.3 for the one-dimensional wave profile, we introduce the following covariance function:

$$\Psi(X, Y, \tau) = \lim_{x_0, y_0, t_0 \to \infty} \frac{1}{x_0 y_0 t_0} \int_{-x_0/2}^{x_0/2} \int_{-y_0/2}^{y_0/2} \int_{-t_0/2}^{t_0/2} \eta(x, y, t)$$
$$\times \eta(x + X, y + Y, t + \tau) \, dx \, dy \, dt \, . \tag{11.54}$$

By substituting Eq. (9.12), which describes the two-dimensional wave profile, into the above, and by introducing the notation $\alpha = (kx \cos \theta + ky \sin \theta - \omega t + \varepsilon)$ with $\omega = 2\pi f$ for abbreviation, we have

$$\Psi(X, Y, \tau) = \lim_{x_0, y_0, t_0 \to \infty} \frac{1}{x_0 y_0 t_0} \int \int \int \sum_{n=1}^{\infty} \sum_{m=1}^{\infty} a_n a_m \cos \alpha_n$$
$$\times \cos(\alpha_m + k_m X \cos \theta_m + k_m Y \sin \theta_m - \omega_m \tau) \, dx \, dy \, dt \, . \tag{11.55}$$

---

[c]As explained in Sec. 9.3, an autocorrelation function with zero mean is called the autocovariance function.

Because the expected value of $\cos \alpha_n \cos \alpha_m$ is 0 for $n \neq m$, the above equation can be shown to become

$$\Psi(X, Y, \tau) = \sum_{n=1}^{\infty} \frac{1}{2} a_n^2 \cos(k_n X \cos \theta_n + k_n Y \sin \theta_n - \omega_n \tau). \qquad (11.56)$$

Combination of Eq. (11.56) with the definition of the directional spectral function, Eq. (9.14), yields the following relation between the covariance function and directional spectrum:

$$\Psi(X, Y, \tau) = \int_0^{\infty} \int_0^{2\pi} S_k(k, \theta) \cos(kX \cos \theta + kY \sin \theta - \omega \tau) \, d\theta \, dk . \quad (11.57)$$

The inverse transform of the above can be obtained by utilizing the theorems of the Fourier transform and its inverse transform (e.g., Ref. 20). For this purpose, the polar coordinates $(k, \theta)$ are transformed to the Cartesian coordinates $(u, v)$, and $\omega$ is treated as an independent variable; that is, the dispersion relationship of Eq. (9.3) is not used at this stage. After some manipulation, we obtain the following relations:

$$\Psi_0(X, Y, \tau) = \int_{-\infty}^{\infty} \int_{-\infty}^{\infty} \int_{-\infty}^{\infty} S_{k_0}(u, v, \omega) e^{i(uX + vY - \omega \tau)} du \, dv \, d\omega , \qquad (11.58)$$

$$S_{k_0}(u, v, \omega) = \frac{1}{(2\pi)^3} \int_{-\infty}^{\infty} \int_{-\infty}^{\infty} \int_{-\infty}^{\infty} \Psi_0(X, Y, \tau) e^{-i(uX + vY - \omega \tau)} dX \, dY \, d\tau ,$$

$$(11.59)$$

where

$$u = k \cos \theta , \quad v = k \sin \theta . \qquad (11.60)$$

In Eqs. (11.58) and (11.59), the angular frequency $\omega$ is treated as varying independently of the wavenumber $k = |u^2 + v^2|^{1/2}$, and thus the directional spectral density is defined in the three-dimensional space of $(u, v, \omega)$. The subscript "0" on $\Psi_0$ and $S_{k_0}$ is added to indicate that the functions attached are defined in the ranges $-\infty < \tau < \infty$ and $-\infty < \omega < \infty$; these functions have values of one half of the functions defined in the range $[0, \infty)$.

If the covariance function of the two-dimensional wave profile with respect to $X, Y$ and $\tau$ is known with sufficient density in the whole domain, the directional wave spectrum can be estimated with Eq. (11.59). Information on $\Psi_0(X, Y, \tau)$ with sufficient density requires that the wave profile data be obtained uniformly in the whole area, $x = -x_0/2 \sim x_0/2$, $y = -y_0/2 \sim y_0/2$ and throughout the time span $t = -t_0/2 \sim t_0/2$. Specifically speaking, it requires several hundred consecutive stereophotographs of the sea surface taken over the same area. This is not impossible, but it is unfeasible in practice because of the excessive cost of operation and analysis. When the U.S. Navy carried out the Stereo Wave Observation Project (SWOP) in 1954, stationarity of the wave field was assumed and the directional spectrum $S_{k_0}(u,v)$ was estimated from the information of $\Psi_0(X,Y,0)$, which was calculated from a contour map of the surface elevation (see Ref. 21, for example). A shortcoming of this technique is the incapability of distinguishing values of $S_{k_0}(u,v)$ and $S_{k_0}(-u,-v)$, that is, two wave components propagating in the directions differing by 180°. Therefore, for the purpose of the analysis of the SWOP data, it was assumed that the directional wave spectrum existed only in the azimuth within $\pm 90°$ from the mean wave direction. The situation is the same for the hologram method. In the use of wave gauge arrays (except for the linear array) and wave buoys, such a problem does not exist and the directional wave spectrum can be estimated in the full directional range ($\theta = 0° \sim 360°$).

### 11.3.2  *Estimate of Directional Spectra with a Wave Gauge Array*

(A) *Direct Fourier transform method*

Stereophotogrametry for obtaining the directional wave spectrum requires a large volume of data analysis, and thus it is not suited for routine wave observation. On the other hand, the measurement of the directional wave spectrum with a wave gauge array requires only the simultaneous recording of wave profiles, and the data analysis is not as voluminous as compared to that for the stereophotogrametric method. Thus, wave gauge arrays are sometimes employed in the measurement of directional wave spectra of sea waves, though they are more popular in seismology.

In the definition of the covariance function, Eq. (11.54), the spatial lags $X$ and $Y$ will be regarded as fixed for the time being. By denoting the covariance function under this condition as $\Psi'$, it is expressed as

$$\Psi'(\tau|X,Y) = \lim_{t_0 \to \infty} \frac{1}{t_0} \int_{-t_0/2}^{t_0/2} \eta(t|x,y)\,\eta(t+\tau|x+X,y+Y)\,dt. \quad (11.61)$$

This function can be obtained from the simultaneous records of wave profiles at any two stations, whose locations are denoted by $(x,y)$ and $(x+X,\,y+Y)$. We consider the function $\Phi_0(f|X,Y)$, which is the Fourier transform of $\Psi'$, similar as in Eq. (9.27) in Sec. 9.3:

$$\Phi_0(f|X,Y) = \int_{-\infty}^{\infty} \Psi'(\tau|X,Y)\, e^{-i2\pi f\tau} d\tau\,, \qquad (11.62)$$

$$\Psi'(\tau|X,Y) = \int_{-\infty}^{\infty} \Phi_0(f|X,Y)\, e^{i2\pi f\tau} df\,, \qquad (11.63)$$

in which $\Phi_0$ is defined in the range $-\infty < f < \infty$.

The function $\Phi_0(f|X,Y)$ is called the cross-spectrum, and it is often expressed in terms of its real and imaginary parts as

$$\Phi_0(f|X,Y) = C_0(f|X,Y) - iQ_0(f|X,Y) : -\infty < f < \infty\,, \qquad (11.64)$$

where

$$C_0(f|X,Y) = \int_{-\infty}^{\infty} \Psi'(\tau|X,Y)\cos 2\pi f\tau\, d\tau\,, \qquad (11.65)$$

$$Q_0(f|X,Y) = \int_{-\infty}^{\infty} \Psi'(\tau|X,Y)\sin 2\pi f\tau\, d\tau\,. \qquad (11.66)$$

The real part is generally called the *co-spectrum*, and the imaginary part is called the *quadrature-spectrum*.

From the definition given in Eq. (11.61), the following property of $\Psi'$ is derived:

$$\Psi'(\tau|X,Y) = \Psi'(-\tau|-X,-Y)\,. \qquad (11.67)$$

In general, $\Psi'(\tau|X,Y)$ is not equal to $\Psi'(-\tau|X,Y)$. For the cross-spectrum, there exist the following relations:

$$C_0(f|X,Y) = C_0(f|-X,-Y) = C_0(-f|-X,-Y) = C_0(-f|X,Y)\,, \qquad (11.68)$$

$$Q_0(f|X,Y) = -Q_0(f|-X,-Y) = Q_0(-f|-X,-Y) = -Q_0(-f|X,Y)\,. \qquad (11.69)$$

We can prove from these relations that the covariance function $\Psi'$ is a real function, even though the cross-spectrum $\Phi_0$ is a complex function.

The covariance function $\Psi'$ corresponds to $\Psi_0$ on the right side of Eq. (11.59) if the triple integration of the inverse Fourier transform is first performed with respect to $\tau$. This correspondence is based on the definitions in Eqs. (11.54) and (11.61). Thus, by performing the integration first with respect to $\tau$ in Eq. (11.59) by utilizing Eq. (11.62), we have the result

$$S_{k_0}(u,v|f_0) = \frac{1}{(2\pi)^2} \int_{-\infty}^{\infty}\int_{-\infty}^{\infty} \Phi_0^*(X,Y|f_0)\, e^{-i(uX+vY)} dX\, dY\,, \qquad (11.70)$$

where $\Phi_0^*$ is the conjugate function of the cross-spectrum $\Phi_0$; i.e., $\Phi_0^* = C_0 + iQ_0$. Note that the frequency $f_0$ is fixed in this integration.

Equation (11.70) indicates that the directional wave spectrum $S_{k_0}(u, v, f)$ can be estimated if wave profiles are simultaneously recorded at a large number of points in an area with sufficient density. From such records, all possible pairs of wave gauges are formed. The distances are denoted as $(0, 0)$, $(X_1, Y_1)$, $(X_2, Y_2)$, .... The cross-spectra between these pairs are calculated, and directional spectrum is estimated by replacing the integrations in Eq. (11.70) with summations. In practice, however, the number of wave gauges is rather limited and the integrations cannot be evaluated accurately.

For this situation, Barber[22] proposed that the integral in Eq. (11.70) be replaced by the following summation as a way of estimating the directional spectrum. He assigned the value of zero to the cross-spectrum at any pair of arbitrary distances $X$ and $Y$ other than those realized among the actual pairs of wave gauges. That is,

$$\hat{S}_{k_0}(u, v | f_0) = \frac{1}{(2\pi)^2} \sum_{n=-M}^{M} \Phi_0^*(X_n, Y_n | f_0) e^{-i(uX_n + vY_n)}, \qquad (11.71)$$

where $M$ denotes the number of pairs of wave gauges and is given by $M = N(N - 1)/2$ for an array of $N$ wave gauges. The above equation can be written in terms of real variables alone by utilizing Eqs. (11.68) and (11.69):

$$\hat{S}_{k_0}(u, v | f_0) = \frac{1}{(2\pi)^2} \left\{ C_0(0, 0 | f_0) + 2 \sum_{n=1}^{M} [C_0(X_n, Y_n | f_0) \cos(uX_n + vY_n) \right.$$
$$\left. + Q_0(X_n, Y_n | f_0) \sin(uX_n + vY_n)] \right\}. \qquad (11.72)$$

The above equation enables estimation of the directional spectral function in the wavenumber domain $(u, v)$ in the frequency range $-\infty < f_0 < \infty$. The directional wave spectrum $S(f, \theta)$ in the frequency and direction domain such as defined by Eq. (9.13) in Sec. 9.2 can be estimated by the following procedure. First, the dispersion relationship, Eq. (9.3), is assumed to hold between $f_0$ and $k = |u^2 + v^2|^{1/2}$. Thus, we have

$$S(f_0, \theta) = \alpha S_{k_0}(u, v | f_0), \qquad (11.73)$$

where $\alpha$ is a proportionality constant, the value of which varies with $f_0$ according to the dispersion relationship. The directional spectrum $S(f, \theta)$ is assumed to be given by the product of the frequency spectrum $S(f)$ and the directional spreading function $G(f; \theta)$, as in Eq. (2.20) in Sec. 2.3.2. That is,

$$S(f_0, \theta) = S(f_0)\, G(\theta | f_0).$$  (11.74)

The co-spectrum $C_0(0, 0 | f_0)$ appearing on the right side of Eq. (11.72) satisfies the relation $C_0(0, 0 | f_0) = S(f_0)/2$, as proved by comparison of Eqs. (11.65) and (9.28). Thus, the estimate of $S(f_0)$, denoted with a caret, is obtained as

$$\hat{S}(f_0) = 2C_0(0, 0 | f_0).$$  (11.75)

The estimate of the directional spreading function $G(\theta | f_0)$ is then given by

$$\hat{G}_1(\theta | f_0) = \alpha' \left\{ 1 + 2 \sum_{n=1}^{M} [C_*(X_n, Y_n | f_0) \cos(k_0 X_n \cos\theta + k_0 Y_n \sin\theta) \right.$$

$$\left. + Q_*(X_n, Y_n | f_0) \sin(k_0 X_n \cos\theta + k_0 Y_n \sin\theta)] \right\},$$  (11.76)

where

$$\left. \begin{array}{l} C_*(X_n, Y_n | f_0) = C_0(X_n, Y_n | f_0)/C_0(0, 0 | f_0), \\ Q_*(X_n, Y_n | f_0) = C_0(X_n, Y_n | f_0)/C_0(0, 0 | f_0). \end{array} \right\}$$  (11.77)

The coefficient of proportionality $\alpha'$ in Eq. (11.76) is determined by the normalization condition that the integral of $\hat{G}$ in the range of $\theta = -\pi \sim \pi$ should be 1. In the normalization of the cross-spectrum by Eq. (11.77), it is recommended that the co-spectrum $C_0(0, 0 | f_0)$ be taken as the geometric mean of the data of the various pairs of wave gauges.

The above method of estimating the directional wave spectrum has been named as the *direct Fourier transform method* by Kinsman,[21] as it is based on the direct Fourier transform of the cross-spectrum between the pairs of wave gauges.

## (B) *Maximum likelihood method*

Although the direct Fourier transform method is straightforward in the logic of its derivation, its performance in the estimate of the directional spectrum is not good, particularly if the number of wave gauges is small. The directional resolution is rather dull and Eq. (11.76) yields negative values of the estimated spreading function in some ranges of direction. This problem of negative spectral density is remedied by rewriting Eq. (11.76) in the following form:

$$\hat{G}_2(\theta|f_0) = \alpha' \left\{ N + 2 \sum_{n=1}^{M} [C_*(X_n, Y_n|f_0) \cos(k_0 X_n \cos\theta + k_0 Y_n \sin\theta) \right.$$
$$\left. + Q_*(X_n, Y_n|f_0) \sin(k_0 X_n \cos\theta + k_0 Y_n \sin\theta)] \right\},$$

$$(11.78)$$

although the directional resolution is degraded.

There are other methods of estimating the directional spectrum from data of the cross-spectra. A fairly common technique is the so-called maximum likelihood method developed by Capon *et al.*[23] for the analysis of seismic waves with a sensor array. It is designed to minimize the variance of the difference between the estimate and the true spectrum under the constraint that the amplitude of unidirectional plane waves with no contamination by noise is passed without bias, as described by Pawka.[24] The formula for the estimation of the directional spectral density by maximum likelihood method is

$$\hat{G}_3(\theta|f_0) = \frac{\alpha'}{\hat{S}(f_0)} \left\{ \sum_{i=1}^{N} \sum_{j=1}^{N} \Phi_{ij}^{-1}(f_0) \right.$$
$$\left. \times \exp\left[-i(kX_{ij}\cos\theta + kY_{ij}\sin\theta)\right] \right\}^{-1}, \qquad (11.79)$$

where $X_{ij} = X_i - X_j$, $Y_{ij} = Y_i - Y_j$, and $\Phi_{ij}^{-1}$ denotes the $(i,j)$ component of the inverse matrix of the complex matrix composed of the conjugate cross-spectrum $\Phi_{ij}^* = C_{ij} + iQ_{ij}$, defined in the range $0 < f < \infty$, and $\hat{S}(f_0)$ is the mean of the frequency spectral density estimated from the records of $N$ gauges.

The maximum likelihood method has a directional resolution much higher than the direct Fourier transform method. Figure 11.4 is an example of the comparison of directional resolution by both methods for unidirectional irregular waves, as determined by four wave gauges arranged in the form of star array.[25] It shows the directional spreading function at the frequency corresponding to the relative gauge spacing $D/L = 0.47$ ($D =$ gauge spacing, $L =$ wavelength). Even though numerical noise with a root-mean-square ratio of 20% to the signal was added to the simulated irregular wave records, the maximum likelihood method result shown with the open circles yielded quite accurate estimates of the input spectrum. The direct Fourier transform method, on the other hand, gave poor spectral estimates, as shown with the crosses.

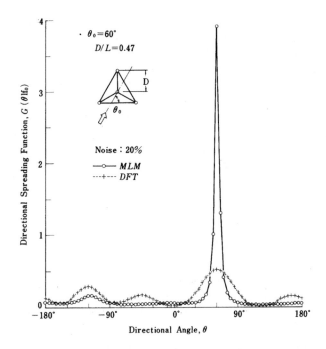

Fig. 11.4 Example of directional resolution of star array to unidirectional irregular waves by the maximum likelihood method and by the direct Fourier transform method.[25]

When the actual directional spectrum is known to have very sharp directional peaks owing to the effect of shadowing by islands in the offshore, for example, a technique for enhancing the directional resolution by the maximum likelihood method is available through an iterative procedure, as presented by Pawka.[24]

A problem associated with the maximum likelihood method is the possibility of splitting of the peak in the directional spreading function, due to the property of high directional resolution of the method. Splitting has occasionally appeared when the method was tested in numerical simulations of random wave profiles for directional spectra with relatively broad peaks, as reported by Yamaguchi and Tsuchiya,[26] for example.

## (C) *Layout of wave gauge arrays*

In the measurement procedure for the directional spectrum, the layout of the wave gauge array should be optimized to yield the best estimate of the directional spectrum with the least number of wave gauges. The problem of the optimum array layout has been investigated in the fields of radio wave detection and seismology. Guidelines for the optimum gauge array may be summarized as follows:

(1) No pair of wave gauges should have the same vector distance between gauges.
(2) The vector distance should be distributed uniformly in as wide a range as possible.
(3) The minimum separation distance between a pair of wave gauges should be less than one half of the smallest length of the component waves for which the directional analysis is to be made.

The star array shown in the inset of Fig. 11.4 satisfies the first requirement. If four gauges are arranged at the corners of a square, the two pairs of gauges at the horizontal sides of the square have the same vector distance, and so do the two pairs at the vertical sides. Thus, the number of independent pairs of wave gauges decreases to only four from six possible combinations, and the directional resolution is decreased accordingly. As discussed earlier, the estimate of the directional spectrum from information of the cross-spectra between pairs of wave gauges is based on the approximation of the integral in Eq. (11.70) by the summation of cross-spectra at a finite number of vector distances between wave gauges, as given by Eq. (11.71). The accuracy of the approximation is improved as the number of wave gauges increases and as they are arranged with a higher density. Therefore, duplication of vector distance should be avoided.

The second requirement is related to the directional resolution. When the frequency spectrum is estimated by the autocorrelation method of Blackman and Tukey,[6] the maximum value of the time lag $\tau$ governs the frequency resolution of the spectrum. Similarly, the directional resolution of the directional wave spectrum increases as the maximum distance between wave gauges increases. A uniform distribution of vector distance, on the other hand, is a requirement for homogeneity in the spectral estimate with respect to direction. A wave gauge array set along a straight line, called a *line array*, has sharp resolution in the direction normal to the line, but the resolution decreases as the direction of the wave incidence departs from the normal.

The third requirement is easily understood by considering the situation that a plane wave is incident to a pair of wave gauges having a separation distance of one half the wavelength along the direction of the axis connecting the two gauges. The wave records from the two gauges will

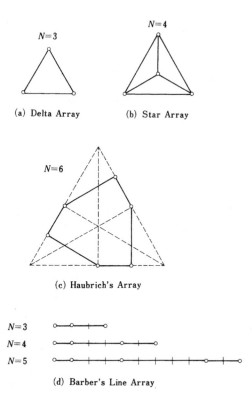

Fig. 11.5   Example of optimum gauge arrays.

show the same profile, but with a phase difference of 180°. The cross-spectrum becomes $|C_*| = 1$ and $Q_* = 0$. For any pair of wave gauges with a separation distance of a multiple of $L/2$, the cross-spectrum becomes $|C_*| = 1$ and $Q_* = 0$. Such a gauge pair cannot detect the side from which the waves are coming. There also appear false peaks in the spectral density, called side lobes, at directions different from the true direction of wave incidence.

In summary, measurement procedures for the directional spectrum with a wave gauge array have various limitations, and accurate and reliable measurements are difficult if the number of gauges is small. Figure 11.5 illustrates several gauge layouts which have been employed in directional measurements. A layout for six gauges was proposed by Haubrich.[27] The line arrays are based on Barber's proposal.[22] Although line arrays cannot distinguish waves approaching from opposite sides of the line, they may be usefully employed in measurements of nearshore waves, since most of wave energy is incident from the offshore side.

### 11.3.3    *Estimate of Directional Wave Spectra with a Directional Buoy and with a Two-Axis Current Meter*

Many directional wave measurements have been made by means of directional wave buoys and two-axis current meters. The number of such directional measurements reported in technical papers surpasses that of wave gauge arrays. An early-developed directional buoy is the *pitch and roll buoy*, reported by Longuet-Higgins *et al.*,[28] which measures the angles of pitch and roll and the heaving acceleration of the buoy. Mitsuyasu *et al.*[29] employed a cloverleaf buoy, which has the capability of measuring the curvature of the water surface in addition to the pitch, roll and heave. On the basis of their detailed measurements of directional wave spectra, and other published data, Mitsuyau *et al.* proposed a standard form for the directional spreading function, as discussed in Sec. 2.3.2. Turning to current meters, Nagata[30] measured the water particle velocity components in two Cartesian coordinates by means of a two-axis electromagnetic current meter, and he estimated the directional wave spectrum from these measurements. Recent directional wave measurements with *two-axis current meters* are accompanied by simultaneous measurement of the surface elevation or pressure fluctuation, so as to enable to distinguish from which side of the half plane the waves are arriving.

The principle of directional wave measurements by means of a buoy was given by Longuet-Higgins *et al.*[28] The quantities measurable by a pitch and roll buoy are

$$
\left.
\begin{aligned}
\xi_1 &= \eta = \sum_{n=1}^{\infty} a_n \cos(k_n x \cos\theta_n + k_n y \sin\theta_n - \omega_n t + \varepsilon_n)\,, \\
\xi_2 &= \frac{\partial \eta}{\partial x} = -\sum_{n=1}^{\infty} k_n a_n \cos\theta_n \sin(k_n x \cos\theta_n + k_n y \sin\theta_n - \omega_n t + \varepsilon_n)\,, \\
\xi_3 &= \frac{\partial \eta}{\partial y} = -\sum_{n=1}^{\infty} k_n a_n \sin\theta_n \sin(k_n x \cos\theta_n + k_n y \sin\theta_n - \omega_n t + \varepsilon_n)\,.
\end{aligned}
\right\}
$$

$$(11.80)$$

The heave displacement is obtained by twice integrating the heave acceleration with respect to time. The covariance functions between these three quantities are calculated as

$$\Psi_{12}(\tau) = \overline{\xi_1(t)\xi_2(t+\tau)} = \sum_{n=1}^{\infty} \frac{1}{2} k_n a_n^2 \cos\theta_n \sin\omega_n \tau \,,$$

$$\Psi_{13}(\tau) = \overline{\xi_1(t)\xi_3(t+\tau)} = \sum_{n=1}^{\infty} \frac{1}{2} k_n a_n^2 \sin\theta_n \sin\omega_n \tau \,, \qquad (11.81)$$

$$\Psi_{23}(\tau) = \overline{\xi_2(t)\xi_3(t+\tau)} = \sum_{n=1}^{\infty} \frac{1}{2} k_n^2 a_n^2 \cos\theta_n \sin\theta_n \cos\omega_n \tau \,.$$

Cross-spectra are calculated from these covariance functions by Eqs. (11.65) and (11.66), and they can be related to the directional wave spectrum as in the following:

$$C_{23}(f) = \int_{-\infty}^{\infty} \Psi_{23}(\tau) \cos 2\pi f\tau \, d\tau = \frac{1}{2} \int_0^{2\pi} S(f,\theta) k^2 \cos\theta \sin\theta \, d\theta \,,$$

$$Q_{12}(f) = \int_{-\infty}^{\infty} \Psi_{12}(\tau) \sin 2\pi f\tau \, d\tau = \frac{1}{2} \int_0^{2\pi} S(f,\theta) k \cos\theta \, d\theta \,,$$

$$Q_{13}(f) = \int_{-\infty}^{\infty} \Psi_{13}(\tau) \sin 2\pi f\tau \, d\tau = \frac{1}{2} \int_0^{2\pi} S(f,\theta) k \sin\theta \, d\theta \,,$$

$$\text{(11.82)}$$

$$C_{12}(f) = C_{13}(f) = Q_{23}(f) = 0 \,. \qquad (11.83)$$

Furthermore, the autocorrelation functions yield the following co-spectra, which are related to the directional spectrum:

$$C_{11}(f) = \int_{-\infty}^{\infty} \Psi_{11}(\tau) \cos 2\pi f\tau \, d\tau = \frac{1}{2} \int_0^{2\pi} S(f,\theta) \, d\theta \,,$$

$$C_{22}(f) = \int_{-\infty}^{\infty} \Psi_{22}(\tau) \cos 2\pi f\tau \, d\tau = \frac{1}{2} \int_0^{2\pi} S(f,\theta) k^2 \cos^2\theta \, d\theta \,,$$

$$C_{33}(f) = \int_{-\infty}^{\infty} \Psi_{33}(\tau) \cos 2\pi f\tau \, d\tau = \frac{1}{2} \int_0^{2\pi} S(f,\theta) k^2 \sin^2\theta \, d\theta \,,$$

$$\text{(11.84)}$$

$$Q_{11}(f) = Q_{22}(f) = Q_{33}(f) = 0 \,. \qquad (11.85)$$

Thus, we can obtain six quantities from the three cross-spectra and three autocorrelation functions, which are related to the integrals of the directional spectrum with respect to direction as given by Eqs. (11.82) and (11.84). To derive an estimate of the directional spectrum, we assume the

$S(f, \theta)$ can be expanded in a Fourier series as follows:

$$S(f, \theta) = \frac{1}{2} A_0(f) + \sum_{n=1}^{\infty} [A_n(f) \cos n\theta + B_n(f) \sin n\theta] . \qquad (11.86)$$

By substituting Eq. (11.86) into Eqs. (11.82) and (11.84), and by carrying out the integrations, Fourier coefficients for $n = 0$, 1 and 2 can be determined as

$$\left. \begin{array}{c} A_0(f) = \dfrac{2}{\pi} C_{11}(f) , \quad A_1(f) = \dfrac{2}{\pi k} Q_{12}(f) , \quad A_2(f) = \dfrac{2}{\pi k^2} [C_{22}(f) - C_{33}(f)] , \\[2mm] B_1(f) = \dfrac{2}{\pi k} Q_{13}(f) , \quad B_2(f) = \dfrac{4}{\pi k^2} C_{23}(f) . \end{array} \right\}$$

$$(11.87)$$

The resultant estimate of the directional spectrum represents the infinite Fourier series of Eq. (11.86) up to the second-order terms only, and thus, it is a biased estimate of the true spectrum. The estimate with the coefficients of Eq. (11.87), denoted as $\hat{S}_1(f, \theta)$, bears the following relationship to the true spectrum:

$$\hat{S}_1(f, \theta) = \frac{1}{2\pi} \int S(f, \theta) W_1(\phi - \theta) \, d\phi , \qquad (11.88)$$

where

$$W_1(\phi) = \sin \frac{5}{2}\phi \Big/ \sin \frac{1}{2}\phi = 1 + 2\cos\phi + 2\cos 2\phi . \qquad (11.89)$$

In order to reduce the extent of the bias caused by the use of the window function $W_1(\phi)$, Longuet-Higgins *et al.*[28] proposed use of the following formula for the estimation of the directional spectrum:

$$\hat{S}_2(f, \theta) = \frac{1}{2} A_0 + \frac{2}{3} (A_1 \cos\theta + B_1 \sin\theta) + \frac{1}{6} (A_2 \cos 2\theta + B_2 \sin 2\theta) .$$

$$(11.90)$$

This is equivalent to application of the window function

$$W_2(\phi) = \frac{8}{3} \cos^4 \frac{1}{2}\phi = 1 + \frac{4}{3}\cos\phi + \frac{1}{3}\cos 2\phi . \qquad (11.91)$$

The above method of making the spectral estimation is characterized by th initial assumption of the spectral form and the subsequent determination of the coefficients from the observation data. In this sense, it can be called a *parametric method*. A similar method is applicable to directional wave measurements with wave gauge arrays. Borgman,[31] and Panicker and Borgman[32] have proposed a parametric method for wave gauge arrays.

In the present section, the cross-spectra have been assumed to take the ensemble mean values. In practice, any value of the cross-spectra calculated form wave records is subject to sampling variability. The author[33] has carried out a numerical simulation study on the sampling variability of directional estimates, and has found the following empirical relationship for the co-spectra between the vertical velocity and the $x$ and $y$ components of the slope of the water surface:

$$\frac{\sigma[C_{ij}(f)]}{[C_{ii}(f)C_{jj}(f)]^{1/2}} = \sqrt{\frac{2}{r}}, \qquad (11.92)$$

where $\sigma$ denotes the standard deviation, $C_{ij}$ is the co-spectrum between the $i$th and $j$th wave components, and $r$ is the number of degrees of freedom. Because of variability in the cross spectral estimates, the directional spectral density estimated from wave records is accompanied by some statistical variability. In fact, Kuik and van Vledder[34] demonstrated by means of the Monte Carlo simulations that the estimate of the mean wave direction from a pitch and roll buoy data is accompanied by sampling variations and that the magnitude of the root-mean-square error is in agreement with the theoretical prediction by Borgman *et al.*[35]

### 11.3.4 *Advanced Theories of Directional Spectrum Estimates*

(A) *Transfer function for wave kinematics*

The estimation methods of the directional spectrum discussed above were developed for either a wave gauge array, a directional buoy, or a two-axis current meter, and they were incapable of dealing with a mixed array consisting of multiple wave gauges and current meters. Furthermore, these methods could not take into account the presence of noises in records and statistical variability of the cross-spectrum. A breakthrough was made by Isobe *et al.*[36] in 1984, who proposed the extended maximum likelihood method for an array of any combination of wave sensors. While Isobe's approach was deterministic without any consideration for noises and sampling variability, Hashimoto *et al.*[37−40] developed the maximum entropy principle method and a method using the Bayesian approach for directional spectrum estimates, based on the probabilistic concept: Hashimoto's methods are summarized in Ref. 41. A survey of various methods of directional wave spectrum estimate in the late 1990s has been made by Benoit *et al.*[42]

Hashimoto's approach is to treat the problem of directional spectrum estimation as a procedure to find the solution $S_k(k, \theta)$ of the integral

Eq. (11.57) from the information of a given set of covariances $\Psi$. Since the covariances are estimated at discrete values of the frequency, the problem is to find the solution of the directional spreading function $G(\theta|f)$ at respective frequencies. Thus, Eq. (11.57) is rewritten as

$$\phi_m(f) = \int_0^{2\pi} H_m(f,\theta)G(\theta|f)\,d\theta \; : \; m = 1, 2, \ldots, M \,, \qquad (11.93)$$

where $M$ denotes the number of cross-spectra formed from $N$ wave sensors of any type $(M = N(N+1)/2)$. The function $\phi_m(f)$ is a normalized form of cross-spectrum defined by

$$\phi_m(f) = \Phi_{ij}(f)/S(f) \,, \qquad (11.94)$$

and $H_m$ is the kernel function defined by

$$\begin{aligned} H_m(f) = H_i(f,\theta)H_j^*(f,\theta)[&\cos(kx_{ij}\cos\theta + ky_{ij}\sin\theta) \\ &-i\sin(kx_{ij}\cos\theta + ky_{ij}\sin\theta)] \,, \end{aligned} \qquad (11.95)$$

where $x_{ij}$ and $y_{ij}$ denote the distances in the $x$ and $y$ directions between the $i$th and $j$th wave sensors, respectively.

The function $H_i(f,\theta)$ is the transfer function from the water surface elevation to the wave property recorded by the $i$th sensor, and $H_j^*(f,\theta)$ is the conjugate complex of the transfer function for the $j$th sensor. The transfer function is hereby defined as

$$H(f,\theta) = K(k,f)\cos^\alpha\theta\sin^\beta\theta \,. \qquad (11.96)$$

According to Isobe et al.,[36] the transfer function is derived by the small amplitude wave theory as listed in Table 11.1.

### (B) Extended maximum likelihood method (EMLM)

By introducing the transfer function of Eq. (11.96), Isobe et al.[36] made it possible to apply the maximum likelihood method to a mixed array of wave sensors. Equation (11.79) is modified for a mixed array as in the following:

$$\hat{G}(\theta|f) = G_0 \left\{ \sum_{i=1}^{N}\sum_{j=1}^{N} \psi_{ij}^{-1} \exp[-i(kx_{ij}\cos\theta + ky_{ij}\sin\theta)]\cos^{\alpha_i+\alpha_j}\theta\sin^{\beta_i+\beta_j}\theta \right\}^{-1} \,, \qquad (11.97)$$

where $\psi_{ij}^{-1}$ is the element of the inverse matrix of a complex matrix composed of the normalized cross-spectra $\psi_{ij}$ of the following:

$$\psi_{ij} = \frac{\Phi_{ij}^*(f)}{K_i(k,f)K_j^*(k,f)\hat{S}(f)} \,. \qquad (11.98)$$

Table 11.1 Transfer function for directional spectral measurements.

| Wave Motion | | $K(k, f)$ | $\alpha$ | $\beta$ |
|---|---|---|---|---|
| Surface elevation | $\eta$ | $1$ | 0 | 0 |
| Vertical acceleration | $\eta_{tt}$ | $-4\pi^2 f^2$ | 0 | 0 |
| Surface slope $(x)$ | $\eta_x$ | $ik$ | 1 | 0 |
| Surface slope $(y)$ | $\eta_y$ | $ik$ | 0 | 1 |
| Orbital velocity $(x)$ | $u$ | $2\pi f \dfrac{\cosh k(h + z)}{\sinh kh}$ | 1 | 0 |
| Orbital velocity $(y)$ | $v$ | $2\pi f \dfrac{\sinh k(h + z)}{\sinh kh}$ | 0 | 1 |
| Pressure variation | $p$ | $\rho g \dfrac{\cosh k(h + z)}{\cosh kh}$ | 0 | 0 |

Note: $z$ is the elevation measured upward from the mean water level, $h$ the water depth, $\rho$ the density of water, and $g$ the acceleration of gravity.

The term $G_0$ on the right side of Eq. (11.97) is a constant to satisfy the normalization condition of Eq. (2.21) in Sec. 2.3.2.

The EMLM has a sharp directional resolution, and it is simple to use because of a straightforward algorithm for programming: it has been widely utilized in directional wave measurements with a combination of a two-axis current meter and a pressure sensor. However, the method sometimes yields a split at the peak of spreading function, and it is inferior to the MEP and BDM methods to be described in the following subsections in resolving multiple directional peaks.

## (C) *Maximum entropy principle method (MEP) and extended maximum entropy method (EMEP)*

Hashimoto and Kobune[37] introduced the concept of entropy for the directional spreading function as defined below,

$$H = - \int_0^{2\pi} G(\theta|f) \ln G(\theta|f) \, d\theta, \qquad (11.99)$$

and proposed to find the solution of $G(\theta|f)$ which will maximize the above entropy under the constraint of Eq. (11.93). The original proposal, called the maximum entropy principle method (MEP),[37] was for a three-element measurement system such as a pitch and roll buoy, but later Hashimoto et al.[40] extended it to a combination of any number of wave sensors and called it the extended maximum entropy principle method (EMEP).

To solve the problem, the directional spreading function is assumed to have the following expression:

$$G(\theta|f) = G_0 \exp\left\{ \sum_{n=1}^{N^*}[a_n(f)\cos n\theta + b_n(f)\sin n\theta] \right\}, \qquad (11.100)$$

where $a_n(f)$ and $b_n(f)$ are unknown parameters and $N^*$ is the order of the model. Equation (11.100) is substituted into Eq. (11.93) to find the solutions for these parameters. However, by considering the existence of errors by noises and sampling variability, it is modified to the following form:

$$\epsilon_m = G_0 \int_0^{2\pi} [\phi_m - H_m(f,\theta)] \exp\left\{ \sum_{n}^{N^*}[a_n(f)\cos n\theta \right.$$

$$\left. + b_n(f)\sin n\theta] \right\} d\theta \ : \ m = 1,2,\ldots,M^*, \qquad (11.101)$$

where $\epsilon_m$ stands for the amount of error associated with the $m$th cross-spectrum, and $M^*$ is the number of independent equations after the elimination of meaningless equations such as those involving zero co-spectra and quadrature-spectra.

The solutions for $a_n(f)$ and $b_n(f)$ are sought for by minimizing the total amount of errors; i.e., $\sum \epsilon_m^2$. It is carried out through a numerical technique of local linearization and iteration. The order of the model $N^*$ is chosen as the optimal one using the minimum Akaike's Information Criterion. For the procedure of numerical computations, refer to Ref. 41. In the MEP for a three-element measurement system, the model order $N^*$ is preset at 2, and the solutions for the unknown parameters are obtained through a numerical computation technique.

The EMEP has the capability to estimate the directional spectrum very close to the true one from a limited number of wave motion records, much better than the EMLM. Its performance is nearly as good as the Bayesian approach, which is the best method of directional spectrum estimation but requires quite a lot of computational works.

## (D) *Bayesian directional spectrum estimation method (BDM)*

The Bayesian approach is a probabilistic procedure to estimate the cause of a phenomenon from the result. To find the directional spreading function $G(\theta|f)$ from a set of the observed cross-spectra $\phi_m(f)$, as stated by

Eq. (11.93), is a typical subject of the Bayesian approach, which was undertaken by Hashimoto *et al.*[38,39] In this approach, the directional spreading function is not pre-assigned any specific form, but it is assumed to be of a step function having constant (positive) values over small directional intervals as expressed by

$$G(\theta|f) \approx \sum_{k=1}^{K} \exp[z_k(f)] I_k(\theta) , \tag{11.102}$$

in which $z_k(f)$ is the unknown constant to be solved by the Bayesian approach, $K$ is the number of division of the directional range $0 \le \theta \le 2\pi$, and $I_k$ is defined as

$$I_k(\theta) = \begin{cases} 1 : (k-1)\Delta\theta \le k\Delta\theta , \\ 0 : \text{otherwise}; \quad k = 1, 2, \ldots, K . \end{cases} \tag{11.103}$$

By substituting Eq. (11.102) into Eq. (11.93), we have the following approximate equation, provided the number of division $K$ is sufficiently large:

$$\phi_m(f) \approx \sum_{k=1}^{K} \alpha_{m,k} \exp[z_k(f)] + \epsilon_m : m = 1, 2, \ldots, M , \tag{11.104}$$

where

$$\alpha_{m,k} = \int_0^{2\pi} H_m(f, \theta) I_k(\theta) d\theta \approx H_m(f, \theta) \Delta\theta , \tag{11.105}$$

and $\epsilon_m$ represents the error term inherent in any measurement data and is assumed to follow the normal distribution with the mean 0 and the variance $\sigma^2$ of unknown magnitude. The solutions for the unknown constant $z_k$ can be obtained by maximizing the likelihood of $z_k$ and $\sigma^2$ for a given set of $\phi_m$. The constraint on $z_k$ is such that its value should vary gradually over consecutive sections, because the directional spreading function is expected to vary smoothly over the directional range.

The above is the basic concept of the BDM by Hashimoto *et al.*[38] More details of the method including computational techniques are described in Ref. 41. Because the BDM does not make any assumption for the functional shape of the directional spreading function, it has a large flexibility to fit well to any shape of spreading function. Through a number of simulation tests,[38,39] the BDM has been shown to perform the best among the existing methods of directional spectrum estimation, when the number of wave sensors is equal to or greater than four. For a three-element measurement system, the evaluation of the error terms contained in the cross-spectra becomes difficult and the BDM does not perform satisfactory: the MEP is recommended instead.

## 11.4   Resolution of Incident and Reflected Waves of Irregular Profiles

### 11.4.1   *Measurement of the Reflection Coefficient in a Wave Flume*

In hydraulic model tests of maritime structures in a wave flume, the first item of measurement is the characteristics of the incident waves, and the second item is the coefficient of reflection of the model structure. Although these measurements are not difficult for a regular train of waves, the same technique cannot be employed for an irregular train of waves, because irregular waves must be generated for a sufficiently long duration to exhibit their full statistical characteristics, and thus re-reflected waves by the wave paddle inevitably contaminate the incident wave train. Concerning this problem, Kajima[43] proposed a method for resolving the incident and reflected wave spectra by using the autocorrelation function. Thornton and Calhoun[44] estimated the coefficients of reflection and transmission of a rubble mound breakwater in the field, using a method slightly different from that by Kajima. Then, Goda and Suzuki[45] developed another method using the fast Fourier transform technique, which has become one of the standard techniques of flume wave tests in the world. This method will be described below.

In a wave flume, with installation of a model structure to be tested, waves reflected by the model travel back to the wave paddle and are re-reflected there. They propagate toward the structure and are reflected again, and the process is repeated many times. Thus, a multi-reflection system of wave trains is formed in the wave flume. The waves propagating toward the structure therefore consist of the superposition of the original incident waves, the first waves re-reflected by the paddle, the second re-reflected waves, and so on. Although they appear complicated, the components of these multi-reflection waves having the same frequency can be synthesized into a single train of waves, because the components all have the same frequency and the phase differences are fixed. A similar expression is possible for waves traveling toward the wave paddle. The synthesized profiles of incident and reflected waves for a specific frequency can be expressed as

$$\left. \begin{array}{l} \eta_I = a_I \cos(kx - \omega t + \varepsilon_I), \\ \eta_R = a_R \cos(kx + \omega t + \varepsilon_R). \end{array} \right\} \qquad (11.106)$$

The subscripts "$I$" and "$R$" denote the incident and reflected waves, respectively. The horizontal coordinate $x$ is taken as positive in the direction from the wave paddle to the model structure.

Suppose wave profiles are recorded at two locations, $x = x_1$ and $x = x_2 = x_1 + \Delta l$, with the separation distance $\Delta l$. The wave profiles are expressed as

$$\left.\begin{aligned}
\eta_1 &= (\eta_I + \eta_R)_{x=x_1} = A_1 \cos \omega t + B_1 \sin \omega t\,, \\
\eta_2 &= (\eta_I + \eta_R)_{x=x_2} = A_2 \cos \omega t + B_2 \sin \omega t\,,
\end{aligned}\right\} \tag{11.107}$$

where

$$\left.\begin{aligned}
A_1 &= a_I \cos \phi_I + a_R \cos \phi_R\,, \\
B_1 &= a_I \sin \phi_I - a_R \sin \phi_R\,, \\
A_2 &= a_I \cos(k\Delta l + \phi_I) + a_R \cos(k\Delta l + \phi_R)\,, \\
B_2 &= a_I \sin(k\Delta l + \phi_I) - a_R \sin(k\Delta l + \phi_R)\,,
\end{aligned}\right\} \tag{11.108}$$

$$\phi_I = kx_1 + \varepsilon_I\,, \quad \phi_R = kx_1 + \varepsilon_R\,. \tag{11.109}$$

In the above, Eq. (11.108) represents a system of four equations containing the four unknown quantities $a_I, a_R, \phi_I$ and $\phi_R$. By eliminating $a_R$ and $\phi_R$ from the expressions for $A_2$ and $B_2$, we have

$$\left.\begin{aligned}
A_2 &= (A_1 \cos k\Delta l + B_1 \sin k\Delta l) - 2a_I \sin k\Delta l \sin \phi_I\,, \\
B_2 &= (-A_1 \sin k\Delta l + B_1 \cos k\Delta l) + 2a_I \sin k\Delta l \cos \phi_I\,.
\end{aligned}\right\} \tag{11.110}$$

The quantity $a_I$ can be obtained by eliminating the terms containing $\phi_I$ from the above, and similarly for $a_R$. The result is

$$\begin{aligned}
a_I &= \frac{1}{2|\sin k\Delta l|}[(A_2 - A_1 \cos k\Delta l - B_1 \sin k\Delta t)^2 \\
&\quad + (B_2 + A_1 \sin k\Delta l - B_1 \cos k\Delta l)^2]^{1/2}\,, \\
a_R &= \frac{1}{2|\sin k\Delta l|}[(A_2 - A_1 \cos k\Delta l + B_1 \sin k\Delta l)^2 \\
&\quad + (B_2 - A_1 \sin k\Delta l - B_1 \cos k\Delta l)^2]^{1/2}\,.
\end{aligned} \tag{11.111}$$

Thus, the amplitudes $a_I$ and $a_R$ of the incident and reflected waves can be calculated from the four amplitudes $A_1, B_1, A_2$ and $B_2$, and the phase lag $k\Delta l$. Furthermore, the phase angle $\phi_I$ can be obtained by eliminating the terms containing $a_I$ in Eq. (11.110), and similarly for $\phi_R$, as demonstrated by Fan.[46] The result is

$$\left.\begin{aligned}
\phi_I &= \tan^{-1}\left[\frac{-A_2 + A_1 \cos k\Delta l + B_1 \sin k\Delta l}{B_2 + A_1 \sin k\Delta l - B_1 \cos k\Delta l}\right]\,, \\
\phi_R &= \tan^{-1}\left[\frac{-A_2 + A_1 \cos k\Delta l - B_1 \sin k\Delta l}{-B_2 + A_1 \sin k\Delta l + B_1 \cos k\Delta l}\right]\,.
\end{aligned}\right\} \tag{11.112}$$

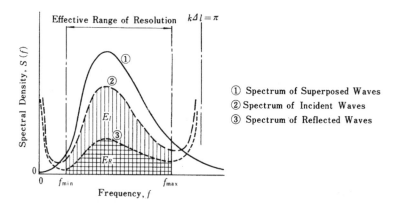

Fig. 11.6   Illustration of the spectral resolution of the incident and reflected waves.

The above method of resolution is applied to each Fourier component of the irregular wave profiles recorded at two locations separated by an appropriate distance, after the records have been decomposed into the Fourier series by using the fast Fourier transform technique. An assumption in this analysis is that the dispersion relationship, Eq. (9.3), is satisfied between the wavenumber $k$ and the frequency $f$ in the range of analysis. The amplitudes $a_I$ and $a_R$, thus estimated for each Fourier component, represent the root-mean-square amplitude $|A_k^2 + B_k^2|^{1/2}$ for the wave profile expressed by Eq. (11.14). The frequency spectra of the incident and reflected waves are then estimated by the procedure described in Sec. 11.2.

The result of a spectral resolution by this method is schematically shown in Fig. 11.6.   It is seen that the spectral estimates diverge in the neighborhood of frequencies satisfying the condition $k\Delta l = n\pi$ for $n = 0, 1, 2, \ldots$, because the factor $|\sin k\Delta l|$ in the denominator on the right side of Eq. (11.111) becomes very small, and errors due to noise are greatly amplified. The spectral estimates are effective in the frequency range outside of the neighborhood of such diverging points. The effective frequency range of resolution can be judged with the following guideline:

$$\left.\begin{array}{l} \text{Upper limit } (f_{\max}) \,:\, \Delta l / L_{\min} \simeq 0.45\,, \\ \text{Lower limit } (f_{\min}) \;\,:\, \Delta l / L_{\max} \simeq 0.05\,. \end{array}\right\} \tag{11.113}$$

The symbols $L_{\min}$ and $L_{\max}$ denote the wavelengths corresponding to the upper $(f_{\max})$ and lower $(f_{\min})$ limits of the effective frequency range, respectively. When the distance between the two wave gauges has been fixed, the effective frequency range of the resolution of the incident and reflected waves can be determined from Eq. (11.113). Working the other way, when the plan of a hydraulic model test is being prepared, the distance $\Delta l$ should be selected such that the major part of the wave energy of the spectrum of the test wave is contained in the effective frequency range $f_{\min}$ to $f_{\max}$.

For the estimation of the reflection coefficient, the energies of the incident and reflected waves, $E_I$ and $E_R$, within the range $f_{\min}$ to $f_{\max}$ need to be calculated. This is accomplished as

$$\left.\begin{array}{c} E_I = \displaystyle\int_{f_{\min}}^{f_{\max}} S_I(f)df = \frac{\Delta t}{2t_0}\sum_{f_{\min}}^{f_{\max}} I_I\,, \\[4mm] E_R = \displaystyle\int_{f_{\min}}^{f_{\max}} S_R(f)df = \frac{\Delta t}{2t_0}\sum_{f_{\min}}^{f_{\max}} I_R\,, \end{array}\right\} \tag{11.114}$$

where $I_I$ and $I_R$ are the periodograms of the incident and reflected waves as equal to $Na_I^2$ and $Na_R^2$, respectively. The calculation is best made with the periodograms in order to eliminate the effect of leakage of wave energy into neighboring frequency bands, which is introduced through the use of a filter in smoothing of the periodograms. Because the energies of the incident and reflected waves must be proportional to the squares of the respective wave heights, the reflection coefficient defined as the ratio of heights can be estimated from

$$K_R = \sqrt{E_R/E_I}\,. \tag{11.115}$$

This reflection coefficient represents that of the wave group as a whole. The incident wave height $H_I$ and the reflected wave height $H_R$ can be estimated with the above coefficient of reflection $K_R$ and the mean value of wave heights at the two locations, which is denoted by $H_S$, as

$$H_I = \frac{1}{(1+K_R^2)^{1/2}}H_S \quad\text{and}\quad H_R = \frac{K_R}{(1+K_R^2)^{1/2}}H_S\,. \tag{11.116}$$

The wave heights $H_I, H_R$ and $H_S$ can refer to any definition such as the significant height, the mean height, or other heights, as long as the same definition is employed in the calculation of Eq. (11.116). This calculation is based on the assumption that the observed height in the flume is equal to the square root of the sum of the squares of the incident and reflected wave heights, as expressed by Eq. (3.51) in Sec. 3.7.3. As shown in Fig. 3.48, however, Eq. (3.51) does not hold in the vicinity of a reflective boundary because of strong phase interference. Therefore, it is recommended that the wave gauges be set at a distance of more than one wavelength of the significant wave from both the model structure and the wave paddle.

The above technique of resolving incident and reflected waves can be directly applied to regular wave trains. The amplitudes $a_I$ and $a_R$ become those of the incident and reflected waves, and the reflection coefficient is directly obtained as the ratio $a_R/a_I$. Furthermore, use of the phase angles $\phi_I$ and $\phi_R$, resolved by Eq. (11.112), together with $a_I$ and $a_R$ makes it possible to retrieve the profiles of the incident and reflected wave trains. By applying this technique for the whole Fourier components of irregular waves (within the effective frequency range of resolution), the profiles of incident and reflected irregular wave trains can be reconstructed.

In several hydraulic laboratories around the world, the technique has been extended to a linear array of three wave gauges (e.g., see Refs. 47–49). The motivation behind this is to circumvent the divergence problem at $k\Delta l = n\pi$. With three gauges, one pair may fail to yield the proper solution at a particular frequency owing to the divergence problem, but the other two pairs can yield nondiverging estimates. By taking the average of the results of the pairs of wave gauges not troubled by the divergence problem at the respective frequencies, a three-gauge array can extend the effective range of wave resolution considerably.

## 11.4.2   *Measurement of the Reflection Coefficient of Prototype Structures*

Measurement of waves reflected from a vertical breakwater or a seawall in the field is rather difficult. We can sometimes recognize the presence of reflected waves visually at the site, on aerial photographs, or on the screens of imaging radars employed for directional wave measurements. But the heights of such reflected waves cannot be measured by these means.

The measurement of reflected waves in the field is fundamentally the same as the measurement of the directional wave spectrum. The directional resolution must be good enough to separate the directional peaks of the incident and reflected waves. A serious problem with the directional wave measurement of a wave reflection system is a phase-locked relation between the incident and reflected wave components. In the analysis of the directional wave spectrum, it is assumed that all the component waves are independent, with random and uniformly distributed phase angles. However, each pair of incident and reflected wave components maintain a fixed phase relation, which depends on the relative distance and the angle of incidence to the reflective boundary. The violation of the condition of random phase angles cause a false estimation of the directional spectral density. This has been confirmed in a numerical simulation study by Goda.[25]

For the correct estimation of directional spectra in the presence of phase-locked incident and reflected waves, Isobe and Kondo[50] proposed a modification of the maximum likelihood method under the assumption that the waves are reflected at the reflective boundary without a phase delay and with the angle of reflection the same as the incident angle. In the modified maximum likelihood method (MMLM), the reflection coefficient which is assumed to be the function of frequency and direction is calculated by

$$
r(f, \theta) = - \left[ \sum_{i=1}^{N} \sum_{j=1}^{N} \Phi_{ij}^{-1}(f) \{ \exp[-ik(R_{ij} \cos \theta + Y_{ij} \sin \theta)] \right.
$$

$$
\left. + \exp[ik(R_{ij} \cos \theta - Y_{ij} \sin \theta)] \} \right]
$$

$$
\bigg/ 2 \sum_{i=1}^{N} \sum_{j=1}^{N} \Phi_{ij}^{-1}(f) \exp[ik(X_{ij} \cos \theta + Y_{ij} \sin \theta)] \} , \quad (11.117)
$$

in which the $y$ axis ($x = 0$) is taken at the reflective boundary, $X_{ij}$ and $Y_{ij}$ are the distances in the $x$ and $y$ directions between the $i$th and $j$th wave sensors, respectively, $R_{ij} = X_i + X_j$ is the sum of the distances from the boundary, and $\Phi_{ij}^{-1}(f)$ denotes the $(i, j)$ component of the inverse matrix of the complex matrix composed of the conjugate cross-spectra.

The directional spreading function of the incident waves is estimated by

$$
\hat{G}(\theta|f) = A \left[ \sum_{i=1}^{N} \sum_{j=1}^{N} \Phi_{ij}^{-1}(f) \{ \exp[ik(X_i \cos \theta + Y_i \sin \theta)] \right.
$$

$$
+ r(f, \theta) \exp[-ik(X_i \cos \theta - Y_i \sin \theta)] \}
$$

$$
\times \{ \exp[-ik(X_j \cos \theta + Y_j \sin \theta)] \}
$$

$$
\left. + r(f, \theta) \exp[ik(X_j \cos \theta - Y_j \sin \theta)] \} \right]^{-1} , \quad (11.118)
$$

where $A$ is the constant for normalization of the directional spreading function by the condition of Eq. (2.21) but has the inverse units of spectral density.

The frequency spectral density in front of a reflective structure fluctuates from place to place because of the presence of a partial standing wave system formed by the superposition of incident and reflected random waves. The observed spectral density $\hat{S}_{ii}$ by the $i$th wave gauge should be modified for the estimation of incident wave spectrum by the following formula:

$$\hat{S}_{ii}(f) = \frac{S_{ii}(f)}{\displaystyle\int_{-\pi/2}^{\pi/2} \hat{G}(f,\theta)[1 + 2r(f,\theta)\cos(2kX_i\cos\theta) + r^2(f,\theta)]d\theta} .$$

$$(11.119)$$

The modified frequency spectra of $N$ wave sensors thus obtained are averaged to yield the estimate of the incident wave spectrum. The estimate of incident directional spectrum is obtained by the product of the above frequency spectrum and the directional spreading function by Eq. (11.118).

The MMLM has been utilized for the estimation of reflection coefficient of actual harbor and coastal structures (e.g., see Refs. 50 and 51, etc.). It is relatively simple to use, but it may not yield a good result when the incident waves have multidirectional peaks. Hashimoto et al.[52] have proposed the modified, extended maximum entropy method (MEMEP) as a more powerful method of directional spectrum estimation for the incident and reflected wave field.

An approximate value of the reflection coefficient of prototype structures may also be estimated by simultaneously measuring the wave height $H_S$ in front of a structure (at a distance of more than one wavelength from the structure) and the incident wave height $H_I$ at a location free from the effect of reflected waves. Then, Eq. (3.51) in Sec. 3.7.3 can be utilized to resolve the heights of the reflected and incident waves, yielding an estimate of the reflected wave height as $H_R \simeq \sqrt{H_S^2 - H_I^2}$. Accurate estimation of the reflection coefficient by this method is difficult, however. Even if the reflected wave height is 50% of the incident height, the resultant height of the superposed waves is only 12% greater than the incident height. Sampling variability of random sea waves adds further difficulty to the estimation of wave reflection. A large amount of data for similar wave conditions should be collected and the mean ratio $H_S/H_I$ should be calculated in order to reduce the effect of sampling variability on the estimate of $H_R/H_I$.

## 11.5  Numerical Simulation of Random Sea Waves and Numerical Filters

### 11.5.1  *Principles of Numerical Simulation*

The properties of random waves are basically described with the directional wave spectrum, but we need more detailed information on wave profiles, wave kinematics, and so on for further examination of wave actions on maritime structures. For such purposes, temporal and spatial profiles and kinematics of random waves are simulated by computers for a given input of wave spectrum. Input signals for random wave generation in laboratories are prepared by the numerical simulation technique as described in Sec. 8.3.

The basic of wave simulation is Eq. (9.12) in Sec. 9.2, which is expressed as an infinite series of component waves. The infinite series is replaced with a finite series with $M$ terms, and an extension is made for simulation of an arbitrary wave-related variable $\zeta_i$ as in the following:

$$\zeta_i(x, y, t) = \sum_{m=1}^{M} K_i(f_m, \theta_m) a_m \cos(k_m x \cos\theta_m$$

$$+ k_m y \sin\theta_m - 2\pi f_m t + \epsilon_m + \psi_i), \qquad (11.120)$$

where $a_m, k_m, f_m, \theta_m$, and $\varepsilon_m$ denote the amplitude, wave number, frequency, azimuth, and random phase of the $m$th component wave, respectively. The terms of $K_i(f, \theta)$ and $\psi_i$ are the transfer function from $\eta$ to $\zeta_i$ and the phase lag, respectively. They are given using the function $H(f, \theta)$ of Eq. (11.96) with the information listed in Table 11.1 in Sec. 11.3.4 as below.

$$\left. \begin{array}{c} K_i(f, \theta) = |H_i(k, f)| \cos^\alpha \theta \sin^\beta \theta, \\ \psi_i = \arg[H_i(k, f)]. \end{array} \right\} \qquad (11.121)$$

For making the control signal of a random wave generator, the transfer function is set as $K(f, \theta) = 1/F(f, h)$ using the wave generating function of $F_1(f, h)$ by Eq. (8.1) for piston type or $F_2(f, h)$ by Eq. (8.2) for flap type.

Equation (11.120) describes the superposition of component waves propagating forward in the direction of $\theta_m$ with the $x$ axis. When the azimuth is defined in the direction of propagation toward an observation point as in the case of directional wave observations, the sign of the term of $2\pi f_m t$ is changed from negative to positive and the phase lag of horizontal orbital velocity is added by $\pi$. The present section deals with the case of surface elevation only, i.e. $K_i(f, \theta) = 1, \psi_i = 0$, for the sake of simplicity, but the simulation method for other kinematics are the same.

### 11.5.2 *Methods of Numerical Simulation*

(A) *Double summation method*

Simulation by means of Eq. (11.120) can be made by using either single series or double series. The former method is called the *single summation method*, while the latter is called the *double summation method*. Initially the double summation method was often employed in wave simulation because the component amplitudes are easily related with the directional spectrum. The fundamental equation for the double summation method for wave profile is as follows:

$$\eta(x, y, t) = \sum_{m=1}^{M} \sum_{n=1}^{N} a_{m,n} \cos(k_m x \cos\theta_n + k_m y \sin\theta_n - 2\pi f_m t + \epsilon_{m,n}).$$

(11.122)

The ranges of frequency and wave angle are divided into $M$ and $N$ segments, respectively, $a_{m,n}$ denotes the amplitude of the component wave in the $m$th frequency and $n$th directional segment, and $\epsilon_{m,n}$ is a random phase uniformly distributed in $[0, 2\pi]$. The amplitude $a_{m,n}$ is determined from a given function of the directional spectrum $S(f, \theta)$ by

$$a_{m,n} = \sqrt{2S(f_m, \theta_n)\Delta f_m \Delta \theta_n},$$

(11.123)

where $\Delta f_m$ and $\Delta \theta_n$ denote the bandwidth of the $m$th frequency and that of the $n$th wave angle, respectively. They are not necessarily constant but may vary from a component to another. Selection of the frequencies $f_m$ is discussed in Clause (C). In the double summation method, the azimuth $\theta_m$ is usually selected with an equal spacing between each other.

The numbers of components $M$ and $N$ should be taken as large as possible within the capacity of the computer at hand. A miminum of $M = 50$ and $N = 30$ is recommended for spatial simulation of wave field. The wave crest pattern of wave refraction in shoaling water shown in Fig. 3.5 in Sec. 3.2.1 was simulated by the author with these numbers of components.

The concept of representing the infinite series of Eq. (9.12) with the finite double series is also employed in the evaluation of the refraction and diffraction of directional random waves, which have been discussed in Secs. 3.2 and 3.3.

(B) *Single summation method*

The double summation method has the merit that the process is easily understandable, but the total number of components becomes large and the simulation may take a long computation time. There is also a risk of phase locking among component waves. Because of such reasons, multidirectional wave generation in laboratory basins is mostly carried out with the single summation method. Simulation is made with the single series of

Eq. (11.120). The amplitudes of component waves are calculated from the frequency spectrum by the following:

$$a_m = \sqrt{2S(f_m)\Delta f_m}\,. \tag{11.124}$$

Each component wave is assigned a single azimuth, $\theta_m$, which is selected by a random number $R_m$ being uniformly distributed between 0 and 1 through the cumulative distribution of the directional distribution function corresponding to the given frequency $f_m$; i.e.,

$$\theta_m = F^{-1}(R_m)\ :\ F(\theta|f_m) = \int_0^\theta G(\theta|f_m)d\theta\,. \tag{11.125}$$

The inverse solution of the cumulative distribution $F(\theta|f_m)$ cannot be obtained analytically, but must be sought for by a numerical procedure. Initially the random number $R_m$ is assigned to each frequency. The spreading function $G(\theta|f_m)$ is first integrated for the full range of wave angle so as to evaluate the normalization constant $G_0$. The integration of the second equation of Eq. (11.125) is made step by step until the value $F(\theta|f_m)$ equals $R_m$, which is assigned to the frequency $f_m$. The angle at this condition gives the result for $\theta_m$.

The number of frequency components may vary from several hundreds to more than one thousand.

### (C) *Selection of component frequencies*

The frequency components are selected by two methods, depending on the technique of the superposition of component waves. If the inverse fast Fourier transform (FFT) method is employed for computation of wave profiles, the frequency components are set as equally spaced with the increment of $1/t_0$, where $t_0$ is the duration of wave record to be simulated. The number $M = 2^J$ determines the maximum frequency for simulation as $f_{max} = 2^J/t_0$, and the number of data points to be simulated is $2^{J+1}$. If the sampling time $\Delta t$ and the number of data point $2M = 2^{J+1}$ are set first, then $t_0 = 2M^{J+1}\Delta t$ and $f_{max} = 1/2\Delta t$. In the author's numerical simulation of wave profiles for various frequency spectra,[53] he used $M = 2^{12} = 4096$.

If the inverse FFT method is not used for computation of wave profiles, the number of frequency components may be limited to several hundreds. In that case, it is better to select the frequency components in such a way that the individual amplitudes would be as nearly equal as possible; by so selecting, every wave component can contribute to the simulation result with equal weight. A convenient formula for selecting the frequency components with nearly equal component amplitudes is given by Goda[53] as

$$f_m = \frac{1.007}{T_{1/3}} \left[ \ln \left( \frac{2M}{2m-1} \right) \right]^{1/4}. \tag{11.126}$$

This formula is derived by dividing the zeroth moment of the Bretschneider–Mitsuyasu spectrum (Eq. (2.10)) into the $2M$ segments of equal amount and by taking the odd number of frequencies at the boundaries of $2M$ segments. For other standard frequency spectra, the condition of equal component amplitude is not satisfied, and the component amplitudes need to be calculated by Eq. (11.123) or (11.124).

(D) *Simulation of waves at a fixed location*

If the objective of simulation is to produce the time variation of wave profiles at a fixed location, Eq. (11.120) is rewritten in the following form[54]:

$$\eta(t|x, y) = \sum_{m=1}^{M} A_m \cos(2\pi f_m t - \psi_m), \tag{11.127}$$

where

$$\left. \begin{aligned}
A_m &= \sqrt{C_m^2 + S_m^2}, \\
\psi_m &= \tan^{-1}[S_m/C_m], \\
C_m &= \sum_{n=1}^{N} \cos(k_m x \cos\theta_n + k_m y \sin\theta_n + \varepsilon_{m,n}), \\
S_m &= \sum_{n=1}^{N} \sin(k_m x \cos\theta_n + k_m y \sin\theta_n + \varepsilon_{m,n}).
\end{aligned} \right\} \tag{11.128}$$

When the author[54] examined the statistical variability of wave statistics in 1977, he evaluated the coefficients of Eq. (11.128) by employing $N = 30$ or 36 and then simulated wave profiles with Eq. (11.127) by employing $M = 200$.

When the number of directional components $N$ is sufficiently large and the phase angle $\varepsilon_{m,n}$ are uniformly and randomly distributed, the coefficients $C_m$ and $S_m$ follow the normal distribution with the mean of 0 and the variance of the following:

$$\text{Var}[C_m] = \text{Var}[S_m] = \sum_{n=1}^{N} \frac{1}{2} a_{m,n}^2 = S(f_m) \Delta f_m. \tag{11.129}$$

The normal distribution of $C_m$ and $S_m$ means that the amplitude $A_m$ is a stochastic variate following the chi-square distribution with the two degrees of freedom. For proper reproduction of the statistical variability of wave properties, the component amplitude of Eq. (11.124) needs to be replaced by the following:

$$a_m = \sqrt{S(f_m)\chi_2^2 \Delta f_m}\,, \tag{11.130}$$

where $\chi_2^2$ denotes the chi-square variate with the two degrees of freedom. Its expected value is $E[\chi_2^2] = 2$, and thus the above equation becomes equal to Eq. (11.125) at its mean value.

### (E) *Simulation of time series data*

When wave simulation is made for a fixed location, it is carried out by using Eq. (11.127). When the inverse FFT method is employed to compute Eq. (11.127), the coefficients $C_m$ and $S_m$ are directly used instead of the amplitude $A_m$. When the unequal frequency spacing is used to reduce the number of frequency components and the simulation is made at a constant time interval, i.e., $t_k = k\Delta t$, the use of either set of the following regression relations leads to a considerable saving in the computation time:

$$\left.\begin{aligned}
\cos(k+1)\Delta t &= \cos k\Delta t \cos \Delta t - \sin k\Delta t \sin \Delta t\,,\\
\sin(k+1)\Delta t &= \sin k\Delta t \cos \Delta t + \cos k\Delta t \sin \Delta t\,,
\end{aligned}\right\} \tag{11.131}$$

or

$$\left.\begin{aligned}
\cos(k+1)\Delta t &= 2\cos k\Delta t \cos \Delta t - \cos(k-1)\Delta t\,,\\
\sin(k+1)\Delta t &= 2\sin k\Delta t \cos \Delta t - \sin(k-1)\Delta t\,.
\end{aligned}\right\} \tag{11.132}$$

The former set has been used in several simulation studies by the author,[54,55] while the latter set is due to Medina et al.[56]

The simulation of a one-dimensional irregular wave profile for a given frequency spectrum can be done by the same principle as above. The basic equation is Eq. (11.122) at the origin of the coordinates, i.e., $x = 0$ and $y = 0$, which is written as

$$\eta(t) = \sum_{m=1}^{M} a_m \cos(2\pi f_m t - \epsilon_m)\,. \tag{11.133}$$

In the case of Eq. (11.128), the coefficients $C_m$ and $S_m$ are normally distributed with the mean 0 and the variance $S(f_m)\Delta f_m$ as described in Clause (D). Thus, the amplitude $a_m$ in Eq. (11.133) should be evaluated with Eq. (11.130) with due consideration of statistical variability.

(F) *Preparation of control signals for multidirectional wave generator*

As discussed in Sec. 8.2, the control signals for multidirectional wave generators are prepared by the single summation method, because the use of the double summation method introduces the problem of phase locking among wave components and violates the condition of homogeneity. According to Takayama and Hiraishi,[57] the motion of a the $r$th segment of a wave paddle is formulated as

$$\eta_r(t) = \sum_{m=1}^{M} \frac{a_m}{F(f_m, h)} \sin(2\pi f_m t - k_m r b \cos\theta_m + \epsilon_m)\sin\theta_m\,, \quad (11.134)$$

where $b$ denotes the width of a wave paddle so that $x = rb$ indicates the distance of the $r$th paddle from the origin of the coordinate. The frequency $f_m$ and the wave amplitude $a_m$ are determined by the methods described in Clauses (B) and (C). The azimuth $\theta_m$ is assigned by Eq. (11.125).

The control signals for wave generators must be fed with a short time interval for smooth operation of the servo system. When Eq. (11.131) or Eq. (11.132) is used for computation of wave profiles, $\Delta t$ can be set as required from the servo system. When the inverse FFT method is employed, the computation is made with a coarse time interval and the resultant output is interpolated to yield the signals in a fine time interval.

### 11.5.3  *Pseudo-Random Number Generating Algorithm*

Selection of the random phase angle $\epsilon_m$ in Eq. (11.120) or $\epsilon_{m,n}$ in Eq. (11.122) is the important step in correct simulation of random waves. It is made by means of a series of random number $R$ uniformly distributed between 0 and 1. There are several computer algorithms for generating random numbers, but the generated ones are not true but pseudo random numbers. A series of the generated random numbers repeat themselves after a certain number of operations. When they are arranged in a two-dimensional matrix, they may show a certain pattern which betrays uniform randomness. Thus a random number generation algorithm needs to be tested for the nonrepeatability and the uniformity of randomness in two-dimensions.

The author has been using the following algorithm to generate pseudo-random numbers by computer. First, select an arbitrary odd integer $X_1$, which is called the seed number. The second integer $X_2$ is generated by

$$X_2 = \mathrm{mod}(Y, b)\,, \quad Y = aX_1\,, \quad (11.135)$$

where $\mathrm{mod}(Y, b)$ denotes a remainder after $Y$ is divided by $b$. Then, the process is repeated by replacing $X_1$ with $X_2$ for any number of cycles, yielding a series of integers $X_i$. The random number assigned between 0 and 1 is obtained by

$$R_i = X_i/q : i = 1, 2, \ldots . \tag{11.136}$$

The constants are given the following specific values:

$$a = 7909, \quad b = 2^{36}, \quad q = 2^{35} - 1. \tag{11.137}$$

The random integer $X_i$ may exceed $2^{35}$ during the generating process. In that case, $X_i$ is replaced by the integer $(2^{36} - X_i)$ and the process is continued. The above algorithm has been tested for the randomness of the results, and it has passed the randomness test for a series of up to $2 \times 10^7$ random numbers.[55]

### 11.5.4 *Numerical Filtering of Wave Record*

The concept of numerical simulation of irregular wave profiles is applicable to the problem of numerical filtering. For example, a record of nearshore waves is often contaminated by the fluctuation of the mean water level due to surf beat. In such a case, the data is decomposed into a finite Fourier series with the aid of the fast Fourier transform algorithm. Then all the Fourier amplitudes outside the frequency range of interest are set to zero, and the operation of the inverse FFT is performed by retaining only the amplitudes within the frequency range of interest.

Figure 11.7 shows an example of such a computation.[58] A wave record consisting of 1800 data points sampled at the rate of $\Delta t = 1$ s was analyzed, and the profiles of surf beat in the frequency range of $f \leq 0.05$ Hz were reconstructed with about 90 pairs of Fourier coefficients as shown with the dashed line; in this computation, the regression formula of Eq. (11.131) was used instead of the inverse FFT method. This example presents a case of extracting the profile of long period components, but the method can

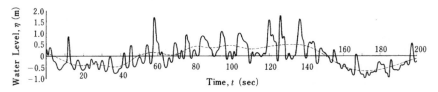

Fig. 11.7　Example of the profile of surf beat extracted by means of a numerical filter.[58]

also be utilized for correction of the fluctuating mean water level in a wave record, by setting the Fourier coefficients of the surf beat frequency to zero and retaining those in the normal frequency range as they are.

## References

1. M. J. Tucker, "Analysis of records of sea waves," *Proc. Inst. Civil Engrg.* **26**(10) (1963), pp. 305–316.
2. Y. Goda, "The observed joint distribution of periods and heights of sea waves," *Proc. 16th Int. Conf. Coastal Engrg.* (Hamburg, ASCE) (1978) pp. 227–246.
3. Y. Goda, "Estimation of wave statistics from spectral information," *Proc. Int. Symp. Ocean Wave Measurement and Analysis (WAVES '74)* ASCE **1** (1974), pp. 320–337.
4. A. Kimura, "Statistical properties of random wave groups," *Proc. 17th Int. Conf. Coastal Engrg.* (Sydney, ASCE) (1980), pp. 2955–2973.
5. L. H. Koopmans, *The Spectral Analysis of Time Series* (Academic Press, 1974) pp. 258–265.
6. R .B. Blackman and J. W. Tukey, *The Measurement of Power Spectra from the Point of View of Communications Engineering* (Dover Pub., 1958) 190p.
7. Y. Goda, "Analysis of wave grouping and spectra of long-travelled swell," *Rept. Port and Harbour Res. Inst.* **22**(1) (1983), pp. 3–41.
8. M. Donelan and W. J. Pierson, Jr., "The sampling variability of estimates of spectra of wind-generated gravity waves," *J. Geophys. Res.* **88**(C7) (1983), pp. 4381–4392.
9. Koopmans, *loc. cit.*, p. 305.
10. K. Rikiishi, "Methods of computing the power spectrum for equally-spaced time serise of finite length," *J. Applied Meteorology* **15**(10) (1976), pp. 1102–1110.
11. Koopmans, *loc. cit.*, pp. 300–302.
12. S. Kuwajima and K. Nagai, "The fast Fourier transform for the sample of arbitrary length and its application to spectral analyses," *Tech. Note of Port and Harbour Res. Inst.* **155** (1973), 33p (*in Japanese*).
13. K. Rikiishi and H. Mitsuyasu, "On the use of windows for the computation of power spectra," *Rept. Res. Inst. Applied Mech., Kyushu Univ.* **XXI**(68) (1973), pp. 53–71.
14. Koopmans, *loc. cit.*, p. 273.
15. H. Rye, "Ocean wave groups," *Dept. Marine Tech. Norwegian Inst. Tech., Rept.* **UR 82–8** (1982), 214p.
16. N. N. Panicker, "Review of techniques for directional wave spectrum," *Proc. Int. Symp. Ocean Wave Measurement and Analysis (WAVES '74)* ASCE **1** (1974), pp. 669–688.
17. D. Stilwell, "Directional energy spectra of the sea from photographs," *J. Geophys. Res.* **74**(8) (1969), pp. 1974–1986.

18. D. B. King and O. H. Shemdin, "Radar observations of hurricane wave directions," *Proc. 16th Int. Conf. Coastal Engrg.* (Hamburg, ASCE) (1978), pp. 209–226.

19. J. F. Vesecky, S. V. Hsiao, C. C. Teague, O. H. Shemdin and S. S. Pawka, "Radar observations of wave transformations in the vicinity of island," *J. Geophys. Res.* **85**(C9) (1980), pp. 4977–4986.

20. I. A. Sneddon, *Fourier Transform* (McGraw-Hill, 1951), p. 44.

21. B. Kinsman, *Wind Waves* (Prentice-Hall, 1965), pp. 460–471.

22. N. P. Barber, "The directional resolving power of an array of wave detectors," *Ocean Wave Spectra* (Prentice-Hall, 1963), pp. 137–150.

23. J. Capon, R. J. Greenfield and R. J. Kolker, "Multidimensional maximum-likelihood processing of a large aperture seismic array," *Proc. IEEE* **55** (1967), pp. 192–211.

24. S. S. Pawka, "Island shadows in wave directional spectra," *J. Geophys. Res.* **88**(C4) (1983), pp. 2579–2591.

25. Y. Goda, "Simulation in examination of directional resolution," *Proc. Conf. on Directional Wave Spectra Applications* (Berkeley, ASCE) (1981), pp. 387–407.

26. M. Yamaguchi and Y. Tsuchiya, "Directional spectra of wind-waves in growing state," *Proc. 27th Japanese Conf. Coastal Engrg.* (1980), pp. 99–103 (*in Japanese*).

27. R. A. Haubrich, "Array design," *Bull. Seismological Soc. America* **58**(3) (1968), pp. 979–991.

28. M. S. Longuet-Higgins, D. E. Cartwright and N. D. Smith, "Observations of the directional spectrum of sea waves using the motions of a floating buoy," *Ocean Wave Spectra* (Prentice-Hall, 1963), pp. 111–136.

29. H. Mitsuyasu, F. Tasai, T. Suhara, S. Mizuno, M. Ohkusu, T. Honda and K. Rikiishi, "Observation of the directional spectrum of ocean waves using a cloverleaf buoy," *Physical Oceanography* **5**(4) (1975), pp. 750–760.

30. Y. Nagata, "The statistical properties of orbital wave motions and their application for the measurement of directional wave spectra," *Oceanogr. Soc. Japan.* **19**(4) (1964), pp. 169–191.

31. L. E. Borgman, "Directional spectral model for design use for surface waves," *Hyd. Engrg. Lab., Univ. Calif., Rept.* **HEL 1–12** (1969), 56p.

32. N. N. Panicker and L. E. Borgman, "Enhancement of directional wave spectrum estimates," *Proc. 14th Int. Conf. Coastal Engrg.* (Copenhagen, ASCE) (1974), pp. 258–279.

33. Y. Goda, "Numerical examination of the measuring technique of wave direction with the 'covariance method'," *Rept. Port and Harbour Res. Inst.* **20**(3) (1981), pp. 53–92 (*in Japanese*).

34. A. J. Kuik and G. Ph. van Vledder, "Proposed method for the routine analysis of pitch-roll buoy data," *Proc. Symp. Description and Modelling of Directional Seas*, Paper No. A-5, Tech. Univ. Denmark (June, 1984), 13p.

35. L. E. Borgman, R. L. Hagan and A. J. Kuik, "Statistical precision of directional spectrum estimation with data from a tilt-and-roll buoy," *Topics in*

*Ocean Physics*, ed. A. R. Osborne and P. Malanotte Rizzoli (Noord-Holland, Amsterdam, 1982), pp. 418–438.

36. M. Isobe, K. Kondo and K. Horikawa, "Extension of MLM for estimating directional wave spectrum," *Proc. Symp. Description and Modelling of Directional Seas*, Tech. Univ. Denmark, Paper No. A-6, (1984), 15p.

37. N. Hashimoto and K. Kobune, "Estimation of directional spectra from the maximum entropy principle," *Rept. Port and Harbour Res. Inst.* **24**(3) (1985), pp. 123–145 (*in Japanese*).

38. N. Hashimoto, K. Kobune and Y. Kameyama, "Estimation of directional spectrum using the Bayesian approach, and its application to field data analysis," *Rept. Port and Harbour Res. Inst.* **26**(5) (1987), pp. 57–100.

39. N. Hashimoto and K. Kobune, "Directional spectrum estimation from a Bayesian approach," *Proc. 21st Int. Conf. Coastal Engrg.* (Malaga, Spain, ASCE) (1988), pp. 62–76.

40. N. Hashimoto, T. Nagai and T. Asai, "Extension of the maximum entropy principle method for directional wave spectrum estimation," *Proc. 24th Int. Conf. Coastal Engrg.* (Kobe, ASCE) (1994), pp. 232–246.

41. N. Hashimoto, "Analysis of the directional wave spectrum from field data," *Advances in Coastal and Ocean Engineering*, ed. P. L.-F. Liu **3** (1997), pp. 103–143.

42. M. Benoit, P. Frogaard and H. A. Schäffer, "Analysing multidirectional wave spectra: A tentative classification of available methods," *Proc. IAHR Seminar on Multidirectional Waves and Their Interaction with Structures* 27th IAHR Congress (San Francisco, National Research Council of Canada) (1997), pp. 159–182.

43. R. Kajima, "Estimation of incident wave spectrum in the sea area influenced by reflection," *Coastal Engineering in Japan* **12** (1969), pp. 9–16.

44. E. B. Thornton and R. J. Calhoun, "Spectral resolution of breakwater reflected waves," *Proc. ASCE* **98** (WW4) (1972), pp. 443–460.

45. Y. Goda and Y. Suzuki, "Estimation of incident and reflected waves in random wave experiments," *Proc. 15th Int. Conf. Coastal Engrg.* (Hawaii, ASCE) (1976), pp. 828–845.

46. Q. Fan, "Separation of time series on incident and reflected waves in model test with irregular waves," *China Ocean Engrg.* **2**(4) (1988), pp. 45–60.

47. W. N. Seelig, "Effect of breakwaters on waves: Laboratory tests of wave transmission by overtopping," *Coastal Structures '79* ASCE (1979), pp. 941–961.

48. E. P. D. Mansard and E. R. Funke, "The measurement of incident and reflected wave spectra using a least squares method," *Proc. 17th Int. Conf. Coastal Engrg.* (Sydney, ASCE) (1980), pp. 154–172.

49. P. Gaillard, H. Gauthier and F. Holly, "Method of analysis of random wave experiments with reflecting coastal structures," *Proc. 17th Int. Conf. Coastal Engrg.* (Sydney, ASCE) (1980), pp. 204–220.

50. M. Isobe and K. Kondo, "Method for estimation of directional wave spectrum in incident and reflected wave field," *Proc. 19th Int. Conf. Coastal Engrg.* (Houston, ASCE) (1984), pp. 467–483.

51. D. A. Huntley, D. J. Simmonds and M. A. Davidson, "Estimation of frequency-dependent reflection coefficients using current and elevation sensors," *Proc. Coastal Dynamics '95,* (Gdańsk, Poland, ASCE, 1995), pp. 57–68.

52. N. Hashimoto, T. Nagai and T. Asai, "Modification of extended maximum entropy principle method for estimating directional spectrum in incident and reflected wave field," *Rept. Port and Harbour Res. Inst.* **32**(4) (1993), pp. 25–47 (*in Japanese*) (see also *Proc. 16th OMAE '97,* **II** pp. 1–7.)

53. Y. Goda, "Statistical variability of sea state parameters as a function of a wave spectrum," *Coastal Engineering in Japan* **31**(1) (1988), pp. 39–52.

54. Y. Goda, "Numerical experiments on statistical variability of ocean waves," *Rept. Port and Harbour Res. Inst.* **16**(2) (1977), pp. 3–26.

55. Y. Goda, "Numerical examination of several statistical parameters of sea waves," *Rept. Port and Harbour Res. Inst.* **24**(4) (1985), pp. 65–102 (*in Japanese*).

56. J. R. Medina, J. Aguilar and J. J. Diez, "Distortions associated with random sea simulators," *Proc. ASCE* **111** (WW4) (1985), pp. 603–628.

57. T. Takayama and T. Hiraishi, "Reproducibility of directional random waves in laboratory wave simulation," *Rept. Port and Harbour Res. Inst.* **28**(4) (1989), pp. 3–24.

58. Y. Goda, "Irregular wave deformation in surf zone," *Coastal Engineering in Japan* **18** (1975), pp. 13–26.

Chapter 12

# 2-D Computation of Wave Transformation with Random Breaking and Nearshore Currents

## 12.1 Overview of Numerical Computation Models for 2-D Wave Transformations

Nowadays numerical computation techniques are widely used in solving various problems related with water waves. Begining with wind field estimation from meteorological data, they include forecasting and hindcasting of wave generation, growth, and decay, transformations of waves during their propagation from deep water to the shore, wave actions on maritime structures, sediment transport near the shore, beach morphology, and many others. The term of *numerical analysis* is hereby used for referring to the numerical computation which solves a given set of differential equations in grid-by-grid and/or in time steps.

The present chapter deals with the problems of wave transformations among various fields of numerical wave analysis. Furthermore, discussions are limited to the two-dimensional (2-D) wave transformations for which the vertical wave kinematics are expressed with the solutions of the small amplitude wave theory. There are several advanced methods of numerical wave analysis which enable detailed analysis of wave kinematics and turbulence structures under waves of large heights. Examples are a numerical wave flume called the *CADMUS-SURF* developed by a group of Japanese coastal engineers,[1,2] which solves the Navier-Stokes equations in combination with the VOF (Volume Of Fluid) method, and the technique with 3-D LES (Large Eddy Simulation).[3] These advanced numerical analysis methods are not dealt with here.

Even limited to the numerical analysis methods for 2-D wave transformations, there have been developed a large number of numerical models since around the 1960s. The author has selected several typical models

Table 12.1   A tentative list of numerical analysis models for 2-D wave transformation.

| Type | Name or Basic Equation | Area | Characteristics | Wave Breaking |
|---|---|---|---|---|
| Phase-averaged | Helmholtz eq. | Harbor basin | Multi-zone divisions (constant depth) | Not considered |
| | Takayama method | Ditto | Multi-diffraction and reflection | Ditto |
| | Energy balance eq. | Very large | Directional spectrum | Energy dissipation term |
| | SWAN model | Ditto | Wave-current interaction | Bore-type model |
| | Mild slope eq. | Medium | Refraction and diffraction | Energy dissipation term |
| | Parabolic eq. | Large | Neglect of reflected waves | Ditto |
| | PEGBIS model | Ditto | Parabolic eq. modified | Gradational breaker index |
| Phase-resolving | Boussinesq eq. | Medium | Weakly nonlinear and weakly dispersive | Judged by pressure gradient |
| | Nonlinear mild slope eq. | Small | Strongly nonlinear and strongly dispersive | Judged by velocity ratio |
| | Nonlinear, multilayer wave eq. | Ditto | Multilayer formulation | not introduced yet |

somewhat subjectively and listed them in Table 12.1 according to their characteristics. These models are capable of application for random waves.

The numerical analysis models are classified into the *phase-averaged type* and the *phase-resolving type*. The phase-averaged type model yields the spatial distribution of wave amplitude and direction only without computing wave profiles. It can solve wave transformation in a large area within a relatively short computation time. The computation grids can be set with the grid spacing at one-tenth of wavelength or larger.

On the other hand, the phase-resolving type model, which is also called the time-evolution model or the time-domain model, computes the propagation of wave profiles in time steps and thus enables to examine minute phenomena such as wave splitting over a shoal and others. However, the computation must be made with the grid spacing being less than a few hundredths of wavelength and the time step being less than a few percent of wave period. The computation load becomes quite heavy with a long computation time, which may need to be counted by the units of days instead of hours or minutes. The size of computation area is limited by the capacity of a computer available. When a wave-related problem is to be solved with a numerical analysis model, selection must be made of the model appropriate for the assigned task in consideration of the characteristics of candidate models and the nature of the solution required.

The numerical model based on the Helmholtz equation was developed in France in the 1960s to compute the spatial distribution of wave heights within a harbor basin having a relatively narrow entrance. It is listed in Table 12.1 as being one of the early models for analysis of harbor tranquility, even though it employs monochromatic waves. Since the late 1970s, harbor tranquility against directional random waves has been analyzed in Japan with the Takayama method,[4,5] which solves the multiple diffraction and reflection problems in harbor basins based on the Sommerfeld equation for wave diffraction and the use of fictitious breakwaters for propagation of waves reflected by structures within a harbor. An example of harbor tranquility analysis was presented by Goda *et al.*[6] in 1978. Both the Helmholtz equation model and the Takayama method do not consider wave decay by breaking, and no further comment is made on the both methods in this chapter, which focuses on numerical models with wave breaking process.

In the wave field where waves are breaking in a certain area of the computation domain, computed results depend largely on modeling of wave breaking process employed in the numerical model rather than the basic equation for wave analysis itself. Figure 3.43 in Sec. 3.6.4 is a demonstration of the influence of wave breaking modeling on cross-shore distribution of wave heights on a planar beach. Difference in wave height distributions brings forth different spatial distribution of radiation stress and yields different prediction of nearshore currents. The author[7] has presented examples of different longshore current profiles resulted from seven models of random wave transformation by breaking on planar beaches for the same offshore wave condition; the peak velocity is not much affected by the models, but the width of longshore current zone differs greatly. When a project of wave field and/or nearshore current computation is undertaken, selection of the computation model should be made from the viewpoint of appropriateness of random wave breaking modeling in addition to the computation performance of the numerical model.

## 12.2	Outline of Phase-Averaged Type Wave Transformation Models

(A) *Energy balance equation and wave action balance equation*

The numerical model based on the *energy balance equation* solves the shoaling and refraction processes of directional random waves. The basic equation has been presented as Eq. (3.19) in Sec. 3.2.3. Karlsson[8] presented this model in 1969 by applying the numerical wave forecasting model of the first generation to the problems of wave transformation in relatively shallow water by restricting wave propagation in the forward direction only. Nagai *et al.*[9] employed this model in 1974 for analyzing wave transformation from the offshore toward the design site of a breakwater in Japan. The energy balance equation can be solved with a large grid size such as several grid points per one wave length, which would be sufficient for computing wave refraction. Because the model allows the use of large grid size, it has been used in many practical applications for wave propagation and transformation from the offshore to the shore.

Another phase-averaged type model for directional random wave is the model based on the so-called *wave action balance equation*, which solves the following equation for the wave action denoted with $N$ and defined as the directional spectral density $S(f, \theta)$ divided by the angular frequency $\omega$, i.e., $N = S(f, \theta)/\omega$:

$$\frac{\partial N}{\partial t} + \frac{\partial (Nc_x)}{\partial x} + \frac{\partial (Nc_y)}{\partial y} + \frac{\partial (Nc_\sigma)}{\partial \sigma} + \frac{\partial (Nc_\theta)}{\partial \theta} = \frac{S}{\omega}, \quad (12.1)$$

where $c$ denotes the vector sum of the wave celerity and the velocity of co-existing currents. The term $S$ at the right side of Eq. (12.1) is a source function expressing the inflow and outflow of the energy of waves in the system. As implied from its formulation, Eq. (12.1) has originally been developed for describing generation and growth of wind waves in shallow water. Holthuijsen *et al.*[10] proposed a numerical computation model of stationary transformation of spectral waves in shallow water by deleting the first term in the left side of Eq. (12.1) for temporal variation of wave action, and they named the model as *SWAN* (Simulating WAves in the Near shore). A detailed explanation of SWAN can be found in the book by Holthuijsen.[11]

Both the energy balance and wave action balance equations do not include the terms for wave diffraction process. Though the wave diffraction by a large topographic barrier such as headland or island is approximately represented by the directional spreading process (refer to Sec. 3.3.4), the refraction-diffraction phenomenon over a shoal and the wave diffraction through a narrow opening between breakwaters cannot be computed by the both models based on these equations. To remedy the deficiency, Booji et al.[12] and Rivero et al.[13] have proposed to introduce the terms which approximately represent the diffraction effect into the wave action balance equation. On the other hand, Mase[14] and Mase et al.[15] formulated the diffraction term based on the parabolic equation and introduced the term explicitly in the energy balance equation.

The energy dissipation by random wave breaking is introduced in the source function $S$ of Eq. (12.1). The rate of energy dissipation in terms of the root-mean-square wave height $H_{\mathrm{rms}}$ is evaluated with the bore-type model by Battjes and Janssen[16] in the case of SWAN model.[17,18] Models based on the energy balance equation introduce a dissipation term proportional to the spectral density, $-\varepsilon_b' S(f, \theta)$, in the right side of Eq. (3.19) in Sec. 3.2.3. Takayama et al.[19] have evaluated the energy dissipation rate $\varepsilon_b'$ with the random wave breaking model by Goda in 1975.[20]

Both the models based on the energy balance and wave action balance equations yield a single characteristic wave height such as $H_{\mathrm{rms}}$ or $H_{1/3}$ only, and cannot provide the information on other characteristic wave heights such as $H_{1/250}$ and $H_{1/20}$, which are needed for design of maritime structures. When the information is required, it is necessary to employ other models such as PEGBIS and Boussinesq equation models.

## (B) *Mild slope equation and parabolic equation*

The basis of the majority of current wave transformation models is the mild slope equation proposed by Berkoff[21] in 1972. According to the lecture note by Nadaoka,[22] the complex-valued wave amplitudes $\hat{\eta}$ at respective coordinates are expressed by the following equation of elliptic type:

$$\nabla \cdot (cc_g \nabla \hat{\eta}) + k^2 cc_g \hat{\eta} = 0, \tag{12.2}$$

where $c$ and $c_g$ denotes the wave celerity and group velocity, respectively, $k$ is the wave number, and $\nabla = (\partial/\partial x, \partial/\partial y)$. Variation of surface elevation by waves is expressed by $\eta(x, y, t) = \hat{\eta}(x, y) \exp[i\omega t]$ with $\omega$ being the angular frequency.

Equation (12.2) of elliptic type requires specification of the boundary conditions along the whole periphery and has some difficulty in numerical computation. Nevertheless, Sato et al.[23] solved the wave height distribution in a harbor basin of arbitrary shape using this equation in 1988. Most of current wave transformation models, however, are based on the equations of parabolic type derived from Eq. (12.2) by separating wave components into those propagating in the principal direction and those propagating in the opposite direction and by deleting the latter components. Radder[24] was the first proposing the parabolic type equation of the following form:

$$
\frac{\partial \phi}{\partial x} = \left\{ ik - \frac{1}{2kcc_g} \frac{\partial}{\partial x}(kcc_g) \right\} \phi + \frac{i}{2kcc_g} \frac{\partial}{\partial y} \left( cc_g \frac{\partial \phi}{\partial y} \right), \qquad (12.3)
$$

where $\phi$ denotes the complex amplitude of the velocity potential of waves at the steady state condition. The amplitude and direction of waves are given by the absolute value and the phase of the velocity potential, respectively.

In derivation of Eq. (12.3), Radder assumed that waves propagate nearly parallel to the $x$ axis at all the points. The assumption induces a certain amount of error for waves of oblique incidence. To remedy this problem, several modifications have been proposed, among which the following is due to Hirakuchi and Maruyama[25]:

$$
\frac{\partial \phi}{\partial x} = \left\{ i \left( k_x + \frac{k_y^2}{2k_x} \right) - \frac{1}{2k_x cc_g} \frac{\partial}{\partial x}(k_x cc_g) \right\} \phi + \frac{i}{2k_x cc_g} \frac{\partial}{\partial y} \left( cc_g \frac{\partial \phi}{\partial y} \right),
$$

$$(12.4)$$

where $k_x$ and $k_y$ denote the wave numbers in the $x$ and $y$ directions, respectively. The $x$ axis is taken positive offshore perpendicularly to the shoreline, while the $y$ axis is parallel to the shoreline.

Hirakuchi and Maruyama have shown that Eq. (12.4) can solve the wave refraction-diffraction over an circular shoal quite correctly to the incident angle of up to 30° and simulate the wave field properly to the incident angle of up to 45° when a small difference from the exact solution is allowed in the wave height distribution. Improvement in the analysis of oblique incident waves is continuing as exemplified by the works by Li[26] and Saied and Tsanis.[27] Both works have introduced weak nonlinearity in the parabolic equations, following the approach by Kirby.[28]

The parabolic equation models can solve the wave diffraction by breakwaters and other barriers quite correctly except for the situation in which the diffracted waves have to penetrate into the sheltered area in the direction opposite to the wave incidence; a case is the breakwater aligned with the angle less than 90° to the $x$ axis. The parabolic equation models have another difficulty in dealing with the steep seabed, because they are based on the assumption of gradual bottom change employed in Eq. (12.3) as indicated in its name of mild slope; Nadaoka[22] suggests the seabed slope to be milder than 1 on 3. When the models are applied for a group of submerged, detached breakwaters with wide crests which have the side slopes of 1 on 2 or so, refraction of waves at the side slopes are overestimated and so is the wave attenuation by them.[29]

Transformation of directional random waves can be solved with the parabolic equation model by computing transformation of individual wave components and linearly superposing the computed results. For validation tests of the parabolic equation models, the laboratory test data of directional random wave refraction, diffraction, and breaking over an elliptic shoal by Vincent and Briggs[30] is often employed. Özkan and Kirby[31] are the first in comparing their computed wave height distribution with the measurement data by Vincent and Briggs, and later followed by Yoon *et al.*[32] and the author.[33] In the parabolic equation models, wave attenuation is simulated by adding a term of energy dissipation which is proportional to the complex amplitude of the velocity potential. The energy dissipation rate in the model of Özkan and Kirby is estimated by the random wave breaking scheme by Thornton and Guza,[34] in the model of Yoon *et al.* by the scheme by Battjes and Janssen[35] and in the author's model by the gradational breaker index scheme.

For the test cases without wave attenuation by breaking, all three models have produced the computed wave heights in good agreement with the measured data, but the models have not succeeded in reproducing the wave height distribution for the test cases with intensive wave breaking. According to the computer analysis by Choi *et al.*[36] using a phase-resolving type model based on the Boussinesq equation, there exists a strong wave-induced jet-like current on the rear side of the shoal flowing toward the down-wave direction. The current defocuses wave concentration by refraction and reduces the wave heights along the central longitudinal axis, thus contributing to unsuccessful predictions of the above three models.

## 12.3   Outline of Phase-Resolving Type Wave Transformation Models

(A) *Time-dependent mild slope equation and nonlinear mild slope equation*

The mild slope equation of Eq. (12.2) is said to be difficult in solving numerically, and thus it is sometimes reformulated in a time-evolution type. For example, Watanabe and Maruyama[37] have proposed the following formulation (after Nadaoka[22]):

$$
\left.
\begin{aligned}
\frac{\partial \eta}{\partial t} + \nabla \cdot \vec{Q} &= 0 , \\
\frac{\partial \vec{Q}}{\partial t} + \frac{1}{n} c^2 \nabla(n\eta) &= 0 ,
\end{aligned}
\right\}
\tag{12.5}
$$

where $\vec{Q}$ denotes the current flux vector and $n$ represents the ratio of group velocity to wave celerity, i.e., $n = c_g/c$.

Equation (12.5) cannot be used for random waves having the components of diverse periods in this form, because the wave celerity $c$ and the velocity ratio $n$ are the function of wave period. Thus, Isobe[38] and Ishii *et al.*[39] have derived the time-dependent mild slope equations which are applicable for waves characterized with frequency spectra.

Nonlinear shoaling phenomenon in which the wave height increases rapidly as waves approach the breaking point cannot be dealt with Eq. (12.5), because it is a set of linear equations. Arikawa and Isobe[40] have derived a nonlinear mild slope equation based on the proposal by Isobe.[41] The equation can deal with wave transformation under the strongly nonlinear and strongly dispersive conditions, though limited to regular waves. Wave breaking is judged by employing the ratio of water orbital velocity to wave celerity as an breaker index, and the energy dissipation by breaking is included in the equation. Application of this model to a prototype bathymetry has been reported by Arikawa and Okayasu.[42]

## (B) *Nonlinear, strongly-dispersive wave equation*

The nonlinear mild slope equation in the above has been derived by approximately representing the velocity potential with a series of multiple vertical distribution functions, and it is one of the multiple coupling formulations after Nadaoka.[22] There have been several such models, which may be classified according to the selection of vertical distribution functions and the coupling method of multiple functions. Most of the papers dealing with nonlinear, strongly-dispersive equations discuss the methodology of theoretical development rather than applications on actual bathymetry. An exception is the work by Kanayama,[43] who computed the transformation of regular waves around a 2-D submerged breakwater by using a set of multi-level wave equations. The water body in the computation area is divided into multiple levels, and the Boussinesq equations are applied for respective levels to solve wave kinematics. Kanayama did not consider wave attenuation by breaking, but its introduction should be possible in principle.

Numerical models based on the nonlinear mild slope equations as well as nonlinear, strongly-dispersive wave equations can compute wave transformation accurately for a water area of small size. Because the computation load is heavy, however, it would take some years before these models are widely used for solving practical problems. Extension of these models for directional random waves would be another task for practical applications.

(C) *Boussinesq equation model*

Among various phase-resolving type wave models, those based on the Boussinesq equation are popular presently. The basis of the Boussinesq equation was presented by Peregrine[44] in 1967 in the following functional form:

$$\frac{\partial \eta}{\partial t} + \nabla[(\eta + h)\vec{u}] = 0\,, \tag{12.6}$$

$$\frac{\partial \vec{u}}{\partial t} + \vec{u}\nabla\vec{u} + g\nabla\eta = \frac{h}{2}\nabla\left[\nabla\cdot\left(h\frac{\partial\vec{u}}{\partial t}\right)\right] - \frac{h^2}{6}\nabla\left[\nabla\cdot\frac{\partial\vec{u}}{\partial t}\right]\,, \tag{12.7}$$

where $\vec{u}$ denotes the horizontal velocity vector averaged over the depth, and $h$ is the water depth.

The Boussinesq equation is weakly nonlinear, and it does not satisfy the dispersion relationship between the angular frequency and wavenumber, because it was originally derived for long waves. Thus it could not be employed for waves in water having relatively large depth-to-wavelength ratio and so remained for some years. In 1991, however, Madsen *et al.*[45] have proposed a formulation for improving the accuracy of approximation to the dispersion relationship, by adding a few modification terms to Eq. (12.7). The modified equation of motion is expressed as in the following:

$$\frac{\partial \vec{u}}{\partial t} + \vec{u}\nabla\vec{u} + g\nabla\eta = Bgh\nabla[\nabla\cdot(h\nabla\eta)]$$
$$+ h\frac{\partial}{\partial t}\left\{\left(\frac{1}{2}+B\right)\nabla[\nabla\cdot(h\vec{u})] - \frac{1}{6}h\nabla(\nabla\cdot\vec{u})\right\}\,, \tag{12.8}$$

where $B$ is a parameter to which Madsen *et al.* assigned the value of $B = 1/15$ as best approximating the dispersion relationship.

There have been proposed several other formulations to improve the accuracy of approximation to the dispersion relationship. Hirayama[46] has examined various formulations from the viewpoints of both the approximation to the dispersion relationship and the accuracy in dealing with secondary interaction waves. He recommends the value of $B = 1/15$ based on his examination results.

For execution of wave field computation with the Boussinesq equation, various computer programs have been developed to solve the problems of wave absorption at the wave generating area and the side boundaries, partial wave reflection from structures within the computation area, and others. Hirayama,[46] for example, discusses the techniques to deal with these problems in detail.

Deformation of wave profiles by depth-controlled breaking can be dealt with the Boussinesq equation. Kabiling and Sato[47] expressed the wave breaking effect by means of the turbulent eddy viscosity. They added a dispersion term to the momentum equation by taking account of the breaker-induced turbulence. The approach was also adopted by Hirayama,[46] who initially employed the wave breaking condition such that the absolute orbital velocity exceeds the speed being 64% of the wave celerity which is approximated by $\sqrt{gh}$. This breaking condition, however, yields the constant breaker height among irregular waves at a given water depth, and it cannot simulate actual situation of random wave breaking in which individual breaker heights vary over a certain range (refer to Fig. 3.31 in Sec. 3.6.2).

To remedy the deficiency, Hara *et al.*[48] introduced the condition of zero vertical pressure gradient at a wave crest, which owes to Nadaoka *et al.*,[49] and evaluated the energy dissipation rate with a pseudo bore model in time-domain. Because the Boussinesq equation is only weakly nonlinear and the numerical computation is carried out in a finite difference scheme, Hirayama and Hiraishi[50] have relaxed the wave breaking condition of zero gradient as in the following:

$$-\frac{1}{\rho g}\frac{\partial p}{\partial z}\bigg|_{z=\eta} = 0.5\,. \tag{12.9}$$

Introduction of this breaker index has enabled to appropriately simulate not only the spatial variation of characteristic wave heights but also the distribution of individual wave heights within the surf zone.

A group of engineers at the Port and Airport Research Institute and other institutions in Japan has developed and been improving a numerical wave model based on the Boussinesq equation under the name of NOWT-PARI. Computation of wave transformations in an area of several kilometers square may take a few days per case presently. Similar development of wave models has been carried out at various institutions in the world, and the computation time per case would be similar.

## 12.4   Wave Transformation Analysis with PEGBIS Model

(A) *Basic equations*

Design of maritime structures needs the information of various characteristic wave heights other than the significant or the root-mean-square wave height. Examples are the maximum wave height $H_{\max}$ or $H_{1/250}$ for composite breakwaters and the highest one-twentieth wave height $H_{1/20}$ for armor units of rubble mound breakwaters. The assumption of the Rayleigh distribution does not hold when waves are affected by the breaking process, and thus the information of $H_{m0}$ or $H_{\mathrm{rms}}$ cannot yield reliable estimates of $H_{1/250}$ or $H_{1/20}$. A phase-resolving type model such as the Boussinesq equation may be run over a sufficiently long duration so that the information of individual wave heights and periods is obtained by the zero-crossing method. However, the computation load would be too heavy for routine design works.

One of the wave transformation models which can produce the spatial variation of wave height distribution within the surf zone is the author's model[20] in 1975. However, it deals with planar beaches only and is difficult to be applied to beaches of arbitrary bathymetry. Thus, the author[29,33] has proposed a phase-averaged wave transformation model based on the parabolic equation and using a gradational breaker index, which is named as the *PEGBIS* (Parabolic Equation with Gradational Breaker Index for Spectral waves). The present section is dedicated to description of this model.

The basic equation of this model is the parabolic equation proposed by Hirakuchi and Maruyama[25] and presented as Eq. (12.4). The energy dissipation by bottom friction and wave breaking is introduced by a term, $-f_D\phi$, in the right side of the equation, i.e.,

$$\frac{\partial \phi}{\partial x} = \left\{ i \left( k_x + \frac{k_y^2}{2k_x} \right) - \frac{1}{2k_x cc_g} \frac{\partial}{\partial x} (k_x cc_g) \right\} \phi$$
$$+ \frac{i}{2k_x cc_g} \frac{\partial}{\partial y} \left( cc_g \frac{\partial \phi}{\partial y} \right) - f_D \phi. \qquad (12.10)$$

The wave celerity $c$ is modified as in the following so as to approximately express the celerity increase by the nonlinearity effect:

$$c = \begin{cases} c_A \left[ 1 + \dfrac{3}{2} \left( \dfrac{a}{h} \right)^2 \right]^{1/2} & : a < h, \\[3mm] c_A \left[ 1 + \dfrac{3}{2} \left( \dfrac{a}{h} \right) \right]^{1/2} & : a \geq h, \end{cases} \tag{12.11}$$

where $c_A$ denotes the wave celerity of small amplitude waves (Airy's waves), $a$ is the wave amplitude as defined by the absolute value of the complex velocity potential $\phi$, and $h$ is the water depth. The formulation of celerity increase is made by referring to the nonlinear wave celerity of the third order Stokes waves. The wave nonlinerity effect is not considered for the group velocity $c_g$, because the transport of wave energy is least affected by wave nonlinearity.

The rate of energy dissipation is assumed to be proportional to $\phi$, and its coefficient $f_D$ is treated as the sum of the coefficient $f_{Df}$ to account for the dissipation due to bottom friction and the coefficient $f_{Db}$ for the energy dissipation by breaking, i.e., $f_D = f_{Df} + f_{Db}$. The frictional dissipation coefficient is derived from calculation of energy dissipation within the turbulent boundary layer at the sea bottom as

$$f_{Df} = \frac{4}{3\pi} C_f \frac{a}{h^2} \frac{k^2 h^2}{\sinh kh \left( \sinh 2kh + 2kh \right)}, \tag{12.12}$$

where $C_f$ denotes the friction coefficient at the sea bottom, which may be taken at $C_f = 0.01$ in general.

### (B) *Coefficient of energy dissipation by breaking*

The coefficient of energy dissipation by breaking has been set as in the following, by referring to the approach by Dally *et al.*[51]:

$$f_{Db} = \begin{cases} 0 & : a \leq \kappa h, \\[3mm] \dfrac{K_b}{2h} \left[ \left( \dfrac{a}{\kappa h} \right)^2 - 1 \right]^{1/2} & : a > \kappa h, \end{cases} \tag{12.13}$$

where $K_b$ is a factor, the value of which depends on the beach slope $s$ as set in the following:

$$K_b = \frac{3}{8}(0.3 + 2.5s). \tag{12.14}$$

The term $\kappa$ in Eq. (12.13) represents a breaker index defined as the ratio of the wave amplitude at the breaking limit to the local water depth. The energy dissipation is set in such a way that it takes place in proportion to the value of $[(a/\kappa h)^2 - 1]^{1/2}$ when the wave amplitude $a$ exceeds the threshold of $\kappa h$. Many models on wave attenuation by breaking assign a constant value to $\kappa$, but the PEGBIS model assigns the value varying with the individual wave height levels. First, the $m$th wave height $H_m$ is defined for $M$ levels with the following formula for the incident offshore significant wave height $(H_{1/3})_0$:

$$H_m = 0.706(H_{1/3})_0 \left[ \ln \frac{2M}{2m-1} \right]^{1/2} : m = 1, 2, \ldots, M. \qquad (12.15)$$

The above wave height represents the odd number level height when the Rayleigh distribution of wave heights is divided by $2M$ with an equal probability of occurrence; $H_1$ represents the highest level, while $H_M$ corresponds to the lowest level.

The gradational breaker index $\kappa_m$ is defined for the $m$th level of wave height as in the following:

$$\kappa_m = \left( C_b \frac{L_0}{h} \left\{ 1 - \exp\left[ -\frac{1.5\pi h}{L_0}(1 + 15s^{2.5}) \right] \right\} \right.$$
$$\left. + \beta_0 \frac{H_m}{h} \left( \frac{H_m}{L_0} \right)^{-0.38} \exp[30s^2] \right) \times \left( \frac{H_m}{H_1} \right)^p . \qquad (12.16)$$

The constants $C_b$, $\beta_0$, and $p$ have been assigned the following values:

$$\left. \begin{array}{l} C_b = 0.080, \beta_0 = 0.016, p = 0.333 \ : \ s > 0 \,, \\ C_b' = 0.070, \beta_0' = 0.016, p' = 0.667 \ : \ s \leq 0 \,. \end{array} \right\} \qquad (12.17)$$

As shown in Eq. (12.17), two sets of constants are used depending on whether the seabed has a positive or non-positive slope. The set of constants for non-positive slopes, $C_0', \beta_0'$, and $p'$, is required for simulating the distribution of individual wave heights in trough zones of bar and trough beaches or on horizontal steps.

Over the sea bottom of complicated bathymetry, the slope $s$ is to be calculated with the depth difference between the measuring point and the offshore location at the distance of $0.1L_0$ in the principal wave direction. Figure 12.1 shows the variation of the gradational breaker index with respect to the wave height level. The upper diagram is the case of the water depth decreasing toward the shore with a uniform slope of $1/20$, while the lower diagram is the case of a horizontal step. The abscissa is the ratio of

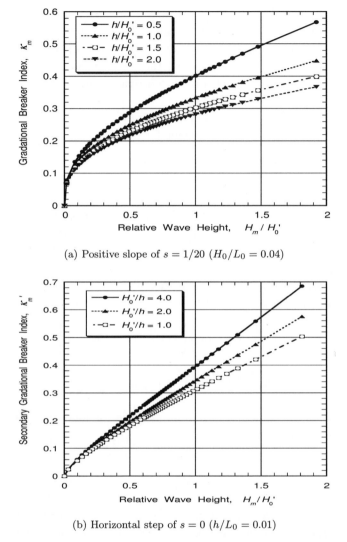

(a) Positive slope of $s = 1/20$ $(H_0/L_0 = 0.04)$

(b) Horizontal step of $s = 0$ $(h/L_0 = 0.01)$

Fig. 12.1   Variation of gradational breaker index with respect to wave height level.[33]

the $m$th level wave height $H_m$ to the equivalent deepwater wave height $H_0'$, which has been introduced in Sec. 3.4. In the upper diagram, the inspection site is shifted from the depth of $h = 2.0H_0'$ to that of $h = 0.5H_0'$. As the water become shallow, the value of gradational breaker index increases. In the lower diagram, the water depth is constant but the incident wave height varies.

The gradational breaker index of Eq. (12.16) has been formulated with the following idea. The first term within the large parantheses in the right-hand side with the constant $C_b$ is a modification of the breaker index of regular waves of Eq. (3.32) in Sec. 3.6.1; the constant value of $C_b = 0.080$ is equivalent to $A = 0.16$. The second term within the large parantheses in the right-hand side with the constant $\beta_0$ is a modification of the coefficient $\beta_0$ in Table 3.6 in Sec. 3.6.3 for the purpose of retaining a finite wave height at the shoreline. The multiplicative term of $(H_m/H_1)^p$ is for adjustment of the wave height distribution. The values of the constants $C_0, \beta_0$, and $p$ for positive slopes have been so selected to yield the variation of characteristic wave heights across the surf zone as close as possible to those shown in Figs. 3.39 to 3.42 in Sec. 3.6.3. The values of the constants $C_0', \beta_0'$, and $p'$ for non-positive slopes have been adjusted with several laboratory data available.

(C) *Synthesis of wave height distribution from multiple level computation*

Wave transformation and deformation are computed with the parabolic equation of Eq. (12.10) at $M$ levels of wave heights. At every grid point in the computation area, the results of $M$ level wave heights are synthesized to produce the distribution of transformed wave heights. Figure 12.2 shows

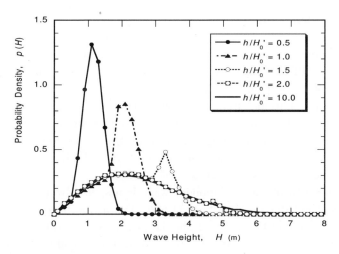

Fig. 12.2 Spatial variation of wave height distribution in PEGBIS model.[33] $(s = 1/100, H_0' = 4.0$ m).

an example of the variation of wave height distribution from the offshore site with $h = 10H_0'$ to a shallow site with $h = 0.5H_0'$ on a uniform planar beach of $s = 1/100$ for the unidirectional random waves of $(H_{1/3})_0 = 4.0$ m and $(L_{1/3})_0 = 100$ m. At the offshore site, the wave height follows the Rayleigh distribution as expected. At the location of $h = 2.0H_0'$, only the portion of waves greater than 5.5 m disappears. As waves propagate toward the shore, however, the right tail of the distribution is gradually removed and there appear a triangular-shaped concentration of wave height at the right side. Though the simulated wave height distribution is not as smooth as that exhibited in Fig. 3.31 in Sec. 3.6.2(B), the spatial variation of characteristic wave height ratios across the surf zone is approximately reproduced as demonstrated in Figs. 3.32 and 3.33.

(D) *Synthesis of wave height from computed directional components*

Computation of 2-D wave field needs to be made with due consideration for the spreading of wave energy in the frequency and directional domains, which is expressed with the directional wave spectrum. The PEGBIS model employs the combination of the JONSWAP spectrum of Eq. (2.12) in Sec. 2.3.1 and the Mitsuyasu-type spreading function of Eqs. (2.22) and (2.25) in Sec. 2.3.2 as the input directional wave spectrum. The frequency spectrum is initially divided into $N_0$ frequencies given by

$$f_n = 1.057 f_p \left[ \ln \frac{2N_0}{2N_0 - 2n + 1} \right]^{-1/4} \quad : n = 1, 2, \ldots, N_0, \qquad (12.18)$$

where $f_p$ represents the frequency at the spectral peak. The frequencies defined by Eq. (12.18) equally divide the wave energy given by the JONSWAP spectrum with the peak enhancement factor of $\gamma = 1$. For the spectrum with $\gamma > 1$, additional frequency components are supplemented around the spectral peak so that all the frequency components are provided with the approximately equal energy. The total number of component waves becomes $N = 1.23N_0$ for the case of $\gamma = 2$, $N = 1.64N_0$ for the case of $\gamma = 3.3$, and $N = 4.3N_0$ for the case of $\gamma = 20$.

The azimuth of each component wave is assigned by the single summation method described in Sec. 11.5.2. The azimuth is limited within the range $\pm 0.55\pi (10/s_{\max})^{0.35}$ from the principal wave direction so as to economize computation works for integrating the spreading function and assigning the azimuth by means of a random number being uniformly distributed between [0, 1]; integration of the spreading function for each frequency is made with one hundred equal divisions of the azimuth range. For oblique incident waves, the azimuth is further limited within the range not exceeding $\pm 90°$ from the $x$ axis. The $N$ component waves often yields somewhat asymmetric distribution of azimuths around the principal wave direction, which leads to asymmetric wave height distribution. To remedy this problem, another set of $N$ component waves are added with the azimuths assigned at the directions symmetrical to those of the original set with respect to the principal wave direction. Thus the computation is made for $2N$ component waves in total.

The flow of computation of the PEGBIS model is shown in Fig. 12.3. Computation for wave transformation of $2N$ spectral components are repeated at $M$ levels of wave height. Because the spectral components are given the equal energy, the arithmetic mean of computed results yields the wave height and other wave statistics at each wave height level, and the result is stored in memory for all grid points in the computation area. Contribution of each wave height level to the radiation stresses is also computed and stored in memory. When computation of the $M$ wave height levels is completed, the highest $(1/n)$th wave heights, the principal directions, the radiation stresses and others are calculated at all grid points and the cycle is closed.

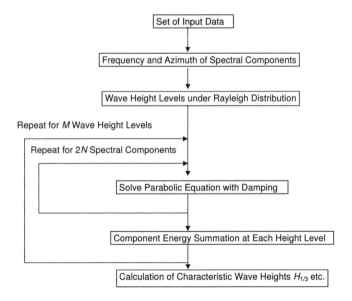

Fig. 12.3   Computation flow of PEGBIS model.

(E) *Setting of computational conditions and computation time*

Computation of solving the parabolic equation is repeated $M \times 2N$ times in total. When a detailed information on wave height variation is sought for, the wave height level $M$ is better set at a large number. The results shown in Fig. 12.2 have been obtained by computation with $M = 831$. When the information on the characteristic wave height such as $H_{1/3}$ or $H_{\mathrm{rms}}$ alone is necessary, the number of wave height level may be much lowered. A simple decrease in the number $M$ while using the breaking constant value of $C_b = 0.08$ makes the gradational breaker index $\kappa_1$ for the highest height $H_1$ at slightly larger than the optimal value, however, It is recommended to reduce the constant value to $C_b = 0.077$ for $M = 101$, $C_b = 0.076$ for $M = 61$, and $C_b = 0.074$ for $M = 31$. No adjustment is required for $M > 300$.

The spectral component number $N$ should be selected in consideration of the objective of analysis. Computation presented in Fig. 12.2 was carried out with $N = 10$ for unidirectional waves without directional spreading. For comparison with the laboratory data on an elliptic shoal by Vincent and Briggs,[30] spectral components with $2N = 148$ were employed at the wave height level of $M = 61$. The computation area was divided in 111 grids in the $x$ direction and 151 grids in the $y$ direction. The computation time by a PC with a Pentium III processor was about 30 minutes. When a wave field in the area of several kilometers square was analyzed with the PEGBIS model, the area was composed of $1,300 \times 700$ grids and computation was executed with $2N = 100$, and $M = 61$. The computation time by a PC with a Pentium IV processor was about 12 hours.

The selection of grid spacing depends on the bathymetry. Where the variation in bathymetry is gradual, several grids per wavelength would suffice. Where the bathymetric variation is large, about twenty grids per wavelength would be recommended. For simulation of linear wave shoaling, the depth difference $\Delta h$ between the adjacent grids needs to be set less than about one-tenth of the minimum water depth to which waves are to propagate. When the wave breaking process during shoaling is taken into account, however, the maximum depth difference can be set at $\Delta h \simeq 0.1(H_{1/3})_0$.

(F) *Comparison of computed cross-shore wave height profiles with field observation and large-scale laboratory data*

As stated before, the constants $C_b$, $\beta_0$, and $p$ of the gradational breaker index $\kappa_m$ have been so adjusted to minimize the difference between the

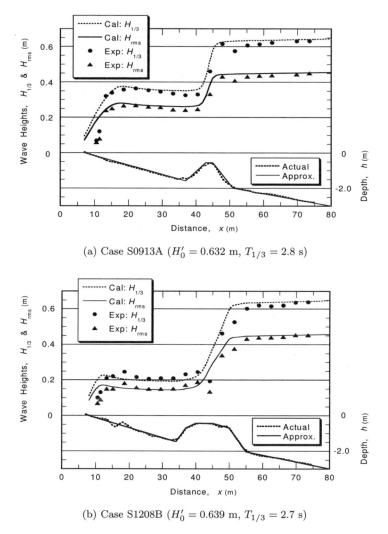

(a) Case S0913A ($H_0' = 0.632$ m, $T_{1/3} = 2.8$ s)

(b) Case S1208B ($H_0' = 0.639$ m, $T_{1/3} = 2.7$ s)

Fig. 12.4   Comparison of cross-shore wave heights computed with PEGBIS model and those measured in laboratory for beaches with bar and trough.[33]

wave heights predicted by the PEGBIS model and the Goda model in 1975. Thus the critical test of the PEGBIS model is its validation against the wave transformation on beaches of complicated bathymetry. One of the data sets to be utilized for validation is the SUPERTANK project reported by Kraus *et al.*,[52] which measured the evolution of sandy beach profiles under irregular waves with the height $H_0' = 0.60 \sim 0.80$ m. Figure 12.4 shows comparison of the computed and measured heights of $H_{1/3}$ and $H_{rms}$ for two cases of distinct bar and trough bathymetries as presented by Goda.[33] Beach profiles are shown in the lower panels. Wave heights drop rapidly on top of bars and then gradually recover in trough zones. The measured cross-shore variations of both $H_{1/3}$ and $H_{rms}$ are well reproduced with the PEGTBIS model for both cases.

As for the field measurement data of cross-shore wave heights on beaches with bars and troughs, several researchers have employed the minute field measurement results at Ajigaura Beach, Ibaragi Prefecture in Japan, reported by Hotta and Mizuguchi[53] in 1980, for validation tests of their numerical wave models. Waves were swell of low steepness ($H_0/L_0 \simeq$ 0.0055). The cross-shore variations of the three characteristic wave heights, $H_{1/10}, H_{1/3}$, and $H_{rms}$, computed with the PEGBIS model are compared with the measured ones in the upper panel of Fig. 12.5. The beach profile is shown in the lower panel. The measured wave heights, especially of $H_{1/10}$, exhibit conspicuous increases at the location of $x = 80$ to $100$ m from the shore, owing to the nonlinear wave shoaling. Linear wave transformation models including PEGBIS cannot reproduce such nonlinear features. Except for this nonlinear shoaling section, the PEGBIS model yields the cross-shore variations of three characteristic wave heights in good agreement with the measured ones.

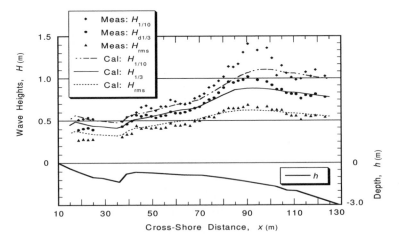

Fig. 12.5  Comparison of cross-shore wave heights computed with PEGBIS model and those measured at Ajigaura Beach, Ibaragi Prefecture, Japan[33]: data by Hotta and Mizuguchi.[53] ($H_0' = 0.70$ m and $T_p = 9.0$ s).

For both the large-scale laboratory data and the field data in Figs. 12.4 and 12.5, the breaking constant $C_b$ was set at the value of 0.08. However, when the PEGBIS model was applied to the field data obtained during DUCK 94 and SandyDuck projects at Duck Beach in Virgina, USA and another set of data at Leadbetter Beach in California, Beach, reported by Thornton and Guza,[54] use of $C_b = 0.08$ produced some overestimation of wave heights within the surf zone. It was necessary to lower the breaking constant value to 0.07 or lower for beaches extending over several hundred meters.[55] Although further tuning of the constant values is necessary with more field measurement data, the following values are recommended for the time being when compution of waves is made at the area where no wave measurement data is available:

$$\left. \begin{array}{l} C_b = 0.070, \, \beta_0 = 0.016, \, p = 0.333 \; : \; s > 0\,, \\ C_b' = 0.061, \, \beta_0' = 0.016, \, p' = 0.667 \; : \; s \leq 0\,. \end{array} \right\} \qquad (12.19)$$

## 12.5  Outline of Numerical Computation of Nearshore Currents

(A) *Basic equations*

As waves propagate from the offshore toward the shore, the height and direction of waves gradually change owing to wave refraction, breaking and other processes depending on the bathymetry. Accordingly, the local mean water level vary from place to place and wave-induced currents are generated in both the alongshore and cross-shore directions. Detailed observations have revealed the presence of the onshore flow (mass transport) near the surface within the surf zone, and the offshore return flow which compensates the mass transport in the middle to the lower layer of water body. Though such 3-D structure of nearshore currents is the subject to be pursued further, the present chapter focuses on the depth-averaged mean flow and the mean water level, disregarding the 3-D current structure.

Let us denote the rise of the local mean water level from the still water level with $\bar{\eta}$, and the $x$ and $y$ components of the depth-averaged mean current velocity with $U$ and $V$ in water of the depth $h$, respectively. The $x$ axis is taken positive offshore from the shoreline and the $y$ axis is on the shoreline. The equations of continuity and momentum are expressed as in the following:

$$\frac{\partial \bar{\eta}}{\partial t} + \frac{\partial (DU)}{\partial x} + \frac{\partial (DV)}{\partial y} = 0 \; : \; D = h + \bar{\eta}\,, \qquad (12.20)$$

$$\left.\begin{aligned}
\frac{\partial U}{\partial t} + U\frac{\partial U}{\partial x} + V\frac{\partial U}{\partial y} + g\frac{\partial \overline{\eta}}{\partial x} + R_x - L_x + F_x = 0, \\
\frac{\partial V}{\partial t} + U\frac{\partial V}{\partial x} + V\frac{\partial V}{\partial y} + g\frac{\partial \overline{\eta}}{\partial y} + R_y - L_y + F_y = 0,
\end{aligned}\right\} \qquad (12.21)$$

where $R_x$ and $R_y$ are the stresses acting on the water body being the sum of the radiation stresses and the stresses due to surface roller, $L_x$ and $L_y$ are the horizontal dispersion terms, and $F_x$ and $F_y$ are the average shear stresses acting on the sea bottom. The subscripts $x$ and $y$ indicate the components in the $x$ and $y$ directions, respectively.

The stresses $R_x$ and $R_y$ are expressed as in the following:

$$\left.\begin{aligned}
R_x = \frac{1}{\rho D}\left[\left(\frac{\partial S_{xx}}{\partial x} + \frac{\partial S_{yx}}{\partial y}\right) + \frac{\partial}{\partial x}(2E_{sr}\cos^2\theta) + \frac{\partial}{\partial y}(E_{sr}\sin 2\theta)\right], \\
R_y = \frac{1}{\rho D}\left[\left(\frac{\partial S_{xy}}{\partial x} + \frac{\partial S_{yy}}{\partial y}\right) + \frac{\partial}{\partial x}(E_{sr}\sin 2\theta) + \frac{\partial}{\partial y}(2E_{sr}\sin^2\theta)\right],
\end{aligned}\right\}$$
$$(12.22)$$

where $S_{xx}, S_{xy}, S_{yx}$, and $S_{yy}$ are the radiation stresses, and $E_{sr}$ denotes the kinetic energy of surface roller, which will be discussed in the next clause. The wave direction $\theta$ is measured as the angle between the $x$ axis and the direction of wave incidence.

The horizontal diffusion terms are expressed as in the following:

$$\left.\begin{aligned}
L_x = \frac{1}{h}\left[\frac{\partial}{\partial x}\left(\nu_t D\frac{\partial U}{\partial x}\right) + \frac{\partial}{\partial y}\left(\nu_t D\frac{\partial U}{\partial y}\right)\right], \\
L_y = \frac{1}{h}\left[\frac{\partial}{\partial x}\left(\nu_t D\frac{\partial V}{\partial x}\right) + \frac{\partial}{\partial y}\left(\nu_t D\frac{\partial V}{\partial y}\right)\right],
\end{aligned}\right\} \qquad (12.23)$$

where $\nu_t$ denotes the turbulent eddy viscosity.

The expressions in Eqs. (12.20) through (12.23) are those employed by Kuriyama and Ozaki,[56] while the terms with the surface roller are introduced by referring to Reniers.[57]

## (B) *Radiation stress and surface roller*

When wave height and direction vary in space, the radiation stress associated with wave momentum flux varies from place to place. The spatial variation of radiation stress exercises the force on the water body, which generates the gradient of mean water level and induces the nearshore currents. The radiation stress is a tensor having the following components:

$$
\left.
\begin{aligned}
S_{xx} &= \frac{1}{8}\rho g H^2 \left[ n(\cos^2 \theta + 1) - \frac{1}{2} \right], \\
S_{yy} &= \frac{1}{8}\rho g H^2 \left[ n(\sin^2 \theta + 1) - \frac{1}{2} \right], \\
S_{xy} &= S_{yx} = \frac{1}{16}\rho g H^2 n \sin 2\theta,
\end{aligned}
\right\}
\qquad (12.24)
$$

where $H$ is the wave height and $n$ denotes the ratio of group velocity to wave celerity. The radiation stress $S_{xx}$ acts in the $x$ direction along the $x$ axis, $S_{yy}$ acts in the the $y$ direction along the $y$ axis, and $S_{xy}$ acts in the $y$ direction along the $x$ axis.

In the PEGBIS model, transformation of directional wave components, which have different frequencies but approximately equal energy, are computed with a given wave height at each height level and the radiation stress is calculated as the arithmetic mean of all directional components at that wave height level. By summing up the contributions from the $M$ wave height levels, the total amount of the radiation stress is estimated. The calculation is made at all the grid points in the computation area.

The surface roller is the concept proposed by Svendsen[58] such that there appears a vortex (roller) with the horizontal axis and the area of $A_{sr}$ per unit width in the frontal part of a breaker. The surface roller has the following kinetic energy $E_{sr}$:

$$
E_{sr} = \frac{\rho A_{sr} c}{2T}, \qquad (12.25)
$$

where $c$ is the wave celerity and $T$ is the wave period.

The surface roller grows initially by absorbing a part of wave energy dissipated by breaking, and then decays gradually as waves propagate through the surf zone. Dally and Brown[59] gave a mathematical formulation to the growth and decay process of surface roller, and Tajima and Madsen[60] extended the idea of Dally and Brown to the following formulation for the steady state condition:

$$
\alpha \left[ \frac{\partial}{\partial x}(E c_g \cos\theta) + \frac{\partial}{\partial y}(E c_g \sin\theta) \right] + \left[ \frac{\partial}{\partial x}(E_{sr} c \cos\theta) + \frac{\partial}{\partial y}(E_{sr} c \sin\theta) \right]
$$
$$
= -\frac{K_{sr}}{h} E_{sr} c, \qquad (12.26)
$$

where $\alpha$ is a factor expressing the ratio of the energy transferred to the surface roller to the wave energy dissipated by breaking and it takes a value between 0 and 1. The term $K_{sr}$ is the energy dissipation rate of the surface roller. It is assumed to be the same as that of energy dissipation factor $K_b$ by breaking as given by Eq. (12.14), by referring to the approach by Tajima and Madsen. The first term in the left-hand side of Eq. (12.26) represents the energy to be transferred to the surface roller and the second term is for its growth. The right-hand side of Eq. (12.26) represents the decay of the surface roller energy.

When the variation in the $y$ direction is negligible as in the case of a beach having nearly uniform bathymetry in the $y$ direction, Eq. (12.26) is rewritten as in the following:

$$\alpha \frac{\partial}{\partial x} \left( \frac{1}{8} \rho g H_{\text{rms}}^2 c_g \cos \theta \right) + \frac{\partial}{\partial x} \left( \frac{\rho A_{sr}}{2T} c^2 \cos \theta \right) = -\frac{K_{sr}}{h} \frac{\rho A_{sr}}{2T} c^2 . \quad (12.27)$$

The rate of energy transfer in the first term in the left-hand side of Eq. (12.27) can be calculated from the difference of wave energy between the adjacent grid points within the surf zone. By setting $A_{sr} = 0$ at the offshore zone and by solving Eq. (12.27) in grid by grid toward the shore with the information of wave deformation by random breaking, the growth and decay of the surface roller is evaluated. In regular waves, a surface roller is associated with a single breaking wave. In random waves, the surface roller solved by means of Eq. (12.27) represents a summation of the energy of individual surface rollers at respective grid points within the surf zone. The representative kinetic energy of the surface roller is calculated by substituting the solution of $A_{sr}$ from Eq. (12.27) into Eq. (12.25), and thus the stresses $R_x$ and $R_y$ are assessed by Eq. (12.22).

(C) *Equations for calculation of mean water level and longshore currents on planar beaches*

Most of previous studies on wave setup and longshore currents have not considered the contribution of surface roller. Because a part of wave energy dissipated by breaking is transferred to the surface roller, the process significantly affects the calculated values of wave setup and longshore currents. On a planar beach which has uniform bathymetry in the $y$ direction, the equation for calculation of the rise of mean water level $\bar{\eta}$ for the steady state condition is derived from the upper equation of Eq. (12.21) as in the following:

$$\frac{\partial \bar{\eta}}{\partial x} = -\frac{1}{\rho g (h + \bar{\eta})} \left[ \frac{\partial S_{xx}}{\partial x} + \frac{\partial}{\partial x} (2 E_{sr} \cos^2 \theta) \right] . \quad (12.28)$$

The radiation stress $S_{xx}$ acting in the $x$ direction along the $x$ axis is usually estimated with $H_{rms}$ for the sake of convenience. As explained in Clause (A), the PEGBIS model calculates the radiation stress as the summation of all contributions from the $M$ wave height levels.

The lower equation of Eq. (12.21) also yields the equation for the longshore current velocity $V$ on a planar beach because of $U = 0$ for the steady state condition. By substituting the upper equations of Eqs. (12.22) and (12.23) into the lower equation of Eq. (12.21), we obtain

$$\frac{\partial S_{xy}}{\partial x} + \frac{\partial}{\partial x}(E_{sr} \sin 2\theta) - \frac{\partial}{\partial x}\left(\rho\nu_t D \frac{\partial V}{\partial x}\right) + \tau_y = 0, \qquad (12.29)$$

where $\tau_y$ denotes the bed shear stress ($= \rho D F_y$). Deletion of the second term in Eq (12.29) leads to the longshore current equation derived by Longuet-Higgins.[61]

He gave the following formulation for the bottom shear stress based on linear theory:

$$\tau_y = \frac{2}{\pi}\rho C_f u_{max} V, \qquad (12.30)$$

where $C_f$ is the friction coefficient of sea bottom which is assigned the value of 0.01 in general, and $u_{max}$ is the amplitude of horizontal orbital velocity, which is calculated as $u_{max} = \pi H/(T \sinh kh)$ by the small amplitude wave theory.

Equation (12.30) has been derived under the assumption that the longshore current velocity is small compared with the amplitude of bottom orbital velocity by waves. When this assumption does not hold, it has been argued that the bed shear stress should be evaluated with the vector sum of the longshore currents and orbital velocity. For example, a number of numerical computation models of nearshore currents employ the approximate nonlinear solution of bed shear stress by Nishimura.[62] Nevertheless, the linear solution by Longuet-Higgins can also yield good estimate of longshore currents by employing some appropriate value of friction coefficient. Larson and Kraus[63] examined the effect of the linear and nonlinear solutions of bed shear stress on the calculated longshore currents on bar and trough beaches. They have indicated that the linear solution may be utilized by raising the friction coefficient up to two times the one selected for the nonlinear solution. Thornton and Guza[54] analyzed the field measurement data of longshore currents at Leadbetter Beach, California, by employing the optimal friction coefficient of $C_f = 0.009 \pm 0.001$ for the linear solution and $C_f = 0.006 \pm 0.0007$ for the nonlinear solution of the bed shear stress.

Calculation of longshore currents induced by directional random waves is influenced by a number of unsettled factors, including formulation of random wave breaking process, assessment of turbulent eddy viscosity to be discussed in the next clause, energy transfer of breaking wave energy to surface roller, and others. These unsettled factors exercise far greater influence on the calculated velocity of longshore currents than the selection of the linear or nonlinear formulation of bed shear stress. Thus, the linear formulation of bed shear stress as expressed in Eq. (12.30) is employed in the present chapter.

## (D) *Formulation of turbulent eddy viscosity*

When Longuet-Higgins[61] proposed a theory of longshore current generation based on the concept of radiation stress, he introduced the turbulent eddy viscosity which has the dimensions of $\rho L U$, where $L$ is a typical length scale and $U$ is a typical velocity. By taking $L \propto x$ and $U \propto (gh)^{1/2}$, he expressed the turbulent eddy viscosity as in the following:

$$\nu_t = N|x|(gh)^{1/2}, \tag{12.31}$$

where $N$ is a constant to which Longuet-Higgins assigned a value of $0 \sim 0.016$ to fit the longshore current velocities to the larboratory data of regular waves. With this formular, the turbulent eddy viscosity increases linearly with the distance from the shore. When longshore currents induced by random waves are calculated with this formula, calculation yields longshore currents extending far outside the surf zone because of excessive horizontal diffusion. Calculation necessitates use of a very small constant value such as $N = 0.0001$ for random waves, but such a small value brings forth a difficulty in dealing with strong diffusion process around trough zones of bar and trough beaches.

To remedy the difficulty, Battjes[64] proposed the following formula by taking account of the physical process involved in horizontal diffusion process:

$$\nu_t = Mh\left(\frac{D_w}{\rho}\right)^{1/3}, \tag{12.32}$$

where $D_w$ denotes the rate of wave energy dissipation per area and $M$ is a constant which takes a value up to around 1. Battjes' formula is physically more rational than Longuet-Higgins' one, and several nearshore current models employ this formula. A difficulty remains in use of this formula, however, such as how to evaluate energy dissipation rate in the area behind a detached breakwater where the wave height decreases by diffraction but no dissipation takes place.

Another proposal on the turbulent eddy viscosity has been made by Larson and Kraus[63] who employed the wave height $H$ as a typical length and the maximum orbital velocity at bottom $u_{max}$ as a typical velocity. Thus,

$$\nu_t = \Lambda u_{max} H , \qquad (12.33)$$

where $\Lambda$ is a constant which is supposed to take a value $0.3 \sim 0.5$. The formula of Eq. (12.33) yields the turbulent eddy viscosity in decrease with the distance from the shore, and the calculated longshore currents do not extend much outside the surf zone. The formula also has the advantage of quickly giving an estimate of the turbulent eddy viscosity once the wave field is computed. Thus, Eq. (12.33) is employed for evaluation of the turbulent eddy viscosity in the present chapter.

(E) *Role of energy transfer factor to surface roller*

The variation of the mean water level within the surf zone and the longshore currents are affected by the degree of energy transfer from the breaking waves to the surface roller as represented by the factor $\alpha$ in Eq. (12.26). Figure 12.6 exhibits the cross-shore variation of the mean water level on a uniform beach with the slope of $1/20$ when waves with the offshore significant wave of 2.0 m are approaching with the offshore incident angle of $30°$,

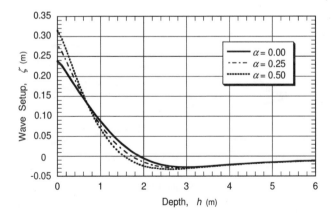

Fig. 12.6 Effect of the factor of energy transfer to surface roller on the cross-shore variation of mean water level.[55] ($H_0' = 2.0$ m, $T_p = 9.1$ s, and $\theta_0 = 30°$).

which have been calculated with the PEGBIS model. The energy transfer factor is set at three levels of $\alpha = 0$, 0.25, and 0.50. The case with $\alpha = 0$ does not introduce the process of energy transfer from breaking waves to surface roller. Compared with this case, the cases with $\alpha = 0.25$ and 0.50 yield a shift of the end point of wave setdown toward the shoreline and a slow start of wave setup. Because of increase in the gradient of mean water level by introduction of energy transfer to surface roller, however, the mean water level rise higer at the shoreline as the value of $\alpha$ becomes large.

Wave-induced longshore currents are also affected by the degree of energy transfer from the breaking waves to the surface roller. With increase in the energy transfer factor, the location of peak velocity shifts toward the shoreline with enhancement of peak velocity. Figure 12.7 shows comparison

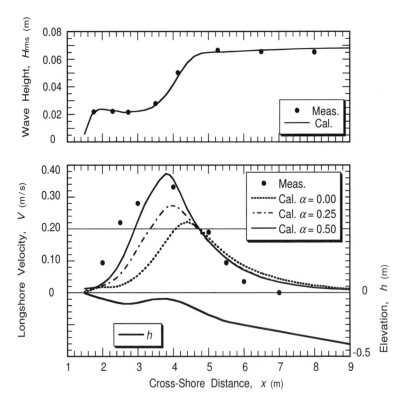

Fig. 12.7   Comparison of wave height and longshore currents measured by Reniers and Battjes[65] and those computed with PEGBIS model.[55] ($H_{\mathrm{rms}} = 0.07$ m, $T_p = 1.2$ s, and $\theta_0 = 30°$).

between the laboratory test data by Reniers and Battjes[65] and the PEGBIS computation. The test was made using a unidirectional train of irregular waves with the input of $H_{\mathrm{rms}} = 0.07$ m, $T_p = 1.2$ s, and $\theta_0 = 30°$ at the offshore area with the depth of 0.55 m. The bar and trough beach was made of fixed bed with mortar. The cross-shore variation of wave height shown in the upper panel is well reproduced by the PEGBIS model, while the computed longshore current profile fits well to the measurement by setting the energy transfer factor at $\alpha = 0.5$. Computation was made by setting the breaking constant at $C_b = 0.08$, the constant for turbulent eddy visocity at $\Lambda = 0.05$, and the friction coefficient at $C_f = 0.025$.

A comparison with field observation data is shown in Fig. 12.8 for the case reported by Kuriyama and Ozaki,[56] who carried out measurements

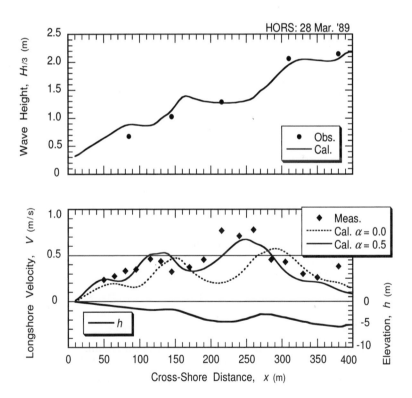

Fig. 12.8   Comparison of wave height and longshore currents in Hazaki Coast observed by Kuriyama and Ozaki[56] and those computed with PEGBIS model.[55] (($H_{1/3})_0 = 2.35$ m, $T_p = 9.75$ s, and $\theta_0 = 25°$).

at Hazaki Coast, Ibaragi Prefecture, Japan. The offshore waves were estimated as $(H_{1/3})_0 = 2.35$ m, $T_p = 9.75$ s, and $\theta_0 = 25°$. Computation of wave field was made by setting the breaking constant at $C_b = 0.06$ for adjustment with measured data. Longshore currents were computed by setting the constant for turbulent eddy visocity at $\Lambda = 1.0$ and the friction coefficient at $C_f = 0.0075$. As shown in the lower panel, use of $\alpha = 0.5$ for the rate of energy transfer to surface roller succeeds in reproducing the current peak around the location $x = 240$ m.

Further comparison with other laboratory and field data has indicated that the value of $\alpha = 0.5$ for the rate of energy transfer from breaking waves to surface roller brings forth good agreement in general. For a few cases of field observation of waves of very low steepness which seem to have produced conspicuous plunging breakers, however, the value of $\alpha = 0.25$ has yielded better agreement with the data than $\alpha = 0.5$. Examination of appropriate value of the energy transfer factor to surface roller would be an important subject of further studies.

## 12.6   Prediction of Wave Setup and Longshore Currents on Planar Beaches

(A) *Wave setup at the shoreline*

The author has prepared a design diagram for prediction of the amount of wave setup at the shoreline based on his random wave breaking model of 1975. The diagram was reproduced in the previous edition of this book as Fig. 3.35. Because the diagram was prepared without due consideration of directional wave characteristics, the PEGBIS model was mobilized for calculation of wave setup for various wave conditions.[66]

Computation was carried out by fixing the deepwater significant wave height at $(H_{1/3})_0 = 2.0$ m and varying the spectral peak period in the range of $T_p = 4.55 \sim 18.19$ s so as to yield the deepwater wave steepness of $H_0/L_0 = 0.005, 0.007, 0.010, 0.014, 0.02, 0.03, 0.04, 0.06$ and $0.08$ (as calculated with the significant wave period). Although waves with steepness of 0.08 rarely appear in the nature as envisaged from Fig. 3.4 in Sec. 3.1.3, they were included in the computation to see the trend of variation with respect to wave steepness. The directional spectral characteristics were specified with the peak enhancement factor $\gamma$ of the JONSWAP sprectrum and the maximum directional spreading parameter of $s_{max}$ of the Mitsuyasu-type spreading function. A combination of $\gamma = 10$ and $s_{max} = 150$ was assigned for waves with $H_0/L_0 = 0.005$. The values of $\gamma$ and $s_{max}$ were gradually lowered with increase of wave steepness to a combination of $\gamma = 1.05$ and $s_{max} = 10$ for waves with $H_0/L_0 = 0.08$. The factor of energy transfer from breaking waves to surface roller was set at $\alpha = 0.20$ for $H_0/L_0 = 0.005$ and assigned a gradually increasing value toward $\alpha = 0.50$ for $H_0/L_0 = 0.08$.

Five linear slopes of $s = 1/10, 1/20, 1/30, 1/50$ and $1/100$ were employed, while the deepwater incident angle was varied at 8 directions of $\theta_0 = 1°, 10°(10°)70°$. The breaking constant of the gradational breaker index was set at $C_b = 0.07$.

Computation was carried out concurrently for the wave setup and longshore current velocity. The result of wave setup at the shoreline has been analyzed and presented in Sec. 3.6.2 (C) in the form of design diagram in Fig. 3.35 and with a set of the prediction formulas of Eqs. (3.41) to (3.43).

### (B) *Fitting of Weibull distribution of longshore current profile*

In 1991, the author and Watanabe[68] computed longshore currents on planar beaches and proposed empirical prediction formulas together with a set of design diagrams. Computation employed the author's random wave breaking model in 1975. New computation of longshore currents has been executed using the PEGBIS model together with wave setup as described above. Different values of the constant for the turbulent eddy viscosity were employed, depending on the slope, i.e., $\Lambda = 0.05, 0.275, 0.40, 0.50$, and $0.60$ for $s = 1/10, 1/20, 1/30, 1/50$, and $1/100$, respectively. It was necessary to use a small constant value for steep slope in order not to induce excessive horizontal diffusion and reduction in the current speed.

The longshore current profiles were diverse in correspondence to wave and slope conditions. However, they had a common feature of single peak, zero speed at the shoreline, and a slightly heavy tail toward offshore. To represent such a feature, a fitting of the following Weibull distribution to the longshore current profiles has been attempted:

$$V = V_0 \left(\frac{z}{a}\right)^{k-1} \exp\left[-\left(\frac{z}{a}\right)^k\right] : z = \frac{h}{H_0}, \qquad (12.34)$$

where $V_0$ is a representative velocity having the dimension of velocity, $k$ is the shape parameter, $a$ is the scale parameter, $h$ is the local water depth, and $H_0$ denotes the deepwater significant wave height. Both $k$ and $a$ are nondimensional parameters.

The Weibull distribution by Eq. (12.34) has a single peak when $k > 1$ at the following location:

$$z_{\mathrm{mod}} = a \left(1 - \frac{1}{k}\right)^{1/k}.$$ 
(12.35)

The peak velocity at $z_{\mathrm{mod}}$ is given by

$$V_{\max} = V_0 \left(1 - \frac{1}{k}\right)^{1-1/k} \exp\left[-\left(1 - \frac{1}{k}\right)\right].$$ 
(12.36)

The location of the center of gravity of the Weibull distribution is also calculated as

$$\overline{z} = \frac{\displaystyle\int_0^\infty V(z)z\,dz}{\displaystyle\int_0^\infty V(z)\,dz} = a\,\Gamma\left(1 + \frac{1}{k}\right),$$ 
(12.37)

where $\Gamma(\cdot)$ denotes the gamma function.

From a computed longshore current profile, the location of peak velocity $z_{\mathrm{mod}}$ is identified and the location of the center of gravity $\overline{z}$ can be calculated easily. Substituting these quantities into Eqs. (12.35) and (12.37) yields the estimate of the shape parameter $k$ and the scale parameter $a$. Figure 12.9 presents some examples of fitting of the Weibull distribution (thin lines) to the numerically computed longshore currents (thick lines). Despite some small differences, the fitted distributions approximately represent the computed results.

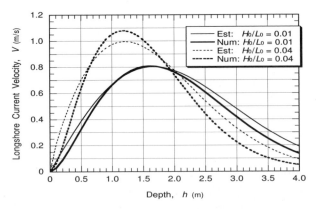

Fig. 12.9  Examples of comparison of numerically computed longshore current profiles and representation with the Weibull distribution.[66] ($s = 1/30$, $\theta_0 = 30°$, and $H_0 = 2.0$ m).

Estimation of the scale and shape parameters has been made for all the computed longshore profiles, and multiple regression analysis has been performed for each parameter. The regression analysis has yielded the following empirical formula for the shape parameter:

$$k = [A_k + B_k \ln H_0/L_0] \, (\cos \theta_0)^{r_k} \,, \tag{12.38}$$

where the constants are given by

$$\left.\begin{aligned}
A_k &= -0.9017 - 1.9486 \ln s - 0.3783 (\ln s)^2 \,, \\
B_k &= -0.6584 - 0.5869 \ln s - 0.1246 (\ln s)^2 \,, \\
r_k &= 0.326 + 0.218 \ln s + 0.0446 (\ln s)^2 \,.
\end{aligned}\right\} \tag{12.39}$$

The empirical formulas for the scale parameter has been obtained as follows:

$$a = [A_a + B_a \ln H_0/L_0 + C_a (\ln H_0/L_0)^2] \, (\cos \theta_0)^{r_a} \,, \tag{12.40}$$

where the constants are given by

$$\left.\begin{aligned}
A_a &= 3.148 + 1.855 \ln s + 0.3631 (\ln s)^2 \,, \\
B_a &= 1.766 + 1.122 \ln s + 0.1929 (\ln s)^2 \,, \\
C_a &= 0.2211 + 0.1194 \ln s + 0.02019 (\ln s)^2 \,, \\
r_a &= -0.751 - 0.327 \ln H_0/L_0 - 0.0228 (\ln H_0/L_0)^2 \,.
\end{aligned}\right\} \tag{12.41}$$

The maximum velocity $V_{\max}$ by Eq. (12.36) is represented as the product of the reference velocity $V_c$ and the peak velocity coefficient $c_{\max}$, i.e., $V_{\max} = c_{\max} V_c$. The reference velocity has been formulated as a function of wave height and others as in the following:

$$V_c = \frac{s\sqrt{gH_0}}{C_f} \sin \theta_0 \cos(0.8\theta_0) \,. \tag{12.42}$$

The value of the peak velocity coefficient has been calculated for the computed maximum velocity of all the data. The calculated results have been subjected to multiple regression analysis, yielding the following empirical formula:

$$c_{\max} = V_{\max}/V_c = A_c + B_c \ln H_0/L_0 + C_c (\ln H_0/L_0)^2 \,, \tag{12.43}$$

where the constants are given by

$$\left.\begin{aligned}
A_c &= 0.8642 + 0.3141 \ln s + 0.02741 (\ln s)^2 \,, \\
B_c &= 0.3292 + 0.1616 \ln s + 0.01616 (\ln s)^2 \,, \\
C_c &= 0.03281 + 0.01856 \ln s + 0.002024 (\ln s)^2 \,.
\end{aligned}\right\} \tag{12.44}$$

The peak velocity coefficient $c_{\max}$ does not vary with the incident angle.

With the values of $c_{\max}$ and $V_c$ thus determined, the representative velocity $V_0$ is obtained by solving Eq. (12.36) with the result of

$$V_0 = c_{\max} V_c \left( 1 - \frac{1}{k} \right)^{-(1-1/k)} \exp\left[ \left( 1 - \frac{1}{k} \right) \right] . \qquad (12.45)$$

Figures 12.10 to 12.12 are the diagrams for estimation of the shape parameter $k$, the scale parameter $a$, and the peak velocity coefficient $c_{\max}$. The former two diagrams are for the cases of normal incidence.

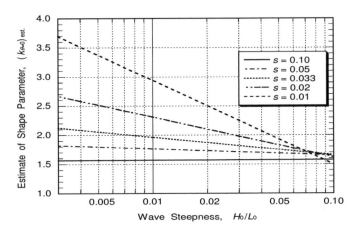

Fig. 12.10   Design diagram of the shape parameter $k_{\theta_0=0}$ for normal incidence.[66]

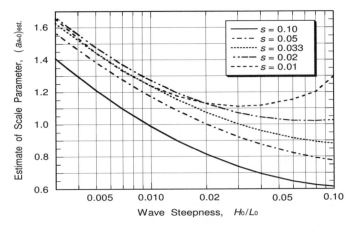

Fig. 12.11   Design diagram of the scale parameter $a_{\theta_0=0}$ for normal incidence.[66]

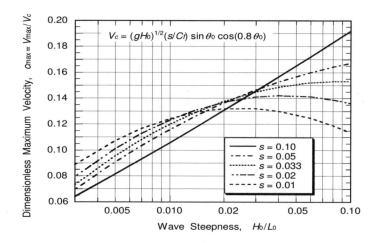

Fig. 12.12   Design diagram for the peak velocity coefficient $c_{\max} = V_{\max}/V_c$.[66]

The value of the peak velocity coefficient remains in the range of $c_{\max} = 0.07 \sim 0.16$ except for very steep waves of $H_0/L_0 \geq 0.05$ on the steep slope $s = 1/10$. By assuming the friction coefficient being $C_f = 0.01$, a formula for approximate estimation of the maximum longshore current velocity is given with the deepwater wave characteristics as in the following:

$$V_{\max} \simeq (7 \sim 16)\, s\sqrt{gH_0}\sin\theta_0\,\cos(0.8\theta_0)\,. \tag{12.46}$$

The formula can also be rewritten with the wave characteristics at breaking as below.

$$V_{\max} = \frac{K_{\max}}{C_f}\, s\sqrt{gH_b}\sin 2\theta_b\,, \tag{12.47}$$

where

$$K_{\max} = \frac{c_{\max}\sqrt{H_0}\sin\theta_0\,\cos(0.8\theta_0)}{\sqrt{H_b}(K_r)_b\sin 2\theta_b}\,. \tag{12.48}$$

Evaluation of the empirical constant $K_{\max}$ has been made with the breaker height $H_b$ at the breaking depth $h_b$, which are estimated by using the concept of incipient breaking presented in Sec. 3.6.2 (A). The refraction coefficient at breaking $(K_r)_b$ and the incident angle of breaking wave $\theta_b$ are calculated by the Snell law at the depth $h_b$. For the range of the beach slope of $s = 1/100 \sim 1/10$ and the wave steepness of $H_0/L_0 = 0.005 \sim 0.05$, the empirical constant takes the value of $K_{\max} = 0.09 \sim 0.13$ with the mean of 0.115. Thus, a formula for approximate estimation of the maximum longshore current velocity in terms of breaker characteristics becomes as follows:

$$V_{max} \simeq \frac{0.115}{C_f} s\sqrt{gH_b} \sin 2\theta_b. \tag{12.49}$$

If the term of $s/C_f$ is set at 4.3 (e.g., $s = 1/23$ and $C_f = 0.01$) together with the relationship of $H_{1/3} \simeq 1.4H_{rms}$, Eq. (12.49) is reduced to the following expression, which is due to Komar[67]:

$$V_{mid} = 0.58 \sqrt{g(H_{rms})_b} \sin 2\theta_b, \tag{12.50}$$

where $V_{mid}$ refers to the longshore current velocity at the mid-surf zone.

**Example 12.1**

Swell with the offshore height of $(H_{1/3})_0 = 2.5$ m and the period of $T_{1/3} = 8.3$ s are acting on a nearly planar beach with the slope of $s = 1/45$ with the offshore incident angle of $\theta_0 = 35°$. Assuming the friction coefficient as being $C_f = 0.01$, estimate the cross-shore variation of longshore current veclocity.

**Solution**

First, the wave steepness is calculated. As the deepwater wavelength is $L_0 = 1.56 \times 8.3^2 = 107.5$ m, the steepness is $H_0/L_0 = 2.5/107.5 = 0.0233$. Thus, we have $\ln H_0/L_0 = \ln(0.0233) = -3.759$ and $\ln s = \ln(1/45) = -0.3807$.

Second, the reference velocity is calculated as below.

$$V_c = 1/45 \times (9.8 \times 2.5)^{1/2}/0.01 \times \sin 35° \times \cos(0.8 \times 35°) = 5.57 \text{ m/s}.$$

The constants of the formula for $c_{max}$ as given by Eq. (12.44) are calculated as

$$A_c = 0.0657, \ B_c = -0.0517, \ C_c = -0.00851.$$

Thus, the maximum velocity is obtained as below.

$$c_{max} = 0.0657 - 0.0517 \times (-3.759) - 0.00851 \times (-3.759)^2 = 0.1398,$$
$$V_{max} = c_{max} V_c = 0.1398 \times 5.57 = 0.779 \text{m/s}.$$

Third, the constants for the shape parameter $k$ are calculated by Eq. (12.39) as below.

$$A_k = 1.034, \ B_k = -0.260, \ r_k = 0.142.$$

Thus, the shape parameter is obtained by Eq. (12.38) as

$$k = [1.034 - 0.260 \times (-3.759)] \times (\cos 35°)^{0.142} = 1.955.$$

Fourth, the constants for the scale parameter $a$ are calculated by Eq. (12.41) as below.

$$A_a = 1.349, \ B_a = 0.290, \ C_a = 0.0592, \ r_a = 0.156.$$

Thus, the scale parameter is obtained by Eq. (12.40) as

$$a = [1.349 + 0.290 \times (-3.759) + 0.0592 \times (-3.759)^2] \times (\cos 35°)^{0.156}$$
$$= 1.062.$$

The location of the maximum longshore velocity as denoted by $h_{mod}$ is estimated by Eq. (12.35) as below.

$$h_{mod} = 1.062 \times 2.5 \times (1 - 1/1.955)^{1/1.955} = 1.84 \text{ m.}$$

The representative velocity $V_0$ is calculated by Eq. (12.45) as below.

$$V_0 = 0.779/(1 - 1/1.955)^{-(1-1/1.955)} \times \exp[1 - 1/1.955] = 1.802 \text{ m/s.}$$

The cross-shore variation of longshore current velocity is calculated with the values of the shape and scale parameters and the representative velocity thus obtained by substituting them into Eq. (12.34). Calculation result is listed in Table 12.2.

Table 12.2   Cross-shore variation of longshore current velocity of Example 12.1.

| $x$ (m) | $h$ (m) | $z/a$ | $V$ (m/s) | $x$ (m) | $h$ (m) | $z/a$ | $V$ (m/s) |
|---------|---------|-------|-----------|---------|---------|-------|-----------|
| 0   | 0    | 0     | 0     | 140 | 3.11 | 1.172 | 0.536 |
| 20  | 0.44 | 0.167 | 0.317 | 160 | 3.56 | 1.339 | 0.406 |
| 40  | 0.89 | 0.335 | 0.563 | 180 | 4.00 | 1.507 | 0.287 |
| 60  | 1.33 | 0.502 | 0.720 | 200 | 4.44 | 1.674 | 0.191 |
| 80  | 1.78 | 0.670 | 0.778 | 220 | 4.89 | 1.841 | 0.119 |
| 100 | 2.22 | 0.837 | 0.750 | 240 | 5.33 | 2.009 | 0.070 |
| 120 | 2.67 | 1.003 | 0.660 | 260 | 5.78 | 2.176 | 0.039 |

(C) *Water depth of peak velocity*

With fitting of the Weibull distribution to the longshore current profiles, we can calculate the longshore current velocity at any location for a given wave and slope condition as indicated in Example 12.1. The location at which the maximum velocity of longshore currents appears is estimated by Eq. (12.35) once the shape and scale parameters are assessed. Figure 12.3 is a design diagram for estimation of the water depth of peak velocity on the planar beach with the slope of $s = 1/20$ and $1/50$.

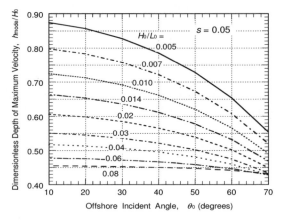

(a) Beach slope of $s = 1/20$

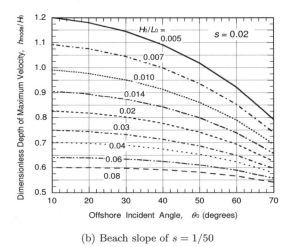

(b) Beach slope of $s = 1/50$

Fig. 12.13   Ratio of water depth of peak velocity to offshore wave height.[33]

The abscissa of this figure is the offshore incident angle $\theta_0$, while the ordinate is the water depth at which the peak velocity appears, $h_{\mathrm{mod}}$, relative to the offshore wave height $(H_{1/3})_0$. The wave steepness $H_0/L_0$ functions as the parameter. The location of the peak velocity of longshore currents gradually shifts offshore as the wave steepness decreases or as the beach slope becomes mild. Increase in the offshore incident angle causes a on-shore shift of the location of the peak velocity. The tendency of the shift of peak velocity location with respect to the wave and beach conditions is the same as the previous study by Goda and Watanabe.[68] Quantitatively, however, the location of peak velocity is shallower as a whole than that in the previous study. The difference mainly owes to the introduction of the surface roller term in Eq. (12.29), and it indicates the importance of surface roller in nearshore current computation.

(D) *Setting of constant values in nearshore current computation*

Computation of wave field and nearshore currents needs the setting of several constant and coefficient values. The PEGBIS model has a tuning parameter of the breaking constants $C_b$ and $C_b'$, the default values of which are $C_b = 0.07$ and $C_b' = 0.0613$; the latter is set at seven eighths of the former. When the wave height data in the field are available, better prediction of wave filed will be obtained by tuning these values in the range of $C_b = 0.06 \sim 0.08$ and $C_b' = 0.0525 \sim 0.07$.

Computation of nearshore currents requires the setting of the friction coefficient, $C_f$, the constant of turbulent eddy viscosity, $\Lambda$, when Eq. (12.33) is employed, and the factor of energy transfer from breaking waves to surface roller, $\alpha$. When the author[55] employed the PEGBIS model for computation of longshore currents for several cases of field and laboratory data, it was necessary to set these parameters in the range of $C_f = 0.0070 \sim 0.025$, $\Lambda = 0.05 \sim 1.0$, and $\alpha = 0.25$ or $0.5$. Even though the computation presented in this subsection was carried out by using $C_f = 0.01$, $\Lambda = 0.05 \sim 0.6$, and $\alpha = 0.2 \sim 0.5$, selection of these values was made somewhat subjectively. A search for more rational selection of the parameter values will be the subject of further studies.

(E) *Estimation of longshore current on beaches of arbitrary profiles*

Empirical formulation of the longshore current velocity presented in the present section has been made for planar beaches of uniform slope.

Representation of longshore current profile with the Weibull distribution is not appropriate for beaches with bars and troughs or varying slope. Nevertheless, the present empirical formulation may be utilized for the order of magnitude estimate of longshore currents on beaches of arbitrary profiles. First thing to do is to assess the mean slope $\bar{s}$ across the surf zone. The shape parameter $k$, the scale parameter $a$, and the reference velocity $V_0$ are assessed for the given wave conditions. The amount of wave setup at the shoreline, $\zeta$, may also need to be assessed with Eqs. (3.41) to (3.43) in Sec. 3.6.2 (C). Because Eq. (12.34) expresses the longshore current velocity as a function of the relative depth $h/H_0$, substitution of respective water depths corresponding to the distance from the shoreline into Eq. (12.34) yields an approximate estimate of cross-shore profile of longshore currents. Ref. 66 illustrates several examples of approximate longshore current estimations for beaches of arbitrary profiles.

## References

1. M. Isobe, X. Yu, K. Umemura and S. Takahashi, "Study on development of Numerical Wave Flume, " *Proc. Coastal Engrg., JSCE* **46** (1999), pp. 36–40 (*in Japanese*).

2. S. Takahashi, Y. Kotake, R. Fujiwara and M. Isobe, "Performance evaluation of perforated-wall caissons by VOF numerical simulations," *Coastal Engineering 2002* (*Proc. 28th Int. Conf.*) (Cardif, Wales, World Scientific) (2003), pp. 1364–1376.

3. Y. Watanabe, M. Yasuhara and H. Saeki, "Evolution of three-dimensional large-scale swirl eddies, oblique eddies and plunging jets," *Proc. Coastal Engrg., JSCE* **46** (1999), pp. 141–145 (*in Japanese*).

4. T. Takayama and Y. Kamiyama, "Diffraction of sea waves by rigid or cushion type breakwaters," *Rept. Port and Harbour Res. Inst.* **16**(3) (1977), pp. 3–37.

5. T. Takayama, "Wave diffraction and wave height distribution inside a harbor," *Tech. Note Port and Harbour Res. Inst.* No. 367 (1981), 140p (*in Japanese*).

6. Y. Goda, T. Takayama and Y. Suzuki, "Diffraction diagrams for directional random waves," *Proc. 16th Int. Cont. Coastal Engrg.* (Hamburg, ASCE, 1978), pp. 628–650.

7. Y. Goda, "Examination of the influence of several factors on longshore current computation with random waves," *Coastal Engineering,* **53**(2-3) (2006), pp. 157–170.

8. T. Karlsson, "Refraction of continuous ocean wave spectra," *Proc. ASCE* **95**(WW4) (1969), pp. 437–448.

9. K. Nagai, T. Horiguchi and T. Takai, "Computation of the propagation of sea waves having directional spectra from offshore to shallow water," *Proc. 21st Japanese Conf. Coastal Engrg.* (1974), pp. 349–253 (*in Japanese*).

10. L. H. Holthuijsen, N. Booji and R. C. Ris, "A spectral wave model for the coastal zone," *Proc. 2nd Int. Symp. on Ocean Wave Measurement and Analysis* (New Orleans, ASCE) (1993), pp. 630–641.

11. L. H. Holthuijsen, *Waves in Oceanic and Coastal Waters* (Cambridge Univ. Press, 2008), 387p.

12. N. Booji, L. H. Holthuijsen, N. Doorn and A. T. M. M. Kieftenburg, "Diffraction in a spectral wave model," *Proc. 3rd Int. Symp. on Ocean Wave Measurement and Analysis* (Virginia Beach, Virginia, ASCE) (1997), pp. 243–255.

13. F. J. Rivero, A. S.-Archilla and E. Carci, "An analysis of diffraction in spectral wave models," *Proc. 3rd Int. Symp. on Ocean Wave Measurement and Analysis* (Virginia Beach, Virginia, ASCE) (1997), pp. 431–445.

14. H. Mase, "Multi-directional random wave transformation based on energy balance equation," *Coastal Engineering Journal* **43**(4) (2001), pp. 339–337.

15. H. Mase, T. Takayama and T. Kitano, "Spectral wave transformation model with wave diffraction effect," *Proc. 4th Int. Symp. on Ocean Wave Measurement and Analysis* (San Francisco, ASCE) (2001), pp. 814–823.

16. J. A. Battjes and J. P. F. M. Janssen, "Energy loss and set-up due to breaking of random waves," *Proc. 16th Int. Conf. Coastal Eng.* (Hamburg, ASCE) (1978), pp. 1–19.

17. N. Booji, R. C. Ris and L. H. Holthuijsen, "A third-generation wave model for coastal regions, Part I: Model description and validation," *J. Geophys. Res* **104**(C4) (1999), pp. 7649–7666.

18. R. C. Ris, N. Booji and L. H. Holthuijsen, "A third-generation wave model for coastal regions, Part II: Verification," *J. Geophys. Res.* **104**(C4) (1999), pp. 7667–7681.

19. T. Takayama, N. Ikeda and T. Hiraishi, "Practical computation method of directional random wave transformation," *Rept. Port and Harbour Res. Inst.* **30**(1) (1991), pp. 21–67.

20. Y. Goda, "Irregular wave deformation in the surf zone," *Coastal Engineering in Japan* **18** (1975), pp. 13–26.

21. J. C. W. Berkoff, "Computation of combined refraction-diffraction," *Proc. 13th Int. Conf. Coastal Eng.* (Vancouver, ASCE) (1972), pp. 471–490.

22. K. Nadaoka, "Wave equations — Theories and numerical simulation," *Lecture Notes of the 35th Summer Seminar on Hydraulic Engrg, JSCE* (1999), pp. B-2-1~B-2-19 (*in Japanese*).

23. N. Sato, M. Isobe and T. Izumita, "A numerical model for calculating wave height distribution in a harbor of arbitrary shape," *Coastal Engineering in Japan* **33** (1988), pp. 119–131.

24. A. C. Radder, "On the parabolic equation method for wave-wave propagation," *J. Fluid Mech.* **95** (1979), pp. 159–176.

25. H. Hirakuchi and K. Maruyama, "An extension of the parabolic equation for application to obliquely incident waves," *Proc. 33rd Japanese Conf. Coastal Engrg.* (1986), pp. 114–118 (*in Japanese*).

26. B. Li, "Parabolic model for water waves," *J. Waterway, Port, Coastal, and Ocean Engrg.* **123**(4) (1997), pp. 192–199.

27. U. M. Saied and I. K. Tsanis, "Improved parabolic water wave transformation models," *Coastal Engineering* **52** (2005), pp. 139–149.

28. J. T. Kirby, "Rational approximation in the parabolic equation method for water waves," *Coastal Engineering* **10** (1986), pp. 355–378.

29. Y. Goda, "Wave field computation around artificial reefs with gradational breaker model," *Coastal Structures 2003* (*Proc. Conf.*, Portland, Oregon, ASCE) (2003), pp. 824–830.

30. C. L. Vincent and M. J. Briggs, "Refraction-diffraction of irregular waves over a mound," *J. Waterway, Port, Coastal, and Ocean Engrg.* **115**(2) (1989), pp. 269–284.

31. H. T. Özkan and J. T. Kirby, "Evolution of breaking directional spectral waves in the nearshore," *Proc. 2nd Int. Symp. on Ocean Wave Measurement and Analysis* (New Orleans, ASCE) (1993), pp. 849–863.

32. S. B. Yoon, J. W. Lee, Y. J. Yeon, B. H. and Cho, "A note on the numerical simulation of wave deformation over a submerged shoal," *Proc. 1st Conf. Asian and Pacific Coastal Engrg.* (Dalian, China) (2001), pp. 315–324.

33. Y. Goda, "A 2-D random wave transformation model with gradational breaker index," *Coastal Engineering Journal* **46**(1) (2004), pp. 1–38.

34. E. B. Thornton and R. T. Guza, "Transformation of wave height distribution," *J. Geophys. Res.* **88**(C10) (1983), pp. 5925–5938.

35. loc. cit. Ref. 16.

36. J. Choi, C. H. Lim, J. I. Lee and S. B. Yoon, "Evolution of currents over a submerged laboratory shoal," *Coastal Engineering* **56** (2009), pp. 297–312.

37. A. Watanabe and K. Maruyama, "Numerical modeling of nearshore wave field under combined refraction, diffraction and breaking," *Coastal Engineering in Japan* **29** (1986), pp. 19–39.

38. M. Isobe, "Time-dependent mild-slope equations for random waves," *Proc. 24th Int. Conf. Coastal Engrg.* (Kobe, ASCE) (1994), pp. 285–299.

39. T. Ishii, M. Isobe and A. Watanabe, "Two-dimensional analysis of wave transformation by rational-approximation-based, time-dependent mild-slope equation for random waves," *Proc. 25th Int. Conf. Coastal Engrg.* (Orlando, Florida, ASCE) (1996), pp. 754–766.

40. T. Arikawa and M. Isobe, "Development of numerical model for breaking and run-up waves with nonlinear mild-slope equation," *Proc. Coastal Engrg., JSCE* **47** (2000), pp. 186–190 (*in Japanese*).

41. M. Isobe, "A proposal of nonlinear mild-slope equation," *Proc. Coastal Engrg., JSCE* **41** (1994), pp. 1–5 (*in Japanese*).

42. T. Arikawa and A. Okayasu, "Development of a 2DH wave breaking model based on nonlinear mild slope equations," *Proc. Coastal Engrg., JSCE* **49** (2002), pp. 26–30 (*in Japanese*).

43. S. Kanayama, "On the applicability of multi-level model for wave deformation analysis on steep bathymetry," *Annual J. Civil Engrg. in the Ocean, JSCE* **22** (2006), pp. 115–120 (*in Japanese*).

44. D. H. Peregrine, "Long waves on a beach," *J. Fluid Mech.* **27** (1967), pp. 815–827.

45. P. A. Madsen, R. Murray, O. R. and Sorensen, "A new form of the Boussinesq equations with improved linear dispersion characteristics," *Coastal Engineering* **15** (1991), pp. 371–388.

46. K. Hirayama, "Utilization of numerical simulation on nonlinear irregular wave for port and harbor design," *Tech. Note Port and Airport Res. Inst.* No. 1036 (2002), 162p (*in Japanese*).

47. M. K. Kabiling and S. Sato, "Two-dimensional nonlinear dispersive wave-current and three-dimensional beach deformation model," *Coastal Engineering in Japan* **36**(2) (1993), pp. 195–212.

48. N. Hara, K. Hirayama and T. Hiraishi, "Application of bore model to nonlinear wave transformation," *Proc. 13th Int. Offshore and Polar Engrg. Conf. (ISOPE)* (2003), pp. 796–801.

49. K. Nadaoka, O. Ono and H. Kurihara, "Analysis of near-crest pressure gradient of irregular water waves as a dynamic criterion of wave breaking," *Proc. 7th Int. Offshore and Polar Engrg. Conf. (ISOPE)* (1997), pp. 170–174.

50. K. Hirayama and T. Hiraishi, "A Boussinesq model for wave breaking and runup in a coastal zone: 1D," *Proc. 5th Int. Symp. on Ocean Wave Measurement and Analysis* (Madrid, ASCE, 2005), Paper 151 (CD-ROM).

51. W. R. Dally, R. G. Dean and R. A. Darlymple, "Wave height variation across beaches of arbitrary profile," *J. Geophys. Res.* **90**(C6) (1985), pp. 11,917–11,927.

52. N. C. Kraus, J. M. Smith and C. K. Sollitt, "SUPERTANK laboratory data collection project," *Proc. 23rd Int. Conf. Coastal Engrg.* (Venice, ASCE) (1992), pp. 2191–2204.

53. S. Hotta and M. Mizuguchi, "A field study of waves in the surf zone," *Coastal Engineering in Japan* **23** (1980), pp. 59–79.

54. E. B. Thornton and R. T. Guza, "Surf zone longshore currents and random waves: field data and models," *J. Phys. Oceanogr.* **16** (1986), pp. 1165–1178.

55. loc. cit. Ref. 7.

56. Y. Kuriyama and Y. Ozaki, "Longshore current distribution on a bar-trough beach — Field measurements at HORF and numerical model," *Rept. Port and Harbour Res. Inst.* **32**(3) (1993), pp. 3–37.

57. A. Reniers, "Longshore current dynamics," *Communications on Hydraulic Eng., Dept. Civil and Geoscience, Delft Univ.* Tech. Rept. No. 99-2 (1999), 132p.

58. I. A. Svendsen, "Wave heights and set-up in a surf zone," *Coastal Engineering* **8** (1984), pp. 303–329.

59. W. R. Dally and C. A. Brown, "A modeling investigation of the breaking wave roller with application to cross-shore currents," *J. Geophys. Res.* **100**(C12) (1995), pp. 24,873–24,883.

60.  Y. Tajima and O. S. Madsen, "Modeling near-shore waves and surface roller," *Proc. 2nd Int. Conf. Asian and Pacific Coasts (APAC 2003)* (Makuhari, Chiba, Japan) (2003), Paper No. 28 (CD-ROM), 12p.

61.  M. S. Longuet-Higgins, "Longshore current generated by obliquely incident sea waves, 1 and 2," *J. Geophys. Res.* **75**(33) (1970), pp. 6779–6801.

62.  H. Nishimura, "Numerical simulation of nearshore circulations," *Proc. 29th Japanese Conf. Coastal Engrg.* (1982), pp. 333–337 (*in Japanese*).

63.  M. Larson and N. C. Kraus, "Numerical model of longshore current for bar and trough beaches," *J. Waterway, Port, Coastal, and Ocean Engrg.* **117**(4), (1991), pp. 326–347.

64.  J. A. Battjes, "Modeling of turbulence in the surf zone," *Proc. Symp. Modeling Techniques* (1975), pp. 1050–1061.

65.  A. J. H. M. Reniers and J. A. Battjes, "A laboratory study of longshore currents over barred and non-barred beaches," *Coastal Engineering* **30** (1997), pp. 1–22.

66.  Y. Goda, "Wave setup and longshore currents induced by directional spectral waves: Prediction formulas based on numerical computation results," *Coastal Engineering Journal* **50**(4) (2008), pp. 397–440.

67.  P. D. Komar, "Beach slope dependency of longshore currents," *J. Waterway, Port, Coastal, and Ocean Engrg.* **105**(WW4), (1979), pp. 460–464.

68.  Y. Goda and N. Watanabe, "A longshore current formula for random breaking waves," *Coastal Engineering in Japan, JSCE* **34**(2) (1991), pp. 159–175.

# Part III

# Statistical Analysis of
# Extreme Waves

# Chapter 13

# Statistical Analysis of Extreme Waves

## 13.1 Introduction

### 13.1.1 Data for Extreme Wave Analysis

(A) *Preparation of sample*

The first step in designing a maritime structure is the selection of design waves. In most cases, storm wave heights which would be exceeded once in a given period of years, say 100 years, are chosen on the basis of statistical analysis of extreme events. Needs for the analysis of extremes arise in many branches of science and technology. Peak river discharges for flood-plain protection, hurricane winds for suspension bridge designs, and storm surge heights for coastal defense works are well-known examples in civil engineering. In this chapter, statistical techniques frequently used in extreme wave analysis are introduced and discussed.

Depending on the method of selecting a set of wave data (which is called a *sample* in statistics), there are three different approaches. One method tries to utilize the whole data of wave heights observed visually or instrumentally during a number of years. The data are analyzed in a form of cumulative distribution to be fitted to some distribution function. Once a best-fitting distribution function is found, the design wave height is estimated by extrapolating the distribution function to the level of probability which corresponds to a given period of years being considered in design process. This method is called the *total sample method*.

The other two methods use only the maxima of wave heights in time series data. The *annual maxima method* picks up the largest significant wave height in each year, whereas the *peaks-over-threshold method* takes the peak heights of storm waves exceeding a certain threshold value.

The three methods have their own proponents. The choice to make among these methods is somewhat subjective. One important requisite for a statistical sample is *independency*. It means that individual data in a sample must be statistically independent of each other; in other words, the correlation coefficient between successive data should nearly be zero. Another important requisite is *homogeneity*. Individual data in a sample must have a common distribution, all belonging to a single group of data, which is called the *population*. The distribution of the data in a population is called the *parent distribution*. Storm waves during the monsoon season and waves during the off-monsoon season would exhibit some difference in their cumulative distributions, and thus they would belong to different populations. A population of waves generated by tropical cyclones would probably be different from that by extratropical cyclones.

The total sample method is not recommended according to the above requisites. Ocean waves have a tendency of being persistent for many hours. The correlation coefficient between wave heights, 24 hours apart, has been found to have a high value of 0.3 to 0.5 according to Goda;[1] thus independency is not satisfied for data sets of the total sample method. Furthermore, the group of small wave heights is likely to constitute a population different from that of the group of large heights. Van Vledder *et al.*[2] reports a case that the total sample method predicts a 100-year wave height 10% larger than the estimation by the peaks-over-threshold method, probably owing to the influence of low wave height data. Thus, no further discussion will be given to the total sample method.

The annual maxima method and peaks-over-threshold method both satisfy the requisite of independency. The annual maxima method is widely used in the analysis of extreme flood discharges and other data of environmental loads. However, existing databases of storm waves in many countries rarely cover a period of more than 30 years as of the late 2000s. Such a short record length of extreme wave data brings forth a problem of low reliability in statistical sense; a small sample size produces a wide range of confidence interval. The peaks-over-threshold method (henceforth abbreviated as POT) can have a relatively large number of data in a sample, and thus have a smaller range of confidence interval. Therefore, the discussion hereinafter is mainly focused on POT. Nevertheless, the techniques of extreme data analysis for the annual maxima method are identical with those of POT, and the following descriptions are also applicable for the use of the annual maxima method.

(B) *Parameters of sample of extreme data*

Two parameters are important in describing the nature of a sample of extreme data. One is the *mean rate* of the extreme events. The mean rate denoted by $\lambda$ is defined with the number of events $N_T$ which have occurred during the period of $K$ years as

$$\lambda = \frac{N_T}{K}.$$ (13.1)

In the annual maxima method, one data is taken from each year so that $\lambda = 1$. In POT, the mean rate may vary from a few to several dozens depending on the threshold value which defines the extreme events. The number of years $K$ need not be an integer but can have a decimal.

Another parameter is related with the censoring process. When a wave hindcasting project is undertaken for the purpose of collecting samples of extreme wave heights, there is a possibility that medium to minor storms have not been detected on weather maps and waves generated by these storms are dropped from the list of data. Thus, it is often recommended to employ the data of large storm waves only, by omitting the data of low wave heights. This is an example of censoring process. Another example is the treatment of downtime of wave measurement system. If the maximum wave during the downtime is known to be below a certain moderate value by information from some other sources, the period of downtime can be included in the effective duration of measurement period $K$, provided that other measured data below that value be omitted from the extreme wave analysis. In these examples, the existence of minor data should be taken into account in extreme analysis so as not to distort the shape of distribution function. For this purpose, the following parameter called the *censoring parameter* denoted by $\nu$ is introduced:

$$\nu = \frac{N}{N_T},$$ (13.2)

where $N$ refers to the number of data used in the analysis and $N_T$ the total number of storm events which would have occurred during the period of extreme wave analysis; $N_T$ need not be accurate, but its approximate estimate suffices.

### 13.1.2 *Distribution Functions for Extreme Data Analysis*

In the extreme data analysis, many distribution functions are employed for fitting to samples. The distribution function, which is an abbreviation of the cumulative distribution function, is defined to represent the probability of a random variable $X$ equal to or less than some specified value $x$, i.e.,

$$F(x) = \Pr[X \leq x], \tag{13.3}$$

where $\Pr[\mathcal{A}]$ denotes the probability of the event $\mathcal{A}$ to take place.

If a random variable $X_{\max}$ represents the maximum value among a large number of independent variables $z$ randomly extracted from a same initial distribution such as $X_{\max} = \max(z_1, z_2, \ldots, z_n)$, the theory of extreme statistics states that the distribution function for $X$ belongs one of the following three types of asymptotic functions for most of initial distributions, e.g., Gumbel[3] and Coles[4]:

(1) Fisher–Tippett type I (abbreviated as FT-I) or Gumbel distribution:

$$F(x) = \exp\left[-\exp\left(-\frac{x-B}{A}\right)\right] : -\infty < x < \infty, \tag{13.4}$$

$$f(x) = \frac{1}{A}\exp\left[-\frac{x-B}{A} - \exp\left(-\frac{x-B}{A}\right)\right], \tag{13.5}$$

where $f(x)$ denotes the probability density function.

(2) Fisher–Tippett type II (abbreviated as FT-II) or Frechét distribution:

$$F(x) = \exp\left[-\left(1 + \frac{x-B}{kA}\right)^{-k}\right] : k > 0, \quad B - kA \leq x < \infty, \tag{13.6}$$

$$f(x) = \frac{1}{A}\left(1 + \frac{x-B}{kA}\right)^{-(1+k)}\exp\left[-\left(1 + \frac{x-B}{kA}\right)^{-k}\right]. \tag{13.7}$$

(3) Fisher–Tippett type III (abbreviated as FT-III) or maximal Weibull distribution:

$$F(x) = \exp\left[-\left(1 - \frac{x-B}{kA}\right)^{k}\right] : k > 0, \quad -\infty < x \leq B + kA, \tag{13.8}$$

$$f(x) = \frac{1}{A}\left(1 - \frac{x-B}{kA}\right)^{k-1}\exp\left[-\left(1 - \frac{x-B}{kA}\right)^{k}\right]. \tag{13.9}$$

Among the parameters used in the above equations, the parameter $A$ is called the *scale parameter* because it governs the linear scale of $x$. The parameter $B$ is called the *location parameter* because it fixes the location of the axis of $x$. The parameter $k$ is called the *shape parameter* because it determines the functional shape of distribution. The parameter $k$ has no dimension, but the parameters $A$ and $B$ have the same units with $x$. The notations for these parameters are not universal; the readers are advised to check the notations when they refer to various literatures. Depending on the textbooks, the functional expressions for the FT-II and FT-III may also differ from the above.

The FT-III distribution has been derived for the extreme data of the maxima in samples and has the upper bound of $B+kA$. Although it is sometimes called the Weibull distribution, it is different from another Weibull distribution to be discussed later. The FT-III distribution may be named here as *maximal Weibull distribution* so as to differentiate it from the later one expressed by Eqs. (13.12) and (13.13).

The three asymptotic distributions of Eqs. (13.4) to (13.9) can be synthesized into the following distribution called the *Generalized extreme-value distribution*[4]:

(4) Generalized extreme-value distribution (abbreviated as GEV):

$$F(x) = \exp\left[-\left(1 + \xi\frac{x-B}{A}\right)^{-1/\xi}\right] : A + \xi(x-B) > 0, \qquad (13.10)$$

$$f(x) = \frac{1}{A}\left(1 + \xi\frac{x-B}{A}\right)^{-(1+1/\xi)} \exp\left[-\left(1 + \xi\frac{x-B}{A}\right)^{-1/\xi}\right]. \qquad (13.11)$$

Depending on the value of the shape parameter $\xi$ being positive, zero, or negative, the GEV distribution becomes the FT-II, FT-I, or FT-III distribution through conversion of the shape parameter by $\xi = 1/k$.

There are other asymptotic distributions for a random variable $X_{min}$ that represents the minimum value among a large number of independent variables $z$ randomly extracted from a same initial distribution such as $X_{min} = \min(z_1, z_2, \ldots, z_n)$. One of such distributions is the following, which is called the Weibull distribution[5]:

(5) Weibull distribution:

$$F(x) = 1 - \exp\left[-\left(\frac{x-B}{A}\right)^k\right] : k > 0, \quad B \le x < \infty, \qquad (13.12)$$

$$f(x) = \frac{k}{A}\left(\frac{x-B}{A}\right)^{k-1} \exp\left[-\left(\frac{x-B}{A}\right)^k\right]. \qquad (13.13)$$

The Weibull distribution was introduced by Weibull in the late 1930s for description of the breaking strength of materials, the lower limit of which was his main concern.[6] In contrast to the maximal Weibull distribution of Eqs. (13.8) and (13.9), this distribution may be called the minimal Weibull distribution. However, the prefix minimal is not used hereinafter, because the maximal Weibull distribution is not going to be discussed further in the present chapter. The Weibull distribution has been used for extreme wave analysis in many occasions. Petruaskas and Aagaard[7] are the early supporters of this distribution function. The Weibull distribution is also employed in the field of hydrology as reported by Greenwood *et al.*[8] and Boes *et al.*[9]

Another distribution function which is often employed in the extreme wave analysis is the lognormal distribution of the following:

(6) Lognormal distribution:

$$F(x) = \int_{-\infty}^{x} f(t)\, dt \ : \ 0 < x < \infty, \tag{13.14}$$

$$f(x) = \frac{1}{\sqrt{2\pi}Ax} \exp\left[ -\frac{(\ln x - B)^2}{2A^2} \right]. \tag{13.15}$$

In this expression, the scale parameter $A$ and the location parameter $B$ have no dimensions.

In the extreme statistical analysis, many more distribution functions are fitted to samples of extreme data. For example, Hosking and Wallis[10] list eleven functions including the Gumbel, Generalized extreme-value, Generalized Pareto, Pearson type III, and other distributions as the candidates for data fitting in the regional frequency analysis.

Although some statisticians promote the use of the asymptotic distributions of Eqs. (13.4) to (13.11) for maximal data, Hosking and Wallis do not give any priority to the asymptotic distributions, arguing that the number of storm events in a year is rarely large enough to justify the extreme-value approximation for the annual maximum streamflow data.[11] The asymptotic distributions may be used only after confirmation of their superiority in the goodness of fit in comparison with other distributions. The POT data are not maximal data but merely constitute a sample randomly taken from an initial distribution, and thus there is no theoretical ground for a sample of POT data to belong to the family of the GEV distributions.

### 13.1.3  *Characteristics of Selected Distribution Functions*

Following the approach by Petruaskas and Aagaard as well as another by Hoking and Wallis, no priority is given to the asymptotic distribution functions in the present chapter. They are simply taken as one of candidates of initial distributions with wide variety of distribution characteristics to fit to a sample of extreme wave data. However, the FT-III or maximal Weibull distribution of Eqs. (13.8) and (13.9) is excluded from candidate distributions, because it assigns an upper bound for the range of extreme data; i.e., $x_{upper} = B + kA$. Exclusion is made because there is no physical ground to believe that extreme wave heights would have an upper limit.

When waves are recorded in relatively shallow water, the height of storm waves may be limited by depth-controlled breaking. In such a case, use of the FT-III may become appropriate. However, wave data recorded in a restricted depth of water cannot be utilized for determination of design waves in deeper water, because we cannot estimate the offshore wave height from the wave data within the surf zone. It has been a practice in wave measurements, at least in Japan, such that the wave recording should be made at the depth being greater than three times the largest possible significant wave height.

The GEV distribution is also excluded from candidate distributions for extreme wave analysis, because it becomes the FT-III distribution when

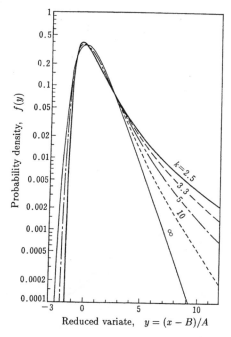

Fig. 13.1 Probability densities of FT-I ($k = \infty$) and FT-II ($k = 2.5$ to $10$) distributions.[12]

the shape parameter takes a negative value. Therefore, the present chapter employs the FT-I, FT-II, Weibull and lognormal distributions as the candidates for fitting to the samples of extreme wave data.

Figure 13.1 exhibits the probability density of the FT-II distribution with the shape parameter $k = 2.5$, 3.3, 5 and 10 together with that of the FT-I distribution, which is designated with $k = \infty$. Designation is so made, because Eq. (13.4) of the FT-I distribution is the asymptote of Eq. (13.6) of the FT-II distribution at the limit of $k \to \infty$.

The abscissa of Fig. 13.1 is a dimensionless variate of $y = (x - B)/A$, which is called the *reduced variate*. As the value of the shape parameter $k$ decreases, the distribution of FT-II becomes broader with heavier (i.e., longer) tails. A broad distribution predicts a very large 100-year wave height, when compared with a light-tail distribution.

Figure 13.2 shows the probability density of the Weibull distribution with the shape parameter $k = 0.75$, 1.0, 1.4 and 2.0. The case with $k = 1$ is the exponential distribution, which is included in the Weibull distribution. With the decrease in the $k$ value, the distribution becomes broader. The lognormal distribution demonstrates behaviors similar to the Weibull distribution with $k = 2$; the two distributions are often fitted to a sample of extreme wave heights with almost the same degree of goodness of fit.[13] Thus, a fitting of lognormal distribution can be replaced to that of the Weibull distribution with $k = 2$.

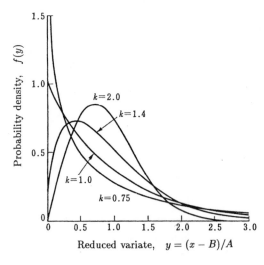

Fig. 13.2   Probability density of Weibull distribution ($k = 0.75$ to 2.0).[13]

Table 13.1  Characteristics of distribution functions for extreme analysis.

| Distribution | Mode | Mean | Standard Deviation |
|---|---|---|---|
| FT-II | $B+kA\left[\left(\dfrac{k}{1+k}\right)^{1/k}-1\right]$ | $B+kA\left[\Gamma\left(1-\dfrac{1}{k}\right)-1\right]$ | $kA\left[\Gamma\left(1-\dfrac{2}{k}\right)-\Gamma^2\left(1-\dfrac{1}{k}\right)\right]^{1/2}$ |
| FT-I | $B$ | $B+A\gamma$ | $\dfrac{\pi}{\sqrt{6}}A$ |
| Weibull | $B+A\left(1-\dfrac{1}{k}\right)^{1/k}$ | $B+A\Gamma\left(1+\dfrac{1}{k}\right)$ | $A\left[\Gamma\left(1+\dfrac{2}{k}\right)-\Gamma^2\left(1+\dfrac{1}{k}\right)\right]^{1/k}$ |
| Lognormal | $\exp(B-A^2)$ | $\exp\left(B+\dfrac{A^2}{2}\right)$ | $\exp\left(B+\dfrac{A^2}{2}\right)(\exp A^2-1)^{1/2}$ |

Note: $\Gamma(\cdot)$ is the gamma function and $\gamma$ is Euler's constant ($=0.5772\ldots$).

The characteristics of the FT-I, FT-II, Weibull, and lognormal distributions are listed in Table 13.1 for their modes, means and standard deviations. When the shape parameter $k$ of the FT-II distribution is less than 2, the distribution becomes too broad and its standard deviation cannot be defined. When the shape parameter $k$ of the Weibull distribution is less than 1, the probability densities diverge as the variate $x$ approaches $B$ and the mode cannot be defined.

## 13.1.4  *Return Period and Return Value*

Extreme statistics have the objective of estimating an expected value of extreme event which would occur once in a long period of time. For this purpose, the concepts of *return period* and *return value* are introduced. The return period is defined as the average duration of time during which extreme events exceeding a certain threshold value would occur once. The return value is the threshold value which defines a given return period. A 100-year wave height is a return value which would be exceeded once in every 100 years on the average.

The return period denoted by $R$ is derived from the distribution function as follows. For the sake of simplicity, the case of annual maxima method is discussed first. The distribution function $F(x)$ is assumed to be known. The probability that the extreme variate $x$ does not exceed a given value $x_u$ in one year is $F(x_u)$ by definition. Suppose that the event of $x \geq x_u$ occurred in one year, the variate $x$ did not exceed $x_u$ during the other $n-1$ years, and it did exceed $x_u$ in the $n$th year. Because the probability of nonexceedance for $n-1$ years is given by $F^{n-1}(x_u)$ and that of exceedance in one year is $1 - F(x_u)$, the probability of the above event is expressed as

$$P_n = F^{n-1}(x_u)[1 - F(x_u)].\qquad(13.16)$$

The above event may occur in the first year of $n = 1$, in the second year of $n = 2$, or in any arbitrary year, but may not occur until $n = \infty$. The expected value of $n$ is the return period by definition, and it is calculated as below.

$$R = E[n] = \sum_{n=1}^{\infty} nP_n = [1 - F(x_u)] \sum_{n=1}^{\infty} nF^{n-1}(x_u) = \frac{1}{1 - F(x_u)} . \quad (13.17)$$

The return value corresponding to the return period $R$ is denoted by $x_R$. It is obtained with the inverse function of the cumulative distribution as

$$x_R = F^{-1}\left(1 - \frac{1}{R}\right) . \quad (13.18)$$

In textbooks of extreme statistics, the technical term of *quantile* is often used. The quantile is the value of variate corresponding to a given probability to be calculated by the inverse function of the distribution function. The return value is a quantile with the nonexceedance probability of $1 - 1/R$ in case of the annual maximum data.

In the case of POT with the mean rate $\lambda$, each year is divided into time segments of $1/\lambda$ year by assuming that each time segment has the same probability of extreme events (seasonal variation of events is neglected). Then, the return period and the return value can be given by the following formulas:

$$R = \frac{1}{\lambda[1 - F(x_u)]} , \quad (13.19)$$

$$x_R = F^{-1}\left(1 - \frac{1}{\lambda R}\right) . \quad (13.20)$$

### 13.1.5 *Spread Parameter of Distribution Functions*

When a sample of extreme wave data, either the annual maximal or POT data, is fitted to candidate distribution functions, judgment of the best fitting one may becomes difficult because of small differences in the degree of goodness of fitting among candidate functions. A parameter which may assists the judgment is the spread parameter to be introduced hereon, which defines the degrees of heaviness of the right-hand tail of a probability density function.

The spread parameter is defined by Goda[14] as the ratio of the 50-year return wave height $H_{50}$ to the 10-year return wave height $H_{10}$ as,

$$\gamma_{50} = \frac{H_{50}}{H_{10}} . \quad (13.21)$$

Because the return wave height is estimated as $H_R = A\,y_R + B$ as will be explained in Sec. 13.2.2(E) where $y_R$ denotes the reduced variate for the return period of $R$, the spread parameter can be calculated by the following equation:

$$\gamma_{50} = 1 + \frac{y_{50} - y_{10}}{y_{10} + B/A}, \tag{13.22}$$

where $y_{50}$ and $y_{10}$ are the reduced variates corresponding to the return period of 50 and 10 years, respectively. Therefore, the value of the spread parameter varies depending on the ratio of the location parameter to the scale parameter even if the distribution function is of the same type and has the same shape parameter.

Though the spread parameter is evaluated only after the distribution function for extreme wave heights is selected at a particular site, the parameter serves as an indicator of the tendency of the occurrence of very large wave heights for a long return period; the larger the spread parameter value, the higher the return wave height or quantile. For extreme waves around the Japanese Islands, the northern part of the Pacific coast has been found to have the value of $\gamma_{50} = 1.20 \sim 1.24$, the southern part of the Pacific coast to have the value of $\gamma_{50} = 1.21 \sim 1.27$, and the coast along the Japan Sea to have the value of $\gamma_{50} = 1.13 \sim 1.14$.[14] The spread parameter of extreme wave data has also been reported by Gencarelli *et al.*[15] in the Meditarrean Sea around Italy, being $1.10 \sim 1.24$.

In the cases of daily rainfalls and wind speeds, the spread parameter takes much larger value. Analysis of the annual maximum daily rainfalls and wind speeds in Japan has yielded the spread parameter in excess of 1.3 at many stations with the largest value of 1.77.[14] The result indicates that the extreme values of daily rainfalls and wind speeds have much larger yearly variations than those of extreme wave heights.

The spread parameter of a given sample is directly related with the coefficient of variation of the sample, $V_x = \sigma/\bar{x}$, where $\sigma$ is the standard deviation $\sigma$ and $\bar{x}$ is the mea. By using the coefficient of variation, the spread parameter is expressed as follows:

$$\gamma_{50} = \frac{1 + (y_{50} - y_{10})V_x/(\alpha\,\kappa)}{1 + (y_{10} - \beta)V_x/(\alpha\,\kappa)}, \tag{13.23}$$

where $\alpha$ is the proportionality coefficient of the standard deviation of the population being expressed as $\sigma = \alpha A$ with the scale parameter of $A$ ($\alpha$ can be obtained from Table 13.1), $\kappa$ is the ratio of the standard deviation of the sample to that of the population which can be found in Table 13.9 to appear later, and $\beta$ is the proportionality coefficient to be multiplied to the scale parameter $A$ when the population mean is expressed as $\bar{x} = B + \beta A$ ($\beta$ can be obtained from Table 13.1).

The spread parameter can also be defined as the ratio of 100-year return wave height to 10-year return wave height, i.e., $\gamma_{100} = H_{100}/H_{10}$. Two definitions of the spread parameter are convertible by the following equation:

$$\gamma_{100} = \frac{H_{100}}{H_{10}} = 1 + \frac{y_{100} - y_{10}}{y_{50} - y_{10}}(\gamma_{50} - 1). \qquad (13.24)$$

A distribution function is uniquely determined by the location, scale, and shape parameters once the functional type is given. Nevertheless, it is difficult to judge the degree of spreading of the right tail of the probability density function with the parameter values alone. In the probabilistic design of breakwaters and other coastal structures, the Monte Carlo simulation method is often employed to examine the safety level of structures under design. The statistical variation of annual maximum wave height is introduced with an extreme distribution function, the parameters of which must be given initially. Unless the distribution function is already specified at the design site, a responsible engineer needs to select an appropriate distribution function with specific parameter values by referring to the regional characteristics of extreme waves. The governing factor is the spread parameter rather than the functional type of extreme distributions, because several functions with the same value of spread parameter behave similarly at their right tails.

It is recommended to select the value of the spread parameter around 1.1 to 1.3 by referring to the regional characteristics as exemplified in the case of the coasts around the Japanese Islands. Then the scale and location parameters for the specified distribution function can be determined by using the following formulas:

$$\left.\begin{array}{l} A = H_{50}\dfrac{1 - 1/\gamma_{50}}{y_{50} - y_{10}}, \\[3mm] B = H_{50}\left[1 - \dfrac{1 - 1/\gamma_{50}}{1 - y_{10}/y_{50}}\right]. \end{array}\right\} \qquad (13.25)$$

**Example 13.1**

Set the scale parameter $A$ and the location parameter $B$ for a site where the return wave height has been determined as $H_{50} = 7.8$ m under the Weibull distribution with $k = 1.0$. The mean rate of storm waves is $\lambda = 8.5$ and the spread parameter is assumed to be $\gamma_{50} = 1.25$.

**Solution**

The reduced variates $y_{50}$ and $y_{10}$ are calculated by Eq. (13.37) to appear in Sec. 13.2.2(E) as follows:

$$y_{50} = [\ln(8.5 \times 50)]^{1.0} = 6.052, \quad y_{10} = [\ln(8.5 \times 10)]^{1.0} = 4.443.$$

Use of Eq. (13.24) yields the following results:

$$A = 7.8 \times \frac{1 - 1/1.25}{6.052 - 4.443} = 7.8 \times \frac{0.2}{1.609} = 0.9695 \text{ m},$$

$$B = 7.8 \times \left[ 1 - \frac{1 - 1/1.24}{1 - 4.443/6.052} \right] = 7.8 \times \left[ 1 - \frac{0.2}{1 - 0.7341} \right] = 1.933 \text{ m}.$$

## 13.2 Estimation of Best-Fitting Distribution Function

### 13.2.1 *Selection of Plotting Position*

(A) *Method of fitting a distribution to a sample*

Once a sample of extreme wave heights is obtained, we need to estimate the parent distribution of the sample. If the type of the parent distribution of storm wave heights is known, estimation of the shape, scale, and location parameters is not so difficult. However, no consensus has ever been made of the parent distributions of waves, winds, floods, and other environmental conditions. Thus, extreme statistical analysis begins from a search of most suitable distribution function for the unknown population.

Recently, the methodology called the *regional frequency analysis* based on $L$-moments has been promoted by Hosking and Wallis,[10] which makes use of multiple samples from a number of stations within a region for enhancement of the capacability of finding the parent distribution. It has been employed mostly in the field of hydrology, but van Gelder *et al.*[16] have used the methodology for analysis of extreme wave heights in the Dutch North Sea. Even in this methodology, analysis begins from the process of distribution fitting.

For fitting a candidate distribution function to a sample, several methods listed below are available.

(i) method of moments,
(ii) method of probability-weighted moments,

(iii) method of $L$-moments,

(iv) maximum likelihood method,

 (v) least squares method,

(vi) extended least squares method.

The method of moments in (i) is to calculate the first, second, and third moments of sample data and compare them with the theoretical mean and standard deviation of a candidate distribution listed in Table 13.1. In the days before use of computers, calculation of moments was within the capacity of manual works using machine calculators and it was favored by many analysts including Gumbel.[3] However, the method yields a large bias for small size samples and is not in use presently.

The method of probability-weighted moments in (ii) was developed in 1979 by Greenwood *et al.*[8] as an improvement over the method of moments. They presented parameter estimation formulas from the probability-weighted moments of a sample. The probability-weighted moments method has been employed in the field of hydrology.

The method of $L$-moments in (iii) was proposed by Hosking[17] in 1990 as further improvement over the probability-weighted moments method. This method is the basis of the regional frequency analysis as described in the book by Hosking and Wallis.[10]

The maximum likelihood method in (iv) is an iterative numerical scheme to find the parameter values which maximize the likelihood function defined as

$$L(x_1, \ldots, x_N; A, B, k) = \prod_{m=1}^{N} f(x_m; A, B, k), \qquad (13.26)$$

where $x_1, \ldots, x_N$ represent the data values and $f$ is the probability density function. The maximum likelihood method is favored by statisticians, because its characteristics can be examined mathematically. However, the theory is not easy to understand and the algorithm of numerical scheme is rather complicated. The variation of the likelihood function around the peak point is rather slow as exhibited in the book by Coles,[4] and the best-fitting distribution may not be so clearly defined.

The least squares method in (v) is a numerical version of graphical fitting techniques, by making regression analysis between the ordered statistics $x_{(m)}$ of extreme data and their reduced variates $y_{(m)}$. The reduced variate is calculated from a chosen distribution function for the non-exceedance probability $F_{(m)}$ assigned to the order number $m$ so that $x_{(m)}$ can be linearly plotted against $y_{(m)}$. Reliability of the least squares method is affected by the selection of the probability $F_{(m)}$, which is called the plotting position; the selection problem is discussed in Clause (C) of this subsection.

The extended least squares method of (vi) has been proposed by Izumiya and Saito.[18] This method improves the efficiency of the least squares method, while computation loads increase little. The method will be described in Sec. 13.2.2(B).

(B) *Selection of distribution fitting method*

A method for fitting a distribution function to a sample of extreme data should be selected on the basis of two criteria called the *unbiasedness* and *efficiency*. *Unbiasedness* refers to the conditions that the values of the parameters and return values estimated from the sample should be the same as those of the population values. As will be discussed in Sec. 13.3.1, a sample of extreme variates represents only a randomly-extracted portion of the population and any statistics calculated from the sample has a certain deviation from the population statistics. Estimates of statistics will differ from one sample to another, but their expected values (ensemble means of many samples) should be in agreement with the population values. This is the definition of the unbiasedness. Even if the expected values of the sample statistics agree with the population values (the unbiasedness condition is satisfied), the sample statistics show a certain degree of variation. *Efficiency* refers to the smallness of the standard deviation of the sample statistics.

Samples of extreme wave data are mostly those censored by the threshold wave height in the POT method. It is desirable for the fitting method being capable of duly dealing with censored samples. The fitting method should also be simple and easy to apply without necessitating any subjective judgment.

According to the above two criteria, the method of moments cannot satisfy the unbiasedness, because the sample moments deviate from the population values as the sample size decreases. Table 13.10 in Sec. 13.3.1 exhibits the effect of sample size on the sample standard deviation. Its efficiency is said to be inferior to the maximum likelihood method and the least squares method. While the method of $L$-moments has a potential of wide application to extreme wave statistics, its application to the POT data begins only recently and its validation remains as a future task.

The maximum likelihood method tends to have a small amount of negative bias for small size samples, but its efficiency seems to be the highest as exemplified in a comparative study with numerically simulated extreme data.[19] However, its application to samples of censored data is difficult because the likelihood functions have not been derived for the case of $\nu < 1$.

The least squares method has often been accused of yielding a positive bias in the return value. However, a positive bias is the resultof the use of inappropriate plotting position formula. With the use of plotting position formulas presented later, it has been verified that the least squares method satisfies the condition of unbiasedness.[13,20] With regards to the efficiency, the least squares method is slightly inferior to the maximum likelihood method, but the extended least squares method has the efficiency as high as that of the maximum likelihood method. Although it is difficult to apply the least squares method to a three parameter distribution directly, the method remains workable through the conversion of a three-parameter distribution to a two-parameter distribution by fixing the shape parameters at several pre-selected values. In this chapter, the least squares method for extreme analysis is discussed in detail because of its simplicity in algorithm and applications.

(C) *Plotting position formulas*

A sample of data arranged in the ascending or descending order belongs to the category of *ordered statistics*. As the present chapter is concerned with the statistics of extremely large wave heights, the descending order is taken and the order number is expressed with $m$. The variate and its nonexceedance probability of the $m$th order are denoted with the subscript $(m)$, i.e., $x_{(m)}$ and $F_{(m)}$. The information of the nonexceedance probability $F_{(m)}$ is sought for when the ordered statistics are compared with the corresponding values estimated from the distribution fitted to the sample and in the process of distribution fitting by means of the least squares method. The formula which assigns the probability to the ordered variate is called the *plotting position formula*.

The best-known plotting position formula is the Weibull formula of the following:

$$\hat{F}_{(m)} = 1 - \frac{m}{N+1}. \qquad (13.27)$$

Equation (13.13) is derived as the expected probability of the $m$th ordered variate in the population; i.e., $E[F(x_{(m)})]$. Gumbel (Ref. 3, Sec 1.2.7) advocated the use of this formula based on somewhat intuitive arguments. But the Weibull plotting position formula always produces a positive bias

Table 13.2  Constants of unbiased plotting position formula.

| Distribution | $\alpha$ | $\beta$ | Authors |
|---|---|---|---|
| FT-II | $0.44 + 0.52/k$ | $0.12 - 0.11/k$ | Goda and Onozawa[12] |
| FT-I | $0.44$ | $0.12$ | Gringorten[21] |
| Weibull | $0.20 + 0.27/\sqrt{k}$ | $0.20 + 0.23/\sqrt{k}$ | Goda[13,20] |
| Normal | $0.375$ | $0.25$ | Blom[22] |
| Lognormal | $0.375$ | $0.25$ | Blom[22] |

in the return value, amounting to several percent when the sample size is less than a few dozens.[13,20]

The unbiased plotting position formula varies depending on the distribution function being applied. According to a numerical simulation study,[13] the Gringorten formula[21] yields almost no bias when applied to the FT-I distribution. For the normal distribution, the Blom formula[22] brings forth little bias. For the Weibull distribution, Petruaskas and Aagaard[7] derived a formula in such a way that it gives the probability corresponding to the expected value of the $m$th ordered variate; i.e., $F\{E[x_{(m)}]\}$. When examined in a numerical simulation study, however, their formula has produced a small amount of negative bias. Then Goda[13,20] proposed a modified version of the Petruaskas and Aagaard formula. For the FT-II distribution, Goda and Onozawa[12] proposed an empirical formula based on another numerical simulation study. The latter two proposals are both based on the Monte Carlo simulations for the sample size ranging from 10 to 200, each size with 10,000 samples.

The unbiased plotting position formula can be expressed in the following general form:

$$\hat{F}_{(m)} = 1 - \frac{m - \alpha}{N_T + \beta}, \quad m = 1, 2, \ldots, N. \tag{13.28}$$

The values of constants $\alpha$ and $\beta$ are given in Table 13.2. The above formula uses the total number $N_T$ instead of the sample size $N$ so that the formula can be applied for both censored and uncensored samples.

## 13.2.2  Estimation of Return Values with the Least Squares Method

### (A) Procedure of parameter estimation

Estimation of the parameters of a distribution function to be fitted to a sample is made in the following steps:

(i) Select a certain number of candidate distributions for fitting. Three-parameter distributions such as the FT-II and Weibull are treated as two-parameter distributions by fixing the value of shape parameter at one of pre-selected values (see Clause (C)).

(ii) Sort the data in a given sample in the descending order of $x_{(m)}$.

(iii) Approximately estimate the total number $N_T$ of storm events during the period of data analysis, and assign the nonexceedance probability $\hat{F}_{(m)}$ by Eq. (13.28) using the constants listed in Table 13.2.

(iv) Calculate the reduced variate $y_{(m)}$, which depends on the candidate distribution, for respective probability $\hat{F}_{(m)}$.

(v) Apply the least squares method by assuming a linear relationship between the ordered statistics $x_{(m)}$ and $y_{(m)}$ of the following expression:

$$x_{(m)} = \hat{B} + \hat{A} y_{(m)} . \tag{13.29}$$

The correlation coefficient $r$ between $x_{(m)}$ and $y_{(m)}$ must be estimated together with $\hat{A}$ and $\hat{B}$. Any numerical algorithm for the least squares method suffices for solving Eq. (13.29). However, attention is called for the expression of Eq. (13.29) which is different from the conventional form of $y = a + bx$.

(B) *Simultaneous equations for obtaining the scale and location parameters*

The parameters $\hat{A}$ and $\hat{B}$ in Eq. (13.29) are estimated by solving the following simultaneous equations:

$$\left.\begin{aligned} A \sum_{m=1}^{N} y_{(m)}^2 + B \sum_{m=1}^{N} y_{(m)} &= \sum_{m=1}^{N} x_{(m)} y_{(m)} , \\ A \sum_{m=1}^{N} y_{(m)} + NB &= \sum_{m=1}^{N} x_{(m)} . \end{aligned}\right\} \tag{13.30}$$

In order to improve the performance of the least squares method, Izumiya and Saito[18] have proposed to apply the weighting function $w_{(m)}^2$ to each term in Eq. (13.30) such that

$$\left.\begin{aligned} A \sum_{m=1}^{N} w_{(m)}^2 y_{(m)}^2 + B \sum_{m=1}^{N} w_{(m)}^2 y_{(m)} &= \sum_{m=1}^{N} w_{(m)}^2 x_{(m)} y_{(m)} , \\ A \sum_{m=1}^{N} w_{(m)}^2 y_{(m)} + B \sum_{m=1}^{N} w_{(m)}^2 &= \sum_{m=1}^{N} w_{(m)}^2 x_{(m)} . \end{aligned}\right\} \tag{13.31}$$

Izumiya and Saito called the method with Eq. (13.31) as the *extended least squares method* and determined the weighting function $w_{(m)}^2$ in consideration of the variance of the ordered statistics under a particular parent distribution. The weighting function has been given for the FT-I distribution as

$$w_{(m)}^2 = \frac{N_T + \alpha + \beta - m}{m - \alpha} \left[ \ln \left( \frac{N_T + \alpha + \beta - m}{m - \alpha} \right) \right]^2 . \qquad (13.32)$$

It is given for the Weibull distribution as

$$w_{(m)}^2 = \frac{m - \alpha}{N_T + \alpha + \beta - m} \left[ - \ln \left( \frac{m - \alpha}{N_T + \beta} \right) \right]^{2(k-1)/k} , \qquad (13.33)$$

where $\alpha$ and $\beta$ are the constants listed in Table 13.2.

The value of the weighting function $w_{(m)}^2$ is smallest for $m = 1$ (largest data) and increases as the order number $m$ becomes large. Because of this characteristic, the influence of an outlier (the largest data for $m = 1$ being far greater than the second largest data and others) is suppressed as minimal and the regression analysis based on Eq. (13.31) seems to yield stabler estimates of the scale parameter $\hat{A}$ and the location parameter $\hat{B}$ than the simple least squares method based on Eq. (13.30). However, the extended least squares method is not applicable for the FT-II distribution, because Izumiya and Saito did not provide the weighting function for this distribution.

Though the extended least squares method is recommended for the extreme wave analysis, the examples of data analysis and various empirical formulas in this chapter have been prepared using the simple least squares method based on Eq. (13.30).

(C) *Candidate distribution functions*

As discussed in Sec. 13.1.3, the FT-I, FT-II and Weibull distributions are considered in this chapter as the candidates of the parent distribution of extreme waves. The scale, location, and shape parameters of these candidate functions are estimated for a given sample by solving either Eq. (13.30) or (13.31), and the goodness of fit to each function is compared for selection of the best-fitting distribution.

The least squares method can yield the best estimate of two parameters in a single operation. As the FT-I distribution has two parameters of $A$ and

$B$, it can be analyzed by the least squares method directly. The FT-II and Weibull distributions have three parameters however, and thus they have to be modified into a form of two parameter functions. In this chapter, the shape parameter $k$ is fixed at one of the following values for the purpose of distribution fitting:

$$\left. \begin{array}{l} \text{FT-II distribution} \quad : k = 2.5, \ 3.33, \ 5.0 \text{ and } 10.0, \\ \text{Weibull distribution} : k = 0.75, \ 1.0, \ 1.4 \text{ and } 2.0. \end{array} \right\} \qquad (13.34)$$

Once the shape parameter is fixed, each distribution becomes an independent candidate function and is competed with other functions for best fitting. Thus, a proposal is hereby made to employ nine cumulative distributions (one FT-I, four FT-II's, and four Weibull's) as the candidate distributions.

The main reason for fixing the shape parameter is the difficulty in predicting the true parent distribution from a sample of small size, say a few dozen to one hundred. Goda[13,20] has demonstrated this difficulty by a Monte Carlo simulation study. The statistical variability of these distributions as well as confidence intervals of parameter estimates and return values have also been analyzed for the above nine distributions, and the results are presented in a form of tables and empirical formulas.

(D) *Calculation of reduced variate and explanatory example*

The next step in the parameter estimation is the preparation of the order statistics $x_{(m)}$ of extreme data in the descending order and the assignment of the nonexceedance probability $\hat{F}_{(m)}$ by Eq. (13.28). Then the reduced variate $y_{(m)}$ for the $m$th ordered data is calculated by the following equation:

$$\left. \begin{array}{l} \text{FT-I distribution} \quad : y_{(m)} = -\ln[-\ln \hat{F}_{(m)}], \\[2mm] \text{FT-II distribution} \quad : y_{(m)} = k\left[\left(-\ln \hat{F}_{(m)}\right)^{-1/k} - 1\right], \\[2mm] \text{Weibull distribution} : y_{(m)} = \left[-\ln\left(1 - \hat{F}_{(m)}\right)\right]^{1/k}. \end{array} \right\} \qquad (13.35)$$

As an explanatory example, the Kodiak data of hindcasted storm waves[2] is analyzed below. Wave hindcasting was carried out by the Coastal

Table 13.3. Peak significant wave heights of Kodiak data set.
($K = 20$ years, $N = N_T = 78$, $\lambda = 3.9$, $\nu = 1$).

| Year | $H_s$ (m) | Year | $H_s$ (m) |
|------|-----------|------|-----------|
| 1956 | 6.2 | 1966 | 7.3, 8.6, 7.4 |
| 1957 | − | 1967 | 7.1, 6.0, 6.3, 6.0, 6.7 |
| 1958 | 8.8, 6.6, 6.9, 7.8, 6.3 | 1968 | 6.6, 6.5, 6.9, 7.7, 8.2, 6.7, 7.4 |
| 1959 | 11.7, 7.2, 7.4 | 1969 | 6.4, 6.1, 7.1, 6.5, 8.5, 8.8, 9.1 |
| 1960 | 9.9, 8.9, 7.5, 7.0, 6.7 | 1970 | 8.0, 6.3, 9.1 |
| 1961 | 9.2, 6.2, 6.3 | 1971 | 6.6 |
| 1962 | 8.1, 6.3, 7.2, 6.3, 6.0 | 1972 | 6.7, 7.2, 10.2, 7.0, 10.1 |
| 1963 | 8.4, 6.8, 9.3, 6.7, 6.5, 7.2, 8.5 | 1973 | 7.8, 6.1, 6.3, 8.6, 7.1, 10.0 |
| 1964 | 6.9. 6.6, 9.4, 8.2 | 1974 | 8.0, 6.1, 8.4 |
| 1965 | 6.3, 7.6 | 1975 | 7.4, 8.2, 8.1 |

Engineering Research Center of the US Army[23] for the North-Eastern Pacific Ocean. The data was retrieved from a grid point called Kodiak off Alaska, located at 57°50′N and 148°78′W. The data set consists of all peak storm waves with the significant height exceeding 6 m, which were generated by 78 storms during a period of 20 years. Table 13.3 lists the Kodiak data set in chronological sequences. As the exact number of storm events in this period was not scrutinized, the data set is treated here as an uncensored sample. The Kodiak data is one of the two extreme wave data sets which were jointly analyzed by a working group on extreme statistics of the Section of Maritime Hydraulics of the International Association of Hydraulic Research (IAHR), as reported by van Vledder *et al.*[2]

The Kodiak data set is rearranged in the descending order according to the magnitude of significant wave height, the nonexceeding probability is assigned, and the reduced variate is calculated for several candidate distributions. Then the least squares method is applied. Table 13.4 lists a part of the results of calculation. Estimates of return waves corresponding to the return period of 10, 20, 50, and 100 years are also listed, toghether with the spread parameter introduced in Sec. 13.1.4.

(E) *Estimation of return value*

With the parameter values estimated for a candidate distribution function, the estimate of return value $\hat{x}_R$ for a given return period $R$ is made using the reduced variate $y_R$ as

$$\hat{x}_R = \hat{A}\,y_R + \hat{B}\,. \tag{13.36}$$

Table 13.4. Results of Kodiak data analysis by the least squares method.
(sample size: $N = 78$, mean: $\bar{x} = 7.501$ m, standard deviation: $\sigma_x = 1.214$ m).

| $m$ | $x_{(m)}$ | FT-II ($k=10$) $\hat{F}_{(m)}$ | $y_{(m)}$ | FT-I $\hat{F}_{(m)}$ | $y_{(m)}$ | Weibull ($k=1.4$) $\hat{F}_{(m)}$ | $y_{(m)}$ | Weibull ($k=2.0$) $\hat{F}_{(m)}$ | $y_{(m)}$ |
|---|---|---|---|---|---|---|---|---|---|
| 1 | 11.7 | 0.9935 | 6.540 | 0.9928 | 4.934 | 0.9927 | 3.121 | 0.9922 | 2.204 |
| 2 | 10.2 | 0.9807 | 4.825 | 0.9800 | 3.903 | 0.9800 | 2.648 | 0.9795 | 1.971 |
| 3 | 10.1 | 0.9679 | 4.081 | 0.9672 | 3.402 | 0.9672 | 2.405 | 0.9667 | 1.845 |
| 4 | 10.0 | 0.9551 | 3.607 | 0.9544 | 3.065 | 0.9544 | 2.238 | 0.9539 | 1.754 |
| 5 | 9.9 | 0.9423 | 3.261 | 0.9416 | 2.811 | 0.9417 | 2.109 | 0.9412 | 1.683 |
| 6 | 9.4 | 0.9295 | 2.990 | 0.9288 | 2.606 | 0.9289 | 2.003 | 0.9284 | 1.624 |
| ⋮ | ⋮ | ⋮ | ⋮ | ⋮ | ⋮ | ⋮ | ⋮ | ⋮ | ⋮ |
| ⋮ | ⋮ | ⋮ | ⋮ | ⋮ | ⋮ | ⋮ | ⋮ | ⋮ | ⋮ |
| 74 | 6.1 | 0.0589 | 0.989 | 0.0584 | −1.044 | 0.0615 | 0.140 | 0.0607 | 0.250 |
| 75 | 6.1 | 0.0461 | 1.063 | 0.0456 | −1.128 | 0.0488 | 0.118 | 0.0479 | 0.222 |
| 76 | 6.0 | 0.0333 | 1.152 | 0.0328 | −1.229 | 0.0360 | 0.094 | 0.0351 | 0.189 |
| 77 | 6.0 | 0.0205 | 1.270 | 0.0200 | −1.364 | 0.0233 | 0.069 | 0.0224 | 0.150 |
| 78 | 6.0 | 0.0077 | 1.464 | 0.0072 | −1.597 | 0.0105 | 0.039 | 0.0096 | 0.098 |
| Parameters | | $\hat{A} = 0.8292$ m $\hat{B} = 6.937$ m | | $\hat{A} = 0.9567$ m $\hat{B} = 6.955$ m | | $\hat{A} = 1.8621$ m $\hat{B} = 5.805$ m | | $\hat{A} = 2.6228$ m $\hat{B} = 5.178$ m | |
| Correlation | | $r = 0.98738$ | | $r = 0.99191$ | | $r = 0.99629$ | | $r = 0.98906$ | |
| $H_R$ (m) $R = 10$ years | | 10.56 | | 10.42 | | 10.51 | | 10.20 | |
| $R = 20$ years | | 11.42 | | 11.09 | | 11.13 | | 10.65 | |
| $R = 50$ years | | 12.65 | | 11.97 | | 11.91 | | 11.20 | |
| $R = 100$ years | | 13.66 | | 12.64 | | 12.47 | | 11.58 | |
| $\gamma_{50}$ | | 1.198 | | 1.149 | | 1.133 | | 1.098 | |

The reduced variate $y_R$ is calculated as a function of the return period $R$ and the mean rate $\lambda$ as follows:

$$
\begin{aligned}
\text{FT-I distribution} \quad &: y_R = -\ln\left\{-\ln\left[1 - \frac{1}{\lambda R}\right]\right\}, \\
\text{FT-II distribution} \quad &: y_R = k\left\{\left[-\ln\left(1 - \frac{1}{\lambda R}\right)\right]^{-1/k} - 1\right\}, \\
\text{Weibull distribution} &: y_R = [\ln(\lambda R)]^{1/k}.
\end{aligned}
\tag{13.37}
$$

In the case of the Kodiak data, the Weibull distribution with $k = 1.4$ has the scale and location parameters of $\hat{A} = 1.8621$ m and $\hat{B} = 5.805$ m. For this distribution, the 100-year wave height is estimated as follows:

$$y_{100} = [\ln(3.9 \times 100)]^{1/1.4} = 3.5815,$$

$$\hat{x}_{100} = 5.805 + 1.8621 \times 3.5815 = 12.47 \, \text{m}.$$

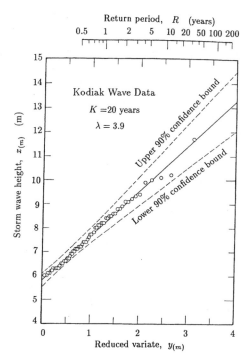

Fig. 13.3 Fitting of the Kodiak storm wave data to the Weibull distribution with $k = 1.4$.

Figure 13.3 shows the fitting of the Kodiak data to the Weibull distribution with $k = 1.4$. The solid line represents the best-fitting line, while the dashed line indicates the 90% confidence interval to be discussed in Sec. 13.3.3. The scale in the upper axis indicates the return period in years.

### 13.2.3 Selection of Most Probable Parent Distribution

(A) *Goodness of fit tests*

As mentioned earlier, the candidate distribution which best fits to the sample is selected as the most probable parent distribution. Nevertheless this does not exclude use of a single candidate distribution based on one's postulation about the parent distribution of storm wave heights. As the database of extreme waves by observations and hindcasting is expanded, there will be many cases of reliable extreme wave analyses. Then it could be possible in the future to establish some parent distribution of extreme wave heights, which would vary from coast to coast.

Goodness of fit is measured by several tests. The Kolmogorov–Smirnov test, the Anderson–Darling test and the chi-square test are often used for this purpose. When the parameter estimate is done by the least squares method, however, the degree of goodness of fit is simply represented with the value of correlation coefficient between the ordered data $x_{(m)}$ and its reduced variate $y_{(m)}$; the nearer the coefficient is toward 1, the better the fitting is. The Kodiak data set has been fitted to nine distributions including the four distributions listed in Table 13.4. Among the candidate functions, the Weibull distribution with $k = 1.4$ yields the correlation coefficient closest to 1, and is judged as the best-fitting one.

### (B) *MIR criterion for goodness of fit test*

The degree of correlation coefficient being near to 1 depends on a candidate distribution. Samples from a distribution with a narrow range of spreading such as the Weibull with $k = 2$ tend to yield the correlation coefficient much closer to 1 compared with samples from a distribution with a broad spreading. To examine the statistical characteristics of correlation coefficient, its residue from 1 is defined here as $\Delta r = 1 - r$. The residue is a statistical variate, the value of which varies from sample to sample. Goda and Kobune[20] reported the results of another Monte Carlo simulation study on extreme statistics.

Figure 13.4 shows the mean value of the residue of correlation coefficient of samples for several distributions. Simulation was done with 10,000 samples for each sample size ranging from 10 to 400 for respective distributions. The symbols of $\triangle$, $\circ$, $+$, $\triangledown$, $\times$, and $\bullet$ in Fig. 13.4 refer to Wibull $k = 0.075$, 1.0, 1.4, and 2.0, FT-I, and Log-normal distributions, respectively. The same is applied to Fig. 13.5. As seen in Fig. 13.4, $\Delta r_{\mathrm{mean}}$ of the Weibull distribution with $k = 0.75$ is larger than that with $k = 2$ at any sample size. Thus, a test of goodness of fit by means of the absolute value of correlation coefficient tends to yield unfavorable results against a broad distribution.

A remedy for the above bias is to use the ratio of the residue of a sample to the mean residue of a fitted distribution. Figure 13.5 shows examples of the cumulative distributions of the ratio $\Delta r / \Delta r_{\mathrm{mean}}$; the sample size is $N = 40$ and the data are censored ones with the censoring parameter $\nu = 0.5$. When the residue of correlation coefficient is normalized with its mean value, differences between various distributions are greatly reduced and a fair comparison of goodness of fit becomes possible. Goda and

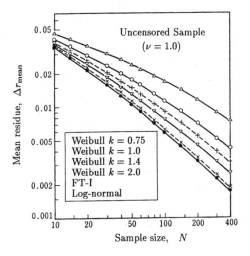

Fig. 13.4   Mean residue $\Delta r_{\text{mean}}$ of various distributions.[20]

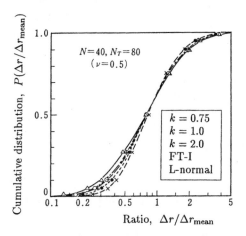

Fig. 13.5   Examples of cumulative distributions of the ratio $\Delta r/\Delta r_{\text{mean}}$.[20]

Kobune[20] proposed to use the MIR (MInimum Ratio of residual correlation coefficient) criterion for judgment of best fitting; a distribution with the smallest ratio is a best fitting one. They derived an empirical formula for estimating the mean residue $\Delta r_{\text{mean}}$ for a given distribution, sample size and censoring parameter from the data of simulation study. The formula is given as

$$\Delta r_{\text{mean}} = \exp[a + b \ln N + c \left(\ln N\right)^2].  \tag{13.38}$$

Table 13.5　Empirical coefficients for $\Delta r_{\mathrm{mean}}$ in the MIR criterion.[20]

| Distribution | $a$ | $b$ | $c$ |
|---|---|---|---|
| FT-II ($k = 2.5$) | $-2.470 + 0.015\nu^{3/2}$ | $-0.1530 - 0.0052\nu^{5/2}$ | $0$ |
| FT-II ($k = 3.33$) | $-2.462 - 0.009\nu^2$ | $-0.1933 - 0.0037\nu^{5/2}$ | $-0.007$ |
| FT-II ($k = 5.0$) | $-2.463$ | $-0.2110 - 0.0131\nu^{5/2}$ | $-0.019$ |
| FT-II ($k = 10.0$) | $-2.437 + 0.028\nu^{5/2}$ | $-0.2280 - 0.0300\nu^{5/2}$ | $-0.033$ |
| FT-I | $-2.364 + 0.54\nu^{5/2}$ | $-0.2665 - 0.0457\nu^{5/2}$ | $-0.044$ |
| Weibull ($k = 0.75$) | $-2.435 - 0.168\nu^{1/2}$ | $-0.2083 + 0.1074\nu^{1/2}$ | $-0.047$ |
| Weibull ($k = 1.0$) | $-2.355$ | $-0.2612$ | $-0.043$ |
| Weibull ($k = 1.4$) | $-2.277 + 0.056\nu^{1/2}$ | $-0.3169 - 0.0499\nu$ | $-0.044$ |
| Weibull ($k = 2.0$) | $-2.160 + 0.113\nu$ | $-0.3788 - 0.0979\nu$ | $-0.041$ |
| Log-normal | $-2.153 + 0.059\nu^2$ | $-0.2627 - 0.1716\nu^{1/4}$ | $-0.045$ |

The coefficients $a$, $b$ and $c$ are formulated for various distributions as listed in Table 13.5. Relative error of Eq. (13.38) in predicting the mean residue value is less than $\pm 3\%$.

For the case of the Kodiak data set with $N = 78$ and $\nu = 1$, $\Delta r_{\mathrm{mean}}$ and $\Delta r/\Delta r_{\mathrm{mean}}$ are calculated for the four distributions of Table 13.4 as follows:

$$\text{FT-II } (k = 10) \quad : \Delta r_{\mathrm{mean}} = 0.01562, \quad \Delta r/\Delta r_{\mathrm{mean}} = 0.808\,,$$

$$\text{FT-I} \quad : \Delta r_{\mathrm{mean}} = 0.01105, \quad \Delta r/\Delta r_{\mathrm{mean}} = 0.732\,,$$

$$\text{Weibull } (k = 1.4) : \Delta r_{\mathrm{mean}} = 0.00952, \quad \Delta r/\Delta r_{\mathrm{mean}} = 0.390\,,$$

$$\text{Weibull } (k = 2.0) : \Delta r_{\mathrm{mean}} = 0.00743, \quad \Delta r/\Delta r_{\mathrm{mean}} = 1.472\,.$$

Thus, the Weibull distribution with $k = 1.4$ satisfies the MIR criterion for best fitting.

The criteria of the largest correlation coefficient and the MIR both give the same judgment of best fitting for the Kodiak data set. However, use of the MIR criterion for the situation where the true parent distribution is unknown requires some caution. In a joint study by several hydraulic institutions on extreme wave statistics,[19] 500 samples of numerically simulated data from the population of the Weibull distribution with $k = 1.4$ were analyzed by various methods. With the least squares method, use of a simple criterion of the largest correlation coefficient produced slightly better results in predicting return values than that of the MIR criterion; further examination by numerical simulation is needed.

### 13.2.4 Rejection of Distribution Function

(A) *Outlier detection by the DOL criterion*

A sample of extreme data sometimes contains a data which exhibits the value much larger than the rest of data. When the sample is tried to fit to a candidate distribution, the particular data would be plotted at a position far above the line of a fitted distribution curve. Such a data is called an *outlier*. In other cases, the largest data $x_{(1)}$ might be only slightly greater than the second largest data $x_{(2)}$. The data $x_{(1)}$ would then be plotted far below the fitted distribution curve. It is also an outlier.

Detection of an outlier can be made with the DOL (Deviation of Outlier) criterion proposed by Goda and Kobune[20] and/or a statistical test by Barnett and Lewis (Ref. 24, pp. 144–150). The DOL criterion uses the following dimensionless deviation $\xi$:

$$\xi = \frac{x_{(1)} - \overline{x}}{s}, \tag{13.39}$$

where $\overline{x}$ is the mean of a sample and $s$ is the standard deviation of a sample defined as $s^2 = \sum_{i=1}^{N}(x_i - \overline{x})^2/N$. For the Kodiak data set, $x_{(1)} = 11.7$ m, $\overline{x} = 7.501$ m, $s = 1.206$ m, and thus $\xi = $ is 3.48.

In the statistical test of the normality of a sample, Thompson's test is used by comparing its mean value with the overall mean of a large number of samples. This test can be modified to yield the theoretical value of $\xi$ having the nonexceedance probability $P$ as below.

$$\xi_P = \left[ \frac{(N-1)\, F(1, N-2:\alpha)}{N-2+F(1, N-2:\alpha)} \right]^{1/2}, \tag{13.40}$$

where $N$ is the sample size and $F(1, N-2:\alpha)$ denotes the $F$ distribution with the $(1, N-2)$ degrees of freedom at the exceedance probability $\alpha$. For the largest data $x_{(1)}$, the probability $\alpha$ is given as $2(1 - P^{1/N})$.

The cumulative distribution of $\xi$ has been calculated by Eq. (13.40) and compared with the simulation data sampled from a population of the normal distribution.[20] As shown in Fig. 13.6, the $\xi$ value by simulation agrees with the theory except for the range of low probability. The agreement supports the validity of $\xi$ as a statistical variate. Then the $\xi$ value can be used to judge whether the largest data $x_{(1)}$ of a sample is an outlier or not.

For example, if the $\xi$ value of a sample exceeds the population value $\xi_{95\%}$ corresponding to the exceedance probability of 0.95, the largest data $x_{(1)}$ is judged as an outlier at the level of significance of 0.05. If the $\xi$ value of a sample is below the population value $\xi_{5\%}$ corresponding to the

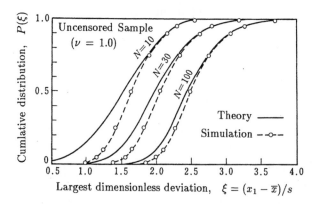

Fig. 13.6 Cumulative distribution of dimensionless deviation of the largest data $\xi$ for samples from the normal distribution.[20]

exceedance probability of 0.05, $x_{(1)}$ is also judged as an outlier at the level of significance of 0.05. The threshold value $\xi_{95\%}$ of the population is called the upper DOL, and the threshold value $\xi_{5\%}$ is called the lower DOL. The upper and lower DOLs of various distributions have been estimated with the simulation data of 10,000 samples for respective conditions, by reading the cumulative distribution curves of simulation data such as those shown in Fig. 13.6 as the function of the sample size $N$ and the censoring parameter $\nu$. The measured $\xi$ values have been approximated by the following empirical formula:

$$\xi_{95\%} \quad \text{and} \quad \xi_{5\%} = a + b \ln N + c \left( \ln N \right)^2 . \tag{13.41}$$

The empirical coefficients $a$, $b$ and $c$ have been formulated as listed in Tables 13.6 and 13.7. Relative error of Eq. (13.41) in predicting $\xi_{95\%}$ and $\xi_{5\%}$ is less than $\pm 2\%$.

The DOL criterion for outlier detection is applicable for any sample, and it is independent of distribution fitting methods being employed. It simply determines whether the largest data of a sample is an outlier of a distribution function fitted to a sample or not. In the case of the Kodiak data set, the largest data of 11.7 m is not an outlier for nine distributions discussed in this chapter. When the largest data of a sample is judged as an outlier, a quality check of the data should be made first. If no error is found in the data acquisition process, then the data should not be removed from a sample but the distribution for fitting to the sample should be eliminated from the candidate distributions instead.

Table 13.6   Empirical coefficients for the upper DOL criterion $\xi_{95\%}$.[20]

| Distribution | $a$ | $b$ | $c$ |
|---|---|---|---|
| FT-II ($k = 2.5$) | $4.653 - 1.076\nu^{1/2}$ | $-2.047 + 0.307\nu^{1/2}$ | 0.635 |
| FT-II ($k = 3.33$) | $3.217 - 1.216\nu^{1/4}$ | $-0.903 + 0.294\nu^{1/4}$ | 0.427 |
| FT-II ($k = 5.0$) | $0.599 - 0.038\nu^2$ | $0.518 - 0.045\nu^2$ | 0.210 |
| FT-II ($k = 10.0$) | $-0.371 + 0.171\nu^2$ | $1.283 - 0.133\nu^2$ | 0.045 |
| FT-I | $-0.579 + 0.468\nu$ | $1.496 - 0.227\nu^2$ | $-0.038$ |
| Weibull ($k = 0.75$) | $-0.256 - 0.632\nu^2$ | $1.269 + 0.254\nu^2$ | 0.037 |
| Weibull ($k = 1.0$) | $-0.682$ | $1.600$ | $-0.045$ |
| Weibull ($k = 1.4$) | $-0.548 + 0.452\nu^{1/2}$ | $1.521 - 0.184\nu$ | $-0.065$ |
| Weibull ($k = 2.0$) | $-0.322 + 0.641\nu^{1/2}$ | $1.414 - 0.326\nu$ | $-0.069$ |
| Log-normal | $0.178 + 0.740\nu$ | $1.148 - 0.480\nu^{3/2}$ | $-0.035$ |

Table 13.7   Empirical coefficients for the lower DOL criterion $\xi_{5\%}$.[20]

| Distribution | $a$ | $b$ | $c$ |
|---|---|---|---|
| FT-II ($k = 2.5$) | $1.481 - 0.126\nu^{1/4}$ | $-0.331 - 0.031\nu^2$ | 0.192 |
| FT-II ($k = 3.33$) | $1.025$ | $-0.077 - 0.050\nu^2$ | 0.143 |
| FT-II ($k = 5.0$) | $0.700 + 0.060\nu^2$ | $0.139 - 0.076\nu^2$ | 0.100 |
| FT-II ($k = 10.0$) | $0.424 + 0.088\nu^2$ | $0.329 - 0.094\nu^2$ | 0.061 |
| FT-I | $0.257 + 0.133\nu^2$ | $0.452 - 0.118\nu^2$ | 0.032 |
| Weibull ($k = 0.75$) | $0.534 - 0.162\nu$ | $0.277 + 0.095\nu$ | 0.065 |
| Weibull ($k = 1.0$) | $0.308$ | $0.423$ | 0.037 |
| Weibull ($k = 1.4$) | $0.192 + 0.126\nu^{3/2}$ | $0.501 - 0.081\nu^{3/2}$ | 0.018 |
| Weibull ($k = 2.0$) | $0.050 + 0.182\nu^{3/2}$ | $0.592 - 0.139\nu^{3/2}$ | 0 |
| Log-normal | $0.042 + 0.270\nu$ | $0.581 - 0.217\nu^{3/2}$ | 0 |

## (B) *REC criterion for rejection of candidate distribution*

Presence of an outlier suggests that a particular distribution is better eliminated from the candidates of parent distributions. When the distribution fitting is made with the least squares method, the value of the correlation coefficient $r$ between the ordered variate $x_{(m)}$ and the reduced variate $y_{(m)}$ can provide another test for rejection of candidate distributions. For this purpose, the residue of correlation coefficient from 1; i.e., $\Delta r = 1 - r$, is employed.

Figure 13.7 shows the cumulative distribution curves of $\Delta r$ for the Weibull distribution with $k = 1$ for the sample size of 10 to 300, which were

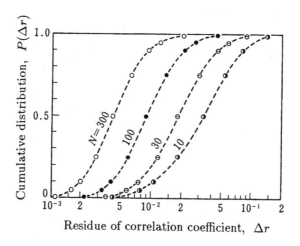

Fig. 13.7   Cumulative distribution of the residue of correlation coefficient $\Delta r$ for uncensored samples from the Weibull distributions with $k = 1.0$.[20]

prepared from simulation data of 10,000 samples for respective conditions.[20] By assuming these cumulative distributions of simulated data being almost the same as those of the population, a criterion for the rejection of candidate function which is called the REC (REsidue of Correlation coefficient) has been prepared. The exceedance probability of 0.95 was set for establishing the threshold value of $\Delta r$ at the level of significance of 0.05. The threshold value $\Delta r_{95\%}$ has been obtained from the simulation data and formulated in the following empirical expression of Eq. (13.42) with the coefficients listed in Table 13.8.[20] Relative error in predicting the threshold value $\Delta r_{95\%}$ is mostly less than $\pm 3\%$.

$$\Delta r_{95\%} = \exp[a + b \ln N + c \, (\ln N)^2]. \qquad (13.42)$$

In the case of the Kodiak data, all the nine candidate distribution functions yield the residual correlation coefficient $\Delta r$ below the threshold value $\Delta r_{95\%}$, and thus they are not rejected from the candidates of the parent distribution.

(C) *Rejection of candidate distribution with DOL and REC criteria*

As discussed in Sec. 13.2.3(A), most of statistical tests for the goodness of fit are not powerful enough for differentiating the performance of various distribution functions for extreme wave data. Instead of finding the best fitting distribution, a rejection test of a candidate distribution can be made

Table 13.8   Empirical Coefficients for $\Delta r_{95\%}$ in the REC criterion.[20]

| Distribution | $a$ | $b$ | $c$ |
|---|---|---|---|
| FT-II ($k = 2.5$) | $-1.122 - 0.037\nu$ | $-0.3298 + 0.0105\nu^{1/4}$ | $0.016$ |
| FT-II ($k = 3.33$) | $-1.306 - 0.105\nu^{3/2}$ | $-0.3001 + 0.0404\nu^{1/2}$ | $0$ |
| FT-II ($k = 5.0$) | $-1.463 - 0.107\nu^{3/2}$ | $-0.2716 + 0.0517\nu^{1/4}$ | $-0.018$ |
| FT-II ($k = 10.0$) | $-1.490 - 0.073\nu$ | $-0.2299 - 0.0099\nu^{5/2}$ | $-0.034$ |
| FT-I | $-1.444$ | $-0.2733 - 0.0414\nu^{5/2}$ | $-0.045$ |
| Weibull ($k = 0.75$) | $-1.473 - 0.049\nu^2$ | $-0.2181 + 0.0505\nu$ | $-0.041$ |
| Weibull ($k = 1.0$) | $-1.433$ | $-0.2679$ | $-0.044$ |
| Weibull ($k = 1.4$) | $-1.312$ | $-0.3356 - 0.0449\nu$ | $-0.045$ |
| Weibull ($k = 2.0$) | $-1.188 + 0.073\nu^{1/2}$ | $-0.4401 - 0.0846\nu^{3/2}$ | $-0.039$ |
| Log-normal | $-1.362 + 0.360\nu^{1/2}$ | $-0.3439 - 0.2185\nu^{1/2}$ | $-0.035$ |

by using the DOL and REC criteria introduced hereinbefore. The both criteria do not have theoretical background, but they perform well when the sample size is large, say one hundred or greater.

The objectives of rejection tests are as follows:

(i) To minimize the danger of misfitting when the true parent distribution is unknown by reducing the number of candidate distributions,

(ii) To narrow the range of distribution functions for the parent distribution by making rejection tests for multiple samples belonging to a certain regional group,

(iii) To judge if the maximum data of a sample is an outlier when the parent distribution is almost known through various studies.

The regional frequency analyis promoted by Hosking and Wallis[10] has the objective of finding a parent distribution which is commonly applicable to the samples of flood discarges and others at many sites withing a certain region. A test for discordancy by means of Monte Carlo simulation has been deviced to determine if a particular site belongs to the group common in the region.

Use of the DOL and REC criteria was made by Goda *et al.*[25] for many samples of extreme wave data by the POT method around Japan. They examined the wave data at 30 stations with the record lengths of 10 to 29 years, which were grouped into three geographical regions. After removing distributions judged as inappropriate by the DOL and REC criteria, distribution fucntions most often cited as best-fitting in respective regions were selcted as the most probable parent distributions. They were the Weibull distribution with the shape parameter of $k = 1$ (exponential distribution) for the northern Pacific coast, the FT-I distribution for the southern Pacific coast, and the Weibull distribution with $k = 1.4$ for the Japan Sea coast.

## 13.3   Confidence Interval of Return Value

### 13.3.1   *Statistical Variability of Samples of Extreme Distributions*

In the ocean environment, we sometimes encounter the event of extremely large storm waves having the return period far exceeding the design condition. It could have been generated by a truly abnormal meteorological condition, but it often results from the situation such that the extreme distribution of storm waves estimated from the previous data set was inappropriate because the sample size was not large enough. This is the problem of statistical variability of samples of extreme statistics.

Figure 13.8 provides an example of variation of samples drawn from an extreme distribution.[19] One hundred samples with the size 100 are numerically simulated from the population of the Weibull distribution with $k = 1.4$, which has the 1-year wave height of 8 m and the 100-year wave height of 13 m at the mean rate $\lambda = 5$; the sample size 100 is equivalent to the period of $K = 20$ years in duration. As each line in Fig. 13.7 represents the plot of $x_{(m)}$ versus $y_{(m)}$ of one sample, the spread of lines indicate the

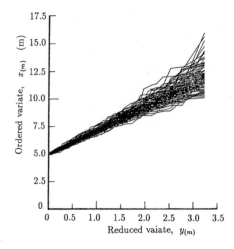

Fig. 13.8   Plot of 100 samples of simulated storm wave height data.[19]

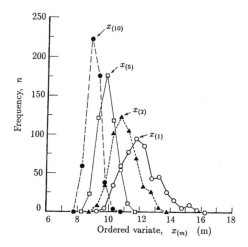

Fig. 13.9   Histograms of the first, second, fifth and tenth largest data of 500 samples.[19]

extent of sampling variability. Figure 13.9 shows the histograms of the first, second, fifth and tenth largest wave heights of 500 samples drawn from the same distribution. The largest wave height in 20 years varies from 9 to 16 m, even though the 100-year wave height of the population is 13 m. The data set was another subject of joint examination by the working group of the Section of Maritime Hydraulics of IAHR.

The extent of sampling variability can be examined through the analysis of the standard deviation of a sample. Although the unbiased variance of a sample defined as $\sigma_x^2 = \sum_{i=1}^{N}(x_i - \overline{x})^2/(N-1)$ has the expected value equal to the population variance, the expected value of $\sigma_x$ itself is smaller than the population standard deviation as listed in Table 13.9. The difference increases as the sample size decreases. Thus, the terminology of *unbiased standard deviation* should be used with some caution.

Analysis of the statistical characteristics of sample (unbiased) standard deviation has been made for various extreme distribution functions by the author (Ref. 13 and an unpublished source) through a Monte Carlo simulation technique. Table 13.9 lists the mean and the coefficient of variation of the unbiased standard deviation of a sample of extreme distributions. The data is based on the analysis of 20,000 simulated samples for each case, and the results have been slightly modified so that a smooth variation with respect to the sample size $N$ would be realized.

Decrease of the sample standard deviation from the population value causes a consistent underestimation of the scale parameter $A$ when the

Table 13.9    Mean and coefficient of variation of the sample standard deviation.

| Size | FT-II | | | | |
| N | $k = 2.5$ | $k = 3.33$ | $k = 5.0$ | $k = 10.0$ | FT-I |
|---|---|---|---|---|---|
| 10 | 0.632 (1.25) | 0.798 (0.79) | 0.882 (0.55) | 0.928 (0.41) | 0.951 (0.32) |
| 14 | 0.662 (1.12) | 0.827 (0.71) | 0.905 (0.49) | 0.946 (0.36) | 0.965 (0.27) |
| 20 | 0.690 (0.99) | 0.851 (0.63) | 0.923 (0.42) | 0.958 (0.31) | 0.975 (0.23) |
| 30 | 0.718 (0.87) | 0.873 (0.54) | 0.940 (0.36) | 0.969 (0.25) | 0.984 (0.19) |
| 40 | 0.736 (0.80) | 0.887 (0.49) | 0.950 (0.32) | 0.975 (0.22) | 0.988 (0.16) |
| 60 | 0.761 (0.74) | 0.906 (0.44) | 0.962 (0.28) | 0.983 (0.18) | 0.992 (0.13) |
| 100 | 0.792 (0.66) | 0.929 (0.38) | 0.974 (0.23) | 0.989 (0.15) | 0.995 (0.10) |
| 140 | 0.810 (0.61) | 0.941 (0.35) | 0.981 (0.20) | 0.992 (0.13) | 0.996 (0.09) |
| 200 | 0.827 (0.56) | 0.951 (0.32) | 0.987 (0.18) | 0.994 (0.11) | 0.998 (0.07) |

| Size | Weibull | | | | |
| N | $k = 0.75$ | $k = 1.00$ | $k = 1.4$ | $k = 2.0$ | Normal |
|---|---|---|---|---|---|
| 10 | 0.870 (0.56) | 0.925 (0.42) | 0.954 (0.31) | 0.971 (0.25) | 0.975 (0.24) |
| 14 | 0.899 (0.49) | 0.942 (0.36) | 0.966 (0.26) | 0.980 (0.20) | 0.982 (0.20) |
| 20 | 0.923 (0.42) | 0.958 (0.30) | 0.977 (0.22) | 0.986 (0.17) | 0.987 (0.17) |
| 30 | 0.944 (0.35) | 0.971 (0.25) | 0.985 (0.18) | 0.991 (0.14) | 0.992 (0.14) |
| 40 | 0.956 (0.31) | 0.977 (0.22) | 0.989 (0.15) | 0.993 (0.12) | 0.994 (0.12) |
| 60 | 0.969 (0.25) | 0.984 (0.18) | 0.992 (0.13) | 0.995 (0.10) | 0.996 (0.09) |
| 100 | 0.980 (0.20) | 0.990 (0.14) | 0.995 (0.10) | 0.997 (0.08) | 0.998 (0.07) |
| 140 | 0.985 (0.18) | 0.993 (0.12) | 0.996 (0.08) | 0.998 (0.06) | 0.998 (0.06) |
| 200 | 0.989 (0.15) | 0.995 (0.10) | 0.997 (0.07) | 0.999 (0.06) | 0.999 (0.05) |

Note: The figures outside the parentheses denote the ratio of the mean of the sample standard deviation to the population value, and the figures inside the parentheses are the coefficient of variation.

method of moments is employed. Gumbel (Ref. 3, Sec. 6.2.3) lists a table of the sample standard deviation for various sample sizes for the FT-I distribution, but the table gives the values smaller than those listed in Table 13.9. As the data source of Gumbel's table is unknown, it is safe to refer to Table 13.9 when necessary.

### 13.3.2    *Confidence Interval of Parameter Estimates*

Scatter of samples around the population such as those shown in Fig. 13.8 suggests a certain scatter of the parameter values fitted to each sample. Thus, the estimates of the parameters of extreme distributions constitute statistical variates having their own distributions. It is important to examine the confidence intervals of the parameter estimates. The magnitude of confidence intervals depends on the method of data fitting to a theoretical distribution. When the method of moments is employed, the variability of scale parameter is the same as that of sample standard deviation. For the case of maximum likelihood method, a few theoretical analyses on the parameter estimates are available. Lawless[26] has given a solution for the estimates of parameters of the FT-I distribution, and Challenor[27] has prepared tables of the confidence intervals of the parameter estimates.

For the case of the least squares method, no theory is available on the parameter estimates. Thus, the author (Ref. 13 and an unpublished source) has analyzed the data of the Monte Carlo simulation studies to estimate the confidence intervals of the scale and location parameters of various distributions. The results are listed in Table 13.10, which is based on the data of 20,000 samples for each case. It is found that the confidence intervals of the parameter estimates by the least squares method for the FT-I distribution are about 20% greater than those by the maximum likelihood method, according to the comparison with Challenor's tables.

**Example 13.2**

Estimates of $\hat{A} = 1.20$ and $\hat{B} = 4.77$ have been obtained from a sample of the FT-I distribution with the size $N = 20$. What are the 95% confidence intervals of these parameters?

**Solution**

By reading the row of $N = 20$ of the FT-I distribution of Table 13.10, we have the following dimensionless values at 2.5% and 97.5% levels:

$$[A/\hat{A}]_{2.5\%} = 0.67\,, \quad [A/\hat{A}]_{97.5\%} = 1.61\,,$$

$$[(\hat{B} - B)/\hat{A}]_{2.5\%} = -0.49\,, \quad [(\hat{B} - B)/\hat{A}]_{97.5\%} = 0.53\,.$$

Thus, the confidence intervals of the population parameters are calculated as follows:

$$A = (0.67 \sim 1.61) \times 1.20 = 0.80 \sim 1.93\,,$$
$$B = 4.77 + (-0.49 \sim 0.53) \times 1.20 = 4.18 \sim 5.41\,.$$

### 13.3.3    *Confidence Interval of Return Value*

(A) *Statistical variability of return value*

A sample of extreme wave data exhibits quite a large magnitude of variability as demonstrated in Figs. 13.8 and 13.9. The return values estimated with the best-fitting distribution function vary considerably around the true

Table 13.10   Confidence intervals of parameter estimates by the least squares method.

| Distribution | N | Scale Parameter $A/\hat{A}$ | | | | | Location Parameter $(\hat{B} - B)/\hat{A}$ | | | | |
|---|---|---|---|---|---|---|---|---|---|---|---|
| | | 2.5% | 25% | 75% | 97.5% | $\sigma$ | 2.5% | 25% | 75% | 97.5% | $\sigma$ |
| FT-II | 10 | 0.30 | 0.89 | 2.04 | 4.05 | 0.98 | −1.27 | −0.31 | 0.39 | 1.42 | 0.66 |
| ($k = 2.5$) | 14 | 0.31 | 0.89 | 1.90 | 3.45 | 0.82 | −0.92 | −0.25 | 0.37 | 1.21 | 0.53 |
| | 20 | 0.35 | 0.89 | 1.78 | 3.05 | 0.70 | −0.66 | −0.20 | 0.35 | 1.05 | 0.44 |
| | 30 | 0.37 | 0.89 | 1.65 | 2.65 | 0.59 | −0.51 | −0.15 | 0.33 | 0.92 | 0.37 |
| | 40 | 0.39 | 0.89 | 1.59 | 2.48 | 0.53 | −0.47 | −0.13 | 0.32 | 0.85 | 0.34 |
| | 60 | 0.42 | 0.89 | 1.51 | 2.25 | 0.47 | −0.44 | −0.10 | 0.30 | 0.75 | 0.30 |
| | 100 | 0.45 | 0.90 | 1.43 | 2.01 | 0.40 | −0.43 | −0.08 | 0.28 | 0.65 | 0.28 |
| | 140 | 0.48 | 0.90 | 1.39 | 1.90 | 0.36 | −0.43 | −0.08 | 0.27 | 0.60 | 0.26 |
| | 200 | 0.49 | 0.90 | 1.35 | 1.80 | 0.33 | −0.43 | −0.07 | 0.27 | 0.56 | 0.25 |
| FT-II | 10 | 0.36 | 0.85 | 1.73 | 3.19 | 0.74 | −1.09 | −0.28 | 0.30 | 1.11 | 0.54 |
| ($k = 3.33$) | 14 | 0.38 | 0.86 | 1.62 | 2.73 | 0.61 | −0.82 | −0.23 | 0.27 | 0.91 | 0.43 |
| | 20 | 0.42 | 0.86 | 1.53 | 2.43 | 0.51 | −0.62 | −0.19 | 0.24 | 0.76 | 0.35 |
| | 30 | 0.46 | 0.87 | 1.43 | 2.14 | 0.43 | −0.47 | −0.15 | 0.21 | 0.65 | 0.28 |
| | 40 | 0.49 | 0.88 | 1.38 | 2.00 | 0.38 | −0.40 | −0.13 | 0.20 | 0.57 | 0.25 |
| | 60 | 0.52 | 0.88 | 1.32 | 1.83 | 0.33 | −0.34 | −0.10 | 0.17 | 0.48 | 0.21 |
| | 100 | 0.56 | 0.89 | 1.26 | 1.65 | 0.28 | −0.30 | −0.09 | 0.15 | 0.39 | 0.18 |
| | 140 | 0.59 | 0.90 | 1.23 | 1.56 | 0.24 | −0.28 | −0.08 | 0.14 | 0.34 | 0.16 |
| | 200 | 0.62 | 0.91 | 1.20 | 1.49 | 0.22 | −0.26 | −0.07 | 0.12 | 0.30 | 0.14 |
| FT-II | 10 | 0.43 | 0.84 | 1.53 | 2.63 | 0.57 | −0.96 | −0.26 | 0.26 | 0.95 | 0.47 |
| ($k = 5.0$) | 14 | 0.46 | 0.85 | 1.44 | 2.29 | 0.47 | −0.74 | −0.21 | 0.22 | 0.76 | 0.37 |
| | 20 | 0.50 | 0.86 | 1.36 | 2.03 | 0.39 | −0.57 | −0.18 | 0.19 | 0.62 | 0.30 |
| | 30 | 0.55 | 0.87 | 1.29 | 1.81 | 0.32 | −0.44 | −0.14 | 0.16 | 0.51 | 0.24 |
| | 40 | 0.59 | 0.88 | 1.25 | 1.70 | 0.28 | −0.37 | −0.12 | 0.15 | 0.44 | 0.21 |
| | 60 | 0.62 | 0.89 | 1.21 | 1.57 | 0.24 | −0.30 | −0.10 | 0.12 | 0.36 | 0.17 |
| | 100 | 0.67 | 0.91 | 1.16 | 1.44 | 0.19 | −0.24 | −0.08 | 0.09 | 0.28 | 0.13 |
| | 140 | 0.71 | 0.91 | 1.14 | 1.37 | 0.17 | −0.21 | −0.07 | 0.08 | 0.24 | 0.11 |
| | 200 | 0.74 | 0.92 | 1.12 | 1.32 | 0.15 | −0.18 | −0.06 | 0.07 | 0.20 | 0.10 |
| FT-II | 10 | 0.51 | 0.84 | 1.39 | 2.26 | 0.46 | −0.85 | −0.24 | 0.25 | 0.86 | 0.42 |
| ($k = 10.0$) | 14 | 0.54 | 0.86 | 1.32 | 1.98 | 0.37 | −0.68 | −0.19 | 0.21 | 0.69 | 0.34 |
| | 20 | 0.59 | 0.87 | 1.26 | 1.77 | 0.30 | −0.53 | −0.16 | 0.18 | 0.55 | 0.27 |
| | 30 | 0.64 | 0.88 | 1.20 | 1.60 | 0.24 | −0.42 | −0.13 | 0.15 | 0.44 | 0.22 |
| | 40 | 0.67 | 0.90 | 1.18 | 1.51 | 0.21 | −0.36 | −0.11 | 0.13 | 0.38 | 0.19 |
| | 60 | 0.71 | 0.91 | 1.14 | 1.41 | 0.18 | −0.28 | −0.09 | 0.10 | 0.31 | 0.15 |
| | 100 | 0.76 | 0.92 | 1.11 | 1.31 | 0.14 | −0.22 | −0.07 | 0.08 | 0.23 | 0.12 |
| | 140 | 0.79 | 0.93 | 1.09 | 1.26 | 0.12 | −0.19 | −0.06 | 0.07 | 0.20 | 0.10 |
| | 200 | 0.82 | 0.94 | 1.08 | 1.22 | 0.10 | −0.16 | −0.05 | 0.06 | 0.16 | 0.08 |
| FT-I | 10 | 0.58 | 0.86 | 1.30 | 2.03 | 0.37 | −0.77 | −0.21 | 0.25 | 0.83 | 0.40 |
| | 14 | 0.62 | 0.87 | 1.24 | 1.78 | 0.30 | −0.61 | −0.18 | 0.21 | 0.66 | 0.32 |
| | 20 | 0.67 | 0.88 | 1.19 | 1.61 | 0.24 | −0.49 | −0.15 | 0.18 | 0.53 | 0.26 |
| | 30 | 0.71 | 0.90 | 1.14 | 1.47 | 0.19 | −0.39 | −0.12 | 0.14 | 0.42 | 0.20 |
| | 40 | 0.74 | 0.91 | 1.13 | 1.39 | 0.17 | −0.34 | −0.11 | 0.12 | 0.36 | 0.18 |
| | 60 | 0.78 | 0.92 | 1.10 | 1.30 | 0.13 | −0.27 | −0.09 | 0.10 | 0.29 | 0.14 |
| | 100 | 0.82 | 0.94 | 1.08 | 1.23 | 0.10 | −0.21 | −0.07 | 0.07 | 0.22 | 0.11 |
| | 140 | 0.85 | 0.95 | 1.06 | 1.19 | 0.09 | −0.18 | −0.06 | 0.06 | 0.19 | 0.09 |
| | 200 | 0.87 | 0.95 | 1.05 | 1.16 | 0.07 | −0.15 | −0.05 | 0.05 | 0.18 | 0.08 |

Table 13.10  (*Continued*)

| Distribution | N | Scale Parameter $A/\hat{A}$ | | | | | Location Parameter $(\hat{B} - B)/\hat{A}$ | | | | |
|---|---|---|---|---|---|---|---|---|---|---|---|
| | | 2.5% | 25% | 75% | 97.5% | $\sigma$ | 2.5% | 25% | 75% | 97.5% | $\sigma$ |
| Weibull | 10 | 0.42 | 0.82 | 1.70 | 3.58 | 0.86 | −0.41 | −0.12 | 0.36 | 1.14 | 0.40 |
| ($k = 0.75$) | 14 | 0.47 | 0.83 | 1.55 | 2.88 | 0.65 | −0.39 | −0.11 | 0.31 | 0.93 | 0.34 |
| | 20 | 0.50 | 0.84 | 1.43 | 2.45 | 0.50 | −0.36 | −0.11 | 0.26 | 0.75 | 0.29 |
| | 30 | 0.56 | 0.85 | 1.34 | 2.09 | 0.40 | −0.33 | −0.10 | 0.22 | 0.61 | 0.24 |
| | 40 | 0.59 | 0.87 | 1.28 | 1.89 | 0.33 | −0.31 | −0.09 | 0.19 | 0.53 | 0.21 |
| | 60 | 0.64 | 0.88 | 1.22 | 1.67 | 0.27 | −0.28 | −0.08 | 0.16 | 0.42 | 0.18 |
| | 100 | 0.70 | 0.90 | 1.17 | 1.50 | 0.20 | −0.24 | −0.07 | 0.12 | 0.33 | 0.15 |
| | 140 | 0.73 | 0.91 | 1.15 | 1.41 | 0.18 | −0.22 | −0.06 | 0.11 | 0.28 | 0.13 |
| | 200 | 0.77 | 0.92 | 1.12 | 1.34 | 0.15 | −0.19 | −0.06 | 0.09 | 0.24 | 0.11 |
| Weibull | 10 | 0.51 | 0.83 | 1.44 | 2.58 | 0.55 | −0.34 | −0.12 | 0.24 | 0.83 | 0.30 |
| ($k = 1.0$) | 14 | 0.55 | 0.84 | 1.36 | 2.22 | 0.43 | −0.31 | −0.11 | 0.21 | 0.66 | 0.25 |
| | 20 | 0.60 | 0.86 | 1.28 | 1.92 | 0.34 | −0.28 | −0.09 | 0.17 | 0.52 | 0.21 |
| | 30 | 0.65 | 0.88 | 1.22 | 1.70 | 0.26 | −0.25 | −0.08 | 0.13 | 0.41 | 0.16 |
| | 40 | 0.69 | 0.89 | 1.18 | 1.58 | 0.23 | −0.22 | −0.08 | 0.12 | 0.35 | 0.15 |
| | 60 | 0.72 | 0.90 | 1.15 | 1.44 | 0.18 | −0.20 | −0.07 | 0.10 | 0.28 | 0.12 |
| | 100 | 0.78 | 0.92 | 1.11 | 1.33 | 0.14 | −0.16 | −0.05 | 0.07 | 0.21 | 0.09 |
| | 140 | 0.80 | 0.93 | 1.09 | 1.27 | 0.12 | −0.14 | −0.05 | 0.06 | 0.18 | 0.08 |
| | 200 | 0.83 | 0.94 | 1.08 | 1.22 | 0.10 | −0.12 | −0.04 | 0.05 | 0.15 | 0.07 |
| Weibull | 10 | 0.60 | 0.86 | 1.30 | 2.05 | 0.38 | −0.30 | −0.12 | 0.17 | 0.66 | 0.24 |
| ($k = 1.4$) | 14 | 0.64 | 0.87 | 1.24 | 1.79 | 0.30 | −0.27 | −0.10 | 0.15 | 0.52 | 0.20 |
| | 20 | 0.69 | 0.88 | 1.19 | 1.61 | 0.24 | −0.24 | −0.09 | 0.11 | 0.38 | 0.15 |
| | 30 | 0.73 | 0.90 | 1.14 | 1.47 | 0.19 | −0.20 | −0.07 | 0.10 | 0.31 | 0.13 |
| | 40 | 0.75 | 0.91 | 1.12 | 1.39 | 0.16 | −0.18 | −0.07 | 0.08 | 0.26 | 0.11 |
| | 60 | 0.79 | 0.92 | 1.10 | 1.29 | 0.13 | −0.15 | −0.06 | 0.07 | 0.21 | 0.09 |
| | 100 | 0.83 | 0.94 | 1.07 | 1.22 | 0.10 | −0.12 | −0.05 | 0.05 | 0.16 | 0.07 |
| | 140 | 0.86 | 0.95 | 1.06 | 1.18 | 0.08 | −0.11 | −0.04 | 0.04 | 0.13 | 0.06 |
| | 200 | 0.88 | 0.96 | 1.05 | 1.15 | 0.07 | −0.09 | −0.03 | 0.04 | 0.11 | 0.05 |
| Weibull | 10 | 0.66 | 0.87 | 1.22 | 1.80 | 0.29 | −0.30 | −0.12 | 0.17 | 0.66 | 0.24 |
| ($k = 2.0$) | 14 | 0.70 | 0.88 | 1.17 | 1.60 | 0.23 | −0.26 | −0.10 | 0.13 | 0.49 | 0.19 |
| | 20 | 0.74 | 0.90 | 1.14 | 1.46 | 0.19 | −0.22 | −0.09 | 0.11 | 0.38 | 0.15 |
| | 30 | 0.78 | 0.92 | 1.11 | 1.36 | 0.15 | −0.19 | −0.07 | 0.08 | 0.29 | 0.12 |
| | 40 | 0.80 | 0.92 | 1.09 | 1.29 | 0.13 | −0.17 | −0.06 | 0.07 | 0.24 | 0.10 |
| | 60 | 0.83 | 0.94 | 1.07 | 1.23 | 0.10 | −0.14 | −0.05 | 0.06 | 0.19 | 0.08 |
| | 100 | 0.87 | 0.95 | 1.05 | 1.17 | 0.08 | −0.11 | −0.04 | 0.04 | 0.14 | 0.06 |
| | 140 | 0.89 | 0.96 | 1.05 | 1.14 | 0.06 | −0.10 | −0.04 | 0.04 | 0.12 | 0.05 |
| | 200 | 0.90 | 0.97 | 1.04 | 1.11 | 0.05 | −0.08 | −0.03 | 0.03 | 0.09 | 0.04 |
| Normal | 10 | 0.66 | 0.87 | 1.21 | 1.78 | 0.29 | −0.68 | −0.22 | 0.22 | 0.71 | 0.35 |
| | 14 | 0.70 | 0.89 | 1.16 | 1.58 | 0.22 | −0.57 | −0.18 | 0.18 | 0.57 | 0.29 |
| | 20 | 0.75 | 0.90 | 1.13 | 1.44 | 0.18 | −0.46 | −0.15 | 0.15 | 0.47 | 0.23 |
| | 30 | 0.79 | 0.92 | 1.10 | 1.33 | 0.14 | −0.37 | −0.12 | 0.12 | 0.37 | 0.19 |
| | 40 | 0.81 | 0.93 | 1.08 | 1.28 | 0.12 | −0.32 | −0.11 | 0.11 | 0.32 | 0.16 |
| | 60 | 0.84 | 0.94 | 1.07 | 1.21 | 0.09 | −0.26 | −0.09 | 0.09 | 0.26 | 0.13 |
| | 100 | 0.87 | 0.95 | 1.05 | 1.16 | 0.07 | −0.20 | −0.07 | 0.07 | 0.20 | 0.10 |
| | 140 | 0.89 | 0.96 | 1.04 | 1.13 | 0.06 | −0.17 | −0.06 | 0.06 | 0.17 | 0.09 |
| | 200 | 0.91 | 0.97 | 1.03 | 1.10 | 0.05 | −0.14 | −0.05 | 0.05 | 0.14 | 0.07 |

Note: $\sigma$ stands for the standard deviation of respective parameter estimates in dimensionless forms.

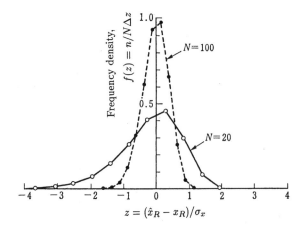

Fig. 13.10   Examples of spread of the return value estimated from uncensored samples of the FT-I distribution: sample sizes $N = 20$ and $N = 100$ for the return period $R = 100$.[19]

values of population. Figure 13.10 shows examples of statistical variability of return values estimated from individual samples. The FT-I distribution is employed as the population, and uncensored samples ($\nu = 1$) are simulated by a Monte Carlo technique. Two cases of the sample size $N = 20$ and $N = 100$ are analyzed. By assuming the mean rate $\lambda = 1$, the sample sizes correspond to the annual maxima data of 20 and 100 years, respectively. For each sample, the scale and location parameters are estimated with the least squares method by assuming the parent distribution being the FT-I. Then the return value corresponding to the period $R = 100$ (years) is calculated for each sample. The estimates of return value are shown in Fig. 13.10 in a form of probability density function. The following reduced variate is introduced to make the results independent of parametric values:

$$z = \frac{\hat{x}_R - x_R}{\sigma_x}, \qquad (13.43)$$

where $x_R$ is the true return value of the population and $\sigma_x$ is the unbiased standard deviation of the sample.

As seen in Fig. 13.10, the 100-year wave height estimated from a sample of annual maximum height of 20 years may have a deviation of more than $\sigma_x$ from the true value. In the real situation, we have only one set of extreme wave heights with no information on the parent distribution. The true return value may be on the larger side or the smaller side of the estimate, but we have no information to judge which side is the correct one. The only procedure we can take is to estimate the confidence interval of the estimated return value within which the true value would be located.

## (B) *Standard deviation of return value when the parent distribution is known*

Estimate of the confidence interval of any statistics requires the information on its cumulative distribution. The data shown in Fig. 13.10 provides such an information. When a sample of extreme data is analyzed, the confidence interval can be estimated by means of a Monte Carlo simulation technique as suggested by Mathiesen *et al.*[28] The distribution fitted to the original sample is assumed to be the parent distribution; a large number of artificial samples (say 2000 samples) are simulated, return values are estimated for these samples, and a cumulative distribution of the estimated return values is established. For practical purposes, however, the normal distribution of the return value is assumed (almost true of the case of $N = 100$ but not for the case of $N = 20$ in Fig. 13.10), and the confidence interval is estimated by means of the standard deviation of the return value. The 90% confidence interval is set at the range of $\pm 1.645$ times the standard deviation around the estimate of the return value.

For the FT-I distribution, Gumbel (Ref. 3, Sec. 6.2.3) has given the following formula for the case when the method of moments is employed:

$$\sigma(\hat{x}_R) = [1 + 0.8885(y_R - \gamma) + 0.6687(y_R - \gamma)^2]^{1/2}\sigma/\sqrt{N}, \qquad (13.44)$$

where $\sigma$ is the standard deviation of the population (which is unknown), $y_R$ stands for the reduced variate for the return period $R$, and $\gamma$ is Euler's constant.

For the case of the least squares method, the author,[12,13,29] has derived the empirical formula for the estimate of the standard deviation of the return value as in the following:

$$\sigma(\hat{x}_R) = \sigma_z \cdot \sigma_x, \qquad (13.45)$$

where $\sigma_z$ is the standard deviation of the reduced variate defined by Eq. (13.43) and is given by

$$\sigma_z = [1.0 + a(y_R - c + \alpha \ln \nu)^2]^{1/2}/\sqrt{N}, \qquad (13.46)$$

Table 13.11　Constants for the standard deviation of return value for FT-I and Weibull distribution.

| Distribution | $a_1$ | $a_2$ | $\kappa$ | $c$ | $\alpha$ |
|---|---|---|---|---|---|
| FT-I | 0.64 | 9.0 | 0.93 | 0 | 1.33 |
| Weibull ($k = 0.75$) | 1.65 | 11.4 | −0.63 | 0 | 1.15 |
| Weibull ($k = 1.0$) | 1.92 | 11.4 | 0 | 0.3 | 0.90 |
| Weibull ($k = 1.4$) | 2.05 | 11.4 | 0.69 | 0.4 | 0.72 |
| Weibull ($k = 2.0$) | 2.24 | 11.4 | 1.34 | 0.5 | 0.54 |

Table 13.12　Constants for the standard deviation of return value for FT-II distribution.

| Shape Parameter | $a_1$ | $a_2$ | $N_0$ | $\kappa$ | $\nu_0$ | $c$ | $\alpha$ |
|---|---|---|---|---|---|---|---|
| $k = 2.5$ | 1.27 | 0.12 | 23 | 0.24 | 1.34 | 0.3 | 2.3 |
| $k = 3.33$ | 1.23 | 0.09 | 25 | 0.36 | 0.66 | 0.2 | 1.9 |
| $k = 5.0$ | 1.34 | 0.07 | 35 | 0.41 | 0.45 | 0.1 | 1.6 |
| $k = 10.0$ | 1.48 | 0.06 | 60 | 0.47 | 0.34 | 0 | 1.4 |

in which the constant $a$ is to be calculated by the following formula:

$$
a = \begin{cases} a_1 \exp[a_2 N^{-1.3} + \kappa(-\ln \nu)^2] & : \text{for FT-I and Weibull} \\ & \quad \text{distributions}, \\ \\ a_1 \exp\{a_2[\ln(N\nu^{0.5}/N_0)]^2 - \kappa[\ln(\nu/\nu_0)]\} & : \text{for FT-II distribution}. \end{cases}
$$

$$(13.47)$$

The constants appearing in Eqs. (13.46) and (13.47) have been set as listed in Tables 13.11 and 13.12 based on the data of Monte Carlo simulation studies, which have been utilized for the derivation of the DOL, REC and MIR criteria.

**Example 13.3**

Estimate the confidence interval of the 100-year wave height of the Kodiak data.

**Solution**

The 100-year wave height of the Kodiak data has been estimated as $x_{100} = 12.47$ m in Sec. 13.2.1(D). The corresponding reduced variate is $y_{100} = 3.5815$, and the sample standard deviation is $\sigma_x = 1.214$ m. The standard deviation of the 100-year wave height is calculated by using Eqs. (13.45) to (13.47) as follows:

$$a = 2.05 \exp[11.4 \times 78^{-1.3} + 0.69 \times (-\ln 1.0)^2] = 2.133 \,,$$

$$\sigma_z = [1.0 + 2.133 \times (3.5815 - 0.4 + 0.72) \times \ln 1.0)^2]^{1/2}/\sqrt{78} = 0.538 \,,$$

$$\sigma(\hat{x}_{100}) = 0.538 \times 1.214 = 0.653 \,\mathrm{m} \,.$$

Thus, the 90% confidence interval of the 100-year wave height is estimated as follows:

$$x_{100} = 12.47 \pm 1.645 \times 0.653 = 12.47 \pm 1.07 = 11.4 \sim 13.5 \,\mathrm{m} \,.$$

The two dashed lines above and below the straight line in Fig. 13.3 for the Kodiak data have been drawn by calculation such as shown in the above. Thus, the confidence interval can easily be calculated for any sample of extreme wave data. However, introduction of the concept of confidence interval of design wave height into design practice is still in its infancy stage. Practitioners are not accustomed with the concept of confidence interval, and they are reluctant in dealing with it. Nevertheless, engineers in charge of designing maritime structures should keep in mind the uncertainty of design wave height and should provide a sufficient margin of safety against the attack of storm waves much severer than the design condition.

### (C) *Standard deviation of return value when the parent distribution is unknown*

The question inherent in extreme wave analysis is such: what is the true parent distribution of extreme storm waves? As explained in Sec. 13.1.2, no answer exists to this question at present and the distribution best fitting to a sample among several candidates is selected as the *most probable* parent distribution. However, there is a possibility to establish the parent distribution of extreme storm waves in the future. If storm wave data are collected at a number of stations in one ocean region where the nature of storm waves can be regarded homogeneous, they will constitute multiple samples drawn from the same population. By applying the DOL and REC criteria for these samples, the least rejectable distribution for the particular ocean region will emerge by such analysis. In fact, Goda *et al.*[25] made such an analysis as described in Sec. 13.2.4(C) and reported that the northern Pacific coast, the southern Pacific coast, and the Japan Sea coast of Japan seem to have the most probable parent distributions of the Weibull $(k = 1)$,

FT-I, and Weibull ($k = 1.4$), respectively. Their results were incomplete because the database employed covered a relatively short span of time duration. With expansion of the storm wave database, more detailed analyses similar with Goda *et al* would indicate the parent distribution of extreme waves along respective ocean regions.

In the present situation in which no parent distribution is established yet, we are always faced with the problem of misfitting a sample to a distribution different from the true parent distribution, because of sampling variability such as shown in Fig. 13.8. In a joint simulation study reported by Goda *et al.*,[19] samples drawn from the Weibull distribution with $k = 1.4$ were fitted to a wide range of distributions including the FT-I and the Weibull with $k$ being less than 1 to greater than 2. The empirical formulas of Eqs. (13.45) to (13.47) for the standard deviation of return values are based on the condition that the samples are fitted to the parent distribution. When a sample is fitted to a distribution different from the parent, the confidence interval of return value would be wider than that estimated with the empirical formulas of Eqs. (13.45) to (13.47). However, there is no way to know the amount of difference. The only thing we can endeavor is to increase the sample size $N$ by extending the duration of database $K$ and by increasing the mean rate $\lambda$. By increasing the sample size, we can diminish the possibility of misfitting and decrease the range of confidence interval.

### (D) *Effect of sample size on confidence interval*

As indicated in Eqs. (13.44) and (13.46), the standard deviation of return value is inversely proportional to the square root of sample size $N$. A trial calculation has been made to demonstrate the effect of sample size on the magnitude of the standard deviation of return value.[30] The FT-I distribution is taken as a model, and the ratio of $\sigma(\hat{x}_{100})$ to $\sigma_x$ is calculated with Eqs. (13.46) and (13.47) for the case of no censoring ($\nu = 1$). The result is shown in Fig. 13.11.

The abscissa is the time span of extreme wave data set $K$ and the mean rate $\lambda$ is taken as the main parameter, while the sample size $N$ is shown as the secondary parameter. For a given value of $\lambda$, the ratio $\sigma(\hat{x}_{100})/\sigma_x$ gradually decreases as the duration $K$ becomes long. In practice, the duration $K$ for a particular site is fixed by the availability of wave data source. Adoption of as many of the storm waves as possible by the peaks-over-threshold (POT) method enables the increase of the mean rate and sample

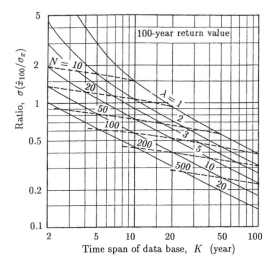

Fig. 13.11 Standard deviation of a 100-year return value of FT-I distribution $\sigma(\hat{x}_{100})$ relative to the sample standard deviation $\sigma_x$ for uncensored samples.[30]

size, thus reducing the magnitude of the standard deviation of return value. Caution should be taken however for the possibility of mixed population to be discussed in Sec. 13.4.1 and Sec. 13.5.1(E).

## 13.4 Several Topics on Extreme Wave Statistics

### 13.4.1 *Treatment of Mixed Populations*

(A) *Cumulative distribution of extreme variate for mixed populations*

As discussed in the beginning of this chapter, homogeneity is one of the important requisites in extreme statistics. Ideally speaking, each storm wave height is to be classified according to the types of wave generation sources, and each group of storm wave heights should be analyzed separately. In practice, the homogeneity check of individual storm waves is a tedious work, and extreme wave analysis is often carried out without due regard to the homogeneity condition. However, storm wave data can be classified by seasons with little difficulty if not by the type of wave generation source. Extremal analysis of such classified storm wave data should be encouraged.

When multiple groups of extreme wave data are analyzed, different distribution functions should be fitted to respective groups. Then it becomes necessary to combine these distributions into a single cumulative distribution. Carter and Challenor[31] made estimates of return values of wave heights and wind speeds around U.K. by using the monthly maximum values. They derived the cumulative distribution of annual maxima as the product of 12 monthly cumulative distributions; i.e.,

$$F(x) = \text{Prob.}(X \le x) = \prod_{j=1}^{12} F_j(x), \qquad (13.48)$$

where $F(x)$ refers to the cumulative distribution of annual maxima and $F_j(x)$ that of monthly maxima.

For POT data, its cumulative distribution is first converted to that of annual maxima by assuming that the number of storm events per year follows the Poisson distribution expressed as

$$P_r = \frac{1}{r!} e^{-\lambda} \lambda^r, \quad r = 0, 1, 2, \ldots, \qquad (13.49)$$

where $r$ denotes the storm number per year. Examination of the Kodiak data in Table 13.3 indicates the assumption of Poisson distribution acceptable. When $r$ peak wave heights occur in one year, the annual maximum is the largest of $r$ data and its nonexceedance probability is given as $F_*^r(x)$, where $F_*(x)$ denotes the nonexceedance probability (i.e., distribution function) of individual storm events. Because the number of storm events $r$ is a statistical variate, the cumulative distribution of annual maxima $F(x)$ is derived as the expected value of the probability from $r = 0$ to $\infty$. Thus,

$$F(x) = \sum_{r=0}^{\infty} P_r F_*^r(x) = \exp\{-\lambda[1 - F_*(x)]\}. \qquad (13.50)$$

When there are $n$ groups of extreme storm waves, the cumulative distribution for the annual maxima can be derived by applying the technique of Eq. (13.48) as follows:

$$F(x) = \prod_{j=1}^{n} \exp\{-\lambda_j[1 - F_j(x)]\} = \exp\left\{ -\sum_{j=1}^{n} \lambda_j[1 - F_j(x)] \right\}, \qquad (13.51)$$

where $n$ is the number of populations of storm waves, and $\lambda_j$ and $F_j(x)$ denote the mean rate and the distribution function of the $j$th group of POT data, respectively.

(B) *Return value and confidence interval of return value of mixed populations*

As Eq. (13.51) defines the cumulative distribution of the annual maxima, the return value for the return period $R$ is derived as $F^{-1}(1 - 1/R)$ by Eq. (3.18). Except for the case that all the distribution function of POT data belong to the FT-I distribution, $F(x)$ is obtained in a tabular form only and the return value must be obtained by interpolation.

The variance of the return value of mixed populations can be assessed by the following formula given by Izumiya and Saito[32]:

$$\sigma^2(\hat{x}_R) = \frac{\sum_{j=1}^{n} \lambda_j^2 f_j^2(\hat{x}_R) \sigma_j^2(\hat{x}_R)}{\left[\sum_{j=1}^{n} \lambda_j f_j(\hat{x}_R)\right]^2}, \qquad (13.52)$$

where $n$ is the number of the groups of extreme storm waves, $\lambda_j$ is the mean rate of storm waves for the $j$th group, $f_j(\hat{x}_R)$ is the value of the frequency distribution of the $j$th group at $x = \hat{x}$, and $\sigma_j^2(\hat{x}_R)$ denotes the variance of the $j$th group. The standard deviation $\sigma$ of the return value of mixed populations is obtained as the square root of the above variance, and the 90% confidence interval may be set as the range of $\pm 1.645\sigma$ around the estimated return value $\hat{x}_R$.

### 13.4.2 Encounter Probability

Estimation of a return value for a given return period is one thing in extreme wave statistics, and specification of the return period is another thing. Borgman[33] has presented the concept of encounter probability for selection of the return period. Let the design working life of a structure be $L$ and the design return period of the structure be $R$. The cumulative distribution of the annual maxima of storm wave heights is assumed to have been established as $F(x)$. The probability that an annual maximum wave height in one year exceeds the design value $x_R$ is $1 - F(x_R)$ by definition, and it is equal to $1/R$ by Eq. (13.17). The probability that the structure will not experience storm waves greater than the design condition in one year is given by $F(x_R) = 1 - 1/R$. The encounter probability that the structure will experience storm waves greater than the design condition during $L$ years is then obtained as

$$P_L = 1 - \left(1 - \frac{1}{R}\right)^L. \qquad (13.53)$$

If the design condition is set with the return period $R = 100$ years for a structure which is planned to serve for $L = 30$ years, there will be a probability of 0.26 that the structure will encounter storm waves exceeding the design condition. Borgman prepared a table of encounter probability for various combinations of the design working life and return period. However, the following approximation to Eq. (13.53) is available when both $L$ and $R$ are sufficiently large:

$$P_L = 1 - \exp\left(-\frac{L}{R}\right) . \tag{13.54}$$

The above equation indicates that if $R$ is set equal to $L$, the encounter probability will be as high as 0.63. If the encounter probability is to be kept below 0.3, the return period should be longer than 2.8 times the design working life.

### 13.4.3 L-Year Maximum Wave Height and Its Confidence Interval

(A) *L-year maximum wave height*

Another approach to the selection of design wave height is to use the expected value of the largest wave height during the design working life of a structure, which is hereby called the *L*-year maximum height. Use of the *L*-year maximum load is a common practice in reliability-based designs called the Load and Resistance Factor Design (e.g., see Refs. 34 and 35).

First, the largest wave height within $L$ years is denoted with $x_L$ and its cumulative distribution with $\Phi(x_L)$. The meaning of $\Phi(x_L)$ is explained as follows. A very long stationary time series of annual maximum wave heights is supposed to exist. The time series is segmented into a large number of samples with the duration of $L$ years each. The value of $x_L$ varies from sample to sample, and a histogram of $x_L$ can be prepared. By summing up the histogram from the lowest to the largest $x_L$, a cumulative distribution of $x_L$ will be constructed. This is an empirical form of $\Phi(x_L)$. In statistical sense, $\Phi(x_L)$ is defined with the cumulative distribution of annual maximum wave height $F(x)$ as in the following:

$$\Phi(x_L) = \text{Prob.}(X \leq x_L) = F^L(x_L) . \tag{13.55}$$

The probability density function of $x_L$ denoted with $\phi(x_L)$ is derived by taking the derivative of Eq. (13.55) as follows:

$$\phi(x_L) = L \, F^{L-1}(x_L) \, f(x_L) . \tag{13.56}$$

The $L$-year maximum wave height is the mean of $x_L$ of many samples; i.e., $\bar{x}_L = E[x_L]$, which is calculated using Eq. (13.56). The magnitude of variation of $x_L$ of individual samples is represented with its standard deviation, and it is derived from the variance $E[(x_L - \bar{x}_L)^2]$. Computation of the mean and standard deviation of $x_L$ generally requires numerical integration for the first and second moments of Eq. (13.56). When the annual maximum wave heights are fitted by the FT-I or FT-II distribution, however, the analytical expression for $\Phi(x_L)$ can be readily derived, and the mean and standard deviation are obtained by referring to Table 13.1. The results are as follows:

### (i) FT-I distribution:

$$\text{Distribution}: \Phi(x_L) = \exp\left\{-\exp\left[-\frac{x - (B + A\ln L)}{A}\right]\right\}, \quad (13.57)$$

$$\text{Mean}: \bar{x}_L = B + A(\ln L + \gamma), \quad (13.58)$$

$$\text{Standard deviation}: \delta(x_L) = \frac{\pi}{\sqrt{6}}A, \quad (13.59)$$

where $\gamma$ is Euler's constant $(= 0.5772\ldots)$.

### (ii) FT-II distribution:

$$\text{Distribution}: \Phi(x_L) = \exp\left\{-\left[1 + \frac{x - (B + kAL^{1/k} - kA)}{kAL^{1/k}}\right]^{-k}\right\},$$
$$(13.60)$$

$$\text{Mean}: \bar{x}_L = B + kA\left[L^{1/k}\Gamma\left(1 - \frac{1}{k}\right) - 1\right], \quad (13.61)$$

$$\text{Standard deviation}: \delta(x_L) = kAL^{1/k}\left[\Gamma\left(1 - \frac{2}{k}\right) - \Gamma^2\left(1 - \frac{1}{k}\right)\right]^{1/2}.$$
$$(13.62)$$

For the Weibull distribution, Izumiya[36] has derived the approximate solution of the distribution function of the $L$-maximum wave height and related statistics by making use of the Taylor expansion of the Weibull distribution function.

## (iii) **Weibull distribution:**

$$\text{Distribution}: \Phi(x_L) = \left\{ 1 - \frac{1}{L} \exp[-a_L(x_L - x_{Lp})] \right\}^N$$

$$\simeq \exp\{-\exp[-a_L(x_L - x_{Lp})]\}, \qquad (13.63)$$

$$\text{Mean}: \overline{x}_L \simeq A \left[ (\ln L)^{1/k} + \frac{\gamma}{k(\ln L)^{1-1/k}} \right] + B, \quad (13.64)$$

$$\text{Standard deviation}: \delta(x_L) \simeq \frac{\pi}{\sqrt{6}} \cdot \frac{A}{k(\ln L)^{1-1/k}}, \qquad (13.65)$$

where

$$x_{Lp} = (\ln L)^{1/k} A + B, \qquad a_L = \frac{k}{A}(\ln L)^{1-1/k}. \qquad (13.66)$$

The distribution function $\Phi(x_L)$ of Eq. (13.63) expresses its asymptotic approach to the FT-I distribution. Thus, the mean and standard deviation of Eqs. (13.64) and (13.65) are derived using the formulas of FT-I listed in Table 3.1.

The return period corresponding to the $L$-year maximum wave height $\overline{x}_L$ for the FT-I distribution is calculated by substituting Eq. (13.58) into Eq. (13.4) and by using Eq. (13.17) as follows:

$$R(\overline{x}_L) = \frac{1}{1 - \exp[-\exp(-\ln L - \gamma)]} \simeq 1.78L + \frac{1}{2}. \qquad (13.67)$$

Thus, the return period of $\overline{x}_L$ is approximately equal to $1.8L$. In the case of the FT-II distribution, $R(\overline{x}_L)$ increases with the decrease in the shape parameter $k$ up to about $2.7L$. The return period of $\overline{x}_L$ for the Weibull distribution is given by[36]

$$R(\overline{x}_L) \simeq 1.78L \exp\left[ \frac{(k-1)\gamma^2}{2k \ln L} \right]. \qquad (13.68)$$

Equations (13.57) to (13.68) apply to the distribution function of annual maximum wave heights. For the POT data of storm wave heights, the distribution fitted to a sample must be converted to that of annual maximum heights by means of Eq. (13.50). Therefore, even if the POT data is fitted to the FT-I distribution, Eqs. (13.57) to (13.59) cannot be applied directly and numerical integration involving $\phi(x_L)$ must be carried out. A compromise is possible, however, by replacing the design working life $L$ with $\lambda L$. As the reliability-based design of maritime structures is at its early stage of development, further examinations of the characteristics of $L$-year maximum wave height will be necessary.

### (B) *Confidence interval of L-year maximum wave height*

The reliability-based design[34,35,37] examines the probability of failure during the design working life of a structure. Examination is made through the analysis of statistical distributions of the magnitudes of the loads to be exerted on structures and the capacity of structures in resisting the loads. (*Load and Resistance Factor Design.*) The concepts of load and resistance factors are thus introduced. The coefficients of variation of these factors constitute the important elements in the reliability-based design. The standard deviation $\delta(x_L)$ in Eqs. (13.59), (13.62), or (13.65) is a measure of dispersion of $x_L$ of individual samples with the duration of $L$ years each. It is evaluated on the assumption that the functions $F(x_L)$ and $f(x_L)$ being employed in Eq. (13.56) represent the true parent distribution. In practice, however, the extreme distribution function has to be estimated from one sample of storm wave data, and there remains the uncertainty in the estimate of return value due to sampling variability as discussed in the preceding sections. This uncertainty should be added in the evaluation of the coefficient of variation of $L$-year maximum wave height. It can be done by taking a sum of the variance of $x_L$ due to the nature of parent distribution itself and that due to statistical variability of a sample of storm wave data. Thus,

$$\text{C.V. } [\overline{x}_L] = \frac{\sqrt{\delta^2(x_L) + \sigma^2(\hat{x}_L)}}{\overline{x}_L}, \tag{13.69}$$

where $\sigma(\hat{x}_L)$ denotes the standard deviation of $\overline{x}_L$ due to sampling variability which can be estimated with Eqs. (13.45) to (13.47), when the distribution fitting is made by means of the least squares method.

### Example 13.4

A design working life of a structure is set at $L = 50$ years. Suppose the extreme wave statistics of the site is almost the same as that of the Kodiak data. Estimate the $L$-year maximum wave height and its coefficient of variation.

### Solution

The most probable parent distribution of the Kodiak data is the Weibull with $k = 1.4$ and the estimated scale and location parameters of $\hat{A} = 1.8621$ m and $\hat{B} = 5.805$ m. The mean and standard deviation of the 50-year maximum wave height are calculated by Eqs. (13.64) and (13.65) as follows:

$$\bar{x}_{L=50} = 1.8621 \times \left[ (\ln 50)^{1/1.4} + \frac{0.5772}{1.4 \times (\ln 50)^{1/1.4}} \right] + 5.805$$

$$= 1.8621 \times \left[ 2.46494 + \frac{0.5772}{1.4 \times 1.4776} \right] + 5.805$$

$$= 5.453 + 5.804 = 11.25 \text{ m},$$

$$\delta(x_{50}) = \frac{\pi}{\sqrt{6}} \cdot \frac{1.8621}{1.4 \times (\ln 50)^{1-1/1.4}} = 1.16 \text{ m}.$$

The uncertainty of $\bar{x}_{50}$ due to sampling variability is estimated in a similar manner as Example 13.2. The reduced variate $y_{50}$ for $\bar{x}_{50} = 11.25$ m is calculated as $y_{50} = (11.25 - 5.804)/1.8621 = 2.925$ m. Use of Eq. (13.46) yields the following standard deviation $\sigma(\bar{x}_{50})$ due to sampling variability as

$$\sigma(\bar{x}_{50}) = 1.214 \times [1.0 + 2.133 \times (2.925 - 0.4)^2]^{1/2}/\sqrt{78} = 0.525 \text{ m}.$$

The coefficient of variation of $\bar{x}_{50}$ is assessed by taking into account both $\delta(x_{50})$ and $\sigma(\bar{x}_{50})$ as

$$\text{C.V. } [\bar{x}_{50}] = \frac{\sqrt{1.16^2 + 0.525^2}}{11.25} = 0.113.$$

The uncertainty due to sampling variability is thus taken into account in the reliability-based design of maritime structures as shown in the above example.

## 13.5  Design Waves and Related Problems

### 13.5.1  *Database of Extreme Waves and Its Analysis*

(A) *Available sources of wave data*

Several sources of wave data are availabe for the extreme wave analysis. They are

(i) long record of instrumentally measured data of wave heights and periods,

(ii)  long record of visually observed data of wave heights and periods,

(iii)  wave hindcasting project for storm waves over a long time span,

(iv)  data bank of world-wide wave forecasting operations.

Among them, instrumentally measured data are the best source, provided that they cover a sufficiently long time span and the downtime is kept minimum. Hindcasted data of storm waves are the second best source, provided that the wave hindcasting method has been well calibrated with several storm wave records in the area of interest. World-wide wave forecasting operations has been carried out since the 1970s by meteorological institutions of many countries, and the forecasted wave data are accumulated in their data banks. Some institutions make the wave data available to the public through consulting companies. Though the accuracy of the forecasted wave information is inferior to instrumental records and hindcasted wave data, the records of forecasted wave data provide a valuable asset for a coastal project in the region where no wave information is available.

Visually observed wave data are the compilations of ship report data, which are presented in a form of a joint distribution (or a scatter diagram) of wave heights and periods, (e.g., see Refs. 38 and 39). Visual observation data should be used as the last recourse, because the accuracy of individual wave data is low. Furthermore, ocean-going ships try to avoid the area of high waves as much as possible, and thus wave data compiled from ship reports tend to have a negative bias in high wave ranges.

(B) *Requirements for good wave database*

When a set of wave data is analyzed for extreme statistics, three requirements must be met as the database of extreme waves. The first requirement is that all the data are free of any error. Malfunctioning of a wave recorder at the peaks of storm waves is a typical error in instrumental measurements; if it did occur, the erroneous data should be replaced with some estimate based on wave hindcasting or information from neighboring stations.

The second requirement is the duration of record. Duration should be as long as possible, preferably more than 30 years and a minimum of 10 years. A short time span of wave data yields only a small number of storm wave data, and thus increases the magnitude of uncertainty due to sampling variability. From this point of view, hindcasted wave data would be preferred to instrumental wave data at present, because the duration of the latter rarely exceeds 30 years.

The third requirement is that all major storms during the period of concern must be included in the database. In the case of instrumental wave data, a check should be made on the depth of water at the measurement site, in order to guarantee that no data is affected by a depth-limited wave breaking process; if affected, the distribution of wave heights exhibits a tendency of saturation on the upper tail. Also, if a wave recorder failed to register a peak height of big storm waves in one year for some reason, and if it was unable to fill the gap by any means, then the particular season or year should be deleted from the effective duration of wave data. In case an extremely large storm event occurred and the peak wave height is undoubtedly the largest in the past, but without any reliable height estimate, the least squares method can deal with such a situation. Just leave the position of the largest data $x_{(1)}$ vacant but include the event in counting the sample size; the largest among the existing wave data is assigned the order number 2, the rest are given the order number 3 onward, and the least squares method is carried out normally.

## (C) *Censoring of wave data*

When a wave hindcasting project is undertaken, a detailed search should be made initially for meteorological disturbances capable of generating medium to large storm waves. Even so, we cannot eliminate a possibility that some undetected meteorological disturbances might have generated storm waves of appreciable heights. Thus, it is recommended to delete the lower one third to one quarter of peak wave heights from the set of wave data so as to guarantee all major storms are included in the data set. In such a situation, the total number $N_T$ is better set to a slightly larger than the number of hindcasted cases, and the data is treated as a censored one. The censoring parameter $\nu$ is calculated as the ratio of the number of adopted storm events to the probable number of total events $N_T$.

There has been raised a concern that wave measurements taken at three to six hour intervals might miss the peak of storm waves, and thus return values estimated on the basis of such data might have a negative bias.[28] A study by Forristall *et al.*[40] has clarified this problem. The significant wave height derived from a record of limited length, say 20 minutes, has a statistical variability with the coefficient of variation of the order of several percent, as discussed in Sec. 10.6. As the peak wave height of storm waves is defined as the largest among the measured data, a negative bias due to a failure to measure the storm peak at regular recording times is mostly canceled by a positive bias due to the statistical variability by a limited record length, although the amount of error depends on the recording interval relative to the storm duration.

(D) *Preparation of a sample of extreme wave data*

As stated in the beginning of this chapter, the total sample method is not recommended for extreme wave analysis. However, when a visually observed wave data is the only available data source, the total sample method is an inevitable choice. In such a case, a search should be made for the average duration $\tau$ of storm waves in the area under study. It provides the unit time of storm event in extreme analysis. During a given return period $R$, there will be $R/\tau$ units of storm events, and the cumulative distribution for the return wave height $x_R$ is given by $F(x_R) = 1 - \tau/R$ (units of time must be the same for $\tau$ and $R$). The same principle is applied for the total sample method using the instrumentally recorded wave data.

In the peaks-over-threshold (POT) method, setting of the threshold is a delicate problem. If it is set too high, the number of peak wave heights becomes too small. If it is set too low, a sequence of several storms might be judged as one storm and the number will be decreased. Even if the threshold is set at a moderate level, a storm wave activity may be slackened midway and one storm event is judged to consist of two events. To prevent such an artificial division, a limitation of minimum time interval between successive peaks should be imposed; for example, Mathiesen *et al.*[28] recommend two to four days. Probably the threshold would be better set to maximize the number of storm events in a year or season; by increasing the sample size, a possibility of misfitting is reduced and the range of confidence interval is narrowed.

There is a criticism against the POT method such that the estimate of return value is lowered as the threshold level is raised because the sample standard deviation decreases. The criticism is correct if the distribution parameters are estimated by the method of moments, but incorrect when the parameters are estimated by the least squares method with due consideration for the rate of data censoring.

(E) *Homogeneity check of wave data*

Another problem associated with the POT method is how to maintain homogeneity of wave data. Examination of each wave source on synoptic charts is a tedious work, even though it is a desirable procedure. When the working group of the Maritime Hydraulics Section of IAHR analyzed the Halkenbanken data off Norway for nine years together with the Kodiak data, the group noted that the estimate of 100-year wave height gradually decreased as the threshold was raised.[2] It was considered that peak wave heights at lower levels might have belonged to a population different from that at higher levels.

A possible way to separate different populations of storm waves would be to divide wave records by season and to make individual analyses for respective sets of wave data. The results of analyses need to be combined as of mixed populations as discussed in Sec. 13.4.1.

Another way is to raise the threshold of wave height in preparation of the POT data set gradually, say every 0.5 m. Extreme wave analysis is to be made for the data sets at respective threshold levels. If the wave data is of mixded populations, a raise of the threshold would yield a downward shift of return wave heights. One should take the lowest threshold beyond which little change appears in the estimate of return wave heights as the optimum threshold.

### 13.5.2   *Selection of Design Wave Height and Period*

(A) *Selection of design wave height*

The foregoing discussion in this chapter concerns with the estimation of a return wave height for a given length of a return period. The return wave height provides the basic information for selection of the design wave height, but the selection is not a straightforward procedure, because several factors await the decision of the engineer in charge of designing a structure.

First, distinction must be made between the two approaches of deterministic and probabilistic designs. In the former approach, the design wave height is selected as a fixed value, and detailed analysis of structural stability and strength is carried out to insure a sufficient margin of safety against the design condition. In the latter approach, the reliability of structure or the probability of failure during the design working life of a structure is examined under a certain set of conditions; the probability of failure is required to be kept below a certain level.

In the deterministic approach, the return period of a design wave height needs to be assigned *a priori*. The encounter probability provides some guidelines for setting a return period, but there is no recommendable rule to set the encounter probability. In many designs of maritime structures, a return period of 50 to 100 years is often employed for evaluation of a design wave height. However, it remains as a conventional practice without a theoretical background. The confidence interval of a return wave height is only utilized for a safety check against the attack of design load greater than the design condition.

The reliability-based design or the probabilistic approach has a theoretical superiority over the deterministic approach, because it takes into account the uncertainties of various factors involved in structural designs. As described in Chapter 6, breakwaters in Japan are now being designed with the partial factor sytem in Level I method of the reliability-based design or with the expected sliding distance method. A crucial factor in the probabilistic design of maritime structure is the assessment of uncertainty involved in the design factors such as wave heights. The $L$-year maximum wave height discussed in Sec. 13.4.3 has its own variability as expressed with the standard deviation under the assigned parent distribution. In addition, there is another source of uncertainty owing to the inherent variability of a wave data set being a statistical sample. Example 13.3 shows a way to combine two sources of uncertainty. Both variabilities must be duly taken into account when the reliability-based design is carried out.

One of the unsolved problems in the extreme wave statistics is the long-term climatic changes in the ocean environment. Tropical cyclones, for example, are known to be under the influence of the El Niño–Southern Oscillations, which have a time scale of several years (e.g., Refs. 41 to 44). In the evaluation of extreme storm surge levels, historical records in the past several hundred years can shed light to the previous abnormal storm surge levels. In the case of storm wave heights, however, measured data exist only after the 1960s and they often cover a relatively short time span at each site. Meteorological information on storms before World War II is not detailed enough to make wave hindcasting reliable. We do not know what wave climatic changes took place in the past and neither do we know what will occur in the future. Selection of design wave heights on the basis of extreme wave statistics is simply made on the wishful postulation that the future wave climate during the lifetime of structure will remain the same as that in the past.

(B) *Wave period associated with design wave height*

Wave loads on structures are mainly determined by the wave height, but it is also dependent on the wave period. Once the design wave height is selected, then the range of wave period for the design condition needs to be fixed. Presently no reliable method has been established for selection of the wave period associated with the design wave height. Conventional practice is one of the following three approaches:

(i) refer to a scatter diagram of the heights and periods of the POT data set,
(ii) apply the mean relationship between significant wave height and period of wind waves,
(iii) construct a bivariate distribution of extreme wave height and period, and select the wave period at a certain exceedance probability.

In the first approach, some trend curve or a regression line is drawn on the scatter diagram for the relationship between wave height and period. In many cases, correlation between wave heights and periods of storm waves is weak and a large degree of ambiguity remains in setting the relationship, because of a relativel small number of peak wave data. In order to increase the number of data, Sekimoto et al.[45] have proposed to add the wave data within six hours from the time of a storm wave peak.

Among many efforts in the second approach, Callaghan et al.[46] empirically expressed the significant wave period as an empirical function of the significant wave height for the data at Narrabeen Beach in the New South Wales, Australia. For simplicity, the relationship of $T_{1/3}(\text{s}) \simeq 3.3(H_{1/3})^{0.63}(\text{m})$ introduced as Eq. 3.6 in Sec. 3.1.1 may be utilized. The relationship has been derived on the basis of Wilson's formulas for wind wave generation so that swell-dominated waves will show longer period. For drawing a regression curve on the scatter diagram between wave height and period, this relationship may provide a guideline with adjustment of the constant of 3.3.

The third approach is to construct a bivariate distribution of wave height and period based on the available data of extreme waves. Sekimoto et al.[45] employed the joint Weibull distribution by introducing a correlation coefficient between wave height and period. Repko et al.[47] analyzed the marginal distributions of wave height and wave period, and combined the two distribution through the parameters of wave steepness. De Waal and van Gelder[48] introduced the concept of copula for the analysis of the bivariate distribution. Galiatsatou and Prinos[49] applied the concept of copula to the wave data at a Dutch station in the North Sea. In the third approach, the question of what exceedance probability of wave period should be taken remains to be scrutinized. It will be one of the future topics to be pursued in maritime structure designs.

# References

1. Y. Goda, "A review on statistical interpretation of wave data," *Rept. Port and Harbour Res. Inst.* **18**(1) (1979), pp. 5–32.
2. G. van Vledder, Y. Goda, P. Hawkes, E. Mansard, M. H. Martin, M. Mathiesen, E. Peltier and E. Thompson, "Case studies of extreme wave analysis: A comparative analysis," *Proc. 2nd Int. Symp. Ocean Wave Measurement and Analysis* (ASCE, New Orleans, 1993), pp. 978–992.
3. E. J. Gumbel, *Statistics of Extremes* (Columbia Univ. Press, New York, 1953).
4. S. Coles, *An Introduction to Statistical Modeling of Extreme Values* (Springer, London, 2001), pp. 45–49.
5. H. A. David and H. N. Nagaraja, *Order Statistics (Third Ed.)* (John Wiley and Sons, 2003), p. 284.
6. W. Weibull, "A statistical theory of strength of materials," *Ing. Vet. Ak. Handl.*, No. 151 (Stockholm, 1939) [after Gumbel, *loc. cit.*, p. 370].
7. C. Petruaskas and P. M. Aagaard, "Extrapolation of historical storm data for estimating design wave heights," *J. Soc. Petroleum Engrg.* **11** (1971), pp. 23–37 (also, Prepr. 2nd OTC, 1970, Paper No. 1190).
8. J. A. Greenwood, J. M. Landwehr and N. C. Matalas, "Probability weighted moments: Definition and relation to parameters of several distributions expressable in inverse form," *Water Res. Res* **15**(5) (1979), pp. 1049–1054.
9. D. C. Boes, J.-H. Heo and J. D. Salas, "Regional flood quantile estimation for a Weibull model," *Water Res. Res* **25**(5) (1989), pp. 979–990.
10. J. R. M. Hosking and J. R. Wallis, *Regional Frequency Analysis: An Approach Based on L-moments* (Cambridge Univ. Press, Cambridge, 1997), pp. 73-78 and 191-209.
11. Hosking and Wallis, *loc. cit.*, p. 77.
12. Y. Goda and M. Onozawa, "Characteristics of the Fisher-Tippett type II distribution and their confidence intervals," *Proc. Japan Soc. Civil Engrg.* (417/II-13) (1990), pp. 289–292 (*in Japanese*).
13. Y. Goda, "Numerical investigations on plotting formulas and confidence intervals of return values in extreme statistics," *Rept. Port and Harbour Res. Inst.* **27**(1) (1988), pp. 31–92 (*in Japanese*), also Y. Goda, "On the methodology of selecting design wave height," *Proc. 21st Int. Conf. Coastal Engrg.* (Malaga, ASCE, 1988), pp. 899–913.
14. Y. Goda, "Spread parameter of extreme wave height distribution for performance-based design of maritime structures," *J. Waterway, Port, Coastal, and Ocean Engrg.* ASCE, **130**(1) (2004) pp. 29-38.
15. R. Gencarelli, G. R. Tomasicchio and P. Veltri, "Wave height long term prediction based on the use of the spread parameter," *Coastal Engineering 2006 (Proc. 30th ICCE)* (San Diego, USA, World Scientific) (2007), pp. 4482–4493.
16. P. H. A. J. M van Gelder, J. De Ronde and N. W. Neykov, "Regional frequency analysis of extreme wave heights: trading space for time," *Coastal Engineering 2000 (Proc. 26th ICCE)* (Sydney, ASCE) (2000), pp. 1099–1112.

17. J. R. M. Hosking, "*L*-moments: Analysis and estimation of distributions using linear combinations of order statistics," *J. Roy. Statistical Soc., Series B* **52** (1990), pp. 105–124.

18. T. Izumiya and M. Saito, "Extended least squares method for estimating parameters for extreme distribution function," *Proc. Coastal Engrg., JSCE* **44** (1997), pp. 181–185 (*in Japanese*).

19. Y. Goda, P. Hawkes, E. Mansard, M. H. Martin, M. Mathiesen, E. Peltier, E. Thompson and G. van Vledder, "Intercomparison of extremal wave analysis method using numerically simulated wave data," *Proc. 2nd Int. Symp. Ocean Wave Measurement and Analysis*, ASCE (1993), pp. 963–977.

20. Y. Goda and K. Kobune, "Distribution function fitting for storm wave data," *Proc. 22nd Int. Conf. Coastal Engrg.* (Delft, 1990), pp. 18–31.

21. I. I. Gringorten, "A plotting rule for extreme probability paper," *J. Geophys. Res.* **68**(3) (1963), pp. 813–814.

22. G. Blom, *Statistical Estimates and Transformed Beta-Variates* (John Wiley and Sons, New York) (1958), Chapter 12.

23. M. Andrew, O. P. Smith and J. M. Mckee, "Extremal analysis of hindcast wind and wave data at Kodiak, Alaska," *Tech. Rept., CERC-85-4*, U.S. Army Corps of Engrs., Waterways Experiment Station (Vicksburg, Miss.) (1985).

24. V. Barnett and T. Lewis, *Outliers in Statistical Data (2nd ed.)* (John Wiley and Sons) (1984), 463p.

25. Y. Goda, O. Konagaya, N. Takeshita, H. Hitomi and T. Nagai, " Population distribution of extreme wave heights estimated through regional analysis," *Proc. 27th Int. Conf. Coastal Engrg.* (Sydney, ASCE) (2000), pp. 1078–1091.

26. J. F. Lawless, *Statistical Models and Methods for Lifetime Data* (John Wiley and Sons, New York) (1982), 580p.

27. P. G. Challenor, "Confidence limits for extreme value statistics," *Inst. Oceanographic Sciences, Rept.* **82** (1979), 27p.

28. M. Mathiesen, Y. Goda, P. Hawkes, E. Mansard, M. H. Martin, E. Peltier, E. Thompson and G. van Vledder, "Recommended practice for extreme wave analysis," *J. Hydraulic Res.*, IAHR **32**(6) (1994), pp. 803–814.

29. Y. Goda, "Uncertainty of design parameters from viewpoint of extreme statistics," *Trans. Amer. Soc. Mech. Engrg.* **114** (1992), pp. 76–82.

30. Y. Goda, "On the uncertainties of wave heights as the design load for maritime structures," *Proc. Int. Workshop on Wave Barriers in Deep Waters*, Port and Harbour Res. Inst. (Yokosuka, Japan) (1994).

31. D. J. T. Carter and P. G. Challenor, "Estimating return values of environmental parameters," *Quart. Jour. Roy. Meteorol. Soc.* **107** (1981), pp. 259–266.

32. T. Izumiya and M. Saito, "Unbiasedness condtions in extreme data analysis and estimation of confidense interval based on asymptotic approximation," *Proc. Coastal Engrg., JSCE* **45** (1998), pp. 206–210 (*in Japanese*).

33. L. E. Borgman, "Risk criteria," *J. Waterway and Harbor Div., Proc. ASCE* **89**(WW3) (1963), pp. 1–35.

34. M. K. C. Ravindra, C. A. Cornell and T. V. Galambos, "Wind and snow load factors for use in LRFD," *J. Structural Engrg., Proc. ASCE* **104**(ST9) (1978), pp. 1443–1457.

35. B. Ellingwood and T. V. Galambos, "General specification for design loads," *Proc. Symp. Probabilistic Methods in Structural Engrg.* (ASCE) (1987), pp. 27–42.

36. T. Izumiya, "Formulation of the N-year maximum value with the Weibull distribution and its applicability" *Proc. Coastal Engrg., JSCE* **46** (1999), pp. 236–240 (*in Japanese*).

37. H. F. Burcharth, "Reliability-based design of coastal structures," *Advances in Coastal and Ocean Engineering* **3** ed. P. L.-F. Liu (World Scientific, Singapore) (1997), pp. 145–214.

38. N. Hogben and F. E. Lumb, *Ocean Wave Statistics* (H.M.S.O., London) (1967).

39. British Maritime Technology, *Global Wave Statistics* (Unwin Brothers Ltd.) (1986).

40. G. Z. Forristall, J. C. Heideman, I. A. Legget, B. Roskam and L. Vanderschuren, "Effect of sampling variability on hindcast and measured wave heights," *J. Waterway, Port, Coastal, and Ocean Engrg.*, ASCE **122**(5) (1996), pp. 216–225.

41. R. J. Seymour, R. R. Strange III, D. R. Cayan and R. A. Nathan, "Influence of El Niños on California's wave climate," *Proc. Int. Conf. Coastal Engrg.* (Houston, ASCE) (1984), pp. 577–592.

42. R. J. Seymour, "Wave climate variability in Southern California," *J. Waterway, Port, Coastal and Ocean Engrg.* **122**(4) (1996), pp. 182–186.

43. P. A. Hastings, "Southern oscillation influences on tropical activity in the Australian/south-west Pacific region," *Int. J. Climatology* **10** (1990), pp. 291–298.

44. D. L. Proh and M. R. Gourlay, "Interannual climate variations and tropical cyclones in the eastern Australian region," *Proc. 13th Australasian Coastal and Ocean Engrg. Conf. and 6th Australasian Port and Harbour Conf.* (Christchurch, 1997), pp. 681–686.

45. T. Sekimoto, M. Hanayama, H. Katayama and T. Shimizu, "On the estimation of design wave period," *Proc. Coastal Engrg., JSCE* **46** (1999), pp. 256–250 (*in Japanese*).

46. D. P. Callaghan, P. Nielsen, A. Short and R. Ranasinghe, "Statistical simulation of wave climate and extreme beach erosion," *Coastal Engineering* **55**(5) (2008), pp. 375–390.

47. A. Repko, P. H. A. J. M. van Gelder, H. G. Voortman and J. K. Vrijling, "Bibariate description of offshore wave conditions with physics-based extreme value statistics," *Applied Ocean Res.* **26** (2004), pp. 162–170.

48. D. J. de Waal and P. H. A. J. M. van Gelder, "Modelling of extreme wave heights and periods through copulas," *Extremes* **8** (2005), pp. 345–356.

49. P. Galiatsatou and P. Prinos, "Bivariate analysis and joint exceedance probabilities of extreme wave heights and periods," *Coastal Engineering 2008* (*Proc. 31th ICCE*) (Hamburg, World Scientific) (2009), pp. 4121–4133.

# Part IV

# Waves and Beach Morphology

## Chapter 14

# Coastline Change and Coastal Reconnaissance

## 14.1 Introduction

This and the following chapters differ from the preceding chapters in their objective of presentation. The intended reader is an engineer or manager who is assigned the responsibility of protection and rehabilitation of a certain extension of sandy beach. He may not have a background in coastal engineering but will be responsible for supervising field surveys, making a project plan and executing the project. Another possible reader is a coastal engineer who has been assigned a field study for beach protection at a foreign coast on which he has little knowledge. The two chapters aim at presenting an overview on morphological beach problems for such a reader to assist him in his works.

Admittedly, the author has only a few experiences in dealing with beach morphological problems in the field. He has published a limited number of papers on sediment transport topics. Nevertheless, he recognizes a need to look at the problems of beach morphology from a standpoint different from the conventional approach employed by most of coastal sediment researchers.

One aspect is the necessity of understanding the history and origin of a sandy coast under study in a millennial time span. This is a call for the attention of people engaged in a specific project in the field. The present chapter begins with an overview of historical coastline changes since some 15,000 years ago, when the sea level was about 100 to 130 m below the present level globally. Examples of coastline advance in the past millennia are presented. Then the focus is given on sandy coasts, and their geological features are discussed. The natural causes of shoreline changes are itemized with brief explanations for them. The estimated rates of littoral sediment transport from various sources are listed in a table to provide a coastal manager and/or engineer with a sense for an order of magnitude of coastal sediment problems. Next the anthropogenic influence on coastal morphology is discussed with several cases of historical events. Construction of a structure on a sandy coast always brings forth certain patterns of shoreline changes, which are classified in ten types. For any coastal management and protection project, coastal reconnaissance is the first step to be undertaken. Its methodology is discussed with guidelines for search of sediment supply and longshore transport direction.

## 14.2   Overview of Historical Coastline Change

### 14.2.1   *Geological View of Coastline Change*

(A) *Holocine rise of sea level*

The coastline marks the boundary between land and sea. When sea level rises, the coastline recedes toward land. Geologically speaking, the position of coastline was not fixed but shifted inward and outward widely. Throughout the history of the earth, even the shape and location of the continents have changed dramatically. They have drifted with tectonic movements of plates, which sometimes trigger earthquakes and generate tsunamis. The Panamanian Isthmus was formed about 2.7 million years ago in the late Pliocene Epoch when the Central American Seaway was closed by volcanic activity.[1] It was in the middle of the Tertiary Period, and the Quaternary Period begins about 1.8 million years ago with the Pleistocene and Holocene Epochs. The Pleistocene Epoch is characterized with a series of Glacial and Interglacial stages. The Holocene (Recent) Epoch refers to the period from 10,000 years ago to the present.

About 15,000 to 18,000 years ago, the Ice Age came to the end and the climate commenced changing to a warmer one. At that time, the sea level is estimated to have been positioned at a level of 100 to 130 m below the present. Under the warming climate, continental glaciers melted gradually and supplied a huge amount of water to the ocean. The sea level began to rise worldwide with an average rate of 1 m per century with several occasions of slowdown and small reversals. Around five thousand years ago, the sea level rise stopped at the height of several meters above than the present level,[a] and then gradually fell to the present level around three thousands year ago. There have been several slight rises and falls of the sea level since then. The technical terms of *transgression* and *regression* are used to denote the advance and recession of the coastline corresponding to the rise and fall of sea level. The amounts of relative sea level fluctuations vary from region to region, and consultation with geologists is recommended for more information.

An evidence of the Holocene transgression associated with sea level rise is the presence of submarine canyons, which are easily recognized in the charts as extension of major rivers. The rivers flowed on then-dried plains and cut the land to form valleys, transporting a large amount of sediments to the sea. Rivers continued to transport sediments even after the end of the Holocene transgression and contributed to formation of deltas and sandy beaches everywhere in the world.

Presently sea level rise due to the global warming is a hot topic of the world. The fourth report of the Intergovernmental Panel on Climate Change (IPCC) published in 2007 estimates that mean sea level has risen by 0.17 m in the twentieth century and the rate of sea level rise between the period from 1993 to 2003 was 3.1 mm per year. IPCC predicts a sea level rise of 0.26 to 0.59 m by 2100 for the scenario that the consumption of fossil fuel continues at the present rate. However, the estimate does not include the effect of possible melting of the Arctic summer sea-ice, the Greenland ice sheet, and the west Antarctic ice sheet (Lenton *et al.*[2]). The IPCC report says that the process of ice melting is poorly understood and a possible sea level rise due to the ice melting could not be evaluated at the stage of the report preparation. Nevertheless, there are strong arguments for inclusion of the ice melting effect in some coastal management and defense plans over long periods such as the planning for the year 2100.

(B) *Historical records of coastline advancement by river sediments*

The Mesopotamian cities such as Ur and Uruk are said to have developed at the estuaries of the Tigris and the Euphrates some 5000 years ago. Presently they are located more than 250 km inland from the head of Persian Gulf. The Holocene transgression at its peak might have brought the sea near

---

[a]Japan has the evidence of transgression with many remains of shell middens in the periphery of Tokyo Bay and Osaka Bay at the elevation up to 5 m. It is called the Jomon transgression. Northern France also has the Flandrian transgression.

to the Mesopotamian cities, but the huge amout of sediments by the flood every year must have advanced the shoreline away from the ancient cities. Although no quantitative or historical records are available for the case of Tigris and Euphrates rivers, we can cite several cases of coastline advancement by sediments brought by rivers.

The Chang Jiang (Yangtze River) is the largest river in China with a length of about 6300 km and a watershed basin area of 1810 thousands km$^2$. The annual volume of river flow is about 100 billion cubic meters or 33,000 cubic meters per second on the average, and the total sediment discharge is estimated about 200 million cubic meters per year.[3] It is estimated that the river estuary must have been located about 250 km upstream 6000 years ago. Shanghai was under the sea before the tenth century. The name of the town of Shanghai appears in documents only after the twelfth century.

China's second largest river of the Huang He (Yellow River) has a length of about 5500 km and a watershed basin area of 750 thousands km$^2$. The annual volume of river flow is about 47 billion cubic meters, but its concentration of suspended loess is 35 g/$l$ as the annual mean.[4] Thus, the annual sediment discharge is about 640 million cubic meters, which is about three times that of the Chang Jiang. Because of the settlement of densely suspended sediment, the river bed often becomes higher than the inland at many places. The levees were raised in pace with the rise of river bed, but they were breached by big floods. The Huang He has switched its course widely over the expanse of more than 650 km. The present delta of the Huang He began to grow after the latest switch of its course in 1855, and the coastline advanced with a rate of about 150 m per year since then.

The Danube is the second largest river in Europe with a length of about 2900 km and a watershed basin area of 820 thousand km$^2$. Before World War II, the Danube carried sediments of 73 million cubic meters annually (Bondar and Panin[5]) which formed a large delta of the Danube; it has been designated as a World Heritage.

The Tevere which flows through the city of Rome is the third largest river in Italy with a length of 405 km. The Emperor Trajanus (98–117 AD) excavated an artificial harbor of hexagon shape next to the river mouth of the Tevere. Each side of the hexagon is 350 m long, and the basin had the size of 700 m by 606 m. The harbor was the entry port of grain and other important goods to the capital of the Roman Empire. Since the collapse of the empire, no civilization took care of the Trajanus's harbor, and sediments carried by the Tevere gradually filled up the harbor and advanced the coastline toward the west by more than 3 km. Presently, we can see on Google Earth the hexagon pool at the south side of the Via del Lago di Traiano next to the Leonard da Vinci Airport of Rome. The rate of shoreline advancement is about 1.6 m per year. It is small compared with the cases of the Chang Jian and the Huang He, but is documented in the historical record.

### 14.2.2   *Geological Features of Sandy Coast*

(A) *Origin of beach sand*

Beaches are composed of materials of various sizes ranging from silt to boulder. Most beaches are made of sand of fine, medium, or coarse particles. Pebble beaches may be found around the mouths of steep rivers, while cobble or boulder beaches may exist at the base of sea cliffs having been attacked by high waves.

Beach sand is either of terrestrial, volcanic, or biogenic origin. Rocks in mountains are eroded by mechanical weathering and become so-called clastic sediments. Mountains are also eroded by direct actions of river currents. While clastic sediments are carried by currents, they are reduced in size by mutual collision and abrasion. Upon reaching the river mouth, they are dispersed into the sea, but a certain portion of them are relocated back to the shore by waves and are carried along the shore by coastal currents.

Beach sand is also produced by erosion of sea cliffs, which had been formed by various geological processes. Cliffs are made of rocks (igneous, sedimentary, or metamorphic), sand deposit layers, and/or loess deposit layers. Cliff sand may be termed as of rock origin mostly.

Volcanic sand is the product of volcanic eruption. Beaches around volcanic islands are made of coarse volcanic ashes as seen in Hawaiian Islands, Aleutian Islands, and elsewhere.

Examples of biogenic sand are fragments of bivalve shells and coral sand. It is called carbonate sand and is mostly found in subtropical and tropical regions. However, the Romanian coast of the Black Sea has 40-km long beach made of shell fragments at the feet of cliffs between Constanta and the frontier border to Bulgaria. Limestone is a sedimentary rock formed under the sea by the remains of foraminifer, calcareous algae, coral, bivalve shell, and others millions of years ago.

## (B) *Mineralogical characteristic of sand*

Minerals contained in beach sand reflect the origin of sand. Carbonate sand is composed of calcium carbonate, which is a compound of calcium oxide (CaO) and carbon dioxide ($CO_2$).

Terrestrial and volcanic sands contain quartz ($SiO_2$), feldspar ($MT_4O_8$), and other minerals. Because quartz is inert to chemical actions, it withstands weathering processes and the quartz content increases as sand is transported over long distance. Feldspars appear in different types of crystals; M stands for alkali elements of Ca, Na, K, or Ba, while T stands for Al or Si. Because feldspars are susceptible to weathering processes, feldspar sand is found on beaches close to sources of igneous and metamorphic rocks.

Sand may also contain heavy minerals in black or reddish color. A number of minerals with specific gravities greater than 2.87[b] may be found in beach sand. If one can identify specific heavy minerals and analyze the spatial variation of mineral contents in sand, it may be possible to trace the origin of the beach sand. Some beaches may contain iron sand (magnetite: $Fe_3O_4$) deposited in thin layers. Iron sand is black in color and heavy on one's palm, and thus it is easy to detect iron sand. However, deposits of iron sand appear sporadically along long stretches of coasts, and thus it is difficult to use it for tracing its origin.

## (C) *Sources of beach sand*

Beach sand is considered to have been supplied from the following sources:

  (i) submarine deposits of sediment,
 (ii) discharge from rivers,
(iii) collapsed cliffs,
 (iv) littoral transport from adjacent beaches, and
 (v) biological production.

Throughout geologic times, land sank in the sea and rose from the sea many times by crustal movements. Rivers continued to erode mountains and to carry sediments to the sea. Most of the coastal plains in the world have emerged through the last upheaval of crustal movements. Long stretches of beaches along coastal plains seem to be the remains of

---

[b]This is the specific gravity of the liquid bromoform used to separate light and heavy minerals.

submarine deposits when the plain was under the sea, and they must have been provided with additional contemporary supply from rivers. Presence of subsurface sand ridges on the continental shelf is reported in many places. The Coastal Engineering Manual (CEM) cites the case of the Middle Atlantic Bight of the United States.[6]

Rivers are the most important source of sand supply to beaches. Sand can be transported over a distance of a few hundred kilometers by coastal currents induced by waves, tides, and winds. Thus, one should look for a wide area for search of substantial rivers that are capable of supplying sand to the beach under study.

Cliffs on the coast are eroded by the actions of waves and currents with rates varying from place to place. Erosion may take a form of sudden collapse of the cliff face. Loess and silt in the collapsed mass of a cliff are carried away by waves, but sand and coarse materials are deposited at the feet of cliffs, thus providing a source of beach sand.

Along the coast composed of a chain of pocket beaches, sediments are easily transported across headlands separating pocket beaches. Unless a headland project into the sea with a very steep slope, wave-induced currents can flow beyond the tip of the headland toward the next pocket beach, while carrying suspended sediment. Thus a pocket beach is an intermediate point of sand movement, and the true source of sand supply needs to be sought elsewhere.

Biogenic sand is produced by living organisms. Under normal circumstances, they produce a certain constant volume of biogenic sand every year. Quantitative estimate of production rate will be difficult, however.

(D) *Typical shapes of sandy coast*

Coastal topography is diverse, and many experts have presented coastal classifications. CEM refers to the classification by Shepard.[7,8] Sandy coast is one of coastal features, and the following is a tentative list of its topographic shapes:

(i) long beach along a coastal plain,
(ii) pocket beach between headlands,
(iii) sand spit,
(iv) barrier beach,
(v) barrier island, and
(vi) beach ridges and swales.

A long beach along a coastal plain is a typical shape of a sandy coast. As explained in Clause (C), sand would be the remains of submarine deposit brought by ancient rivers in millions of years ago, which emerged together with the coastal plain by crustal movements. The length of such a beach may extend from a few tens of kilometers to more than one hundred kilometers. Waves obliquely incident to the shore induce alongshore currents in the surf zone and transport sediments in suspension. Depending on the season, a large quantity of sediments may move along the beach.

A *pocket beach* of concaved arc shape is developed between two headlands of a rocky coast or an undulating cliff coast. It is also called an embayment beach or an arcuate beach. Not only the shoreline in a shape of concaved arc, but also the depth contours in shallow waters are parallel to the shoreline. Waves incident to a pocket beach are refracted by such bathymetry and break near the shore almost in parallel with the shoreline. Only a weak alongshore currents are induced and sediment transport within a pocket beach is small. Thus, a pocket beach maintains its stable shape against waves coming from any direction. However, the orientation of pocket beaches oscillates in response to seasonal changes in wave direction. Pocket beaches usually appear in a series, and Silvester and Hsu[9] applied the term of *crenulate shaped bays* to a chain of pocket beaches.

In places where a large amount of sediments are transported alongshore in one direction, sediments may form a spit beyond a promontory. It is called a *sand spit*. Cape Skagen at northern Denmark and Cape Cod in the East Coast of the United States are examples of sand spits. Sand spits can be found at many coasts of the world.

As the sediments transported alongshore come at one corner of a bay entrance, sediments build a spit and the spit gradually increases in length. With elongation of a sand spit, it may come to the stage of closing the bay, and then it is called a *barrier beach*. Along the northern Black Sea coast of Romania, there were several Hellenic and Roman colonies, the ships of which freely sailed across the Black Sea. Some colonial towns such as Historia and Ovidiu were depedent on sea ports located within embayments. Over the period of two to three thousand years, terrestrial sediments moving southwest from the Danube gradually closed bay mouths, and embayments became seashore lakes as nowadays. The United States have an innumerable number of barrier beaches along its East Coast and the Gulf Coast.

A water body enclosed by a barrier beach may create an opening by itself to respond to tidal exchange of water. A barrier beach may be cut by overwash of storm waves or rise of the water level of interior waters due to heavy rainfall. When a part of barrier beach is detached from the rest, it is called a *barrier island*. Many of barrier beaches and barrier islands of the United States change their shapes by sediment transport, formation of natural new opening, and/or man-made inlets for access to harbors and marinas behind barrier islands.

Beach ridges and swales are a system of shore-parallel long stretches of slightly elevated (*beach ridge*) and depressed (*swale*) topography. They are found at the edge of coastal plains, the head of sand spits, and others. The number of rows of ridges and swales may amount to a few hundreds, and they may extend over several tens of kilometers as reported by Anthony.[10] They seem to have developed in response to many cycles of small fluctuations of the mean sea level after the Holocene transgression in the coast where abundant sediment supply is available (Tanner[11]). The longevity of beach ridge and swale systems was attested by Jankaew *et al.*,[12] who searched for an evidence of historical tsunamis before the Great Indian Ocean Tsunami of 2004. They dug a pit in one of swales at Phra Thong Island (Latitude 9°08′ N and Longitude 98°16′ E) in Thailand. They found a sand sheet of about 0.1 m thick at the depth of about 0.3 m below the surface, which must have been brought by the preceding tsunami uprush. Radio carbon dating revealed that the sand sheet was deposited in the thirteenth or fourteenth century. There were two more sand sheets underneath which indicate two more great tsunamis in the past milleniums.

### 14.2.3 *Natural Process of Shoreline Change*

(A) *Overview*

The shoreline of a sandy beach changes its position gradually due to natural processes and can also be affected by human activities to a great extent. In this clause, changes of shoreline position under natural processes are discussed.

The causes of natural shoreline changes of sandy coast may be listed as follows:

A. Causes for shoreline advancement:

    A-1. Uplift of the ground by isostatic adjustment or tectonic movements associated with earthquakes,

    A-2. Fall of sea level,

    A-3. Supply of sediments from rivers,

    A-4. Supply of sediments by collapse of a neighboring cliff.

B. Causes for shoreline recession:

B-1. Subsidence of the ground by tectonic movement associated with earthquakes,

B-2. Rise of sea level,

B-3. Erosion of cliffs,

B-4. Erosive actions of waves and currents.

The *isostatic adjustment* refers to the process of the crustal movement toward the gravitational equilibrium when a superposed load is changed. A typical example is the depression of land mass under the weight of glaciers in the Ice Age, and its rebound after melting of glaciers in the Holocene Epoch. The lands of Alaska, Scandinavia, and an eastern part of England, which were covered with thick ice sheets during the Ice Age, continue rising even today. The Coastal Engineering Manual cites a diagram of yearly mean sea level changes at Juneau (58°22′ N and 134°36′ W), Alaska from 1936 to 1986.[13] It shows a linear trend of isostatic rebound with a rate of about 1.3 cm per year.

Uplift or subsidence of the ground by a major earthquake may amount to a few meters. The Great Kanto Earthquake in 1923 with a magnitude of $M = 7.9$ caused an uplift of 1.8 m at the tip of Boso Penisula (35°55′ N and 139°54′ E), Chiba Prefecure, Japan.

The change of mean sea level has been discussed in Sec. 14.2.1(A), and examples of shoreline advance by river sediments have been presented in Sec. 14.2.1(B). Erosive actions of waves and currents on natural coasts are evident and cases of beach erosions are found at many places.

(B) *Examples of historical cliff and beach erosion*

Erosion of sea cliffs by waves and currents has been noticed since early days and the extent of erosion can be assessed by comparison of old documents/maps and present ones. Erosion of cliffs is a rather common feature of coastal morphology in the world. Zenkovich[14] states that a part of the cliffed coast in eastern England consisting of unconsolidated Quaternary deposits is receding at an unusually rapid rate averaging between 2.1 and 4.5 m per year.

In the United States, Dean[15] cites several estimates of historical erosion rate of the Outer Cape Cod cliffs in Massachussetts, which varies from 0.66 to 0.97 m per year. The Chatham Lighthouse (41°40′ N and 69°57′ W) located at the elbow of Cape Cod experienced severe recession of the fronting shoreline after the barrier beach was breached and waves began to attack the beach directly. The beach had the width of 69 m in 1847, but it was lost completely by 1880, and the lighthouse tower slid over the bank (Dean[16]).

The Cape Hatteras Lighthouse (35°14′ N and 75°33′ W) in North Carolina, standing at the outmost junction point of two long barrier beaches (Outer Banks), has been providing indispensable aids to navigation since it was built in 1870. At that time, the lighthouse had a frontal beach of 490 m wide, but the beach was eroded by a rate of 7.6 m per year with the result of the beach having a width of only 150 m remaining in 1919. Various protection works including beach nourishment were executed, but the beach erosion was unstoppable. To save the Lighthouse from the sea, it was decided to move it inland by 490 m. The Lighthouse was lifted on a steel plate and slowly moved upon iron rails to the new site in July 1999 (Dean[17]).

As for cliff coasts in Japan, Horikawa[18] refers to Sunamura's work on cliff erosion, who compiled the mean erosion rate of cliffs in Japan, ranging from 0.3 to 2.2 m per year. Being one of the cliff coasts in Japan, Byobu-ga-Ura in Chiba Prefecture has a most famous erosional history. It is about 10 km long with a height of 40 to 50 m, located between Iioka Town (presently a part of Ashahi City, 35°41′ N and 140°33′ W) and Choshi City (35°42′ N and 140°50′ E). The lower part of the cliff is tuff of the Tertiary Period, the middle is sand layers, and the upper layer is loess. According to documents in the thirteenth century, the coastline of Byobu-ga-Ura was 2000 to 6000 m offshore. An old map in 1794 indicates the coastline at a distance of about 800 m from the present. Thus the rate of cliff erosion is estimated as 4 to 6 m per year (Toyoshima[19]).

### 14.2.4 Examples of Estimated Rates of Littoral Sediment Transport

(A) *Alongshore transport rate*

Sediments supplied from rivers and collapsed cliffs are transported offshore and alongshore by actions of waves and currents. Sediments at beaches are also carried away by waves and currents. Sediment transport along the shore is usually expressed as the total volume of sediments moving across

Table 14.1 Examples of longshore sediment transport rates in the world.

(units of transport rate: thousdands cubic meters/year)

| Country | Place | Direction-wise Rate | | Net Rate | Waves | Refrences and Remarks |
|---|---|---|---|---|---|---|
| | | (S) | (N) | | | |
| Barbados | Southwest coast | — | — | 70 | C | Smith et al.[20] (rocky coast) |
| Brazil | Fortaleza Port | — | — | 600 ~ 860 | B | Onaka et al.[21] (impoundment) |
| Costa Rica | Caldera Port | 0 | 120 | 120 | C – D | Rodriguez and Katoh[22] (estimate by dredging volume) |
| India (East coast) | Madras | 680 | 1,030 | 350 | B | Chandramohan et al.[23] |
| | Manamelkudi | 640 | 1,430 | 790 | B | (estimate by CERC formula) |
| India (West coast) | Trivandrum | 1,630 | 620 | 1,010 | B | ditto |
| | Cochin | 980 | 690 | 290 | B | ditto |
| Israel | Haifa | 200 | 80 | 120 | C | Perlin and Kit[24] |
| Israel | Ashdod | 250 | 50 | 200 | C | (estimate by LITPAK package) |
| Japan | Kashima Port | 640 | 600 | 40 | B | Sato and Tanaka[25] (submerged impoundment) |
| Romania | Mamaia Beach | 140 | 160 | 20 | C – D | Kuroki et al.[26] (CERC formula) |
| Sri Lanka (East coast) | Trinconamale | 330 | 240 | 90 | B | Chandramohan et al.[23] |
| Sri Lanka (West Coast) | Dondra | 280 | 1,820 | 1,540 | B | (estimate by CERC formula) |
| | Columbo | 1,060 | 680 | 380 | B – C | ditto |
| | Kalpitiya | 650 | 800 | 150 | B – C | ditto |
| U.S.A. (Atlantic) | Long Island, NY | — | — | ~200 | B – C | Rosati et al.[27] |
| | Sandy Hook, NJ | — | — | 350 | B | CEM Table III-2-1 |
| | Ocean City, NJ | — | — | 310 | B | ditto |
| U.S.A. (Pacific) | Santa Barbara, CA | — | — | 210 | B | ditto |
| | Port Hueneme, CA | — | — | 380 | B | ditto |
| U.S.A. | Great Lakes area | — | — | 10 ~ 70 | D | ditto |

Note: (1) The direction-wise rates with (S) and (N) refer to the southward and northward gross rates.

(2) The symbols of B, C, D in the column of waves refer to the magnitude of the storm significant wave height such as from 5 to 8 m (period of 8 to 12 s), 3 to 5 m (period of 6 to 9 s), and less than 3 m (period less than 7 s), respectively, by subjective judgment of the present author.

(3) CEM stands for the Coastal Engineering Manual by the US Army Corps of Engineers.

a vertical section normal to the shoreline every year on the average, which is called the *longshore sediment transport rate*.

There have been many efforts for estimating the longshore transport rate. The *impoundment method* is a favored estimation technique. The volume of sediments accumulated at one side of a jetty or breakwater is estimated by topographic and bathymetric surveys, and the volume is divided by the number of years of the period in which sediment accumulation took place to yield transport rate. Some empirical formulas have also been used to estimate the longshore sediment transport rate based on wave climate data, though they may have a large extent of deviation from the true rate because of the uncertainty involved in the empirical formulas. The volume of maintenance dredging at an entrance of a harbor can also provide information of sediment transport rate. Table 14.1 is a compilation of available data of longshore transport rate as a reference for the order of magnitude of sediment transport in the field. Because the direction of sediment transport may differ by season, Table 14.1 lists the transport rates in both directions (e.g., southward and northward, or left and right) wherever the data available and the net transport rate which is the difference between the rates in both directions.

As seen in Table 14.1, the rate of alongshore transport of sediments varies widely depending on the locality. Because the transport rate is a function of the longshore component of wave energy flux as will be discussed in Sec. 15.3.1, the azimuth of the coastline relative to the predominant wave direction and the wave height are most influential for the longshore sediment transport rate. At many coasts, predominant direction of waves shifts by season, and sediments are transported to both directions of coasts. The net transport rate in Table 14.1 is the difference between the transport rates in one direction and the other. Where no information is given on direction-wise transportation rates, the reported transport rate is listed as a net rate.

When the development of Kashima Port in Japan was started in 1961 by excavating channels and basins amid a long-stretch of sandy coast, the coast was estimated to have southward and northward transport rates of 640 and 600 thousand cubic meters per year, respectively, with the net southward transport of 40 thousand cubic meters per year (Sato and Tanaka[25]). Later field investigations, however, have indicated a slight dominance of the northward sediment transport. The incidence testifies the difficulty in quantitative prediction of sediment transport rates.

(B) *Offshore transport rate*

When a beach is eroded, a certain amount of sediments are carried away offshore and the rest is moved alongshore. Sediments moved offshore during a storm often build longshore bars, which function as energy dissipaters by forcing incoming waves to break. After a storm, longshore bars migrate toward the shore being pushed by waves of low steepness. The process is repeated whenever storm waves attack a beach, but at the end of one year cycle the beach usually stays near its initial position. Exceptions are coasts with submarine canyons reaching near the coast. Sediments transported alongshore fall into the canyon and are lost forever.

Some beaches without nearby canyons, however, are suffering from slow but continuous erosive processes. Historical beach erosion at the Cape Hatteras Lighthouse in Sec. 14.2.3(B) is such an example. Analysis of sediment budget over a long stretch of coast can yield an estimate of the offshore transport rate. A few examples of such estimates are presented here. Kuroki *et al.*[28] analyzed the sediment budget of a 75-km stretch of the northern coast of Niigata Prefecture (from $37°47'$ N and $137°49'$ E to $38°41'$ N and $139°27'$ W) over the period of 1947 to 1975. They estimated that the coast has lost 9 cubic meters of sediment per meter per year on the average (total of 675 thousand cubic meters per year). This coastal stretch includes the Niigata coast around Niigatga Port, which suffered from severe beach erosion after the opening of a flood diversion channel of the Shinano River outside this coastal stretch in 1922. The Niigata coast had advanced over years with the sediment discharge of the Shinano River, and without further supply of sediments the sea is taking back its share.

At the apex of the Suruga Bay facing the Pacific in central Japan, the seabed drops to depths of more than 1000 m with a steep slope of 1 on 7. The Fuji River flows into the Suruga Bay at the location of $35°08'$ N and $138°39'$ E. A number of flood control projects were executed in the upstream of the river after World War II. Ogura *et al.*[29] analyzed the sediment budget of a 12-km long coast at both sides of the mouth of the Fuji River in the period from 1961 to 1988. The annual sediment loss is 410 thousand cubic meters, which is equivalent to the rate of 34 cubic meters per meter per year. Most of sediments must have gone into the depths of the Suruga Bay.

The sediment discharge of the Ooi River flowing into the western coast of the Suruga Bay (the river mouth located at $34°46'$ N and $138°13'$ E) was much reduced by construction of a number of large hydropower dams after World War II. The river still discharges sediments amounting to about

500 thousands cubic meters per year, but the volume of 294 thousands cubic meters per year is carried away offshore, according to the estimate by Kunieda *et al.*[30] Because the coast directly affected by the sediment from the Ooi River is about 14 km long, the annual sediment loss is calculated at 21 cubic meters per meter per year.

These offshore sediment losses are manifestations of the erosive power of waves acting on the coast.

## 14.3    Anthropogenic Influence on Coastal Morphology

### 14.3.1    *Outline*

With progress of civilization, people transformed the natural environment by various means at many places in the world. Cutting forests to make agricultural fields is an example, and coastal reclamation is another. Human activities have exercised large influences on coastal morphology over many years. Anthropogenic influences on coasts may be classified as in the following:

A.  Accretive influence:

   A-1. Increase of river sediment discharge through deforestation,
   A-2. Impoundment of sediment at the up-drift side of jetty or breakwater,
   A-3. Formation of tombolo or salient by construction of detached breakwater, offshore island, etc.

B.  Erosive influence:

   B-1. Decrease of river sediment discharge by erosion control projects in mountain areas, construction of dams and flood diversion channels, sand mining, etc.,
   B-2. Obstruction of alongshore sediment transport by construction of jetties, breakwaters, etc.,
   B-3. Sediment attraction behind a shadow area of offshore structure from adjacent beaches,
   B-4. Reduction of sediment supply by coastal protection works to adjacent coasts,
   B-5. Acceleration of ground subsidence through pumping of ground water.

The item A-1 of coastal accretion has occurred in some early days of civilization, though the morphological change of coasts must have remained unnoticed by people in those days. On the other hand, people of the present days are trying to reduce the peak flood discharge by means of forestation in mountain areas, dam construction, flood diversion channels, and others. In the steep mountain streams, rock and sand arrestation works are executed to minimize the damage by debris flow. All these measures are the sources of the item B-1. Reduction of the sediment delivery to the coast immediately induces coastal erosion. Some examples will be presented in the next subsection.

Aggregate mining from river beds for concrete production had been a practice in civil construction works. As the adverse effect of aggregate mining on coastal erosion has been recognized, legal restriction against aggregate mining is being enforced in many places.

The items A-2 and B-2 are a pair of events induced by a single construction work. The items A-3 and B-3 are the same. These pairs of influence will be discussed in Sec. 14.3.3.

The item B-4 refers to the situation in which a civil works to protect one area against erosion might give rise to the erosion of adjacent beaches if the eroding coast has been supplying sediment to these areas. Construction of vertical or steeply-sloped seawalls often disperses sand in frontal beaches offshore by the waves reflected from seawalls. Adjacent beaches may suffer from erosion through disappearance of a buffer beach and decrease of alongshore sediment supply.

The item B-5 applies to the cities developed in coastal alluvial plains. The subterranean ground is made of relatively young deposits in the Quaternary Period, and is subject to slow consolidation processes. When the subterranean water is pumped up for industrial, municipal and other purposes, the ground water level is lowered and the ground consolidation is accelerated. A number of coastal metropolises in the world have experienced such problems of ground subsidence. Beaches of such cities need to appropriately monitor the ground subsidence.

### 14.3.2   *Typical Cases of Significant Shoreline Recession by Anthropogenic Influence*

(A) *Influence of dams*

A dam built in the upstream of a river stores a large amount of water, and sediments are deposited upon the bottom of the dam lake. Water is discharged for irrigation, hydropower generation and others, whenever requested. However, the outflow of sediments, especially the coarse fraction, is greatly restricted. The sediment discharge to the coast is reduced accordingly, and the coast suffers from severe erosion. There are many examples of coastal erosion after construction of dams.

A typical case is the recession of the Rosetta Promontory (31°29′ N and 30°23′ E) at the mouth of the Rosetta branch of the Nile after construction of the Aswan Low Dam in 1902 and the High Dam in 1970. By referring to Frihy et al.,[31] Komar[32] introduces the case that the promontory receded by about 2200 m over the period from 1955 to 1983. Ismail and El-Sayed[33] describe further progress of coastal recession up to 2005 and discuss the interaction of seawall construction and beaches.

Another example is the Danube, where the sediment discharge of 73 million cubic meters per year was halved to 36 million cubic meters per year as of 1990 due to the construction of many dams in the main stream and tributaries after World War II (Bondar and Panin[5]). The Northern Romanian Black Sea shore is receding at a rate of several meters per year with the highest rate of 19 m per year over a stretch of more than 100 km.[26,34]

### (B) *Influence of flood control projects*

The historical coastal accretion of the Tevere in Italy introduced in Sec. 14.2.1(B) was turned into a case of coastal recession in the late twentieth century because of the decrease of sediment delivery owing to river control works and excessive dredging on the river bed (Toti et al.[35]). The coast of Ostia has receded at an average rate of about 1.7 m per year, which is almost the same as the millennial averaged rate of accretion. To cope with the erosion problem, Italian engineers carried out a beach nourishment project over the beach length of 3 km with continuous submerged dikes at the offshore edge (perched beach).

The Kaike Coast (35°27′ N and 133°22′ E), Tottori Prefecture, Japan was supplied with large quantities of sediments since the eighteenth century by iron sand mining for iron making in the upstream of the Hino River (77 km long). The iron making with iron sand was ceased by 1921, and the Kaike Coast began to show the tendency of erosion since then. Erosion control works in the mountainous area of the Hino River beginning in the 1930s accelerated the pace of beach erosion. The amount of erosion is estimated as up to 300 m by 1960 (Toyoshima[36]). Since 1969, the coast has been provided with a chain of detached breakwaters which have succeeded in formation of tombolos behind them. At the down-drift coast of the group of detached breakwaters, coastal erosion is continuing however.

(C) *Influence of flood diversion channels*

The Niigata Coast ($37°57'$ N and $139°04'$ E) can be cited as a typical case of significant anthropogenic influence. The coastal plain of Niigata was the estuary of the Shinano River (length of 367 km and watershed basin area of 12 thousand $km^2$) four hundred years ago. With the sediment discharge of 12 million ton per year (a contemporary estimate with many dams constructed upstream in Toyoshima[37]), the Shinano River filled swamps in the estuary and advanced the shoreline seaward. The estuarine port of Niigata prospered since the seventeenth century as one of hub ports of coastal routes around Japan.

The Shinano at the same time has brought frequent flood disasters upon people and rice fields. In response to the eager, repeated petitions by local people for flood control works, the Government of Japan undertook the project of constructing a flood diversion channel of 10 km long at a location 58 km upstream of the river mouth. Construction of the channel began in 1909 and was completed in 1922. The diversion channel (named the New Shinano) virtually eliminated the flood disaster from Niigata and converted the Niigata Plain as one of the largest rice production areas in Japan. At the outlet of the diversion channel, a new beach has been expanding with a great speed to the extent of 6 km long and 1 km wide presently.

While the New Shinano was a great success from the viewpoint of flood control, it caused very severe erosion of the Niigata coast adjacent to the river mouth of the Shinano. By 1945, the coast lost more than 200 m of beach and dunes, and a number of seaside houses were threatened of destruction by waves. Since then a massive coastal protection work has been undertaken, including recent beach nourishment together with construction of offshore low-crested structures and jetties. Many Japanese papers and textbooks have been published dealing with the erosion of the Niigata Coast.

(D) *Influence of longshore sediment transport obstruction by breakwaters*

A spectacular example of sediment impoundment by construction of a breakwater is found in the case of Madras Port (presently Chennai Port) in the last quarter of the nineteenth century. According to Madras Port Trust,[38] the Port began its operation as an open roadstead with passengers and cargo carried by boats pulled out from the sandy beach. An open pier on iron screw piles was built by 1861 with a length of 334 m to a depth of 9 m. To accommodate the rapidly increasing amount of cargo, the Government of the day decided to build a harbor sheltered by two breakwater arms and began the breakwater construction in March 1877. Figure 14.1(a) shows the situation before the start of breakwater construction. A small beach at the left of the center pier seems to indicate beach accretion by a preparatory work for the harbor construction works.

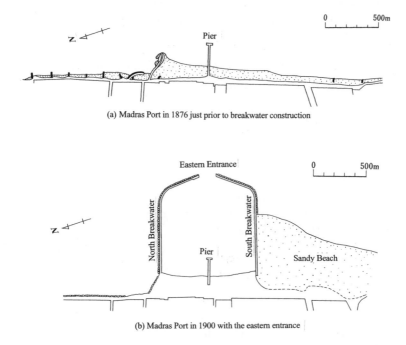

(a) Madras Port in 1876 just prior to breakwater construction

(b) Madras Port in 1900 with the eastern entrance

Fig. 14.1  Shoreline change by construction of breakwaters in Madras Port (adapted from Madras Port Trust.[38])

Extension of the south breakwater was followed by shoreline advancement with almost the same speed. Construction of the north and south breakwaters was completed in 1881 with the eastern entrance of 167 m wide. The southern beach continued to advance at a rate of 21 m per year, however. Figure 14.1(b) shows the completed harbor in 1900 with a wide sandy beach at the south of the south breakwater. Because the continuous northward progression of sediments threatened ship navigation through the eastern entrance, the Port in 1902 decided to close the eastern entrance, to build the east breakwater by 360 m toward the north, and to remove a part of the north breakwater for opening the northern entrance. Presently the east breakwater has been extended by more than 900 m to avoid the intrusion of sediments into the main harbor at the northern part, which serves for large container ships and oil carriers.

The northward sediment transport is estimated as 1.03 million cubic meters per year in Table 14.1. Blocking of the sediment transport by Madras (Chennai) Port has brought forth extensive erosion in the coast north of the port. Judging from the topographic map, recession of the northern coast extends more than 6 km with the largest recession of about 2 km.

### 14.3.3   *Patterns of Shoreline Changes Caused by Structure Construction on Sandy Coast*

Among the anthropogenic influences on coastal morphology, construction of structures such as jetties and breakwaters on sandy coast induces morphological changes within the time span of several months to a few years. Tanaka[39] collected a number of aerial photographs of sandy coasts before and after construction of structures at some 120 sites in Japan and examined the changes of shoreline positions due to structure construction. He found out that there are certain patterns of morphological changes and he classified them in ten types as shown in Fig. 14.2.

These patterns of morphological changes and their causes could be better understood by means of the following three axioms:

AXIOM A:  In the coast where the littoral transport of sediment is predominant in one direction, a new structure extended from a beach always causes accretion in the up-drift side and erosion at the down-drift side.

AXIOM B:  A natural beach has an equilibrium form of concave shape between two promontories. When a new jetty, breakwater, or mole is built, the beach regards the structure as a new promontory and tries to reform itself into a new equilibrium state of concave shape.

AXIOM C:  Sediments are attracted toward the center of a sheltered area, where sediment pickup by waves is weak and slow currents encourage sediment settlement there.

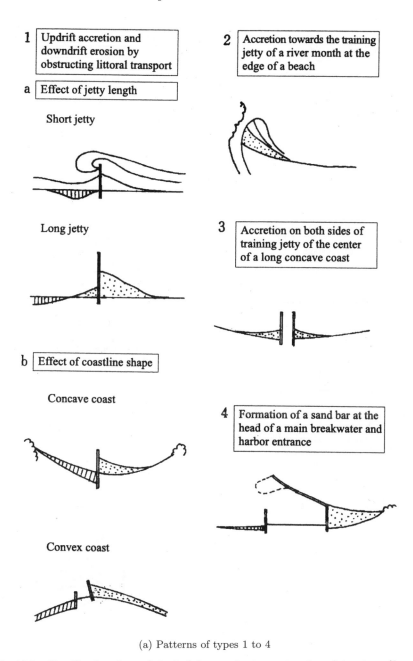

(a) Patterns of types 1 to 4

Fig. 14.2   Classification of morphological changes due to construction of structures (from Tanaka[39]). The direction of net sediment transport is leftward.

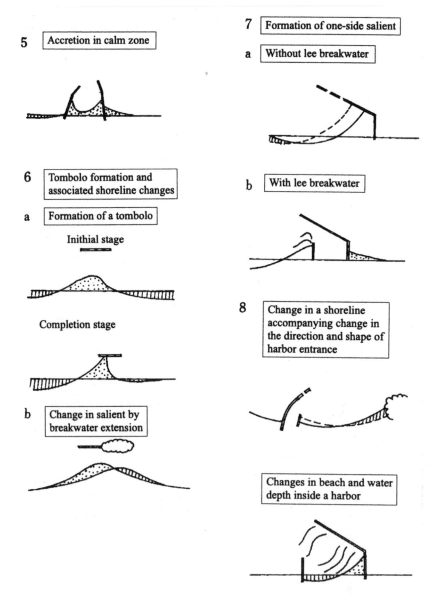

(b) Patterns of types 5 to 8

Fig. 14.2 (*Continued*)

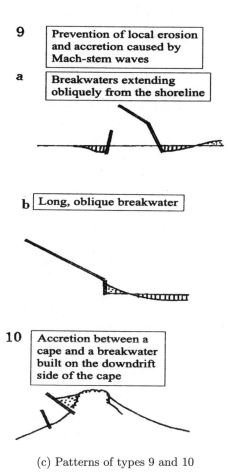

**9** | Prevention of local erosion and accretion caused by Mach-stem waves

**a** | Breakwaters extending obliquely from the shoreline

**b** | Long, oblique breakwater

**10** | Accretion between a cape and a breakwater built on the downdrift side of the cape

(c) Patterns of types 9 and 10

Fig. 14.2 (*Continued*)

Characteristics of the ten types of morphological changes due to structure construction are explained below with identification of the axiom applicable to them. In Fig. 14.2, accreted portions are shown with dots, while eroded portions are hatched.

Type 1(a): A jetty blocks the longshore sediment transport. The extent of erosion at the down-drift side depends on the length of jetty (AXIOM A).

Type 1(b): If the coastline is concave in plan view, the shoreline advances on the up-drift side and recedes on the down-drift side so that those shorelines maintain approximately the same direction on each other. If the shoreline is convex, on the other hand, the shoreline on the up-drift side advances almost to the tip of the jetty and erosion extends very far along the down-drift coast (AXIOM A).

Type 2: If a jetty is located at the mouth of a river next to a headland at one end of a beach, a flat wide beach appears near the jetty (AXIOM B).

Type 3: If a pair of long jetties is built at the center of a long and concave beach, distinct accumulation develops on both sides of the jetties. This type involves a larger area of accumulation and little erosion, as compared to Type 1 (AXIOM B).

Type 4: Accumulation of sediment takes place immediately at the foot of a breakwater by impoundment of longshore sediment transport. With the increase in accumulation, a submerged sand spit is formed from the breakwater tip and causes shoaling of a basin (AXIOM A).

Type 5: Accumulation of sediment takes place in the interior of a sheltered basin without much erosion outside (AXIOM C).

Type 6(a): A salient or cuspate beach develops in the lee of a detached breakwater (AXIOM C). When a salient develops into a tombolo attached to the breakwater, the tombolo functions as a barrier to longshore sediment transport, causing erosion on the down-drift coast (AXIOM A).

Type 6(b): If a sheltered area behind an island is expanded by construction of a breakwater from the island, a salient already in existence shifts its position toward the new center of the sheltered area (AXIOM C). This causes erosion in one side of the tombolo and accretion in the other side.

Type 7(a): Accretion takes place in the sheltered area of a hook-shaped breakwater with the sediment taken from the down-drift side of the sheltered area, thus causing severe erosion there (AXIOM C). Accretion and erosion grow as the breakwater is extended further. This type of beach deformation occurs quite often in the ordinary coast of stable beaches.

Type 7(b): Installation of a short jetty at the down-drift side of a hook-shaped breakwater can reduce the accretion in the sheltered basin and the erosion in the down-drift beach (AXIOM C). If the tip of the hook-shaped breakwater is extended much further, however, noticeable accretion on the down-drift side of the jetty and beach erosion on the further down-drift side take place and may induce shoaling of the sheltered basin.

Type 8: Extension of a jetty or breakwater results in a newly stabilized concave shoreline configuration, which brings forth accretion and erosion along the original shoreline (AXIOM B). If the width or position of the entrance of a harbor is changed by extension of breakwaters, the shoreline inside the harbor changes its orientation and curvature to reach a new stabilized form.

Type 9(a): If breakwaters of upright type are extended obliquely from the shoreline, local erosion takes place at the toe of the breakwaters owing to waves reflected by the breakwaters.

Type 9(b): Similar erosion will take place if the shore-normal part of the breakwater is short.

Type 10: If a harbor is constructed at the down-drift side of a headland of the coast, accumulation of sediment occurs rapidly between the breakwater and the headland (AXIOM B).

Type 1 is a typical case of obstruction of longshore sediment transport. Type 6(b) is an interesting case such that a salient behind a natural island has moved behind the center of a combined barrier of the island and a breakwater when the latter was extended from the island to create a marina. Type 7(a) exhibits a standard pattern of morphological change by construction of a hook-shaped breakwater. As sketched in Fig. 14.3, accretion occurs in the area sheltered against the prevailing wave direction and erosion takes place in the area further away from the breakwater. The boundary between accretion and erosion is approximately located at the junction point of the shoreline and a straight line from the tip of breakwater with an angle of 30° to the prevailing wave direction.

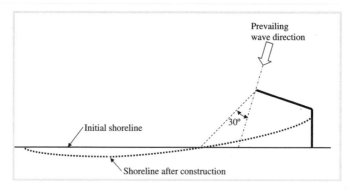

Fig. 14.3   Zones of accretion and erosion induced by construction of a single breakwater.

By applying the three axioms of shoreline change induced by structure construction and referring to the ten types of morphological patterns shown in Fig. 14.2, one can make qualitative prediction of possible shoreline change when a new coastal project is planned on a sandy coast, without mobilizing any numerical model. As exemplified in Fig. 14.2, construction of a new structure or extension of existing one always changes the nearshore current and wave exposure systems and brings forth a certain amount of shoreline changes. In coastal zone management, one should expect that a sandy beach would change its shape whenever a new man-made structure is built there. He should make his best efforts in estimating the extent of morphological changes before implementing a new project.

## 14.4   Coastal Reconnaissance for Beach Protection Project

### 14.4.1   *Collection and Analysis of Documents Before Coastal Reconnaissance*

(A)  *Outline*

The first task of a manager or an engineer who has been assigned a coastal project at a new site is to visit the site, walk along the coast, and grasp the general characteristics of the coast. It may be called *coastal reconnaissance*. Before visiting the site, any available data and documents should be collected and analyzed. When the study site is within a developed country where a number of coastal studies have been carried out, many data and documents may be obtained without much difficulty. In a developing country, relevant data and documents may be difficult to obtain. The present subsection mainly deals with the latter case.

The documentary survey includes the following items:

(1) Satellite images,
(2) Topographic maps and aerial photographs,
(3) Tide tables and nautical charts of small and large scales,
(4) Meteorological and hydrological data,
(5) Worldwide wave data,
(6) Technical reports of any project at the site and/or in the neighborhood,
(7) Historical documents and old newspapers.

These items are briefly discussed in the subsequent clauses.

(B) *Satelite images*

Since 2005, *satellite images* of the earth's surface at any place have become accessible through the Google Earth via internet. One can zoom the images to a resolution less than 1 m, though it may depend on the area. Coastal topography and conditions of vegetation are clearly visible on them together with the latitude and longitude information. A distance between two arbitrary locations can be indicated on the monitor screen. For examples, multiple rows of beach ridges at Cape Skagen and Phra Thong Island are easily viewed on Google Earth.

Satellite images provide both the overall picture of the coast and the local beach conditions as the first input data of documentary survey. Presence of large rivers capable of supplying sediments to the coast should be examined on the satellite images of small scale. They will also tell us how many dams have been built along the river.

(C) *Topographic maps and aerial photographs*

*Topographic maps* with scales of 1:25,000 to 1:200,000 are made by the offices of geodetic survey of respective countries, which often belong to the army. Availability of such maps differs from country to country. Even though they might be classified to the public, permission to release them might be granted to the officers in charge of coastal projects of the government. The same applies for the *aerial photographs* of coastal areas, which have been taken for preparation of topographic maps. The information gained from the satellite images of Google Earth should be confirmed with topographic maps.

Topographic maps, aerial photographs, and satellite images are used to describe the general features of coastal areas and to determine indications of the direction of longshore sediment transport. Evidence of morphological changes is sometimes obtained by comparing the data taken several years apart. Search of old aerial photographs at the archives of geodetic offices is recommended for analysis of shoreline changes. One technical problem in such analyses is the variation of the shoreline location with the tides. The tide level at the time of taking photographs needs to be calculated with the information of tidal constituents, and the shoreline at the mean water level is to be estimated with some assumption of the foreshore slope.

Sometimes aerial photographs reveal submarine topographic features such as the presence of longshore bars. They appear as zones or strips of light, brownish color, because of shallow depths over them. If longshore bars are detected on an aerial photograph, searches should be made on other photographs of the same location, taken at different seasons and/or years, to examine how they undergo changes from time to time.

### (D)  *Tide tables and nautical charts*

*Tide tables* are a compilation of the predicted heights and hours of daily high and low tides for one year. Tide tables of regional ports in the world can be purchased at hydrographic offices of respective countries. Tide tables of major ports in the world are also found in the publications of the British Admiralty, U.S. Navy Hydrographic Office, and other hydrographic agencies of major countries, because the tide information is vital for the navigational safety of international vessels, which call at various ports in the world.

Tide tables are indispensable for knowing the tide level of the site at the time when a coastal reconnaissance team visits a location of interest. *Tide curves* or the temporal variation of tide level at a particular date of reconnaissance must be prepared beforehand by plotting the heights of flood and ebb tides at the prescribed hours and drawing a smooth curve connecting them with the aid of personal computer. A book of tide tables is usually provided with additional information on the difference of the heights and hours of tides between the standard ports listed in the tide tables and adjacent harbors, for which tide predictions are not given. By using these tide differences, it is not difficult to construct a tide curve of the site for which a tide table is not available. Tide tables also give us the information on the spring and neap tidal ranges, even if they could not be obtained from other sources.

*Nautical charts* of small scales covering the area of several hundreds to a few thousands of kilometers are available on almost all ocean coasts at the hydrographic offices of respective countries and/or those of major countries. Charts of large scales covering an area of a few tens of kilometers have been prepared for many harbor and anchorage areas for the purpose of safe navigation, but they cover only a limited portion of coastlines.

Charts are analyzed for four items. The first is the general features of the coastlines. Together with analysis of topographic maps, we can get an idea on the topographic condition of the coastal area in concern before we actually inspect it in the field. The second is the general features of *submarine topography*. The mean beach (seabed) gradient can be calculated by reading the sounding (depth) data on the charts. The third is the *classification of sediment* on the seabed, which is marked at a number of locations on the charts for aiding ship captains to judge anchoring conditions; sediment classification is made with marks of mud, silt, fine sand, coarse sand, gravel, rock, etc. The fourth is the *tidal information*. The flood and ebb currents at spring tides are often shown with arrows for the directions and figures for their strength on the charts. The high and low water levels of spring and neap tides are listed as explanatory notes on the charts of large scales. The tidal ranges are calculated from these tide levels.

### (E) *Meteorological and hydrological data*

*Meteorological data* are not of absolute necessity for the analysis of coastal morphology, but they provide us with the background data on the area under study. *Temperature, rainfall,* and *winds* are the major elements to be reviewed for coastal reconnaissance. Meteorological disturbances such as storms and cyclones would directly affect the coastal area by generating large waves and storm surges. If these disturbances often affect the area under study, efforts must be made to collect the information related to them.

*Hydrological data* are related with those of rivers that supply the sediment to the coast under study. It is not easy to estimate the source of littoral sediment, but large rivers nearby are the first candidates of sediment supply. Because sediment can be transported alongshore over a distance of a few hundred kilometers, rivers at a far distance should not be omitted from examination if they are medium to large ones. The data most needed in the analysis of coastal morphology is the annual quantity of *sediment load* carried by a river to the sea. However, it will be difficult to obtain a reliable estimate of sediment quantity. *Flood discharge* is also needed for estimation of sediment quantity. The catchment area of the river, the maximum daily rainfall of the area, the gradient of riverbed and others are the basic data for estimation of flood discharge. Assistance of experts in hydrology should be sought for to obtain the hydrological information.

(F) *Worldwide wave data*

Waves are the principal driving force of sediment transport and beach deformation. However, waves are the environmental conditions that are most difficult to assess, because they are constantly changing in their shape, magnitude, and directions. Even a judgment of wave height at a beach is not easy if tried with bare eyes. The best data is from the records of wave measurements with special instruments such as wave buoys, bottom-mounted pressure gauges, ultrasonic wave meters (inverted echo-sounders), and others. However, such data usually become available only after a project is started and wave gauges are installed at the site. In most cases, a crude estimate must be made initially, based on scant information.

A promising source of wave data at a project site without any record is the databank of daily wave forecasts over the oceans undertaken by the meteorological or naval offices of several countries. Wave forecasts are made by means of directional spectral analysis over the grid spacing of 100 km or so. The spectral wave forecasting by the U.S. Navy Fleet Numerical Oceanographical Center was the earliest operation on global scales which began in the early 1970s. The European Centre for Medium-range Weather Forecast (ECMWF) has been active since the 1980s in making medium-range forecasting of weather, winds, and waves over the world. A databank of such forecasted waves can be utilized as a substitute to the instrumental wave records measured in the field, although the accuracy is inferior to the actual measurements. The databanks of forecasted waves can be accessed through the hands of meteorological and/or oceanographic consulting companies (Barstow *et al.*[40]). For beach morphology study, wave data every six hours over ten years would be sufficient to yield reliable wave climate information.

The merit of the databank of forecasted waves is the inclusion of wave direction, which enables calculation of direction-wise wave energy flux and estimation of longshore sediment transport rate after evaluation of wave transformation from the forecasted point to the project site at shore. Nevertheless, care should be exercised in accepting longshore sediment transport calculation based on forecasted wave data because of their inferior accuracy compared with instrumental wave data.

A check is better made for the capability of predicting long-traveled swell across the equator. Some of previous wave forecasting models in the 1990s could not make reliable prediction of long-traveled swell, though most recent versions must have solved the problem.

### (G) *Historical documents and old newspapers*

The usefulness of technical reports of any project at the site and/or in the neighborhood is so evident and it is unnecessary to be discussed here. If the project includes coastal disaster mitigation, then a search of historical documents and old newspaper articles becomes necessary.

Detailed information of coastal disasters does not survive over a time span of several decades. Only a slight memory of the past disasters may have been kept by senior persons among local residents. The events of great storms, storm surges, and/or tsunamis in the past are the subjects of investigation for any project in the coastal area. The frequencies and intensities of such events are the important design factors, because the safety of people living in the coastal area is threatened by these events. A search should be made on historical documents and old newspapers if such events have ever occurred at the locality and its neighborhood. A record of the calamity in the sea with a loss of many ships and lives onboard often indicates a passage of a great storm. An old newspaper article on the extent of seawater penetration onto the land by a storm or tsunami may reveal the height of storm surge or tsunami run-up, if the particular locations mentioned in the article are identified on the present topographic map.

### 14.4.2   *Field Inspection Works*

### (A) *Works and objectives of field inspection*

An expedition of field inspection is a major event of coastal reconnaissance. The items of works during an expedition of field inspection are as follows:

(1) Visual observation and sketching of overland and beach topography,
(2) Judgment of wave intensity with the elevation of vegetation and other features,
(3) Finding some clues on the direction of longshore sediment transport and its source,

(4) Sampling of sediment for sieve analysis in the laboratory, and
(5) Interview with local people for information on winds, waves, storm surges, and tsunamis.

A reconnaissance team is strongly advised to visit the coastal area under study by flying over in a small airplane or helicopter. The team will then have the best overall view of topographic features of the area. Sea and swell conditions may be judged by observing breakers near the shore. The volume of river discharge at the time of visit may be guessed by a spread of color of effluents. The flight should cover several hundred kilometers of coastline in both the left and right directions from the site of interest.

After a flight over the area, the team must traverse the coastline. A four-wheel drive car will make it easy to move along the beach if the beach face is hard enough. Inspection from the sea by renting a boat is also necessary, because sediment sampling from the seabed in the nearshore and the offshore is required in most cases. A field visit also provides the team with a handy opportunity to experience the traffic condition between the project site and major cities. It will help a planning of construction schedules of the project.

(B) *Tools and instruments of field inspection*

A field inspection team is advised to bring the following tools and instruments with them:

(a) Documents: maps, charts, tide curves specially prepared for the date of inspection, and results of documentary study.
(b) Survey tools: GPS receiver, surveyor's hand-level and poles, carpenter's rule, tape measure, clinometer with compass, etc.
(c) Instruments on ship: GPS receiver, a lead line for sounding (or portable echo sounder), small bottom sampler, etc.
(d) Recording equipment: digital camera, stationeries, surveyor's note-book, etc.
(e) Sampling tools: small shovels, several scores of small plastic bags, felt-tipped marking pens, etc.
(f) Others: a few pairs of binoculars, etc.

During a field inspection, a distance over land is better measured by counting the number of pairs of strides; timing of single steps does not match with the pace of counting, but counting the number of pairs of strides makes a harmonized rhythm with human breathing. One should make a practice of walking with equal stride and calibrate his stride by walking a known distance. For example, the author walks the distance of 1.5 m by his pair of strides.

### (C) *Inspection along the shoreline*

The reconnaissance team must visit many beaches along the coastline under study. Roads often run in the inland away from the shoreline, and the team may have to find out some side roads toward the shore. When there are hills and/or headlands overlooking the shore, their summits will provide the team with best places to overview the shoreline features. Members of a reconnaissance team are asked to take notes of the topographic features of beaches, such as their plan shapes, beach slopes, sediment classification, presence of sand dunes with their sizes, elevation of vegetations at the coast, and others.

Figure 14.4 is a definition sketch of beach profile, which was reproduced from Horikawa.[41] The *foreshore* or *beach face* is the zone from the mean low water to the upper limit of normal wave wash at high tide. Beyond the upper limit of wave wash, somewhat a flat portion may exist: it is called the berm. The zone between the end of foreshore and the coastline is called the *backshore*. The *coastline* is the boundary between the shore and the coastal land. Permanent vegetation, sand dunes, cliffs, etc. marks the edge of coastline. The foreshore is washed up and down by ordinary

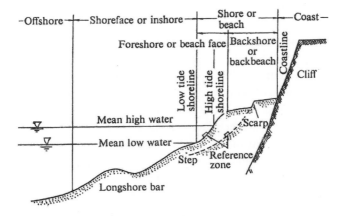

Fig. 14.4   Beach profile and definition of terminology (reproduced from Horikawa[41]).

waves, whereas the backshore is washed over only in the occasions of storm waves. The backshore may have multiple berms, a higher one having been formed by severe storm waves. A scarp or low steep slope may exist at edges of higher berms, having been cut by erosive power of large waves. The *inshore* or *shoreface* is the zone extending from the low water shoreline toward the outermost breaker. It is equivalent to the surf zone. Another term of *nearshore zone* refers to a zone from the shoreline well beyond the surf zone. The width of inshore varies with wave conditions and its outer edge is not easily defined because waves are random. The findings of field inspection must be recorded in a notebook. Table 14.2 is a sample format to be filled in during a field inspection.

During a field inspection, rough measurements of various elevations listed in Table 14.2 should be carried out by using a surveyor's hand-level or a GPS receiver. They are measured with the reference at the present shoreline. These elevations are to be converted to the heights from the datum level of the charts or the geodetic datum, by adding the tide level at the time of measurements. The tide curves prepared before the field inspection is used for this purpose.

### (D) *Sediment sampling during field inspection*

Samples of sediment are collected along the shoreline and in the inshore zone. They are later put to sieve analysis for examination of grain size distribution. The volume of each sample is about 200 to 400 cubic centimeters, or about 500 grams in mass. Each sediment sample is put into a small plastic bag, which is sealed and marked with the location number, the date of collection, and other information. Needless to say, sketches of beaches with marking of the sampling locations are an essential part of the sediment sampling process.

Along the shoreline, sediment at the beach face slightly above the waterline is to be sampled at an interval of several hundred meters. A "reference zone" marked in Fig. 14.4 is the place at which sediment samples are collected. Sediment in the beach face is sometimes deposited in thin layers of different grain sizes. On such a beach, samples from these layers are taken together by inserting a shovel into the beach face and collecting a vertical core of sediment with the thickness of a few tens of centimeters. If cusps are developed along a beach, sampling points must be fixed at either the apexes or the midpoint of cusps, because the sediment is relatively coarse at the former and fine at the latter.

Table 14.2. Sample format for recording observations during field inspection.

Beach: _____; Coast: _____;
Area: _____.
Date (day/month/year): _____/_____/_____/; Hour: _____;
Estimated tide level: _____ m.
Recorded by: _____

| Item | Remarks |
|---|---|
| Beach configuration | Shoreline shapes (tick appropriate boxes): Extension of beach: [ ] short, [ ] medium, or [ ] long. Curvature: [ ] much curved, [ ] slightly curved, or [ ] nearly straight.<br><br>Planar features (tick appropriate boxes):<br>[ ] cusps present: mean distance between cusps _____ m.<br>[ ] tombolo or salient present:<br>   Shoreline advanced at: [ ] left, [ ] right, or [ ] no difference.<br>[ ] sand bars / spit present at river or lagoon mouth:<br>   Shoreline advanced at: [ ] left, [ ] right, or [ ] no difference.<br>[ ] groins and/or jetties present:<br>   Shoreline advanced at: [ ] left, [ ] right, or [ ] no difference.<br>[ ] others (describe below)<br><br>_____<br>_____ |
| Beach profile | Tick appropriate boxes or fill in:<br>Beach is composed of: [ ] sand, [ ] shingles, [ ] rocks, or [ ] others<br>Sand size of beach face:<br>   At the surface:          [ ] fine, [ ] medium, or [ ] coarse.<br>   10 to 20 cm below the surface: [ ] fine, [ ] medium, or [ ] coarse.<br>Beach face slope: _____.<br>Elevation of crest of foreshore: _____ m.<br>Width of foreshore:          _____ m.<br>Width of backshore:          _____ m.<br>Deposit of driftwood or other drifted remnants: [ ] yes; [ ] no.<br>   If yes; distance from shoreline: _____ m; elevation: _____ m. |
| Coastal land features behind beach | Tick appropriate boxes or fill in:<br>[ ] cliff: elevation of its foot; _____ m.<br>[ ] trees and/or bushes: elevation of their seaward edge; _____ m.<br>[ ] sand dunes: foot elevation; _____ m, crest elevation; _____ m.<br>[ ] others (describe below):<br><br>_____<br>_____ |

In the inshore zone, sampling is made from a boat. Several traverse lines normal to the shoreline are set by erecting two poles per line on the backshore, and sediment samples are collected along these lines. A small bottom sampler of grab type is used, but a diver can also do it. The sampling locations are surveyed with a GPS receiver, and the water depths are measured with a sounding lead line. Furthermore, a few lateral survey lines following certain isobaths are preferably set for sediment sampling.

The boat ride also provides the reconnaissance team with an opportunity to view the coastal area from the sea. A pair of binoculars is a necessity for such inspection from the sea.

A sieve analysis of each sediment sample is made after the reconnaissance is finished and the samples are brought to a soil test laboratory. The *mean diameter* $d_m$ or the *median diameter* $d_{50}$ of sediment grains is calculated from the *cumulative grain size distribution*. A *uniformity coefficient* (or sorting coefficient) is also calculated as $S_0 = \sqrt{d_{75}/d_{25}}$, where $d_{75}$ and $d_{25}$ denote the grain diameter at the accumulation levels of 75% and 25%, respectively. An alternative method of grain size analysis is the use of the *phi scale*, which is converted from the particle diameter $d_{mm}$ in millimeters by the following equation:

$$\phi = -\log_2(d_{mm}) \qquad (14.1)$$

The mean diameter, the standard deviation of grain size, the skewness, and the kurtosis of the grain size distribution can be easily calculated with the phi scale. Larson *et al.*[42] as well as Dean and Dalrymple[43] provide detailed explanation on the use of the phi scale as well as sediment sampling and geotechnical methods.

In the coast where sediment is supplied from a fixed source such as a river mouth or a cliff being eroded, the mean or median diameter of sediment grains gradually decreases as the sediments are transported from the source, and the uniformity coefficient decreases toward the value of 1.0 (the grain size being uniform). By examining the variations in the diameter and the uniformity of sediment grains, we can sometimes determine the source of sediment supply and direction of sediment transport.

(E) *Interview with local people*

People living in the seaside and/or getting the livelihood on the sea have the best information on the general wind and wave conditions of the locality. The reconnaissance team should try to make appointments for interview with local fishermen who fish in small boats, crews of coastal lighters and barges, yachtsmen living there, and/or senior persons having been living for many years in the area. General conditions of winds and waves, shoreline changes by season and/or decades, and experience of coastal disasters are main items of questions.

Answers to the interview may not be consistent from one interviewee to another, and they may not be sufficiently quantitative. Nevertheless, the answers from local residents provide the reconnaissance team with valuable information in making an initial judgment on the environmental conditions of the coastal area under study.

### 14.4.3   Guidelines for Search of Sediment Supply and Longshore Transport Direction

(A) *Source of sediment supply*

One of the important objectives of coastal reconnaissance is to identify the source of sediment supply to the beach under study and to assess any change in the amount of sediment supply. A sandy beach keeps its position and profile under a dynamic balance of erosive and accretive forces as discussed in Sec. 14.2.3. Exception may be a pocket beach or embayment beach, which is at a dynamically equilibrium condition.

As listed in Sec. 14.2.2(C), there are five possible sources of sediment supply to a beach. *Submarine deposits* of sand must have contributed to formation of a long stretch of coastal plain when the land was emerged by the upheaval of seabed by crustal movements. Komar[44] cites the findings by Bowen and Inman[45] that the sediments of 76,000 cubic meters per year come from offshore sources to the littoral cell of Grover Beach at Arroyo Grande in the southern California. However, it seems to be a rare report of sediment supply from offshore. It is doubtful if offshore sand will be continuously transported to the shore under natural process of waves and currents in a decadal or centenary time span.

*Sediment discharge from rivers* is the primary source of sand to the majority of beaches as mentioned earlier. Sediment supply could have been continuing for a millennial time span. A search should be made of big rivers that have the capability of supplying a sufficient amount of sediments to the beach under study. The prevailing wave direction can indicate the direction in which a search of rivers is to be made. In the case of the Danube, the sediments from its mouth were transported southwestward over more than 180 km and formed barrier beaches, thus closing the entrance channels to Hellenic and Roman colonial ports as explained in Sec. 14.2.2(D).

Often several rivers supply sediments to a beach in concern. Alongshore distributions of the sediment size and the uniformity coefficient may provide a clue to the question which river is a main supplier of beach sand. Identification of the types of heavy minerals and their contents in the sediments may also reveal the sources of beach sand. A fluorescence X-ray analysis is also effective in revealing quantitative contents of various compounds such as $SiO_2$, $CaO$, $TiO_2$, and $MnO$. The analysis succeeded in differentiating terrestrial sand from the Danube and carbonate sand of biogenic origin[26,34] in the case of the Southern Romanian Black Sea shore.

*Collapse of cliffs* produces sediments to adjacent beaches. Prevailing wave direction will indicate the direction to which sediments from cliffs will be transported. The cliff retreat speed and the composition of cliff deposite layers are important in assessing the contribution of cliffs to sediment supply to beaches.

An adjacent beach across a headland can also supply sediments by littoral transport passing around the headland. A section of coast bounded by headlands or long jetties (or breakwaters) is called a *littoral cell*. The Coastal Engineering Manual[46] defines it as "A reach of the coast that is isolated sedimentologically from adjacent coastal reaches and that features its own sources and sinks. Isolation is typically caused by protruding headlands, submarine canyons, inlets, and some river mouths that prevent sediment from one cell to pass into the next." It is also called the *coastal sediment cell* by European Commission.[47] Large headlands may block littoral transport of sediments, but a certain portion of sediments can be transported beyond a headland. Thus, a search of sediment supply should be made beyond a littoral cell.

### (B) *Direction of longshore sediment transport*

Judgment of the *direction of longshore sediment transport* is sometimes easy because of persistent waves coming from a certain direction, but it is not an easy task in general. One reason is the fact that the direction may change from season to season depending on the prevailing direction of incident waves. Several observations at different seasons must be made before the direction of longshore sediment transport is judged. Collection and analysis of as many aerial photographs as possible will help to avoid the pitfall of judging the direction from the data of one season only.

Certain shoreline and coastline features can be utilized in judging the direction of longshore sediment transport. Goda and Tanaka[48] have prepared several diagrams for guiding judgment of the direction of longshore sediment transport. Figure 14.5 shows three patterns of micro-scale topography, which are clearly formed by a consistent longshore sediment transport. The open arrows indicate the direction of prevailing longshore sediment transport. For the meso-scale shoreline shapes, Fig. 14.6 will assist

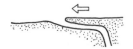

(a) Direction of a mouth of small river
or a tidal inlet

(b) Difference between the shoreline locations
at the both sides of a jetty or promontory

(c) Difference between the shoreline locations
at the both sides of a tombolo

Fig. 14.5   Judgment of longshore sediment transport direction based on micro-scale topography.[48]

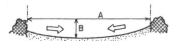

(a) Long beach with shallow depression facing
an open sea ( A >> B )

(b) Short beach at the head of deep cove facing
an open sea ( A ≤ B )

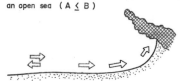

(c) Beach adjoining a long headland

(d) Beach adjoining a sand spit

Fig. 14.6   Judgment of longshore sediment transport direction based on meso-scale topography.[48]

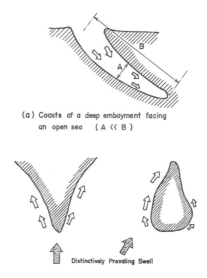

(a) Coasts of a deep embayment facing
an open sea    ( A ≪ B )

(b) Coasts with prevaling swell from one direction

Fig. 14.7   Judgment of longshore sediment transport direction based on macro-scale topography.[48]

one's judgment. A littoral cell of long beach shown in the upper panel (a) has a tendency of sediment moving toward the center, while a littoral cell of short beach shown in the second panel (b) indicates sediment movement away from the center. On the macro-scale of geography, a certain coast is subject to the intensive longshore sediment transport in one direction. Figure 14.7 exhibits such examples for which no explanation will be necessary.

## References

1. http://en.wikipedia/wiki/Central_American_Seaway.
2. T. M. Lenton, H. Held, E. Kreigler, J. W. Hall, W. Lucht, S. Rahmstoff and H. J. Schellnhuber, "Tipping elements in the Earth's climate system," PNAS (*Proc. Nat. Acad. Sci. USA*) **105**(6) (Feb. 12, 2008), pp. 1786–1793.
3. Chang Jiang Watershed Region Planning Division, *Hisotry of Water Management of Chang Jiang River*, (Kokin Shoten, Tokyo) (1992), 324p (*translated into Japanese from Chinese*).
4. Huang He Water Management Committee, *Flood Control and Development of Huang He River*, (Kokin Shoten, Tokyo) (1989), 254p (*translated into Japanese from Chinese*).

5. C. Bondar and N. Panin, "The Danube delta hydrologic database and modelling," *GEO-ECO-MARINA* No. (5-6)/(2000-2001), pp. 5–52.
6. A. Morang, L. T. Gorman, D. B. King and E. Meisburger, "Coastal classification and morphology," Chapter IV-2 of *Coastal Engineering Manual* EM 1110-2-1100, U.S. Corps of Engineers (2002), pp. IV-2-63–64.
7. A. Morang, L. T. Gorman, D. B. King and E. Meisburger, *loc. cit.*, pp. IV-2-1–3.
8. F. P. Shepard, *Submarine Geology* (3rd ed.) (Harper & Row) (1973).
9. R. Silvester and J. R. C. Hsu, *Coastal Stabilization* (World Scientific, Singapore) (1997), p. 206.
10. E. J. Anthony, "Beach-ridge development and sediment supply: examples from West Africa," *Marine Geology* **129** (1995), pp. 175–186.
11. W. F. Tanner, "Origin of beach ridges and swales," *Marine Geology* **129** (1995), pp. 149–161.
12. K. Jankaew, B. F. Atwater, Y. Sawai, M. Choowong, T. Charoentitirat, M. E. Martin and A. Predergast, "Medieval forewarning of the 2004 Indian Ocean tsunami in Thailand," *Nature* **455** (30 October 2008), pp. 1228–1231.
13. A. Morang, "Coastal terminology and geologic environments," Chapter IV-1 of *Coastal Engineering Manual* EM 1110-2-1100, U.S. Corps of Engineers (2002), pp. IV-1-30.
14. V. P. Zenkovich, *Process of Coastal Development* (English translation edited by J.A. Steers, Oliver and Boyd, Edinburgh and London) (1967), p. 164.
15. C. Dean, *Against the Tide* (Columbia Univ. Press, New York) (1999), p. 18.
16. C. Dean, *loc. cit.* 15, p. 33.
17. C. Dean, *loc. cit.* 15, p. 181–185.
18. K. Horikawa, *Coastal Engineering* (University of Tokyo Press, Tokyo) (1978), pp. 239–242.
19. O. Toyoshima, *Coastal Engineering for Field Engineers: Volume of Erosional Problems* (Morikita Pub., Tokyo) (1972), pp. 43–51 (*in Japanese*).
20. D. A. Smith, P. S. Warner, R. M. Sorensen and L. A. Nurse, "Development of a sediment budget for the west and southwest coasts of Barbados," *Proc. Coastal Sediments '99* (Long Island, N.Y., ASCE) (1999), pp. 818–827.
21. S. Onaka, T. Uda, T. San-nami and K. Kobune, "Coastal sediment balance under condition of existence of large-scale sand dunes in Northeastern Brazil," *Annual J. Civil Engrg. in the Ocean, JSCE* **22** (2006), pp. 457–462 (*in Japanese*).
22. J. G. P. Rodriguez and K. Katoh, "Control of littoral drift in Caldera Port, Costa Rica," *Proc. Hyro-Port '94* (Port and Harbour Res. Inst., Yokosuka) (1994), pp. 1019–1040.
23. P. Chandramohan, B. U. Nayak and V. S. Raju, "Longshore-transport model for South India and Sri Lankan Coasts," *J. Waterway, Port, Coastal and Ocean Engrg.* **116**(4) (1990), pp. 408–424.
24. A. Perlin and E. Kit, "Longshore sediment transport on Mediterranean Coast of Israel," *J. Waterway, Port, Coastal and Ocean Engrg.* **125**(2) (1999), pp. 80–87.

25. S. Sato and N. Tanaka, "Field investigation on sand drift at Port Kashima facing the Pacific Ocean," *Proc. 10th Int. Conf. Coastal Engrg.* (Tokyo, ASCE) (1966).

26. K. Kuroki, Y. Goda, N. Panin, A. Stanica, D. I. Diaconeasa and G. Babu, "Beach erosion and coastal protection plan along the Southern Romanian Black Sea shore," *Coastal Engineering 2006 (Proc. 30th Int. Conf.*, San Diego, California) (2007) pp. 3788–3799.

27. J. D. Rosati, M. B. Gravens and W. G. Smith, "Regional sediment budget for Fire Island to Montauk Point, New York, USA," *Proc. Coastal Sediments '99* (Long Island, N.Y., ASCE) (1999), pp. 802–817.

28. K. Kuroki, D. Obata, K. Chikagawa and T. Takano, "Research about a littoral transport trend of North Niigata Coast," *Proc. Coastal Engrg., JSCE* **49** (2002), pp. 536–540 (*in Japanese*).

29. S. Ogura, K. Uno, N. Sugiyama, J. Kikuchi, A. Katano and M. Hattori, "Budget of littoral sand along the Suruga-Bay Coast — Aerial photograph analyses," *Proc. Coastal Engrg., JSCE* **49** (2002), pp. 546–550 (*in Japanese*).

30. J. Kunieda, M. Iino, Y. Oishi, H. Sasaki, M. Sakurada and Y. Kurata, "Sedimentation and sediment balance of Suruga Coast," *Proc. Coastal Engrg., JSCE* **49** (2002), pp. 551–555 (*in Japanese*).

31. O. E. Frihy, A. M. Fanos, A. A. Khafagy and P. D. Komar, "Patterns of nearshore sediment transport along the Nile Delta, Egypt," *Coastal Engineering* **15** (1991), pp. 409–429.

32. P. D. Komar, *Beach Processes and Sedimentation*(Second Ed.) (Prentice Hall, Inc.) (1997), p. 71 and 433–435.

33. N. M. Ismail and W. El-Sayed, "Coastal processes at Rosetta Headland and seawall–beach interaction," *Coastal Structures 2007 (Proc. 5th Int. Conf.*, Venice, World Scientific) (2009), pp. 490–501.

34. ECOH CORPORATION, *The Study on Protection and Rehabilitation of the Southern Romanian Black Sea Shore in Romania* (Japan International Co-operation Agency (JICA) Report 07-030, Project ID 7241013F0) (2007), **I**, pp. 4-10∼4-14.

35. M. Toti, P. Cuccioletta and A. Ferrante, "Beach nourishment at Lido di Oscia (Rome)," *Proc. 27th Int. Navigation Congress* (Tokyo, PIANC) (1990), S.II–2-1, pp. 23–28.

36. O. Toyoshima, *loc. cit.* 19, pp. 18–23.

37. O. Toyoshima, *loc. cit.* 19, p. 9.

38. Madras Port Trust, *The Port of Madras: Past, Present and Future* (March 1967), 52p.

39. N. Tanaka, "A study on characteristics of littoral drift along the coast of Japan and topographic change related from construction of harbors on sandy beaches," *Tech. Note of Port and Harbour Res. Institute* No. 453 (1983), 148p. (*in Japanese*).

40. S. Barstow, G. Mork, L. Lonseth, P. Schjolberg, U. Machado, G. Athanassoulis, K. Belibassakis, T. Gerostathis and G. Spaan, "World waves: High quality coastal and offshore wave data within minutes for any global site," *Book of Abstracts, COPEDEC 2003* (Colombo, Sri Lanka) (2003), Paper No. 102 (CD-ROM).

41. K. Horikawa, *Coastal Engineering* (University of Tokyo Press, Tokyo) (1978), 402p.
42. R. Larson, A. Morang and L. Gorman, "Monitoring the coastal environment; Part II: Sediment sampling and geotechnical methods," *J. Coastal Res.* **13**(2) (1997), pp. 308–330.
43. R. G. Dean and R. A. Dalrymple, *Coastal Processes with Engineering Applications* (Cambridge Univ. Press) (2002), pp. 21–27.
44. P. D. Komar, *loc. cit.* 32, pp. 70–71.
45. A. J. Bowen and D. L. Inman, "Budget of littoral sands in the vicinity of Point Arguello, California," *U.S. Army Coastal Engineering Research Center Tech. Memo* No. 19 (1966).
46. A. Morang and A. Szuwalski (editors), "Glossary of coastal terminology," Appendix A of *Coastal Engineering Manual* EM 1110-2-1100, U.S. Corps of Engineers (2002), pp. A-45.
47. DGENV European Commission, "Development of a Guidance Document on Strategic Environment Assessment (SEA) and Coastal Erosion," (2004), p. 14.
48. Y. Goda and N. Tanaka, "Guidelines for coastal reconnaissance for shore protection and coastal development," *Proc. 4th Coastal and Port Engineering for Developing Countries* (*COPEDC IV*, Rio de Janeiro, Brazil) (1995), pp. 1834-1849.

## Chapter 15

# Prediction and Control of Shoreline Evolution

## 15.1 Introduction

The process of wave transformation and deformation as well as the wave actions on structures can be analyzed for design purposes with an error range of ±25% or so. It may be said that they have reached to a saturation level of reliability. Prediction of beach morphological changes, however, has not attained such a level of reliability. Difference by one order of magnitude between prediction and reality is not a rare case. Thus, the methodology of beach morphological prediction is at the level of the state-of-the-art presently.

There are many reasons, which will be discussed in Sec. 15.2.2. The main reason from the author's viewpoint is the deficiency in our understanding of the process of sediment suspension by strong turbulence associated with wave breaking within the surf zone. Because of the complexity of this process, most of coastal engineering researchers have avoided tackling the problem of sediment suspension by breaking waves.

The present chapter begins with an overview of the studies on sediment motion, which began in the late 1940s. Then the problems inherent to beach morphological prediction are discussed for explanation of its low reliability. Next viewed is the present state of prediction of sediment suspension rate in the surf zone, though it is at an infancy stage yet. An overview is also given on various prediction formulas of longshore sediment transport and beach morphology models. The purpose of presenting an overview is not to recommend them for engineering applications but rather to indicate their limitations so that a coastal manager or engineer will not rely upon them blindingly. An example of approach for incorporation of wave randomness and wave climate variability in beach morphological prediction is presented for a quantitative assessment of sediment impoundment by a groin.

Although many beach morphology models need fine tuning of several parameters by specialists, the so-called one-line model has been a working tool of coastal engineers. A case study with the one-line model is presented to emphasize the importance of field calibration data for its appropriate applications. Lastly various shore protection facilities are described with some comments on them. Their design aspects are not dealt with in this chapter, because they belong within the category of engineering practice.

The present chapter intentionally omits discussion of the cross-shore sediment transport and seasonal changes in beach profiles. They have been the subjects of active research works over years and are certainly interesting topics. Excluding special circumstances, however, cross-shore changes are seasonal and natural beaches return to their equilibrium state on the annual basis. What contributes to the voluminous accretion and/or erosion of a beach is the longshore sediment transport. The present chapter focuses on a medium- to long-term shoreline changes, thus without dealing with the cross-shore morphological changes.

## 15.2   State of the Art of Beach Morphological Prediction

### 15.2.1   *Overview of Studies on Sediment Movement*

(A) *Early studies with the methodology of river bedload transport*

The problem of beach erosion occurs at many places in the world from time to time, but American people seem to have the keenest concern on conservation of beaches for summer recreation. Upon the demand of many people, the U.S. Army Corps of Engineers created the Beach Erosion Board in 1930 to deal with severe beach erosion problems along the American coastline. After World War II, the U.S. Navy also found interests in wave-related problems. For one of the projects commissioned to the University of California at Berkeley, Einstein[1] gave a suggestion for the direction of approach to be taken for the study on beach sand movements by waves, saying that the concept of river sediment transport can be extended to the study of beach sand movements even though a river flows in one direction while beach sands are exposed to oscillatory flows of wave orbital motions.

Einstein had established his famous formula on bedload transport in rivers in 1942. His formula was based on a statistical concept with dimensional analysis. Later Kalinske[2] augmented the Einstein formula by giving more physical reasoning. Both of them had recognized the necessity to include the suspended load in the total transport, but they analyzed the bed load transport only, probably because of the difficulty in deriving any generalized conclusion on the amount of suspended load.

In the U.K., Bagnold[3] initiated investigations on the motion of sand by waves and presented a paper in 1946; he had been famous for his experiments on impulsive breaking wave force on a vertical wall as well as his pioneering work on aeolian sand transport. He immersed a swinging, arc-shaped cradle, hanged from the ceiling, in a water tank. By changing the amplitude and frequency of swinging motion of the cradle, he created oscillating water flows relative to the surface of cradle. Bagnold then placed a layer of sand on the cradle and observed the motion of sand particles by oscillating water flows. The threshold velocities for initiation of particle motions, formation of sand ripples, and suspension of sand particles from sand ripples were thus measured and recorded. Another method of creating oscillating flows relative to a sand layer is to reciprocate a horizontal plate with sand on top of it in a water tank through a piston-motion mechanism. This method was used by Manohar[4] in 1955 for the study of threshold velocities of sand motions in a manner similar to Bagnold's study.

An early laboratory study of sand movement by waves was reported by Scott[5] in 1954, who carried out many tests at the University of California at Berkeley. A few years later, a group of researchers at the Massachusetts Institute of Technology began a series of detailed measurements on the wave-induced motion of particles on slopes. Eagleson and Dean[6] summarized their findings on the vertical distribution of net velocity near the bed, the drag coefficient of sediment particles on bed, and other aspects.

## (B) *Incipient motions of sediment particles by oscillating flows*

Studies on sand particle motions in the early stage stimulated interests of researchers on the incipient motions of sediment particles by oscillating flows. To facilitate the laboratory experiments, Lundgren and Sorensen[7] invented a pulsating water tunnel in 1957, which later became favorite equipment in many coastal engineering laboratories in the world.

The threshold of incipient motions of sediment particles is mostly expressed in terms of the Shields number, which normalizes the critical shear stress for driving particles in reciprocating motions. The so-called Shields curve, i.e., plotting of the threshold Shields number against the Reynolds number (defined by the shear velocity and the particle diameter), has been applied by many researchers to their laboratory data. The *Shields number* is defined below.

$$\Psi = \frac{\tau_b}{(\rho_s - \rho)gd} = \frac{1}{2}\frac{fu_b^2}{(\rho_s/\rho - 1)gd}, \tag{15.1}$$

where $\tau_b$ denotes the bottom shear stress, $f$ is the friction coefficient, $u_b$ stands for the flow velocity at bottom, $\rho_s$ and $\rho$ are the density of sediment and water, respectively, and $d$ is the diameter of sediment grain.

The concept of Shields curve does not have unanimous support, however. Hallermeier[8] analyzed ten sets of laboratory data by various investigators in 1980 and concluded that the Shields curve is not suitable for describing the threshold of incipient sediment motions by oscillating flows. Instead, he proposed two empirical formulas for the critical orbital velocity, one for fine grains under high-frequency oscillations and the other for coarse grains under relatively low-frequency oscillations. In 2000, You[9] also discussed the threshold wave orbital velocity for initiation of sand particle motions in a similar fashion.

### (C) *Sand ripples and sheet flow*

As waves propagate onshore, orbital velocities of water particles by waves gradually increases and sand ripples appear on the seabed owing to enhanced bottom shear stress. With further increase in orbital velocities and shear stress, sediment particles are put into suspension from the crests of sand ripples. Suspended sediment particles are lifted up by turbulence and then settle down by gravity. Sediment concentration is densest near the seabed and thinnest near the surface. The vertical distribution of the time-averaged sediment concentration is determined by the sediment pick-up rate (or the reference concentration at the bottom), the strength of turbulent eddy viscosity, and the fall velocity of sediment. As waves enter the surf zone, intensive turbulence by breaking generates large-scale suspension of sediments and sand ripples disappear. Sediments of a certain thickness near the bed move as a whole by wave actions; the phenomenon is called *sheet flow*. Figure 15.1 is an overall sketch of wave motion and sediment movement from the offshore to the foreshore.

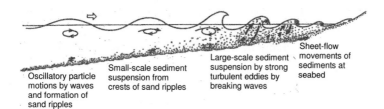

Sheet-flow
movements of
Large-scale sediment \ sediments at
Small-scale sediment suspension by strong seabed
Oscillatory particle suspension from turbulent eddies by
motions by waves crests of sand ripples breaking waves
and formation of
sand ripples

Fig. 15.1   Schematic representation of wave motion and sediment movement from off-shore to foreshore.

(D) *Suspension of sediment particles*

There have been many studies on the concentration of sediment suspended by wave actions; some made detailed experiments in oscillating water tunnels as well as in wave flumes, while others measured sediment concentrations in the sea. In 1972, Das[10] gave a review of early studies on suspended sediment, the amount of which was mostly measured from water samples taken by suction. Sleath[11] is one of earlier investigators who made use of phototransistors to measure the concentration of sediment in 1982. Nowadays, many optical backscatter sensors and acoustic sensors are employed to monitor temporal variations of sediment concentrations, even though some researchers favor the method of water sampling by suction.

Outside the surf zone, the suspended sediment concentration near the bed is in the order of 1 $g/l$ and decreases rapidly to the order of 0.001 $g/l$ as the height from the bed increases. Nielsen[12] gave a formula for the concentration at the sea bottom as being proportional to the cube of a Shields parameter in 1986. The vertical distribution of suspended sediment concentration in waves can be calculated by solving the diffusion equation. The solution generally yields an exponential decay function toward the water surface.

Field measurements of suspended sediment concentration have been made in many countries since the 1950s. Among them, Fairchild[13] succeeded in obtaining more than 800 data in the surf zone at various elevations from the seabed to mid-depth. He moved a tractor-mounted pump sampler along a pier extending from the shore, utilizing one pier at Ventnor in New Jersey and another at Nags Head in North Carolina. The maximum concentrations ranged up to 2.6 $g/l$ at Ventnor and 4.0 $g/l$ at Nags Head. The median diameter of suspended sediment was in the range of 0.12 to 0.20 mm. While the data by Fairchild gave an overall view of suspended sediment in the surf zone, they were not well correlated with wave data and no simultaneous current measurements were taken.

Since the 1970s, many field campaigns have been conducted in a more systematic way. In 1980, Kana and Ward[14] reported the first systematic measurements of sediment concentrations and current velocities along the CERC pier at Duck. They commented that "the sediment concentrations ranged over 3-1/2 orders of magnitude from approximately 0.05 g/$l$ to over 10.0 g/$l$ with highest concentrations in the inner surf zone and near the bed." With introduction of optical backscatter sensors, it became clear that bursts of suspended sediment occur intermittently with the peak concentration more than 20 g/$l$. In 1988, Beach and Sternberg[15] reported that such suspension events are correlated with infragravity waves of 30 to 300 s in period. In the same year, Nadaoka *et al.*[16] attributed suspension events to the three-dimensional large-scale eddies that were produced by plunging breakers. Nadaoka *et al.*[17] further reported field measurements of such eddies and their relationship with sediment suspension. Jaffe and Sallenger[18] observed that suspension events occurred every 1 to 2 minutes with the concentrations being one order of magnitude greater than the rest of time. Although the suspension events occupied less than 10% of the observation time, they contributed to the increase of mean concentration by 15% to 55%. Black *et al.*[19] presented several records of suspended sediment of high-density clouds (on the order of 10 g/$l$) with a description of how they moved back and forth around the sensors in relation with the wave motions. Miller[20] also demonstrated records of suspension events with high-concentration sediment reaching near the water surface; some individual suspension event showed concentrations of 180 g/$l$.

### 15.2.2 *Problems Inherent to Beach Morphological Prediction*

Maritime structures such as breakwaters and seawalls are designed to withstand a certain set of predetermined environmental conditions such as the 100-year storm wave height and surge level and to protect the facilities behind them. Even though determination of the design conditions is an intricate procedure by itself, design is made for only one or a few storm situations. Design methodology has usually been established and little ambiguity remains in the design process.

Contrary to structural design, prediction of beach morphological change involves many inherent problems as listed below.

(i) Not a single event but accumulated outcome of successive beach deformations over a long time span needs to be taken into account,
(ii) Necessity to deal with ever-changing wave conditions,
(iii) Difficulty in incorporating random wave concepts,
(iv) Limitation of laboratory test data because of scale effects,
(v) Difficulty in obtaining high quality field data, and
(vi) Dominance of suspended load during storm events.

*Accumulated outcome of beach deformations*: Beach deformation by a single severe storm may be problematic, but we are usually concerned with a long-term beach deformation, which is the accumulated outcome of beach deformations by a number of waves and swell events. Because a beach is subject to temporary accretions and erosions depending on season and wave conditions, it is difficult to specify a single wave condition for beach morphological prediction. This is the first problem.

*Ever-changing wave conditions*: The above problem is associated with the nature of wave conditions that change continually. Long-term statistics of waves at a given site is called the *wave climate*. The joint frequency tables of significant wave height, period, and directions are the main components of the wave climate statistics. Prediction of beach morphology must be based on the analysis of wave climate for morphological prediction. In principle, daily wave conditions should be input into a beach morphology model and daily changes in beach profiles and bathymetry should be computed. In reality, however, such a procedure requires too excessive computation and is not usually adopted. The prediction performance of morphological models has not reached to the level high enough to yield reliable outcome for such a procedure. The currently favored method is to evaluate the *energy averaged wave heights and periods* for several representative directions and use them as the input for a beach morphological model. An example will be presented in Sec. 15.4.

*Incorporation of random wave concept*: As seen in the overview of the studies on sediment movement in Sec. 15.2.1, beach morphological studies have been pursued with the monochromatic wave concept. Many laboratory tests were conducted with regular waves and theoretical analyses were based on the oscillatory motions of water particles by regular waves. The results have rarely been extended to the level of random waves. The field measurement data were often correlated with single breaker height and period, which were assessed by visual observations. A random wave approach would yield a prediction different from the regular wave approach, but few researchers have tried to clarify the difference between the predictions by regular and random waves. One reason might be the low reliability of quantitative prediction of sediment transport presently; one-half order of difference (0.3 to 3.0 times)[21] between prediction and observation is common as will be discussed in Sec. 15.2.3.

*Limitation of laboratory test data*: Laboratory model tests on sediment transport are prone to scale effects. If medium sand of 40 $\mu$m is scaled down with the length scale of 1/10, the model particle has the diameter of 4 $\mu$m which is silt rather than sand. One of the governing parameters on sediment motion is the *Dean number*[22] of the following:

$$D = \frac{H_b}{w_f \, T} \, ,\qquad\qquad (15.2)$$

where $H_b$ is the breaking wave height, $T$ is the wave period, and $w_f$ denotes the fall velocity of the sediment particle. The sediment fall velocity increases approximately linearly with the increase in diameter as shown in Fig. 15.2 in Sec. 15.2.4. Unless some light-density grains are used as model sediments, the equality of the Dean number cannot be maintained between the prototype and the model tests in small to medium size wave basins. Distortion in the Dean number in the model causes enhancement of bedload and suppression of suspended load. Thus laboratory test results on sediment transport are mostly biased toward a priority to the role of bedload.

*High quality field data*: Field measurement data should replace inadequate laboratory test data, but acquirement of high quality field data is very difficult because of the enormous requirements for human resources, instrumentation, and expenditure. A number of field campaigns have been carried out around the CERC pier (Field Research Facility) at Duck, North Carolina in the United States, but the scope of the data is still limited. There have been many field data measured on nearshore currents and sediment transport accumulated since the 1950s, but most of the data are sporadic observations with a few visual observations of waves and currents. When field data are compared with various formulas concerning sediment transport, they generally exhibit wide ranges of scatter and are incapable of differentiating between these formulas for their performance.

*Dominance of suspended transport*: Sediment transport takes two modes of bedload and suspended load. Komar[23] makes a detailed review of field measurement reports and provides a table of the ratio of suspended

Table 15.1   Tentative assessment of the roles of longshore sediment transport modes.

| Mode | Description | Present State of Understanding | Total Contribution |
|---|---|---|---|
| Bedload | Particles | Almost known | Small |
| | Sheet flow | Somewhat known | Medium |
| Suspension | From crests of sand ripples | Almost known | Little |
| | By turbulence of breakers | Unknown | Dominant |

transport to the total longshore sediment transport. The ratio varies from 0.07 to 1.0 depending on researchers who analyzed the data. Komar states that "the question of the relative role of suspension versus bedload transport on the beaches must remain open and thus in need of additional research."

However, the present author believes in the dominance of suspended transport in storm conditions in which major beach deformations take place. It is based on his personal observation at the Hazaki Oceangraphical Research Station (HORS) in Ibaraki Prefecture, Japan. HORS is an open pier extending 400 m from the beach of Hazaki. While standing on the pier, one can observe a burst of sediment cloud rising up in the water after a plunging breaker of 2 to 3 m high moves out toward the beach. The sediment cloud slowly moves alongshore within the surf zone and does not disappear for several tens of seconds. In another occasion, the author watched from an airplane a spread of brownish color throughout the surf zone over a length of a few kilometers along a straight coast in Espírito Santo State, Brazil. The zone of brownish color moved along a long breakwater at the down-drift side of the coast and spread beyond its head, indicating down-drift transport of suspended sediments. Table 15.1 is the author's subjective assessment of the roles of various longshore transport modes of sediment.

The driving force of sediment suspension by turbulence of breaking waves is the three-dimensional large-scale eddies as revealed by Nadaoka *et al.*[16,17] The mechanism is qualitatively understood, but its quantitative description and prediction remain as the subjects of further study.

Table 15.1 is for the longshore sediment transport. For the cross-shore sediment transport, the bedload will have a certain share in the total transport as exemplified in the occasion of beach profile recovery by waves of low steepness after a storm event.

### 15.2.3    *Sediment Suspension Rate in Surf Zone*

(A) *Role of Shields parameter in the surf zone*

In coastal engineering studies it has been customary to relate sediment motions with the Shields number as a governing parameter. As expressed in Eq. (15.1), the Shields number represents a nondimensional bottom shear stress and is proportional to the square of the flow velocity at bottom. The tradition goes back to the 1950s when the incipient motion of sediment was a focus of discussion. The Shields parameter has been employed to describe the threshold condition of incipient sediment motion. The bedload transport rate has often been expressed as a function of the excess Shields number, i.e., the difference between the Shields number under a given wave and current condition and the threshold value for incipient motion.

The Shields parameter may be appropriate in describing sediment suspension from the crests of sand ripples, because the process is governed by the flow velocity over the ripple crests. However, the horizontal flow velocity near the seabed contributes little to sediment suspension by breaking waves. Kos'yan et al.[24] made a series of field measurements of suspended sediment concentration at Norderney Coast in the German North Sea and other places. In their report in 2001 they stated that "in the surf zone the intensive suspension of sediment from the bottom is determined by the macro-scale turbulence which is generated by breaking waves" and "a high correlation between the sediment concentration and the turbulent kinetic energy is observed." They also said that "it is doubtful to use a definition of the concept of the reference concentration,[a] because the sediment concentration is weakly coherent to the time varying cross-shore velocity."

Weak correlation of suspended sediment concentration with the flow drag or shear stress near the seabed has also been confirmed by van Thiel de Vries et al.[25,26] through large-scale flume experiments with waves of 1.5 m in the significant height in 2008. They found strong correlation between the time and depth averaged sediment concentrations and the maximum surface slope multiplied with an additional factor representing turbulence energy decay over depth.

Another indication of inappropriateness of the Shields parameter for quantifying sediment suspension by breaking waves is its cross-shore variation. According to a model computation by Shimizu et al.,[27] the Shields

---

[a]This refers to the sediment concentration near the bed in the sediment concentration profile, which exponentially decays toward the surface. The reference concentration is usually expressed as a function of the Shields parameter.

number takes the largest value at the water depth of 1.5 to 2.5 times the offshore significant wave height when random waves with a period of 10 s are incident on a planar beach with a slope of 1/100 and composed of sand grains with a diameter of 0.25 mm. As indicated in Fig. 3.42 in Sec. 3.6.3(B), the above water depth corresponds to the outer edge of the surf zone where only several percent of high waves begin to break. Intensive wave breaking occurs in much shallower waters where the Shields number becomes small.

From these reports and other studies, it can be said that any formula for suspended sediment transport within the surf zone expressed with the Shields parameter does not have physical background and its success in application to a prototype situation solely depends on empirical tuning of parameters to respective conditions.

## (B) *Approaches to sediment suspension rate in the surf zone*

A heuristic approach to the problem of sediment suspension is to correlate it with the rate of wave energy dissipation in the surf zone. As one of early efforts, the work of Katayama and Goda[28-30] can be cited. They proposed a 2DH model of beach morphological change with the depth averaged sediment concentration $\overline{C}$. Suspended sediments settle down with the fall velocity inherent to their grain size. A steady state concentration can only be maintained with new suspension of the sediments toward the water surface, the amount of which is equal to that of sediments settling down to the bottom per unit time. The work required to elevate sediments to the surface is expressed as

$$dW = (\rho_s - \rho)\,g w_f\,\overline{C}(x)h dx\,, \qquad (15.3)$$

where $\rho_s$ and $\rho$ is the density of sediment and water, respectively, $g$ is the acceleration of gravity, $w_f$ is the fall velocity, $\overline{C}(x)$ is the average sediment concentration at the distance $x$ fron the shoreline, and $h$ is the water depth. The rate of sediment suspension is equal to the rate of sediment settlement, i.e., $w_f\,\overline{C}(x)$.

The work for sediment suspension must be borne by breaking waves. By assuming that a part of the dissipating rate of wave energy flux by breaking is converted to the sediment suspension work, the average sediment concentration is given by

$$\overline{C}(x) = \frac{\beta_s}{8\,(s_r - 1)\,w_f\,h}\frac{\partial}{\partial x}\left(H_{\text{rms}}^2\,c_g\right) \quad : \quad s_r = \rho_s/\rho\,, \qquad (15.4)$$

where $\beta_s$ is called a *suspension coefficient*, $H_{rms}$ is the root-mean-square wave height, and $c_g$ is the group velocity. Upon calibration with the field data then available, the constant $\beta_s$ was found to have the value of 0.01 to 0.10. The dissipation rate of wave energy flux is represented by the spatial gradient of $H_{rms}^2 c_g$, which can be predicted by any random wave deformation model such as PEGBIS.[31]

For quantifying the constant $\beta_s$, Katayama and Goda assumed that the CERC formula of longshore sediment transport (to be discussed in Sec. 15.3.1) wholly represents the suspended sediment transport and they made use of the formulas of longshore current cross-shore profiles by Goda and Watanabe.[32] Katayama and Goda derived an empirical expression as below.

$$\beta_s \simeq 0.76 \frac{w_f K}{g T_{1/3}} \left(s^{-1} + 18 s^{0.4}\right) (H_0/L_0)^{-0.43}, \qquad (15.5)$$

where $K$ is the constant of the CERC formula, $T_{1/3}$ is the significant wave period, $s$ is the beach slope, and $H_0/L_0$ stands for the deepwater wave steepness.

For developing a 1DH morphological model, Kobayashi and Johnson[33] assumed that the sediment suspension rate $S$ is proportional to the local wave energy dissipation rates $D_B$ and $D_f$ per unit horizontal area due to wave breaking and bottom friction, respectively. Their formulation for suspension rate is as follows:

$$S = \frac{e_B D_B + e_f D_f}{\rho g (s - 1) h}, \qquad (15.6)$$

where $e_B$ and $e_f$ are the empirical suspension efficiencies for $D_B$ and $D_f$, respectively. They used the value of $e_B = 0.002$, 0.005 and 0.01 and $e_f = 0.01$ when they calibrated their model with regular wave data. In a further development by Kobayashi *et al.*,[34] the suspended sediment volume $V_s$ per unit horizontal area is estimated by $V_s = P_s S h / w_f$, where $P_s$ is the probability of sediment movement.

A more detailed analysis of sediment suspension model has recently been presented by van Thiel de Vries.[25] He first solves a 2D random wave deformation model with breaking and surface rollers, which is a time-dependent version of the wave action balance equation (cf. Eq. (12.1) in Sec. 12.2). Then the wave averaged turbulence energy $\overline{k_s}$ is assessed by

$$\overline{k_s} = (D_r/\rho)^{2/3}, \qquad (15.7)$$

where $D_r$ denotes the dissipation rate of roller energy, which is obtained from the wave transformation model. The near-bed turbulence intensity $k_b(t)$ is obtained by applying an exponential turbulence decay model as

$$k_b(t) = \frac{k_s(t)}{\exp[h/L_{\text{mix}}] - 1} \ : \ L_{\text{mix}} \simeq 0.95 H_{\text{rms}}. \tag{15.8}$$

The near-bed turbulence intensity $k_b(t)$ is either treated as the wave averaged value of $\overline{k_s}$ or the wave varying one.

The effect of wave-breaking-induced turbulence is incorporated in the near-bed orbital velocity by the following formula:

$$u_{\text{rms},2} = \sqrt{u_{\text{rms}}^2 + 1.45 k_b}, \tag{15.9}$$

where $u_{\text{rms}}$ denotes the root-mean-square horizontal orbital velocity at the bed.

With this modified horizontal velocity, van Thiel de Vries applies the equivalent sediment concentration formulation by Van Rijn[35] and calculates the depth averaged sediment concentrations. His simulation results agreed well with the measured concentrations which varied from about 0.2 to 40 g/l in water of 1.4 to 0.2 m deep.

### 15.2.4 Fall Velocity and Equilibrium Beach Profile

(A) *Fall velocity*

In dealing with beach morphology, several fundamental quantities and concepts are useful. The Shields number in Eq. (15.1) and the Dean number in Eq. (15.2) are examples. The fall velocity of sediment also appears in various formulations as in Eq. (15.3). It was first formulated by Rubey[36] and has been used by many engineers. Ahrens[37] rewrote it in the following expression:

$$w_f = \frac{\nu}{d}\left[\sqrt{6^2 + \frac{2}{3}A} - 6\right] \ : \ A = \frac{(s_r - 1)g d^3}{\nu^2}, \tag{15.10}$$

where $\nu$ denotes the kinematic viscosity of water, $d$ is the characteristic sediment particle diameter, and $s_r$ is the density ratio of $\rho_s/\rho$.

After comparison with various settling test data and prediction formulas, Ahrens[37] recommends the following:

$$w_f = \frac{\nu}{d}\left[\sqrt{3.61^2 + 1.18 A^{1/1.53}} - 3.61\right]^{1.53}. \tag{15.11}$$

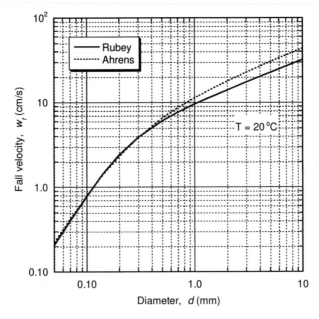

Fig. 15.2   Fall velocity of sand particles by Rubey's and Ahrens' formulas (temperature of $\breve{T} = 20°\mathrm{C}$ and sand density of $\rho_s = 2.65$ g/cm$^3$).

The kinematic viscosity and density of seawater vary with temperature. Ahrens[37,38] has given the following approximate calculation formulas:

$$\rho = 1.028043 - 0.0000721\breve{T} - 0.00000471\breve{T}^2 \;:\; [\mathrm{g/cm}^3], \quad (15.12)$$

$$\nu = 0.0182 - 0.000529\breve{T} + 0.0000069\breve{T}^2 \;:\; [\mathrm{cm}^2/\mathrm{s}], \quad (15.13)$$

where $\breve{T}$ denotes the temperature in degrees Celsius. For example, the density and kinematic viscosity of seawater at the temperature $\breve{T} = 20°\mathrm{C}$ are $\rho = 1.0247$ g/cm$^3$ and $\nu = 0.0104$ cm$^2$/s.

The fall velocity calculated by Rubey's formula (Eq. 15.9) is compared with that by Ahrens' one (Eq. 15.10) in Fig. 15.2. For a sediment particle with a diameter of $d = 0.2$ mm, Rubey's and Ahrens' give the fall velocity of 2.4 and 2.3 cm/s, respectively. For very fine sand of $d = 0.1$ mm, they give $w_f = 0.78$ and 0.79 cm/s, respectively. For medium sand of $d = 0.4$ mm, Ahrens' formula gives $w_f = 5.2$ cm/s which is slightly faster than Rubey's result of $w_f = 5.1$ cm/s. Both formulas have been derived for spherical particles. Actual fall velocity is affected by the shape of sediment grains, but the shape effect has not been scrutinized to date, probably because of various uncertainties involved in beach morphological problems.

## (B) *Equilibrium beach profile*

Cross-shore profiles of sandy beaches vary widely within a year, sometimes with bars and troughs and other times without them. However, the mean profile averaged over one year usually shows a smooth concave curve. Dean[39] called such mean profile as the *equilibrium beach profile* and expressed it with the following function:

$$h(x) = A\,x^{2/3}\,,$$  (15.14)

where $h(x)$ is the water depth at the distance $x$ from the shoreline, both expressed in the units of m, and $A$ is a *sediment scale parameter* or *profile scale factor* having the dimension of $m^{1/3}$. Dean[40] correlated the parameter $A$ with the sediment fall velocity $w_f$ in the units of cm/s as

$$A = 0.067w_f^{0.44}\,.$$  (15.15)

If the beach is composed of fine sand with $d = 0.2$ mm, the fall velocity is $w_f \simeq 2.3$ cm/s and the sediment scale parameter becomes $A = 0.097$ $m^{1/3}$. As the sand becomes coarse or the diameter increases, the value of $A$ increases and the beach becomes steep. For easy applications, Houston[41] simplified Eq. (15.15) by replacing it with $A = 0.21d^{0.46}$ in the temperature range of $15° \sim 25°$C. The readers are referred to the book by Dean and Dalrymple[42] for detailed discussion on the equilibrium beach profiles and some applications.

The profile expressed by Eq. (15.14) has an infinite slope at the shoreline. When an equivalent beach profile is introduced in a numerical model, it will be necessary to modify the initial depth at a few grids next to the shoreline so as to relax an excessively steep slope there.

## 15.3 Overview of Beach Morphology Models

### 15.3.1 *Prediction Formulas for Longshore Sediment Transport*

#### (A) *Total longshore sediment transport formulas*

As explained in Sec. 12.2.4(A), the longshore sediment transport rate refers to the total volume of sediments moving across a vertical section normal to the shoreline per unit time. It was correlated to the incident wave energy flux by Inman and Bagnold[43] in 1963 and Inman and Komar[44] gave an explicit functional form in 1977. The US Army Coastal Engineering Research Center (CERC) adopted it in the 1984 version of the *Shore*

*Protection Manual*[45] in the following form, which has been known as the CERC formula:

$$Q = \frac{K\,(Ec_g)_b}{(\rho_s - \rho)\,g\,(1 - \lambda)}\sin\alpha_b\cos\alpha_b\,, \tag{15.16}$$

where $K$ is a constant characterizing the CERC formula, $E$ is the wave energy density, $c_g$ is the group velocity, $\alpha$ is the incident angle, the subscript $b$ indicates the quantity at breaking point, $\rho_s$ and $\rho$ are the density of sediment and water, respectively, and $\lambda$ is the in-situ porosity of sediment which is about 0.4.

The wave energy density is evaluated by the following equation by assuming the Rayleigh distribution of individual wave heights:

$$E = \frac{1}{8}\rho g H_{\mathrm{rms}}^2 = \frac{1}{16}\rho g H_{1/3}^2\,. \tag{15.17}$$

The constant $K$ was initially given the average value of 0.77, but it varies over a wide range depending on field sites. In general, there has been observed a tendency of decrease with increase in the sediment grain size. By referring to various formulations of the relationship between the $K$ value and sediment diameter $d$, King[46] has proposed the following simple form through a large scatter of data:

$$K = \frac{0.1}{d_{50}\,(\text{in mm})}\,. \tag{15.18}$$

The data which King analyzed are the volume of sediments impounded at the up-drift side of jetties and other structures. When the CERC formula of Eq. (15.16) is employed in the shoreline change model, the constant $K$ becomes the sole tuning parameter. It is necessary to adjust the $K$ value over a wide range to establish the optimum for simulation of the past shoreline change records. Quite often, a value different from 0.77 or prediction by Eq. (15.18) yield optimum simulation results.

There are several other formulas for estimation of longshore sediment transport rate. One is due to Kamphuis[47] based on a series of 3D wave basin tests. It is expressed as

$$Q = 2.27(H_s)_b^2\,T_p^{1.5}\,s_b^{0.75}\,d_{50}^{-2.5}\,\sin^{0.6}(2\alpha_b)\,, \tag{15.19}$$

where $H_s$ denotes the significant wave height, $T_p$ is the spectral peak period, and $s_b$ is the beach slope out to the wave break point.

A unique longshore sediment transport rate formula has been proposed by Bayram *et al.*[48], who have correlated the longshore transport rate to the mean longshore velocity $\overline{V}$ as follows:

$$Q = \frac{\varepsilon}{(\rho_s - \rho)\,g\,w_f\,(1 - \lambda)}(Ec_g\cos\alpha)_b\overline{V}\,, \tag{15.20}$$

where $\varepsilon$ is called the *transport coefficient*. Bayram *et al.* calibrated Eq. (15.20) with six high-quality data sets of field measurements covering 180 data and proposed the following empirical formula for the transport coefficient:[b]

$$\varepsilon = \left( 4.0 + 9.0 \frac{(H_s)_b}{w_f \, T_p} \right) \times 10^{-5}. \tag{15.21}$$

The formula by Bayram *et al.* assumes that the majority of longshore transport is carried by suspended sediments and a part of incident wave energy flux is consumed for the work of bringing sediments into suspension from the seabed. Thus, the sediment fall velocity $w_f$ enters into the formulation. They say that the new formula may include any type of current. However, the assessment of the mean velocity of wave induced longshore currents will require a numerical computation of random wave fields or use of empirical formulas such as introduced in Sec. 12.6(B).

According to Bayram *et al.*, 62% of predicted transport rates by their formula were within a range of 1/2 to 2 times the measured data and 88% of predictions were in a range of 1/4 to 4 times the measurements. The CERC formula with a constant value of $K = 0.39$[c] yielded over-prediction by a factor of four or so (the present author's judgment on a diagram presented in their paper). Individual prediction values are scattered in the range of 0.8 to 80 times the measurements. The Kamphuis formula also showed a tendency of overprediction with the mean ratio of prediction to measurement around 2 and a scatter of the ratio from 0.5 to 80.

### (B) *Local sediment transport rate formulas*

Sediment transport occurs at any place in the nearshore zone, and many efforts have been made to quantify the local sediment transport rate under a given wave condition. There have been proposed scores of prediction formulas to be applicable at an arbitrary site. Most are based on the concept of the shear stress near the bed and employ the Shields parameter explicitly or implicitly. As a representative of such formulas, the equation due to Bailard[49,50] is shown below.

$$q = 0.5 \rho f_w u_0^3 \frac{\varepsilon_b}{(\rho_s - \rho) \, g \tan \gamma} \left( \frac{\delta_V}{2} + \delta_V^3 \right) + 0.5 \rho f_w u_0^4 \frac{\varepsilon_s}{(\rho_s - \rho) \, g w_f} \delta_V u_3^*, \tag{15.22}$$

---

[b]The coefficients of 4.0 and 9.0 are exchanged after the comment by Dr. Yoshiaki Kuriyama and the correction has been acknowledged by Dr. Atila Bayram.

[c]Bayram *et al.* seemed to have employed the wave energy density equation of $E = (1/8)\rho g H_{1/3}^2$ which yields twice the value by Eq. (15.17).

where $\varepsilon_b$ and $\varepsilon_s$ are the efficiency factors related to bedload and suspended load, respectively, $\tan\gamma$ is a dynamic friction factor being given a value of 0.63, and $\delta_V$ and $u_3^*$ are defined below.

$$\delta_V = \frac{V}{u_0}, \quad u_3^* = \frac{\langle |U_t'|^3 \rangle}{u_0},$$

where $V$ is the mean longshore current velocity, $u_0$ is the maximum wave orbital velocity near the bed and $U_t'$ is the instantaneous velocity vector near the bed (wave and current). The efficiency factors $\varepsilon_b$ and $\varepsilon_s$ need to be adjusted to yield best simulation results. Bailard[49] initially gave the values of $\varepsilon_b = 0.21$ and $\varepsilon_s = 0.025$, but he used another set of $\varepsilon_b = 0.13$ and $\varepsilon_s = 0.032$ later.[50] Bayram *et al.*[51] employed the values of $\varepsilon_b = 0.10$ and $\varepsilon_s = 0.025$ for comparison with field measurement data.

The expression of Eq. (15.22) is due to Bayram *et al.* They evaluated five more formulas in addition to Bailard's for comparison with field measurement data for the formulas' predictability of cross-shore distribution of longshore sediment transport rate. The field data consisted of those of the DUCK and the SUPERDUCK surf zone sand transport experiments, both of which were carried out at the CERC pier at Duck, North Carolina in the United States. The percentage that the prediction was within the range of 1/5 to 5 times the measurement varied from 62% to 96%. The standard deviation of $\log(q_{\mathrm{pred}}/q_{\mathrm{meas}})$ varied from 0.35 to 0.87. Bayram *et al.* conclude their paper by saying that "At the present time, there is no well-established transport formula that takes into account all the different factors that control longshore sediment transport in the surf zone, although the Van Rijn (formula) evidently accounts for many of those factors."

Another comparison of prediction formulas with field data has been reported by Davies *et al.*[52] They selected seven research models and five practical models for prediction of longshore sediment transport rate. Field data were suspended sediment concentrations and transport rates at five sites with different hydrodynamic and sediment characteristics. The percentage that the predicted sediment concentration was within the range of 1/10 to 10 times the measurement at four sites varied from 70% to 83%. Comparison of the sediment transport rate was made for the data at one site, and the percentage that the predicted transport rate was within the range of 1/3 to 3 times the measurements varied from 33% to 88%. Davies *et al.* conclude their paper by saying that "Evidently, the state-of-the-art in sand transport research still requires some knowledge of conditions on site, allowing the user to carry out model validation and/or tuning and, hence, make an informal judgment about the optimum choice of model for use in sand transport computations."

### (C) *Suspended sediment rate formulas*

Inadequacy of the present sediment transport models seems to originate from the neglect of sediment suspension by strong turbulence associated with wave breaking in the surf zone. For example, the Bailard formula of Eq. (15.22) has a term for suspended load, but it is correlated with the wave orbital velocity near the bed. It may be said that presently available prediction models are based on the concept of sediment suspension from crests of sand ripples and they are incapable of taking into account the heavy sediment suspension by breaking waves inside the surf zone. The comment also applies to various commercialized softwares developed up to now. In a paper in 2009, van Maanen *et al.*[53] demonstrate that the Bailard and Van Rijn models under-predict their estimates of the depth-averaged longshore flux of suspended sediments, saying that "both models are lacking physics related to wave breaking, such as an increase in bed shear stress due to turbulence injected from the surface during the passage of a surf zone bore."

A crude approach of Eqs. (15.4) and (15.5) by Katayama and Goda[28−30] or a sophisticated approach of Eqs. (15.7) to (15.9) by van Thiel de Vries[25] would lead to development of better models when they are calibrated with many of field measurement data sets. As Davies *et al.*[52] state, "At the present state of research the availability of some field measurements is a necessary requirement for higher accuracy predictions."

### 15.3.2 *Shoreline Change Models*

### (A) *One-line model for shoreline change*

Prediction of beach morphological changes for practical applications has mostly been carried out using shoreline change models. These models deal with the advance and/or retreat of the shoreline position without considering local bathymetric evolution. The typical model is so called the *one-line model*, which assumes the invariability of cross-shore beach profile even when the shoreline moves. Figure 15.3 illustrates the concept of the one-line model.

Let us consider a section of the beach profile with alongshore distance $\Delta y$. During the time $\Delta t$, there is an inflow of the longshore sediment $Q\Delta t$

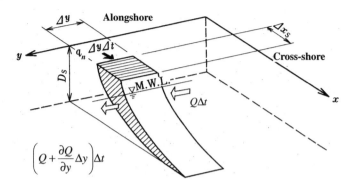

Fig. 15.3   Sketch of sediment budget and shoreline change in the one-line model.

from the right and an outflow of $(Q + \partial Q/\partial y) \times \Delta t$ toward the left. There will also be a net flux of cross-shore sediment into the volume of $q_n \Delta y \Delta t$. Applying the sediment budget requirement, the shoreline will advance by $\Delta x_s$. The fundamental equation for the shoreline change is expressed as

$$\frac{\partial x_s}{\partial t} + \frac{1}{D_s}\left(\frac{\partial Q}{\partial y} - q_n\right) = 0 , \qquad (15.23)$$

where $D_s$ is a vertical distance over which sediment movement is active. It is measured from the seabed at which no significant depth change takes place to the top of the backshore berm denoted by $Y_s$. The former is denoted by $h_c$ and is called as the *depth of closure*. Thus, $D_s = h_c + Y_s$. Both the depth of closure and the berm height will be discussed in the subsequent clauses.

The longshore sediment transport $Q$ has the units of $\mathrm{m}^3/\mathrm{s}$ for example, while the net cross-shore sediment transport rate $q_n$ has the units of $\mathrm{m}^3/\mathrm{m/s}$. The net cross-shore sediment transport can occur due to sediment supply at a river mouth (positive) and/or the continuous loss toward the offshore (negative). The longshore sediment transport $Q$ is usually estimated with the CERC formula of Eq. (15.16), which is for sediment transport on natural beaches. In cases where breakwaters and other structures have been built, the incident wave heights vary locally due to the influence of structures. In such cases, the following modified version due to Ozasa and Brampton[54] is employed:

$$Q = \frac{(Ec_g)_b}{(\rho_s - \rho)\,g(1-\lambda)}\left(K_1 \sin\alpha_b \cos\alpha_b - \frac{K_2}{s}\cos\alpha_b \frac{\partial H_b}{\partial y}\right) , \qquad (15.24)$$

where $s$ is the beach slope.

The constant $K_1$ is taken as the same value in the CERC formula. The constant $K_2$ is usually set at a value slightly smaller than $K_1$. As discussed in Sec. 15.3.1(A), the value of $K_1 = K$ cannot be determined a priori but should be adjusted through validation with field data at a study site.

The one-line model cannot predict local evolution of beach profiles because of its fundamental assumption of invariable profiles of beaches. To remedy this shortcoming, several models have been proposed using multiple-lines of isobaths. They require a certain amount of bathymetric survey data over years for the model validation and several parameters of the models need to be adjusted by experience. It may not be easy for engineers to make good use of the models, except for the persons who developed the models.

## (B) *Depth of closure*

The *depth of closure* refers to the water depth beyond which no significant depth changes occur over several years. It is a site-specific quantity and needs to be assessed through comparison of a number of bathymetric survey data sets.

A correlation of the depth of closure to the wave height in wave climate was first proposed by Hallermeier[55] as in the following:

$$h_c = 2.28 H_{\text{eff}} - 68.5 \left( \frac{H_{\text{eff}}^2}{g T_{\text{eff}}^2} \right) = (2.28 - 10.9 H_{\text{eff}}/L_0) H_{\text{eff}}, \qquad (15.25)$$

where $h_c$ is the depth of closure, $H_{\text{eff}}$ is later called an effective wave height as defined to have the exceedance probability of 0.137% (expected to exceed for 12 hours per year),[d] $T_{\text{eff}}$ is the effective wave period corresponding to $H_{\text{eff}}$, and $L_0$ is the deepwater wavelength that is equal to $g T_{\text{eff}}^2/(2\pi)$. Hallermeier gave an estimation formula of $H_{\text{eff}} = \overline{H} + 5.6\sigma_H$, where $\overline{H}$ and $\sigma_H$ denote the annual mean and the standard deviation of significant wave height at site, for the case of the exponential wave height distribution. Use of the offshore wave steepness of $H_0/L_0 = 0.035$ by referring to Fig. 3.1 in Sec. 3.1 yields an approximate estimation of $h_c \simeq 1.9 H_{\text{eff}}$. Houston[41] suggests another approximation of $h_c \simeq 6.75(H_s)_{\text{aveg}}$ with $(H_s)_{\text{aveg}}$ denoting the annual average of significant wave height.

---

[d]His proposal of the 12-hour-per-year wave height with exceedance of 0.137% for estimation of the depth of closure corresponds to twice wave measurements per day, though he did not explicitly state so.

By analyzing the available data since then, Birkemeier[56] presented a modified version in 1985 as

$$h_c = 1.75 H_{\text{eff}} - 57.9 \left( \frac{H_{\text{eff}}^2}{g T_{\text{eff}}^2} \right) = (1.75 - 9.2 H_{\text{eff}}/L_0) H_{\text{eff}} \,. \qquad (15.26)$$

The above depth of closure may be approximated with $h_c \simeq 1.4 H_{\text{eff}}$ by assuming $H_0/L_0 = 0.035$.

Although Hallermeier defined the effective wave height having the exceedance probability of 0.137%, Kuriyama[57] recommends using the significant height with the exceedance probability of $1 \sim 3\%$ (6 to 13 times a year) for Eq. (15.25) and with the exceedance probability of $0.12 \sim 0.8\%$ (2 to 4 times a year) for Eq. (15.26), based on his examination of the depth of closure at Hazaki Beach, Ibaraki Prefecture, Japan based on a large databank of waves and beach profiles accumulated there.

## (C) *Height of backshore berm*

The backshore is the zone in which waves run up during storms. It is often marked with a berm formed by storm waves. It is necessary to estimate the berm height $Y_s$ for assessing the vertical distance $D_s$ in the one-line model. Two estimation formulas for $Y_s$ based on laboratory tests have been proposed by Rector[58] and Swart,[59] and another by Sunamura[60] based on field data. They are listed in that order in the following:

$$\frac{Y_s}{L_0} = \begin{cases} 0.024 & : \ H_0/L_0 \geq 0.018 \,, \\ 0.18 (H_0/L_0)^{0.5} & : \ H_0/L_0 < 0.018 \,, \end{cases} \qquad (15.27)$$

$$\frac{Y_s}{d_{50}} = 7644 - 7706 \exp \left[ -0.000143 \frac{H_0^{0.488} \, T^{0.93}}{d_{50}^{0.786}} \right] , \qquad (15.28)$$

$$\frac{Y_s}{H_0} = 0.173 \, (H_0/L_0)^{-0.5} , \qquad (15.29)$$

where $d_{50}$ denotes the median diameter of sediment and the berm height $Y_s$ is measured from the still water level.

The offshore wave height $H_0$ to be substituted in Eqs. (15.27) to (15.29) is not specified by respective authors. Kuriyama[61] recommends using the wave height with an exceedance probability of 2% as a guideline. Estimation of the berm height by these equations will yield somewhat different results. Selection should be made through comparison of the field data at site.

### 15.3.3  *Numerical Models for 3-D Beach Deformation*

(A) *Overview*

Prediction of local beach morphological changes requires detailed analysis of spatial and temporal sediment transport and local depth change. Numerical models for such prediction are called as quasi-3-D models, because the depth-averaged nearshore currents and sediment concentrations are employed. Hereinafter, they are called the *3-D beach deformation models* for simplicity. Numerical models for 3-D beach deformation are generally composed of four computational modules as shown in Fig. 15.4.

The *first module* for wave field is to calculate the transformation and deformation of random waves from the offshore to the shoreline, including wave diffraction effects. The module is executed for steady state conditions and yields the spatial distribution of wave heights and directions, energy densities, surface rollers, and radiation stresses. Any wave transformation model may be used in this module, but incorporation of directional spectral waves is essential for realistic simulation of the wave field. As exemplified in Fig. 3.43 in Sec. 3.6.4, several models of random wave breaking processes yield quite different cross-shore variations of wave heights in the surf zone.

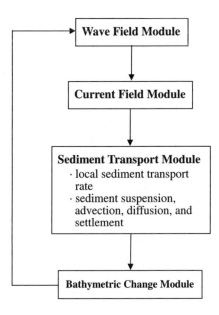

Fig. 15.4  Composition of a 3-D beach deformation model.

A cautious selection of appropriate random wave breaking model is necessary. Energy transfer from breaking waves to surface rollers is also required as exemplified in Figs. 12.6 to 12.8 in Sec. 12.5.

The *second module* is for analysis of nearshore current fields with the input from the wave field module. One of the difficulties in the current computation is a reproduction of flow separation from a head of a jetty or a corner of a structure. A jetty extended from a bank of uniform flow is known to cause flow separation at its head, and the streamline of separation reattaches itself at the bank at the downstream distance of 10 to 15 times the jetty length. Immediately behind the jetty, a large area of counter-circulation flow appears. Flow separation is one of characteristic features of viscous flow. Introduction of some turbulence models is required to simulate flow separation.

The performance of the *third module* for sediment transport depends entirely on the sediment transport rate formula employed. One type of module uses the formula based on the shear stress near the bed, while other type uses the suspended sediment concentration formula.

The *fourth module* is to calculate the temporal variation of local water depth from the imbalance of sediment influx and outflux. After a certain elapsed time with appreciable amount of depth change, the first module is run again with the new bathymetry and the cycle is repeated.

(B) *Beach deformation models based on bed shear stress*

Beach deformation models employing sediment transport rate formulas based on the bed shear stress calculate the sediment transport rate at each grid point without considering advection and diffusion of sediments. The depth change is calculated by solving an equation such as shown below, which is due to Kuroiwa *et al.*[62] presented in 2000.

$$\frac{\partial h}{\partial t} = \frac{1}{1-\lambda}\frac{\partial}{\partial x}\left(q_x + E_s|q_x|\frac{\partial h}{\partial x}\right) + \frac{1}{1-\lambda}\frac{\partial}{\partial y}\left(q_y + E_s|q_y|\frac{\partial h}{\partial y}\right) , \quad (15.30)$$

where $\lambda$ denotes the porosity of sediment, $q_x$ and $q_y$ are the sediment transport rates in the cross-shore and alongshore directions, respectively, and $E_s$ represents a coefficient for the effect of local bottom slope on the sediment transport.

There have been proposed so many models in this category that no specific description to respective ones is given here.

(C) *Beach deformation models based on sediment concentrations*

A beach deformation model mainly focusing on suspended sediment transport must solve the advection and diffusion equation of suspended sediments. Sawaragi *et al.*[63] presented in 1984 the following equation for the depth-average concentration $\overline{C}$:

$$\frac{\partial(h\overline{C})}{\partial t} + \frac{\partial(\overline{C}Uh)}{\partial x} + \frac{\partial(\overline{C}Vh)}{\partial y} - \frac{\partial}{\partial x}\left(K_x h\frac{\partial\overline{C}}{\partial x}\right) - \left(K_y h\frac{\partial\overline{C}}{\partial y}\right) + \overline{C}w_f = q_p,$$
(15.31)

where $K_x$ and $K_y$ are the cross-shore and alongshore components of the depth-averaged horizontal diffusion coefficients, respectively, and $q_p$ represents the sediment pick-up rate from the bottom. Katayama and Goda[30] used a simplified version of Eq. (15.31), while Kuroiwa *et al.*[64] developed its 3D version.

As for the sediment pick-up rate, Kuroiwa *et al.* used a formula based on the Shields parameter, but Katayama and Goda used a formula based on the energy dissipation rate by wave breaking as given by Eq. (15.4). Selection of appropriate formula for sediment pick-up rate is a critical factor for successful development of a beach morphology model based on sediment concentrations. The rate of depth change is given by the following equation:

$$\frac{\partial h}{\partial t} = \frac{1}{1-\lambda}\left[-q_p + w_f\overline{C} + \left(\frac{\partial q_{bx}}{\partial x} + \frac{\partial q_{by}}{\partial y}\right)\right],$$
(15.32)

where $q_{bx}$ and $q_{by}$ are the cross-shore and alongshore components of the bedload transport rate, respectively.

### 15.3.4 *Quantitative Assessment of Sediment Impoundment by Groin*

As discussed in Sec. 15.2.2, the concept of random waves has not been established yet in coastal morphological studies. Many formulas are based on model tests using regular waves and/or regular wave theory, and field data are presented with a single height of visually observed waves. In order to illustrate a method of incorporating random wave characteristics and wave climate statistics into sediment transport problems, derivation of a formula for quantitative assessment of sediment impoundment efficiency by a groin is explained below.

Derivation is due to Nakamura and Goda[65] in 2007. First, they made numerical analysis of nearshore currents around a groin for regular waves by taking into account the interaction between waves and nearshore currents. The offshore wave height was varied from 0.5 to 4.0 m for a groin extended to the water depth of 1, 2, and 3 m. The wave period, incident angle, and beach slope were varied at three steps, respectively. The bedload transport rate was computed with a formula based on the Shields parameter with the combined velocity of nearshore currents and wave orbital velocity.

Comparison of longshore sediment transport rates with and without a groin enabled to derive the following empirical formula for the sediment impoundment efficiency for steady state condtions:

$$\Phi_{\text{reg}}(\xi_0) = 1 - (1 + \xi_0^2) \exp[-\xi_0^2] \ : \ \xi_0 = d_t/H_0 \,, \tag{15.33}$$

where $d_t$ denotes the water depth at the head of a groin and $H_0$ is the deepwater wave height. The impoundment efficiency $\Phi$ ranges between 0 and 1; 0 for no impoundment without a groin and 1 for full impoundment by a groin of immense length. Equation (15.33) was confirmed to function regardless of the incidence angle up to $32°$, the wave period of 6 to 12 s, and the beach slope of 1/20 to 1/60.

The expected impoundment efficiency $\Phi_{\text{exp}}$ for random waves is calculated as an ensemble average for individual waves, the heights of which are assumed to follow the Rayleight distribution. The calculation is made by

$$\Phi_{\text{exp}}(\xi_s) = \frac{\displaystyle\int_0^\infty \Phi_{\text{reg}}(\xi)Q(H)p(H)dH}{\displaystyle\int_0^\infty Q(H)p(H)dH} \ : \ \xi_s = \frac{d_t}{H_s} \,, \tag{15.34}$$

where $Q(H)$ is the longshore sediment transport for regular wave with the height $H$ and $p(H)$ is the probability density function of the wave height, which is expressed by Eq. (10.16) with $a = 1.416$ for $H_* = H_s$ in Sec. 12.1.2(A).

The longshore sediment transport by the CERC formula expressed with the breaking wave characteristics can be rewritten in terms of the offshore wave characteristics as in the following:

$$Q \propto H_0^{2.4} T^{0.2} \alpha_0 \,, \tag{15.35}$$

where $\alpha_0$ denotes the offshore wave incident angle. The above equation is derived with the shallow water wave approximation.

Computation of the expected impoundment efficiency has been made by substituting Eq. (15.35) into Eq. (15.34) together with the probability density function of the Rayleigh distribution. The computation result was expressed approximately with the following empirical formula:

$$\Phi_{\exp} = 1 - \exp[-0.5\xi_s^2] \; : \; \xi_s = d_t/H_s \,. \tag{15.36}$$

Equation (15.36) provides a quick estimate of the impoundment efficiency of a groin. If one wishes to stop the longshore sediment transport by 60% or $\Phi_{\exp} = 0.6$, the equation yields $\xi_s \geq 1.35$ or $d_t \geq 1.35 H_s$ for a given significant wave height.

We may be interested in a single storm event, but the impoundment efficiency over a year is of much concern. When the marginal distribution of the significant wave height over a long duration is established at a site of interest, the overall impoundment efficiency is assessed by the following equation:

$$\Phi_{\text{overall}} = \frac{\displaystyle\int_0^\infty \Phi_{\exp} H_s^{2.4} f(H_s) dH_s}{\displaystyle\int_0^\infty H_s^{2.4} f(H_s) dH_s} , \tag{15.37}$$

where $f(H_s)$ denotes the probability density of the marginal distribution of significant wave height in the wave climate. The approximate relationship of Eq. (15.36) is utilized in the above assessment. The wave period term is omitted as being negligibly small and the incident wave angle term is dropped, being assumed as constant.

Depending on the locality, the wave climate may have the exponential distribution of the significant wave height such as

$$f(H_s) dH_s = \frac{1}{(H_s)_{\text{mean}}} \exp\left[-\frac{H_s}{(H_s)_{\text{mean}}}\right] , \tag{15.38}$$

where $(H_s)_{\text{mean}}$ denotes the annual mean of the significant wave height. Results of computation by substitution of Eq. (15.38) into Eq. (15.37) have been expressed in the following empirical formula for the overall impoundment efficiency:

$$\Phi_{\text{overall}}(\xi_{\text{mean}}) = 1 - \exp[-0.17\xi_{\text{mean}}^{1.5}] \; : \; \xi_{\text{mean}} = d_t/(H_s)_{\text{mean}} \,. \tag{15.39}$$

The overall impoundment efficiency is much smaller than the expected impoundment efficiency of Eq. (15.36) because high storm waves transport large amount of longshore sediments. According to Eq. (15.39), a groin with the depth at its head being 2, 4, and 6 times the mean significant wave height is estimated to have the overall impoundment efficiency of 38%, 74%, and 92%, respectively, at a site where the wave climate is described with Eq. (15.38) for the significant wave height. Conversely, a groin needs to be extended to the depth of 1.4, 2.6, and 4.5 times the mean significant wave height to hold the overall impoundment efficiency of 25%, 50%, and 80%, respectively. Even though the derivation of the impoundment efficiency formulas does not consider the longshore transport of suspended sediments, Eq. (15.39) will serve for a quantitative planning of groins and jetties.

## 15.4     A Case Study of Shoreline Change Prediction by One-Line Model

### (A)  *Outline of study site*

An application of the one-line model for shoreline change has been made for a beach of 11 km long at Mamaia in the City of Constanta, Romania, as a part of the shore protection and rehabilitation study commissioned by Japan International Cooperation Agency (JICA).[66,67] Some features of the study are introduced here as a case study of shoreline change prediction.

Figure 15.5 shows a location map of Mamaia at the western coast of the Black Sea. The map covers 190 km in the north to south and 180 km in the east to west. The Danube Delta is located the upper right corner, and Sulina below the delta is the entrance of the Danube International Channel. Sediments from the Danube have been transported southwestward along the northern shore of Romanian Black Sea and have nourished the seashore down to the Port of Constanta. Lagoons at the north of the coastline were embayments several thousand years ago, but they were gradually closed by development of barrier beaches. Mamaia Beach is located between Navodari and Constanta in the lower center of the map. Just behind the beach there is a fresh water lake of about 8 km long, which has been separated from the sea by the growth of a barrier beach about two thousand years ago.

Mamaia Beach used to have a broad shore of 100 m wide and accommodated a large number of summer tourists from European countries since the early twentieth century. However, the beach experienced a rapid recession of the shoreline since the late 1970s. The cause of the recession is considered to be the extension of a breakwater of Midia Port located next to Navodari from a depth of 5 m toward a depth of 10 m. Terrestrial sediments from the Danube carried by wave-induced currents were impounded by the breakwater and could not reach Mamaia Beach. By the end of 1990, the beach was eroded by up to 80 m and some of seashore hotels were threatened of collapse.

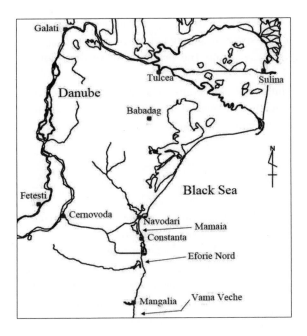

Fig. 15.5    Location map of the northern and southern shore of Romanian Black Sea.

The Government of Romania pumped up about 500 thousand cubic meters of fine sand from the lake behind the beach and nourished the beach with construction of six offshore breakwaters with the extension of 250 m each and the opening of 250 m between them along the isobath of 5 m. Beach-filled sand soon moved toward the offshore breakwaters to form the equilibrium profile corresponding to its small diameter. The speed of beach erosion was slowed down, but it had the rate of 2.3 m per year at the severest location. The study commissioned by JICA had the objectives of clarifying the processes of beach erosion and preparing the rehabilitation plan of the seashore from Navodari to Vama Veche, including Mamaia.

### (B) *Calculation of representative wave dimensions*

The JICA study employed the one-line model for simulation of the shoreline change, because of its long extension and necessity of making prediction over 20 years into the future. As mentioned in Sec. 15.2.2, a beach morphological model is run with some energy averaged waves. The study made use of the wave forecast data offshore of Constanta from 1991 to 2002 by the European Center for Medium-range Weather Forecast (ECMWF). The wave data list the spectral significant wave height $H_{m0}$, the spectral significant period $T_{m-1,0}$, and the principal wave direction $\theta_p$ at six hours intervals.

The wave data were divided into the northern and southern groups, and the representative height, period, and direction were computed for the two groups by the following equations:

$$H_{\rm rep} = \sqrt{\frac{\sum_{i=1}^{N} H_i^2 T_i}{\sum_{i=1}^{N} T_i}}, \quad T_{\rm rep} = \frac{\sum_{i=1}^{N} T_i}{N}, \quad \theta_{\rm rep} = \frac{\sum_{i=1}^{N} \theta_i H_i^2 T_i}{\sum_{i=1}^{N} H_i^2 T_i},$$

$$(15.40)$$

where $N$ is the number of data in the northern or southern groups. Calculation yielded the northern representative waves as $H_{1/3} = 1.65$ m, $T_{1/3} = 6.2$ s, and $\theta_p =$ N64.0°E and the southern representative waves as $H_{1/3} = 1.11$ m, $T_{1/3} = 6.2$ s, and $\theta_p =$ N115.2°E.

For the two representative waves, the wave transformation was analyzed over the area of 120 km in the north to south and 65 km in the east to west by means of the energy balance equation of Eq. (3.19) in Sec. 3.2.3. The computational grids were set at a spacing of 250 m each. Wave breaking was assumed to take place at the offshore-most grids satisfying the condition of $H_{1/3} \geq 0.8h$, and wave refraction in the area shallower than the wave breaking grid was estimated with the Snell's law. Wave diffraction by the offshore breakwaters was assessed by means of the angular spreading method in Sec. 3.3.4.

Seasonal variation of wave climate was introduced by adjusting the duration of the northern and southern representative waves for each month. For this purpose, the mean wave energy fluxes of the northern and southern wave groups were calculated for each month from January to December and the duration of representative waves in each month was assigned in proportion to the ratio of their energy fluxes.

For evaluation of the depth of closure, the effective wave height and period were calculated from the ECMWF data as $H_{\rm eff} = 5.0$ m and $T_{\rm eff} = 9.1$ for exceedance of 12 hours per year. The depth of closure was assessed as $h_t = 9.3$ m by Eq. (15.25).

(C) *Simulation of shoreline changes*

For the southern shore of the Romanian Black Sea, the National Institute for Marine Research and Development of Romania has been carrying out beach profile surveys at a number of baselines for more than 30 years. The Water Resources Administration Agency of Romania has also executed intermittent shoreline surveys since 1976. Four shoreline maps of 1976, 1980, 1995, and 1997 were available for calibration data of the shoreline simulation by the one-line model.

The one-line model was applied to an extension of 20 km from Navadori, to Constanta with a grid spacing of 20 m. Trial simulations were made by setting the constant $K_1$ of the CERC formula at three stages of 0.077, 0.154, and 0.308, among which $K_1 = 0.154$ gave the best agreement with the field survey data and thus it was adopted. The second constant $K_2$ was set at the level of $K_2 = 0.81K_1$ according to previous experiences of Kuroki, the first author of Ref. 66, and thus $K_2 = 0.125$ was used. Mamaia Beach is composed of fine sand with the median diameter of 0.18 to 0.27 mm. Use of empirical formulas such as Eq. (15.18) yields an estimate of $K = 0.56$ to 0.37, but use of such empirical value would have yielded simulation results quite different from what had occurred in Mamaia.

Beach fill of 500 thousand cubic meters over the longshore distance of 1000 m was introduced in the model in the period of 1989 to 1990. The rate of offshore sediment flux was set at $q_n = -3$ m$^3$/m/year after some trials. Figure 15.6 shows the results of simulation of the shoreline change since 1976 over the extension of 20 km in comparison with the field survey data.

The top panel shows the plan shape of the study area. Six offshore breakwaters appear at the area at the alongshore distance of 9,300 to 12,200 m. The second panel exhibits the shoreline change between 1976 and 1980. Beach erosion began to appear in the area at the alongshore distance of 10,000 to 12,700 m. In the period of 1976 and 1990 shown in the third panel, the effect of beach nourishment is clearly visible. However, the benefit of beach nourishment was lost by 1995 and the shoreline continued to recede as seen in the fourth and fifth panels. The area at the alongshore distance of 8,000 to 9,500 m at the north of the offshore breakwaters began to recede after 1990.

Computation by the one-line model yielded the simulation results quite close to the field survey data in the area at the alongshore distance of 6,000 to 12,700 m. In the area at the alongshore distance of 13,500 to 17,000 m, no field survey data was available. In the area at the alongshore distance of 700 to 6,000 m, the model indicates appreciable accretion while the field survey data shows no accretion. There seems to exist some outflow of sediments along the jetty of Midia Port at the left end of the panel, but no particular

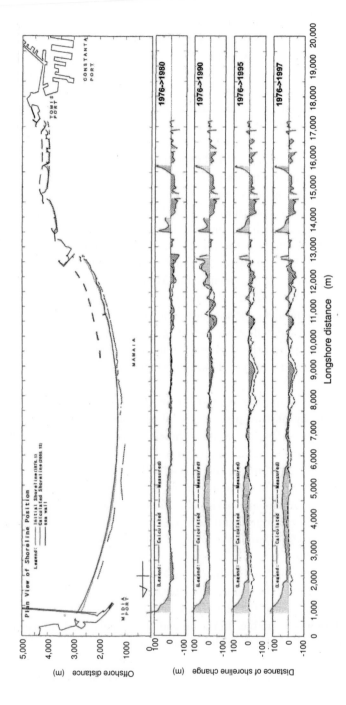

Fig. 15.6　Comparison of the shoreline changes between field survey data and simulation results in the period from 1976 to 1997.

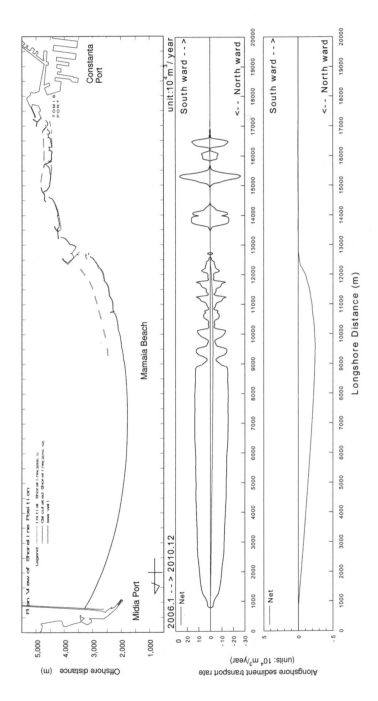

Fig. 15.7 Longshore sediment transport rate in Mamaia Beach.

adjustment of the offshore sediment flux rate was made because the beach at this area is not threatened by erosion and it was located beyond the target area of study.

Computation by the one-line model also yields the estimate of longshore sediment transport rate. The northern representative waves produce the southward transport, while the southern representative waves produce the northward transport. Figure 15.7 shows the estimated rates of southward and northward sediment transport and the net transport rate, which was computed for the prediction period of five years from 2006 to 2010. The longshore sediment transport rate varies considerably with location because of the presence of various structures. The southward transport reaches 140 thousand cubic meters per year, while the northward transport rate has a maximum of 160 thousand cubic meters per year shown in the middle panel. As the result, the net transport is northward with a maximum rate of 20 thousand cubic meters per year as shown in the bottom panel.

As exemplified in this case study, availability of the field data for calibration is an essential requirement for successful prediction of shoreline change by means of the one-line model. Qualitative analysis of shoreline change is possible any time, but quantitative prediction without field calibration data is futile. Coastal managers should collect beach morphology data for the period as long as possible and keep good records of such data.

## 15.5   Overview of Shore Protection Facilities

Beach erosion is unwelcome phenomenon, and people have tried to stop it since early days. *Coastal Structures 2007*[68] reproduces an old printing (1528) of Venice with piled rock groins for shore protection at Lido di Venizia, Italy. Various types of structures have been built to mitigate beach erosion. Artificial nourishment of beaches (beach fill) with sand brought from outside was executed in 1922 for a length of 1.1 km at Coney Island, New York in the United States and beach fill projects became popular in the twentieth century. In this subsection, several measures against beach erosion are overviewed with some comments. Because the topics cover many aspects of planning, design, and monitoring which cannot be dealt with here, the readers are asked to refer to relevant technical books and engineering manuals.

## (A) Seawalls and coastal dikes

Seawalls and coastal dikes are the facilities to protect the land behind them from the attacks of high waves, storm surges, and/or tsunamis. When these structures are built with broad beaches in front of them, their protective function is high and beaches may not be subject to erosive threat by waves. However, if the frontal beach is not wide enough, waves reflected by seawalls and coastal dikes often take away beach sand and beaches may disappear. The main parameter is the reflection coefficient of these facilities. Equation (3.50) in Sec. 3.7.1 provides an estimate of reflection coefficient.

Erosion of cliff coasts can be stopped by constructing seawalls and/or wave dissipating structures at their feet. It will be a good remedy for the cliffs themselves, but the measure may cause the erosion of neighboring beaches if the eroding cliffs have been supplying sediments to the neighboring beaches. Decision of seawall construction should be made by paying due considerations to the merit of protecting cliffs and the threat of beach erosion.

## (B) Groins

A groin is a structure jutting from the shoreline of a sandy beach to trap and retain sediments moving alongshore. It is built either singly or in a group of a dozen or more to deal with beach erosion. It is usually extended midway in the inshore and thus relatively short. A jetty is a longer structure extended offshore to prevent shoaling of a channel by littoral materials and to direct and confine the stream of tidal flow, built at the river mouths or tidal inlets, according to the Coastal Engineering Manual.[69] However, a groin may be called a jetty when it is extended to the offshore zone.

A successful example of sediment control by a groin system is that built at Westhampton Beach, Long Island, New York in the United States.[70,71] To prevent the beach erosion at the eastern south shore of Long Island, eleven groins were built in 1966 and four groins were added in 1970. The groins have a nominal length of 146 m and a spacing of approximately 365 m.[72] They have been functioning in protecting a coastal section of 5.6 km with trapping of sand exceeding 3 million cubic meters in total. However, the coast immediately at the down-drift (west) side has been suffering from a deficit of longshore sediment supply and the shoreline has been receding. Hanson *et al.*[72] propose to shorten the length of groins by 30.4 m (100 ft) so as to enhance the sediment transport to the down-drift side on the basis of their numerical prediction of shoreline change.

Groin design has been made mostly on experience, by referring to successful cases of groin constructions. Little discussion has been made on the impoundment efficiency of groins. Equation (15.39) presented in Sec. 15.3.4 will hopefully stimulate discussions on quantitative performance of groin systems. When using Eq. (15.39), it is suggested to employ the mean tide level for definition of the depth at the head of groin, because the equation has been derived as the performance on the annual average.

(C) *Submerged groin system*

Against a conventional system of emergent groins, there has been proposed a system of submerged groins made of geotextile tubes filled with fresh concrete[73,74] or sand-filled geotextile bags.[75,76] Though Holmberg[73] presented several field cases of successful sand accretion by installation of submerged groin systems, he gave no physical explanation for the mechanism of sand accretion. Goudas *et al.*[74] tried to analyze the flow circulation near the bed between submerged groins. Aminti *et al.*[76] report monitoring results of shoreline changes after installation of four submerged groins (150 to 240 m long) with spacings of 230 and 500 m. They make a numerical analysis of wave and current field around the groin system, but provide no explicit explanation on the function of a submerged groin system.

According to the numerical analysis by Goda *et al.*,[77] submerged groins yield fluid resistance to longshore currents owing to flow separation at their crests. Goda *et al.* have simulated the fluid resistance by artificially increasing the frictional resistance at three grids in the downstream of each submerged groin. Nine submerged groins with a height of 0.5 m, a length of 65 m to a depth of 2 m, and a separation of 48 m effectively decreased the longshore current velocities within the groin field against waves of 2 m in the significant height. Shoreline changes were analyzed with the suspension-based 2HD model by Katayama and Goda.[28-30] The analysis demonstrated a certain amount of sediment impoundment at the upstream side, little deposition or erosion within the groin field, and quick recovery of sediment transport at the downstream side. A submerged groin system seems to be a promising countermeasure to beach erosion, though more detailed analysis is needed.

## (D) *Low-crested structures*

Shore-parallel breakwaters with low crests have been built at many coasts of the world. They have been given various names such as offshore breakwaters and detached breakwaters. In a recent joint study among European countries, the structures are given the terminology of *low-crested structures* with the acronym of LCS (Lamberti *et al.*[78]). They include both emergent and submerged ones.

A submerged LCS is favored from the aesthetic viewpoint and the water circulation behind LCS. However, a submerged LCS allows higher wave transmission than an emergent one. A low degree of energy dissipation by a submerged LCS is compensated by an increase in the crest width. Such a submerged LCS with a wide crest is sometimes called an *artificial reef*, though the same term is also used to refer to aquatic structures for aquaculture. A group of LCSs may be laid perpendicular to the shoreline, which was named by Goda[79] as a longitudinal reef system against a conventional layout which was called a lateral reef system. Numerical computation against random waves has shown the performance of the former more efficient than the latter.[80]

There are some guidelines for planning LCSs for shore protection, but they are mostly empirical or those based on regular wave tests. No criterion has been established for the target wave transmission coefficient of LCS. Equations (3.59) through (3.63) would serve for quantitative assessment of the wave transmission coefficient of LCS of any dimension.

Low-crested structures may require periodic repairs and maintenance because of frequent wave overtopping. There may be occasions of subsidence due to a loss of sand beneath them. Placement of geotextile sheets and gravel mat beneath LCS is a necessity. Scour behind LCS and in the seabed at the opening between LCSs is another concern in design.

## (E) *Artificial headlands*

Artificial headlands are the concept promoted by Silvester and Hsu[81] under the name of headland control. The name of artificial headlands appear in the book by Dean and Dalrymple.[82] A headland introduced in the book by Silvester and Hsu is a short shore-parallel structure at each tip of crenulate shaped bays, and it is intended to function similar as a natural headland. In a beach fill project at a coast to which predominant waves are obliquely incident, a number of artificial headlands are placed at some distance from the original shoreline, and beach fill is so made to have the new shoreline formed in concave curves between the artificial headlands. Such a beach will maintain an equilibrium planform against obliquely incident waves.

The concept has been expanded to a system of medium-size T-type groins in Japan, which are constructed in a coast of normal wave incidence (Saito *et al.*[83]). Depending on seasons, waves may come from the left or the right obliquely and the shoreline will respond to the seasonal waves to form an equilibrium plan shape. So long as the sediments between two headlands do not move out and the temporary recession of the shoreline does not interfere the use of the coastal land, the headland system will maintain a stabilized beach on the annual basis. Most of so-called headlands built in Japan, however, have a length of 100 to 150 m. Thus a large amount of sediments move across the tip of a headland. Tsukiyama *et al.*[84] reports an impoundment efficiency of 23% for a 100-m long headland built at the Sendai Coast in Japan based on field surveys with fluorescence-dyed sand as tracers. If the information were given on the annual mean significant wave height and the depth at the tip of this headland, the impoundment efficiency could have been estimated by Eq. (15.39) in Sec. 15.3.4.

### (F) *Beach fill with protective facilities*

The philosophy of beach fill projects differs between the United States and other countries. American projects are mostly executed by dumping a large quantity of offshore sand without any protective structures such as groins and low-crested structures. A focal point of project planning is the prediction of longevity of beach fill or the period of time until the next nourishment will be required. Dean and Dalrymple[85] provide detailed discussion on the planning of beach fill without protective facilities.

In Europe, a number of beach fill projects have been carried out with protective facilities. In Japan, beach fill without protective facilities is a rare case. Groins are extended at the sides of a filled beach and emergent or submerged low-crested breakwaters are built at the offshore periphery. An offshore sill or toe structure is sometimes built at a certain distance from the shore to limit the volume of fill sand. Beach fill is made close to the crest of the offshore sill. Filled beach with the offshore sill is called a *perched beach*. The beach fill project of 3 km long at Ostia (Rome) was protected by a continous LCS (Lamberti *et al.*[78] and Toti *et al.*[86]). The LCS is connected to the shoreline at the both ends. Another beach fill project in Italy at Lido di Dante is also protected by a continuous LCS of 770 m long with the crest of 12 m wide and 0.5 m below the mean sea level.[78,87,88] The LCS is located at a distance of 180 m from the shoreline, and two groins (one-half length of deeper portions being submerged) connect the LCS to the shore. Zanuttigh[87] reports the morphological response of beach through numerical modeling.

Japanese beach fill projects are mostly of small scale; the majority has the fill amount less than 50 thousand cubic meters (Nishi *et al.*[89]). Japanese projects favor protection by two long groins at the both sides of beach fill areas and offshore breakwaters distributed along the project boundary. Open spaces are provided between the breakwaters and the groins for enhancement of water circulation. Groins are extended as long as the cost permits to minimize the loss of filled sand.

Our capability of numerically predicting the morphological changes of a filled beach protected by structures seems to be limited presently, especially for a new site where the information of beach morphology is limited. Zanuttigh calls attention to the correct representation of wave transmission, especially in presence of emergent structures. Reliable simulation of sediment suspension by breaking waves is another task for improvement of numerical models for beach fill projects. The present author is looking forward to seeing good progress in the near future.

# References

1. H. A. Einstein, "Movement of beach sands by water waves," *Trans. Amer. Geophys. Union* **29**(5) (1948), pp. 653–655.
2. A. A. Kalinske, "Movement of sediment as bed load in rivers," *Trans. Amer. Geophy. Union* **28**(4) (1947), pp. 615–620.
3. R. A. Bagnold, "Motion of waves in shallow water: interaction between waves and sand bottom," *Proc. Roy. Soc. London* **187** (1946), pp. 1–15.
4. M. Manohar, "Mechanics of bottom sediment movement due to wave action," *Beach Erosion Board, Tech. Memo.* No. 75 (1955), 121p.
5. T. Scott, "Sand movement by waves," *Beach Erosion Board Tech. Memo.* No. 48 (1954), 37p.
6. P. S. Eagleson and R. G. Dean, "Wave-induced motion of bottom sediment particles," *Trans. ASCE* **126**(I) (1961), pp. 1162–1189.
7. H. Lundgren and T. Sorensen, "A pulsating water tunnel," *Proc. 6th Conf. Coastal Engrg.* (Florida) (1957), pp. 356–358.
8. R. J. Hallermeier, "Sand motion initiation by water waves: two asymptotes," *J. Waterway, Port, Coastal, and Ocean Engrg.* **106**(WW3) (1980), pp. 299–318.
9. Z.-J. You, "A simple model of sediment initiation under waves," *Coastal Engineering* **41** (2000), pp. 399–412.

10. M. M. Das, "Suspended sediment and longshore sediment transport data review," *Proc. 13th Int. Conf. Coastal Engrg.* (Vancouver, ASCE) (1972), pp. 1027–1048.

11. J. F. A. Sleath, "The suspension of sand by waves," *J. Hydraulic Res.* **20**(5) (1982), pp. 439–452.

12. P. Nielsen, "Suspended sediment concentrations under waves," *Coastal Engineering* **10** (1986), pp. 23–31.

13. J. C. Fairchild, "Longshore transport of suspended sediment," *Proc. 13th Int. Conf. Coastal Engrg.* (Vancouver, ASCE) (1972), pp. 1069–1088.

14. T. W. Kana and L. G. Ward, "Nearshore suspended sediment load during storm and post-storm conditions," *Proc. 17th Int. Conf. on Coastal Engrg.* (Sydney, ASCE) (1980), pp. 1158–1173.

15. R. A. Beach and R. W. Sternberg, "Suspended sediment transport in the surf zone: response to cross-shore infragravity motion," *Marine Geology* **80** (1988), pp. 61–79.

16. K. Nadaoka, S. Ueno and T. Igarashi, "Sediment suspension due to large scale eddies in the surf zone," *Proc. 21st Int. Conf. on Coastal Engrg.* (Malaga, Spain, ASCE) (1988), pp. 1646–1660.

17. K. Nadaoka, S. Ueno and T. Igarashi, "Field observation of three-dimensional large-scale eddies and sediment suspension in the surf zone," *Coastal Engineering in Japan* **31**(2) (1988), pp. 277–287.

18. B. E. Jaffe and A. H. Sallenger, Jr., "The contribution of suspension events to sediment transport in the surf zone," *Proc. 23rd Int. Conf. on Coastal Engrg.* (Venice, ASCE) (1992) pp. 2680–2693.

19. K. P. Black, R. M. Gorman and G. Symonds, "Sediment transport near the break point associated with cross-shore gradient in vertical eddy diffusivity," *Coastal Engineering* **26** (1995), pp. 153–175.

20. H. C. Miller, "Field measurements of longshore sediment transport during storms," *Coastal Engineering* **36** (1999), pp. 301–321.

21. Y. Goda, "Call for engineering judgment in coastal engineering research," *Coastal Structures 2007* (*Proc. 5th Int. Conf.*, Venice, World Scientific) (2009), p. 3–16.

22. R. G. Dean and R. A. Dalrymple, *Coastal Processes with Engineering Applications* (Cambridge Univ. Press) (2002), p. 44.

23. P. D. Komar, *Beach Processes and Sedimentation* (Second Ed.) (Prentice Hall, Inc.) (1997), pp. 406–416.

24. R. Kos'yan, H. Kunz, N. Pykhov, S. Kuznetsov, I. Podymov and P. Vorobyez, "Physical regularities for the suspension and transport of sand under irregular waves," *Die Künste* **64** (2001), pp. 161–200.

25. J. S. M. van Thiel de Vries, "Dune erosion during storm surges," *Deltares Select Series* **3** (IOS Press, Amsterdam) (2009), 202p.

26. J. S. M. van Thiel de Vries, M. R. A. Gent, D. J. R. Walstra and A. J. H. M. Reniers, "Analysis of dune erosion processes in large-scale flume experiments," *Coastal Engineering* **55** (2008) pp. 1028–1040.

27. T. Shimizu, K. Kondo and A. Watanabe, "Study on applicability of local sediment transport formulas to field conditions," *Proc. Coastal Engrg. JSCE* **37** (1990), pp. 274–278 (*in Japanese*).

28. H. Katayama and Y. Goda, "Sediment suspension by random breaking waves evaluated from the CERC formula," *Proc. Coastal Sediments '99* (Long Island, New York, ASCE) (1999), pp. 1019–1033.

29. H. Katayama and Y. Goda, "A sediment pickup rate formula based on energy dissipation rate by random breaking," *Proc. 27th Int. Conf. Coastal Engrg.* (Sydney, ASCE) (2000), pp. 2859–2872.

30. H. Katayama and Y. Goda, "2DH beach changes due to suspended sediment picked-up by random breaking waves," *Coastal Engineering 2002 (Proc. 28th Int. Conf.*, Cardiff, Wales, World Scientific) (2003) pp. 2767–2779.

31. Y. Goda, "A 2-D random wave transformation model with gradational breaker index," *Coastal Engineering Journal* **46**(1) (2004), pp. 1–38.

32. Y. Goda and N. Watanabe, "A longshore current formula for random breaking waves," *Coastal Engineering in Japan* **34**(2) (1991), pp. 159–175.

33. N. Kobayashi and B. D. Johnson, "Sand suspension, storage, advection, and settling in surf and swash zones," *J. Geophys. Res.* **106**(C5) (2001), pp. 9363–9376.

34. N. Kobayashi, M. Buck, A. Payo and B. D. Johnson, "Berm and dune erosion during a storm," *J. Waterway, Port, Coastal, and Ocean Engrg.* **135**(1) (2009), pp. 1–10.

35. L. C. Van Rijn, "Unified view of sediment transport by currents and waves: Parts I, II, III and IV," *J. Hyraulic Engrg.* **133** (6, 7) (2007), pp. 649–689 (Parts I and II) and pp. 761–793 (Parts III and IV).

36. W. W. Rubey, "Settling velocity of gravel, sand, and silt," *American J. Science* **25**(148) (1933), pp. 325–338.

37. J. P. Ahrens, "Simple equations to calculate fall velocity and sediment scale parameter," *J. Waterway, Port, Coastal, and Ocean Engrg.* **129**(3) (2003), pp. 146–150.

38. J. P. Ahrens, "A fall velocity equation," *J. Waterway, Port, Coastal, and Ocean Engrg.* **126**(2) (2000), pp. 99–102.

39. R. G. Dean, " Equilibrium beach profiles: Principles and applications," *J. Coastal Res.* **7**(1) (1991), pp. 53–84.

40. R. G. Dean, "Coastal sediment processes: Toward engineering solutions," *Proc. Coastal Sediments* (ASCE) (1987), pp. 1–24.

41. J. R. Houston, "Simplified Dean's method for beach-fill design," *J. Waterway, Port, Coastal and Ocean Engrg.* **122**(3) (1996), pp. 143–148.

42. R. G. Dean and R. A. Dalrymple, *loc. cit.* 22, pp. 162–203.

43. D. L. Inman and R. A. Bagnold, "Littoral Processes," *The Sea* (ed. by M.N. Hill, Interscience, New York) (1963), pp. 3529–3533.

44. P. D. Komar and D. L. Inman, "Longshore sand transport on beaches," *J. Geophys. Res.* **76**(3) (1977), pp. 713–721.

45. U.S. Army Corps of Engrs., Coastal Engineering Research Center, *Shore Protection Manual* (4th Ed., U.S. Gov. Print. Office) (1984).

46. D. B. King, "Dependence of the CERC formula $K$ coefficient on grain size," *Coastal Engineering 2006 (Proc. 30th Int. Conf.*, San Diego, California, World Scientific) (2007), pp. 3381–3390.

47. J. W. Kamphuis, "Alongshore sediment transport rate," *J. Waterway, Port, Coastal and Ocean Engrg.* **117**(6) (1991), pp. 624–640.

48. A. Bayram, M. Larson and H. Hanson, "A new formula for the total longshore sediment transport rate," *Coastal Engineering* **54**(9) (2007), pp. 700–710.

49. J. A Bailard, "An energetic total load sediment transport model for a plane sloping beach," *J. Geophys. Res.* **86**(C11) (1981), pp. 10,938–10,954.

50. J. A. Bailard, "A simplified model for longshore sediment transport," *Proc. 19th Int. Conf. on Coastal Engrg.* (Houston, ASCE) (1984), pp. 1454–1470.

51. A. Bayram, M. Larson, H. C. Miller, and N. C Kraus, "Cross-shore distribution of longshore sediment transport: comparison between predictive formulas and field measurements," *Coastal Engineering* **44** (2001), pp. 79–99.

52. A. G. Davies, L. C. van Rijn, J. S. Damgaard, J. van de Graaff and J. S. Ribberink, " Intercomparison of research and practical sand transport models," *Coastal Engineering* **46** (2002), pp. 1–23.

53. B. van Maanen, P. J. de Ruiter and B. G. Ruessink, "An evaluation of two alongshore transport equation with field measurements models," *Coastal Engineering* **56** (2009), pp. 313–319.

54. H. Ozasa and A. H. Brampton, "Mathematical modeling of beaches backed by seawalls," *Coastal Engineering* **4** (1980), pp. 47–63.

55. R. J. Hallermeier, "Uses for a calculated limit depth of beach erosion," *Proc. 16th Int. Conf. Coastal Engrg.* (Hamburg, ASCE) (1978), pp. 1483–1512.

56. W. A. Birkemeier, "Field data on seaward limit of profile change," *J. Waterway, Port, Coastal and Ocean Engrg.* **111**(3) (1985), pp. 598–602.

57. Y. Kuriyama, *Beach Deformation — Features, Prediction, and Prevention* (Gihodo Pub., Tokyo) (2006), pp. 65–66 (*in Japanese*).

58. R. L. Rector, "Laboratory study of equilibrium profiles of beach," *Beach Erosion Board Tech. Memo.* No. 41 (1954), 38p.

59. D. H. Swart, "A schematization of onshore-offshore transport," *Proc. 14th Int. Conf. Coastal Engrg.* (Copenhagen, ASCE) (1974), pp. 884–900.

60. T. Sunamura, "'Static' relationship among beach slope, sand size, and wave properties," *Geographical Review of Japan* **48**(7) (1975).

61. Y. Kuriyama, *loc cit.* 57, pp. 67–68.

62. M. Kuroiwa, H. Noda, C.-B. Son, K. Kato and S. Taniguchi, "Numerical prediction of bottom topographical change using quasi-3D nearshore current model," *Coastal Engineering 2000* (*Proc. 27th Int. Conf.*, Sydney, ASCE) (2001), pp. 2914–2927.

63. T. Sawaragi, S.-I. Lee and I. Deguchi, "Study on the model of nearshore currents and beach deformation around a river mouth," *Proc. 31st Japanese Conf. Coastal Engrg.* (1984), pp. 411–415 (*in Japanese*).

64. M. Kuroiwa, Y. Matsubara, T. Kuchiishi, K. Katoh, H. Noda and C.-B. Son, "A morphodynamic model based on Q-3D nearshore current model and application to barred beach," *Coastal Engineering 2002* (*Proc. 28th Int. Conf.*, Cardiff, Wales, World Scientific), (2003), pp. 3409–3421.

65. S. Nakamura and Y. Goda, "Quantitative evaluation of the blockage efficiency of a groin against alongshore bedload transport in consideration of random wave features and wave climate," *J. Japan Soc. Civil Engrs.* **63-B**(4) (2007), pp. 272–281 (*in Japanese*).

66. K. Kuroki, Y. Goda, N. Panin, A. Stanica, D. I. Diaconeasa and G. Babu, "Beach erosion and coastal protection plan along the Southern Romanian Black Sea shore," *Coastal Engineering 2006* (*Proc. 30th Int. Conf.*, San Diego, California, World Scientific) (2007), pp. 3788–3799.

67. ECOH CORPORATION, *The Study on Protection and Rehabilitation of the Southern Romanian Black Sea Shore in Romania* (Japan International Co-operation Agency (JICA) Report 07-030, Project ID 7241013F0) (2007), **I**, pp. 4-10–4-14.

68. L. Franco, G. R. Tomasicchio, and A. Lamberti (editors), *Coastal Structures 2007* (*Proc. 5th Int. Conf.*, Venice, World Scientific) (2009), p. 1.

69. A. Morang and A. Szuwalski (editors), "Glossary of coastal terminology," Appendix A of *Coastal Engineering Manual* EM 1110-2-1100, U. S. Corps of Engineers (2002), pp. A-42.

70. G. K. Nersesian, N. C. Kraus and F. C. Carson, "Functioning of groins at Westhampton Beach, Long Island, New York," *Proc. 23rd Int. Conf. on Coastal Engrg.* (Venice, ASCE) (1992) pp. 3357–3370.

71. R. G. Dean and R. A. Dalrymple, *loc. cit.* 22, p. 387.

72. H. Hanson, L. Bocamazo, M. Larson and N. C. Kraus, "Long-term beach response to groin shortening Westhampton Beach, Long Island, New York," *Coastal Engineering 2008* (*Proc. 31th Int. Conf.*, Hamburg, World Scientific) (2009) pp. 1927–1939.

73. D. Holmberg, "Alternative to traditional ways of treating shoreline erosion," *Proc. 1st Conf. Soft Shore Protection* (Patras, Greece) (2001), pp. 139–150.

74. C. L. Goudas, G. A. Katsiaris, G. Labeas, G. Karahalios and G. Pnevmatikos, "Soft protection using submerged groin arrangements — Dynamic analysis of system stability and review of application impacts," *Proc. 1st Conf. Soft Shore Protection* (Patras, Greece) (2001), pp. 167–186.

75. P. L. Aminti, C. Cammelli, L. E. Cipriani and E. Pranzini, "Evaluating the effectiveness of a submerged groin as soft shore protection," *Proc. 1st Conf. Soft Shore Protection* (Patras, Greece) (2001), pp. 151–158.

76. P. L. Aminti, L. Cappietti, C. D'Elso and E. Mori, "Numerical simulation of an experimental submerged groin system," *Coastal Structures 2007* (*Proc. 5th Int. Conf.*, Venice, World Scientific) (2009), pp. 1511–1519.

77. Y. Goda, N. Ono and Y. Uno, "Examination of efficacy of submerged groin system against beach erosion through numerical simulation," *Coastal Dynamics 2009* (*Proc. 6th Int. Conf.*, Tokyo) (2009), Paper No. 41 (CD-ROM).

78. A. Lamberti, R. Archetti, M. Kramer, D. Paphitis, C. Mosso and M. Di Risio, "European experience of low crested structures for coastal management," *Coastal Engineering,* **52** (2005), pp. 841–866.

79. Y. Goda, "Wave damping characteristics of longitudinal reef systems," *Advances in Coastal Structures and Breakwaters* (Ed. by J. E. Clifford, Thomas Telford, London) (1996), pp. 192–203.

80. Y. Goda, "Wave field computation around artificial reefs with gradational breaker model," *Coastal Structures 2003* (*Proc. Conf.*, Portland, Oregon, ASCE) (2004), pp. 824–836.

81. R. Silvester and J. R. C. Hsu, *Coastal Stabilization* (World Scientific) (1993), pp. 340–385.

82. R. G. Dean and R. A. Dalrymple, *loc. cit.* 22, pp. 400–401.
83. K. Saito, T. Uda, K. Yokota, S. Ohara, K. Kawanakajima and K. Uchida, "Observations of nearshore currents and beach changes around headlands built on the Kashimanada Coast, Japan," *Proc. 25th Int. Conf. Coastal Engrg.* (Orland, Florida, ASCE) (1996), pp. 4000–4013.
84. T. Tsukiyama, A. Kimura, T. Takagi and S. Hashimoto, "Study on the sand trapping effect of headland control works in the southern coast of Sendai Bay," *Proc. Coastal Engrg., JSCE* **50** (2003), pp. 521–525 (*in Japanese*).
85. R. G. Dean and R. A. Dalrymple, *loc. cit.* 22, p. 343–374.
86. M. Toti and, P. Cuccioletta and A. Ferrante "Beach nourishment at Lido di Oscia (Rome)," *Proc. 27th Int. Navigation Congress* (Tokyo, PIANC) (1990), S.II-2-1, pp. 23–28.
87. B. Zanuttigh, "Numerical modelling of the morphological response induced by low-crested structures in Lido di Dante, Italy," *Coastal Engineering* **54** (2007), pp. 31–47.
88. A. Lamberti, B. Zanuttigh and L. Martinelli, "On the predictability of nourishment performance by numerical models: A prototype case in Emilia Romagna, Italy," *Coastal Engineering 2008* (*Proc. 31st Int. Conf.*, Hamburg, World Scientific) (2009), pp. 2519–2531.
89. R. Nishi, R. D. Dean and R. Tanaka, "Beach nourishment projects in Japan in terms of its size and cost," *Annual J. Civil Engrg. Ocean, JSCE* **21** (2005), pp. 355–360 (*in Japanese*).

Appendix

# List of Wavelength and Celerity for a Given Wave Period and Water Depth

Table A.1   List of wavelength and celerity for a given wave period and water depth $(g = 9.8 \text{ m/s}^2)$.

| Wave period (s) Water depth (m) | 2.0 Wavelength (m) | Celerity (m/s) | 2.5 Wavelength (m) | Celerity (m/s) | 3.0 Wavelength (m) | Celerity (m/s) | 4.0 Wavelength (m) | Celerity (m/s) | 5.0 Wavelength (m) | Celerity (m/s) |
|---|---|---|---|---|---|---|---|---|---|---|
| 0.1 | 1.97 | 0.97 | 2.45 | 0.98 | 2.95 | 0.98 | 3.94 | 0.99 | 4.94 | 0.99 |
| 0.2 | 2.71 | 1.35 | 3.42 | 1.37 | 4.14 | 1.38 | 5.55 | 1.39 | 6.96 | 1.39 |
| 0.3 | 3.26 | 1.63 | 4.15 | 1.66 | 5.03 | 1.68 | 6.77 | 1.69 | 8.50 | 1.70 |
| 0.4 | 3.69 | 1.85 | 4.74 | 1.89 | 5.76 | 1.92 | 7.79 | 1.95 | 9.79 | 1.96 |
| 0.5 | 4.05 | 2.03 | 5.24 | 2.09 | 6.39 | 2.13 | 8.67 | 2.17 | 10.92 | 2.18 |
| 0.6 | 4.36 | 2.18 | 5.67 | 2.27 | 6.95 | 2.32 | 9.45 | 2.36 | 11.93 | 2.39 |
| 0.7 | 4.62 | 2.31 | 6.05 | 2.42 | 7.45 | 2.48 | 10.17 | 2.54 | 12.85 | 2.57 |
| 0.8 | 4.85 | 2.42 | 6.40 | 2.56 | 7.90 | 2.63 | 10.82 | 2.71 | 13.70 | 2.74 |
| 0.9 | 5.04 | 2.52 | 6.70 | 2.68 | 8.31 | 2.77 | 11.43 | 2.86 | 14.49 | 2.90 |
| 1.0 | 5.21 | 2.61 | 6.98 | 2.79 | 8.69 | 2.90 | 11.99 | 3.00 | 15.23 | 3.05 |
| 1.1 | 5.36 | 2.68 | 7.23 | 2.89 | 9.04 | 3.01 | 12.52 | 3.13 | 15.93 | 3.19 |
| 1.2 | 5.49 | 2.74 | 7.46 | 2.99 | 9.36 | 3.12 | 13.02 | 3.26 | 16.59 | 3.32 |
| 1.3 | 5.60 | 2.80 | 7.67 | 3.07 | 9.66 | 3.22 | 13.50 | 3.37 | 17.22 | 3.44 |
| 1.4 | 5.70 | 2.85 | 7.87 | 3.15 | 9.95 | 3.32 | 13.94 | 3.49 | 17.82 | 3.56 |
| 1.5 | 5.78 | 2.89 | 8.04 | 3.22 | 10.21 | 3.40 | 14.37 | 3.59 | 18.40 | 3.68 |
| 1.6 | 5.85 | 2.93 | 8.20 | 3.28 | 10.46 | 3.49 | 14.77 | 3.69 | 18.95 | 3.79 |
| 1.8 | 5.96 | 2.98 | 8.48 | 3.39 | 10.90 | 3.63 | 15.53 | 3.88 | 19.98 | 4.00 |
| 2.0 | 6.05 | 3.02 | 8.72 | 3.49 | 11.30 | 3.77 | 16.22 | 4.05 | 20.94 | 4.19 |
| 2.2 | 6.11 | 3.05 | 8.91 | 3.56 | 11.65 | 3.88 | 16.85 | 4.21 | 21.84 | 4.37 |
| 2.5 | 6.16 | 3.08 | 9.14 | 3.66 | 12.09 | 4.03 | 17.71 | 4.43 | 23.08 | 4.62 |
| 3.0 | 6.21 | 3.11 | 9.40 | 3.76 | 12.67 | 4.22 | 18.95 | 4.74 | 24.92 | 4.98 |
| 3.5 | 6.23 | 3.11 | 9.55 | 3.82 | 13.09 | 4.36 | 19.98 | 5.00 | 26.52 | 5.30 |
| 4.0 | 6.23 | 3.12 | 9.64 | 3.86 | 13.39 | 4.46 | 20.85 | 5.21 | 27.93 | 5.59 |
| 4.5 | 6.24 | 3.12 | 9.69 | 3.88 | 13.60 | 4.53 | 21.57 | 5.39 | 29.18 | 5.84 |
| 5.0 | 6.24 | 3.12 | 9.72 | 3.89 | 13.75 | 4.58 | 22.18 | 5.55 | 30.29 | 6.06 |
| 6.0 | 6.24 | 3.12 | 9.74 | 3.90 | 13.91 | 4.64 | 23.11 | 5.78 | 32.17 | 6.43 |
| 7.0 | 6.24 | 3.12 | 9.75 | 3.90 | 13.99 | 4.66 | 23.75 | 5.94 | 33.67 | 6.73 |
| 8.0 | 6.24 | 3.12 | 9.75 | 3.90 | 14.02 | 4.67 | 24.19 | 6.05 | 34.86 | 6.97 |
| 9.0 | 6.24 | 3.12 | 9.75 | 3.90 | 14.03 | 4.68 | 24.47 | 6.12 | 35.81 | 7.16 |
| 10.0 | 6.24 | 3.12 | 9.75 | 3.90 | 14.03 | 4.68 | 24.65 | 6.16 | 36.56 | 7.31 |
| 11.0 | 6.24 | 3.12 | 9.75 | 3.90 | 14.04 | 4.68 | 24.77 | 6.19 | 37.15 | 7.43 |
| 12.0 | 6.24 | 3.12 | 9.75 | 3.90 | 14.04 | 4.68 | 24.84 | 6.21 | 37.60 | 7.52 |
| 13.0 | 6.24 | 3.12 | 9.75 | 3.90 | 14.04 | 4.68 | 24.89 | 6.22 | 37.95 | 7.59 |
| 14.0 | 6.24 | 3.12 | 9.75 | 3.90 | 14.04 | 4.68 | 24.91 | 6.23 | 38.22 | 7.64 |
| 15.0 | 6.24 | 3.12 | 9.75 | 3.90 | 14.04 | 4.68 | 24.93 | 6.23 | 38.42 | 7.68 |
| 16.0 | 6.24 | 3.12 | 9.75 | 3.90 | 14.04 | 4.68 | 24.94 | 6.23 | 38.57 | 7.71 |
| 17.0 | 6.24 | 3.12 | 9.75 | 3.90 | 14.04 | 4.68 | 24.95 | 6.24 | 38.68 | 7.74 |
| 18.0 | 6.24 | 3.12 | 9.75 | 3.90 | 14.04 | 4.68 | 24.95 | 6.24 | 38.77 | 7.75 |
| 19.0 | 6.24 | 3.12 | 9.75 | 3.90 | 14.04 | 4.68 | 24.95 | 6.24 | 38.83 | 7.77 |
| 20.0 | 6.24 | 3.12 | 9.75 | 3.90 | 14.04 | 4.68 | 24.95 | 6.24 | 38.87 | 7.77 |
| Deepwater waves | 6.24 | 3.12 | 9.75 | 3.90 | 14.04 | 4.68 | 24.96 | 6.24 | 38.99 | 7.80 |

Table A.2  List of wavelength and celerity for a given wave period and water depth $(g = 9.8 \text{ m/s}^2)$.

| Wave period (s) Water depth (m) | 6.0 Wave-length (m) | Cel-erity (m/s) | 7.0 Wave-length (m) | Cel-erity (m/s) | 8.0 Wave-length (m) | Cel-erity (m/s) | 9.0 Wave-length (m) | Cel-erity (m/s) | 10.0 Wave-length (m) | Cel-erity (m/s) |
|---|---|---|---|---|---|---|---|---|---|---|
| 0.5 | 13.16 | 2.19 | 15.39 | 2.20 | 17.62 | 2.20 | 19.84 | 2.20 | 22.06 | 2.21 |
| 1.0 | 18.43 | 3.07 | 21.61 | 3.09 | 24.78 | 3.10 | 27.94 | 3.10 | 31.09 | 3.11 |
| 1.5 | 22.36 | 3.73 | 26.29 | 3.76 | 30.19 | 3.77 | 34.08 | 3.79 | 37.95 | 3.80 |
| 2.0 | 25.57 | 4.26 | 30.14 | 4.31 | 34.67 | 4.33 | 39.18 | 4.35 | 43.68 | 4.37 |
| 2.5 | 28.31 | 4.72 | 33.46 | 4.78 | 38.56 | 4.82 | 43.62 | 4.85 | 48.67 | 4.87 |
| 3.0 | 30.71 | 5.12 | 36.39 | 5.20 | 42.01 | 5.25 | 47.58 | 5.29 | 53.13 | 5.31 |
| 3.5 | 32.84 | 5.47 | 39.02 | 5.57 | 45.13 | 5.64 | 51.18 | 5.69 | 57.19 | 5.72 |
| 4.0 | 34.75 | 5.79 | 41.42 | 5.92 | 47.98 | 6.00 | 54.48 | 6.05 | 60.92 | 6.09 |
| 4.5 | 36.49 | 6.08 | 43.61 | 6.23 | 50.61 | 6.33 | 57.53 | 6.39 | 64.40 | 6.44 |
| 5.0 | 38.07 | 6.34 | 45.63 | 6.52 | 53.05 | 6.63 | 60.38 | 6.71 | 67.64 | 6.75 |
| 6.0 | 40.84 | 6.81 | 49.24 | 7.03 | 57.47 | 7.18 | 65.57 | 7.29 | 73.58 | 7.36 |
| 7.0 | 43.19 | 7.20 | 52.39 | 7.48 | 61.37 | 7.67 | 70.20 | 7.80 | 78.92 | 7.89 |
| 8.0 | 45.19 | 7.53 | 55.16 | 7.88 | 64.86 | 8.11 | 74.38 | 8.26 | 83.77 | 8.38 |
| 9.0 | 46.91 | 7.82 | 57.61 | 8.23 | 68.01 | 8.50 | 78.19 | 8.69 | 88.22 | 8.82 |
| 10.0 | 48.37 | 8.06 | 59.78 | 8.54 | 70.85 | 8.86 | 81.68 | 9.08 | 92.32 | 9.23 |
| 11.0 | 49.62 | 8.27 | 61.72 | 8.82 | 73.44 | 9.18 | 84.89 | 9.43 | 96.12 | 9.61 |
| 12.0 | 50.69 | 8.45 | 63.44 | 9.06 | 75.80 | 9.48 | 87.85 | 9.76 | 99.67 | 9.97 |
| 13.0 | 51.60 | 8.60 | 64.98 | 9.28 | 77.96 | 9.74 | 90.59 | 10.07 | 102.98 | 10.30 |
| 14.0 | 52.38 | 8.73 | 66.35 | 9.48 | 79.93 | 9.99 | 93.14 | 10.35 | 106.07 | 10.61 |
| 15.0 | 53.03 | 8.84 | 67.58 | 9.65 | 81.73 | 10.22 | 95.51 | 10.61 | 108.98 | 10.90 |
| 16.0 | 53.58 | 8.93 | 68.66 | 9.81 | 83.39 | 10.42 | 97.71 | 10.86 | 111.71 | 11.17 |
| 17.0 | 54.04 | 9.01 | 69.63 | 9.95 | 84.90 | 10.61 | 99.77 | 11.09 | 114.29 | 11.43 |
| 18.0 | 54.42 | 9.07 | 70.49 | 10.07 | 86.29 | 10.79 | 101.68 | 11.30 | 116.71 | 11.67 |
| 19.0 | 54.74 | 9.12 | 71.25 | 10.18 | 87.56 | 10.95 | 103.47 | 11.50 | 119.00 | 11.90 |
| 20.0 | 55.00 | 9.17 | 71.92 | 10.27 | 88.72 | 11.09 | 105.14 | 11.68 | 121.16 | 12.21 |
| 22.0 | 55.39 | 9.23 | 73.03 | 10.43 | 90.76 | 11.35 | 108.14 | 12.02 | 125.12 | 12.51 |
| 24.0 | 55.65 | 9.28 | 73.89 | 10.56 | 92.46 | 11.56 | 110.76 | 12.31 | 128.66 | 12.87 |
| 26.0 | 55.83 | 9.30 | 74.54 | 10.65 | 93.86 | 11.73 | 113.04 | 12.56 | 131.83 | 13.18 |
| 28.0 | 55.94 | 9.32 | 75.03 | 10.72 | 95.02 | 11.88 | 115.01 | 12.78 | 134.66 | 13.47 |
| 30.0 | 56.02 | 9.34 | 75.40 | 10.77 | 95.97 | 12.00 | 116.72 | 12.97 | 137.19 | 13.72 |
| 35.0 | 56.11 | 9.35 | 75.96 | 10.85 | 97.64 | 12.20 | 120.03 | 13.34 | 142.38 | 14.24 |
| 40.0 | 56.14 | 9.36 | 76.22 | 10.89 | 98.61 | 12.33 | 122.26 | 13.58 | 146.25 | 14.63 |
| 45.0 | 56.15 | 9.36 | 76.33 | 10.90 | 99.16 | 12.39 | 123.75 | 13.75 | 149.10 | 14.91 |
| 50.0 | 56.15 | 9.36 | 76.39 | 10.91 | 99.46 | 12.43 | 124.71 | 13.86 | 151.16 | 15.12 |
| 55.0 | 56.15 | 9.36 | 76.41 | 10.92 | 99.63 | 12.45 | 125.32 | 13.92 | 152.64 | 15.26 |
| 60.0 | 56.15 | 9.36 | 76.42 | 10.92 | 99.72 | 12.46 | 125.71 | 13.97 | 153.68 | 15.37 |
| 65.0 | 56.15 | 9.36 | 76.42 | 10.92 | 99.77 | 12.47 | 125.95 | 13.99 | 154.41 | 15.44 |
| 70.0 | 56.15 | 9.36 | 76.42 | 10.92 | 99.97 | 12.47 | 126.10 | 14.01 | 154.91 | 15.49 |
| 75.0 | 56.15 | 9.36 | 76.43 | 10.92 | 99.81 | 12.48 | 126.19 | 14.02 | 155.25 | 15.53 |
| 80.0 | 56.15 | 9.36 | 76.43 | 10.92 | 99.81 | 12.48 | 126.25 | 14.03 | 155.49 | 15.55 |
| Deepwater waves | 56.15 | 9.36 | 76.43 | 10.92 | 99.82 | 12.48 | 126.34 | 14.04 | 155.97 | 15.60 |

Table A.3  List of wavelength and celerity for a given wave period and water depth $(g = 9.8 \text{ m/s}^2)$.

| Wave period (s) Water depth (m) | 11.0 Wavelength (m) | Celerity (m/s) | 12.0 Wavelength (m) | Celerity (m/s) | 13.0 Wavelength (m) | Celerity (m/s) | 14.0 Wavelength (m) | Celerity (m/s) | 15.0 Wavelength (m) | Celerity (m/s) |
|---|---|---|---|---|---|---|---|---|---|---|
| 1.0 | 34.2 | 3.11 | 37.4 | 3.12 | 40.5 | 3.12 | 43.7 | 3.12 | 46.8 | 3.12 |
| 2.0 | 48.2 | 4.38 | 52.6 | 4.39 | 57.1 | 4.39 | 61.6 | 4.40 | 66.0 | 4.40 |
| 3.0 | 58.6 | 5.33 | 64.2 | 5.35 | 69.6 | 5.36 | 75.1 | 5.37 | 80.6 | 5.37 |
| 4.0 | 67.3 | 6.12 | 73.7 | 6.14 | 80.1 | 6.16 | 86.5 | 6.18 | 92.8 | 6.19 |
| 5.0 | 74.9 | 6.81 | 82.0 | 6.84 | 89.2 | 6.86 | 96.3 | 6.88 | 103.4 | 6.90 |
| 6.0 | 81.5 | 7.41 | 89.4 | 7.45 | 97.3 | 7.48 | 105.1 | 7.51 | 113.0 | 7.53 |
| 7.0 | 87.6 | 7.96 | 96.1 | 8.01 | 104.7 | 8.05 | 113.2 | 8.08 | 121.6 | 8.11 |
| 8.0 | 93.1 | 8.46 | 102.3 | 8.52 | 111.4 | 8.57 | 120.6 | 8.61 | 129.6 | 8.64 |
| 9.0 | 98.1 | 8.92 | 108.0 | 9.00 | 117.7 | 9.05 | 127.4 | 9.10 | 137.1 | 9.14 |
| 10.0 | 102.8 | 9.35 | 113.2 | 9.44 | 123.6 | 9.50 | 133.8 | 9.56 | 144.1 | 9.60 |
| 11.0 | 107.2 | 9.75 | 118.2 | 9.85 | 129.1 | 9.93 | 139.9 | 9.99 | 150.6 | 10.04 |
| 12.0 | 111.3 | 10.12 | 122.8 | 10.24 | 134.2 | 10.33 | 145.6 | 10.40 | 156.8 | 10.45 |
| 13.0 | 115.2 | 10.47 | 127.2 | 10.60 | 139.1 | 10.70 | 151.0 | 10.78 | 162.7 | 10.85 |
| 14.0 | 118.8 | 10.80 | 131.3 | 10.95 | 143.8 | 11.06 | 156.1 | 11.15 | 168.3 | 11.22 |
| 15.0 | 122.2 | 11.11 | 135.3 | 11.27 | 148.2 | 11.40 | 161.0 | 11.50 | 173.7 | 11.58 |
| 16.0 | 125.5 | 11.41 | 139.0 | 11.58 | 152.4 | 11.72 | 165.7 | 11.83 | 178.8 | 11.92 |
| 17.0 | 128.5 | 11.68 | 142.6 | 11.88 | 156.4 | 12.03 | 170.1 | 12.15 | 183.8 | 12.25 |
| 18.0 | 131.4 | 11.95 | 145.9 | 12.16 | 160.3 | 12.33 | 174.4 | 12.46 | 188.5 | 12.57 |
| 19.0 | 134.2 | 12.20 | 149.2 | 12.43 | 163.9 | 12.61 | 178.6 | 12.75 | 193.0 | 12.87 |
| 20.0 | 136.8 | 12.44 | 152.3 | 12.69 | 167.5 | 12.88 | 182.5 | 13.04 | 197.4 | 13.16 |
| 22.0 | 141.7 | 12.89 | 158.1 | 13.17 | 174.1 | 13.39 | 190.0 | 13.57 | 205.7 | 13.72 |
| 24.0 | 146.2 | 13.29 | 163.4 | 13.61 | 180.3 | 13.87 | 197.0 | 14.07 | 213.5 | 14.23 |
| 26.0 | 150.2 | 13.66 | 168.3 | 14.02 | 186.0 | 14.31 | 203.5 | 14.53 | 220.8 | 14.72 |
| 28.0 | 153.9 | 13.99 | 172.8 | 14.40 | 191.3 | 14.72 | 209.6 | 14.97 | 227.6 | 15.17 |
| 30.0 | 157.3 | 14.30 | 176.9 | 14.74 | 196.2 | 15.10 | 215.3 | 15.38 | 234.1 | 15.60 |
| 35.0 | 164.4 | 14.95 | 186.0 | 15.50 | 207.2 | 15.94 | 228.1 | 16.29 | 248.7 | 16.58 |
| 40.0 | 170.1 | 15.46 | 193.5 | 16.12 | 216.5 | 16.65 | 239.1 | 17.08 | 261.4 | 17.43 |
| 45.0 | 174.5 | 15.86 | 199.6 | 16.64 | 224.4 | 17.26 | 248.7 | 17.76 | 272.6 | 18.17 |
| 50.0 | 178.0 | 16.18 | 204.7 | 17.06 | 231.0 | 17.77 | 256.9 | 18.35 | 282.5 | 18.83 |
| 55.0 | 180.7 | 16.42 | 208.8 | 17.40 | 236.6 | 18.20 | 264.1 | 18.86 | 291.1 | 19.41 |
| 60.0 | 182.7 | 16.61 | 212.1 | 17.68 | 241.4 | 18.57 | 270.3 | 19.31 | 298.8 | 19.92 |
| 70.0 | 185.5 | 16.86 | 216.9 | 18.08 | 248.7 | 19.13 | 280.3 | 20.02 | 311.6 | 20.77 |
| 80.0 | 187.0 | 17.00 | 220.0 | 18.33 | 253.7 | 19.52 | 287.7 | 20.55 | 321.5 | 21.43 |
| 90.0 | 187.8 | 17.07 | 221.9 | 18.49 | 257.2 | 19.78 | 293.1 | 20.93 | 329.1 | 21.94 |
| 100.0 | 188.3 | 17.11 | 233.0 | 18.58 | 259.5 | 19.96 | 297.0 | 21.21 | 334.9 | 22.32 |
| 120.0 | 188.6 | 17.15 | 224.1 | 18.67 | 261.9 | 20.15 | 301.6 | 21.54 | 342.5 | 22.83 |
| 140.0 | 188.7 | 17.15 | 244.4 | 18.70 | 262.9 | 20.23 | 303.8 | 21.70 | 346.6 | 23.11 |
| 160.0 | 188.7 | 17.16 | 224.5 | 18.71 | 263.3 | 20.26 | 304.9 | 21.78 | 348.7 | 23.25 |
| 180.0 | 188.7 | 17.16 | 224.6 | 18.72 | 263.5 | 20.27 | 305.3 | 21.81 | 349.8 | 23.32 |
| 200.0 | 188.7 | 17.16 | 224.6 | 18.72 | 263.6 | 20.27 | 305.5 | 21.82 | 350.4 | 23.36 |
| Deepwater waves | 188.7 | 17.16 | 224.6 | 18.72 | 263.6 | 20.28 | 305.7 | 21.84 | 350.9 | 23.40 |

Table A.4 List of wavelength and celerity for a given wave period and water depth ($g = 9.8$ m/s$^2$).

| Wave period (s) Water depth (m) | 16.0 Wave-length (m) | Cel-erity (m/s) | 17.0 Wave-length (m) | Cel-erity (m/s) | 18.0 Wave-length (m) | Cel-erity (m/s) | 19.0 Wave-length (m) | Cel-erity (m/s) | 20.0 Wave-length (m) | Cel-erity (m/s) |
|---|---|---|---|---|---|---|---|---|---|---|
| 1.0 | 50.0 | 3.12 | 53.1 | 3.12 | 56.2 | 3.12 | 59.4 | 3.12 | 62.5 | 3.13 |
| 2.0 | 70.5 | 4.40 | 74.9 | 4.41 | 79.4 | 4.41 | 83.8 | 4.41 | 88.2 | 4.41 |
| 3.0 | 86.1 | 5.38 | 91.5 | 5.38 | 97.0 | 5.39 | 102.4 | 5.39 | 107.9 | 5.39 |
| 4.0 | 99.1 | 6.20 | 105.4 | 6.20 | 111.8 | 6.21 | 118.1 | 6.21 | 124.4 | 6.22 |
| 5.0 | 110.5 | 6.91 | 117.6 | 6.92 | 124.7 | 6.93 | 131.8 | 6.93 | 138.8 | 6.94 |
| 6.0 | 120.8 | 7.55 | 128.5 | 7.56 | 136.3 | 7.57 | 144.1 | 7.58 | 151.8 | 7.59 |
| 7.0 | 130.1 | 8.13 | 138.5 | 8.15 | 146.9 | 8.16 | 155.3 | 8.17 | 163.7 | 8.19 |
| 8.0 | 138.7 | 8.67 | 147.7 | 8.69 | 156.7 | 8.71 | 165.7 | 8.72 | 174.7 | 8.74 |
| 9.0 | 146.7 | 9.17 | 156.3 | 9.19 | 165.9 | 9.22 | 175.4 | 9.23 | 185.0 | 9.25 |
| 10.0 | 154.2 | 9.64 | 164.4 | 9.67 | 174.6 | 9.69 | 184.6 | 9.72 | 194.7 | 9.73 |
| 11.0 | 161.3 | 10.08 | 172.0 | 10.12 | 182.6 | 10.15 | 193.2 | 10.17 | 203.8 | 10.19 |
| 12.0 | 168.0 | 10.50 | 179.2 | 10.54 | 190.3 | 10.57 | 201.4 | 10.60 | 212.5 | 10.63 |
| 13.0 | 174.4 | 10.90 | 186.1 | 10.95 | 197.7 | 10.98 | 209.3 | 11.01 | 220.8 | 11.04 |
| 14.0 | 180.5 | 11.28 | 192.6 | 11.33 | 204.7 | 11.37 | 216.7 | 11.41 | 228.7 | 11.44 |
| 15.0 | 186.3 | 11.65 | 198.9 | 11.70 | 211.4 | 11.75 | 223.9 | 11.79 | 236.4 | 11.82 |
| 16.0 | 191.9 | 11.99 | 204.9 | 12.06 | 217.9 | 12.11 | 230.8 | 12.15 | 243.7 | 12.18 |
| 17.0 | 197.3 | 12.33 | 210.7 | 12.40 | 224.1 | 12.45 | 237.5 | 12.50 | 250.8 | 12.54 |
| 18.0 | 202.4 | 12.65 | 216.3 | 12.72 | 230.1 | 12.78 | 243.9 | 12.84 | 257.6 | 12.88 |
| 19.0 | 207.4 | 12.96 | 221.7 | 13.04 | 235.9 | 13.11 | 250.1 | 13.16 | 264.2 | 13.21 |
| 20.0 | 212.2 | 13.26 | 226.9 | 13.35 | 241.5 | 13.42 | 256.1 | 13.48 | 270.6 | 13.53 |
| 22.0 | 221.3 | 13.83 | 236.8 | 13.93 | 252.2 | 14.01 | 267.5 | 14.08 | 282.8 | 14.14 |
| 24.0 | 229.9 | 14.37 | 246.1 | 14.48 | 262.3 | 14.57 | 278.3 | 14.65 | 294.3 | 14.72 |
| 26.0 | 237.9 | 14.87 | 254.9 | 14.99 | 271.8 | 15.10 | 288.6 | 15.19 | 305.3 | 15.26 |
| 28.0 | 245.5 | 15.34 | 263.2 | 15.48 | 280.8 | 15.60 | 298.3 | 15.70 | 315.7 | 15.78 |
| 30.0 | 252.7 | 15.79 | 271.1 | 15.95 | 289.4 | 16.08 | 307.5 | 16.19 | 325.6 | 16.28 |
| 35.0 | 269.0 | 16.81 | 289.1 | 17.01 | 309.1 | 17.17 | 328.9 | 17.31 | 348.6 | 17.43 |
| 40.0 | 283.4 | 17.71 | 305.2 | 17.95 | 326.7 | 18.15 | 348.1 | 18.32 | 369.3 | 18.46 |
| 45.0 | 296.2 | 18.51 | 319.5 | 18.80 | 342.6 | 19.03 | 365.5 | 19.23 | 388.1 | 19.41 |
| 50.0 | 307.6 | 19.23 | 332.4 | 19.56 | 357.0 | 19.83 | 381.3 | 20.07 | 405.4 | 20.27 |
| 55.0 | 317.8 | 19.86 | 344.1 | 20.24 | 370.1 | 20.56 | 395.8 | 20.83 | 421.3 | 21.06 |
| 60.0 | 326.9 | 20.43 | 354.7 | 20.86 | 382.0 | 21.22 | 409.1 | 21.53 | 435.9 | 21.80 |
| 70.0 | 342.4 | 21.40 | 372.9 | 21.94 | 403.0 | 22.39 | 432.7 | 22.77 | 462.1 | 23.10 |
| 80.0 | 354.9 | 22.18 | 387.9 | 22.82 | 420.5 | 23.36 | 452.8 | 23.83 | 484.6 | 24.23 |
| 90.0 | 364.9 | 22.80 | 400.3 | 23.55 | 435.3 | 24.19 | 470.0 | 24.73 | 504.2 | 25.21 |
| 100.0 | 372.8 | 23.30 | 410.4 | 24.14 | 447.8 | 24.88 | 484.7 | 25.51 | 521.2 | 26.06 |
| 120.0 | 383.9 | 23.99 | 425.4 | 25.03 | 466.9 | 25.94 | 508.0 | 16.74 | 548.8 | 27.44 |
| 140.0 | 390.6 | 24.41 | 435.2 | 25.60 | 480.1 | 26.67 | 524.9 | 27.63 | 569.6 | 28.48 |
| 160.0 | 394.4 | 24.65 | 441.4 | 25.96 | 489.1 | 27.17 | 537.0 | 28.26 | 585.0 | 26.25 |
| 180.0 | 396.6 | 24.79 | 445.2 | 26.19 | 495.0 | 27.50 | 545.5 | 28.71 | 596.4 | 29.82 |
| 200.0 | 397.8 | 24.87 | 447.5 | 26.32 | 498.8 | 27.71 | 551.4 | 29.02 | 604.6 | 30.23 |
| Deepwater waves | 399.3 | 24.96 | 450.8 | 26.52 | 505.3 | 28.07 | 563.1 | 29.63 | 623.9 | 31.19 |

# Author Index

Aagaard, P.M., 542, 553
Abbott, M.B., 298
Ahrens, J.P., 132, 149, 150, 655
Airy, Sir G.B., 4
Alberti, P., 146
Allsop, N.W.H., 240, 325
Aminti, P.L., 678
Anthony, E.J., 607
Arikawa, T., 498
Arthur, R.W., 45

Bagnold, R.A., 169, 180, 181, 645, 657
Bailard, J.A., 659
Barailler, L., 298
Barber, N.P., 458, 463
Barstow, S., 628
Barthel, V., 346
Battjes, J.A., 8, 129, 383, 495, 516, 518
Bayram, A., 658, 660
Beach, R.A., 648
Benoit, M., 467
Berkoff, J.C.W., 495
Biésel, F., 340
Birkemeier, W.A., 664
Black, K.P., 648
Blackman, R.B., 444, 462
Blenkarn, K.A., 320
Blom, G., 553
Boes, D.C., 542
Bondar, C., 602, 615

Booji, N., 495
Borgman, L.E., 45, 466, 467, 581
Boussinesq, J., 4
Bouws, E., 36
Bowen, A.J., 635
Bowers, E.C., 57, 346
Brampton, A.H., 662
Bretschneider, C.L., 34, 67, 97, 399
Briggs, M.J., 7, 77, 89, 497, 508
Brown, C.A., 513
Bruun, P., 323
Burcharth, H.F., 334

Calhoun, R.J., 472
Capon, J., 460
Cartwright, D.E., 373, 403
Cavanié, A.G., 402, 426
Challenor, P.G., 571
Chandramohan, P., 611
Choi, J., 497
Christensen, M., 344
Coles, S., 540
Collin, J.I., 7, 129
Crammer, H., 378

Daemrich, K.F., 234
Dally, W.R., 129, 502, 513
Dalrymple, R.A., 634, 657, 679, 680
Das, M.M., 647
Davenport, A.G., 320, 377
Davies, A.G., 660, 661
De Rouck, J., 128, 229, 237

de Vries, G., 4
de Waal, D.J., 592
de Waal, J.P., 148, 335
Dean, C., 608, 609
Dean, R.G., 420, 634, 645, 657, 679, 680
Donelan, M., 447

Eagleson, P.S., 645
Ebersole, B.A., 111, 410
Einstein, H.A., 644
El-Sayed, W., 615
Endo, T., 240
Esaki, K., 197
Ewans, K.C., 46

Fairchild, J.C., 647
Fan, Q.-J., 473
Forristall, G.Z., 49, 375, 379, 588
Freyer, D.K., 342
Frigaard, P., 344
Frihy, O.E., 615
Fuji-ike, T., 205
Fujihata, S., 320
Fukuda, N., 235, 240
Fukumori, T., 169, 258
Funke, E.R., 340, 344, 347, 382

Gaillard, B., 168
Gaillard, P., 298
Galiatsatou, P., 592
Gencarelli, R., 547
Gerstner, F.J.V., 4
Goda, Y., 7, 9, 10, 20, 24, 37, 39, 46, 64, 78, 82, 100, 101, 105, 106, 109, 113, 129, 144, 149, 166, 218, 245, 278, 283, 333, 334, 374, 380, 381, 387, 401, 407, 410, 414, 421, 426, 439, 446, 472, 476, 482, 493, 510, 529, 538, 546, 553, 556, 560, 563, 567, 577, 578, 636, 653, 661, 667, 678, 679
Goudas, C.L., 678
Greenwood, J.A., 542, 550
Gringorten, I.I., 553

Gumbel, E.J., 540, 552, 570, 575
Guza, R.T., 129, 511, 515

Hallermeier, R.J., 646, 663
Hamada, T., 420
Hamm, L., 109
Hanslow, D., 116
Hanson, H., 678
Hara, N., 500
Hashimoto, N., 420, 423, 467, 469, 471, 478
Hasselmann, K., 34, 35, 420
Hattori, M., 149
Haubrich, R.A., 463
Hibino, T., 55
Hiraishi, T., 58, 234, 327, 347, 484, 500
Hirakuchi, H., 496, 501
Hirayama, K., 500
Hiroi, I., 162, 168
Holman, R.A., 108
Holmberg, D., 678
Holthuijsen, L.H., 7, 494
Hom-ma, M., 93
Horikawa, K., 631
Hosking, J.R.M., 542, 549, 550, 567
Hotta, S., 111, 410, 510
Houston, J.R., 657
Hsu, F.H., 320
Hsu, J.R.C., 606, 679
Huang, N.E., 36, 409
Hubble, E.N., 34, 37
Hudspeth, R.T., 420, 423
Hughes, S.A., 111, 410
Hunt, J.N., 359

Ikeda, N., 257, 280
Ikeno, K., 327
Inman, D.L., 635, 657
Inoue, M., 246
Irie, I., 88
Ishii, T., 498
Ismail, N.M., 615
Isobe, M., 467, 468, 477, 498
Ito, Y., 77, 139, 141, 161, 168, 169, 173, 186, 348

Iwagaki, Y., 100, 101
Iwasaki, T., 148
Izumiya, T., 246, 551, 554, 581, 583

Jaffe, B.E., 648
Jankaew, K., 607
Janssen, J.P.F.M., 8, 129, 495
Jeffreys, E.R., 344
Jensen, O., 323
Johnson, B.D., 654

Kabiling, M.K., 500
Kajima, R., 237, 334, 472
Kajiura, K., 55
Kakizaki, S., 169
Kalinske, A.A., 645
Kamphuis, J.W., 104, 658
Kana, T.W., 648
Kanayama, S., 498
Karlsson, T., 6, 75, 494
Kasugai, Y., 258
Katayama, H., 653, 661, 667, 678
Katayama, T., 149
Kato, H., 58, 421, 423
Katoh, K., 115, 611
Kawai, H., 271, 278
Kendall, M.G., 425
Kim, T.M., 280
Kimura, A., 246, 383, 388, 421, 439
Kimura, K., 202–204, 212, 240
King, D.B., 454, 658
Kinsman, B., 459
Kirby, J.T., 496
Kishira, Y., 245
Kit, E., 611
Kjeldsen, S.P., 375, 413
Klopman, G., 348
Kobayashi, N., 654
Kobune, K., 142, 469, 560, 563
Komar, P.D., 526, 615, 650, 657
Komen, G.J., 63
Kondo, H., 147
Kondo, K., 477
Koopmans, L.H., 363, 442, 449–451
Koreteweg, D.J., 4
Kortenhaus, A., 237

Kos'yan, R., 652
Kraus, N.C., 103, 129, 509, 515
Kudaka, M., 374
Kuik, A.J., 467
Kunieda, J., 613
Kuo, S.T., 129
Kuriyama, Y., 512, 519, 664
Kuroiwa, M., 666, 667
Kuroki, K., 611, 612, 673
Kuwajima, S., 449, 450
Kweon, H.M., 101

Lamb, H., 4
Lamberti, A., 148, 182, 679, 680
Larras, J., 161
Larson, M., 129, 515
Larson, R., 634
Lawless, J.F., 571
Lenton, T.M., 601
Li, B., 496
Li, Y.C., 105
Long, S.R., 410
Longuet-Higgins, M.S., 6, 26, 27, 113,
    360, 371, 373, 377, 383, 393, 395,
    396, 399, 402, 403, 413, 415, 464,
    466, 515
Lundgren, H., 645
Lykke Andersen, T., 334

Madsen, O.S., 513
Madsen, P., 499
Manohar, M., 645
Mansard, E.P.D., 340, 347, 382
Martinelli, L., 182
Maruo, H., 317
Maruyama, K., 496, 497, 501
Mase, H., 134, 214, 495
Masuda, A., 420
Mathiesen, M., 575, 589
Matsumi, M., 348
Matsumoto, A., 344
Medina, J.R., 483
Miles, M.D., 344
Miligram, J.H., 344
Miller, H.C., 648
Minami, K., 258

Minikin, R.R., 169
Mitsui, H., 141
Mitsuyasu, H., 7, 34, 38, 420, 449, 464
Mizuguchi, M., 111, 410, 510
Mori, N., 376, 410
Morihira, M., 140
Morinobu, K., 334
Moskowitz, L., 34
Munk, W.H., 4, 55
Murakami, H., 141
Myrhaug, D., 375, 413

Nadaoka, K., 495, 498, 500, 648, 651
Naess, A., 375
Nagai, K., 6, 24, 72, 79, 407, 449,
    450, 494
Nagao, T., 257, 271
Nagase, K., 327
Nagata, Y., 464
Nakai, K., 58
Nakamura, S., 667
Newman, H., 317
Nielsen, P., 116, 647
Nishi, R., 681
Nishimura, H., 515
Nobuoka, H., 58
Numata, A., 148, 149

O'Reilly, W.C., 78
Ochi, M.K., 34, 37
Ogura, S., 612
Ohno, K., 235
Ojima, R., 335
Okayasu, A., 498
Okihiro, M., 421
Okuyama, Y., 140
Onaka, S., 611
Onozawa, M., 553
Osaki, N., 184
Osato, M., 142
Ottesen Hansen, N.E., 346
Owen, N.W., 9
Ozaki, A., 310, 313
Ozaki, Y., 512, 519
Ozasa, H., 662
Özkan, H.T., 497

Palmer, H.S., 162
Panicker, N.N., 453, 466
Panin, N., 602, 615
Pawka, S.S., 460
Penney, W.G., 4
Peregrine, D.H., 499
Perlin, A., 611
Pernnard, C., 109
Petruaskas, C., 542, 553
Pierre, R., 206
Pierson, W.J., Jr., 5, 22, 34, 37, 44,
    45, 447
Pinkster, J.A., 320
Price, A.T., 4
Prinos, P., 592
Putz, R.R., 5

Radder, A.C., 496
Rattanapikikon, W., 103
Rayleigh, Lord, 4
Rector, R.L., 664
Reid, R.O., 97
Reniers, A.J.H.M., 512, 518
Reptiko, A., 592
Rice, S.O., 51, 369, 383, 391, 393, 403
Rikiishi, K., 447, 449
Rivero, F.J., 495
Rodriguez, J.G.P., 611
Rosati, J.D., 611
Rubey, W.W., 655
Russel, R.C.H., 291
Russell, J.S., 4
Rye, H., 452

Saeki, H., 325
Saied, U.M., 496
Sainflou, G., 4, 168
Saito, K., 680
Saito, M., 551, 554, 581
Sakai, T., 100
Sakakiyama, T., 237, 334
Sakamoto, T., 149
Sallenger, A.H., 108, 648
Salter, S.H., 10, 339, 344, 347
Samejima, S., 162
Sand, S.E., 346, 421

Sato, H., 323
Sato, I., 147
Sato, N., 325, 495
Sato, S., 500, 611
Sawaragi, T., 667
Schäffer, H.A., 348
Scott, T., 645
Seelig, W.N., 132
Sekimoto, T., 56, 246, 592
Severdrup, H.U., 4
Sharma, J.N., 420
Shemdin, O.H., 454
Shibayama, T., 103
Shimada, M., 335
Shimizu, T., 652
Shimosako, K., 184, 186, 188, 257, 276, 281, 283
Shiraishi, S., 320, 323
Shuto, N., 100
Silvester, R., 606, 679
Sleath, J.F.A., 647
Smith, D.W., 611
Smith, E.R., 103
Smith, G., 109
Sorensen, T., 645
Sternberg, R.W., 648
Stevenson, Th., 168
Stive, M.J.F., 129
Stokes, Sir G.G., 4
Stuart, A.T., 425
Sunamura, T., 664
Suquet, F., 340
Suzuki, Y., 7, 10, 39, 78, 472
Svendsen, I.A., 513
Sverdrup, H.U., 4
Swart, D.H., 664

Tada, T., 281
Tajima, Y., 513
Takagi, H., 278, 283
Takahashi, S., 166, 181, 182, 184, 186, 188, 199, 203, 257, 276, 281
Takayama, T., 182, 257, 280, 306, 484, 493, 495
Tamada, T., 225
Tanaka, N., 150, 611, 618, 636

Tanimoto, K., 133, 134, 139, 141, 166, 169, 181, 184, 186, 190, 191, 197, 202, 298, 335
Tanner, W.F., 607
Tayfun, M.A., 375, 414
Thomsen, A.L., 335
Thornton, E.B., 129, 472, 511, 515
Tick, L.J., 420
Ting, F.C.K., 109
Tominaga, M., 149, 239
Tomita, T., 235
Tonomo, K., 247
Toti, M., 615, 680
Toyoshima, O., 609, 615
Tsanis, I.K., 496
Tsuchiya, Y., 461
Tsuda, M., 182
Tsukiyama, T., 680
Tsuruta, S., 9
Tsuruya, H., 346
Tuah, H., 420, 423
Tucker, M.J., 36, 55, 426, 434
Tukey, J.W., 444, 462

Ueda, S., 320, 323

van der Meer, J.W., 132, 148, 229
van Gelder, P.H.A.J.M., 549, 592
van Gent, M.R.A., 214
van Maanen, B., 661
van Rijn, L.C., 655
van Thiel de Vries, J.S.M., 652, 654, 661
van Vledder, G.Ph., 383, 467, 538, 557
Venezian, G., 359
Vesecky, J.F., 454
Vincent, C.L., 7, 77, 497, 508

Wallis, J.R., 542, 549, 567
Wamsley, T.V., 150
Ward, L.G., 648
Watanabe, A., 497
Watanabe, N., 521, 529, 654
Weibull, W., 541

Wiegel, R.L., 134
Wilson, B.W., 41, 64

Yamaguchi, M., 461
Yamashiro, M., 235
Yanagishima, S., 115
Yasuda, T., 410
Yoon, S.B., 497

Yoshida, K., 55
Yoshimura, T., 144
Yoshioka, T., 271, 285
You, Z.J., 646

Zanuttigh, B., 132, 681
Zenkovich, F.P., 608

# General Index

accumulated sliding distance, 276, 281

added mass, 316

aerial photograph, 625

Akita Port, 88, 296

Alger Port, 206

aliasing, 448

allowable stress, 253

analog wave data, analysis of, 433

angular spreading method, 92, 136

annual maxima method, 537

anthropogenic influence on coast, 613–618

artificial headland, 679

artificial reef, 679

Aswan Dam, 615

asymmetry of wave profiles, 413

atiltness parameter, 414

attenuation coefficient, of wave height, 97

autocorrelation function, 363

autocorrelation method, 444

autocovariance function, 365

backshore, 631

bandwidth of the spectral resolution, 452

barrier beach, 606

barrier island, 607

Bayesian directional spectrum estimation method (BDM), 470

beach deformation model, 665

Beach Erosion Board, 97, 644

beach face, 631

beach fill, 671, 676

beach ridge, 607

beach sand, 603–605

bearing capacity, 189

bed shear stress, 515

biogenic sand, 603, 606

Blackman–Tukey method, 367

block-mound seawall, 221

bore model by Batthes and Janssen, 8, 129, 495

Boussinesq equation, 499

breaker height, 103

breaker index, 103

breakwater opening, 84

breakwater, optimum design of, 282

breakwater, performance-based design of, 274

breakwater, reliability-based design of, 260

Bretschneider–Mitsuyasu spectrum, 34, 78, 82

Bretschneider–Mitsuyasu spectrum modified, 35

buoyancy, 173

Byobu-ga-Ura, 609

CADMUS-SURF, 491

caisson breakwater, 166

caisson, optimum width of, 283

Caldera Port, 380, 387, 389, 421

Cape Cod, 606, 608
Cape Hatteras, 609
Cape Skagen, 606, 625
carbonate sand, 603, 604, 636
censoring parameter, 539
CERC formula, 658, 673
CERC pier, 648, 650, 660
characteristic ratio, 256
Chatham Lighthouse, 608
chi-square distribution, 443
CLASH project, 128, 229
clastic sediments, 603
cliffs, collapse of, 636
climatic changes, 591
co-spectrum, 457
coastal accretion, 614
coastal dike, 9, 214, 677
Coastal Engineering Manual, 605,
    616, 636, 677
coastal erosion, 614–616
coastal reconnaissance, 624
coastal sediment cell, 636
coastline, 631
coefficient of friction between
    concrete and rubble stones, 189
coefficient of nonlinear damping force,
    316
composite breakwater, 8, 161
composite breakwater, wave
    transmission of, 146
concrete caissons, design of, 200
conditional run of wave heights, 381
Coney Island, 676
confidence interval of parameter
    estimates, 570
confidence interval, effect of sample
    size on, 578
Constanta, 670
correction for parabolic water level
    change, 437
correlation coefficient between
    successive wave heights, 383
Costa Rica, 66
covariance function, 454
crenulate shaped bay, 606, 679

crest elevation of block-mound
    seawall, 242
crest elevation of breakwater, 190
crest elevation of vertical seawall, 242
crest elevation, effect of crown width
    of, 244
crest elevations of sloped seawall, 245
cross-spectrum, 457
cumulative distribution function, 540
cumulative distribution of relative
    wave energy, 44
cumulative grain size distribution,
    634

damage ratio, 351
Danube, 602, 615, 635, 636, 670
data length of wave record, 435
data sampling interval, 448
data window, 448
Dean number, 650
degree of goodness of fit, 560
delta array, 463
depth of closure, 662, 663
design point, 263
design values, uncertainty of, 253
design wave for wave pressure
    calculation, 170
design wave height, selection of, 590
design working life, 581
deterministic design method, 253
diffraction, 12, 80
diffraction coefficient of random
    waves, 81
diffraction diagram, 80
diffraction diagram, regular waves, 96
diffraction diagrams of random
    waves, 82
digital wave data, analysis of, 435
direct Fourier transform method, 456,
    459
directional adjustment, 171
directional buoy, 464
directional spreading function, 38
directional wave spectrum, 32, 37, 360
directional wave spectrum, measuring
    techniques of, 453

dispersion relationship, 358
distorted model, 332
distribution functions for extreme
  data analysis, 539
distribution functions, characteristics
  of, 542
DOL criterion, 563
double summation method, 343, 479,
  480
dynamic similarity, 331

ECMWF, 628, 672
efficiency of extreme wave analysis,
  551
EMLM, 469
encounter probability, 581
energy averaged waves, 649, 671
energy balance equation, 6, 75, 494
energy dissipation by adverse wind,
  137
envelope amplitude, 370
equilibrium beach profile, 657
equivalent deepwater wave, 12, 96
ergodicity, 364
Euler's constant, 378
EurOtop Manual, 214, 229, 234
expected sliding distance method,
  276, 591
extended least squares method, 555
extended maximum entropy method
  (EMEP), 469
extreme wave analysis, 537
extreme wave data, preparation of a
  sample of, 589
extreme wave height distribution, 280
extreme waves, database of, 586

feldspar, 604
fender, 318
fetch, 63
fetch diagram, 63
FFT method, 444
filter, 451
First District Port Construction
  Bureau, 88, 113, 120

Fisher–Tippett type I distribution
  (FT-I), 540
Fisher–Tippett type II distribution
  (FT-II), 540
Fisher–Tippett type III distribution
  (FT-III), 540
flap type wave paddle motion, 340
floating body, motions of, 315
folding frequency, 448
foot-protection block, 167, 202
foreshore, 631
FORM, 261
Fourier coefficients, 440
Fourier coefficients, computation of,
  450
Fourier series, 440
freak waves, 376
Frechét distribution, 540
frequency spectral density function,
  33
frequency spectrum, 33, 440
Fresnel integrals, 139
friction coefficient at sea bottom, 502,
  515
frictional dissipation coefficient, 502
Froude law, 332
FT-I distribution, 553, 556, 558, 583
FT-II distribution, 553, 556, 558, 583
Fuji River, 612

Gaussian distribution, 364
Gaussian process, 364
Generalized extreme-value
  distribution, 541
geometric similarity, 331
Goda formulas, accuracy of, 177
Goda formulas, wave pressure by, 8,
  170
goodness of fit tests, 559
GPS receiver, 630, 632, 633
gradational breaker index, 503, 505
Great Indian Ocean Tsunami, 607
Great Kanto Earthquake, 608
Grenander uncertainty principle, 447
groin, 676
ground subsidence, 614

group velocity, 75, 99
groupiness factor, 382
Gumbel distribution, 540

harbor agitation by reflected waves,
    311
harbor tranquility, 291
harbor tranquility, elements of, 293
harbor tranquility, graphical solution
    for, 305
harbor tranquility, improvement of,
    309
Hatteras Lighthouse, 609
Hazaki, 115, 520, 651, 664
headland control, 679
heaving, 316
Helmholtz equation, 493
HF doppler radar, 454
highest maximum of irregular wave
    profile, 407
highest one-tenth wave, 23, 27, 31,
    373
highest wave, 23, 30, 170
highest wave height, 23, 29, 30, 170,
    377
highest wave period, 23, 31, 170
highest wave representation method,
    14
Himekawa Port, 240
hindcasted wave data, 587
Hino River, 615
Hiroi's formula, 168, 177
Holocere Epoch, 600, 608
hologram method, 454
homogeneity, 538
homogeneity chek of wave data, 589
horizontal diffusion, 512
horizontally-composite breakwater,
    183, 199
HORS, 519, 651
Hosojima Port, 166
hydraulic model test, 331
hydraulic model tests, necessity of,
    335

IAHR working group, 49, 557, 590
impoundment method, 610
impulsive breaking wave pressure, 180
impulsive breaking wave pressure
    coefficient, 184
impulsive pressure, 181
impulsive pressure coefficient, 183,
    184
impulsive pressure, effect on
    breakwater stability, 182
impulsive pressure, generating
    conditions of, 182
impulsive pressure, mechanism of, 180
incident wave height, 475
incident waves, resolution of, 472
incipient breaker index, 106
incipient breaking, 104
independency of statistical data, 538
inertia matrix, 316
infragravity waves, 55
input signal to wave generator, 340
inshore, 632
instrumentally measured data, 587
intermediate-depth water waves, 11
inward corner of reflective structures,
    141
IPCC, 601
Iribarren number, 213, 215
iron sand, 604, 616
irregular standing wave height, 138
irregular wave profile envelope, 369
irregular wave test, 10
irregular wave test method, 15
island breakwater, 144
ISO 21650, 254, 257
ISO 2394, 254
isostatic adjustment, 608

jetty, 677
JICA, 670
joint distribution of wave heights and
    periods, 393, 399
JONSWAP spectrum, 35, 51, 427,
    506, 520

$K$ value, 658, 673
Kaike Coast, 615
Kamaishi Port, 166, 197
Kashima Port, 611
kinematic similarity, 331
Kodiak data, 557, 558
kurtosis, 409, 412

$L$-year maximum wave height, 259, 582
$L$-year maximum wave height, confidence interval of, 585
largest wave height, 29, 376
LCS, 148, 679, 680
least squares method, 437, 550, 555
Level I method, 261
Level I method, modified, 285
Level II method, 260
Level III method, 260
limit state function, 261
line array, 462
littoral cell, 636
Load and Resistance Factor Design, 585
location parameter, 540
Lognormal distribution, 542
Long Island, 677
long period waves, 55, 57, 326, 346
long waves, 11
long waves, free, 56
long waves, group-bounded, 55, 421
longshore current equation, 515
longshore current profile, 521
longshore sediment transport rate, 610, 657, 676
longshore sediment transport, direction of, 636
low-crested structures (LCS), 679
low-crested structures, wave transmission of, 148

Mach-stem reflection, 134
Madras Port, 617
Maldive, 66
Mamaia beach, 670

maxima of irregular wave profile, 402, 405
maximum entropy principle method (MEP), 469
maximum likelihood method, 459, 550
mean diameter, 634
mean frequency, 369
mean period, 391
mean period, coefficient of variation of, 426
mean rate of the extreme events, 539
mean water level, 514
mean water level, correction for linear change, 437
mean wave height, 24, 27, 371
mean wave period, 24, 31, 371
meandering damage, 141
median diameter, 634
method of $L$-moments, 550
method of moments, 550
method of probability-weighted moments, 550
Midia Port, 670
mild slope equation, 495, 498
Minikin's formula, 177
MIR criterion, 560, 561
mirror-image method, 305
Mitsuyasu-type spreading function, 7, 38, 506
mixed population, return value of, 581
mixed populations, 579
model scale, 332
modified maximum likelihood method (MMLM), 477
modified, extended maximum entropy method (MEMEP), 478
mooring force, 316
mooring lines, 318
mooring system, 318
mooring system, natural frequency of, 319
mound breakwater, 9
multidirectional random wave generator, 339, 347

multidirectional random wave
    generator, input signals to, 343
multidirectional wave generator,
    control signals for, 484

Nagoya Port, 81
narrow-band spectrum, 369
nautical chart, 627
nearshore current equation, 511
nearshore zone, 632
New Zealand, 66
Niigata Coast, 241, 615
non-reflective wave generator, 344
nonlinear spectral component, 420
nonlinearity of surface elevation, 407
Nordency Coast, 652
NOWT-PARI, 501
number of degrees of freedom, 445,
    451
numerical analysis, 491
numerical filtering of wave record, 485
numerical simulation of random sea
    waves, 478
Nyquist frequency, 448

obliquely incident waves, 90
offshore breakwater, 671, 679
Ofunato Port, 166
one-line model, 661, 673
Ooi River, 612
ordered statistics, 552
Otaru Port, 162
outlier, 563

P–N–J method, 5
parabolic equation, 496, 501
parabolic filter, 452
parametric method, 466
parent distribution, 538
parent distribution, most probable,
    559
partial factor, 270, 591
peak enhancement factor, 35, 51
peak frequency, 34
peaks-over-threshold method, 537
PEGBIS model, 109, 114, 501

perched beach, 615, 680
perforated-wall caisson breakwater,
    133
performance-based design, 254
periodogram, 444
periodogram, calculation of, 450
periodograms, smoothing of, 450
Petruaskas and Aagaard formula, 553
phase-delayed wave generation signal,
    342
phi scale, 634
photopole method, 111
Phra Thong Island, 607, 625
PIANC, 323
Pierson–Neumann–James method, 93
piston type wave paddle motion, 340
pitch and roll buoy, 464
pitching, 316
plotting position formula, 552
pocket beach, 606
population, 538
POT, 538
pressure correction factor, 199
probabilistic design method, 254
probability calculation method, 15
profile scale factor, 657
pseudorandom number generating
    algorithm, 484
pulsating water tunnel, 645

quadrature-spectrum, 457
quantile, 546
quartz, 604
quay walls of the energy-dissipating
    type, 314

radiation stress, 112, 513
random sea waves, description of, 360
random signals for driving the wave
    paddle, 342
random wave breaking, 12, 102
random wave diffraction diagrams, 88
random wave generation, 340
random wave generator, 339
random wave refraction analysis, 74
rational return period, 283

Rayleigh distribution, 6, 49, 117, 371
REC criterion, 565
rectangular filter, 451
reduced variate, 544, 556
reference concentration, 642, 652
reflected wave height, 475
reflected waves propagation, 133
reflected waves, absorption of, 348
reflected waves, fictitious
  breakwaters, 135
reflected waves, resolution of, 472
reflection coefficient, 132, 475
reflection coefficient of prototype
  structures, 476
reflective waterfront, 305, 310
refraction, 12
refraction coefficient, 70
refraction coefficient of random
  waves, 71
refraction diagram, 70
regional frequency analysis, 549
regression, 601
regular wave diffraction diagrams, 96
regular wave generator, 337
rejection of candidate distribution,
  566
relative energy of component waves,
  72
relative water depth, 100
reliability index, 262
reliability-based design, 254
representative frequency, 72
resolution of incident and reflected
  waves, 472
return period, 545
return value, 545
return value, confidence interval of,
  575
return value, sample size effect on,
  578
return value, standard deviation of,
  575
return value, statistical variability of,
  571
revetment, 214
rms surface elevation, 49

Rock Manual, 214, 229, 234
rolling, 316
root-mean-square wave height, 108,
  417
Rosetta Promontory, 615
rubble mould foundation, armor unit
  of, 202
rubble mound foundation, design of,
  201
run length, 380
run length, probability distribution
  of, 382
run length theory, 388
run length, variance of, 425
run of high wave heights, 380
run-up height, 211

S–M–B method, 4, 63
Sado Island, 93
safety factor, 189, 253
Sainflou's formula, 177
Sakata Port, 120, 387
sample variability, 569
sampling interval for wave profile, 435
sampling variability, 424
sampling variability of directional
  estimates, 467
sand spit, 606
satellite image, 625
scale effect, 333, 650
scale effects in model tests, 237, 333,
  334
scale parameter, 540
sea level rise, 600
seabed scouring, 206
seawall, 9, 214, 677
seawall design, overtopping-based,
  239
seawall design, runup-based, 239
seawall, crest elevation of, 238, 241
secondary wave crest, 345
sediment discharge from rivers, 635
sediment scale parameter, 657
sediment supply source, 604, 635
sediment suspension rate, 652
seed number, 342

semi-infinite breakwater, 83
semi-infinite structure, 139
sensitivity factor, 263
shallow water waves, 11
Shanghai, 602
shape parameter, 540
sheet flow, 646
Shields number, 646, 652
Shinano River, 612, 616
ship mooring, 317, 321
ship mooring, dynamic analysis of, 319
ship motions, 314
ship motions, modes of, 316
ship movement, allowable amplitude, 322
shoaling, 12
shoaling coefficient, 99
shoreface, 632
shoreline advancement, 607
shoreline change pattern, 618–624
shoreline recession, 608, 614
Shore Protection Manual, 97, 658
short-crested waves, 21
significant wave, 4
significant wave height, 27, 49, 137, 373, 415, 419
significant wave height, exceedance probability of, 296
significant wave height, spectral, 49
significant wave height, statistical, 49
significant wave period, 23
significant wave period, correlation with wave height, 65
significant wave period, spectral, 53
significant wave representation method, 14
similarity laws, 331
simulation of a one-dimensional irregular wave profile, 483
Sines Port, 8
single breaking wave, generation of, 344
single summation method, 344, 480, 506
SIWEH, 382

skewness, 409, 411
sliding distance, 188, 276
sliding of vertical breakwater, 185
sloping-top caisson breakwater, 200
smooth slope, run-up height on, 212
smoothed instantaneous wave energy history, 382
smoothed periodogram method, 444, 447
smoothing function, 451
SORM, 266
spectral analysis, theory of, 440
spectral bandwidth, 373
spectral bandwidth, effect of, 375
spectral calculation method, 14
spectral estimate, reliability of, 445
spectral moments, 369
spectral peak frequency, 34
spectral peakedness parameter, 391
spectral resolution, 447, 452
spectral shape parameter, 51, 387
spectral variability, 447
spectral width parameter, 373, 395, 402, 406
spherical shoal, 76
spread parameter, 281, 286, 546
spreading function, 38
spreading parameter, 40
spreading parameter, refraction effect on, 43
stability number, 202
star array, 461
stationarity, 363
statistical variability of samples of extreme distributions, 568
Stereo Wave Observation Project (SWOP), 456
stochastic process, 362
Stokes waves, 502
submarine deposit, 635
submerged groin system, 678
superposition of incident and reflected waves, 137
SUPERTANK project, 509
surf beat, 55, 116
surf beat amplitude, 116

surf beat, profiles of, 485
surf zone, 7, 104, 121
surf zone, hydrodynamics of, 104
surface elevation root-mean-square,
    48
surface elevation, coefficient of
    variation of the variance of, 426
surface roller, 513, 517
surface roller, energy transfer factor,
    513, 517
surging, 316
Suruga Bay, 612
suspension coefficient, 654
swale, 607
SWAN model, 7, 494, 495
swaying, 316
swell decay, 66
swell spectrum, 37
SWOP, 43, 45

Takayama method, 493
taper, 449
Technical Standards for Port and
    Harbour Facilities in Japan, 270
terrestrial sand, 603, 604
Tevere, 602, 615
tide curve, 626
tide table, 626
TMA spectrum, 36
topographic map, 625
total run of wave heights, 381
total sample method, 537, 589
total sliding distance, 276
total uplift pressure, 175
total wave pressure, 174
Trajanus, 602
transfer function for directional
    spectral measurements, 468
transfer function for wave generation,
    341
transfer function for wave kinematics,
    467
transgression, 601
transmitted waves, propagation of,
    153
transport coefficient, 659

triangular filter, 452
tsunami breakwater, 166
Tucker method, 434
turbulent eddy viscosity, 512, 516
two-axis current meters, 464

U.S. Army Corps of Engineers, 644
unbiased standard deviation, 569
unbiasedness, 551
undistorted model, 332
uniformity coefficient, 634
uplift pressure, 173
upright section, stability condition
    for, 188, 194, 197, 198
upright section, stable width of, 190
upright sections, design of, 188, 194

variance spectral density function,
    367
variance wave spectrum density, 362
vertical breakwater, 161
visually observed wave data, 587
volcanic sand, 603

Wallops spectrum, 36, 51, 427
water level correction, 437
wave action balance equation, 494
wave angle for simulation, 481
wave attenuation by an adverse wind,
    137
wave attenuation, of short period
    waves, 333
wave breaking point, 103
wave celerity, 75
wave climate, 296, 649
wave concentration at an inward
    corner, 142
wave damping coefficient, 316
wave data analysis, 433, 435
wave diffraction, 80
wave direction, 296
wave drift force, 317
wave force variation, spatial, 144
wave forecasting, 63, 628
wave frequency for simulation, 481
wave gauge array, 456

wave gauge arrays, layout of, 461
wave generation, transfer function of, 341
wave generator, 337
wave grouping, 379
wave height distribution, 5, 24, 371
wave height in surf zone, 122, 126
wave height variation, 138, 139
wave height, planar beaches, 117
wave heights, 375
wave heights, variability of, 427
wave hindcasting, 296
wave nonlinearity, 376
wave nonlinearity effect, 414, 502
wave nonlinearity parameter, 410
wave overtopping, 9
wave overtopping rate, 216
wave overtopping rate, of block-mound seawall, 223
wave overtopping rate, of uniform slope, 226–228
wave overtopping rate, of vertical seawall, 219
wave overtopping rate, range of variation, 225, 231
wave overtopping rate, tolerable limit of, 240
wave overtopping, expected rate of, 217
wave overtopping, mean rate of, 218
wave period associated with design wave height, 591
wave period, distribution of, 30, 391
wave period, standard deviation of, 399, 400
wave period, marginal distribution of, 397
wave periods, variability of, 428
wave pressure coefficient, 174
wave pressure formulas for upright sections, 168
wave pressure under wave crest, 170
wave pressure under wave trough, 195
wave reflection, 131

wave refraction, 68
wave refraction diagram, 69
wave refraction on a coast with straight, parallel depth-contours, 78
wave run-up, 211
wave setdown, 8, 113
wave setup, 8, 113, 514, 518, 520
wave shoaling, 99
wave spectrum, 31
wave spectrum, nonlinear components of, 419
wave statistics, standard deviations of, 425
wave steepness, 67
wave transformation, 6
wave transformation model, phase-averaged type, 492
wave transformation model, phase-resolving type, 492
wave transmission, 146
wave transmission coefficient, 146
wave-absorbing structures, 311
wavelength, 358
Weibull distribution, 375, 379, 521, 541, 556, 558
Weibull distribution, maximal, 540
Weibull formula, 552
weight function, 451
Wiener–Khintchine relations, 366
Wilson's formulas, 63
wind duration, 63
wind spectrum, 322
wind waves, 63

Yangtze River, 602
yawing, 316
Yellow River, 602
Yokohama Port, 162
Yoshihara Coast, 238

zero-downcrossing method, 22, 414
zero-upcrossing method, 22, 414
zero-upcrossing point, 438